McGraw-Hill
Dictionary of
CHEMISTRY

McGraw-Hill
Dictionary of
CHEMISTRY

Sybil P. Parker

Editor in Chief

McGraw-Hill Book Company

New York   St. Louis   San Francisco

Auckland   Bogotá   Guatemala   Hamburg
Johannesburg   Lisbon   London   Madrid   Mexico
Montreal   New Delhi   Panama   Paris   San Juan
São Paulo   Singapore   Sydney   Tokyo   Toronto

1 2 3 4 5 6 7 8 9 0   DODO   8 9 1 0 9 8 7 6 5 4

ISBN 0-07-045420-5

Library of Congress Cataloging in Publication Data

McGraw-Hill dictionary of chemistry.

1. Chemistry—Dictionaries. I. Parker, Sybil P.
II. McGraw-Hill Book Company. III. Title: Dictionary
of chemistry.
QD5.M357   1984   540'.3'21   84-12205
ISBN 0-07-045420-5

Preface

The *McGraw-Hill Dictionary of Chemistry* is a unique addition to the chemical literature in that it focuses on the vocabulary of theoretical and applied chemistry rather than on chemicals and materials, and includes the specialized terminology of atomic and nuclear physics.

The more than 9000 terms that are defined in the Dictionary represent 11 fields, including analytical chemistry, biochemistry, chemical engineering, crystallography, geochemistry, inorganic, organic, and physical chemistry, and spectroscopy, as well as related fields of physics. This multidisciplinary approach will prove to be invaluable for most professional or pedagogical needs.

Terms and definitions for this Dictionary were selected from the *McGraw-Hill Dictionary of Scientific and Technical Terms* (3d ed., 1984). Synonyms, acronyms, and abbreviations are given with the definitions and are also listed in the alphabetical sequence as cross-references to the defining term. Empirical or line formulas accompany all entries for compounds.

The *McGraw-Hill Dictionary of Chemistry* will serve as an indispensable reference tool for chemists and other scientists, chemical engineers, educators, students, and writers needing clear, concise definitions of chemical terminology.

Sybil P. Parker
EDITOR IN CHIEF

Editorial Staff

Sybil P. Parker, Editor in Chief

Jonathan Weil, Editor
Betty Richman, Editor

Edward J. Fox, Art director

Ann D. Bonardi, Art production supervisor

Joe Faulk, Editing manager

Ann Jacobs, Editing supervisor
Frank Kotowski, Jr., Editing supervisor
Patricia W. Albers, Senior editing assistant

Consulting and Contributing Editors

from the McGraw-Hill Dictionary of Scientific and Technical Terms

Prof. George S. Bonn—Formerly, Graduate School of Library Science, University of Illinois. LIBRARY CONSULTANT.

Robert L. Davidson—Formerly, Editor in Chief, Business Books and Services, Professional and Reference Division. McGraw-Hill Book Company. CHEMICAL ENGINEERING.

Prof. Roland H. Good, Jr.—Department of Physics, Pennsylvania State University. PHYSICS.

Dr. N. Karle Mottet—Professor of Pathology, University of Washington School of Medicine. BIOCHEMISTRY.

Dr. Charles Oviatt—State Department of Education of Missouri. CHEMISTRY.

Dr. Leonard Spero—Walter Reed Hospital Unit, Fort Dietrick, Maryland. CHEMISTRY.

Dr. C. N. Touart—Senior Scientist, Air Force Geophysics Laboratory. GEOCHEMISTRY.

How to Use the Dictionary

I. ALPHABETIZATION

The terms in the *McGraw-Hill Dictionary of Chemistry* are alphabetized on a letter-by-letter basis; word spacing, hyphen, comma, and prime in a term are ignored in the sequencing. Also ignored in the sequencing of terms (usually chemical compounds) are italic elements, numbers, small capitals, and Greek letters. For example, the following terms appear within alphabet letter "A":

> 1-aminoadamantane
> amino alcohol
> *para*-aminobenzoic acid
> γ-aminobutyric acid
> 2-aminoisovaleric acid
> 1-amino-2-propanol
> 3′,5′-AMP

II. CROSS-REFERENCING

A cross-reference entry directs the user to the defining entry. For example, the user looking up "amyl carbinol" finds:

> **amyl carbinol** *See* hexyl alcohol.

The user then turns to the "H" terms for the definition.

Cross-references are also made from variant spellings, acronyms, abbreviations, and symbols.

> **AES** *See* Auger electron spectroscopy.
> **aluminium** *See* aluminum.
> **at. wt** *See* atomic weight
> **Au** *See* gold.

The user turning directly to a defining entry will find the above type of information included, introduced by "Also known as . . . ," "Also spelled . . . ," "Abbreviated . . . ," "Symbolized . . . ," "Derived from . . . ."

III. CHEMICAL FORMULAS

Chemistry definitions may include either an empirical formula (say, for abscisic acid, $C_{16}H_{20}O_4$) or a line formula (for acrylonitrile, CH_2CHCN), whichever is appropriate.

McGraw-Hill
Dictionary of
CHEMISTRY

A

a axis One of the crystallographic axes used as reference in crystal description, usually oriented horizontally, front to back.

abalyn A liquid rosin that is a methyl ester of abietic acid; prepared by treating rosin with methyl alcohol; used as a plasticizer.

Abegg's rule An empirical rule, holding for a large number of elements, that the sum of the maximum positive and negative valencies of an element equals eight.

Abel tester A laboratory instrument used in testing the flash point of kerosine and other volatile oils having flash points below 120°F (49°C); the oil is contained in a closed cup which is heated by a fixed flame below and a movable flame above.

abietic acid $C_{20}H_{30}O_2$ A tricyclic, crystalline acid obtained from rosin; used in making esters for plasticizers.

ABS *See* acrylonitrile butadiene styrene.

abscisic acid $C_{16}H_{20}O_4$ A plant hormone produced by fruits and leaves that promotes abscission and dormancy and retards vegetative growth. Formerly known as abscisin.

abscisin *See* abscisic acid.

absolute alcohol Ethyl alcohol that contains no more than 1% water. Also known as anhydrous alcohol.

absolute boiling point The boiling point of a substance expressed in the unit of an absolute temperature scale.

absolute density *See* absolute gravity.

absolute gravity Density or specific gravity of a fluid reduced to standard conditions; for example, with gases, to 760 mmHg pressure and 0°C temperature. Also known as absolute density.

absolute reaction rate The rate of a chemical reaction as calculated by means of the (statistical-mechanics) theory of absolute reaction rates.

absorb To take up a substance in bulk.

absorbance The common logarithm of the reciprocal of the transmittance of a pure solvent. Also known as absorbancy; extinction.

absorbancy *See* absorbance.

absorbency Penetration of one substance into another.

absorbency index *See* absorptivity.

absorber Equipment in which a gas is absorbed by contact with a liquid.

absorptiometer **1.** An instrument equipped with a filter system or other simple dispersing system to measure the absorption of nearly monochromatic radiation in the visible range by a gas or a liquid, and so determine the concentration of the absorbing constituents in the gas or liquid. **2.** A device for regulating the thickness of a liquid in spectrophotometry.

absorptiometric analysis Chemical analysis of a gas or a liquid by measurement of the peak electromagnetic absorption wavelengths that are unique to a specific material or element.

absorption The taking up of matter in bulk by other matter, as in dissolving of a gas by a liquid.

absorption constant *See* absorptivity.

absorption edge The wavelength corresponding to a discontinuity in the variation of the absorption coefficient of a substance with the wavelength of the radiation. Also known as absorption limit.

absorption limit *See* absorption edge.

absorption line A minute range of wavelength or frequency in the electromagnetic spectrum within which radiant energy is absorbed by the medium through which it is passing.

absorption peak A wavelength of maximum electromagnetic absorption by a chemical sample; used to identify specific elements, radicals, or compounds.

absorption plant A facility to recover the condensable portion of natural or refinery gas.

absorption process A method in which light oil is introduced into an absorption tower so that it absorbs the gasoline in the rising wet gas; the light oil is then distilled to separate the gasoline.

absorption spectrophotometer An instrument used to measure the relative intensity of absorption spectral lines and bands. Also known as difference spectrophotometer.

absorption spectroscopy The study of spectra obtained by the passage of radiant energy from a continuous source through a cooler, selectively absorbing medium.

absorption spectrum The array of absorption lines and absorption bands which results from the passage of radiant energy from a continuous source through a cooler, selectively absorbing medium.

absorption tube A tube filled with a solid absorbent and used to absorb gases and vapors.

absorptive power *See* absorptivity.

absorptivity The constant a in the Beer's law relation $A = abc$, where A is the absorbance, b the path length, and c the concentration of solution. Also known as absorptive power. Formerly known as absorbency index; absorption constant; extinction coefficient.

ABS resin *See* acrylonitrile butadiene styrene resin.

acaroid resin   A gum resin from aloelike trees of the genus *Xanthorrhoea* in Australia and Tasmania; used in varnishes and inks. Also known as gum accroides; yacca gum.

acceleration globulin   A globulin that acts to accelerate the conversion of prothrombin to thrombin in blood clotting; found in blood plasma in an inactive form.

accelerator   In the manufacture of rubber, a substance that acts as a catalyst of vulcanization but undergoes a chemical change in the process.

accelofilter   A filtration device that uses a vacuum or pressure to draw or force the liquid through the filter to increase the rate of filtration.

acceptor   1. A chemical whose reaction rate with another chemical increases because the other substance undergoes another reaction. 2. A species that accepts electrons, protons, electron pairs, or molecules such as dyes.

accessory element   *See* trace element.

accumulator   An auxiliary ram extruder on blow-molding equipment used to store melted material between deliveries.

acediamine hydrochloride   *See* acetamidine hydrochloride.

acenaphthene $C_{12}H_{10}$ An unsaturated hydrocarbon whose colorless crystals melt at 92°C; insoluble in water; used as a dye intermediate and as an agent for inducing polyploidy.

1,2-acenaphthenedione   *See* acenaphthequinone.

acenaphthequinone $C_{10}H_6(CO)_2$ A three-ring hydrocarbon in the form of yellow needles melting at 261–263°C; insoluble in water and soluble in alcohol; used in dye synthesis. Also known as 1,2-acenaphthenedione.

acene   Any condensed polycyclic compound with fused rings in a linear arrangement; for example, anthracene.

acenocoumarin   *See* acenocoumarol.

acenocoumarol $C_{19}H_{15}NO_6$ A tasteless, odorless, white, crystalline powder with a melting point of 197°C; slightly soluble in water and organic solvents; used as an anticoagulant. Also known as acenocoumarin.

acephate $C_4H_{10}NO_3PS$ A white solid with a melting point of 72–80°C; very soluble in water; used as an insecticide for a wide range of aphids and foliage pests. Also known as *O*,*S*-dimethyl acetylphosphoramidothioate.

acephatemet $CH_3OCH_3SPONH_2$ A white, crystalline solid with a melting point of 39–41°C; limited solubility in water; used as an insecticide to control cutworms and borers on vegetables. Also known as *O*,*S*-dimethylphosphoroamidothioate; methamidophos.

acetal   1. $CH_3CH(OC_2H_5)_2$ A colorless, flammable, volatile liquid used as a solvent and in manufacture of perfumes. Also known as 1,1-diethoxyethane. 2. Any one of a class of compounds formed by the addition of alcohols to aldehydes.

acetaldehydase   An enzyme that catalyzes the oxidation of acetaldehyde to acetic acid.

acetaldehyde C_2H_4O A colorless, flammable liquid used chiefly to manufacture acetic acid. Also known as ethanal.

para-acetaldehyde   *See* paraldehyde.

acetaldehyde ammonia   *See* aldehyde ammonia.

acetaldehyde cyanohydrin   *See* lactonitrile.

acetal resins   Linear, synthetic resins produced by the polymerization of formaldehyde (acetal homopolymers) or of formaldehyde with trioxane (acetal copolymers); hard, tough plastics used as substitutes for metals. Also known as polyacetals.

acetamide CH_3CONH_2 The crystalline, colorless amide of acetic acid, used in organic synthesis and as a solvent. Also known as ethanamide.

acetamidine hydrochloride $C_2H_6N_2 \cdot HCl$ Deliquescent crystals that are long prisms with a melting point reported as either 174°C or 164–166°C; soluble in water and alcohol; used in the synthesis of imidazoles, pyrimidines, and triazines. Also known as acediamine hydrochloride; α-amino-α-iminoethane hydrochloride; ethanamidine hydrochloride; ethenylamidine hydrochloride.

acetamidoacetic acid   *See* aceturic acid.

para-acetamidobenzenesulfonyl chloride   *See* *N*-acetylsulfanilyl chloride.

2-acetamido-4-mercaptobutyric acid-γ-thiolactone   *See* *N*-acetylhomocysteinethiolactone.

l-α-acetamido-β-mercaptopropionic acid   *See* acetylcysteine.

acetamidophenol   *See* acetaminophen.

α-acetamido-γ-thiobutyrolactone   *See* *N*-acetylhomocysteinethiolactone.

acetaminophen $C_8H_9O_2N$ Large monoclinic prisms with a melting point of 169–170°C; soluble in organic solvents such as methanol and ethanol; used in the manufacture of azo dyes and photographic chemicals, and as an analgesic and antipyretic. Also known as acetamidophenol; acetaminophenol; *N*-acetyl-*para*-aminophenol (APAP); *para*-hydroxyacetanilide.

acetaminophenol   *See* acetaminophen.

acetanilide   An odorless compound in the form of white, shining, crystalline leaflets or a white, crystalline powder with a melting point of 114–116°C; soluble in hot water, alcohol, ether, chloroform, acetone, glycerol, and benzene; used as a rubber accelerator, in the manufacture of dyestuffs and intermediates, as a precursor in penicillin manufacture, and as a painkiller. Also known as *N*-phenylacetamide.

acetanisidine   *See* methacetin.

acetarsone   *See* 3-acetamido-4-hydroxybenzenearsonic acid.

acetate   One of two species derived from acetic acid, CH_3COOH; one type is the acetate ion, CH_3COO^-; the second type is a compound whose structure contains the acetate ion, such as ethyl acetate.

acetate C-8   *See* *n*-octyl acetate.

acetate dye   1. Any of a group of water-insoluble azo or anthroquinone dyes used for dyeing acetate fibers. 2. Any of a group of water-insoluble amino azo dyes that are treated with formaldehyde and bisulfate to make them water-soluble.

acetate of lime   Calcium acetate made from pyroligneous acid and a water suspension of calcium hydroxide.

acetate process   Acetylation of cellulose (wood pulp or cotton linters) with acetic acid or acetic anhydride and sulfuric acid catalyst to make cellulose acetate resin or fiber.

acetate rayon   *See* acetate.

acetenyl   *See* ethinyl.

acetic acid CH_3COOH **1.** A clear, colorless liquid or crystalline mass with a pungent odor, miscible with water or alcohol; crystallizes in deliquescent needles; a component of vinegar. Also known as ethanoic acid.   **2.** A mixture of the normal and acetic salts; used as a mordant in the dyeing of wool.

acetic anhydride $(CH_3CO)_2O$ A liquid with a pungent odor that combines with water to form acetic acid; used as an acetylating agent.

acetic ester   *See* ethyl acetate.

acetic ether   *See* ethyl acetate.

acetic thiokinase   An enzyme that catalyzes the formation of acetyl coenzyme A from acetate and adenosinetriphosphate.

acetidin   *See* ethyl acetate.

acetin $C_3H_5(OH)_2OOCCH_3$ A thick, colorless, hygroscopic liquid with a boiling point of 158°C, made by heating glycerol and strong acetic acid; soluble in water and alcohol; used in tanning, as a dye solvent and food additive, and in explosives. Also spelled acetine. Also known as glyceryl monoacetate; monoacetin.

acetine   *See* acetin.

acetoacetanilide   *See* acetoacetic acid.

acetoacetate   A salt which contains the CH_3COCH_2COO radical; derived from acetoacetic acid.

acetoacetic acid CH_3COCH_2COOH A colorless liquid miscible with water; derived from β-hydroxybutyric acid in the body. Also known as acetoacetanilide; α-acetylacetanilide; β-ketobutyranilide.

acetoacetic ester   *See* ethyl acetoacetate.

acetoacetyl coenzyme A $C_{25}H_{41}O_{18}N_7P_3S$ An intermediate product in the oxidation of fatty acids.

acetoamidoacetic acid   *See* aceturic acid.

acetobromamide   *See* *N*-bromoacetamide.

α-acetobutyrolactone   *See* α-acetylbutyrolactone.

acetocinnamone   *See* benzylideneacetone.

acetoin $CH_3COCHOHCH_3$ A slightly yellow liquid, melting point 15°C, used as an aroma carrier in the preparation of flavors and essences; produced by fermentation or from diacetyl by partial reduction with zinc and acid. Also known as acetylmethyl carbinol; dimethylketol.

acetol CH_3COCH_2OH A colorless liquid soluble in water; a reducing agent. Also known as 1-hydroxy-2-propanone.

acetolactic acid $C_5H_8O_4$ A monocarboxylic acid formed as an intermediate in the synthesis of the amino acid valine.

acetolysis Decomposition of an organic molecule through the action of acetic acid or acetic anhydride.

acetone CH_3COCH_3 A colorless, volatile, extremely flammable liquid, miscible with water; used as a solvent and reagent. Also known as 2-propanone.

acetone-benzol process A dewaxing process in petroleum refining, with acetone and benzol used as solvents.

acetone body See ketone body.

acetone chloroform See chlorobutanol.

acetone cyanohydrin $(CH_3)_2COHCN$ A colorless liquid obtained from condensation of acetone with hydrocyanic acid; used as an insecticide or as an organic chemical intermediate.

acetone glucose See acetone sugar.

acetone glycerol See 2,2-dimethyl-1,3-dioxolane-4-methanol.

acetone number A ratio used to estimate the degree of polymerization of materials such as drying oils; it is the weight in grams of acetone added to 100 grams of a drying oil to cause an insoluble phase to form.

acetone oxime See acetoxime.

acetone pyrolysis Thermal decomposition of acetone into ketene.

acetone-sodium bisulfite $(CH_3)_2C(OH)SO_3Na$ Crystals that have a slight sulfur dioxide odor and slightly fatty feel; freely soluble in water, decomposed by acids; used in photography and in textile dyeing and printing. Also known as acetone sulfite; 2-hydroxy-2 propanesulfonic acid sodium salt; sodium-acetone bisulfite.

acetone sugar Any reducing sugar that contains acetone; examples are 1,2-monoacetone-D-glucofuranose and 1,2-5,6-diacetone-D-glucofuranose. Also known as acetone glucose.

acetone sulfite See acetone-sodium bisulfite.

acetonitrile CH_3CN A colorless liquid soluble in water; used in organic synthesis. Also known as ethane nitrile; methyl cyanide.

acetonylacetone $CH_3COCH_2CH_2COCH_3$ A colorless liquid with a boiling point of 192.2°C; soluble in water; used as a solvent and as an intermediate for pharmaceuticals and photographic chemicals.

3-(α-acetonylbenzyl)-4-hydroxycoumarin See warfarin.

3-(α-acetonyl-4-chlorobenzyl)-4-hydroxy-coumarin See coumachlor.

acetophenetidide See acetophenetidin.

acetophenone $C_6H_5COCH_3$ Colorless crystals with a melting point of 19.6°C and a specific gravity of 1.028; used as a chemical intermediate. Also known as phenyl methyl ketone.

acetostearin A general term for monoglycerides of stearic acid acetylated with acetic anhydride; used as a protective food coating and as plasticizers for waxes and synthetic resins to improve low-temperature characteristics.

acetoxime $(CH_3)_2CNOH$ Colorless crystals with a chlorallike odor and a melting point of 61°C; soluble in alcohol, ethers, and water; used in organic synthesis and as a solvent for cellulose ethers. Also known as acetone oxime; 2-propanone oxime.

(2-acetoxypropyl)trimethylammonium chloride *See* methacholine chloride.

acetozone *See* acetyl benzoyl peroxide.

aceturic acid $CH_3CONHCHCH_2COOH$ Long, needlelike crystals with a melting point of 206–208°C; soluble in water and alcohol; forms stable salts with organic bases; used in medicine. Also known as acetamidoacetic acid; acetylaminoacetic acid; N-acetylglycine; acetylglycocoll; ethanoylaminoethanoic acid.

acetyl $CH_3CO—$ A two-carbon organic radical containing a methyl group and a carbonyl group.

acetylacetanilide *See* acetoacetanilide.

α-acetylacetanilide *See* acetoacetic acid.

acetylacetone $CH_3COCH_2OCCH_3$ A colorless liquid with a pleasant odor and a boiling point of 140.5°C; soluble in water; used as a solvent, lubricant additive, paint drier, and pesticide. Also known as diacetylmethane; 2,4–pentanedione.

acetylaminoacetic acid *See* aceturic acid.

***para*-acetylaminobenzenesulfonyl chloride** *See* N-acetylsulfonilyl chloride.

acetylaminohydroxyphenylarsonic acid *See* 3-acetamido-4-hydroxybenzenearsonic acid.

N-acetyl-*para*-aminophenol *See* acetaminophen.

***para*-acetylaminophenol** *See* acetaminophen.

acetylase Any enzyme that catalyzes the formation of acetyl esters.

acetylating agent A reagent, such as acetic anhydride, capable of bonding an acetyl group onto an organic molecule.

acetylation The process of bonding an acetyl group onto an organic molecule.

acetyl benzoyl peroxide $C_6H_5CO\cdot O_2\cdot OCCH_3$ White crystals with a melting point of 36.6°C; moderately soluble in ether, chloroform, carbon tetrachloride, and water; used as a germicide and disinfectant. Also known as acetozone; benzozone.

acetyl bromide CH_3COBr A colorless, fuming liquid with a boiling point of 81°C; soluble in ether, chloroform, and benzene; used in organic synthesis and dye manufacture.

α-acetylbutyrolactone $C_6H_8O_3$ A liquid with an esterlike odor; soluble in water; used in the synthesis of 3,4-disubstituted pyridines. Also known as α-acetobutyrolactone; α-acetyl-γ-hydroxybutyric acid γ-lactone; α-(2-hydroxyethyl)acetoacetic acid γ-lactone.

acetyl chloride CH_3COCl A colorless, fuming liquid with a boiling point of 51–52°C; soluble in ether, acetone, and acetic acid; used in organic synthesis, and in the manufacture of dyestuffs and pharmaceuticals. Also known as ethanoyl chloride.

acetylcholine $C_7H_{17}O_3N$ A compound released from certain autonomic nerve endings which acts in the transmission of nerve impulses to excitable membranes.

acetylcholinesterase   An enzyme found in excitable membranes that inactivates acetylcholine.

acetyl coenzyme A $C_{23}H_{39}O_{17}N_7P_3S$ A coenzyme, derived principally from the metabolism of glucose and fatty acids, that takes part in many biological acetylation reactions; oxidized in the Krebs cycle.

acetylene C_2H_2 A colorless, highly flammable gas that is explosive when compressed; the simplest compound containing a triple bond; used in organic synthesis and as a welding fuel. Also known as ethyne.

acetylene black   A form of carbon with high electrical conductivity; made by decomposing acetylene by heat.

acetylene dichloride   See sym-dichloroethylene.

acetylene series   A series of unsaturated aliphatic hydrocarbons, each containing at least one triple bond and having the general formula C_nH_{2n-2}.

acetylene tetrabromide $CHBr_2CHBr_2$ A yellowish liquid with a boiling point of 239–242°C; soluble in alcohol and ether; used for separating minerals and as a solvent. Also known as sym-tetrabromoethane.

acetylene tetrachloride   See sym-tetrachloroethane.

acetylenic   Pertaining to acetylene or being like acetylene, such as having a triple bond.

acetylenyl   See ethinyl.

N-acetylethanolamine $CH_3CONHC_2H_4OH$ A brown, viscous liquid with a boiling range of 150–152°C; soluble in alcohol, ether, and water; used as a plasticizer, humectant, high-boiling solvent, and textile conditioner. Also known as hydroxyethylacetamide.

acetyl formic acid   See pyruvic acid.

N-acetylglycine   See aceturic acid.

acetylglycocoll   See aceturic acid.

N-acetylhomocysteinethiolactone $C_6H_9NO_2S$ Needlelike crystals that melt at 111.5–112.5°C; used as a photographic antifogging agent and a reagent for insolubilizing antibodies. Abbreviated AHCTL. Also known as 2-acetamido-4-mercaptobutyric acid γ-thiolactone; α-acetamido-γ-thiobutyrolactone; citiolone; N-(tetrahydro-2-oxo-3-thienyl)acetamide.

α-acetyl-γ-hydroxybutyric acid   See α-acetylbutyrolactone.

acetylide   A compound formed from acetylene with the H atoms replaced by metals, as in cuprous acetylide (Cu_2C_2).

acetyl iodide CH_3COI A colorless, transparent, fuming liquid with a boiling point of 105–108°C; soluble in ether and benzene; used in organic synthesis.

acetylisoeugenol $C_6H_3(CHCHCH_3)(OCH_3)(OCOCH_3)$ White crystals with a clovelike odor and a congealing point of 77°C; used in perfumery and flavoring. Also known as isoeugenol acetate.

acetyl ketene   See diketene.

acetylmethylcarbinol *See* acetoin.

O-acetyl-β-methylcholine *See* methacholine chloride.

acetyl-β-methylcholine chloride *See* methacholine chloride.

acetyl number A measure of free hydroxyl groups in fats or oils determined by the amount of potassium hydroxide used to neutralize the acetic acid formed by saponification of acetylated fat or oil.

acetyl peroxide $(CH_3CO)_2O_2$ Colorless crystals with a melting point of 30°C; soluble in alcohol and ether; used as an initiator and catalyst for resins. Also known as diacetyl peroxide.

acetyl phosphate $C_2H_5O_5P$ The anhydride of acetic and phosphoric acids occurring in the metabolism of pyruvic acid by some bacteria; phosphate is used by some microorganisms, in place of ATP, for the phosphorylation of hexose sugars.

acetylpropionic acid *See* levulinic acid.

acetyl propionyl $CH_3COCOCH_2CH_3$ A yellow liquid with a boiling point of 106–110°C; used in butterscotch- and chocolate-type flavors. Also known as methyl ethyl diketone; methyl ethyl glyoxal; 2,3-pentanedione.

acetylsalicylic acid $CH_3COOC_6H_4COOH$ A white, crystalline, weakly acidic substance, with melting point 137°C; slightly soluble in water; used medicinally as an antipyretic. Also known by trade name aspirin.

N-acetylsulfanilyl chloride $C_8H_8ClNO_3S$ Thick, light tan prisms ranging to brown powder or fine crystals with a melting point of 149°C; soluble in benzene, chloroform, and ether; used as an intermediate in the preparation of sulfanilamide and its derivatives. Abbreviated ASC. Also known as *para*-acetamidobenzenesulfonyl chloride; *para*-acetylaminobenzenesulfonyl chloride.

acetylurea $CH_3CONHCONH_2$ Crystals that are colorless and are slightly soluble in water.

acetyl valeryl $CH_3COCOC_4H_9$ A yellow liquid used for cheese, butter, and other flavors. Also known as heptadione-2,3.

acfm *See* actual cubic feet per minute.

a-c fracture A type of tension fracture lying parallel to the a-c fabric plane and normal to plane b in a crystal.

achiral molecules Molecules which are superposable to their mirror images.

achroglobin A colorless respiratory pigment present in some mollusks and urochordates.

acid **1.** Any of a class of chemical compounds whose aqueous solutions turn blue litmus paper red, react with and dissolve certain metals to form salts, and react with bases to form salts. **2.** A compound capable of transferring a hydrogen ion in solution. **3.** A substance that ionizes in solution to yield the positive ion of the solvent. **4.** A molecule or ion that combines with another molecule or ion by forming a covalent bond with two electrons from the other species.

1,2,4-acid *See* 1-amino-2-naphthol-4-sulfonic acid.

π-acid An acid that readily forms stable complexes with aromatic systems.

acid acceptor A stabilizer compound added to plastic and resin polymers to combine with trace amounts of acids formed by decomposition of the polymers.

acid alcohol A compound containing both a carboxyl group (—COOH) and an alcohol group (—CH$_2$OH, =CHOH, or ≡COH).

acid ammonium tartrate *See* ammonium bitartrate.

acid anhydride An acid with one or more molecules of water removed; for example, SO$_3$ is the acid anhydride of H$_2$SO$_4$, sulfuric acid.

acid azide **1.** A compound in which the hydroxy group of a carboxylic acid is replaced by the azido group (—NH$_3$). **2.** An acyl or aroyl derivative of hydrazoic acid. Also known as acyl azide.

acid-base catalysis The increase in speed of certain chemical reactions due to the presence of acids and bases.

acid-base equilibrium The condition when acidic and basic ions in a solution exactly neutralize each other; that is, the pH is 7.

acid-base indicator A substance that reveals, through characteristic color changes, the degree of acidity or basicity of solutions.

acid-base pair A concept in the Brönsted theory of acids and bases; the pair consists of the source of the proton (acid) and the base generated by the transfer of the proton.

acid-base titration A titration in which an acid of known concentration is added to a solution of base of unknown concentration, or the converse.

acid blowcase *See* blowcase.

acid calcium phosphate *See* calcium phosphate.

acid cell An electrolytic cell whose electrolyte is an acid.

acid chloride A compound containing the radical —COCl; an example is benzoyl chloride.

acid conductor A vessel designed for refortification of hydrolyzed acid by heating and evaporation of water, or sometimes by distillation of water under partial vacuum.

acid disproportionation The self-oxidation of a sample of an oxidized element to the next higher oxidation state and then a corresponding reduction to lower oxidation states.

acid dye Any of a group of sodium salts of sulfonic and carboxylic acids used to dye natural and synthetic fibers, leather, and paper.

acid egg *See* blowcase.

acid electrolyte A compound, such as sulfuric acid, that dissociates into ions when dissolved, forming an acidic solution that conducts an electric current.

acid gases The hydrogen sulfide and carbon dioxide found in natural and refinery gases which, when combined with moisture, form corrosive acids; known as sour gases when hydrogen sulfide and mercaptans are present.

acid halide A compound of the type RCOX, where R is an alkyl or aryl radical and X is a halogen.

acid heat test   The determination of degree of unsaturation of organic compounds by reacting with sulfuric acid and measuring the heat of reaction.

acidic   1. Pertaining to an acid or to its properties. 2. Forming an acid during a chemical process.

acidic group   The radical COOH, present in organic acids.

acidic oxide   An oxygen compound of a nonmetal, for example, SO_2 or P_2O_5, which yields an oxyacid with water.

acidic sodium aluminum phosphate   *See* sodium aluminum phosphate.

acidic titrant   An acid solution of known concentration used to determine the basicity of another solution by titration.

acidification   Addition of an acid to a solution until the pH falls below 7.

acidimeter   An apparatus or a standard solution used to determine the amount of acid in a sample.

acidimetry   The titration of an acid with a standard solution of base.

acidity   The state of being acid.

acidity coefficient   The ratio of the oxygen content of the bases in a rock to the oxygen content in the silica. Also known as oxygen ratio.

acid magnesium citrate   *See* dibasic magnesium citrate.

acid manganous phosphate $MnHPO_4 \cdot 3H_2O$ Acid-soluble pink powder. Also known as manganese hydrogen phosphate; secondary manganous phosphate.

acid number   *See* acid value.

acidolysis   A chemical reaction involving the decomposition of a molecule, with the addition of the elements of an acid to the molecule; the reaction is comparable to hydrolysis or alcoholysis, in which water or alcohol, respectively, is used in place of the acid. Also known as acyl exchange.

acid phosphatase   An enzyme in blood which catalyzes the release of phosphate from phosphate esters; optimum activity at pH 5.

acid phosphate   A mono- or dihydric phosphate; for example, M_2HPO_4 or MH_2PO_4, where M represents a metal atom.

acid phthalic anhydride   *See* phthalic anhydride.

acid potassium phthalate   *See* potassium biphthalate.

acid potassium sulfate   *See* potassium bisulfate.

acid reaction   A chemical reaction produced by an acid.

acid recovery plant   In some refineries, a facility for separating sludge acid into acid oil, tar, and weak sulfuric acid, with provision for later reconcentration.

acid salt   A compound derived from an acid and base in which only a part of the hydrogen is replaced by a basic radical; for example, the acid sulfate $NaHSO_4$.

acid sludge   The residue left after treating petroleum oil with sulfuric acid for the removal of impurities.

acid sodium tartrate   *See* sodium bitartrate.

acid solution An aqueous solution containing more hydrogen ions than hydroxyl ions.

acid treating A refining process in which unfinished petroleum products, such as gasoline, kerosine, and diesel oil, are contacted with sulfuric acid to improve their color, odor, and other properties.

acid value The acidity of a solution expressed in terms of normality. Also known as acid number.

aconitase An enzyme involved in the Krebs citric acid cycle that catalyzes the breakdown of citric acid to *cis*-aconitic and isocitric acids.

aconitic acid $C_6H_6O_6$ A white, crystalline organic acid found in sugarcane and sugarbeet; obtained during manufacture of sugar.

a-c plane A plane at right angles to the surface of movement in a crystal.

acrasin The chemotactic substance thought to be secreted by, and to effect aggregation of, myxamebas during their fruiting phase.

Acree's reaction A test for protein in which a violet ring appears when concentrated sulfuric acid is introduced below a mixture of the unknown solution and a formaldehyde solution containing a trace of ferric chloride.

acridine $(C_6H_4)_2NCH$ A typical member of a group of organic heterocyclic compounds containing benzene rings fused to the 2,3 and 5,6 positions of pyridine; derivatives include dyes and medicines.

acridine orange A dye with an affinity for nucleic acids; the complexes of nucleic acid and dye fluoresce orange with RNA and green with DNA when observed in the fluorescence microscope.

acriflavine $C_{14}H_{14}N_3Cl$ A yellow acridine dye obtained from proflavine by methylation in the form of red crystals; used as an antiseptic in solution.

acrolein CH_2—$CHCHO$ A colorless to yellow liquid with a pungent odor and a boiling point of 52.7°C; soluble in water, alcohol, and ether; used in organic synthesis, pharmaceuticals manufacture, and as an herbicide and tear gas. Also known as acrylaldehyde; acrylic aldehyde; 2-propenal.

acrolein cyanohydrin CH_2:$CHCH(OH)CN$ A liquid soluble in water and boiling at 165°C; copolymerizes with ethylene and acrylonitrile; used to modify synthetic resins.

acrolein dimer $C_6H_8O_2$ A flammable, water-soluble liquid used as an intermediate for resins, dyestuffs, and pharmaceuticals. Also known as 2-formyl-3,4-dihydro-2*H*-pyran.

acrolein test A test for the presence of glycerin or fats; a sample is heated with potassium bisulfate, and acrolein is released if the test is positive.

acrosin A proteolytic enzyme located in the acrosome of a spermatozoon; thought to be involved in penetration of the egg.

acrylaldehyde *See* acrolein.

acrylamide $CH_2CHCONH_2$ Colorless, odorless crystals with a melting point of 84.5°C; soluble in water, alcohol, and acetone; used in organic synthesis, polymerization, sewage treatment, ore processing, and permanent press fabrics.

acrylamide copolymer A thermosetting resin formed of acrylamide with other resins, such as the acrylic resins.

acrylate   1. A salt or ester of acrylic acid. 2. *See* acrylate resin.

acrylate resin   Acrylic acid or ester polymer with a —CH_2—CH(COOR)— structure; used in paints, sizings and finishes for paper and textiles, adhesives, and plastics. Also known as acrylate.

acrylic acid $CH_2CHCOOH$ An easily polymerized, colorless, corrosive liquid used as a monomer for acrylate resins.

acrylic aldehyde   *See* acrolein.

acrylic ester   An ester of acrylic acid.

acrylic resin   A thermoplastic synthetic organic polymer made by the polymerization of acrylic derivatives such as acrylic acid, methacrylic acid, ethyl acrylate, and methyl acrylate; used for adhesives, protective coatings, and finishes.

acrylic rubber   Synthetic rubber containing acrylonitrile; for example, nitrile rubber.

acrylonitrile CH_2CHCN A colorless liquid compound used in the manufacture of acrylic rubber and fibers. Also known as vinylcyanide.

acrylonitrile butadiene styrene resin   A polymer made by blending acrylonitrile-styrene copolymer with a butadiene-acrylonitrile rubber or by interpolymerizing polybutadiene with styrene and acrylonitrile; combines the advantages of hardness and strength of the vinyl resin component with the toughness and impact resistance of the rubbery component. Abbreviated ABS resin.

acrylonitrile copolymer   Oil-resistant synthetic rubber made by polymerization of acrylonitrile with compounds such as butadiene or acrylic acid.

ACTH   *See* adrenocorticotropic hormone.

actin   A muscle protein that is the chief constituent of the Z-band myofilaments of each sarcomere.

actinide series   The group of elements of atomic number 89 through 103. Also known as actinoid elements.

actinism   The production of chemical changes in a substance upon which electromagnetic radiation is incident.

actinium   A radioactive element, symbol Ac, atomic number 89; its longest-lived isotope is ^{227}Ac with a half-life of 21.7 years; the element is trivalent; chief use is, in equilibrium with its decay products, as a source of alpha rays.

actinium emanation   *See* actinon.

actinochemistry   A branch of chemistry concerned with chemical reactions produced by light or other radiation.

actinochitin   A form of birefringent or anisotropic chitin found in the seta of certain mites.

actinon   A radioactive isotope of radon, symbol An, atomic number 86, atomic weight 219, belonging to the actinium series. Also known as actinium emanation (Ac-Em).

actinouranium   A naturally occurring radioactive isotope of the actinium series, emitting only alpha decay; symbol AcU; atomic number 92; mass number 235; half-life 7.1 × 10^8 years; isotopic symbol ^{235}U.

activated complex   An energetically excited state which is intermediate between reactants and products in a chemical reaction.

activation   Treatment of a substance by heat, radiation, or activating reagent to produce a more complete or rapid chemical or physical change.

activation energy   The energy, in excess over the ground state, which must be added to an atomic or molecular system to allow a particular process to take place.

activator   1. A substance that increases the effectiveness of a rubber vulcanization accelerator; for example, zinc oxide or litharge. 2. A trace quantity of a substance that imparts luminescence to crystals; for example, silver or copper in zinc sulfide or cadmium sulfide pigments.

active amyl alcohol   *See* 2-methyl-1-butanol.

active element   A chemical element which has one or more radioactive isotopes.

activity   1. The intensity of a radioactive source. Also known as radioactivity. 2. A thermodynamic function that correlates changes in the chemical potential with changes in experimentally measurable quantities, such as concentrations or partial pressures, through relations formally equivalent to those for ideal systems.

activity coefficient   A characteristic of a quantity expressing the deviation of a solution from ideal thermodynamic behavior; often used in connection with electrolytes.

actol   *See* silver lactate.

actomyosin   A protein complex consisting of myosin and actin; the major constituent of a contracting muscle fibril.

actual cubic feet per minute   A measure of the volume of gas at operating temperature and pressure, as distinct from volume of gas at standard temperature and pressure. Abbreviated acfm.

AcU   *See* actinouranium.

acyclic compound   A chemical compound with an open-chain molecular structure rather than a ring-shaped structure; for example, the alkane series.

acyl   A radical formed from an organic acid by removal of a hydroxyl group; the general formula is RCO, where R may be aliphatic, alicyclic, or aromatic.

acylation   Any process whereby the acyl group is incorporated into a molecule by substitution.

acyl azide   *See* acid azide.

acylcarbene   A carbene radical in which at least one of the groups attached to the divalent carbon is an acyl group; for example, acetylcarbene.

acyl carnitine   *See* fatty acyl carnitine.

acyl carrier protein   A protein in fatty acid synthesis that picks up aceytl and malonyl groups from acetyl coenzyme A and malonyl coenzyme A and links them by condensation to form β-keto acid acyl carrier protein, releasing carbon dioxide and the sulfhydryl form of acyl carrier protein. Abbreviated ACP.

acyl-coenzyme A   *See* fatty acyl-coenzyme A.

acyl exchange *See* acidolysis.

acyl halide One of a large group of organic substances containing the halocarbonyl group; for example, acyl fluoride.

acylnitrene A nitrene in which the nitrogen is covalently bonded to an acyl group.

acyloin An organic compound that may be synthesized by condensation of aldehydes; an example is benzoin, $C_6H_5COCHOHC_6H_5$.

acyloin condensation The reaction of an aliphatic ester with metallic sodium to form intermediates converted by hydrolysis into aliphatic α-hydroxyketones called acyloins.

adamantane A $C_{10}H_{16}$ alicyclic hydrocarbon whose structure has the same arrangement of carbon atoms as does the basic unit of the diamond lattice.

Adam's catalyst Finely divided platinum(IV) oxide, made by fusing hexachloroplatinic(IV) acid with $NaNO_3$.

adamsite *See* phenarsazine chloride.

adatom An atom adsorbed on a surface so that it will migrate over the surface.

addition agent A substance added to a plating solution to change characteristics of the deposited substances.

addition polymer A polymer formed by the chain addition of unsaturated monomer molecules, such as olefins, with one another without the formation of a by-product, as water; examples are polyethylene, polypropylene, and polystyrene. Also known as addition resin.

addition reaction A type of reaction of unsaturated hydrocarbons with hydrogen, halogens, halogen acids, and other reagents, so that no change in valency is observed and the organic compound forms a more complex one.

addition resin *See* addition polymer.

addition solid solution Random addition of atoms or ions in the interstices within a crystal structure.

adduct 1. A chemical compound that forms from chemical addition of two species; for example, reaction of butadiene with styrene forms an adduct, 4-phenyl-1-cyclohexene. 2. The complex compound formed by association of an inclusion complex.

adenase An enzyme that catalyzes the hydrolysis of adenine to hypoxanthine and ammonia.

adenine $C_5H_5N_5$ A purine base, 6-aminopurine, occurring in ribonucleic acid and deoxyribonucleic acid and as a component of adenosinetriphosphate.

adenosine $C_{10}H_{13}N_5O_4$ A nucleoside composed of adenine and D-ribose.

adenosine 3′,5′-cyclic monophosphate *See* cyclic adenylic acid.

adenosine 3′,5′-cyclic phosphate *See* cyclic adenylic acid.

adenosinediphosphatase An enzyme that catalyzes the hydrolysis of adenosinediphosphate. Abbreviated ADPase.

adenosinediphosphate $C_{10}H_{15}N_5O_{10}P_2$ A coenzyme composed of adenosine and two molecules of phosphoric acid that is important in intermediate cellular metabolism. Abbreviated ADP.

adenosinemonophosphate *See* adenylic acid.

adenosinetriphosphatase An enzyme that catalyzes the hydrolysis of adenosinetriphosphate. Abbreviated ATPase.

adenosinetriphosphate $C_{10}H_{16}N_5O_{12}P_3$ A coenzyme composed of adenosinediphosphate with an additional phosphate group; an important energy compound in metabolism. Abbreviated ATP.

adenylic acid 1. A generic term for a group of isomeric nucleotides. 2. The phosphoric acid ester of adenosine. Also known as adenosinemonophosphate (AMP).

adiabatic calorimeter An instrument used to study chemical reactions which have a minimum loss of heat.

adiabatic flame temperature The highest possible temperature of combustion obtained under the conditions that the burning occurs in an adiabatic vessel, that it is complete, and that dissociation does not occur.

adipate Salt produced by reaction of adipic acid with a basic compound.

adipic acid $HOOC(CH_2)_4COOH$ A colorless crystalline dicarboxylic acid, sparingly soluble in water; used in nylon manufacture. Also known as 1,4-butanedicarboxylic acid; hexanedioic acid.

adipinketone *See* cyclopentanone.

adipocellulose A type of cellulose found in the cell walls of cork tissue.

adiponitrile $NC(CH_2)_4CN$ The high-boiling liquid dinitrile of adipic acid; used to make nylon intermediates. Also known as hexanedinitrile.

adjective dye Any dye that needs a mordant.

adonite *See* adonitol.

adonitol $C_5H_{12}O_5$ A pentitol from the dicotyledenous plant *Adonis vernalis;* large crystals that are optically inactive and melt at 102°C; it does not reduce Fehling's solution, and is freely soluble in water and hot alcohol. Also known as adonite; ribitol.

ADP *See* adenosinediphosphate

ADPase *See* adenosinediphosphatase.

adrenal cortex hormone Any of the steroids produced by the adrenal cortex. Also known as adrenocortical hormone; corticoid.

adrenaline *See* epinephrine.

adrenergic blocking agent Any substance that blocks the action of epinephrine or an epinephrine-like substance.

adrenochrome $C_9H_9O_3N$ A brick-red oxidation product of epinephrine which can convert hemoglobin into methemoglobin.

adrenocortical hormone *See* adrenal cortex hormone.

adrenocorticotropic hormone The chemical secretion of the adenohypophysis that stimulates the adrenal cortex. Abbreviated ACTH. Also known as adrenotropic hormone.

adrenotropic hormone *See* adrenocorticotropic hormone.

adsorbate A solid, liquid, or gas which is adsorbed as molecules, atoms, or ions by such substances as charcoal, silica, metals, water, and mercury.

adsorbent A solid or liquid that adsorbs other substances; for example, charcoal, silica, metals, water, and mercury.

adsorption The surface retention of solid, liquid, or gas molecules, atoms, or ions by a solid or liquid, as opposed to absorbtion, the penetration of substances into the bulk of the solid or liquid.

adsorption chromatography Separation of a chemical mixture (gas or liquid) by passing it over an adsorbent bed which adsorbs different compounds at different rates.

adsorption indicator An indicator used in solutions to detect slight excess of a substance or ion; precipitate becomes colored when the indicator is adsorbed. An example is fluorescein.

adsorption isobar A graph showing how adsorption varies with some parameter, such as temperature, while holding pressure constant.

adsorption isotherm The relationship between the gas pressure p and the amount w, in grams, of a gas or vapor taken up per gram of solid at a constant temperature.

aeration cell An electrolytic cell whose electromotive force is due to electrodes of the same material located in different concentrations of dissolved air. Also known as oxygen cell.

aerobic digestion Digestion of matter suspended or dissolved in waste by microorganisms under favorable conditions of oxygenation.

aerogel A porous solid formed from a gel by replacing the liquid with a gas with little change in volume so that the solid is highly porous.

aerosol A gaseous suspension of ultramicroscopic particles of a liquid or a solid.

AES *See* Auger electron spectroscopy.

AET *See* S-(2-aminoethyl))isothiuronium bromide hydrobromide.

affinity chromatography A chromatographic technique that utilizes the ability of biological molecules to bend to certain ligands specifically and reversibly; used in protein biochemistry.

affinity labeling A method for introducing a label into the active site of an enzyme by relying on the tight binding between the enzyme and its substrate (or cofactors).

aflatoxin The toxin produced by some strains of the fungus *Aspergillus flavus*, the most potent carcinogen yet discovered.

afterglow *See* phosphorescence.

Ag *See* silver.

agaric acid $C_{19}H_{36}(OH)(COOH)_3$ An acid with melting point 141°C; soluble in water, insoluble in benzene; used as an irritant. Also known as agaricin.

agaricin *See* agaric acid.

agavose $C_{12}H_{22}O_{11}$ A sugar found in the juice of the agave tree; used in medicine as a diuretic and laxative.

aggregate recoil The ejection of atoms from the surface of a sample as a result of their being attached to one atom that is recoiling as the result of α-particle emission.

aggressive carbon dioxide The carbon dioxide dissolved in water in excess of the amount required to precipitate a specified concentration of calcium ions as calcium carbonate; used as a measure of the corrosivity and scaling properties of water.

aging All irreversible structural changes that occur in a precipitate after it has formed.

aglycon The nonsugar compound resulting from the hydrolysis of glycosides; an example is 3,5,7,3′,4′-pentahydroxyflavylium, or cyanidin.

agmatine $C_5H_{14}N_4$ Needlelike crystals with a melting point of 231°C; soluble in water; a product of the enzymatic decarboxylation of argenine. Also known as 4-(aminobutyl)guanidine.

air A predominantly mechanical mixture of a variety of individual gases forming the earth's enveloping atmosphere.

air-break switch *See* air switch.

Airco-Hoover sweetening Removal of mercaptans from gasoline by caustic and water washes, then heating the dried gasoline and passing it with some oxygen through a reactor containing a slurry of diatomaceous earth impregnated with copper chloride; the oxygen regenerates the catalyst.

air-cure To vulcanize at ordinary room temperatures, or without the aid of heat.

air deficiency Insufficient air in an air-fuel mixture causing either incomplete fuel oxidation or lack of ignition.

air-fuel ratio The ratio of air to fuel by weight or volume which is significant for proper oxidative combustion of the fuel.

Airlift Thermofor catalytic cracking A continuous catalytic process for the conversion of heavy gas oils into lighter products; utilizes a moving bed of small pellets or beads of catalyst traveling continuously through the oil-cracking vessel and then through a regeneration kiln.

air line Lines in a spectrum due to the excitation of air molecules by spark discharges, and not ordinarily present in arc discharges.

air shower *See* cosmic-ray shower.

air-slaked Having the property of a substance, such as lime, that has been at least partially converted to a carbonate by exposure to air.

air-suspension encapsulation A technique for microencapsulation of various types of solid particles; the particles undergo a series of cycles in which they are first suspended by a vertical current of air while they are sprayed with a solution of coating material, and are then moved by the airstream into a region where they undergo a drying treatment. Also known as Wurster process.

air sweetening A process in which air or oxygen is used to oxidize lead mercaptides to disulfides instead of using elemental sulfur.

ajmaline $C_{20}H_{26}N_2O_2$ An amber, crystalline alkaloid obtained from *Rauwolfia* plants, especially *R. serpentina*.

alanine $C_3H_7NO_2$ A white, crystalline, nonessential amino acid of the pyruvic acid family.

alanyl   The radical CH_3CHNH_2CO—;   occurs in, for example, alanyl alanine, a dipeptide.

Alberger process   A method of manufacturing salt by heating brine at high pressure and passing it to a graveler which removes calcium sulfate; the salt crystallizes as the pressure is reduced and thus is separated from the brine.

albite law   A rule specifying the orientation of alternating lamellae in multiple twin feldspar crystals; the twinning plane is brachypinacoid and is common in albite.

albumin   Any of a group of plant and animal proteins which are soluble in water, dilute salt solutions, and 50% saturated ammonium sulfate.

albumin-globulin ratio   The ratio of the concentrations of albumin to globulin in blood serum.

albumose   A protein derivative formed by the action of a hydrolytic enzyme, such as pepsin.

alchemy   A speculative chemical system having as its central aims the transmutation of base metals to gold and the discovery of the philosopher's stone.

alcogel   A gel formed by an alcosol.

alcohol   1. C_2H_5OH A colorless, volatile liquid; boiling point of pure liquid is 78.3°C; it is soluble in water, chloroform, and methyl alcohol; used as solvent and in manufacture of many chemicals and medicines. Also known as ethanol; ethyl alcohol; grain alcohol. 2. Any of a class of organic compounds containing the hydroxyl group, OH.

alcoholate   A compound formed by the reaction of an alcohol with an alkali metal. Also known as alkoxide.

alcohol dehydrogenase   The enzyme that catalyzes the oxidation of ethanol to acetaldehyde.

alcoholysis   The breaking of a carbon-to-carbon bond by addition of an alcohol.

alcosol   Mixture of an alcohol and a colloid.

aldehyde   One of a class of organic compounds containing the CHO radical.

aldehyde ammonia $CH_3CHOHNH_2$ A white, crystalline solid with a melting point of 97°C; soluble in water and alcohol; used in organic synthesis and as a vulcanization accelerator. Also known as acetaldehyde ammonia; 1-aminoethanol.

aldehyde C–9   *See* nonanal.

aldehyde dehydrogenase   An enzyme that catalyzes the conversion of an aldehyde to its corresponding acid.

aldehyde lyase   Any enzyme that catalyzes the nonhydrolytic cleavage of an aldehyde.

aldehyde polymer   Any of the plastics based on aldehydes, such as formaldehyde, acetaldehyde, butyraldehyde, or acrylic aldehyde (acrolein).

aldicarb $C_7H_{14}N_2O_2S$ A colorless, crystalline compound with a melting point of 100°C; used as an insecticide, miticide, and nematicide to treat soil for cotton, sugarbeets, potatoes, peanuts, and ornamentals. Also known as 2-methyl-2(methylthio)propionaldehyde *O*-(methylcarbamoyl)oxime.

aldohexose A hexose, such as glucose or mannose, containing the aldehyde group.

aldol $CH_3CH(OH)CH_2CHO$ A colorless, thick liquid with a boiling point of 83°C; used in manufacturing rubber age resistors, accelerators, and vulcanizers. Also known as 3-hydroxybutanal.

aldolase An enzyme in anaerobic glycolysis that catalyzes the cleavage of fructose 1,6-diphosphate to glyceraldehyde 3-phosphate; used also in the reverse reaction.

aldol condensation Formation of a β-hydroxycarbonyl compound by the condensation of an aldehyde or a ketone in the presence of an acid or base catalyst. Also known as aldol reaction.

aldol reaction *See* aldol condensation.

aldose A class of monosaccharide sugars; the molecule contains an aldehyde group.

aldosterone $C_{21}H_{28}O_5$ A steroid hormone extracted from the adrenal cortex that functions chiefly in regulating sodium and potassium metabolism.

Aldrin $C_{12}H_8Cl_6$ Trade name for a water-insoluble, white, crystalline compound, consisting mainly of chlorinated dimethanonaphthalene; used as a pesticide.

alfin catalyst A catalyst derived from reaction of an alkali alcoholate with an olefin halide; used to convert olefins (for example, ethylene, propylene, or butylenes) into polyolefin polymers.

algin *See* alginic acid, sodium alginate.

alginate One of a class of salts of algin, such as sodium alginate.

algin fibers A polymer of *D*-mannuronic acid.

alginic acid $(C_6H_8O_6)_n$ An insoluble colloidal acid obtained from brown marine algae; it is hard when dry and absorbent when moist. Also known as algin.

alicyclic 1. Having the properties of both aliphatic and cyclic substances. 2. Referring to a class of organic compounds containing only carbon and hydrogen atoms joined to form one or more rings. 3. Any one of the compounds of the alicyclic class. Also known as cyclane.

aliesterase Any one of the lipases or nonspecific esterases.

alignment A population $p(m)$ of the $2I + 1$ orientational substates of a nucleus; $m = -I$ to $+I$, such that $p(m) = p(-m)$.

aliphatic Of or pertaining to any organic compound of hydrogen and carbon characterized by a straight chain of the carbon atoms; three subgroups of such compounds are alkanes, alkenes, and alkynes.

aliphatic acid Any organic acid derived from aliphatic hydrocarbons.

aliphatic acid ester Any organic ester derived from aliphatic acids.

aliphatic polycyclic hydrocarbon A hydrocarbon compound in which at least two of the aliphatic structures are cyclic or closed.

aliphatic polyene compound Any unsaturated aliphatic or alicyclic compound with more than four carbons in the chain and with at least two double bonds; for example, hexadiene.

aliphatic series A series of open-chained carbon-hydrogen compounds; the two major classes are the series with saturated bonds and with the unsaturated.

alizarin $C_{14}H_6O_2(OH)_2$ An orange crystalline compound, insoluble in cold water; made synthetically from anthraquinone; used in the manufacture of dyes and red pigments. Also known as 1,2-dihydroxyanthraquinone.

alizarin dye Sodium salts of sulfonic acids derived from alizarin.

alizarin red Any of several red dyes derived from anthraquinone.

alkalescence The property of a substance that is alkaline, that is, having a pH greater than 7.

alkali Any compound having highly basic qualities.

alkali-aggregate reaction The chemical reaction of an aggregate with the alkali in a cement, resulting in a weakening of the concrete.

alkali alcoholate A compound formed from an alcohol and an alkali metal base; the alkali metal replaces the hydrogen in the hydroxyl group.

alkali blue The sodium salt of triphenylrosanilinesulfonic acid; used as an indicator.

alkalide A member of a class of crystalline salts with an alkali metal atom.

alkali metal Any of the elements of group Ia in the periodic table: lithium, sodium, potassium, rubidium, cesium, and francium.

alkalimeter 1. An apparatus for measuring the quantity of alkali in a solid or liquid. 2. An apparatus for measuring the quantity of carbon dioxide formed in a reaction.

alkalimetry Quantitative measurement of the concentration of bases or the quantity of one free base in a solution; techniques include titration and other analytical methods.

alkaline 1. Having properties of an alkali. 2. Having a pH greater than 7.

alkaline earth An oxide of an element of group IIa in the periodic table, such as barium, calcium, and strontium. Also known as alkaline-earth oxide.

alkaline-earth metals The heaviest members of group IIa in the periodic table; usually calcium, strontium, magnesium, and barium.

alkaline-earth oxide *See* alkaline earth.

alkaline phosphatase A phosphatase active in alkaline media.

alkaline wash The removal of impurities from kerosine, used for illuminating purposes, by caustic soda solution.

alkalinity The property of having excess hydroxide ions in solution.

alkaloid One of a group of nitrogenous bases of plant origin, such as nicotine, cocaine, and morphine.

alkalometry The measurement of the quantity of alkaloids present in a substance.

alkamine A compound that has both the alcohol and amino groups. Also known as amino alcohol.

alkane A member of a series of saturated aliphatic hydrocarbons having the empirical formula C_nH_{2n+2}.

alkannin $C_{16}H_{16}O_5$ A red powder, the coloring ingredient of alkanet; soluble in alcohol, benzene, ether, and oils; used as a coloring agent for fats and oils, wines, and wax.

alkanolamine One of a group of viscous, water-soluble amino alcohols of the aliphatic series.

Alkar process Catalytic alkylation of aromatic hydrocarbons with olefins to produce alkylaromatics; for example, production of ethylbenzene from benzene and ethylene.

alkene One of a class of unsaturated aliphatic hydrocarbons containing one or more carbon-to-carbon double bonds.

alkoxide *See* alcholate.

alkoxy An alkyl radical attached to a molecule by oxygen, such as the ethoxy radical.

alkyd resin A class of adhesive resins made from unsaturated acids and glycerol.

alkyl A monovalent radical, C_nH_{2n+1}, which may be considered to be formed by loss of a hydrogen atom from an alkane; usually designated by R.

alkylamine A compound consisting of an alkyl group attached to the nitrogen of an amine; an example is ethylamine, $C_2H_5NH_2$.

alkylaryl sulfonates General name for alkylbenzene sulfonates.

alkylate A product of the alkylation process in petroleum refining.

alkylate bottom Residue from fractionation of total alkylate which boils at a higher temperature than aviation gasolines.

alkylation 1. A refinery process for chemically combining isoparaffin with olefin hydrocarbons. 2. A chemical process in which an alkyl radical is introduced into an organic compound by substitution or addition.

alkylbenzene sulfonates Widely used nonbiodegradable detergents, commonly dodecylbenzene or tridecylbenzene sulfonates.

alkyldimethylbenzylammoniumchloride *See* benzalkonium chloride.

alkylene An organic radical formed from an unsaturated aliphatic hydrocarbon; for example, the ethylene radical C_2H_3—.

alkyl halide A compound consisting of an alkyl group and a halogen; an example is ethylbromide.

alkyne One of a group of organic compounds containing a carbon-to-carbon triple bond.

allantoic acid $C_4H_8N_4O_4$ A crystalline acid obtained by hydrolysis of allantoin; intermediate product in nucleic acid metabolism.

allantoin $C_4H_6N_4O_3$ A crystallizable oxidation product of uric acid found in allantoic and amniotic fluids and in fetal urine.

allantoinase An enzyme, occurring in nonmammalian vertebrates, that catalyzes the hydrolysis of allantoin.

allantoxanic acid $C_4H_3N_3O_4$ An acid formed by oxidation of uric acid or allantoin.

allanturic acid $C_3H_4N_2O_3$ An acid formed principally by the oxidation of allantoin.

allelochemistry The science of compounds synthesized by one organism that stimulate or inhibit other organisms.

allene C_3H_4 An unsaturated aliphatic hydrocarbon with two double bonds. Also known as propadiene.

Allen-Moore diaphragm cell A rectangular electrolyte diaphragm used early in the history of chlorine manufacture.

allethrin An insecticide, a synthetic pyrethroid, more effective than pyrethrin.

allidochlor $C_8H_{12}NOCl$ An amber liquid with slight solubility in water; used as a preemergence herbicide for vegetable crops, soybeans, sorghum, and ornamentals. Also known as N,N-diallyl-2-chloroacetamide.

Allihn condenser A condenser whose condensing surface is a series of interconnected glass bulbs.

allo- Prefix applied to the stabler form of two isomers.

allobar A form of an element differing in its atomic weight from the naturally occurring form and hence being of different isotopic composition.

allochromatic crystal A crystal having photoconductive properties due to the presence of small particles within it.

allomerism A constancy in crystal form in spite of a variation in chemical composition.

allophanamide *See* biuret.

allosteric effector A small molecule that reacts either with a nonbinding site of an enzyme molecule, or with a protein molecule, and causes a change in the function of the molecule. Also known as allosteric modulator.

allosteric enzyme Any of the regulatory bacterial enzymes, such as those involved in end-product inhibition.

allosteric modulator *See* allosteric effector.

allosteric transition A reversible exchange of one base pair for another on a protein molecule that alters the properties of the active site and changes the biological activity of the protein.

allostery The property of an enzyme able to shift reversibly between an active and an inactive configuration.

allotrope A form of an element or other substance showing allotropy.

allotropy The assumption by an element or other substance of two or more different forms or structures which are most frequently stable in different temperature ranges, such as different crystalline forms of carbon as charcoal, graphite, or diamond. Also known as allotriomorphism; allotropism.

alloxan $C_4H_2N_2O_4$ Crystalline oxidation product of uric acid; induces diabetes experimentally by selective destruction of pancreatic beta cells. Also known as mesoxalylurea.

allulose $CH_2OHCO(CHOH)_3CH_2OH$ A constituent of cane sugar molasses; it is nonfermentable. Also known as D-piscose; D-ribo-2-ketohexose.

allycarbamide   *See* allylurea.

allyl-   A prefix used in names of compounds whose structure contains an allyl cation.

allylacetone $CH_2CHCH_2CH_2COCH_3$ A colorless liquid, soluble in water and organic solvents; used in pharmaceutical synthesis, perfumes, fungicides, and insecticides.

allyl alcohol CH_2CHCH_2OH Colorless, pungent liquid, boiling at 96°C; soluble in water; made from allyl chloride by hydrolysis. Also known as 2-propen-1-ol.

allylamine $CH_2CHCH_2NH_2$ A yellow oil that is miscible with water; boils at 58°C; prepared from mustard oil. Also known as 2-propen-1-amine.

allyl bromide C_3H_5Br A colorless to light yellow, irritating toxic liquid with a boiling point of 71.3°C; soluble in organic solvents; used in organic synthesis and for the manufacture of synthetic perfumes. Also known as bromoallylene; 3-bromo-1-propene.

allyl cation   A carbonium cation with a structure usually represented as $CH_2{=}CH{-}CH_2{}^+$; attachment site is the saturated carbon atom.

allyl chloride CH_2CHCH_2Cl A volatile, pungent, toxic, flammable, colorless liquid, boiling at 46°C; insoluble in water; made by chlorination of propylene at high temperatures. Also known as 3-chloro-1-propene.

allyl cyanide C_4H_5N A liquid with an onionlike odor and a boiling point of 119°C; slightly soluble in water; used as a cross-linking agent in polymerization. Also known as 3-butenenitrile; vinylacetonitrile.

allylene $CH_3C{:}CH$ An acetylenic, three-carbon hydrocarbon; a colorless gas boiling at -24°C; soluble in ether. Also known as methyl acetylene; propine; propyne.

allylic rearrangement   In a three-carbon molecule, the shifting of a double bond from the 1,2 carbon position to the 2,3 position, with the accompanying migration of an entering substituent or substituent group from the third carbon to the first.

allyl isosulfocyanate   *See* allyl isothiocyanate.

allyl isothiocyanate $CH_2CH{:}CH_2NCS$ A pungent, colorless to pale-yellow liquid; soluble in alcohol, slightly soluble in water; irritating odor; boiling point 152°C; used as a fumigant and as a poison gas. Also known as allyl isosulfocyanate; mustard oil.

allyl mercaptan CH_2CHCH_2SH A colorless liquid with a boiling point of 67–68°C; soluble in ether and alcohol; used as intermediate in pharmaceutical manufacture. Also known as allyl thiol; 2-propene-1-thiol.

allyl plastic   *See* allyl resin.

allyl resin   Any of a class of thermosetting synthetic resins derived from esters of allyl alcohol or allyl chloride; used in making cast and laminated products. Also known as allyl plastic.

allyl sulfide $(CH_2CHCH_2)_2S$ A colorless liquid with a garliclike odor and a boiling point of 139°C; used in synthetic oil of garlic. Also known as diallyl sulfide; oil garlic; thioallyl ether.

allylsulfocarbamide *See* allylthiourea.

allyl sulfourea *See* allylthiourea.

allyl thiol *See* allyl mercaptan.

allylthiourea $C_3H_5NHCSNH_2$ A white, crystalline solid that melts at 78°C; soluble in water; used as a corrosion inhibitor. Also known as allylsulfocarbamide; allyl sulfourea; thiosinamine.

allyltrichlorosilane $CH_2CHCH_2SiCl_3$ A pungent, colorless liquid with a boiling point of 117.5°C; used as an intermediate for silicones.

allylurea $C_4H_8N_2O$ Crystals with a melting point of 85°C; freely soluble in water and alcohol; used to manufacture allylthiourea and other corrosion inhibitors. Also known as allylcarbamide.

allyxycarb $C_{16}H_{22}N_2O_2$ A yellow, crystalline compound used as an insecticide for fruit orchards, vegetable crops, rice, and citrus. Also known as 4-diallylamino-3,5-xylyl-*N*-methylcarbamate.

alpha cellulose A highly refined, insoluble cellulose from which sugars, pectin, and other soluble materials have been removed. Also known as chemical cellulose.

alpha decay A radioactive transformation in which an alpha particle is emitted by a nuclide.

alpha emission Ejection of alpha particles from the atom's nucleus.

alpha globulin A heterogeneous fraction of serum globulins containing the proteins of greatest electrophoretic mobility.

alpha olefin An olefin where the unsaturation (double bond) is at the alpha position, that is, between the two end carbons of the carbon chain.

alpha particle A positively charged particle consisting of two protons and two neutrons, identical with the nucleus of the helium atom; emitted by several radioactive substances.

alpha-particle scattering Deviation at various angles of a stream of alpha particles passing through a foil of material.

alpha position In chemical nomenclature, the position of a substituting group of atoms in the main group of a molecule; for example, in a straight-chain compound such as α-hydroxypropionic acid ($CH_3CHOHCOOH$), the hydroxyl radical is in the alpha position.

alternating copolymer A polymer formed of two different monomer molecules that alternate in sequence in the polymer chain.

alternation of multiplicities law The law that the periodic table arranges the elements in such a sequence that their number of orbital electrons, and hence their multiplicities, alternates between even and odd numbers.

alum 1. Any of a group of double sulfates of trivalent metals such as aluminum, chromium, or iron and a univalent metal such as potassium or sodium. 2. *See* aluminum sulfate; ammonium aluminum sulfate; potassium aluminum sulfate.

alumina Al_2O_3 The native form of aluminum oxide occurring as corundum or in hydrated forms, as a powder or crystalline substance.

aluminate   A negative ion usually assigned the formula AlO_2^- and derived from aluminum hydroxide.

alumina trihydrate $Al_2O_3 \cdot 3H_2O$, or $Al(OH)_3$ A white powder; insoluble in water, soluble in hydrochloric or sulfuric acid or sodium hydroxide; used in the manufacture of ceramic glasses and in paper coating. Also known as aluminum hydrate; aluminum hydroxide; hydrated alumina; hydrated aluminum oxide.

aluminium   *See* aluminum.

aluminon $C_{22}H_{23}N_3O_9$ A yellowish-brown, glassy powder that is freely soluble in water; used for the detection and colorimetric estimation of aluminum in foods, water, and tissues, and as a pharyngeal aerosol spray. Also known as aurintricarboxylic acid triammonium salt.

aluminosilicate $3Al_2O_3 \cdot 2SiO_2$ A colorless, crystalline combination of silicate and aluminate in the form of rhombic crystals.

aluminum   A chemical element, symbol Al, atomic number 13, and atomic weight 26.9815. Also spelled aluminium.

aluminum acetate $Al(CH_3COO)_3$ A white, amorphous powder that is soluble in water; used in aqueous solution as an antiseptic.

aluminum ammonium sulfate   *See* ammonium aluminum sulfate.

aluminum borohydride $Al(BH_4)_3$ A volatile liquid with a boiling point of 44.5°C; used in organic synthesis and as a jet fuel additive.

aluminum chloride $AlCl_3$ or Al_2Cl_6 A deliquescent compound in the form of white to colorless hexagonal crystals; fumes in air and reacts explosively with water; used as a catalyst.

aluminum fluoride $AlF_3 \cdot 3\frac{1}{2}H_2O$ A white, crystalline powder, insoluble in cold water.

aluminum fluosilicate $Al_2(SiF_6)_3$ A white powder that is soluble in hot water; used for artificial gems, enamels, and glass. Also known as aluminum silicofluoride.

aluminum halide   A compound of aluminum with a halogen element, such as aluminum chloride.

aluminum hydrate   *See* alumina trihydrate.

aluminum hydroxide   *See* alumina trihydrate.

aluminum monostearate $Al(OH)_2[OOC(CH_2)_{16}CH_3]$ A white to yellowish-white powder with a melting point of 155°C; used in the manufacture of medicine, paint, and ink, in waterproofing, and as a plastics stabilizer.

aluminum nitrate $Al(NO_3)_3 \cdot 9H_2O$ White, deliquescent crystals with a melting point of 73°C; soluble in alcohol and acetone; used as a mordant for textiles, in leather tanning, and as a catalyst in petroleum refining.

aluminum oleate   A soaplike compound of aluminum and oleic acid, used in lubricating oils and greases to improve their viscosity.

aluminum orthophosphate $AlPO_4$ White crystals, melting above 1500°C; insoluble in water, soluble in acids and bases; useful in ceramics, paints, pulp, and paper. Also known as aluminum phosphate.

aluminum oxide Al_2O_3 A compound in the form of a white powder or colorless hexagonal crystals; melts at 2020°C; insoluble in water; used in aluminum production, paper, spark plugs, absorbing gases, light bulbs, artificial gems, and manufacture of abrasives, refractories, ceramics, and electrical insulators.

aluminum palmitate $Al(C_{16}H_{31}O_2)\cdot H_2O$ An aluminum soap used in waterproofing fabrics, paper, and leather and as a drier in paints.

aluminum phosphate *See* aluminum orthophosphate.

aluminum potassium sulfate *See* potassium aluminum sulfate.

aluminum silicate $Al_2(SiO_3)_3$ A white solid that is insoluble in water; used as a refractory in glassmaking.

aluminum silicofluoride *See* aluminum fluosilicate.

aluminum soap Any of various salts of higher carboxylic acids and aluminum that are insoluble in water and soluble in oils; used in lubricating greases, paints, varnishes, and waterproofing substances.

aluminum sodium sulfate $AlNa(SO_4)_2\cdot 12H_2O$ Colorless crystals with an astringent taste and a melting point of 61°C; soluble in water; used as a mordant and for waterproofing textiles, as a food additive, and for matches, tanning, ceramics, engraving, and water purification. Abbreviated SAS. Also known as porous alum; soda alum; sodium aluminum sulfate.

aluminum stearate $Al(C_{17}H_{35}COO)_3$ An aluminum soap in the form of a white powder that is insoluble in water and soluble in oils; used for waterproofing fabrics and concrete and as a drier in paints and varnishes.

aluminum sulfate $Al_2(SO_4)_3\cdot 18H_2O$ A colorless salt in the form of monoclinic crystals that decompose in heat and are soluble in water; used in papermaking, water purification, and tanning, and as a mordant in dyeing. Also known as alum.

aluminum triacetate $Al(C_2H_3O_2)_3$ A white solid, very slightly soluble in cold water.

Am *See* americium; ammonium.

ambident A reagent or substrate which can have two or more attacking sites.

American melting point A temperature 3°F (1.7°C) higher than the ASTM Method D87 paraffin-wax melting point.

americium A chemical element, symbol Am, atomic number 95; the mass number of the isotope with the longest half-life is 243.

americyl ion A dioxo monocation of americium, with the formula $(AmO_2)^-$.

amicron A particle having a size of 10^{-7} centimeter or less, which is a size in a system of classification of particle sizes in colloid chemistry.

amidase Any enzyme that catalyzes the hydrolysis of nonpeptide C=N linkages.

amidation The process of forming an amide; for example, in the laboratory benzyl reacts with methyl amine to form *N*-methylbenzamide.

amide One of a class of organic compounds containing the $CONH_2$ radical.

amidine A compound which contains the radical $CNHNH_2$.

amido Indicating the NH_2 radical when it is present in a molecule with the CO radical.

amidocarbonic acid *See* carbamic acid.

amidohydrolase An enzyme that catalyzes deamination.

Amidol $C_6H_3(NH_2)_2OH \cdot HCl$ A grayish-white crystalline salt; soluble in water, slightly soluble in alcohol; used as a developer in photography and as an analytical reagent. Also known as 2,4-diaminophenol hydrochloride.

amin- *See* amino-.

amination 1. The preparation of amines. 2. A process in which the amino group ($=NH_2$) is introduced into organic molecules.

amine One of a class of organic compounds which can be considered to be derived from ammonia by replacement of one or more hydrogens by organic radicals.

amine oxidase An enzyme that catalyzes the oxidation of tyramine and tryptamine to aldehyde.

amino- Having the property of a compound in which the group NH_2 is attached to a radical other than an acid radical. Also spelled amin-.

aminoacetic acid *See* glycine.

amino acid Any of the organic compounds that contain one or more basic amino groups and one or more acidic carboxyl groups and that are polymerized to form peptides and proteins; only 20 of the more than 80 amino acids found in nature serve as building blocks for proteins; examples are tyrosine and lysine.

amino alcohol *See* alkamine.

1-aminoanthraquinone $C_{14}H_9NO_2$ Ruby-red crystals with a melting point of 250°C; freely soluble in alcohol, benzene, chloroform, ether, glacial acetic acid, and hydrochloric acid; used in the manufacture of dyes and pharmaceuticals.

1-aminobutane *See* n-butyl amine.

2-amino-1-butanol $CH_3CH_2CH(NH_2)\,CH_2OH$ A liquid miscible with water, soluble in alcohols; used in the synthesis of surface-active agents, vulcanizing accelerators, and pharmaceuticals, and as an emulsifying agent for such products as cosmetic creams and lotions. Also known as 2-amino-n-butyl alcohol.

2-amino-n-butyl alcohol *See* 2-amino-1-butanol.

4-(aminobutyl)guanidine *See* agmatine.

α-aminobutyric acid $H_2NCH_2CH_2\,CH_2COOH$ Crystals which are either leaflets or needles, with a melting point of 202°C; thought to be a central nervous system postsynaptic inhibitory transmitter. Abbreviated GABA. Also known as γ-amino-n-butyric acid; piperidic acid.

γ-amino-n-butyric acid *See* γ-aminobutyric acid.

α-aminocaproic acid $C_6H_{13}NO_2$ Crystals with a melting point of 204–206°C; freely soluble in water; used as an antifibrinolytic agent and a spacer for affinity chromatography. Also known as 6-aminohexanoic acid; epsilcapramin.

aminocarb $C_{11}H_{16}N_2O_2$ A tan, crystalline compound with a melting point of 93–94°C; slightly soluble in water; used as an insecticide for control of forest insects and pests of cotton, tomatoes, tobacco, and fruit crops. Also known as 4-(dimethylamino)-*meta*-tolyl-methylcarbamate.

aminodiborane Any compound derived from diborane (B_2H_6) in which one H of the bridge has been replaced by NH_2.

3-amino-2,5-dichlorobenzoic acid $C_7H_5 O_2NCl_2$ A white solid with a melting point of 200–201°C; solubility in water is 700 parts per million at 20°C; used as a preemergence herbicide for soybeans, corn, and sweet potatoes. Also known as chloramben.

***para*-aminodiphenyl** See *para*-biphenylamine.

1-aminoethanol See aldehyde ammonia.

S-(2-aminoethyl)isothiuronium bromide hydrobromide $C_3H_{11}Br_2N_3S$ Hygroscopic crystals with a melting point of 194–195°C; used as a radioprotective agent. Abbreviated AET.

aminoformic acid See carbamic acid.

6-aminohexanoic acid See ε-aminocaproic acid.

α-aminohydrocinnamic acid See phenylalanine.

α-amino-α-iminoethane hydrochloride See acetamidine hydrochloride.

3-aminoisonaphthoic acid See 3-amino-2-naphthoic acid.

2-aminoisovaleric acid See valine.

α-aminoisovaleric acid See valine.

aminomercuric chloride See ammoniated mercury.

aminomethane See methylamine.

2-amino-3-methylbutyric acid See valine.

2-amino-2-methyl-1,3-propanediol $HOCH_2C(CH_3)(NH_2)CH_2OH$ Crystals with a melting point of 109–111°C; soluble in water and alcohol; used in the synthesis of surface-active agents, pharmaceuticals, and vulcanizers, and as an emulsifying agent for cosmetics, leather dressings, polishes, and cleaning compounds.

3-amino-2-naphthoic acid $H_2NC_{10}H_6 COOH$ Yellow crystals in the shape of scales with a melting point of 214°C; soluble in alcohol and ether; used in the determination of copper, nickel, and cobalt. Also known as 3-aminoisonaphthoic acid.

1-amino-2-naphthol-4-sulfonic acid $H_2NC_{10}H_5(OH)SO_3H$ White or gray, needlelike crystals; soluble in hot sodium bisulfite solutions; used in the manufacture of azo dyes. Also known as 1,2,4-acid.

2-amino-5-naphthol-7-sulfonic acid $C_{10}H_5NH_2OHSO_3H$ Gray or white needles that are soluble in hot water; used as a dye intermediate. Also known as 6-amino-1-naphthol-3-sulfonic acid; J acid.

2-amino-8-naphthol-6-sulfonic acid See gamma acid.

6-amino-1-naphthol-3-sulfonic acid   *See* 2-amino-5-naphthol-7-sulfonic acid.

7-amino-1-naphthol-3-sulfonic acid   *See* gamma acid.

amino nitrogen   Nitrogen combined with hydrogen in the amino group. Also known as ammonia nitrogen.

1-aminopentane   *See* n-amylamine.

aminopeptidase   An enzyme which catalyzes the liberation of an amino acid from the end of a peptide having a free amino group.

aminophenol   A type of compound containing the NH_2 and OH groups joined to the benzene ring; examples are *para*-aminophenol and *ortho*-hydroxylaniline.

para-aminophenol p-$HOC_6H_4NO_2$ A phenol in which an amino (—NH_2) group is located on the benzene ring of carbon atoms para (p) to the hydroxyl (—OH) group; used as a photographic developer and as an intermediate in dye manufacture.

α-amino-β-phenylpropionic acid   *See* phenylalanine.

aminopolypeptidase   A proteolytic enzyme that cleaves polypeptides containing either a free amino group or a basic nitrogen atom having at least one hydrogen atom.

aminoprotease   An enzyme that hydrolyzes a protein and unites with its free amino group.

3-aminopyridine   *See* β-aminopyridine.

4-aminopyridine $C_5H_6N_2$ White crystals with a melting point of 158.9°C; soluble in water; used as a repellent for birds. Abbreviated 4-AP.

β-aminopyridine $C_5H_6N_2$ Crystals with a melting point of 64°C; soluble in water, alcohol, and benzene; used in drug and dye manufacture. Also known as 3-aminopyridine.

amino resin   A type of resin prepared by condensation polymerization, with an aldehyde, of a compound containing an amino group.

amino sugar   A monosaccharide in which a nonglycosidic hydroxyl group is replaced by an amino or substituted amino group; an example is D-glucosamine.

amino terminal   The end part of a polypeptide chain which contains a free α-amino group.

2-aminothiazole $C_3H_4N_2S$ Pale-yellow crystals that melt at 92°C; soluble in cold water, slightly soluble in ethyl alcohol; used as an intermediate in the synthesis of sulfathiazole.

aminotoluene   *See* benzylamine.

aminotransferase   *See* transaminase.

aminotriazole $C_2H_4N_4$ Crystals with a melting point of 159°C; soluble in water, methanol, chloroform, and ethanol; used as an herbicide, cotton plant defoliant, and growth regulator for annual grasses and broadleaf and aquatic weeds. Abbreviated ATA. Also known as 3-amino-1H-1,2,4-triazole; amitrole.

3-amino-1H-1,2,4-triazole   *See* aminotriazole.

amitrole   *See* aminotriazole.

ammine   One of a group of complex compounds formed by coordination of ammonia molecules with metal ions.

ammonation   A reaction in which ammonia is added to other molecules or ions by covalent bond formation utilizing the unshared pair of electrons on the nitrogen atom, or through ion-dipole electrostatic interactions.

ammonia NH_3 A colorless gaseous alkaline compound that is very soluble in water, has a characteristic pungent odor, is lighter than air, and is formed as a result of the decomposition of most nitrogenous organic material; used as a fertilizer and as a chemical intermediate.

ammoniac   *See* ammoniacal.

ammoniacal   Pertaining to ammonia or its properties. Also known as ammoniac.

ammonia dynamite   Dynamite with part of the nitroglycerin replaced by ammonium nitrate.

ammonia gum   *See* ammonium aluminum sulfate.

ammonia liquor   Water solution of ammonia, ammonium compounds, and impurities, obtained from destructive distillation of bituminous coal.

ammonia nitrogen   *See* amino nitrogen; imino nitrogen.

ammonia synthesis   Chemical combination of nitrogen and hydrogen gases at high temperature and pressure in the presence of a catalyst to form ammonia.

ammoniated mercuric chloride   *See* ammoniated mercury.

ammoniated mercury $HgNH_2Cl$ A white powder that darkens on light exposure; insoluble in water and alcohol, soluble in ammonium carbonate solutions and in warm acids; used in pharmaceuticals and as a local anti-infective in medicine. Also known as aminomercuric chloride; ammoniated mercuric chloride; ammoniated mercury chloride; ammonobasic mercuric chloride; mercury cosmetic.

ammoniated mercury chloride   *See* ammoniated mercury.

ammoniated ruthenium oxychloride   *See* ruthenium red.

ammoniated superphosphate   A fertilizer containing 5 parts of ammonia to 100 parts of superphosphate.

ammoniation   Treating or combining with ammonia.

ammonia water   A water solution of ammonia; a clear colorless liquid that is basic because of dissociation of NH_4OH to produce hydroxide ions; used as a reagent, solvent, and neutralizing agent.

ammonification   Addition of ammonia or ammonia compounds, especially to the soil.

ammonium   The radical NH_4^+. Abbreviated Am.

ammonium acetate   1. CH_3COONH_4 A normal salt formed by the neutralization of acetic acid with ammonium hydroxide; a white, crystalline, deliquescent material used in solution for the standardization of electrodes for hydrogen ions. 2. $CH_3COONH_4 \cdot CH_3COOH$ An acid salt resulting from the distillation of the neutral salt or from its solution in hot acetic acid; crystallizes in deliquescent needles. **3.** A mixture of the normal and acetic salts; used as a mordant in the dyeing of wool.

ammonium acid carbonate *See* ammonium bicarbonate.

ammonium acid fluoride *See* ammonium bifluoride.

ammonium acid tartrate *See* ammonium betartrate.

ammonium alginate $(C_6H_7O_6 \cdot NH_4)_n$ A high-molecular-weight, hydrophilic colloid; used as a thickening agent/stabilizer in ice cream, cheese, canned fruits, and other food products.

ammonium alum *See* ammonium aluminum sulfate.

ammonium aluminum sulfate $NH_4Al(SO_4)_2 \cdot 12H_2O$ Colorless, odorless crystals that are soluble in water; used in manufacturing medicines and baking powder, dyeing, papermaking, and tanning. Also known as alum; aluminum ammonium sulfate; ammonia alum; ammonium alum.

ammonium benzoate $NH_4C_7H_5O_2$ A salt of benzoic acid prepared as a coarse, white powder; used as a preservative in certain adhesives and rubber latex.

ammonium bicarbonate NH_4HCO_3 White, crystalline, water-soluble salt; used in baking powders and in fire-extinguishing mixtures. Also known as ammonium acid carbonate; ammonium hydrogen carbonate.

ammonium bichromate *See* ammonium dichromate.

ammonium bifluoride $NH_4F \cdot HF$ A salt that crystallizes in the orthorhombic system and is soluble in water; prepared in the form of white flakes from ammonia treated with hydrogen fluoride; used in solution as a fungicide and wood preservative. Also known as ammonium acid fluoride; ammonium hydrogen fluoride.

ammonium bitartrate $NH_4HC_4H_4O_6$ Colorless crystals that are soluble in water; used to make baking powder and to detect calcium. Also known as acid ammonium tartrate; ammonium acid tartrate; ammonium hydrogen tartrate; monoammonium tartrate.

ammonium borate NH_4BO_3 A white, crystalline, water-soluble salt which decomposes at 198°C; used as a fire retardant on fabrics.

ammonium bromide NH_4Br An ammonium halide that crystallizes in the cubic system; made by the reaction of ammonia with hydrobromic acid or bromine; used in photography and for pharmaceutical preparations (sedatives).

ammonium carbamate $NH_4NH_2CO_2$ A salt that forms colorless, rhombic crystals, which are very soluble in cold water; an important, unstable intermediate in the manufacture of urea; found in commercial ammonium carbonate.

ammonium carbonate **1.** $(NH_4)_2CO_3$ The normal ammonium salt of carbonic acid, prepared by passing gaseous carbon dioxide into an aqueous solution of ammonia and allowing the vapors (ammonia, carbon dioxide, water) to crystallize. **2.** $NH_4HCO_3 \cdot NH_2COONH_4$ A white, crystalline double salt of ammonium bicarbonate and ammonium carbamate obtained commercially; the principal ingredient of smelling salts.

ammonium chloride NH_4Cl A white crystalline salt that occurs naturally as a sublimation product of volcanic action or is manufactured; used as an electrolyte in dry cells, as a flux for soldering, tinning, and galvanizing, and as an expectorant.

ammonium chromate $(NH_4)_2CrO_4$ A salt that forms yellow, monoclinic crystals; made from ammonium hydroxide and ammonium dichromate; used in photography as a sensitizer for gelatin coatings.

ammonium citrate $(NH_4)_2HC_6H_5O_7$ White, granular material; used as a reagent.

ammonium dichromate $(NH_4)_2Cr_2O_7$ A salt that forms orange, monoclinic crystals; made from ammonium sulfate and sodium dichromate; soluble in water and alcohol; ignites readily; used in photography, lithography, pyrotechnics, and dyeing. Also known as ammonium bichromate.

ammonium fluoride NH_4F A white, unstable, crystalline salt with a strong odor of ammonia; soluble in cold water; used in analytical chemistry, glass etching, and wood preservation, and as a textile mordant.

ammonium fluosilicate $(NH_4)_2SiF_6$ A toxic, white, crystalline powder; soluble in alcohol and water; used for mothproofing, glass etching, and electroplating. Also known as ammonium silicofluoride.

ammonium formate HCO_2NH_4 Deliquescent crystals or granules with a melting point of 116°C; soluble in water and alcohol; used in analytical chemistry to precipitate base metals from salts of the noble metals.

ammonium gluconate $NH_4C_6H_{11}O_7$ A white, crystalline powder made from gluconic acid and ammonia; soluble in water; used as an emulsifier for cheese and salad dressing and as a catalyst in textile printing.

ammonium halide A compound with the ammonium ion bonded to an ion formed from one of the halogen elements.

ammonium hydrogen carbonate *See* ammonium bicarbonate.

ammonium hydrogen flouride *See* ammonium bifluoride.

ammonium hydrogen tartrate *See* ammonium bitartrate.

ammonium hydroxide NH_4OH A hydrate of ammonia, crystalline below $-79°C$; it is a weak base known only in solution as ammonia water. Also known as aqua ammonia.

ammonium iodide NH_4I A salt prepared from ammonia and hydrogen iodide or iodine; it forms colorless, regular crystals which sublime when heated; used in photography and for pharmaceutical preparations.

ammonium lactate $NH_4C_3H_5O_3$ A yellow, syrupy liquid used in finishing leather.

ammonium lineolate $C_{17}H_{31}COONH_4$ A soft, pasty material used as an emulsifying agent in various industrial applications.

ammonium metatungstate $(NH_4)_6H_2W_{12}O_{40}$ A white powder, soluble in water, used for electroplating.

ammonium molybdate $(NH_4)_2MoO_4$ White, crystalline salt used as an analytic reagent, as a precipitant of phosphoric acid, and in pigments.

ammonium nickel sulfate *See* nickel ammonium sulfate.

ammonium nitrate NH_4NO_3 A colorless crystalline salt; very insensitive and stable high explosive; also used as a fertilizer.

ammonium oxalate $(NH_4)_2C_2O_4 \cdot H_2O$ A salt in the form of colorless, rhombic crystals.

ammonium perchlorate NH_4ClO_4 A salt that forms colorless or white rhombic and regular crystals, which are soluble in water; it decomposes at 150°C, and the reaction is explosive at higher temperatures.

ammonium persulfate $(NH_4)_2S_2O_8$ White crystals which decompose on melting; soluble in water; used as an oxidizing agent and bleaching agent, and in etching, electroplating, food preservation, and aniline dyes.

ammonium phosphate $(NH_4)_2HPO_4$ A salt of ammonia and phosphoric acid that forms white monoclinic crystals, which are soluble in water; used as a fertilizer and fire retardant.

ammonium picrate $NH_4C_6H_2O(NO_2)_3$ Compound with stable yellow and metastable red forms of orthorhombic crystals; used as a military explosive for armor-piercing shells. Also known as ammonium trinitrophenolate; explosive D.

ammonium salt   A product of a reaction between ammonia and various acids; examples are ammonium chloride and ammonium nitrate.

ammonium silicofluoride   *See* ammonium fluosilicate.

ammonium soap   A product from reaction of a fatty acid with ammonium hydroxide; used in toiletry preparations such as soaps and in emulsions.

ammonium stearate $C_{17}H_{35}COONH_4$ A tan, waxlike substance with a melting point of 73–75°C; used in cosmetics and for waterproofing cements, paper, textiles, and other materials.

ammonium sulfamate $NH_4OSO_2NH_2$ White crystals with a melting point of 130°C; soluble in water; used for flameproofing textiles, in electroplating, and as an herbicide to control woody plant species.

ammonium sulfate $(NH_4)_2SO_4$ Colorless, rhombic crystals which melt at 140°C and are soluble in water.

ammonium sulfide $(NH_4)_2S$ Yellow crystals, stable only when dry and below 0°C; decomposes on melting; soluble in water and alcohol; used in photographic developers and for coloring brasses and bronzes.

ammonium sulfocyanate   *See* ammonium thiocyanate.

ammonium tartrate $C_4H_{12}N_2O_6$ Colorless, monoclinic crystals; used in textiles and in medicine.

ammonium thiocyanate NH_4SCN Colorless, deliquescent crystals with a melting point of 149.6°C; soluble in water, acetone, alcohol, and ammonia; used in analytical chemistry, freezing solutions, fabric dyeing, electroplating, photography, and steel pickling. Also known as ammonium sulfocyanate; ammonium sulfocyanide.

ammonium trinitrophenolate   *See* ammonium picrate.

ammonium vanadate NH_4VO_3 A white to yellow, water-soluble, crystalline powder; used in inks and as a paint drier and textile mordant.

ammonobasic mercuric chloride   *See* ammoniated mercury.

ammonolysis   **1.** A dissociation reaction of the ammonia molecule producing H^+ and NH_2^- species. **2.** Breaking of a bond by addition of ammonia.

ammonoxidation   *See* ammoxidation.

ammoxidation A process in which mixtures of propylene, ammonia, and oxygen are converted in the presence of a catalyst, with acrylonitrile as the primary product. Also known as ammonoxidation; oxyamination.

para-**amoxyphenol** *See para*-pentyloxyphenol.

AMP *See* adenylic acid.

3′,5′-AMP *See* cyclic adenylic acid.

amperometric titration A titration that involves measuring an electric current or changes in current during the course of the titration.

amperometry Chemical analysis by techniques which involve measuring electric currents.

amphibolic pathway A microbial biosynthetic and energy-producing pathway, such as the glycolytic pathway.

amphipatic molecule A molecule having both hydrophilic and hydrophobic groups; examples are wetting agents and membrane lipids such as phosphoglycerides.

amphiphile A molecule which has a polar head attached to a long hydrophobic tail.

amphiprotic *See* amphoteric.

ampholyte An amphoteric electrolyte.

ampholytic detergent A detergent that is cationic in acidic solutions and anionic in basic solutions.

amphoteric Having both acidic and basic characteristics. Also known as amphiprotic.

amphoterism The property of being able to react either as an acid or a base.

ampyrone *See* 4-aminoantipyrine.

amygdalin $C_6H_5CH(CN)OC_{12}H_{21}O_{10}$ A glucoside occurring in the kernels of certain plants of the genus *Prunus*.

amyl Any of the eight isomeric arrangements of the radical C_5H_{11} or a mixture of them. Also known as pentyl.

amyl acetate $CH_3COO(CH_2)_2CH(CH_3)_2$ A colorless liquid, boiling at 142°C; soluble in alcohol and ether, slightly soluble in water; used in flavors and perfumes. Also known as banana oil; isoamyl acetate.

sec-n-**amyl alcohol** *See* diethyl carbinol.

amyl alcohol 1. A colorless liquid that is a mixture of isomeric alcohols. 2. An optically active liquid composed of isopentyl alcohol and active amyl alcohol.

amyl aldehyde *See n*-valeraldehyde.

n-**amylamine** $C_5H_{11}NH_2$ A colorless liquid with a boiling point of 104.4°C; soluble in water, alcohol, and ether; used in dyestuffs, insecticides, synthetic detergents, corrosion inhibitors, and pharmaceuticals, and as a gasoline additive. Also known as 1-aminopentane; pentylamine.

amylase An enzyme that hydrolyzes reserve carbohydrates, starch in plants and glycogen in animals.

amyl benzoate *See* isoamyl benzoate.

amyl carbinol *See* hexyl alcohol.

amylene C_5H_{10} A highly flammable liquid with a low boiling point, 37.5–38.5°C; often a component of petroleum. Also known as 2-methyl-2-butene.

amyl ether **1.** Either of two isomeric compounds, n-amyl ether or isoamyl ether; both may be represented by the formula $(C_5H_{11})_2O$. **2.** A mixture mainly of isoamyl ether and n-amyl ether formed in preparation of amyl alcohols from amyl chloride; very slightly soluble in water; used mainly as a solvent.

amyl hydrosulfide *See* amyl mercaptan.

amyl mercaptan $C_5H_{11}SH$ A colorless to light yellow liquid with a boiling range of 104–130°C; soluble in alcohol; used in odorant for detecting gas line leaks. Also known as amyl hydrosulfide; pentanethiol.

amyl nitrate $C_5H_{11}ONO_2$ An ester of amyl alcohol to diesel fuel to raise the cetane number.

amyl nitrite $(CH_3)_2CH(CH_2)_2NO_2$ A yellow liquid; soluble in alcohol, very slightly soluble in water; fruity odor; it is flammable and the vapor is explosive; used in medicine and perfumes. Also known as isoamyl nitrite.

amylolysis The enzyme-catalyzed hydrolysis of starch to soluble products.

amylolytic enzyme A type of enzyme capable of denaturing starch molecules; used in textile manufacture to remove starch added to slash sizing agents.

amylopectin A highly branched, high-molecular-weight carbohydrate polymer composed of about 80% corn starch.

amylopsin An enzyme in pancreatic juice that acts to hydrolyze starch into maltose.

amylose A linear starch polymer.

amyl propionate $CH_3CH_2COOC_5H_{11}$ A colorless liquid with an applelike odor and a distillation range of 135–175°C; used in perfumes, lacquers, and flavors.

amyl salicylate $C_6H_4OHCOOC_5H_{11}$ A clear liquid that occasionally has a yellow tinge; boils at 280°C; soluble in alcohol, insoluble in water; used in soap and perfumes. Also known as isoamyl salicylate.

amyl xanthate A salt formed by replacing the hydrogen attached to the sulfur in amylxanthic acid by a metal; used as collector agent in the flotation of certain minerals.

An *See* actinon.

anabasine A colorless, liquid alkaloid extracted from the plants *Anabasis aphylla* and *Nicotiana glauca*; boiling point is 105°C; soluble in alcohol and ether; used as an insecticide. Also known as neonicotine.

anabolic Pertaining to anabolism.

anabolism A part of metabolism involving the union of smaller molecules into larger molecules; the method of synthesis of tissue structure.

anaerobic glycolysis A metabolic pathway in plants by which, in the absence of oxygen, hexose is broken down to lactic acid and ethanol with some adenosinetriphosphate synthesis.

anakinesis   A process in living organisms by which energy-rich molecules, such as adenosinetriphosphate, are formed.

analysis   The determination of the composition of a substance.

analyte   The sample being analyzed.

analytical chemistry   The branch of chemistry dealing with techniques which yield any type of information about chemical systems.

analytical distillation   Precise resolution of a volatile liquid mixture into its components; the mixture is vaporized by heat or vacuum, and the vaporized components are recondensed into liquids at their respective boiling points.

analytical extraction   Precise transfer of one or more components of a mixture (liquid to liquid, gas to liquid, solid to liquid) by contacting the mixture with a solvent in which the component of interest is preferentially soluble.

analyzing power   In a nuclear scattering process, a measure of the effect on scattering cross sections of changes in the polarization of the beam or target nuclei.

anaphoresis   Upon application of an electric field, the movement of positively charged colloidal particles or macromolecules suspended in a liquid toward the anode.

anchimeric assistance   The participation by a neighboring group in the rate-determining step of a reaction; most often encountered in reactions of carbocation intermediates.

androgen   A class of steroid hormones produced in the testis and adrenal cortex which act to regulate masculine secondary sexual characteristics.

androsin $C_{15}H_{20}O_8$ A glucoside found in the herb *Apocynum androsaemifolium;* yields glucose and acetovanillone on hydrolysis.

androstane $C_{19}H_{32}$ The parent steroid hydrocarbon for all androgen hormones. Also known as etioallocholane.

androstenedione $C_{19}H_{26}O_2$ Any one of three isomeric androgens produced by the adrenal cortex.

androsterone $C_{19}H_{30}O_2$ An androgenic hormone occurring as a hydroxy ketone in the urine of men and women.

anethole $C_{10}H_{12}O$ White crystals that melt at 22.5°C; very slightly soluble in water; affected by light; odor resembles oil of anise; used in perfumes and flavors, and as a sensitizer in color-bleaching processes in color photography. Also known as anise camphor; 1-methoxy-4-propenyl benzene; *para*-methoxypropenylbenzene; *para*-propenylanisole.

aneurine   *See* thiamine.

angiotensin   A decapeptide hormone that influences blood vessel constriction and aldosterone secretion by the adrenal cortex. Also known as hypertensin.

angle-resolved photoelectron spectroscopy   A type of photoelectron spectroscopy which measures the kinetic energies of photoelectrons emitted from a solid surface and the angles at which they are emitted relative to the surface. Abbreviated ARPES.

anhydrase   Any enzyme that catalyzes the removal of water from a substrate.

anhydride A compound formed from an acid by removal of water.

anhydroglucochloral *See* chloralose.

anhydrous Being without water, especially water of crystallization.

anhydrous alcohol *See* absolute alcohol.

anhydrous ammonia Liquid ammonia, a colorless liquid boiling at $-33.3°C$.

anhydrous ferric chloride *See* ferric chloride.

anhydrous hydrogen chloride HCl Hazardous, toxic, colorless gas used in polymerization, isomerization, alkylation, nitration, and chlorination reactions; becomes hydrochloric acid in aqueous solutions.

anhydrous phosphoric acid *See* phosphoric anhydride.

anhydrous plumbic acid *See* lead dioxide.

anhydrous sodium carbonate *See* soda ash.

anhydrous sodium sulfate Na_2SO_4 Water-soluble, white crystals with bitter, salty taste; melts at 888°C; used in the manufacture of glass, paper, pharmaceuticals, and textiles, and as an analytical reagent.

anhydrous wolframic acid *See* tungstic oxide.

anilazine $C_9H_5Cl_3N_4$ A tan solid with a melting point of 159–160°C; used for fungal diseases of lawns, turf, and vegetable crops. Also known as 4,6-dichloro-*N*-(2-chlorophenyl)-1,3,5-triazin-2-amine.

anilide A compound that has the $C_6H_5NH_2-$ group; an example is benzanilide, $C_6H_5NHCOC_6H_5$.

aniline $C_6H_5NH_2$ An aromatic amine compound that is a pale brown liquid at room temperature; used in the dye, pharmaceutical, and rubber industries.

aniline black A black dye produced on certain textiles, such as cotton, by oxidizing aniline or aniline hydrochloride.

aniline chloride *See* aniline hydrochloride.

aniline *N,N*-dimethyl *See* *N,N*-dimethyl aniline.

aniline dye A dye derived from aniline.

aniline-formaldehyde resin A thermoplastic resin made by polymerizing aniline and formaldehyde.

aniline hydrochloride $C_6H_5NH_2\cdot HCl$ White crystals, although sometimes the commercial variety has a greenish tinge; melting point 198°C; soluble in water and ethanol; used in dye manufacture, dyeing, and printing. Also known as aniline chloride; aniline salt.

aniline point The minimum temperature for a complete mixing of aniline and materials such as gasoline; used in some specifications to indicate the aromatic content of oils and to calculate approximate heat of combustion.

aniline salt *See* aniline hydrochloride.

***para*-anilinesulfonic acid** *See* sulfanilic acid.

anilinoacetic acid *See* *N*-phenylglycine.

2-anilinobenzoic acid *See* *N*-phenylanthranilic acid.

animal black Finely divided carbon made by calcination of animal bones or ivory; used for pigments, decolorizers, and purifying agents; varieties include bone black and ivory black.

animal charcoal Charcoal obtained by the destructive distillation of animal matter at high temperatures; used to adsorb organic coloring matter.

anion An ion that is negatively charged.

anion exchange A type of ion exchange in which the immobilized functional groups on the solid resin are positive.

anionic polymerization A type of polymerization in which Lewis bases, such as alkali metals and metallic alkyls, act as catalysts.

anionotropy The breaking off of an ion such as hydroxyl or bromide from a molecule so that a positive ion remains in a state of dynamic equilibrium.

anisaldehyde $C_6H_4(OCH_3)CHO$ A compound with melting point 2.5°C, boiling point 249.5°C; insoluble in water, soluble in alcohol and ether; used in perfumery and flavoring, and as an intermediate in production of antihistamines. Also known as *para*-anisic aldehyde; aubepine; 4-methoxybenzaldehyde; *para*-methoxybenzaldehyde.

anise alcohol *See* anisic alcohol.

anise camphor *See* anethole.

anisic acid $CH_3OC_6H_4COOH$ White crystals or powder with a melting point of 184°C; soluble in alcohol and ether; used in medicine and as an insect repellent and ovicide. Also known as *para*-methoxybenzoic acid.

anisic alcohol $C_8H_{10}O_2$ A colorless liquid that boils in the range 255–265°C; it is obtained by reduction of anisic aldehyde; used in perfumery, and as an intermediate in the manufacture of pharmaceuticals. Also known as anise alcohol; anisyl alcohol; *para*-methoxybenzyl alcohol.

***para*-anisic aldehyde** *See* anisaldehyde.

anisole $C_6H_5OCH_3$ A colorless liquid that is soluble in ether and alcohol, insoluble in water; boiling point is 155°C; vapors are highly toxic; used as a solvent and in perfumery. Also known as methoxybenzene; methylphenyl ether.

anisotropic membrane An ultrafiltration membrane which has a thin skin at the separating surface and is supported by a spongy sublayer of membrane material.

anisylacetone *See* 4-(*para*-methoxyphenyl)-2-butanone.

anisyl alcohol *See* anisic alcohol.

annular atoms The atoms in a cyclic compound that are members of the ring.

annulene One of a group of monocyclic conjugated hydrocarbons having the general formula [—CH=CH—]$_n$.

anode The positive terminal of an electrolytic cell.

anode corrosion efficiency The ratio of actual weight loss of an anode due to corrosion to the theoretical loss as calculated by Faraday's law.

anode effect A condition produced by polarization of the anode in the electrolysis of fused salts and characterized by a sudden increase in voltage and a corresponding decrease in amperage.

anode film   The portion of solution in immediate contact with the anode.

anodic polarization   The change in potential of an anode caused by current flow.

anolyte   The part of the electrolyte at or near the anode that is changed in composition by the reactions at the anode.

anomalon   A nuclear fragment, produced in the collision of a projectile nucleus at relativistic energy with a target nucleus at rest, that has an anomalously short mean free path, comparable to that of a uranium nucleus.

anomalous Zeeman effect   A type of splitting of spectral lines of a light source in a magnetic field which occurs for any line arising from a combination of terms of multiplicity greater than one; due to a nonclassical magnetic behavior of the electron spin.

anomer   One of a pair of isomers of cyclic carbohydrates; resulting from creation of a new point of symmetry when a rearrangement of the atoms occurs at the aldehyde or ketone position.

anonymous dimensionless group 1–4   Four of the dimensionless groups, used to solve problems in transfer processes, gas absorption in wetted-wall columns, and laminar boundary-layer flow.

anorthic crystal   *See* triclinic crystal.

antacid   Any substance that counteracts or neutralizes acidity.

antenna chlorophyll   Chlorophyll molecules which collect light quanta.

antheraxanthin   A neutral yellow plant pigment unique to the Euglenophyta.

anthocyanidin   Any of the colored aglycone plant pigments obtained by hydrolysis of anthocyanins.

anthocyanin   Any of the intensely colored, sap-soluble glycoside plant pigments responsible for most scarlet, purple, mauve, and blue coloring in higher plants.

anthracene $C_{14}H_{10}$ A crystalline tricyclic aromatic hydrocarbon, colorless when pure, melting at 218°C and boiling at 342°C; obtained in the distillation of coal tar; used as an important source of dyestuffs, and in coating applications.

anthraciferous coal   Anthracite-hard coal containing or yielding anthracene.

anthranilic acid   *ortho*-$NH_2C_6H_4COOH$ A white or pale yellow, crystalline acid melting at 146°C; used as an intermediate in the manufacture of dyes, pharmaceuticals, and perfumes.

anthrapurpurin $C_6H_3OH(CO)_2C_6H_2(OH)_2$ Orange-yellow, crystalline needles with a melting point of 369°C; soluble in alcohol and alkalies; used in dyeing. Also known as isopurpurin; purpurin red.

anthraquinone $C_6H_4(CO)_2C_6H_4$ Yellow crystalline diketone that is insoluble in water; used in the manufacture of dyes. Also known as dihydrodiketoanthracene.

anthraquinone pigments   Coloring materials which occur in plants, fungi, lichens, and insects; consists of about 50 derivatives of the parent compound, anthraquinone.

anthrone $C_{14}H_{10}O$ Colorless needles with a melting point of 156°C; soluble in alcohol, benzene, and hot sodium hydroxide; used as a reagent for carbohydrates.

anthropodesoxycholic acid *See* chenodeoxycholic acid.

antiatom An atom made up of antiprotons, antineutrons, and positrons in the same way that an ordinary atom is made up of protons, neutrons, and electrons.

antibaryon One of a class of antiparticles, including the antinucleons and the antihyperons, with strong interactions, baryon number -1, and hypercharge and charge opposite to those for the particles.

anticatalyst A material that slows down the action of a catalyst; an example is lead, which inhibits the action of platinum.

antichlor A chemical used in the manufacture of paper or textiles to remove excess chlorine or bleaching solution.

anticholinesterase Any agent, such as a nerve gas, that inhibits the action of cholinesterase and thereby destroys or interferes with nerve conduction.

anticoincidence The occurrence of an event at one place without a simultaneous event at another place.

antideuteron The antiparticle to the deuteron, composed of an antineutron and an antiproton.

antidiuretic hormone *See* vasopressin.

antienzyme An agent that selectively inhibits the action of an enzyme.

antifertilizin An immunologically specific substance produced by animal sperm to implement attraction by the egg before fertilization.

antifibrinolysin Any substance that inhibits the proteolytic action of fibrinolysin.

antifoaming agent A substance, such as silicones, organic phosphates, and alcohols, that inhibits the formation of bubbles in a liquid during its agitation by reducing its surface tension.

antifreeze A substance added to a liquid to lower its freezing point; the principal automotive antifreeze component is ethylene glycol.

antihemophilic factor A soluble protein clotting factor in mammalian blood. Also known as factor VIII; thromboplastinogen.

anti-infective vitamin *See* vitamin A.

antimolecule A molecule made up of antiprotons, antineutrons, and positrons in the same way that an ordinary molecule is made up of protons, neutrons, and electrons.

antimonate The negative radical $[Sb(OH)_6]^-$ in salts derived from antimony pentoxide, Sb_4O_{10}, and bases.

antimonic Derived from or pertaining to pentavalent antimony.

antimonide A binary compound of antimony with a more positive compound, for example, H_5Sb. Also known as stibide.

antimonous Pertaining to antinomy, especially trivalent antimony.

antimonous chloride *See* antimony trichloride.

antimonous sulfide *See* antimony trisulfide.

antimony A chemical element, symbol Sb, atomic number 51, atomic weight 121.75.

antimony-124 Radioactive antimony with mass number of 124; 60-day half-life; used as tracer in solid-state and pipeline flow studies.

antimony black *See* antimony trisulfide.

antimony chloride *See* antimony trichloride.

antimonyl The inorganic radical SbO^- .

antimonyl potassium tartrate *See* tartar emetic.

antimony needles *See* antimony trisulfide.

antimony orange *See* antimony trisulfide.

antimony(III) oxide Sb_2O_3 Colorless, rhombic crystals, melting at 656°C; insoluble in water; powerful reducing agent.

antimony pentachloride $SbCl_5$ A reddish-yellow, oily liquid; hygroscopic, it solidifies after moisture is absorbed and decomposes in excess water; soluble in hydrochloric acid and chloroform; used in analytical testing for cesium and alkaloids, for dyeing, and as an intermediary in synthesis. Also known as antimony perchloride.

antimony pentafluoride SbF_5 A corrosive, hygroscopic, moderately viscous fluid; reacts violently with water; forms a clear solution with glacial acetic acid; used in the fluorination of organic compounds.

antimony pentasulfide Sb_2S_5 An orange-yellow powder; soluble in alkali, soluble in concentrated hydrochloric acid, with hydrogen sulfide as a by-product, and insoluble in water; used as a red pigment. Also known as antimony persulfide; antimony red; golden antimony sulfide.

antimony perchloride *See* antimony pentachloride.

antimony persulfide *See* antimony pentasulfide.

antimony red *See* antimony pentasulfide.

antimony sodiate *See* sodium antimonate.

antimony sulfate $Sb_2(SO_4)_3$ Antimony(III) sulfate, a white, deliquescent powder; soluble in acids.

antimony sulfide *See* antimony trisulfide.

antimony trichloride $SbCl_3$ Hygroscopic, colorless, crystalline mass; fumes slightly in air, is soluble in alcohol and acetone, and forms antimony oxychloride in water; used as a mordant, as a chlorinating agent, and in fireproofing textiles. Also known as antimonous chloride; antimony chloride; caustic antimony.

antimony trisulfide Sb_2S_3 Black and orange-red rhombic crystals; soluble in concentrated hydrochloric acid and sulfide solutions, insoluble in water; melting point 546°C; used as a pigment, and in matches and pyrotechnics. Also known as antimonous sulfide; antimony black; antimony needles; antimony orange; antimony sulfide; black antimony.

antimony yellow *See* lead antimonite.

antinucleus A nucleus made up of antineutrons and antiprotons in the same way that an ordinary nucleus is made up of neutrons and protons.

antioxidant An inhibitor, such as ascorbic acid, effective in preventing oxidation by molecular oxygen.

antiozonant A protective agent which can be added to rubber during processing to diminish the deteriorating effects of ozone.

antiprotonic atom An atom consisting of an ordinary nucleus with an orbiting antiproton.

antirachitic vitamin *See* vitamin D.

antirad An inhibitor incorporated into rubber during manufacturing to reduce the degrading effects of radiation.

anti-Stokes lines Lines of radiated frequencies which are higher than the frequency of the exciting incident light.

antitermination factor Protein that interferes with normal termination of ribonucleic acid synthesis.

antithrombin A substance in blood plasma that inactivates thrombin.

antivitamin Any substance that prevents a vitamin from normal metabolic functioning.

antixerophthalmic vitamin *See* vitamin A.

ANTU *See* 1-(1-naphthyl)-2-thiourea.

4-AP *See* 4-aminopyridine.

6-APA *See* 6-aminopenicillanic acid.

APAP *See* acetaminophen.

API scale The American Petroleum Institute hydrometer scale for the measurement of the specific gravity of liquids; used primarily in the American petroleum industry.

aplysiatoxin A bislactone toxin produced by the blue-green alga *Lyngbya majuscula.*

apo- A prefix that denotes formation from or relationship to another chemical compound.

apoatropine $C_{17}H_{21}NO_2$ An alkaloid melting at 61°C with decomposition of the compound; highly toxic; obtained by dehydrating atropine.

apoenzyme The protein moiety of an enzyme; determines the specificity of the enzyme reaction.

apoferritin A protein found in intestinal mucosa cells that has the ability to combine with ferric ion.

apomyoglobin Myoglobin that lacks its heme group.

apoprotein The protein portion of a conjugated protein exclusive of the prosthetic group.

apozymase The protein component of a zymase.

apple essence *See* isoamyl valerate.

apple oil *See* isoamyl valerate.

aprotic solvent A solvent that does not yield or accept a proton.

apyrase Any enzyme that hydrolyzes adenosinetriphosphate, with liberation of phosphate and energy, and that is believed to be associated with actomyosin activity.

aqua   Latin for water.

aqua ammonia   *See* ammonium hydroxide.

aquafortis   *See* nitric acid.

aquametry   Analytical processes to measure the water present in materials; methods include Karl Fischer titration, reactions with acid chlorides and anhydrides, oven drying, distillation, and chromatography.

aqua regia   A fuming, highly corrosive, volatile liquid with a suffocating odor made by mixing 1 part concentrated nitric acid and 3 parts concentrated hydrochloric acid; reacts with all metals, including silver and gold. Also known as chloroazotic acid; chloronitrous acid; nitrohydrochloric acid; nitromuriatic acid.

aquation   Formation of a complex that contains water by replacement of other coordinated groups in the complex.

aqueous electron   *See* hydrated electron.

aqueous solution   A solution with the solvent as water.

aquo ion   Any ion containing one or more water molecules.

Ar   *See* argon.

arabinose $C_5IH_{10}O_5$ A pentose sugar obtained in crystalline form from plant polysaccharides such as gums, hemicelluloses, and some glycosides.

arabite   *See* arabitol.

arabitol $CH_2OH(CHOH)_3CH_2OH$ An alcohol that is derived from arabinose; a sweet, colorless crystalline material present in D and L forms; soluble in water; melts at 103°C. Also known as arabite.

arachic acid   *See* eicosanoic acid.

arachidic acid   *See* eicosanoic acid.

aralkyl   A radical in which an aryl group is substituted for an alkyl H atom. Derived from arylated alkyl.

arbutin $C_{12}H_{16}O_7$ A bitter glycoside from the bearberry and certain other plants; sometimes used as a urinary antiseptic.

arc excitation   Use of electric-arc energy to move electrons into higher energy orbits.

archen   *See* emodin.

arc process   A former process that used electric arcs for fixation (oxidation) of atmospheric nitrogen to manufacture nitric acid.

arc spectrum   The spectrum of a neutral atom, as opposed to that of a molecule or an ion; it is usually produced by vaporizing the substance in an electric arc; designated by the roman numeral I following the symbol for the element, for example, HeI.

arecaidine methyl ester   *See* arecoline.

arecoline $C_8H_{13}O_2N$ An alkaloid from the betel nut; an oily, colorless liquid with a boiling point of 209°C; soluble in water, ethanol, and ether; combustible; used as a medicine. Also known as arecaidine methyl ester; methyl-1,2,5,6-tetrahydro-1-methylnicotinate.

arene *See* aromatic hydrocarbon.

argentic Relating to or containing silver.

argentic oxide *See* silver suboxide.

argentocyanides Complexes formed, for example, in the cyanidation of silver ores and in electroplating, when silver cyanide reacts with solutions of soluble metal cyanides. Also known as dicyanoargentates.

argentometry A volumetric analysis that employs precipitation of insoluble silver salts; the salts may be chromates or chlorides.

argentum Latin for silver.

arginase An enzyme that catalyzes the splitting of urea from the amino acid arginine.

arginine $C_6H_{14}N_4O_2$ A colorless, crystalline, water-soluble, essential amino acid of the α-ketoglutaric acid family.

argon A chemical element, symbol Ar, atomic number 18, atomic weight 39.998.

aristolochic acid $C_{17}H_{11}NO_7$ Crystals in the form of shiny brown leaflets that decompose at 281–286°C; soluble in alcohol, chloroform, acetone, ether, acetic acid, and aniline; used as an aromatic bitter. Also known as aristolochine.

aristolochine *See* aristolochic acid.

Armstrong's acid *See* naphthalene-1,5-disulfonic acid.

Arndt-Eistert synthesis A method of increasing the length of an aliphatic acid by one carbon by reacting diazomethane with acid chloride.

aromatic 1. Pertaining to or characterized by the presence of at least one benzene ring. 2. Describing those compounds having physical and chemical properties resembling those of benzene.

aromatic alcohol Any of the compounds containing the hydroxyl group in a side chain to a benzene ring, such as benzyl alcohol.

aromatic aldehyde An aromatic compound containing the CHO radical, such as benzaldehyde.

aromatic amine An organic compound that contains one or more amino groups joined to an aromatic structure.

aromatic amino acid An organic acid containing at least one amino group and one or more aromatic groups; for example, phenylalanine, one of the essential amino acids.

aromatic hydrocarbon A member of the class of hydrocarbons, of which benzene is the first member, consisting of assemblages of cyclic conjugated carbon atoms and characterized by large resonance energies. Also known as arene.

aromatic ketone An aromatic compound containing the —CO radical, such as acetophenone.

aromatic nucleus The six-carbon ring characteristic of benzene and related series, or condensed six-carbon rings of naphthalene, anthracene, and so forth.

aromatization Conversion of any nonaromatic hydrocarbon structure to aromatic hydrocarbon, particularly petroleum.

arostat process A process in which aromatic molecules are saturated by catalytic hydrogenation to produce high-quality jet fuels, low-aromatic-content solvents, and high-purity cyclohexane from benzene.

aroyl The radical RCO, where R is an aromatic (benzoyl, napthoyl) group.

aroylation A reaction in which the aroyl group is incorporated into a molecule by substitution.

ARPES *See* angle-resolved photoelectron spectroscopy.

Arrhenius equation The relationship that the specific reaction rate constant k equals the frequency factor constant s times exp ($-\delta H_{act}/RT$), where δH_{act} is the heat of activation, R the gas constant, and T the absolute temperature.

arsenate 1. AsO_4^{-3} A negative ion derived from orthoarsenic acid, $H_3AsO_4 \cdot \frac{1}{2}H_2O$. 2. A salt or ester of arsenic acid.

arsenic A chemical element, symbol As, atomic number 33, atomic weight 74.9216.

arsenic acid $H_3AsO_4 \cdot \frac{1}{2}H_2O$ White, poisonous crystals, soluble in water and alcohol; used in manufacturing insecticides, glass, and arsenates and as a defoliant. Also known as orthoarsenic acid.

arsenical 1. Pertaining to arsenic. 2. A compound that contains arsenic.

arsenic disulfide As_2S_2 Red, orange, or black monoclinic crystals, insoluble in water; used in fireworks; occurs naturally as realgar.

arsenic oxide 1. An oxide of arsenic. 2. *See* arsenic pentoxide. 3. *See* arsenic trioxide.

arsenic pentasulfide As_2S_5 Yellow crystals that are insoluble in water and readily decompose to the trisulfide and sulfur; used as a pigment.

arsenic pentoxide As_2O_5 A white, deliquescent compound that decomposes by heat and is soluble in water. Also known as arsenic oxide.

arsenic trichloride $AsCl_3$ An oily, colorless liquid that dissolves in water; used in ceramics, organic chemical syntheses, and in the preparation of pharmaceuticals.

arsenic trioxide As_2O_3 A toxic compound, slightly soluble in water; octahedral crystals change to the monoclinic form by heating at 200°C; occurs naturally as arsenolite and claudetite; used in small quantities in some medicinal preparations. Also known as arsenic oxide; arsenious acid.

arsenic trisulfide As_2S_3 An acidic compound in the form of yellow or red monoclinic crystals with a melting point at 300°C; occurs as the mineral orpiment; used as a pigment.

arsenide A binary compound of negative, trivalent arsenic; for example, H_3As or GaAs.

arsenin A heterocyclic organic compound composed of a six-membered ring system in which the carbon atoms are unsaturated and the unique heteroatom is arsenic, with no nitrogen atoms present.

arsenious acid *See* arsenic trioxide.

arsenite 1. AsO_3^{-3} A negative ion derived from aqueous solutions of As_4O_6. 2. A salt or ester of arsenious acid.

arsenobenzene $C_6H_5As:AsC_6H_5$ White needles that melt at 212°C; insoluble in cold water, soluble in benzene; derivatives have some use in medicine.

arseno compound A compound containing an As-As bond with the general formula $(RAs)_n$, where R represents a functional group; structures are cyclic or long-chain polymers.

arsine H_3As A colorless, highly poisonous gas with an unpleasant odor.

arsinic acid An acid of general formula R_2AsO_2H; derived from trivalent arsenic; an example is cacodylic acid, or dimethylarsinic acid, $(CH_3)_2AsO_2H$.

arsonic acid An acid derived from orthoarsenic acid, $OAs(OH)_3$; the type formula is generally considered to be $RAsO(OH)_2$; an example is *para*-aminobenzenearsonic acid, $NH_2C_6H_4AsO(OH)_2$.

arsonium AsH_4 A radical; it may be considered analogous to the ammonium radical in that a compound such as AsH_4OH may form.

arthropodin A water-soluble protein which forms part of the endocuticle of insects.

artificial atmosphere A mixture of gases used in industrial operations in place of air; classified as an active, or process, atmosphere, or an inactive, or protective, atmosphere.

artificial camphor *See* terpene hydrochloride.

artificial gold *See* stannic sulfide.

artificial malachite *See* copper carbonate.

artificial scheelite *See* calcium tungstate.

aryl An organic radical derived from an aromatic hydrocarbon by removal of one hydrogen.

aryl acid An organic acid that has an aryl radical.

arylamine An organic compound formed from an aromatic hydrocarbon that has at least one amine group joined to it, such as aniline.

arylated alkyl *See* aralkyl.

aryl compound Molecules with the six-carbon aromatic ring structure characteristic of benzene or compounds derived from aromatics.

aryl diazo compound A diazo compound bonded to the ring structure characteristic of benzene or any other aromatic derivative.

arylene A radical that is bivalent and formed by removal of hydrogen from two carbon sites on an aromatic nucleus.

aryl halide An aromatic derivative in which a ring hydrogen has been replaced by a halide atom.

arylide A compound formed from a metal and an aryl radical, for example, PbR_4, where R is the aryl radical.

aryloxy compound One of a group of compounds useful as organic weed killers, such as 2,4-dichlorophenoxyacetic acid (2,4-D).

aryne An aromatic species in which two adjacent atoms of a ring lack substituents, with two orbitals each missing an electron. Also known as benzyne.

As *See* arsenic.

asarone $C_{12}H_{16}O_3$ A crystalline substance with melting point 67°C; insoluble in water, soluble in alcohol; found in plants of the genus *Asarum;* used as a constituent in essential oils such as calamus oil. Also known as 2,4,5,-trimethoxy-1-propenyl benzene.

ASC *See* N-acetylsulfanilyl chloride.

ascaridole $C_{10}H_{16}O_2$ A terpene peroxide, explosive when heated; used as an initiator in polymerization.

ascending chromatography   A technique for the analysis of mixtures of two or more compounds in which the mobile phase (sample and carrier) rises through the fixed phase.

ascorbic acid $C_6H_8O_6$ A white, crystalline, water-soluble vitamin found in many plant materials, especially citrus fruit. Also known as vitamin C.

ash   The incombustible matter remaining after a substance has been incinerated.

ashing   An analytical process in which the chemical material being analyzed is oven-heated to leave only noncombustible ash.

asparaginase   An enzyme that catalyzes the hydrolysis of asparagine to asparaginic acid and ammonia.

asparagine $C_4H_8N_2O_3$ A white, crystalline amino acid found in many plant seeds.

aspartase   A bacterial enzyme that catalyzes the deamination of aspartic acid to fumaric acid and ammonia.

aspartate   A compound that is an ester or salt of aspartic acid.

aspartic acid $C_4H_7NO_4$ A nonessential, crystalline dicarboxylic amino acid found in plants and animals, especially in molasses from young sugarcane and sugarbeet.

aspartokinase   An enzyme that catalyzes the reaction of aspartic acid with adenosinetriphosphate to give aspartyl phosphate.

aspartoyl $-COCH_2CH(NH_2)CO-$ A bivalent radical derived from aspartic acid.

aspartyl phosphate $H_2O_3POOCH_2CHNH_2COOH$ An intermediate in the biosynthesis of pyrimidines.

aspergillic acid $C_{12}H_{20}O_2N_2$ A diketopiperazine-like antifungal antibiotic produced by certain strains of *Aspergillus flavus*.

aspirin *See* acetylsalicylic acid.

assay   Qualitative or quantitative determination of the components of a material, as an ore or a drug.

association   Combination or correlation of substances or functions.

astacin $C_{40}H_{48}O_4$ A red carotenoid ketone pigment found in crustaceans, as in the shell of a boiled lobster.

A stage   An early stage in a thermosetting resin reaction characterized by linear structure, solubility, and fusibility of the material.

astatine A radioactive chemical element, symbol At, atomic number 85, the heaviest of the halogen elements.

astaxanthin $C_{40}H_{52}O_4$ A violet carotenoid pigment found in combined form in certain crustacean shells and bird feathers.

asterism A star-shaped pattern sometimes seen in x-ray spectrophotographs.

ASTM gum test 1. Determination of the amount of existing gum in a gasoline by evaporating a sample from a glass dish or an elevated temperature bath with the aid of circulating air. 2. Any gum test carried out in accordance with an ASTM gum test procedure.

ASTM melting point The temperature at which wax first shows a minimum rate of temperature change.

astron A proposed thermonuclear device in which a deuterium plasma is confined by an axial magnetic field produced by a shell of relativistic electrons.

astronomical spectrograph An instrument used to photograph spectra of stars.

astronomical spectroscopy The use of spectrographs in conjunction with telescopes to obtain observational data on the velocities and physical conditions of astronomical objects.

asulam *See* methyl-4-aminobenzene sulfonylcarbamate.

asymmetric carbon atom A carbon atom with four different radicals or atoms bonded to it.

asymmetric synthesis Chemical synthesis of a pure enantiomer, or of an enantiomorphic mixture in which one enantiomer predominates, without the use of resolution.

asymmetry The geometrical design of a molecule, atom, or ion that cannot be divided into like portions by one or more hypothetical planes. Also known as molecular asymmetry.

At *See* astatine.

ATA *See* aminotriazole.

atactic Of the configuration for a polymer, having the opposite steric configurations for the carbon atoms of the polymer chain occur in equal frequency and more or less at random.

atmospheric distillation Distillation operation conducted at atmospheric pressure, in contrast to vacuum distillation or pressure distillation.

atom The individual structure which constitutes the basic unit of any chemical element.

atomic charge The electric charge of an ion, equal to the number of electrons the atom has gained or lost in its ionization multiplied by the charge on one electron.

atomic diamagnetism Diamagnetic ionic susceptibility, important in providing correction factors for measured magnetic susceptibilities; calculated theoretically by considering electron density distributions summed for each electron shell.

atomic energy level A definite value of energy possible for an atom, either in the ground state or an excited condition.

atomic fission   *See* fission.

atomic fusion   *See* fusion.

atomic ground state   The state of lowest energy in which an atom can exist. Also known as atomic unexcited state.

atomic heat capacity   The heat capacity of a gram-atomic weight of an element.

atomic hydrogen   Gaseous hydrogen whose molecules are dissociated into atoms.

atomic magnet   An atom which possesses a magnetic moment either in the ground state or in an excited state.

atomic magnetic moment   A magnetic moment, permanent or temporary, associated with an atom, measured in magnetons.

atomic nucleus   *See* nucleus.

atomic number   The number of protons in an atomic nucleus. Also known as proton number.

atomic orbital   The space-dependent part of a wave function describing an electron in an atom.

atomic particle   One of the particles of which an atom is constituted, as an electron, neutron, or proton.

atomic percent   The number of atoms of an element in 100 atoms representative of a substance.

atomic photoelectric effect   *See* photoionization.

atomic polarization   Polarization of a material arising from the change in dipole moment accompanying the stretching of chemical bonds between unlike atoms in molecules.

atomic radius   1. Half the distance of closest approach of two like atoms not united by a bond. 2. The experimentally determined radius of an atom in a covalently bonded compound.

atomic spectroscopy   The branch of physics concerned with the production, measurement, and interpretation of spectra arising from either emission or absorption of electromagnetic radiation by atoms.

atomic spectrum   The spectrum of radiations due to transitions between energy levels in an atom, either absorption or emission.

atomic structure   The arrangement of the parts of an atom, which consists of a massive, positively charged nucleus surrounded by a cloud of electrons arranged in orbits describable in terms of quantum mechanics.

atomic theory   The assumption that matter is composed of particles called atoms and that these are the limit to which matter can be subdivided.

atomic units   *See* Hartree units.

atomic vibration   Periodic, nearly harmonic changes in position of the atoms in a molecule giving rise to many properties of matter, including molecular spectra, heat capacity, and heat conduction.

atomic volume   The volume occupied by 1 gram-atom of an element in the solid state.

atomic weight The relative mass of an atom based on a scale in which a specific carbon atom (carbon-12) is assigned a mass value of 12. Abbreviated at. wt.

ATPase *See* adenosinetriphosphatase.

ATR *See* attenuated total reflectance.

atrazine *See* 2-chloro-4-ethylamino-6-isopropylamino-*s*-triazine.

atropisomer One of two conformations of a molecule whose interconversion is slow enough to allow separation and isolation under predetermined conditions.

attenuated total reflectance A method of spectrophotometric analysis based on the reflection of energy at the interface of two media which have different refractive indices and are in optical contact with each other. Abbreviated ATR. Also known as frustrated internal reflectance; internal reflectance spectroscopy.

at. wt *See* atomic weight.

Au *See* gold.

aubepine *See* anisaldehyde.

Aufbau principle A description of the building up of the elements in which the structure of each in sequence is obtained by simultaneously adding one positive charge (proton) to the nucleus of the atom and one negative charge (electron) to an atomic orbital.

Auger coefficient The ratio of the number of Auger electrons to the number of ejected x-ray photons.

Auger effect The radiationless transition of an electron in an atom from a discrete electronic level to an ionized continuous level of the same energy. Also known as autoionization.

Auger electron An electron that is expelled from an atom in the Auger effect.

Auger electron spectroscopy The energy analysis of Auger electrons produced when an excited atom relaxes by a radiationless process after ionization by a high-energy electron, ion, or x-ray beam. Abbreviated AES.

Auger recombination Recombination of an electron and a hole in which no electromagnetic radiation is emitted, and the excess energy and momentum of the recombining electron and hole are given up to another electron or hole.

augmentation distance The extrapolation distance, which is the distance between the time boundary of a nuclear reactor and its boundary calculated by extrapolation.

auramine hydrochloride $C_{17}H_{22}ClN_3 \cdot H_2O$ A compound melting at 267°C; very soluble in water, soluble in ethanol; used as a dye and an antiseptic. Also known as yellow pyoktanin.

aurantia $C_{12}H_8N_8O_{12}$ An orange aniline dye, used in stains in biology and in some photographic filters.

aureusidin $C_{15}H_{11}O_5$ A yellow flavonoid pigment found typically in the yellow snapdragon. Also known as 4,6,3′,4′-tetrahydroxyaurone.

auric oxide *See* gold oxide.

aurin $C_{19}H_{14}O_3$ A derivative of triphenylmethane; solid with red-brown color with green luster; melting point about 220°C; insoluble in water; used as a dye intermediate. Also known as pararosolic acid; rosolic acid.

aurintricarboxylic acid triammonium salt *See* aluminon.

auroral line A prominent green line in the spectrum of the aurora at a wavelength of 5577 angstroms, resulting from a certain forbidden transition of oxygen.

autoacceleration The increase in polymerization rate and molecular weight of certain vinyl monomers during bulk polymerization.

autocatalysis A catalytic reaction started by the products of a reaction that was itself catalytic.

Autofining process A fixed-bed catalytic process for desulfurizing distillates; the pelleted catalyst is cobalt molybdate on alumina and may be regenerated in place.

autogenous ignition temperature *See* ignition temperature.

autoignition *See* spontaneous ignition.

autoionization *See* Auger effect.

autoluminescence Luminescence of a material (such as a radioactive substance) resulting from energy originating within the material itself.

autolysis Return of a substance to solution, as of phosphate removed from seawater by plankton and returned when these organisms die and decay.

autolytic enzyme A bacterial enzyme, located in the cell wall, that causes disintegration of the cell following injury or death.

automatic titrator 1. Titration with quantitative reaction and measured flow of reactant. 2. Electrically generated reactant with potentiometric, am"pherometric, or colorimetric end-point or null-point determination.

autooxidation 1. Oxidation caused by the atmosphere. 2. An oxidation reaction that is self-catalyzed and spontaneous. 3. An oxidation reaction begun only by an inductor.

autoprotolysis Transfer of a proton from one molecule to another of the same substance.

autoprotolysis constant A constant denoting the equilibrium condition for the autoprotolysis reaction.

auxiliary electrode An electrode in an electrochemical cell used for transfer of electric current to the test electrode.

auximone Any of certain growth-promoting substances occurring principally in sphagnum peat decomposed by nitrogen bacteria.

auxin Any organic compound which promotes plant growth along the longitudinal axis when applied to shoots free from indigenous growth-promoting substances.

auxochrome Any substituent group such as $-NH_2$ and $-OH$ which, by affecting the spectral regions of strong absorption in chromophores, enhance the ability of the chromogen to act as a dye.

available chlorine   The quantity of chlorine released by a bleaching powder when treated with acid.

avenin   The glutelin of oats.

average molecular weight   The calculated number to average the molecular weights of the varying-length polymer chains present in a polymer mixture.

avidin   A protein constituting 0.2% of the total protein in egg white; molecular weight is 70,000 daltons; combines firmly with biotin but loses this ability when subjected to heat.

axial angle   1. The acute angle between the two optic axes of a biaxial crystal. Also known as optic angle; optic-axial angle. 2. In air, the larger angle between the optic axes after refraction on leaving the crystal.

axial element   The lengths, length ratios, and angles which define a crystal's unit cell.

axial plane   1. A plane that includes two of the crystallographic axes. 2. The plane of the optic axis of an optically biaxial crystal.

axial ratio   The ratio obtained by comparing the length of a crystallographic axis with one of the lateral axes taken as unity.

9-azafluorene   *See* carbazole.

6-azauracil riboside   *See* 6-azauradine.

azelaic acid $HOOC(CH_2)_7CCOH$ Colorless leaflets; melting point 106.5°C; a dicarboxylic acid useful in lacquers, alkyd salts, organic synthesis, and formation of polyamides. Also known as 1,7-heptanedicarboxylic acid; nonanedioic acid.

azelate   A salt of azelaic acid, for example, sodium azelate.

azeotrope   *See* azeotropic mixture.

azeotropic mixture   A solution of two or more liquids, the composition of which does not change upon distillation. Also known as azeotrope.

2-azetidinecarboxylic acid $C_4H_7NO_2$ Crystals which discolor at 200°C and darken until 310°C; soluble in water; a specific antagonist of l-proline; used in the production of abnormally high-molecular-weight polypeptides.

azide   One of several types of compounds containing the —N_3 group and derived from hydrazoic acid, HN_3.

azimethane   *See* diazomethane.

azimethylene   *See* diazomethane.

azimidobenzene   *See* 1,2,3-benzotriazole.

azimino   *See* diazoamine.

azimuthal quantum number   The orbital angular momentum quantum number l, such that the eigenvalue of L^2 is $l(l + 1)$.

azine   A compound of six atoms in a ring; at least one of the atoms is nitrogen, and the ring structure resembles benzene; an example is pyridine.

azine dyes   Benzene-type dyes derived from phenazine; members of the group, such as nigrosines and safranines, are quite varied in application.

azlactone   A compound that is an anhydride of α-acylamino acid; the basic ring structure is the 5-oxazolone type.

azo-   A prefix indicating the radical —N=N—.

azobenzene $C_6H_5NO{=}N{-}C_6H_5$ A compound existing in cis and trans geometric isomers; the cis form melts at 71°C; the trans form comprises orange-red leaflets, melting at 68.5°C; used in manufacture of dyes and accelerators for rubbers. Also known as azobenzol; benzeneazobenzene; diphenyldiazene.

azobenzol   *See* azobenzene.

2,2′-azobisisobutyronitrile $C_8H_{12}N_4$ Crystals that decompose at 107°C; soluble in methanol and in ethanol; used as an initiator of free radical reactions and as a blowing agent for plastics and elastomers.

azo dyes   Widely used commercial dyestuffs derived from amino compounds, with the —N=N— chromophore group; can be made as acid, basic, direct, or mordant dyes.

azoic dye   A water-insoluble azo dye that is formed by coupling of the components on a fiber. Also known as ice color; ingrain color.

azoimide   *See* hydrazoic acid.

azole   One of a class of organic compounds with a five-membered N-heterocycle containing two double bonds; an example is 1,2,4-triazole.

azophenylene   *See* phenazine.

azotometer   *See* nitrometer.

azoxybenzene $C_6H_5NO{=}N{-}C_6H_5$ A compound existing in cis and trans forms; the cis form melts at 87°C; the trans form comprises yellow crystals, melting at 36°C, insoluble in water, soluble in ethanol.

azulene $C_{16}H_{26}O$ The blue coloring matter of wormwood and other essential oils; an oily, blue liquid, boiling at 170°C; insoluble in water; used in cosmetics.

B

b *See* barn.

Babo's law A law stating that the relative lowering of a solvent's vapor pressure by a solute is the same at all temperatures.

backbending A discontinuity in the rotational levels of some rare-earth nuclei around spin 20 (where the same character is Planck's constant divided by 2π), which appears as a backbend on a graph that plots the moment of inertia versus the square of the rotational frequency.

backflash Rapid combustion of a material occurring in an area that the reaction was not intended for.

back mixing The tendency of reacted chemicals to intermingle with unreacted feed in reactors, such as stirred tanks, packed towers, and baffled tanks.

back-reflection photography A method of studying crystalline structure by x-ray diffraction in which the photographic film is placed between the source of x-rays and the crystal specimen.

back-run process A process for manufacturing water gas in which part of the run is made down, by passing steam through the superheater, thence up through the carburetor, down through the generator, and direct to the scrubbers.

back sweetening The controlled addition of commercial-grade mercaptans to a petroleum stock having excess free sulfur in order to reduce free sulfur by forming a disulfide.

back titration A titration to return to the end point which was passed.

backwash 1. In an ion-exchange resin system, an upward flow of water through a resin bed that cleans and reclassifies the resin particles after exhaustion. 2. *See* blowback.

bacterial methanogenesis *See* methanogenesis.

bacteriochlorophyll $C_{52}H_{70}O_6N_4Mg$ A tetrahydroporphyrin chlorophyll compound occurring in the forms *a* and *b* in photosynthetic bacteria; there is no evidence that *b* has the empirical formula given.

Badger's rule An empirical relationship between the stretching force constant for a molecular bond and the bond length.

baeckeol $C_{13}H_{18}O_4$ A phenolic ketone that is crystalline and pale yellow; found in oils from plants of species of the myrtle family.

Baeyer strain theory The theory that the relative stability of penta- and hexamethylene ring compounds is caused by a propitious bond angle between carbons and a lack of bond strain.

baking soda *See* sodium bicarbonate.

balance To bring a chemical equation into balance so that reaction substances and reaction products obey the laws of conservation of mass and charge.

Balmer lines Lines in the hydrogen spectrum, produced by transitions between $n = 2$ and $n > 2$ levels either in emission or absorption; here n is the principal quantum number.

Balmer series The set of Balmer lines.

Bamberger's formula A structural formula for naphthalene that shows the valencies of the benzene rings pointing toward the centers.

banana oil **1.** A solution of nitrocellulose in amyl acetate having a bananalike odor. **2.** *See* amyl acetate.

band **1.** The position and spread of a solute within a series of tubes in a liquid-liquid extraction procedure. Also known as zone. **2.** *See* band spectrum.

band head A location on the spectrogram of a molecule at which the lines of a band pile up.

band spectrum A spectrum consisting of groups or bands of closely spaced lines in emission or absorption, characteristic of molecular gases and chemical compounds. Also known as band.

barban $C_{11}H_9O_2NCl_2$ A white, crystalline compound with a melting point of 75–76°C; used as a postemergence herbicide of wild oats in barley, flax, lentil, mustard, and peas. Also known as 4-chloro-2-butynyl-*meta*-chlorocarbanilate.

barbital $C_8H_{12}N_2O_3$ A compound crystallizing in needlelike form from water; has a faintly bitter taste; melting point 188–192°C; used to make sodium barbital, a long-duration hypnotic and sedative. Also known as diethylbarbituric acid; diethylmalonylurea.

barbituric acid $C_4H_4O_3N_2$ 2,4,6-Trioxypyrimidine, the parent compound of the barbiturates; colorless crystals melting at 245°C, slightly soluble in water. Also known as malonyl urea.

Barfoed's test A test for monosaccharides conducted in an acid solution; cupric acetate is reduced to cuprous oxide, a red precipitate.

Bari-Sol process Removal of waxes from liquid hydrocarbons by extraction of the wax with a mixed ethylene dichloride–benzene solvent, followed by separation from the hydrocarbon in a centrifuge.

barium A chemical element, symbol Ba, with atomic number 56 and atomic weight of 137.34.

barium-140 A radioactive isotope of barium with atomic mass 140; the half-life is 12.8 days, and the decay is by negative beta-particle emission.

barium acetate $Ba(C_2H_3O_2)_2 \cdot H_2O$ A barium salt made by treating barium sulfide or barium carbonate with acetic acids; it forms colorless, triclinic crystals that decompose upon heating; used as a reagent for sulfates and chromates.

barium azide $Ba(N_3)_2$ A crystalline compound soluble in water; used in high explosives.

barium binoxide *See* barium peroxide.

barium bromate $Ba(BrO_3)_2 \cdot H_2O$ A poisonous compound that forms colorless, monoclinic crystals, decomposing at 260°C; used for preparing other bromates.

barium bromide $BaBr_2 \cdot 2H_2O$ Colorless crystals soluble in water and alcohol; used in photographic compounds.

barium carbonate $BaCO_3$ A white powder with a melting point of 174°C; soluble in acids (except sulfuric acid); used in rodenticides, ceramic flux, optical glass, and television picture tubes.

barium chlorate $Ba(ClO_3)_2 \cdot H_2O$ A salt prepared by the reaction of barium chloride and sodium chlorate; it forms colorless, monoclinic crystals, soluble in water; used in pyrotechnics.

barium chloride $BaCl_2$ A toxic salt obtained as colorless, water-soluble cubic crystals, melting at 963°C; used as a rat poison, in metal surface treatment, and as a laboratory reagent.

barium chromate $BaCrO_4$ A toxic salt that forms yellow, rhombic crystals, insoluble in water; used as a pigment in overglazes.

barium citrate $Ba_3(C_6H_5O_7)_2 \cdot 2H_2O$ A grayish-white, toxic, crystalline powder; used as a stabilizer for latex paints.

barium cyanide $Ba(CN)_2$ A white, crystalline powder; soluble in water and alcohol; used in metallurgy and electroplating.

barium dioxide *See* barium peroxide.

barium fluoride BaF_2 Colorless, cubic crystals, slightly soluble in water; used in enamels.

barium fluosilicate $BaSiF_6H$ A white, crystalline powder; insoluble in water; used in ceramics and insecticides. Also known as barium silicofluoride.

barium hydroxide $Ba(OH)_2 \cdot 8H_2O$ Colorless, monoclinic crystals, melting at 78°C; soluble in water, insoluble in acetone; used for fat saponification and fusing of silicates.

barium hyposulfite *See* barium thiosulfate.

barium manganate $BaMnO_4$ A toxic, emerald-green powder which is used as a paint pigment. Also known as Cassel green; manganese green.

barium mercury iodide *See* mercuric barium iodide.

barium molybdate $BaMoO_4$ A toxic, white powder with a melting point of approximately 1600°C; used in electronic and optical equipment and as a paint pigment.

barium monosulfide BaS A colorless, cubic crystal that is soluble in water; used in pigments.

barium monoxide *See* barium oxide.

barium nitrate $Ba(NO_3)_2$ A toxic salt occurring as colorless, cubic crystals, melting at 592°C, and soluble in water; used as a reagent, in explosives, and in pyrotechnics. Also known as nitrobarite.

barium oxide BaO A white to yellow powder that melts at 1923°C; it forms the hydroxide with water; may be used as a dehydrating agent. Also known as barium monoxide; barium protoxide.

barium perchlorate $Ba(ClO_4)_2 \cdot 4H_2O$ Tetrahydrate variety which forms colorless hexagons; used in pyrotechnics.

barium permanganate $Ba(MnO_4)_2$ Brownish-violet, toxic crystals; soluble in water; used as a disinfectant.

barium peroxide BaO_2 A compound formed as white toxic powder, insoluble in water; used as a bleach and in the glass industry. Also known as barium binoxide; barium dioxide; barium superoxide.

barium protoxide *See* barium oxide.

barium rhodanide *See* barium thiocyanate.

barium silicide $BaSi_2$ A compound that has the appearance of metal-gray lumps; melts at white heat; used in metallurgy to deoxidize steel.

barium silicofluoride *See* barium fluosilicate.

barium stearate $Ba(C_{18}H_{35}O_2)_2$ A white, crystalline solid; melting point 160°C; used as a lubricant in manufacturing plastics and rubbers, in greases, and in plastics as a stabilizer against deterioration caused by heat and light.

barium sulfate $BaSO_4$ A salt occurring in the form of white, rhombic crystals, insoluble in water; used as a white pigment, as an opaque contrast medium for roentgenographic processes, and as an antidiarrheal.

barium sulfite $BaSO_3$ A toxic, white powder; soluble in dilute hydrochloric acid; used in paper manufacturing.

barium sulfocyanate *See* barium thiocyanate.

barium sulfocyanide *See* barium thiocyanate.

barium superoxide *See* barium peroxide.

barium tetrasulfide $BaS_4 \cdot H_2O$ Red or yellow, rhombic crystals, soluble in water.

barium thiocyanate $Ba(SCN) \cdot 2H_2O$ White crystals that deliquesce; used in dyeing and in photography. Also known as barium rhodanide; barium sulfocyanate; barium sulfocyanide.

barium thiosulfate $BaS_2O_3 \cdot H_2O$ A white powder that decomposes upon heating; used to make explosives and in matches. Also known as barium hyposulfite.

barium titanate $BaTiO_3$ A grayish powder that is insoluble in water but soluble in concentrated sulfuric acid; used as a ferroelectric ceramic.

barium tungstate $BaWO_4$ A toxic, white powder used as a pigment and in x-ray photography. Also known as barium white; barium wolframate; tungstate white; wolfram white.

barium white *See* barium tungstate.

barium wolframate *See* barium tungstate.

Barker method A method utilizing a number of convenient rules which allow two observers to choose the same reference system to describe the same noncubic crystal.

barkometer A hydrometer calibrated to test the strength of tanning liquors used in tanning leather.

Barlow's rule The rule that the volume occupied by the atoms in a given molecule is proportional to the valences of the atoms, using the lowest valency values.

barn A unit of area equal to 10^{-24} square centimeter; used in specifying nuclear cross sections. Symbolized b.

barrels per calendar day A unit measuring the average rate of oil processing in a petroleum refinery, with allowances for downtime over a period of time. Abbreviated BCD.

barrels per day A unit measuring the rate at which petroleum is produced at the refinery. Abbreviated BD; bpd.

barrels per month A unit measuring the rate at which petroleum is produced at the refinery. Abbreviated BM; bpm.

barrels per stream day A measurement used to denote rate of oil or oil-product flow while a fluid-processing unit is in continuous operation. Abbreviated BSD.

barren liquor Liquid (liquor) from filter-cake washing in which there is little or no recovery value; for example, barren cyanide liquor from washing of gold cake slimes.

barrier separation The separation of a two-component gaseous mixture by selective diffusion of one component through a separative barrier (microporous metal or nonporous polymeric).

Bartlett force A force between nucleons in which spin is exchanged.

Bart reaction Formation of an aryl arsonic acid by treating the aryl diazo compound with trivalent arsenic compounds, such as sodium arsenite.

baryta water A solution of barium hydroxide.

basal cleavage Cleavage parallel to the base of the crystal structure or to the lattice plane which is normal to one of the lattice axes.

basal orientation A crystal orientation in which the surface is parallel to the base of the lattice or to the lattice plane which is normal to one of the lattice axes.

basal plane The plane perpendicular to the long, or c, axis in all crystals except those of the isometric system.

base 1. Any chemical species, ionic or molecular, capable of accepting or receiving a proton (hydrogen ion) from another substance; the other substance acts as an acid in giving of the proton; the hydroxyl ion is a base. 2. The primary substance in solution in crude oil, and remaining after distillation.

base-centered lattice A space lattice in which each unit cell has lattice points at the centers of each of two opposite faces as well as at the vertices; in a monoclinic crystal, they are the faces normal to one of the lattice axes.

base exchange Replacement of certain ions by others in clay.

base-line technique A method for measurement of absorption peaks for quantitative analysis of chemical compounds in which a base line is drawn tangent to the spectrum background; the distance from the base line to the absorption peak is the absorbence due to the sample under study.

base metal Any of the metals on the lower end of the electrochemical series.

BASF process Production of acetylene by controlled burning of light hydrocarbons (for example, natural gas) with oxygen.

basic Of a chemical species that has the properties of a base.

basic bismuth carbonate *See* bismuth subcarbonate.

basic bismuth gallate *See* bismuth subgallate.

basic bismuth nitrate *See* bismuth subnitrate.

basic bismuth pyrogallate *See* bismuth pyrogallate.

basic bismuth salicylate *See* bismuth subsalicylate.

basic copper carbonate *See* copper carbonate.

basic group A chemical group (for example, OH^-) which, when freed by ionization in solution, produces a pH greater than 7.

basic oxide A metallic oxide that is a base, or that forms a hydroxide when combined with water, such as sodium oxide to sodium hydroxide.

basic salt A compound that is a base and a salt because it contains elements of both, for example, $Cu_2(OH)_2CO_3$.

basic titanium sulfate *See* titanium sulfate.

basic titrant A standard solution of a base used for titration.

basic zirconium chloride *See* zirconium oxychloride.

basic zirconium phosphate *See* zirconium phosphate.

basket strainer A porous-sided or screen-covered vessel used to screen solid particles out of liquid or gas streams.

batch distillation Distillation where the entire batch of liquid feed is placed into the still at the beginning of the operation, in contrast to continuous distillation, where liquid is fed continuously into the still.

batch rectification Batch distillation in which the boiled-off vapor is re-condensed into liquid form and refluxed back into the still to make contact with the rising vapors.

bathochromatic shift The shift of the fluorescence of a compound toward the red part of the spectrum due to the presence of a bathochrome radical in the molecule.

bathophenanthroline *See* phenanthroline indicator.

bating Cleaning of depilated leather hides by the action of tryptic enzymes.

battery A series of distillation columns or other processing equipment operated as a single unit.

battery depolarizer *See* depolarizer.

battery electrolyte A liquid, paste, or other conducting medium in a battery, in which the flow of electric current takes place by migration of ions.

battery limits An area in a refinery or chemical plant encompassing a processing unit or battery of units along with their related utilities and services.

battery manganese *See* manganese dioxide.

Baumé hydrometer scale A calibration scale for liquids that is reducible to specific gravity by the following formulas: for liquids heavier than water, specific gravity = $145 \div (145 - n)$ (at 60°F); for liquids lighter than water, specific gravity = 140 ÷ (130 + n) (at 60°F); n is the reading on the Baumé scale, in degrees Baumé. Baumé is abbreviated Bé.

bauxite treating A catalytic petroleum process in which a vaporized petroleum fraction is passed through beds of bauxite; conversion of many different sulfur compounds, particularly mercaptans into hydrogen sulfide, takes place.

Baveno twin law An uncommon twin law applicable in feldspar, in which the twin plane and composition surface are (021); a Baveno twin usually consists of two individuals.

b axis A crystallographic axis that is oriented horizontally, right to left.

BBC *See* bromobenzylcyanide.

B-B fraction A mixture of butanes and butenes distilled from a solution of light liquid hydrocarbons.

BCD *See* barrels per calendar day.

BD *See* barrels per day.

Be *See* beryllium.

Bé *See* Baumé hydrometer scale.

bead test In mineral identification, a test in which borax is fused to a transparent bead, by heating in a blowpipe flame, in a small loop formed by platinum wire; when suitable minerals are melted in this bead, characteristic glassy colors are produced in an oxidizing or reducing flame and serve to identify elements.

beam attenuator An attachment to the spectrophotometer that reduces reference to beam energy to accommodate undersized chemical samples.

beam-condensing unit An attachment to the spectrophotometer that condenses and remagnifies the beam to provide reduced radiation at the sample.

beam-foil spectroscopy A method of studying the structure of atoms and ions in which a beam of ions energized in a particle accelerator passes through a thin carbon foil from which the ions emerge with various numbers of electrons removed and in various excited energy levels; the light or Auger electrons emitted in the deexcitation of these levels are then observed by various spectroscopic techniques. Abbreviated BFS.

beamhouse A place where the initial wet operations of tanning, involving soaking in water and solutions of alkali, are carried out.

bebeerine $C_{36}H_{38}N_2O_6$ An alkaloid derived from the bark of the tropical tree *Nectandra rodiaei*; the dextro form is soluble in acetone, the levo form is soluble in benzene and is an antipyretic; the dextro form is also known as chondrodendrin; the levo, as curine.

Béchamp reduction Reduction of nitro groups to amino groups by the use of ferrous salts or iron and dilute acid.

Beckmann rearrangement An intramolecular change of a ketoxime into its isomeric amide when treated with phosphorus pentachloride.

Becquerel rays Formerly, radiation emitted by radioactive substances; later renamed alpha, beta, and gamma rays.

bed The ion-exchange resin contained in the column in an ion-exchange system.

Beer-Lambert-Bouguer law   *See* Bouguer-Lambert-Beer law.

Beer's law   The law which states that the absorption of light by a solution changes exponentially with the concentration, all else remaining the same.

behenic acid   *See* docosanoic acid.

behenyl alcohol $CH_3(CH_2)_{20}CH_2OH$ A saturated fatty alcohol; colorless, waxy solid with a melting point of 71°C; soluble in ethanol and chloroform; used for synthetic fibers and lubricants. Also known as 1-docosanol.

bell cap   A hemispherical or triangular metal casting used on distillation-column trays to force upflowing vapors to bubble through layers of downcoming liquid.

bempa $C_6H_{18}N_3PO$ A white solid soluble in water; used as chemosterilant for insects. Also known as hexamethylphosphorictriamide.

Bender process   Treatment of petroleum liquids with lead sulfide catalyst to sweeten (to convert mercaptans to disulfides) by oxidation.

bendiocarb   *See* 2,2-dimethyl-1,3-benzodioxol-4-yl-*N*-methylcarbamate.

Benedict equation of state   An empirical equation relating pressures, temperatures, and volumes for gases and gas mixtures; superseded by the Benedict-Webb-Rubin equation of state.

Benedict's solution   A solution of potassium and sodium tartrates, copper sulfate, and sodium carbonate; used to detect reducing sugars.

benefin   *See* *N*-butyl-*N*-ethyl-α,α,α-trifluoro-2,6-dinitro-*para*-toluidine.

benequinox $C_{13}H_{11}N_3O_2$ A yellow-brown powder that decomposes at 195°C; used as a fungicide for grain seeds and seedlings. Also known as 1,4-benzoquinone-*N'*-benzoylhydrazone oxime.

bensulide $C_{14}H_{24}O_4NPS_3$ An *S*-(*O,O*-diisopropyl phosphorodithioate) ester of *N*-(2-mercaptoethyl)-benzenesulfonamide; an amber liquid slightly soluble in water; melting point is 34.4°C; used as a preemergent herbicide for annual grasses and for broadleaf weeds in lawns and vegetable and cotton crops.

benthiocarb $C_{12}H_{16}NOCl$ An amber liquid with a boiling point of 126–129°C; slightly soluble in water; used as an herbicide to control aquatic weeds in rice crops. Also known as *S*-(4-chlorobenzyl)-*N,N*-diethylthiocarbamate.

bentranil   *See* 2-phenyl-3,1-benzoxazinone-(4).

benzadox $C_6H_5CONHOCH_2COOH$ White crystals with a melting point of 140°C; soluble in water; used as an herbicide to control kochia in sugarbeets. Also known as benzamidoxyacetic acid.

benzalacetaldehyde   *See* cinnamaldehyde.

benzalacetic acid   *See* cinnamic acid.

benzalacetone   *See* benzylideneacetone.

benzal chloride $C_6H_5CHCl_2$ A colorless liquid that is refractive and fumes in air; boiling point 207°C; used to make benzaldehyde and cinnamic acid. Also known as benzylidene chloride; chlorobenzal.

benzaldehyde C_6H_5CHO A colorless, liquid aldehyde, boiling at 170°C and possessing the odor of bitter almonds; used as a flavoring agent and an intermediate in chemical syntheses.

benzaldehyde cyanohydrin *See* mandelonitrile.

benzaldoxime C_6H_5CHNOH An oxime of benzaldehyde; the antiisomeric form melts at 130°C, the syn form at 34°C; both forms are soluble in ethyl alcohol and ether; used in synthesis of other organic compounds.

benzalkonium $C_6H_5CH_2N(CH_3)_2R^+$ An organic radical in which R may range from C_8H_{17} to $C_{18}H_{37}$; found in surfactants, as the chloride salt.

benzalkonium chloride $C_6H_5CH_2(CH_3)_2NRCl$ A yellow-white powder soluble in water; used as a fungicide and bactericide; the R is a mixture of alkyls from C_8H_{17} to $C_{18}H_{37}$. Also known as alkyldimethylbenzylammonium chloride.

benzamide $C_6H_5CONH_2$ A compound with melting point 132.5° to 133.5°C; slightly soluble in water, soluble in ethyl alcohol and carbon tetrachloride; used in chemical synthesis.

benzamidoxyacetic acid *See* benzadox.

benzanilide $C_6H_5CONHC_6H_5$ Leaflet crystals with a melting point of 163°C; soluble in alcohol; used to manufacture dyes and perfumes. Also known as N-benzoylaniline.

1,2-benzanthracene *See* benzanthracene.

benzanthracene $C_{18}H_{14}$ A weakly carcinogenic material that is isomeric with naphthacene; melting point 162°C; insoluble in water, soluble in benzene. Also known as 1,2-benzanthracene; 2,3-benzophenanthrene.

benzanthrone $C_{17}H_{10}O$ A compound with melting point 170°C; insoluble in water; used in dye manufacture.

benzene C_6H_6 A colorless, liquid, flammable, aromatic hydrocarbon that boils at 80.1°C and freezes at 5.4–5.5°C; used to manufacture styrene and phenol. Also known as benzol.

benzene azimide *See* 1,2,3-benzotriazole.

benzeneazoanilide *See* diazoaminobenzene.

benzeneazobenzene *See* azobenzene.

benzene carbonitrile *See* benzonitrile.

benzenediazonium chloride $C_6H_5N(N)Cl$ An ionic salt soluble in water; used as a dye intermediate.

1,4-benzenedicarbonyl chloride *See* terephthaloyl chloride.

benzene-*ortho*-dicarboxylic acid *See* phthalic acid.

benzene-*para*-dicarboxylic acid *See* terephthalic acid.

benzene hexachloride *See* 1,2,3,4,5,6-hexachlorocyclohexane.

benzenephosphonic acid $C_6H_5H_2PO_3$ Colorless, combustible crystals with a melting point of 158°C; soluble in water, alcohol, ether, and acetone; used in antifouling paints. Also known as phenylphosphonic acid.

benzenephosphorus dichloride $C_6H_5PCl_2$ An irritating, colorless liquid with a boiling point of 224.6°C; soluble in inert organic solvents; used in organic synthesis and oil additives.

benzene ring The six-carbon ring structure found in benzene, C_6H_6, and in organic compounds formed from benzene by replacement of one or more hydrogen atoms by other chemical atoms or radicals.

benzene series   A series of carbon-hydrogen compounds based on the benzene ring, with the general formula C_nH_{2n-6}, where n is 6 or more; examples are benzene, C_6H_6; toluene, C_7H_8; and xylene, C_8H_{10}.

benzenesulfonate   Any salt or ester of benzenesulfonic acid.

benzenesulfonic acid $C_6H_5SO_3H$ An organosulfur compound, strongly acidic, water soluble, nonvolatile, and hygroscopic; used in the manufacture of detergents and phenols.

1,2,3,4,5-benzenetetracarboxylic acid   *See* pyromellitic acid.

benzene-1,2,3-tricarboxylic acid   *See* hemimellitic acid.

1,2,4-benzenetricarboxylic acid $C_6H_3(COOH)_3$ Crystals with a melting point of 218–220°C; crystallizes from acetic acid or from dilute alcohol; used as an intermediate in the preparation of adhesives, plasticizers, dyes, inks, and resins. Also known as trimellitic acid.

1,2,4-benzenetriol $C_6H_3(OH)_3$ Monoprismatic leaflets with a melting point of 141°C; freely soluble in water, ether, alcohol, and ethylacetate; used in gas analysis. Also known as hydroxyhydroquinone; hydroxyquinol.

benzenoid   Any substance which has the electronic character of benzene.

benzenyl trichloride   *See* benzotrichloride.

benzhydrol $(C_6H_5)_2CHOH$ Colorless needles; melting point 69°C; slightly soluble in water, very soluble in ethanol and ether; used in preparation of other organic compounds including antihistamines. Also known as diphenylcarbinol.

benzidine $NH_2C_6H_4C_6H_4NH_2$ An aromatic amine with a melting point of 128°C; used as an intermediate in syntheses of direct dyes for cotton.

benzil $C_6H_5COCOC_6H_5$ A yellow powder; melting point 95°C; insoluble in water, soluble in ethanol, ether, and benzene; used in organic synthesis. Also known as dibenzoyl; diphenylglyoxal.

benzilic acid $(C_6H_5)_2C(OH)CO_2H$ A white, crystalline acid, synthesized by heating benzil with alcohol and potassium hydroxide; used in organic synthesis.

benzimidazole $C_7H_6N_2$ Colorless crystals; melting point 170°C; slightly soluble in water, soluble in ethanol; used in organic synthesis. Also known as 1,3-benzodiazole; benzoglyoxaline.

benzin   *See* petroleum benzin.

benzindene   *See* fluorene.

benzine   *See* petroleum benzin.

benzoate   A salt or ester of benzoic acid, formed by replacing the acidic hydrogen of the carboxyl group with a metal or organic radical.

benzocaine   *See* ethyl-*para*-amino benzoate.

1,4-benzodiazine   *See* quinoxaline.

2,3-benzodiazine   *See* phthalazine.

benzo-*para*-diazine   *See* quinoxaline.

β-benzo-*ortho*-diazine   *See* phthalazine.

1,3-benzodiazole *See* benzimidazole.

benzodihydropyrone $C_9H_8O_2$ A white to light yellow, oily liquid having a sweet odor; soluble in alcohol, chloroform, and ether; used in perfumery. Also known as dihydrocoumarin.

benzofuran *See* coumarone.

benzoglycolic acid *See* mandelic acid.

benzoglyoxaline *See* benzimidazole.

benzoic acid C_6H_5COOH An aromatic carboxylic acid that melts at 122.4°C, boils at 250°C, and is slightly soluble in water and relatively soluble in alcohol and ether; derivatives are valuable in industry, commerce, and medicine.

benzoic anhydride $(C_6H_5CO)_2O$ An acid anhydride that melts at 42°C, boils at 360°C, and crystallizes in colorless prisms; used in synthesis of a variety of organic chemicals, including some dyes.

benzoin $C_{14}H_{12}O_2I$ An optically active compound; white or yellowish crystals, melting point 137°C; soluble in acetone, slightly soluble in water; used in organic synthesis. Also known as benzoylphenyl carbinol; 2-hydroxy-2-phenyl acetophenone; phenyl benzoyl carbinol.

benzoinam *See* benzoin.

α-benzoin oxime $C_6H_5CH(OH)C(NOH)C_6H_5$ Crystallizes as prisms from benzene; melting point is 151–152°C; soluble in alcohol and in aqueous ammonium hydroxide solution; used in the detection and determination of copper, molybdenum, and tungsten.

benzol *See* benzene.

benzol-acetone process A solvent dewaxing process in which a mixture of the solvent and oil containing wax is cooled until the wax solidifies and is then removed by filtration.

benzoline *See* normal benzine.

benzomate $C_{18}H_{18}O_5N$ A white solid that melts at 71.5–73°C; used as a wettable powder as a miticide. Also known as 3-chloro-N-ethoxy-2,6-dimethoxybenzimidic acid anhydride with benzoic acid.

benzonitrile C_6H_5CN A colorless liquid with an almond odor; made by heating benzoic acid with lead thiocyanate and used in the synthesis of organic chemicals. Also known as benzene carbonitrile; phenyl cyanide.

2,3-benzophenanthrene *See* benzanthracene.

benzophenone $C_6H_5COC_6H_6$ A diphenyl ketone, boiling point 305.9°C, occurring in four polymorphic forms (α, β, γ, and δ) each with different melting point; used as a constituent of synthetic perfumes and as a chemical intermediate. Also known as diphenyl ketone; phenyl ketone.

benzopyrene $C_{20}H_{12}$ A five-ring aromatic hydrocarbon found in coal tar, in cigarette smoke, and as a product of incomplete combustion; yellow crystals with a melting point of 179°C; soluble in benzene, toluene, and xylene.

1,2-benzopyrone *See* coumarin.

2,3-benzopyrrole *See* indole.

benzo[*f*]quinoline *See* 5,6-benzoquinoline.

5,6-benzoquinoline $C_{13}H_9N$ Crystals which are soluble in dilute acids, alcohol, ether, or benzene; melting point is 93°C; used as a reagent for the determination of cadmium. Also known as benzo[*f*]quinoline; naphthopyridine; β-naphthoquinoline.

benzoquinone *See* quinone.

1,4-benzoquinone-*N'*-benzoylhydrazone oxime *See* benequinox.

benzoresorcinol $C_{13}H_{10}O_3$ A compound crystallizing as needles from hot-water solution; used in paints and plastics as an ultraviolet light absorber. Also known as resbenzophenone.

benzosulfimide *See* saccharin.

benzothiazole C_6H_4SCHN A thiazole fused to a benzene ring; can be made by ring closure from *o*-amino thiophenols and acid chlorides; derivatives are important industrial products.

4-benzothienyl-*N*-methylcarbamate $C_{10}H_9NO_2S$ A white powder compound with a melting point of 128°C; used as an insecticide for crop insects.

benzothiofuran *See* thianaphthene.

1,2,3-benzotriazole $C_6H_5N_3$ A compound with melting point 98.5°C; soluble in ethanol, insoluble in water; derivatives are ultraviolet absorbers; used as a chemical intermediate. Also known as azimidobenzene; benzene azimide.

benzotrichloride $C_6H_5CCl_3$ A colorless to yellow liquid that fumes upon exposure to air; has penetrating odor; insoluble in water, soluble in ethanol and ether; used to make dyes. Also known as benzenyl trichloride; phenylchloroform; toluene trichloride.

benzotrifluoride A colorless liquid with a boiling point of 102.1°C; used for dyes and pharmaceuticals, as a solvent and vulcanizing agent, and in insecticides. Also known as toluene trifluoride; trifluoromethyl benzene.

benzoyl The radical $C_6H_5ICO^-$ found, for example, in benzoyl chloride.

***N*-benzoylaniline** *See* benzanilide.

benzoylation Introduction of the aryl radical (C_6H_5CO) into a molecule.

benzoyl chloride C_6H_5COCl Colorless liquid whose vapor induces tears; soluble in ether, decomposes in water; used as an intermediate in chemical synthesis.

benzoyl chloride 2,4,6-trichlorophenylhydrazone $C_6H_5CClN_2HC_6H_2Cl_3$ A white to yellow solid with a melting point of 96.5–98°C; insoluble in water; used as an anthelminthic for citrus.

benzoylformaldoxime *See* isonitrosoacetophenone.

benzoyl peroxide $(C_6H_5CO)_2O_2$ A white, crystalline solid; melting point 103–105°C; explodes when heated above 105°C; slightly soluble in water, soluble in organic solvents; used as a bleaching and drying agent and a polymerization catalyst. Also known as dibenzoyl peroxide.

benzoylphenyl carbinol *See* benzoin.

benzoylpropethyl $C_{18}H_{17}Cl_2NO_3$ An off-white, crystalline compound with a melting point of 72°C; used as a preemergence herbicide for control of wild oats. Also known as ethyl-*N*-benzoyl-*N*-(3,4-dichlorophenyl)-2-aminopropionate.

benzozone   *See* acetyl benzoyl peroxide.

3,4-benzpyrene $C_{20}H_{12}$ A polycyclic hydrocarbon; a chemical carcinogen that will cause skin cancer in many species when applied in low dosage.

benzthiazuron $C_9H_9N_3SO$ A white powder that decomposes at 287°C; slightly soluble in water; used as a preemergent herbicide for sugarbeets and fodder beet crops. Also known as gatinon; 1-methyl-3-(2-benzothiazolyl)-urea.

benzyl   The radical $C_6H_5CH_2{}^-$ found, for example, in benzyl alcohol, $C_6H_5CH_2OH$.

benzyl acetate $C_6H_5CH_2OOCCH_3$ A colorless liquid with a flowery odor; used in perfumes and flavorings and as a solvent for plastics and resins, inks, and polishes. Also known as phenylmethyl acetate.

benzylacetone $C_6H_5(CH_2)_2COCH_3$ A liquid with a melting point of 233–234°C; used as an attractant to trap melon flies.

benzyl alcohol $C_6H_5CH_2OH$ An alcohol that melts at 15.3°C, boils at 205.8°C, and is soluble in water and readily soluble in alcohol and ether; valued for the esters it forms with acetic, benzoic, and sebacic acids and used in the soap, perfume, and flavor industries. Also known as phenylmethanol.

benzylamine $C_6H_5CH_2NH_2$ A liquid that is soluble in water, ethanol, and ether; boils at 185°C (770 mmHg) and at 84°C (24 mmHg); it is toxic; used as a chemical intermediate in dye production. Also known as aminotoluene.

benzylbenzene   *See* diphenyl methane.

benzyl benzoate $C_6H_5COOCH_2C_6H_5$ An oily, colorless liquid ester; used as an antispasmodic drug and as a scabicide.

benzyl bromide $C_6H_5CH_2Br$ A toxic, irritating, corrosive clear liquid with a boiling point of 198–199°C; acts as a lacrimator; soluble in alcohol, benzene, and ether; used to make foaming and frothing agents. Also known as α-bromotoluene.

benzyl carbinol   *See* phenethyl alcohol.

benzylcarbonylchloride   *See* benzyl chloroformate.

benzyl chloride $C_6H_5CH_2Cl$ A colorless liquid with a pungent odor produced by the chlorination of toluene.

benzyl chloroformate $C_8H_7ClO_2$ An oily liquid with an acrid odor which causes eyes to tear; boiling point is 103°C (20 mmHg pressure); used to block the amino group in peptide synthesis. Also known as benzylcarbonyl chloride; carbobenzoxy chloride; chloroformic acid benzyl ester.

benzyl cinnamate $C_8H_7COOCH_2C_6H_5$ White crystals; melting point 39°C; insoluble in water, soluble in ethanol; used in perfumery. Also known as cinnamein.

benzyl cyanide $C_6H_5CH_2CN$ A toxic, colorless liquid; insoluble in water, soluble in alcohol and ethanol; boils at 234°C; used in organic synthesis. Also known as phenylacetonitrile; α-tolunitrile.

benzyl dichloride   *See* benzal chloride.

***N*-benzyl-*N*-(3,4-dichlorophenyl)-*N'*,*N'*-dimethylurea** $C_{14}H_{18}ONCl_2$ A 50% wettable powder; used as a preemergent herbicide for rice, cotton, fruit trees, vineyards, soybeans, and flax.

benzyl disulfide *See* dibenzyl disulfide.

α-(benzyldithio)toluene *See* dibenzyl disulfide.

benzyleneglycol *See* hydrobenzoin.

benzyl ether $(C_6H_5CH_2)_2O$ A liquid unstable at room temperature; boiling point 295–298°C; used in perfumes and as a plasticizer for nitrocellulose. Also known as dibenzyl ether.

1-benzyl-3-ethyl-6,7-dimethoxy isoquinoline *See* eupavarine.

benzyl ethyl ether $C_6H_5CH_2OC_2H_5$ A colorless, oily, combustible liquid with a boiling point of 185°C; used in organic synthesis and as a flavoring.

benzyl fluoride $C_6H_5CH_2F$ A toxic, irritating, colorless liquid with a boiling point of 139.8°C at 753 mmHg; used in organic synthesis.

benzyl formate $C_6H_5CH_2OOCH$ A colorless liquid with a fruity-spicy odor and a boiling point of 203°C; used in perfumes and as a flavoring.

benzylhydroperoxide *See* perbenzoic acid.

benzylideneacetone $C_6H_5CH—CHCOCH_3$ A crystalline compound soluble in alcohol, benzene, chloroform, and ether; melting point is 41–45°C; used in perfume manufacture and in organic synthesis. Also known as acetocinnamone; benzalacetone; cinnamyl methyl ketone; methyl styryl ketone.

3-benzylideneamino-4-4-phenylthiazoline-2-thione *See* fentiazon.

benzylidene chloride *See* benzal chloride.

benzyl isoeugenol $CH_3CHCHC_6H_3(OCH_3)OCH_2C_6H_5$ A white, crystalline compound with a floral odor; soluble in alcohol and ether; used in perfumery.

benzyl mercaptan $C_6H_5CH_2SH$ A colorless liquid with a boiling point of 195°C; soluble in alcohol and carbon disulfide; used as an odorant and for flavoring.

benzyl penicillinic acid $C_{16}H_{18}N_2O_4S$ An amorphous white powder extracted with ether or chloroform from an acidified aqueous solution of benzyl penicillin.

ortho-**benzylphenol** *See* 2-(hydroxydiphenyl)methane.

benzyl propionate $C_2H_5COOCH_2C_6H_5$ A combustible liquid with a sweet odor and a boiling point of 220°C; used in perfumes and for flavoring.

benzyl salicylate $C_{14}H_{12}O_3$ A thick liquid with a slight, pleasant odor; used as a fixer in perfumery and in sunburn preparations.

benzyne C_6H_4 A chemical species whose structure consists of an aromatic ring in which four carbon atoms are bonded to hydrogen atoms and two adjacent carbon atoms lack substitutents; a member of a class of compounds known as arynes.

berbamine $C_{37}H_{40}N_2O_6$ An alkaloid; melting point 170°C; slightly soluble in water, soluble in alcohol and ether.

berberine $C_{20}H_{19}NO_5$ A toxic compound; melting point 145°C; the anhydrous form is insoluble in water, soluble in alcohol and ether; an alkaloid whose sulfate or hydrochloride might have some use in medicine.

Bergius process Treatment of carbonaceous matter, such as coal or cellulosic materials, with hydrogen at elevated pressures and temperatures in the presence of a catalyst, to form an oil similar to crude petroleum. Also known as coal hydrogenation.

Berg's diver method *See* diver method.

berkelium A radioactive element, symbol Bk, atomic number 97, the eighth member of the actinide series; properties resemble those of the rare-earth cerium.

Berl saddle A type of column packing used in distillation columns.

berrylate A salt such as potassium beryllate, formed by action of a strong base, such as potassium hydroxide, and beryllium oxide.

Berthelot equation A form of the equation of state which relates the temperature, pressure, and volume of a gas with the gas constant.

Berthelot-Thomsen principle The principle that of all chemical reactions possible, the one developing the greatest amount of heat will take place, with certain obvious exceptions such as changes of state.

berthollide A compound whose solid phase exhibits a range of composition.

beryllia *See* beryllium oxide.

beryllide A chemical combination of beryllium with a metal, such as zirconium or tantalum.

beryllium A chemical element, symbol Be, atomic number 4, atomic weight 9.0122.

beryllium fluoride BeF_2 A hygroscopic, amorphous solid with a melting point of 800°C; soluble in water; used in beryllium metallurgy.

beryllium nitrate $Be(NO_3)_2 \cdot 3H_2O$ A compound that forms colorless, deliquescent crystals that are soluble in water; used to introduce beryllium oxide into materials used in incandescent mantles.

beryllium nitride Be_3N_2 Refractory, white crystals with a melting point of $2200 \pm 40°C$; used in the manufacture of radioactive carbon-14 and in experimental rocket fuels.

beryllium oxide BeO An amorphous white powder, insoluble in water; used to make beryllium salts and as a refractory. Also known as beryllia.

beta carotene $C_{40}H_{56}$ A carotenoid hydrocarbon pigment found widely in nature, always associated with chlorophylls; converted to vitamin A in the liver of many animals.

betacyanin A group of purple plant pigments found in leaves, flowers, and roots of members of the order Caryophyllales.

beta decay Radioactive transformation of a nuclide in which the atomic number increases or decreases by unity with no change in mass number; the nucleus emits or absorbs a beta particle (electron or positron). Also known as beta disintegration.

beta decay spectrum The distribution in energy or momentum of the beta particles arising from a nuclear disintegration process.

beta disintegration *See* beta decay.

beta emitter A radionuclide that disintegrates by emission of a negative or positive electron.

beta globulin A heterogeneous fraction of serum globulins containing transferrin and various complement components.

betaine $C_5H_{11}O_2N$ An alkaloid; very soluble in water, soluble in ethyl alcohol and methanol; the hydrochloride is used as a source of hydrogen chloride and in medicine. Also known as lycine; oxyneurine.

betalain The name for a group of 35 red or yellow compounds found only in plants of the family Caryophyllales, including red beets, red chard, and cactus fruits.

betanin An anthocyanin that contains nitrogen and constitutes the principal pigment of garden beets.

beta particle An electron or positron emitted from a nucleus during beta decay.

beta ray A stream of beta particles.

beta-ray spectrometer An instrument used to determine the energy distribution of beta particles and secondary electrons. Also known as beta spectrometer.

beta spectrometer *See* beta-ray spectrometer.

betaxanthin The name given to any of the yellow pigments found only in plants of the family Caryophyllales; they always occur with betacyanins.

BET equation *See* Brunauer-Emmett-Teller equation.

Betterton-Kroll process A method for obtaining pure bismuth from softened and desilverized lead.

betula oil *See* methyl salicylate.

betulinic acid $C_{30}H_{48}O_3$ A dibasic acid, slightly soluble in water, ethyl alcohol, and acetone.

BFE *See* bromotrifluoroethylene.

BFS *See* beam-foil spectroscopy.

BHA *See* butylated hydroxyanisole.

BHC *See* 1,2,3,4,5,6-hexachlorocyclohexane.

BHT *See* butylated hydroxytoluene.

Bi *See* bismuth.

biacetyl *See* diacetyl.

biamperometry Amperometric titration that uses two polarizing or indicating electrodes to detect the end point of a redox reaction between the substance being titrated and the titrant.

biaxial crystal A crystal of low symmetry in which the index ellipsoid has three unequal axes.

biaxial indicatrix An ellipsoid whose three axes at right angles to each other are proportional to the refractive indices of a biaxial crystal.

Biazzi process A continuous-flow process for the nitration of glycerin to nitroglycerin; also used to produce glycol dinitrate and diethylene glycol nitrate.

bibenzyl $C_{14}H_{14}$ A hydrocarbon consisting of two benzene rings attached to ethane. Also known as dibenzyl.

bicarbonate A salt obtained by the neutralization of one hydrogen in carbonic acid.

bichloride of mercury *See* mercuric chloride.

bichromate of soda *See* sodium dichromate.

bicyclic compound A compound having two rings which share a pair of bridge-head carbon atoms.

bidentate ligand A chelating agent having two groups capable of attachment to a metal ion.

Biedenharn identity A relationship among the six-*j* symbols of Wigner.

bifenox $C_{14}H_9Cl_2NO_5$ A tan, crystalline compound with a melting point of 84–86°C; insoluble in water; used as a preemergence herbicide for weed control in soybeans, corn, and sorghum, and as a pre- and postemergence herbicide in rice and small greens. Also known as methyl 5-(2′,4′-dichlorophenoxy)-2-nitrobenzoate.

bifluoride An acid fluoride whose formula has the form MHF_2, such as sodium bifluoride, $NaHF_2$.

bilayer A layer two molecules thick, such as that formed on the surface of the aqueous phase by phospholipids in aqueous solution.

bile acid Any of the liver-produced steroid acids, such as taurocholic acid and glycocholic acid, that appear in the bile as sodium salts.

bile pigment Either of two colored organic compounds found in bile: bilirubin and biliverdin.

bile salt The sodium salt of glycocholic and taurocholic acids found in bile.

bilicyanin A blue pigment found in gallstones; an oxidation product of biliverdin or bilirubin.

biliprotein The generic name for the organic compounds in certain algae that are composed of phycobilin and a conjugated protein.

bilirubin $C_{33}H_{36}N_4O_6$ An orange, crystalline pigment occurring in bile; the major metabolic breakdown product of heme.

biliverdin $C_{33}H_{34}N_4O_6$ A green, crystalline pigment occurring in the bile of amphibians, birds, and humans; oxidation product of bilirubin in humans.

bimolecular Referring to two molecules.

bimolecular reaction A chemical transformation or change involving two molecules.

binapacryl $C_{15}H_{18}O_6N_2$ A light tan solid with a melting point of 68–69°C; insoluble in water; used for powdery mildew and for mites on fruits.

binary compound A compound that has two elements; it may contain two or more atoms; examples are KCl and $AlCl_3$.

binary encounter approximation An approximation for predicting the probability that an incident proton will eject an inner shell electron from an atom; it uses a semiclassical treatment of momentum transfer from the incident proton to the ejected electron.

binary separation Separation by distillation or solvent extraction of a fully miscible liquid mixture of two chemical compounds.

bioassay A method for quantitatively determining the concentration of a substance by its effect on the growth of a suitable animal, plant, or microorganism under controlled conditions.

bioautography A bioassay based upon the ability of some compounds (for example, vitamin B_{12}) to enhance the growth of some organisms or compounds and to repress the growth of others; used to assay certain antibiotics.

biocatalyst A biochemical catalyst, especially an enzyme.

biochemistry The study of chemical substances occurring in living organisms and the reactions and methods for identifying these substances.

biochemorphology The science dealing with the chemical structure of foods and drugs and their reactions on living organisms.

biochrome Any naturally occurring plant or animal pigment.

biocytin $C_{16}H_{28}N_4O_4S$ Crystals with a melting point of 241–243°C; obtained from dilute methanol or acetone solutions; characterized by its utilization by *Lactobacillus casei* and *L. delbrückii* LD5 as a biotin source, and by its unavailability as a biotin source to *L. arabinosus*. Also known as biotin complex of yeast.

biocytinase An enzyme present in the blood and liver which hydrolyzes biocytin into biotin and lysine.

bioenergetics The branch of biology dealing with energy transformations in living organisms.

bioflavonoid A group of compounds obtained from the rinds of citrus fruits and involved with the homeostasis of the walls of small blood vessels; in guinea pigs a marked reduction of bioflavonoids results in increased fragility and permeability of the capillaries; used to decrease permeability and fragility in capillaries in certain conditions. Also known as citrus flavonoid compound; vitamin P complex.

biogeochemical cycle The chemical interactions that exist between the atmosphere, hydrosphere, lithosphere, and biosphere.

biogeochemical prospecting A prospecting technique for subsurface ore deposits based on interpretation of the growth of certain plants which reflect subsoil concentrations of some elements.

biogeochemistry A branch of geochemistry that is concerned with biologic materials and their relation to earth chemicals in an area.

biological oxidation Energy-producing reactions in living cells involving the transfer of hydrogen atoms or electrons from one molecule to another.

biological value A measurement of the efficiency of the protein content in a food for the maintenance and growth of the body tissues of an individual.

biologic artifact An organic compound with a chemical structure that demonstrates the compound's derivation from living matter.

biophile Any element concentrated or found in the bodies of living organisms and organic matter; examples are carbon, nitrogen, and oxygen.

biopolymer A biological macromolecule such as a protein or nucleic acid.

biopotency Capacity of a chemical substance, as a hormone, to function in a biological system.

biosynthesis Production, by synthesis or degradation, of a chemical compound by a living organism.

biotin $C_{10}H_{16}N_2O_3S$ A colorless, crystalline vitamin of the vitamin B complex occurring widely in nature, mainly in bound form.

biotin carboxylase An enzyme which condenses bicarbonate with biotin to form carboxybiotin.

biotin complex of yeast *See* biocytin.

biphenyl $C_{12}H_{10}$ A white or slightly yellow crystalline hydrocarbon, melting point 70.0°C, boiling point 255.9°C, and density 1.9896, which gives plates or monoclinic prismatic crystals; used as a heat-transfer medium and as a raw material for chlorinated diphenyls. Also known as diphenyl; phenylbenzene.

***para*-biphenylamine** $C_{12}H_{11}N$ Leaflets with a melting point of 53°C; readily soluble in hot water, alcohol, and chloroform; used in the detection of sulfates and also as a carcinogen in cancer research. Also known as *para*-aminodiphenyl; xenylamine.

2,2′-bipyridine *See* 2,2′-dipyridyl.

biradical A chemical species having two independent odd-electron sites.

Birkeland-Eyde process An arc process of nitrogen fixation in which air passes through an alternating-current arc flattened by a magnetic field to form about 1% nitric oxide.

bis- A prefix indicating doubled or twice.

***O,O*-bis(*para*-chlorophenyl)acetimidoylphosphoramidothioate** *See* phosazetim.

2,2-bis(*para*-chlorophenyl)-1,1-dichloroethane $C_{14}H_{10}Cl_4$ A colorless, crystalline compound with a melting point of 109-111°C; insoluble in water; used as an insecticide on fruits and vegetables. Also known as DDD; TDE.

1,1-bis(4-chlorophenyl)ethanol *See* chlorofenethol.

1,1-bis(*para*-chlorophenyl)methyl carbinol *See* p,p′-dichlorodiphenylmethyl carbinol.

α,α-bis(*para*-chlorophenyl)-3-pyridine methanol *See* parinol.

1,1-bis(*para*-chlorophenyl)-2,2,2-trichloroethanol *See* dicofol.

bis(diethylthiocarbamyl)disulfide *See* tetraethylthiuram disulfide.

bis(dimethylthiocarbamoyl)disulfide *See* thiram.

bis(dimethylthiocarbamoylthio)methylarsine *See* urbacid.

bis(dimethylthiocarbamyl)sulfide *See* tetramethylthiuram monosulfide.

bisectrix A line that is the bisector of the angle between the optic axes of a biaxial crystal.

bishydroxycoumarin *See* dicoumarin.

3,3-bis(hydroxymethyl)pentane *See* 2,2-diethyl-1,3-propanediol.

1,3-bishydroxymethylurea *See* dimethylolurea.

2,4-bis(isopropylamino)-6-methoxy-S-triazine *See* prometon.

2,4-bis(isopropylamino)-6-(methylthio)-S-triazine   *See* prometryn.

1,4-bis(methanesulfonoxy)butane   *See* busulfan.

bismuth   A metallic element, symbol Bi, of atomic number 83 and atomic weight 208.980.

bismuthate   A compound of bismuth in which the bismuth has a valence of $+5$; an example is sodium bismuthate, $NaBiO_3$.

bismuth carbonate   *See* bismuth subcarbonate.

bismuth chloride $BiCl_3$ A deliquescent material that melts at 230–232°C and decomposes in water to form the oxychloride; used to make bismuth salts. Also known as bismuth trichloride.

bismuth chromate $Bi_2O_3 \cdot Cr_2O_3$ An orange-red powder, soluble in alkalies and acids; used as a pigment.

bismuth citrate $BiC_6H_5O_7$ A salt of citric acid that forms white crystals, insoluble in water; used as an astringent.

bismuth hydroxide $Bi(OH)_3$ A water-insoluble, white powder; precipitated by hydroxyl ion from bismuth salt solutions.

bismuth iodide BiI_3 A bismuth halide that sublimes in grayish-black hexagonal crystals melting at 408°C, insoluble in water; used in analytical chemistry.

bismuth nitrate $Bi(NO_3)_3 \cdot 5H_2O$ White, triclinic crystals that decompose in water; used as an astringent and antiseptic.

bismuth oleate $Bi(C_{17}H_{33}COO)_3$ A salt of oleic acid obtained as yellow granules; used in medicines to treat skin diseases.

bismuth oxide   *See* bismuth trioxide.

bismuth oxycarbonate   *See* bismuth subcarbonate.

bismuth oxychloride $BiOCl$ A white powder; insoluble in water, soluble in acid; a toxic material if ingested; used in pigments and cosmetics.

bismuth phenate $C_6H_5O \cdot Bi(OH)_2$ An odorless, tasteless, gray-white powder; used in medicine. Also known as bismuth phenolate; bismuth phenylate.

bismuth phenolate   *See* bismuth phenate.

bismuth phenylate   *See* bismuth phenate.

bismuth potassium tartrate   *See* potassium bismuth tartrate.

bismuth pyrogallate $Bi(OH)C_6H_3(OH)O_2$ An odorless, tasteless, yellowish-green, amorphous powder; used in medicine as intestinal antiseptic and dusting powder. Also known as basic bismuth pyrogallate; helcosol.

bismuth spar   *See* bismutite.

bismuth subcarbonate $(BiO)_2CO_3$ or $Bi_2O_3 \cdot CO_2 \cdot \frac{1}{2}H_2O$ A white powder; dissolves in hydrochloric or nitric acid, insoluble in alcohol and water; used as opacifier in x-ray diagnosis, in ceramic glass, and in enamel fluxes. Also known as basic bismuth carbonate; bismuth carbonate; bismuth oxycarbonate.

bismuth subgallate $C_6H_2(OH)_3COOBi(OH)_2$ A yellow powder; dissolves in dilute alkali solutions, but is insoluble in water, ether, and alcohol; used in medicine. Also known as basic bismuth gallate.

bismuth subnitrate $4BiNO_3(OH)_2 \cdot BiO(OH)$ A white, hygroscopic powder; used in bismuth salts, perfumes, cosmetics, ceramic enamels, pharmaceuticals, and analytical chemistry. Also known as basic bismuth nitrate; Spanish white.

bismuth subsalicylate $Bi(C_7H_5O)_3Bi_2O_3$ A white powder that is insoluble in ethanol and water; used in medicine and as a fungicide for tobacco crops. Also known as basic bismuth salicylate.

bismuth telluride Bi_2Te_3 Gray, hexagonal platelets with a melting point of 573°C; used for semiconductors, thermoelectric cooling, and power generation applications.

bismuth trichloride *See* bismuth chloride.

bismuth trioxide Bi_2O_3 A yellow powder; melting point 820°C; insoluble in water, dissolves in acid; used to make enamels and to color ceramics. Also known as bismuth oxide; bismuth yellow.

bismuth yellow *See* bismuth trioxide.

bisphenoid A form apparently consisting of two sphenoids placed together symmetrically.

bisphenol A $(CH_3)_2C(C_6H_5OH)_2$ Brown crystals that are insoluble in water; used in the production of phenolic and epoxy resins. Also known as *p,p'*-dihydroxydiphenyldimethylene.

N,N-bis(phosphonomethyl)glycine *See* glyphosine.

bistable system A chemical system with two relatively stable states which permits an oscillation between domination by one of these states to domination by the other.

1,3-bis(trimethylamino)-2-propanol diiodide *See* propiodal.

bisulfate A compound that has the HSO_4^- radical; derived from sulfuric acid.

bitartrate A salt with the radical $HC_4H_4O_6^-$. Also known as acid tartrate.

bithionol A halogenated form of bisphenol used as an ingredient in germicidal soaps and as a medicine in the treatment of clonorchiases.

bittern Concentrated sea water or brine containing the bromides and magnesium and calcium salts left in solution after sodium chloride has been removed by crystallization.

biuret $NH_2CONHCONH_2$ Colorless needles that are soluble in hot water and decompose at 190°C; a condensation product of urea. Also known as allophanamide; carbamylurea; dicarbamylamine.

bivalent Possessing a valence of two.

bixin $C_{25}H_{30}O_4$ A carotenoid acid occurring in the seeds of *Bixa orellano;* used as a fat and food coloring agent.

Bk *See* berkelium.

black Fine particles of impure carbon that are made by the incomplete burning of carbon compounds, such as natural gas, naphthas, acetylene, bones, ivory, and vegetables.

black antimony *See* antimony trisulfide.

black cyanide   *See* calcium cyanide.

black mercury sulfide   *See* mercuric sulfide.

Blagden's law   The law that the lowering of a solution's freezing point is proportional to the amount of dissolved substance.

blanc fixe $BaSO_4$ A commercial name for barium sulfate, with some use in pure form in the paint, paper, and pigment industries as a pigment extender.

Blanc rule   The rule that glutaric and succinic acids yield cyclic anhydrides on pyrolysis, while adipic and pimelic acids yield cyclic ketones; there are certain exceptions.

blasticidin-S   A compound with a melting point of 235–236°C; soluble in water; used as a fungicide for rice crops. Also known as blasticidin-*S*-benzylaminobenzenesulfonate.

blasticidin-*S*-benzylaminobenzenesulfonate   *See* blasticidin-S.

bleaching agent   An oxidizing or reducing chemical such as sodium hypochlorite, sulfur dioxide, sodium acid sulfite, or hydrogen peroxide.

bleed   Diffusion of coloring matter from a substance.

blending stock   Any substance used for compounding gasoline, including natural gasoline, catalytically reformed products, and additives.

blocked operation   The use of a single chemical or refinery process unit alternately in more than one operation; for example, a catalytic reactor will first produce a chemical product and then will be blocked from the main process stream during catalyst regeneration.

blocking   Undesired adhesion of granular particles; often occurs with damp powders or plastic pellets in storage bins or during movement through conduits.

blocking group   In peptide synthesis, a group that is reacted with a free amino or carboxyl group on an amino acid to prevent its taking part in subsequent formation of peptide bonds.

block polymer   A copolymer whose chain is composed of alternating sequences of identical monomer units.

blood sugar   The carbohydrate, principally glucose, of the blood.

blotter press   A plate-and-frame filter in which the filter medium is blotting paper.

blowback   **1.** A continuous stream of liquid or gas bled through air lines from instruments and to the process line being monitored; prevents process fluid from backing up and contacting the instrument. **2.** Reverse flow of fluid through a filter medium to remove caked solids. Also known as backwash.

blowcase   A cylindrical or spherical corrosion- and pressure-resistant container from which acid is forced by compressed air to the agitator; used in manufacture of acids but largely superseded by centrifugal pumps. Also known as acid blowcase; acid egg.

blowdown   Removal of liquids or solids from a process vessel or storage vessel or a line by the use of pressure.

blowdown line   A large conduit to receive and confine fluids forced by pressure from process vessels.

blowdown stack A vertical stack or chimney into which the contents of a chemical or petroleum process unit are emptied in case of an operational emergency.

blowing The introduction of compressed air near the bottom of a tank or other container in order to agitate the liquid therein.

blowing still A still or process column in which blown or oxidized asphalt is made.

blowpipe reaction analysis A method of analysis in which a blowpipe is used to heat and decompose a compound or mineral; a characteristic color appears in the flame or a colored crust appears on charcoal.

blowpit *See* blowtank.

blowtank A tank or pit, used in papermaking, into which the contents of a digester are blown upon completion of a cook. Also known as blowpit.

blue tetrazolium $C_{40}H_{32}Cl_2N_8O_2$ Lemon yellow crystals that decompose at 242–245°C; soluble in chloroform, ethanol, and methanol; used in seed germination research, as a stain for molds and bacteria, and in histochemical studies. Also known as dimethoxy neotetrazolium; ditetrazolium chloride; tetrazolium blue.

BM *See* barrels per month.

BNOA *See* β-naphthoxyacetic acid.

boat A platinum or ceramic vessel for holding a substance for analysis by combustion.

boat conformation A boat-shaped conformation in space which can be assumed by cyclohexane or similar compounds; a relatively unstable form.

body-centered lattice A space lattice in which the point at the intersection of the body diagonals is identical to the points at the corners of the unit cell.

Boettger's test A test for the presence of saccharides, uilizing the reduction of bismuth subnitrate to metallic bismuth, a precipitate.

Bohr atom An atomic model having the structure postulated in the Bohr theory.

Bohr-Breit-Wigner theory *See* Breit-Wigner theory.

Bohr effect The effect of carbon dioxide and pH on the oxygen equilibrium of hemoglobin; increase in carbon dioxide prevents an increase in the release of oxygen from oxyhemoglobin.

Bohr magneton The amount $he/4\pi mc$ of magnetic moment, where h is Planck's constant, e and m are the charge and mass of the electron, and c is the speed of light.

Bohr orbit One of the electron paths about the nucleus in Bohr's model of the hydrogen atom.

Bohr radius The radius of the ground-state orbit of the hydrogen atom in the Bohr theory.

Bohr-Sommerfeld theory A modification of the Bohr theory in which elliptical as well as circular orbits are allowed.

Bohr theory A theory of atomic structure postulating an electron moving in one of certain discrete circular orbits about a nucleus with emission or absorption of electromagnetic radiation necessarily accompanied by transitions of the electron between the allowed orbits.

Bohr-Wheeler theory of fission   A theory accounting for the stability of a nucleus against fission by treating it as a droplet of incompressible and uniformly charged liquid endowed with surface tension.

boiler compound   Any chemical used to treat boiler water to prevent corrosion, the fouling of heat-absorbing surfaces, foaming, and the contamination of steam.

boiler scale   Deposits from silica and other contaminants in boiler water that form on the internal surfaces of heat-absorbing components, increase metal temperatures, and result in eventual failure of the pressure parts because of overheating.

boiling   The transition of a substance from the liquid to the gaseous phase, taking place at a single temperature in pure substances and over a range of temperatures in mixtures.

boiling point   Abbreviated bp. **1.** The temperature at which the transition from the liquid to the gaseous phase occurs in a pure substance at fixed pressure. **2.** *See* bubble point.

boiling point elevation   The raising of the normal boiling point of a pure liquid compound by the presence of a dissolved substance, the elevation being in direct relation to the dissolved substance's molecular weight.

boiling range   The temperature range of a laboratory distillation of an oil from start until evaporation is complete.

bond   The strong attractive force that holds together atoms in molecules and crystalline salts. Also known as chemical bond.

bond angle   The angle between bonds sharing a common atom. Also known as valence angle.

bond distance   The distance separating the two nuclei of two atoms bonded to each other in a molecule. Also known as bond length.

bonded-phase chromatography   A type of high-pressure liquid chromatography which employs a stable, chemically bonded stationary phase.

bond energy   The heat of formation of a molecule from its constituent atoms.

bond hybridization   The linear combination of two or more simple atomic orbitals.

bonding   The joining together of atoms to form molecules or crystalline salts.

bonding electron   An electron whose orbit spans the entire molecule and so assists in holding it together.

bonding orbital   A molecular orbital formed by a bonding electron whose energy decreases as the nuclei are brought closer together, resulting in a net attraction and chemical bonding.

bond length   *See* bond distance.

bone ash   A white ash consisting primarily of tribasic calcium phosphate obtained by burning bones in air; used in cleaning jewelry and in some pottery.

Boord synthesis   A method of producing alpha olefins by the reduction of alpha bromo ethers with zinc.

boracic acid   *See* boric acid.

borane   **1.** A class of binary compounds of boron and hydrogen; boranes are used as fuels. Also known as boron hydride.   **2.** A substance which may be considered a derivative of a boron-hydrogen compound, such as BCl_3 and $B_{10}H_{12}I_2$.

borate   **1.** A generic term referring to salts or esters of boric acid.   **2.** Related to boric oxide, B_2O_3, or commonly to only the salts of orthoboric acid, H_3BO_3.

borazole $B_3N_3H_6$ A colorless liquid boiling at 53°C; with water it hydrolyzes to form boron hydrides; the borazole molecule is the inorganic analog of the benzene molecule.

borazon   A form of boron nitride with a zinc blende structure produced by subjecting the ordinary form to high pressure and temperature.

boric acid H_3BO_3 An acid derived from boric oxide in the form of white, triclinic crystals, melting at 185°C, soluble in water. Also known as boracic acid; orthoboric acid.

boric acid ester   Any compound readily hydrolyzed to yield boric acid and the respective alcohol; for example, trimethyl borate hydrolyzes to boric acid and methyl alcohol.

boric oxide B_2O_3 A trioxide of boron obtained as rhombic crystals melting at 460°C; used as an intermediate in the production of boron halides and metallic borides and as a thermal neutron absorber in nuclear engineering. Also known as boron oxide.

boride   A binary compound of boron and a metal formed by heating a mixture of the two elements.

borneol $C_{10}H_{17}OH$ White lumps with camphor odor; insoluble in water, soluble in alcohol; melting point 203°C; used in perfumes, medicine, and chemical synthesis. Also known as bornyl alcohol; 2-camphanol; 2-hydroxycamphane.

Born equation   An equation for determining the free energy of solvation of an ion in terms of the Avogadro number, the ionic valency, the ion's electronic charge, the dielectric constant of the electrolytic, and the ionic radius.

Born-Oppenheimer method   A method for calculating the force constants between atoms by assuming that the electron motion is so fast compared with the nuclear motions that the electrons follow the motions of the nuclei adiabatically.

bornyl acetate $C_{10}H_{17}OOCCH_3$ A colorless liquid that forms crystals at 10°C; has characteristic piny-camphoraceous odor; used in perfumes and for flavoring.

bornyl alcohol   *See* borneol.

bornylane   *See* camphane.

bornyl isovalerate $C_{10}H_{17}OOC_5H_9$ An aromatic fluid with a boiling point of 255–260°C; soluble in alcohol and ether; used in medicine and as a flavoring.

boroethane   *See* diborane.

boron   A chemical element, symbol B, atomic number 5, atomic weight 10.811; it has three valence electrons and is nonmetallic.

boron-10   A nonradioactive isotope of boron with a mass number of 10; it is a good absorber for slow neutrons, simultaneously emitting high-energy alpha particles, and is used as a radiation shield in Geiger counters.

boron carbide   Any compound of boron and carbon, especially B_4C (used as an abrasive, alloying agent, and neutron absorber).

boron fiber   Fiber produced by vapor-deposition methods; used in various composite materials to impart a balance of strength and stiffness. Also known as boron filament.

boron filament   *See* boron fiber.

boron fluoride BF_3 A colorless pungent gas in a dry atmosphere; used in industry as an acidic catalyst for polymerizations, esterifications, and alkylations. Also known as boron trifluoride.

boron fluoride etherate   *See* boron trifluoride etherate.

boron hydride   *See* borane.

boron nitride   BN   A binary compound of boron and nitrogen, especially a white, fluffy powder with high chemical and thermal stability and high electrical resistance.

boron nitride fiber   Inorganic, high-strength fiber, made of boron nitride, that is resistant to chemicals and electricity but susceptible to oxidation above 1600°F (870°C); used in composite structures for yarns, fibers, and woven products.

boron oxide   *See* boric oxide.

boron polymer   Macromolecules formed by polymerization of compounds containing, for example, boron-nitrogen, boron-phosphorus, or boron-arsenic bonds.

boron trichloride BCl_3 A colorless liquid used as a catalyst and in refining of aluminum, magnesium, zinc, and copper.

boron triethoxide   *See* ethyl borate.

boron triethyl   *See* triethylborane.

boron trifluoride   *See* boron fluoride.

boron trifluoride etherate $C_4H_{10}BF_3O$ A fuming liquid hydrolyzed by air immediately; boiling point is 125.7°C; used as a catalyst in reactions involving condensation, dehydration, polymerization, alkylation, and acetylation. Also known as boron fluoride etherate; ethyl ether–boron trifluoride complex.

bottoms   *See* residue.

bottom steam   Steam piped into the bottom of the still during oil distillation.

boturon $C_{12}H_{13}N_2OCl$ A white solid with a melting point of 145–146°C; used as pre- and postemergence herbicide in cereals, orchards, and vineyards. Also known as butyron.

Bouguer-Lambert-Beer law   The intensity of a beam of monochromatic radiation in an absorbing medium decreases exponentially with penetration distance. Also known as Beer-Lambert-Bouguer law; Lambert-Beer law.

Bouguer-Lambert law   The law that the change in intensity of light transmitted through an absorbing substance is related exponentially to the thickness of the absorbing medium and a constant which depends on the sample and the wavelength of the light. Also known as Lambert's law.

boule   A pure crystal, such as silicon, having the atomic structure of a single crystal, formed synthetically by rotating a small seed crystal while pulling it slowly out of molten material in a special furnace.

boundary line   On a phase diagram, the line along which any two phase areas adjoin in a binary system, or the line along which any two liquidus surfaces intersect in a ternary system.

boundary value component   *See* perfectly mobile component.

bound electron   An electron whose wave function is negligible except in the vicinity of an atom.

bound level   An energy level in a nucleus so close to the ground state that it can only decay by gamma emission.

bound water   Water that is a portion of a system such as tissues or soil and does not form ice crystals until the material's temperature is lowered to about $-20°C$.

Bouvealt-Blanc method   A laboratory method for preparing alcohols by reduction of esters utilizing sodium dissolved in alcohol.

bowl classifier   A shallow bowl with a concave bottom so that a liquid-solid suspension can be fed to the center; coarse particles fall to the bottom, where they are raked to a central discharge point, and liquid and fine particles overflow the edges and are collected.

bp   *See* boiling point.

bpd   *See* barrels per day.

bpm   *See* barrels per month.

BPMC   *See* 2-*sec*-butyl phenyl-*N*-methyl carbamate.

Br   *See* bromine.

brachyaxis   The shorter lateral axis, usually the *a* axis, of an orthorhombic or triclinic crystal. Also known as brachydiagonal.

brachydiagonal   *See* brachyaxis.

Brackett series   A series of lines in the infrared spectrum of atomic hydrogen whose wave numbers are given by $R_H[(1/16)-(1/n^2)]$, where R_H is the Rydberg constant for hydrogen and n is any integer greater than 4.

bradykinin $C_{50}H_{73}N_{15}O_{11}$ A polypeptide kinin; forms an amorphous precipitate in glacial acetic acid; released from plasma precursors by plasmin; acts as a vasodilator. Also known as callideic I; kallidin I.

Bragg curve   **1.** A curve showing the average number of ions per unit distance along a beam of initially monoenergetic ionizing particles, usually alpha particles, passing through a gas. Also known as Bragg ionization curve. **2.** A curve showing the average specific ionization of an ionizing particle of a particular kind as a function of its kinetic energy, velocity, or residual range.

Bragg-Kleeman rule   *See* Bragg rule.

Bragg rule   An empirical rule according to which the mass stopping power of an element for alpha particles is inversely proportional to the square root of the atomic weight. Also known as Bragg-Kleeman rule.

branch   **1.** A product resulting from one mode of decay of a radioactive nuclide that has two or more modes of decay.   **2.** A carbon side chain attached to a molecule's main carbon chain.

branching The occurrence of two or more modes by which a radionuclide can undergo radioactive decay. Also known as multiple decay; multiple disintegration.

branching fraction That fraction of the total number of atoms involved which follows a particular branch of the disintegration scheme; usually expressed as a percentage.

branching ratio The ratio of the number of parent atoms or particles decaying by one mode to the number decaying by another mode; the ratio of two specified branching fractions.

brandy A potable alcoholic beverage distilled from wine or fermented fruit juice, usually after the aging of the wine in wooden casks; cognac is a brandy distilled from wines made from grapes from the Cognac region of France.

brassin Any of a class of plant hormones characterized as long-chain fatty-acid esters; brassins act to induce both cell elongation and cell division in leaves and stems.

Bravais indices A modification of the Miller indices; frequently used for hexagonal and trigonal crystalline systems; they refer to four axes: the c-axis and three others at 120° angles in the basal plane.

Bravais lattice One of the 14 possible arrangements of lattice points in space such that the arrangement of points about any chosen point is identical with that about any other point.

breakpoint *See* breakthrough.

breakthrough **1.** A localized break in a filter cake or precoat that permits fluid to pass through without being filtered. Also known as breakpoint. **2.** In an ion-exchange system, the first appearance of unadsorbed ions of the type which deplete the activity of the resin bed; this indicates that the bed must be regenerated.

Breit-Wigner formula A formula which relates the cross section of a particular nuclear reaction with the energy of the incident particle, when the energy is near that required to form a discrete resonance level of the component nucleus.

Breit-Wigner theory A theory of nuclear reactions from which the Breit-Wigner formula is derived. Also known as Bohr-Breit-Wigner theory.

brevitoxin One of several icthyotoxins produced by the dinoflagellate *Ptychodiscus brevis.*

Brewster process Concentration of dilute acetic acid by use of an extraction solvent (for example, isopropyl ether), followed by distillation.

bridging ligand A ligand in which an atom or molecular species which is able to exist independently is simultaneously bonded to two or more metal atoms.

bright-line spectrum An emission spectrum made up of bright lines on a dark background.

Brix degree A unit of the Brix scale.

Brix scale A hydrometer scale for sugar solutions indicating the percentage by weight of sugar in the solution at a specified temperature.

broadening of spectral line A widening of spectral lines by collision or pressure broadening, or possibly by Doppler effect.

Broenner's acid *See* Brönner's acid.

bromacetone $CH_2BrCOCH_3$ A colorless liquid which is a powerful irritant and lacrimator; used as tear gas and to make other chemicals.

bromacil 5-Bromo-3-*sec*-butyl-6-methyluracil, a soil sterilant; general at high dosage and selective at low.

bromate **1.** BrO_3^- A negative ion derived from bromic acid, $HBrO_3$ **2.** A salt of bromic acid. **3.** $C_9H_9ClO_3$ A light brown solid with a melting point of 118–119°C; used as an herbicide to control weeds in crops such as flax, cereals, and legumes. Also known as 2-methyl-4-chlorophenoxyacetic acid.

bromcresol green *See* bromocresol green.

bromcresol purple *See* bromocresol purple.

bromelain An enzyme that digests protein and clots milk; prepared by precipitation by acetone from pineapple juice; used to tenderize meat, to chill-proof beer, and to make protein hydrolysates. Also spelled bromelin.

bromelin *See* bromelain.

bromeosin *See* eosin.

bromic acid $HBrO_3$ A liquid, colorless to slightly yellow; boils with decomposition at 100°C; used in dyes and as a chemical intermediate.

bromic ether *See* ethyl bromide.

bromide A compound derived from hydrobromic acid, HBr, with the bromine atom in the -1 oxidation state.

brominating agent A compound capable of introducing bromine into a molecule; examples are phosphorus tribromide, bromine chloride, and aluminum tribromide.

bromination The process of introducing bromine into a molecule.

bromine A chemical element, symbol Br, atomic number 35, atomic weight 79.904; used to make dibromide ethylene and in organic synthesis and plastics.

bromine number The amount of bromine absorbed by a fatty oil; indicates the purity of the oil and degree of unsaturation.

bromine test A laboratory test in which the unsaturated hydrocarbons present in a crude oil are determined by mixing a sample with bromine; the lower the rate of bromine absorption, the more paraffinic the test sample.

bromine trifluoride BrF_3 A liquid with a boiling point of 135°C.

bromine value An expression representing the number of centigrams of bromine absorbed by 1 gram of oil under test conditions; an indication of the degree of unsaturation of a given oil.

bromine water An aqueous saturated solution of bromine used as a reagent wherever a dilute solution of bromine is needed.

bromo- A prefix that indicates the presence of bromine in a molecule.

N-bromoacetamide $CH_3CONHBr$ Needlelike crystals with a melting point of 102–105°C; soluble in warm water and cold ether; used as a brominating agent and in the oxidation of primary and secondary alcohols. Also known as acetobromamide.

para-bromoacetanilide Crystals with a melting point of 168°C; soluble in benzene, chloroform, and ethyl acetate; insoluble in cold water; used as an analgesic and antipyretic. Also known as bromoanilide; monobromoacetanilide.

bromoacetone $BrCH_2COCH_3$ A colorless liquid used as a lacrimatory agent.

bromo acid *See* eosin.

bromoalkane An aliphatic hydrocarbon with bromine bonded to it. Also known as bromohydrocarbon.

bromoallylene *See* allyl bromide.

bromoanilide *See para*-bromoacetanilide.

4-bromoaniline *See para*-bromoaniline.

para-bromoanisole C_7H_7BrO Crystals which melt at 9–10°C; used in disinfectants.

bromobenzene C_6H_5Br A heavy, colorless liquid with a pleasant odor; used as a solvent, in motor fuels and top-cylinder compounds, and to make other chemicals.

para-bromobenzyl bromide $BrC_6H_4CH_2Br$ Crystals with an aromatic odor and a melting point of 61°C; soluble in cold and hot alcohol, water, and ether; used to identify aromatic carboxylic acids. Also known as *para*,α-dibromotoluene.

bromobenzyl cyanide $C_6H_5CHBrCN$ A light yellow oily compound used as a tear gas for training and for riot control. Abbreviated BBC.

bromochloromethane $BrCH_2Cl$ A clear, colorless liquid with a boiling point of 67°C; volatile, soluble in organic solvents, with a chloroformlike odor; used in fire extinguishers. Also known as methylene chlorobromide.

3-(4-bromo-3-chlorophenyl)-1-methoxy-1-methylurea *See* chlorbromuron.

bromochloroprene $CHCl{=}CHCH_2Br$ A compound used as a nematicide and soil fumigant. Also known as 3-bromo-1-chloropropene-1.

3-bromo-1-chloropropene-1 *See* bromochloroprene.

bromocresol green Tetrabromo-*m*-cresol sulfonphthalein, a gray powder soluble in water or alcohol; used as an indicator between pH 4.5 (yellow) and 5.5 (blue). Also known as bromcresol green.

bromocresol purple Dibromo-*o*-cresol sulfonphthalein, a yellow powder soluble in water; used as an indicator between pH 5.2 (yellow) and 6.8 (purple). Also known as bromcresol purple.

bromocyclen $C_8H_5BrCl_6$ A compound used as an insecticide for wheat crops.

0-(4-bromo-2,5-dichlorophenyl)0-methyl phenylphosphorothioate *See* leptophos.

bromoethane *See* ethyl bromide.

bromofenoxim $C_{13}H_7N_3O_6Br_2$ A cream-colored powder with melting point 196–197°C; slightly soluble in water; used as herbicide to control weeds in cereal crops. Also known as 3,5-dibromo-4-hydroxybenzaldehyde O-(2',4'-dinitrophenyl)oxime.

bromoform $CHBr_3$ A colorless liquid, slightly soluble in water; used in the separation of minerals. Also known as tribromomethane.

bromohydrocarbon   *See* bromoalkane.

bromomethane   *See* methyl bromide.

1-bromo-3-methylbutane   *See* isoamyl bromide.

1-bromonaphthalene $C_{10}H_7Br$ An oily liquid that is slightly soluble in water and miscible with chloroform, benzene, ether, and alcohol; used in the determination of index of refraction of crystals and for refractometric fat determination. Also known as α-bromonaphthalene.

α-bromonaphthalene   *See* 1-bromanaphthalene.

1-bromooctane $CH_3(CH_2)_6CH_2Br$ Colorless liquid that is miscible with ether and alcohol; boiling point is 198–200°C; used in organic synthesis. Also known as n-octyl bromide.

***para*-bromophenacyl bromide** $C_8H_6Br_2O$ Crystals with a melting point of 109–110°C; soluble in warm alcohol; used in the identification of carboxylic acids and as a protecting reagent for acids and phenols. Also known as 2,4'-dibromoacetophenone; *para*,α-dibromoacetophenone.

3-(*para*-bromophenyl)-1-methoxy-1-methylurea   *See* metabromuron.

bromophos $C_8H_8SPBrCl_2O_3$ A yellow, crystalline compound with a melting point of 54°C; used as an insecticide and miticide for livestock, household insects, flies, and lice.

bromophosgene   *See* carbonyl bromide.

bromopicrin CBr_3NO_2 Prismatic crystals with a melting point of 103°C; soluble in alcohol, benzene, and ether; used for military poison gas. Also known as nitrobromoform; tribromonitromethane.

3-bromo-1-propene   *See* allyl bromide.

bromosulfophthalein   *See* sulfobromophthalein sodium.

***N*-bromosuccinimide** $C_4H_4BrNO_2$ Orthorhombic bisphenoidal crystals with a melting point of 173–175°C; used in the bromination of olefins. Also known as succinbromimide.

α-bromotoluene   *See* benzyl bromide.

bromotrifluoroethylene $BrFC:CF_2$ A colorless gas with a freezing point of $-168°C$ and a boiling point of $-58°C$; soluble in chloroform; used as a refrigerant, in hardening of metals, and as a low-toxicity fire extinguisher. Abbreviated BFE.

bromotrifluoromethane $CBrF_3$ Fluorine compound that has a molecular weight of 148.93, melting point $-180°C$, boiling point $-59°C$; used as a fire-extinguishing agent.

bromouracil $C_4H_3N_2O_2Br$ 5-Bromouracil, an analog of thymine that can react with deoxyribonucleic acid to produce a polymer with increased susceptibility to mutation.

α-bromo-*meta*-xylene $CH_3C_6H_4CH_2Br$ A liquid that is a powerful lacrimator; soluble in alcohol and ether; used in organic synthesis and chemical warfare. Also known as *meta*-methylbenzyl bromide; *meta*-xylyl bromide.

bromoxynil $C_7H_3OBr_2N$ A colorless solid with a melting point of 194–195°C; slightly soluble in water; used as a herbicide in wheat, barley, oats, rye, and seeded turf. Also known as bronate; 3,5-dibromo-4-hydroxybenzonitrile.

bromoxynil octanoate $C_{15}H_{17}Br_2NO_2$ A pale brown liquid, insoluble in water; melting point is 45–46°C; used to control broadleaf weeds. Also known as 3,5-dibromo-4-octanoyloxybenzonitrile.

bromthymol blue An acid-base indicator in the pH range 6.0 to 7.6; color change is yellow to blue. Also known as dibromothymolsulfonphthalein.

bronate *See* bromoxynil.

Brönner's acid $C_{10}H_6(NH_2)SO_3H$ A colorless, water-soluble naphthylamine sulfonic acid that forms needle crystals; used in dyes. Also spelled Broenner's acid. Also known as 2-napthylamine-6-sulfonic acid.

Brönsted acid A chemical species which can act as a source of protons. Also known as proton acid; protonic acid.

Brönsted-Lowry theory A theory that all acid-base reactions consist simply of the transfer of a proton from one base to another. Also known as Brönsted theory.

Brönsted theory *See* Brönsted-Lowry theory.

brown acid Oil-soluble petroleum sulfonate found in sludge following sulfuric acid treatment of petroleum products.

brown lead oxide *See* lead dioxide.

brown-ring test A common qualitative test for the nitrate ion; a brown ring forms at the juncture of a dilute ferrous sulfate solution layered on top of concentrated sulfuric acid if the upper layer contains nitrate ion.

broxyquinoline $C_9H_5Br_2NO$ Crystals with a melting point of 196°C; soluble in acetic acid, chloroform, benzene, and alcohol; used as a reagent for copper, iron, and other metals. Also known as 5,7-dibromo-8-hydroxyquinoline.

brucellergen A nucleoprotein fraction of brucellae used in skin tests to detect the presence of *Brucella* infections.

brucine $C_{23}H_{26}N_2O_4$ A poisonous alkaloid from the seeds of plant species such as *Nux vomica;* used in alcohol as a denaturant. Also known as dimethoxy strychnine.

Brunauer-Emmett-Teller equation An extension of the Langmuir isotherm equation in the study of sorption; used for surface area determinations by computing the monolayer area. Abbreviated BET equation.

BSD *See* barrels per stream day.

B stage An intermediate stage in a thermosetting resin reaction in which the plastic softens but does not fuse when heated, and swells but does not dissolve in contact with certain liquids.

bubble cap A metal cap covering a hole in the plate within a distillation tower; designed to permit vapors to rise from below the plate, pass through the cap, and make contact with liquid on the plate.

bubble-cap plate   One of the devices in large-diameter fractional distillation columns that are designed to produce a bubbling action to exchange the vapor bubbles flowing up the column.

bubble point   In a solution of two or more components, the temperature at which the first bubbles of gas appear. Also known as boiling point.

bubble tower   A plate tower used in distillation, with plates containing bubble caps.

bubble tray   A perforated, circular plate placed within a distillation tower at specific places to collect the fractions of petroleum produced in fractional distillation.

bubble tray column   A fractionating column whose plates are formed from bubble caps.

Bucherer reaction   A method of preparation of polynuclear primary aromatic amines; for example, α-naphthylamine is obtained by heating β-naphthol in an autoclave with a solution of ammonia and ammonium sulfite.

Buerger precession method   The recording on film of a single level of the reciprocal lattice of an individual crystal, by means of x-ray diffraction, for the purpose of determining unit cell dimensions and space groups.

buffer   A solution selected or prepared to minimize changes in hydrogen ion concentration which would otherwise occur as a result of a chemical reaction. Also known as buffer solution.

buffer capacity   The relative ability of a buffer solution to resist pH change upon addition of an acid or a base.

buffer solution   *See* buffer.

bufotenine $C_{12}H_{16}N_2O$ An active pressor agent found in the skin of the common toad; a toxic alkaloid with epinephrinelike biological activity.

buksamin   *See* 4-amino-3-hydroxybutyric acid.

bulab-37 $C_{14}H_{19}N_3O_5$ A yellow solid with a melting point of 70–75°C; used as a preemergence herbicide that is placed in the soil for the control of germinating and seeding weed grasses and broadleaf weeds. Also known as 3′,5′-dinitro-4′-(di-N-propyl)aminoacetophenone.

bulking value   The relative ability of a pigment or other substance to increase the volume of paint.

bumping   Uneven boiling of a liquid caused by irregular rapid escape of large bubbles of highly volatile components as the liquid mixture is heated.

α-bungarotoxin   A neurotoxin found in snake venom which blocks neuromuscular transmission by binding with acetylcholine receptors on motor end plates.

Bunsen-Kirchhoff law   The law that every element has a characteristic emission spectrum of bright lines and absorption spectrum of dark lines.

buret   A graduated glass tube used to deliver variable volumes of liquid; usually equipped with a stopcock to control the liquid flow.

Burgers vector   A translation vector of a crystal lattice representing the displacement of the material to create a dislocation.

burner A furnace where sulfur or sulfide ore are burned to produce sulfur dioxide and other gases.

burning velocity The normal velocity of the region of combustion reaction (reaction zone) relative to nonturbulent unburned gas, in the combustion of a flammable mixture.

burnt lime *See* calcium oxide.

busulphan *See* busulfan.

butachlor $C_{17}H_{26}ClNO_2$ An amber-colored liquid with slight solubility in water; used as a preemergence herbicide for weeds in rice. Also known as 2-chloro-2′,6′-diethyl-*N*-(butoxymethyl)acetanilide.

1,3-butadiene C_4H_6 A colorless gas, boiling point $-4.41°C$, a major product of the petrochemical industry; used in the manufacture of synthetic rubber, latex paints, and nylon.

butadiene dimer C_8H_{12} The third ingredient in ethylene-propylene-terpolymer (EPT) synthetic rubbers; isomers include 3-methyl-1,4,6-heptatriene, vinylcyclohexene, and cyclooctadiene.

butadiene rubber *See* polybutadiene.

butaldehyde *See* butyraldehyde.

α-butalene *See* butene-1.

Butamer process A method of isomerizing normal butane into isobutane in the presence of hydrogen and a solid, noble-metal catalyst; used to prepare raw material in a gasoline alkylation process.

***n*-butanal** *See* butyraldehyde.

butane C_4H_{10} An alkane of which there are two isomers, *n* and isobutane; occurs in natural gas and is produced by cracking petroleum.

butane dehydrogenation A process to remove hydrogen from butane to produce butene or butadiene.

1,4-butanedicarboxylic acid *See* adipic acid.

butanedioic acid *See* succinic acid.

butanedioic anhydride *See* succinic anhydride.

1,3-butanediol *See* 1,3-butylene glycol.

1,4-butanediol *See* 1,4-butylene glycol.

2,3-butanediol $CH_3CHOHCHOHCH_3$ A major fermentation product of several species of bacteria. Also known as 2,3-butylene glycol.

2,3-butanedione *See* diacetyl.

butanenitrile *See* butyronitrile.

butane vapor-phase isomerization A process to isomerize normal butane into isobutane in the presence of aluminum chloride catalyst and hydrogen chloride promoter.

butanoic acid *See* butyric acid.

butanol Any one of four isomeric alcohols having the formula C_4H_9OH; colorless, toxic liquids soluble in most organic liquids. Also known as butyl alcohol.

2-butanone *See* methyl ethyl ketone.

butazolidine *See* phenylbutazone.

butene-1 $CH_3CH_2CHCH_2$ A colorless, highly flammable gas; insoluble in water, soluble in organic solvents; used to produce polybutenes, butadiene aldehydes, and other organic derivatives. Also known as α-butalene; ethylethylene.

butene-2 $CH_3CHCHCH_3$ A colorless, highly flammable gas, used to make butadiene and in the synthesis of four- and five-carbon organic molecules; the cis form, boiling point 3.7°C, is insoluble in water, soluble in organic solvents, and is also known as high-boiling butene-2; the trans form, boiling point 0.88°C, is insoluble in water, soluble in most organic solvents, and is also known as low-boiling butene-2.

3-butenenitrile *See* allyl cyanide.

α-butenic acid *See* crotonic acid.

***trans*-2-butenoic acid** *See* ethyl crotonate.

1-butine *See* ethyl acetylene.

butopyronoxyl $C_{12}H_{18}O_4$ A yellow to amber liquid with a boiling point of 256–260°C; miscible with ether, glacial acetic acid, alcohol, and chloroform; used as an insect repellent for skin and clothing. Also known as butyl 3,4-dihydro-2,2-dimethyl-4-oxo-2H-pyran-6-carboxylate; butyl mesityl oxide oxalate.

2-butoxyethanol $HOCH_2CH_2OC_4H_9$ A liquid with a boiling point of 171–172°C; soluble in most organic solvents and water; used in dry cleaning as a solvent for nitrocellulose, albumin, resins, oil, and grease. Also known as ethylene glycol monobutyl ether.

butyl Any of the four variations of the hydrocarbon radical C_4H_9: $CH_3CH_2CH_2CH_2$—, $(CH_3)_2CHCH_2$—, $CH_3CH_2CHCH_3$—, and $(CH_3)_3C$—.

butyl acetate $CH_3COOC_4H_9$ A colorless liquid slightly soluble in water; used as a solvent.

butylacetic acid *See* caproic acid.

butyl acetoacetate $C_8H_{14}O_3$ A colorless liquid with a boiling point of 213.9°C; soluble in alcohol and ether; used for synthesis of dyestuffs and pharmaceuticals.

butyl acrylate $CH_2CHCOOC_4H_9$ A colorless liquid that is nearly insoluble in water and polymerizes readily upon heating; used as an intermediate for organic synthesis, polymers, and copolymers.

butyl alcohol *See* butanol.

butyl aldehyde *See* butyraldehyde.

***n*-butyl aldehyde** *See* butyraldehyde.

***n*-butylamine** $C_4H_9NH_2$ A colorless, flammable liquid; miscible with water and ethanol; used as an intermediate in organic synthesis and to make insecticides, emulsifying agents, and pharmaceuticals. Also known as 1-aminobutane.

***sec*-butylamine** $CH_3CHNH_2C_2H_5$ A flammable, colorless liquid; boils in the range 63–68°C; may be used as an intermediate in organic synthesis. Also known as 2-aminobutane.

tert-butylamine $(CH_3)_3CNH_2$ A flammable liquid; boiling range 63–68°C; may be used in organic synthesis as an intermediate.

2-*tert*-butylamino-4-chloro-6-ethylamino-*S*-triazine *See* terbutylhylazine.

2-*sec*-butylamino-4-ethylamino-6-methoxy-*S*-triazine *See* secbumeton.

2-*tert*-butylamino-4-ethylamino-6-methoxy-*S*-triazine *See* terbumeton.

2-*tert*-butylamino-4-ethylamino-6-methylthio-*S*-triazine *See* terbutryn.

butylate $C_{11}H_{23}NOS$ A colorless liquid used as an herbicide for preplant control of weeds in corn. Also known as S-ethyl diisobutylthiocarbamate.

butylated hydroxyanisole $(CH_3)_3CC_6H_3OH(OCH_3)$ An antioxidant consisting chiefly of a mixture of 2- and 3-*tert*-butyl-4-hydroxyanisole and used to control rancidity of lard and animal fats in foods. Abbreviated BHA.

butylated hydroxytoluene $[(CH_3)_3C]_2C_6H_2(CH_3)OH$ Crystals with a melting point of 72°C; soluble in toluene, methanol, and ethanol; used as an antioxidant in foods, in petroleum products, and for synthetic rubbers. Abbreviated BHT.

butylbenzene $C_6H_5C_4H_9$ A colorless liquid used as a raw material for organic synthesis, especially for insecticides; forms are normal (1-phenylbutane), secondary (2-phenylbutane), and tertiary (2-methyl-2-phenylpropane).

butyl borate *See* tributyl borate.

N-*sec*-butyl-4-*tert*-butyl-2,6-dinitroaniline $C_{14}H_{21}N_3O_4$ Orange crystals with a melting point of 60–61°C; solubility in water is 1.0 part per million at 24°C; used as a preemergence herbicide. Also known as amchem; dibutalin.

butyl carbinol $(CH_3)_3CCH_2OH$ Colorless crystals that melt at 52°C; slightly soluble in water.

butyl chloride C_4H_9Cl A colorless liquid used as an alkylating agent in organic synthesis, as a solvent, and as an anthelminthic; forms are normal (1-chlorobutane), secondary, and iso or tertiary.

tert-butyl chloroacetate $ClCH_2COOC(CH_3)_3$ A liquid with a boiling point of 155°C; hydrolyzes to *tert*-butyl alcohol and chloroacetic acid; used in glycidic ester condensation. Also known as chloroacetic acid *tert*-butyl ester.

3-*tert*-butyl-5-chloro-6-methyluracil *See* terbacil.

4-*tert*-butyl-2-chlorophenyl-*ortho*-methyl methylphosphoramidate *See* crufomate.

butyl citrate $C_3H_5O(COOC_4H_9)_3$ A colorless, odorless nonvolatile liquid, almost insoluble in water; used as a plasticizer, solvent for cellulose nitrate, and antifoam agent. Also known as tributyl citrate.

2-*tert*-butyl-4-(2,4-dichloro-5-isopropoxyphenyl)-δ-2-1-1,3,4-oxadiazalin-5-one *See* oxadiazon.

1-*n*-butyl-3-(3,4-dichlorophenyl)-1-methylurea *See* neburon.

butyl diglycol carbonate $(C_4H_9OCO_2CH_2CH_2)_2O$ A colorless, combustible liquid with a boiling range of 164–166°C; used as a plasticizer and solvent and in pharmaceuticals and lubricants manufacture.

butyl 3,4-dihydro-2,2-dimethyl-4-oxo-2*H*-pyran-6-carboxylate *See* butopyronoxyl.

2-(*sec*-butyl)-4,6-dinitrophenol *See* dinoseb.

2-*tert*-butyl-4,6-dinitrophenyl acetate *See* dinoterb acetate.

2-*sec*-butyl-4,6-dinitrophenylisopropyl carbonate   *See* dinobuton.

4-butyl-1,2-diphenyl-3,5-pyrazolidinedione   *See* phenylbutazone.

butylene   Any of three isomeric alkene hydrocarbons with the formula C_4H_8; all are flammable and easily liquefied gases.

1,3-butylene glycol $HOCH_2CH_2CH(OH)CH_3$ A viscous, colorless, hygroscopic liquid; soluble in water and alcohol; used as a solvent, food additive, and flavoring, and for plasticizers and polyurethanes. Also known as 1,3-butanediol.

1,4-butylene glycol $HOCH_2CH_2CH_2CH_2OH$ A colorless, combustible, oily liquid with a boiling point of 230°C; soluble in alcohol; used as a solvent and humectant, and in plastics and pharmaceuticals manufacture. Also known as 1,4-butanediol; tetramethylene glycol.

2,3-butylene glycol   *See* 2,3-butanedio.

1,2-butylene oxide $H_2COCHCH_2CH_3$ A colorless, water-soluble liquid with a boiling point of 63°C; used as an intermediate for various polymers. Also known as 1,2-epoxybutane.

butyl ether $C_8H_{18}O$ A colorless liquid, boiling at 142°C, and almost insoluble in water; used as an extracting agent, as a medium for Grignard and other reactions, and for purifying other solvents.

butyl ethylacetic acid   *See* 2-ethyl hexoic acid.

5-*n*-butyl-2-ethylamino-4-hydroxy-6-methylpyrimidine   *See* ethirimol.

***N*-butyl-*N*-ethyl-α,α,α-trifluoro-2,6-dinitro-*para*-toluidine** $C_{13}H_{16}O_4NF_3$ Yellowish-orange crystals with a melting point of 65–66.5°C; used as a preemergence herbicide for annual grasses and broadleaf weeds in crops. Also known as benefin.

butyl formate $HCOOC_4H_9$ An ester of formic acid and butyl alcohol.

***tert*-butylhydroperoxide** $(CH_3)_3COOH$ A liquid soluble in organic solvents; used as a catalyst in polymerization reactions, to introduce the peroxy group into organic molecules.

***n*-butyl-9-hydroxyfluorene-9-carboxylate**   *See* flurenol.

6-*tert*-butyl-3-isopropylisothiazolo-(3,4-*d*)pyrimidin-4(5*H*)one $C_{12}H_{17}N_3OS$ A white solid with a melting point of 198–199°C; insoluble in water; used as an herbicide for field corn, sweet corn, and sorghum.

6-*tert*-butyl-3-isopropylisoxazolo-(5,4-*d*)pyrimidin-4(5*H*)one $C_{12}H_{17}N_3O_2$ A white solid with a melting point of 227–229°C; insoluble in water; used as an herbicide for preemergence on sorghum, flax, sugarcane, and field and sweet corn.

butyl lactate $CH_3CHOHCOOC_4H_9$ A stable liquid, water-white and nontoxic, miscible with many solvents; used as a solvent for resins and gums, in lacquers and varnishes, and as a chemical intermediate.

butyl mercaptan C_4H_9SH A colorless, odorous liquid, a component of skunk secretion; used commercially as a gas-odorizing agent.

butyl mesityl oxide oxalate   *See* butopyronoxyl.

2-*tert*-butyl-5-methyl-4,6-dinitrophenyl acetate   *See* medinoterb acetate.

butyl octadecanoate   *See* butyl stearate.

butyl oleate $C_{22}H_{42}O_2$ A butyl ester of oleic acid; used as a plasticizer.

butylphen   *See para-tert*-butylphenol.

***para-tert*-butylphenol** $(CH_3)_3CC_6H_4OH$ Needlelike crystals with a melting point of 98°C; soluble in alcohol and ether; used as an intermediate in production of varnish and lacquer resins, an additive in motor oil, and an ingredient in deemulsifiers in oil fields. Also known as butylphen.

***para-tert*(butyl phenoxy)cyclohexyl 2-propynyl sulfite**   *See* propargite.

2-*sec*-butyl phenyl-*N*-methyl carbamate $C_{12}H_{17}NO_2$ A pale yellow or pale red liquid, insoluble in water; used as an insecticide for pests of rice and cotton. Also known as BPMC.

butyl propionate $C_2H_5COOC_4H_9$ A colorless aromatic liquid; used in fruit essences.

6-*tert*-butyl-3-propylisoxazolo-(5,4-*d*)pyrimidin-4(5*H*)one $C_{12}H_{17}N_3O_2$ A white solid with a melting point of 217–218°C; insoluble in water; used as an herbicide on field corn, sweet corn, and sorghum.

butyl stearate $C_{17}H_{35}COOC_4H_9$ A liquid that solidifies at approximately 19°C; mixes with vegetable oils and is soluble in alcohol and ethers but insoluble in water; used as a lubricant, in polishes, as a plasticizer, and as a dye solvent. Also known as butyl octadecanoate.

butyl titanate   *See* tetrabutyl titanate.

1-butyne   *See* ethyl acetylene.

butynedial $HOCH_2C:CCH_2OH$ White crystals with a melting point of 58°C; soluble in water, aqueous acids, alcohol, and acetone; used as a corrosion inhibitor, defoliant, electroplating brightener, and polymerization accelerator.

butyraldehyde $CH_3(CH_2)_2CHO$ A colorless liquid boiling at 75.7°C; soluble in ether and alcohol, insoluble in water; derived from the oxo process. Also known as butaldehyde; *n*-butanal; *n*-butyl aldehyde; butyric aldehyde.

butyrate   An ester or salt of butyric acid containing the $C_4H_7O_2$ radical.

butyric acid $CH_3CH_2CH_2COOH$ A colorless, combustible liquid with boiling point 163.5°C (757 mm Hg); soluble in water, alcohol, and ether; used in synthesis of flavors, in pharmaceuticals, and in emulsifying agents. Also known as butanoic acid; *n*-butyric acid; ethyl acetic acid; propylformic acid.

***n*-butyric acid**   *See* butyric acid.

butyric aldehyde   *See* butyraldehyde.

butyric anhydride $C_8H_{14}O_3$ A colorless liquid that decomposes in water to form butyric acid; exists in two isomeric forms.

butyric ether   *See* ethyl butyrate.

butyric fermentation   Fermentation in which butyric acid is produced by certain anaerobic bacteria acting on organic substances, such as butter; occurs in putrefaction and in digestion in herbivorous mammals.

butyrinase   An enzyme that hydrolyzes butyrin, found in the blood serum.

butyrolactone $C_4H_6O_2$ A liquid, the anhydride of butyric acid; used as a solvent in the manufacture of plastics.

butyron   *See* boturon.

butyrone   *See* 4-heptanone.

butyronitrile $CH_3(CH_2)_2CN$ A toxic, colorless liquid with a boiling point of 116–117.7°C; soluble in alcohol and ether; used in industrial, chemical, and pharmaceutical products, and in poultry medicines. Also known as butanenitrile; propyl cyanide.

C

C *See* carbon.

Ca *See* calcium.

⁴⁵Ca *See* calcium-45.

Cabannes' factor An equational factor to correct for the depolarization effect of the horizontal components of scattered light during the determination of molecular weight by optical methods.

cacodyl $(CH_3)_2As^-$ A radical found in, for example, cacodylic acid, $(CH_3)_2AsOOH$.

cacodylate Any salt of cacodylic acid.

cacodylic acid $(CH_3)_2AsOOH$ Colorless crystals that melt at 200°C; soluble in alcohol and water; used as an herbicide. Also known as dimethylarsinic acid.

cacotheline $C_{20}H_{22}N_2O_5(NO_2)_2$ An azoic compound used as a metal indicator in chelometric titrations.

cadalene $C_{15}H_{18}$ A colorless liquid which boils at 291–292°C (720 mmHg; 95,990 newtons per square meter) and which is a substituted naphthalene. Also known as 1,6-dimethyl-4-isopropyl naphthalene.

cadaverine $C_5H_{14}N_2$ A nontoxic, organic base produced as a result of the decarboxylation of lysine by the action of putrefactive bacteria on flesh.

cadinene $C_{15}H_{24}$ A colorless liquid that boils at 274.5°C, and is a terpene derived from cubeb oil, cade oil, juniper berry oil, and other essential oils.

cadmium A chemical element, symbol Cd, atomic number 48, atomic weight 112.40.

cadmium acetate $Cd(OOCCH_3)_2 \cdot 3H_2O$ A compound that forms colorless monoclinic crystals, soluble in water and in alcohol; used for chemical testing for sulfides, selenides, and tellurides and for producing iridescent effects on porcelain.

cadmium bromate $Cd(BrO_3)_2$ Colorless powder, soluble in water; used as an analytical reagent.

cadmium bromide $CdBr_2$ A compound produced as a yellow crystalline powder, soluble in water and alcohol; used in photography, process engraving, and lithography.

cadmium carbonate $CdCO_3$ A white crystalline powder, insoluble in water, soluble in acids and potassium cyanide; used as a starting compound for other cadmium salts.

cadmium chlorate $CdClO_3$ White crystals, soluble in water; a highly toxic material.

cadmium chloride $CdCl_2$ A cadmium halide in the form of colorless crystals, soluble in water, methanol, and ethanol; used in photography, in dyeing and calico printing, and as a solution to precipitate sulfides.

cadmium fluoride CdF_2 A crystalline compound with a melting point of 1110°C; soluble in water and acids; used for electronic and optical applications and as a starting material for laser crystals.

cadmium hydroxide $Cd(OH)_2$ A white powder, soluble in dilute acids; used to prepare negative electrodes for cadmium-nickel storage batteries.

cadmium iodide CdI_2 A cadmium halide that forms lustrous, white, hexagonal scales, consisting of two water-soluble allotropes; used in photography, in process engraving, and formerly as an antiseptic.

cadmium nitrate $Cd(NO_3)_2 \cdot 4H_2O$ White, hygroscopic crystals, soluble in water, alcohol, and liquid ammonia; used to give a reddish-yellow luster to glass and porcelain ware.

cadmium oxide CdO In the cubic form, a brown, amorphous powder, insoluble in water, soluble in acids and ammonia salts; used for cadmium plating baths and in the manufacture of paint pigments.

cadmium potassium iodide *See* potassium tetraiodocadmate.

cadmium sulfate $CdSO_4$ A compound that forms colorless, efflorescent crystals, soluble in water; used as an antiseptic and astringent, in the treatment of syphilis, gonorrhea, and rheumatism, and as a detector of hydrogen sulfide and fumaric acid.

cadmium sulfide CdS A compound with two forms: orange, insoluble in water, used as a pigment, and also known as orange cadmium; light yellow, hexagonal crystals, insoluble in water, and also known as cadmium yellow.

cadmium telluride $CdTe$ Brownish-black, cubic crystals with a melting point of 1090°C; soluble, with decomposition, in nitric acid; used for semiconductors.

cadmium tungstate $CdWO_4$ White or yellow crystals or powder; soluble in ammonium hydroxide and alkali cyanides; used in fluorescent paint, x-ray screens, and scintillation counters.

cadmium yellow *See* cadmium sulfide.

caffeic acid $C_9H_8O_4$ A yellow crystalline acid that melts at 223–225°C with decomposition; soluble in water and alcohol. Also known as 3,4-dihydroxy-cinnamic acid.

caffeine $C_8H_{10}O_2N_4 \cdot H_2O$ An alkaloid found in a large number of plants, such as tea, coffee, cola, and mate.

3-caffeoylquinic acid *See* chlorogenic acid.

cage A void occurring in a crystal structure capable of trapping one or more foreign atoms.

Cailletet and Mathias law The law that describes the relationship between the mean density of a liquid and its saturated vapor at that temperature as being a linear function of the temperature.

cake *See* residue.

calabarine *See* physostigmine.

calandria One of the tubes through which the heating fluid circulates in an evaporator.

calandria evaporator *See* short-tube vertical evaporator.

calcification Any process of soil formation in which the soil colloids are saturated to a high degree with exchangeable calcium, thus rendering them relatively immobile and nearly neutral in reaction.

calcined gypsum *See* plaster of Paris.

calcined soda *See* soda ash.

calcitonin A polypeptide, calcium-regulating hormone produced by the ultimobranchial bodies in vertebrates. Also known as thyrocalcitonin.

calcium A chemical element, symbol Ca, atomic number 20, atomic weight 40.08; used in metallurgy as an alloying agent for aluminum-bearing metal, as an aid in removing bismuth from lead, and as a deoxidizer in steel manufacture, and also used as a cathode coating in some types of photo tubes.

calcium-45 A radioisotope of calcium having a mass number of 45, often used as a radioactive tracer in studying calcium metabolism in humans and other organisms; half-life is 165 days. Designated ^{45}Ca.

calcium acetate $Ca(C_2H_3O_2)_2$ A compound that crystallizes as colorless needles that are soluble in water; formerly used as an important source of acetone and acetic acid; now used as a mordant and as a stabilizer of plastics.

calcium acrylate $(CH_2CHCOO)_2Ca$ Free-flowing, water-soluble white powder used for soil stabilization, oil-well sealing, and ion exchange and as a binder for clay products and foundry molds.

calcium arsenate $Ca_3(AsO_4)_2$ An arsenic compound used as an insecticide to control cotton pests.

calcium arsenite $Ca_3(AsO_3)_2$ White granules that are soluble in water; used as an insecticide.

calcium bisulfite $Ca(HSO_3)_2$ A white powder, used as an antiseptic and in the sulfite pulping process.

calcium bromide $CaBr_2$ A deliquescent salt in the form of colorless hexagonal crystals that are soluble in water and absolute alcohol.

calcium carbide CaC_2 An alkaline earth carbide obtained in the pure form as transparent crystals that decompose in water; used to make acetylene gas.

calcium carbonate $CaCO_3$ White rhombohedrons or a white powder; occurs naturally as calcite; used in paint manufacture, as a dentifrice, as an anticaking medium for table salt, and in manufacture of rubber tires.

calcium chlorate $Ca(ClO_3)_2 \cdot 2H_2O$ White monoclinic crystals, decomposed by heating.

calcium chloride $CaCl_2$ A colorless, deliquescent powder that is soluble in water and ethanol; used as an antifreeze and as an antidust agent.

calcium chromate $CaCrO_4 \cdot 2H_2O$ Yellow, monoclinic crystals that are slightly soluble in water; used to make other pigments.

calcium cyanamide $CaCN_2$ In pure form, colorless rhombohedral crystals, the commercial form being a gray material containing 55–70% $CaCN_2$; used as a fertilizer, weed killer, and defoliant.

calcium cyanide $Ca(CN)_2$ In pure form, a white powder that gives off hydrogen cyanide in air at normal humidity; prepared commercially in impure black or gray flakes; used as an insecticide and rodenticide. Also known as black cyanide.

calcium cyclamate $C_{12}H_{24}O_6N_2S_2Ca_2H_2O$ White crystals with a very sweet taste, soluble in water; has been used as a low-calorie sweetening agent.

calcium dihydrogen phosphate *See* calcium phosphate.

calcium fluoride CaF_2 Colorless, cubic crystals that are slightly soluble in water and soluble in ammonium salt solutions; used in etching glass and preparing hydrofluoric acid.

calcium gluconate $Ca(C_6H_{11}O_7)_2\cdot H_2O$ White powder that loses water at 120°C; soluble in hot water but less soluble in cold water, insoluble in acetic acid and alcohol; used in medicine, as a foaming agent, and as a buffer in foods.

calcium hardness Presence of calcium ions in water, from dissolved carbonates and bicarbonates; treated in boiler water by introducing sodium phosphate.

calcium hydride CaH_2 In pure form, white crystals that are insoluble in water; used in the production of chromium, titanium, and zirconium in the Hydromet process.

calcium hydrogen phosphate *See* calcium phosphate.

calcium hydroxide $Ca(OH)_2$ White crystals, slightly soluble in water; used in cement, mortar, and manufacture of calcium salts. Also known as hydrated lime.

calcium hypochlorite $Ca(OCl)_2\cdot 4H_2O$ A white powder, used as a bleaching agent and disinfectant for swimming pools.

calcium iodide CaI_2 A yellow, hygroscopic powder that is very soluble in water; used in photography.

calcium iodobehenate $Ca(OOCC_{21}H_{42}I)_2$ A yellowish powder that is soluble in warm chloroform; used in feed additives.

calcium lactate $Ca(C_3H_5O_3)_2\cdot 5H_2O$ A salt of lactic acid in the form of white crystals that are soluble in water; used in calcium therapy and as a blood coagulant.

calcium metabolism Biochemical and physiological processes involved in maintaining the concentration of calcium in plasma at a constant level and providing a sufficient supply of calcium for bone mineralization.

calcium naphthenate Calcium derivative of cycloparaffin hydrocarbon (generally cyclopentane or cyclohexane base) that is a light, sticky, water-insoluble mass; used as a hardening agent in plastic compounds, in waterproofing, adhesives, wood fillers, and varnishes.

calcium nitrate $Ca(NO_3)_2\cdot 4H_2O$ Colorless, monoclinic crystals that are soluble in water; the anhydrous salt is very deliquescent; used as a fertilizer and in explosives. Also known as nitrocalcite.

calcium orthoarsenate $Ca_3(AsO_4)_2$ A white powder, insoluble in water; used as a preemergence insecticide and herbicide for turf.

calcium oxalate $CaC_2O_4 \cdot H_2O$ A salt of oxalic acid in the form of white crystals that are insoluble in water.

calcium oxide CaO A caustic white solid sparingly soluble in water; the commercial form is prepared by roasting calcium carbonate limestone in kilns until all the carbon dioxide is driven off; used as a refractory, in pulp and paper manufacture, and as a flux in manufacture of steel. Also known as burnt lime; calx; caustic lime.

calcium pantothenate $(C_9H_{16}NO_5)_2Ca$ White slightly hygroscopic powder; soluble in water, insoluble in chloroform and ether; melts at 170–172°C; found in either the dextro or levo form or in racemic mixtures; used in nutrition and in animal feed.

calcium peroxide CaO_2 A cream-colored powder that decomposes in water; used as an antiseptic and a detergent.

calcium phosphate 1. Any phosphate of calcium. 2. Any of the following three calcium orthophosphates, all of which are white or colorless in pure form: $Ca(H_2PO_4)_2$ is used as a fertilizer, as a plastics stabilizer, and in baking powder, and is also known as acid calcium phosphate, calcium dihydrogen phosphate, monobasic calcium phosphate, monocalcium phosphate; $CaHPO_4$ is used in pharmaceuticals, animal feeds, and toothpastes, and is also known as calcium hydrogen phosphate, dibasic calcium phosphate, dicalcium orthophosphate, dicalcium phosphate; $Ca_3(PO_4)_2$ is used as a fertilizer, and is also known as tribasic calcium phosphate, tricalcium phosphate.

calcium plumbate $Ca(PbO_3)_2$ Orange crystals that are insoluble in cold water but decompose in hot water; used as an oxidizer in the manufacture of glass and matches.

calcium plumbite $CaPbO_2$ Colorless crystals that are slightly soluble in water.

calcium pyrophosphate $Ca_2P_2O_7$ White, abrasive powder, used in dentifrice polishes, in metal polishes, and as a food supplement.

calcium resinate Yellowish white, amorphous powder that is soluble in acid, insoluble in water; made by boiling rosin with calcium hydroxide and filtering, or by fusion of melted rosin with hydrated lime; used for waterproofing, leather tanning, and the manufacture of paint driers and enamels. Also known as limed rosin.

calcium reversal lines Narrow calcium emission lines that appear as bright lines in the center of broad calcium absorption bands in the spectra of certain stars.

calcium silicate Any of three silicates of calcium: tricalcium silicate, Ca_3SiO_5; dicalcium silicate, Ca_2SiO_4; calcium metasilicate, $CaSiO_3$.

calcium stearate $Ca(C_{18}H_{35}O_2)_2$ A metallic soap produced as a white powder that is insoluble in water but slightly soluble in petroleum, benzene, and toluene.

calcium sulfate 1. $CaSO_4$ A white crystalline salt, insoluble in water; used in Keene's cement, in pigments, as a paper filler, and as a drying agent. 2. Either of two hydrated forms of the salt: the dihydrate, $CaSO_4 \cdot 2H_2O$, and the hemihydrate, $CaSO_4 \cdot \frac{1}{2}H_2O$.

calcium sulfide CaS In pure form, white cubic crystals, slightly soluble in water; used as a base for luminescent materials. Also known as hepar calcies; sulfurated lime.

calcium sulfite $CaSO_3 \cdot 2H_2O$ A white powder that is soluble in dilute sulfurous acid; may be dehydrated at 150°C to the anhydrous salt; used in the sulfite process for the manufacture of wood pulp.

calcium tungstate $CaWO_4$ White, tetragonal crystals, slightly soluble in water; used in manufacture of luminous paints. Also known as artificial scheelite; calcium wolframate.

calcium wolframate *See* calcium tungstate.

calibration reference Any of the standards of various types that indicate whether an analytical instrument or procedure is working within prescribed limits; examples are test solutions used with pH meters, and solutions with known concentrations (standard solutions) used with spectrophotometers.

California polymerization A polymerization process for converting C_3—C_4 olefins to motor fuel by utilizing a catalyst of phosphoric acid on quartz chips.

californium A chemical element, symbol Cf, atomic number 98; all isotopes are radioactive.

callideic I *See* bradykinin.

callose A carbohydrate component of plant cell walls; associated with sieve plates where calluses are formed.

calmagite $C_{17}H_{14}N_2O_5S$ A compound crystallizing from acetone as red crystals that are soluble in water; used as an indicator in the titration of calcium or magnesium with EDTA.

calomel electrode A reference electrode of known potential consisting of mercury, mercury chloride (calomel), and potassium chloride solution; used to measure pH and electromotive force. Also known as calomel half-cell; calomel reference electrode.

calomel half-cell *See* calomel electrode.

calomel reference electrode *See* calomel electrode.

calorimetric titration *See* thermometric titration.

calsequestrin An acidic protein, with a molecular weight of 44,000 daltons, which binds calcium inside the sarcoplasmic reticulum.

calx *See* calcium oxide.

cAMP *See* cyclic adenylic acid.

camphane $C_{10}H_{18}$ An alicyclic hydrocarbon; white crystals, soluble in alcohol, with a melting point of 158–159°C. Also known as bornylane; dihydrocamphane.

2-camphanol *See* borneol.

camphene $C_{10}H_{16}$ A bicyclic terpene used as raw material in the synthesis of insecticides such as toxaphene and camphor. Also known as 3,3-dimethyl-2-methylenenorcamphane.

camphor $C_{10}H_{16}O$ A bicyclic saturated terpene ketone that exists in optically active dextro and levo forms and as a racemic mixture of these forms; the dextro form is obtained from the wood and bark of the camphor tree, the levo form is found in some essential oils, and the inactive form is obtained from an Asiatic chrysanthemum or made synthetically from certain terpenes.

10-camphorsulfonic acid *See* d-camphorsulfonic acid.

d-camphorsulfonic acid $C_{10}H_{16}O_4S$ A compound crystallizing as prisms from ethyl acetate or glacial acetic acid; slightly soluble in glacial acetic acid and in ethyl acetate; used in the resolution of optically active isomers. Also known as 10-camphorsulfonic acid; camphostyl; camsylate; Reychler's acid.

camphostyl *See* d-camphorsulfonic acid.

camsylate *See* d-camphorsulfonic acid.

canal ray The name given in early gaseous discharge experiments to the particles passing through a hole or canal in the cathode; the ray comprises positive ions of the gas being used in the discharge.

canavanine $C_5H_{12}O_3N_4$ An amino acid found in the jack bean.

cane sugar Sucrose derived from sugarcane.

cannabidiol $C_{21}H_{28}(OH)_2$ A constituent of cannabis which, upon isomerization to a tetrahydrocannabinol, has some of the physiologic activity of marijuana.

cannabinol $C_{21}H_{26}O_2$ A physiologically inactive phenol formed by spontaneous dehydrogenation of tetrahydrocannabinol from cannabis.

Cannizzaro reaction The reaction in which aldehydes that do not have a hydrogen attached to the carbon adjacent to the carbonyl group, upon encountering strong alkali, readily form an alcohol and an acid salt.

canonical form A resonance structure for a cyclic compound in which the bonds do not intersect.

cantharides camphor *See* cantharidin.

cantharidin $C_{10}H_{12}O_4$ Colorless crystals that melt at 218°C; slightly soluble in acetone, chloroform, alcohol, and water; used in veterinary medicine. Also known as cantharides camphor.

capacity In chromatography, a measurement used in ion-exchange systems to express the adsorption ability of the ion-exchange materials.

capillary condensation Condensation of an adsorbed vapor within the pores of the adsorbate.

capillary gas chromatography A highly efficient type of gas chromatography in which the gaseous sample passes through capillary tubes with internal diameters between 0.2 and 0.5 millimeter and lengths up to 100 meters, and adsorption takes place on a medium that is spread on the inner walls of these tubes.

capraldehyde *See* decyl aldehyde.

caprate Any of the salts of capric acid, containing the group $C_9H_{19}COO$—.

capric acid $CH_3(CH_2)_8COOH$ A fatty acid found in oils and animal fats. Also known as n-decanoic acid; decatoic acid; n-decoic acid; decyclic acid; octylacetic acid.

capric aldehyde *See* decyl aldehyde.

capric anhydride $(CH_3(CH_2)_8CO)_2O$ White crystals that are insoluble in water; used as a chemical intermediate.

caprilydene *See* octyne.

capristor *See* rescap.

caproaldehyde *See* n-hexaldehyde.

caproamide $CH_3(CH_2)_4CONH_2$ An amide, melting point 100–101°C; used as a chemical intermediate.

caproic acid $CH_3(CH_2)_4COOH$ A colorless liquid fatty acid found in oils and animal fats; used in synthesizing pharmaceuticals and flavors. Also known as butylacetic acid; hexanoic acid; n-hexoic acid; n-hexylic acid; pentylformic acid.

caproic aldehyde *See* n-hexaldehyde.

caproic anhydride $[CH_3(CH_2)_4COO]_2$ White crystals that are insoluble in water, melting point -40.6°C, boiling point 241–243°C.

caprolactam $(CH_2)_5NH\cdot CO$ White flakes, melting point 68–69°C, made from cyclohexanone; used to make synthetic fiber, particularly nylon-6.

ε-caprolactone $CH_2(CH_2)_4NHCO$ White crystals, used to make synthetic fibers, plastics, films, coatings, and plasticizers; its vapors or fine crystals are respiratory irritants.

capryaldehyde *See* octyl aldehyde.

caprylamide $CH_3(CH_2)_6CONH_2$ An amine, melting point 105–110°C; decomposes above 200°C; used as a chemical intermediate.

capryl compounds A misnomer for octyl compounds; that is, the term octyl halide is preferred for caprylic halides, and octanoic acid for caprylic acid.

1-caprylene *See* 1-octene.

caprylic acid $C_8H_{16}O_2$ A liquid fatty acid occurring in butter, coconut oil, and other fats and oils. Also known as hexylacetic acid; n-octanoic acid; octylic acid.

caprylic anhydride $[CH_3(CH_2)_6CO]_2O$ A white solid that melts at -1°C; used as a chemical intermediate.

capsaicin $C_{18}H_{27}O_3N$ A toxic material extracted from capsicum.

capsanthin $C_{40}H_{58}IO_3$ Carmine-red carotenoid pigment occurring in paprika.

captan $C_9H_8O_2NSCl_3$ A buff to white solid with a melting point of 175°C; used as a fungicide for diseases of fruits, vegetables, and flowers. Also known as *cis*-N(trichloromethylthio)-4-cyclohexene-1,2-dicarboximide.

capture In a crystal structure, the substitution of a trace element for a lower-valence common element.

capture cross section The cross section that is effective for radiative capture.

carbamic acid NH_2COOH An amino acid known for its salts, such as urea and carbamide. Also known as amidocarbonic acid; aminoformic acid.

carbamic acid β-hydroxyphenethyl ester *See* styramate.

carbamidine *See* guanidine.

carbamino A compound formed by the combination of carbon dioxide with a free amino group in an amino acid or a protein.

carbamoyl The radical NH_2CO, formed from carbamic acid.

carbamylcholine chloride *See* carbachol.

carbamylhydrazine hydrochloride *See* semicarbazide hydrochloride.

carbamyl phosphate $NH_2COPO_4H_2$ The ester formed from reaction of phosphoric acid and carbamyl acid.

carbamylurea *See* biuret.

carbanilide $(NHC_6H_5)CO(NHC_6H_5)$ Colorless crystals that are very slightly soluble in water, and dissolve in ether and alcohol; used in organic synthesis. Also known as diphenylurea.

carbanion One of the charged fragments which arise on heterolytic cleavage of a covalent bond involving carbon; the fragment carries an unshared pair of electrons and bears a negative charge.

carbaryl $C_{12}H_{11}NO_2$ A colorless, crystalline compound with a melting point of 142°C; used as an insecticide for crops, forests, lawns, poultry, and pets. Also known as 1-naphthylmethyl carbamate.

carbazide *See* carbodihydrazide.

carbazole One of a group of organic heterocyclic compounds containing a dibenzopyrrole system. Also known as 9-azafluorene.

carbazotic acid *See* picric acid.

carbendazim *See* 2-(methoxycarbonylamino)benzimidazole.

carbene A compound of carbon which exhibits two valences to a carbon atom; the two valence electrons are distributed in the same valence; an example is CH_2.

carbenium ion A cation in which the charged atom is carbon; for example, R_2C^+, where R is an organic group.

carbenoid species A species that is not a free carbene but has the characteristics of a carbene when participating in a chemical reaction.

carbide A binary compound of carbon with an element more electropositive than carbon; carbon-hydrogen compounds are excluded.

carbinol 1. A primary alcohol with general formula RCH_2OH. 2. The radical CH_2OH of primary alcohols. 3. An alcohol derived from methanol.

carbinyl *See* methyl.

carbobenzoxychloride *See* benzyl chloroformate.

carbocation A positively charged ion whose charge resides, at least in part, on a carbon atom or group of carbon atoms.

carbocyclic compound A compound with a homocyclic ring in which all the ring atoms are carbon, for example, benzene.

carbodicyclohexylimide *See* dicyclohexylcarbodiimide.

carbodihydrazide $CO(NHNH_2)_2$ Colorless crystals that melt at 154°C; very soluble in alcohol and water; used in photographic chemicals. Also known as carbazide.

carbodiimide 1. $HN{=}C{=}NH$ An unstable tautomer of cyanamide. 2. Any compound with the general formula $RN{=}C{=}NR$ which is a formal derivative of carbodiimide.

carbofuran $C_{12}H_{15}NO_3$ A white solid with a melting point of 150–152°C; soluble in water; used as an insecticide, miticide, and nematicide in many crops. Also known as 3,3-dihydro-2,2-dimethyl-7-benzofuranyl methylcarbamate.

carbohydrase Any enzyme that catalyzes the hydrolysis of disaccharides and more complex carbohydrates.

carbohydrate Any of the group of organic compounds composed of carbon, hydrogen, and oxygen, including sugars, starches, and celluloses.

carbohydrate gum A polysaccharide which produces a gel of a viscous solution when it is dispersed in water at low concentrations; examples are agar, guar gum, xanthan gum, gum arabic, and sodium carboxymethyl cellulose.

carbohydrate metabolism The sum of the biochemical and physiological processes involved in the breakdown and synthesis of simple sugars, oligosaccharides, and polysaccharides and in the transport of sugar across cell membranes.

carbolic acid *See* phenol.

carbon A nonmetallic chemical element, symbol C, atomic number 6, atomic weight 12.01115; occurs freely as diamond, graphite, and coal.

carbon-12 A stable isotope of carbon with mass number of 12, forming about 98.9% of natural carbon; used as the basis of the newer scale of atomic masses, having an atomic mass of exactly 12 u (relative nuclidic mass unit) by definition.

carbon-13 A heavy isotope of carbon having a mass number of 13.

carbon-14 A naturally occurring radioisotope of carbon having a mass number of 14 and half-life of 5780 years; used in radiocarbon dating and in the elucidation of the metabolic path of carbon in photosynthesis. Also known as radiocarbon.

carbonate 1. An ester or salt of carbonic acid. 2. A compound containing the carbonate $(CO_3{}^{2-})$ radical. 3. Containing carbonates.

carbonate cycle The biogeochemical carbonate pathways, involving the conversion of carbonate to CO_2 and HCO_3, the solution and deposition of carbonate, and the metabolism and regeneration of it in biological systems.

carbonation 1. The process by which a fluid, especially a beverage, is impregnated with carbon dioxide. 2. A process of chemical weathering whereby minerals that contain soda, lime, potash, or basic oxides are changed to carbonates by the carbonic acid in air or water. 3. Conversion to a carbonate.

carbon black 1. An amorphous form of carbon produced commercially by thermal or oxidative decomposition of hydrocarbons and used principally in rubber goods, pigments, and printer's ink. 2. *See* gas black.

carbon burning The synthesis of nuclei in stars through reactions involving the fusion of two carbon-12 nuclei at temperatures of about 5×10^8 K.

carbon burning rate The weight of carbon burned per unit time from the catalytic-cracking catalyst in the regenerator.

carbon cycle 1. The cycle of carbon in the biosphere, in which plants convert carbon dioxide to organic compounds that are consumed by plants and animals, and the carbon is returned to the biosphere in the form of inorganic compounds by processes of respiration and decay. 2. *See* carbon-nitrogen cycle.

carbon dioxide CO_2 A colorless, odorless, tasteless gas about 1.5 times as dense as air.

carbon dioxide absorption tube An absorbent-packed tube used to capture the carbon dioxide formed during the microdetermination of carbon-hydrogen by the Pragl combustion procedure.

carbon dioxide fire extinguisher A type of chemical fire extinguisher in which the extinguishing agent is liquid carbon dioxide, stored under 800–900 pounds per square inch (5.5–6.2 megapascals) at normal room temperature.

carbon disulfide CS_2 A sulfide, used as a solvent for oils, fats, and rubbers and in paint removers.

carbon film Carbon deposited by evaporation onto a specimen to protect and prepare it for electron microscopy.

carbon-hydrogen analyzer A device used in the quantitative analysis of the carbon and hydrogen content of organic compounds.

carbonic acid H_2CO_3 The acid formed by combination of carbon dioxide and water.

carbonic anhydrase An enzyme which aids carbon dioxide transport and release by catalyzing the synthesis, and the dehydration, of carbonic acid from, and to, carbon dioxide and water.

carbonium ion A carbocation which has a positively charged carbon with a co-ordination number greater than 3.

carbonization 1. The conversion of a carbon-containing substance to carbon or a carbon residue as the destructive distillation of coal by heat in the absence of air, yielding a solid residue with a higher percentage of carbon than the original coal; carried on for the production of coke and of fuel gas. 2. In the coalification process, the accumulation of residual carbon by changes in organic material and their decomposition products. 3. Deposition of a thin film of carbon by slow decay of organic matter underwater. 4. A process of converting a carbonaceous material to carbon by removal of other components.

carbon monoxide CO A colorless, odorless gas resulting from the incomplete oxidation of carbon; found, for example, in mines and automobile exhaust; poisonous to animals.

carbonmonoxyhemoglobin A stable combination of carbon monoxide and hemoglobin formed in the blood when carbon monoxide is inhaled. Also known as carbonylhemoglobin and carboxyhemoglobin.

carbon-nitrogen cycle A series of thermonuclear reactions, with release of energy, which presumably occurs in the sun and other stars; the net accomplishment is the synthesis of four hydrogen atoms into a helium atom, the emission of two positrons and much energy, and restoration of a carbon-12 atom with which the cycle began. Also known as carbon cycle; nitrogen cycle.

carbon number The number of carbon atoms in a material under analysis; plotted against chromatographic retention volume for compound identification.

carbon replication A faithful carbon-film, mold of a specimen surface (for example, powders, bones, or crystals) which is thin enough to be studied by electron microscopy.

carbon residue The quantity of carbon produced from a lubricating oil heated in a closed container under standard conditions.

carbon residue test   A destructive-distillation method for estimation of carbon residues in fuels and lubricating oils. Also known as Conradson carbon test.

carbon suboxide C_3O_2 A colorless lacrimatory gas having an unpleasant odor with a boiling point of $-6.8°C$.

carbon tetrachloride CCl_4 Colorless dense liquid, specific gravity 1.595, slightly soluble in water; used as a dry-cleaning agent. Also known as tetrachloromethane.

carbon tetrafluoride CF_4 A colorless gas with a boiling point of $-126°C$; used as a refrigerant. Also known as tetrafluoromethane.

carbon trichloride   *See* hexachloroethane.

carbonyl   A radical (CO) that is made up of one atom of carbon and one atom of oxygen connected by a double bond; found, for example, in aldehydes and ketones. Also known as carbonyl group.

carbonylation   Introduction of a carbonyl radical into a molecule.

1,1′-carbonyl-bis-1-*H*-imidazole   *See* N,N'-carbonyldiimidazole.

carbonyl bromide $COBr_2$ A poisonous liquid boiling at 187.83°C; may be used by the military as a toxic suffocant. Also known as bromophosgene.

carbonyl chloride   *See* phosgene.

carbonyl compound   A compound containing the carbonyl radical (CO).

N,N′-carbonyldiimidazole $C_7H_6N_4O$ Crystals with a melting point of 115.5–116°C; hydrolyzed by water very quickly; used in the synthesis of peptides. Also known as 1,1′-carbonyl-bis-1-*H*-imidazole.

carbonyl fluoride COF_2 A colorless gas that is soluble in water; used in organic synthesis.

carbonyl group   *See* carbonyl.

carbonylhemoglobin   *See* carbonmonoxyhemoglobin.

carbophenothion $C_{11}H_{16}ClO_2PS_3$ An amber liquid used to control pests on fruits, nuts, vegetables, and fiber crops. Also known as S-{[(*para*-chlorophenyl)thio]methyl}-O,O-diethyl phosphorodithioate.

carborane   1. Any of a class of compounds containing boron, carbon, and hydrogen. 2. $B_{10}C_2H_{12}$ A specific member of the class.

carboxin $C_{12}H_{13}NO_2S$ An off-white solid with a melting point of 91.5–92.5°C; used to treat seeds of barley, oats, wheat, corn, and cotton for fungus diseases. Also known as DCMO; 5,6-dihydro-2-methyl-1,4-oxathiin-3-carboxanilide.

carboxy group   —COOH   The functional group of carboxylic acid. Also known as carboxyl group.

carboxyhemoglobin   *See* carbonmonoxyhemoglobin.

carboxylase   Any enzyme that catalyzes a carboxylation or decarboxylation reaction.

carboxylate anion   An anion with the general formula $(RCO_2)^-$, which is formed when the hydrogen attached to the carboxyl group of a carboxylic acid is removed.

carboxylation   Addition of a carboxyl group into a molecule.

carboxylic Having chemical properties resembling those of carboxylic acid.

carboxylic acid Any of a family of organic acids characterized by the presence of one or more carboxyl groups.

carboxyl terminal The end of a polypeptide chain with a free carboxyl group.

carboxymethyl cellulose An acid ether derivative of cellulose used as a sodium salt; a white, odorless, bulky solid used as a stabilizer and emulsifier; negatively charged resin used in ion-exchange chromatography as a cation exchanger. Also known as cellulose gum; CM-cellulose; sodium carboxymethylcellulose; sodium cellulose glycolate.

carboxypeptidase Any enzyme that catalyzes the hydrolysis of a peptide at the end containing the free carboxyl group.

carburetion The process of enriching a gas by adding volatile carbon compounds, such as hydrocarbons, to it, as in the manufacture of carbureted water gas.

carburetor An apparatus for vaporizing, cracking, and enriching oils in the manufacture of carbureted water gas.

carbyne Elemental carbon in a triply bonded form.

cardiolipin A complex phospholipid found in the ether alcohol extract of powdered beef heart; mixed with leutin and cholesterol, it functions as the antigen in the Wassermann complement-fixation test for syphilis. Also known as diphosphatidyl glycerol.

δ-3-carene $C_{10}H_{16}$ A clear, colorless, combustible terpene liquid, stable to about 250°C; used as a solvent and in chemical synthesis. Also known as 3,7,7-trimethylbicyclo-[4.1.0]-hept-3-ene.

Carius method A procedure used to analyze organic compounds for sulfur, halogens, and phosphorus that involves heating the sample with fuming nitric acid in a sealed tube.

Carlsbad law A feldspar twin law in which the twinning axis is the c axis, the operation is rotation of 180°, and the contact surface is parallel to the side pinacoid.

Carlsbad turn A twin crystal in the monoclinic system with the vertical axis as the turning axis.

carminic acid $C_{22}H_{20}O_{13}$ A glucosidal hydroxyanthrapurin that is derived from cochineal; a red crystalline dye used as a stain for biological materials. Also known as cochinilin.

carnaubic acid $C_{24}H_{48}O_2$ An acid found in carnauba wax and beef kidney.

carnitine $C_7H_{15}NO_3$ α-Amino-β-hydroxybutyric acid trimethylbetaine; a constituent of striated muscle and liver, identical with vitamin B_T (the L form of vitamin B).

carnosine $C_9H_{14}N_4O_3$ A colorless, crystalline dipeptide occurring in the muscle tissue of vertebrates.

Carnot's reagent A solution of sodium bismuth thiosulfate in alcohol used for determining potassium.

Caro's acid H_2SO_5 A white solid melting at about 45°C, formed during the acid hydrolysis of peroxydisulfates. Also known as peroxymonosulfuric acid.

carotenase An enzyme that effects the hydrolysis of carotenoid compounds, used in bleaching of flour.

carotene $C_{40}H_{56}$ Any of several red, crystalline, carotenoid hydrocarbon pigments occurring widely in nature, convertible in the animal body to vitamin A, and characterized by preferential solubility in petroleum ether. Also known as carotin.

carotenoid A class of labile, easily oxidizable, yellow, orange, red, or purple pigments that are widely distributed in plants and animals and are preferentially soluble in fats and fat solvents.

carotin *See* carotene.

carotol $C_{15}H_{25}OH$ A sesquiterpenoid alcohol in carrots.

carrageenin A colloid extracted from carrageen and used chiefly as an emulsifying and stabalizing agent in foods, pharmaceuticals, and cosmetics.

carrene *See* methylene chloride.

carry-over Unwanted liquid or solid material carried by the overhead effluent from a fractionating column, absorber, or reaction vessel.

cartridge silk *See* powder silk.

carvacrol $(CH_3)_2CHC_6H_3(CH_3)OH$ A colorless liquid, boiling at 237°C; used in perfumes, flavorings, and fungicides. Also known as 2-hydroxy-*para*-cymene; isopropyl-*ortho*-cresol; 2-methyl-5-isopropyl phenol.

carvone $C_{10}H_{14}O$ A liquid ketone that boils at 231°C; soluble in water and alcohol; it is optically active and occurs naturally in both dextro and levo forms; used in flavorings and perfumery. Also known as carvol.

caryophyllene $C_{15}H_{24}$ A liquid sesquiterpene that is found in some essential oils, particularly clove oil.

caryophyllin $C_{30}H_{48}O_3$ A ketone, soluble in alcohol, extracted from oil of cloves.

Casale process A process that employs promoted iron oxide catalyst for synthesis of ammonia from nitrogen and hydrogen.

cascade cooler Fluid-cooling device through which the fluid flows in a series of horizontal tubes, one above the other; cooling water from a trough drips over each tube, then to a drain. Also known as serpentine cooler; trickle cooler.

cascade gamma emission The emission by a nucleus of two or more gamma rays in succession.

cascade mixer-settler Series of liquid-holding vessels with stirrers, each connected to an unstirred vessel in which solids or heavy immiscible liquids settle out of suspension; light liquid moves through the mixer-settler units, counterflowing to heavy material, in such a manner that fresh liquid contacts treated heavy material, and spent (used) liquid contacts fresh (untreated) heavy material.

cascade tray A fractionating apparatus that consists of a series of parallel troughs arranged in stairstep fashion.

casein The protein of milk; a white solid soluble in acids.

casein-formaldehyde A modified natural polymer.

Cassel green *See* barium manganate.

Castle's intrinsic factor *See* intrinsic factor.

Castner cell A type of mercury cell used in the commercial production of chlorine and sodium.

Castner process A process used industrially to make high-test sodium cyanide by reacting sodium, glowed charcoal, and dry ammonia gas to form sodamide, which is converted to cyanamide immediately; the cyanamide is converted to cyanide with charcoal.

castor oil acid *See* ricinoleic acid.

catabolism That part of metabolism concerned with the breakdown of large protoplasmic molecules and tissues, often with the liberation of energy.

catabolite Any product of catabolism.

catabolite repression An intracellular regulatory mechanism in bacteria whereby glucose, or any other carbon source that is an intermediate in catabolism, prevents formation of inducible enzymes.

cata-condensed polycyclic An aromatic compound in which no more than two rings have a single carbon atom in common.

catalase An enzyme that catalyzes the decomposition of hydrogen peroxide into molecular oxygen and water.

catalysis A phenomenon in which a relatively small amount of substance augments the rate of a chemical reaction without itself being consumed.

catalyst Substance that alters the velocity of a chemical reaction and may be recovered essentially unaltered in form and amount at the end of the reaction.

catalyst carrier A neutral material used to support a catalyst, such as activated carbon, diatomaceous earth, or activated alumina.

catalyst selectivity 1. The relative activity of a catalyst in reference to a particular compound in a mixture. 2. The relative rate of a single reactant in competing reactions.

catalyst stripping Introduction of steam to remove hydrocarbons retained on the catalyst; the steam is introduced where the spent catalyst leaves the reactor.

catalytic activity The ratio of the space velocity of a catalyst being tested, to the space velocity required for a standard catalyst to give the same conversion as the catalyst under test.

catalytic converter A cracker which utilizes catalysts.

catalytic cracking Conversion of high-boiling hydrocarbons into lower-boiling types by a catalyst.

catalytic hydrogenation Hydrogenating by means of catalysts such as nickel or palladium.

catalytic polymerization Polymerization of monomers to form high-molecular-weight molecules in the presence of catalysts.

catalytic reforming Rearranging of hydrocarbon molecules in a gasoline boiling-range feedstock to form hydrocarbons having a higher antiknock quality. Abbreviated CR.

cataphoresis *See* electrophoresis.

cat cracker A refinery unit where catalytic cracking is done.

catechin See catechol.

catechol One of a group of three isomeric dihydroxy benzenes in which the two hydroxyl groups are ortho to each other. Also known as catechin; pyrocatechol.

catecholamine Any one of a group of sympathomimetic amines containing a catechol moiety, including especially epinephrine, norepinephrine (levarterenol), and dopamine.

catenane A compound whose structure has at least two interlocking rings, similar to links in a chain, where the two links remain free of conventional chemical bonds, yet are still fitted together mechanically.

catenation The property of an element to link to itself to form molecules as with carbon.

catforming A naphtha-reforming process with a catalyst of platinum-silica-alumina which results in very high hydrogen purity.

cathepsin Any of several proteolytic enzymes occurring in animal tissue that hydrolyze high-molecular-weight proteins to proteoses and peptones.

cathode The electrode at which reduction takes place in an electrochemical cell, that is, a cell through which electrons are being forced.

cathode efficiency The proportion of current used for completion of a given process at the cathode.

cathodic inhibitor A compound, such as calcium bicarbonate or sodium phosphate, which is deposited on a metal surface in a thin film that operates at the cathodes to provide physical protection over the entire surface against corrosive attack in a conducting medium.

cathodic polarization Portion of electric cell polarization occurring at the cathode.

catholyte Electrolyte adjacent to the cathode in an electrolytic cell.

cation A positively charged atom or group of atoms, or a radical which moves to the negative pole (cathode) during electrolysis.

cation analysis Qualitative analysis for cations in aqueous solution.

cation exchange A chemical reaction in which hydrated cations of a solid are exchanged, equivalent for equivalent, for cations of like charge in solution.

cation exchange resin A highly polymerized synthetic organic compound consisting of a large, nondiffusible anion and a simple, diffusible cation, which later can be exchanged for a cation in the medium in which the resin is placed.

cationic detergent A member of a group of detergents that have molecules containing a quaternary ammonium salt cation with a group of 12 to 24 carbon atoms attached to the nitrogen atom in the cation; an example is alkyltrimethyl ammonium bromide.

cationic hetero atom A positively charged atom, other than carbon, in an otherwise carbon atomic chain or ring.

cationic polymerization A type of polymerization in which Lewis acids act as catalysts.

cationic reagent   A surface-active agent with active positive ions used for ore beneficiation (flotation via flocculation); an example of a cationic reagent is cetyl trimethyl ammonium bromide.

cationtrophy   The breaking off of an ion, such as a hydrogen ion or metal ion, from a molecule so that a negative ion remains in equilibrium.

caustic   **1.** Burning or corrosive.   **2.** A hydroxide of a light metal.

caustic alcohol   *See* sodium ethylate.

caustic antimony   *See* antimony trichloride.

causticity   The property of being caustic.

caustic lime   *See* calcium oxide.

caustic potash   *See* potassium hydroxide.

caustic soda   *See* sodium hydroxide.

caustic wash   **1.** Treating a product with a solution of caustic soda to remove impurities.   **2.** The solution itself.

cavitation   Emulsification produced by disruption of a liquid into a liquid-gas two-phase system, when the hydrodynamic pressure of the liquid is reduced to the vapor pressure.

c axis   A vertically oriented crystal axis, usually the principal axis; the unique symmetry axis in tetragonal and hexagonal crystals.

CCR process   *See* cyclic catalytic reforming process.

Cd   *See* cadmium.

CDEC   *See* 2-chloroallyl diethyldithiocarbamate.

Ce   *See* cerium.

cell   A cup, jar, or vessel containing electrolyte solutions and metal electrodes to produce an electric current (conductiometric or potentiometric) or for electrolysis (electrolytic).

cell constant   The ratio of distance between conductance-titration electrodes to the area of the electrodes, measured from the determined resistance of a solution of known specific conductance.

cellobiose $C_{12}H_{22}O_{11}$ A disaccharide which does not occur freely in nature or as a glucoside; a unit of cellulose and lichenin; crystallizes as minute water-soluble crystals from alcohol. Also known as cellose.

cellosolve $C_2H_5OCH_2CH_2OH$ An important industrial chemical used in varnish removers, in cleaning solutions, and as a solvent for paints, varnishes, and plastics. Also known as 2-ethoxyethanol.

cellulase   Any of a group of extracellular enzymes, produced by various fungi, bacteria, insects, and other lower animals, that hydrolyze cellulose.

cellulose $(C_6H_{10}O_5)_n$ The main polysaccharide in living plants, forming the skeletal structure of the plant cell wall; a polymer of β-D-glucose units linked together, with the elimination of water, to form chains comprising 2000–4000 units.

α-cellulose   *See* alpha cellulose.

cellulose acetate An acetic acid ester of cellulose; a tough, flexible, slow-burning, and longlasting thermoplastic material used as the base for magnetic tape and movie film, in acetate rayon, as a plastic film in food packaging, in lacquers, and for molded receiver cabinets.

cellulose acetate butyrate An ester of cellulose formed by the action of a mixture of acetic acid and butyric acid and their anhydrides on purified cellulose; has high impact resistance, clarity, and weatherability; used in making plastic film, lacquer, lenses, and outdoor signs.

cellulose diacetate The ester formed by esterification of two hydroxyl groups of a cellulose molecule with acetic acid.

cellulose ester Cellulose in which the free hydroxyl groups have been replaced wholly or in part by acidic groups.

cellulose ether The product of the partial or complete etherification of the hydroxyl groups in a cellulose molecule.

cellulose fiber Any fiber based on esters or ethers of cellulose.

cellulose gum *See* carboxymethyl cellulose.

cellulose methyl ether *See* methylcellulose.

cellulose nitrate Any of several esters of nitric acid, produced by treating cotton or some other form of cellulose with a mixture of nitric and sulfuric acids; used as explosive and propellant. Also known as nitrocellulose; nitrocotton.

cellulose propionate An ester of cellulose and propionic acid.

cellulose triacetate A cellulose resin formed by the complete esterification of the cellulose by acetic acid; used as a base in protective coatings.

cellulose xanthate A compound formed by reaction of soda cellulose (prepared by treating cellulose with strong sodium hydroxide solution) with carbon disulfide.

cellulosic Any of the derivatives of cellulose, such as cellulose acetate.

cellulosic resin Any resin based on cellulose compounds such as esters and ethers.

cementation The setting of a plastic material.

centered lattice A crystal lattice in which the axes have been chosen according to the rules for the crystal system, and in which there are lattice points at the centers of certain planes as well as at the corners.

center of inversion A point in a crystal lattice such that the lattice is left invariant by an inversion in the point.

centrifugal extractor A device for separating components of a liquid solution, consisting of a series of perforated concentric rings in a cylindrical drum that rotates at 2000–5000 revolutions per minute around a cylindrical shaft; liquids enter and leave through the shaft; they flow radially and concurrently in the rotating drum.

centrifugal molecular still A device used for molecular distillation; material is fed to the center of a hot, rapidly rotating cone housed in a chamber at a high vacuum; centrifugal force spreads the material rapidly over the hot surface, where the evaporable material goes off as a vapor to the condenser.

centrifugal sedimentation Removing solids from liquids by causing particles to settle through the liquid radially toward or away from the center of rotation (depending on the solid-liquid relative densities) by use of a centrifuge.

centrifugal settler Spinning container that separates solid particles from liquids; centrifugal force causes suspended solids to move toward or away from the center of rotation, thus concentrating them in one area for removal.

centrifuge refining The use of centrifuges for liquids processing, such as separation of solids or immiscible droplets from liquid carriers, or for liquid-liquid solvent extraction.

centrifuge tube Calibrated, tube-shaped glass container used with laboratory centrifuges for volumetric analysis of separable (solid-liquid or immiscible liquid) samples.

CEPHA *See* ethephon.

cephaeline $C_{14}H_{19}O_2N$ An alkaloid, slightly soluble in water, extracted from the root of ipecac; used as an emetic.

cephalin Any of several acidic phosphatides whose composition is similar to that of lecithin but having ethanolamine, serine, and inositol instead of choline; found in many living tissues, especially nervous tissue of the brain.

cerate A metallic salt or soap made from lard.

cerebrose *See* galactose.

cerebroside Any of a complex group of glycosides found in nerve tissue, consisting of a hexose, a nitrogenous base, and a fatty acid. Also known as galactolipid.

cerelose *See* glucose.

ceria *See* ceric oxide.

ceric oxide CeO_2 A pale-yellow to white powder; soluble in sulfuric acid, insoluble in dilute acid and water; used in ceramics and as a polish for optical glass. Also known as ceria; cerium dioxide; cerium oxide.

ceric sulfate $Ce(SO_4)_2 \cdot 4H_2O$ Yellow needles forming a basic salt with excess water; used in waterproofing, mildew-proofing, and in dyeing and printing textiles.

cerinic acid *See* cerotic acid.

cerium A chemical element, symbol Ce, atomic number 58, atomic weight 140.12; a rare-earth metal, used as a getter in the metal industry, as an opacifier and polisher in the glass industry, in Welsbach gas mantles, in cored carbon arcs, and as a liquid-liquid extraction agent to remove fission products from spent uranium fuel.

cerium-140 An isotope of cerium with atomic mass number of 140, 88.48% of the known amount of the naturally occurring element.

cerium-142 A radioactive isotope of cerium with atomic mass number of 142; emits α-particles and has a half-life of 5×10^{15} years.

cerium-144 A radioactive isotope of the element cerium with atomic mass number of 144; a beta emitter with a half-life of 285 days.

cerium dioxide *See* ceric oxide.

cerium fluoride CeF_3 White hexagonal crystals, melting point 1460°C; used in arc carbons to increase the brilliance of carbon-arc lamps.

cerium oxide *See* ceric oxide.

cerium stearate $Ce(C_{18}H_{35}O_2)_2$ White, waxy, inert powder, melting point 100–110°C; used in waterproofing compounds.

cerotic acid $CH_3(CH_2)_{24}COOH$ A fatty acid derived from carnauba wax or beeswax; melts at 87.7°C. Also known as cerinic acid; hexacosanoic acid.

ceruloplasmin The copper-binding serum protein in human blood.

ceryl alcohol $C_{26}H_{53}OH$ An alcohol derived from Chinese wax, melting at 79°C and insoluble in water.

cesium A chemical element, symbol Cs, atomic number 55, atomic weight 132.905.

cesium-134 An isotope of cesium, atomic mass number of 134; emits negative beta particles and has a half-life of 2.19 years; used in photoelectric cells and in ion propulsion systems under development.

cesium-137 An isotope of cesium with atomic mass number of 137; emits negative beta particles and has a half-life of 30 years; offers promise as an encapsulated radiation source for therapeutic and other purposes. Also known as radiocesium.

cesium bromide CsBr A colorless, crystalline powder with a melting point of 636°C; soluble in water; used in medicine, for infrared spectroscopy, and in scintillation counters.

cesium carbonate Cs_2CO_3 A white, hygroscopic, crystalline powder; soluble in water; used in specialty glasses.

cesium chloride CsCl Colorless cuboid crystals, melting point 646°C; used in filaments of radio tubes to increase sensitivity, in photoelectric cells, and for photosensitive deposit on cathodes.

cesium fluoride CsF Toxic, irritating, deliquescent crystals with a melting point of 682°C; soluble in water and methanol; used in medicine, mineral water, and brewing.

cesium hydroxide CsOH Colorless or yellow, fused crystalline mass with a melting point of 272.3°C; soluble in water; used as electrolyte in alkaline storage batteries at subzero temperatures.

cesium iodide CsI A colorless, deliquescent, crystalline powder with a melting point of 621°C; soluble in water and alcohol; crystals used for infrared spectroscopy.

cesium perchlorate $CsClO_4$ A crystalline solid with a melting point of 250°C; soluble in water; used in optics and for specialty glasses.

cesium sulfate Cs_2SO_4 Colorless crystals with a melting point of 1010°C; soluble in water; used for brewing and in mineral waters.

cetane *See* n-hexadecane.

cetane index An empirical method for finding the cetane number of a fuel based on API gravity and the mid boiling point.

cetane number The percentage by volume of cetane (cetane number 100) in a blend with α-methylnaphthalene (cetane number 0); indicates the ability of a fuel to ignite quickly after being injected into the cylinder of an engine.

cetane number improver   A chemical which has the effect of increasing a diesel fuel's cetane number; examples are nitrates, nitroalkanes, nitrocarbonates, and peroxides.

cetene   *See* 1-hexadecene.

cetin $C_{15}H_{31}COOC_{16}H_{33}$ A white, crystalline, waxy substance with a melting point of 50°C; soluble in alcohol and ether; used as a base for ointments and emulsions and in the manufacture of soaps and candles.

cetol   *See* cetyl alcohol.

cetrimonium bromide $CH_3(CH_2)_{15}N(CH_3)_3Br$ Crystals with a melting point of 237–243°C; soluble in alcohol, water, and sparingly in acetone; used as a cationic detergent, antiseptic, and precipitant for nucleic acids and mucopolysaccharides. Also known as cetyltrimethylammonium bromide.

cetyl   The radical represented as $C_{16}H_{33}$—.

cetyl alcohol $C_{15}H_{33}OH$ A colorless wax, insoluble in water although a solution in kerosine forms an insoluble film on water. Also known as cetol; ethal; 1-hexadecanol.

cetylene   *See* 1-hexadecene.

cetylic acid   *See* palmitic acid.

cetyltrimethylammonium bromide   *See* cetrimonium bromide.

cetyl vinyl ether $C_{16}H_{33}OCO:CH_2$ A colorless liquid with a boiling point of 142°C; may be copolymerized with unsaturated monomers to make internally plasticized resins.

Cf   *See* californium.

Chadwick-Goldhaber effect   *See* photodisintegration.

chain   A structure in which similar atoms are linked by bonds.

chain balance   An analytical balance with one end of a fine gold chain suspended from the beam and the other fastened to a device which moves over a graduated vernier scale.

chain decay   *See* series disintegration.

chain disintegration   *See* series disintegration.

chain fission yield   The sum of the independent fission yields for all isobars of a particular mass number.

chain isomerism   A type of molecular isomerism seen in carbon compounds; as the number of carbon atoms in the molecule increases, the linkage between the atoms may be a straight chain or branched chains producing isomers that differ from each other by possessing different carbon skeletons.

chain reaction   A chemical reaction in which many molecules undergo chemical reaction after one molecule becomes activated.

chain scission   The cleavage of polymer chains, as in natural rubber as a result of heating.

chair form   A particular nonplanar conformation of a cyclic molecule with more than five atoms in the ring; for example, in the chair form of cyclohexane, the hydrogens are staggered and directed perpendicularly to the mean plane of the carbons (axial conformation, a) or equatorially to the center of the mean plane (equatorial conformation, e).

chalcogen One of the elements that form group VIa of the periodic table; included are oxygen, sulfur, selenium, tellurium, and polonium.

chalcogenide A binary compound containing a chalcogen and a more electropositive element or radial.

chalking 1. Treating with chalk. 2. Forming a powder which is easily rubbed off.

chalones Substances thought to be molecules of the protein-polypeptide class that are produced as part of the growth-control systems of tissues; known to inhibit cell division by acting on several phases in the mitotic cycle.

chamber acid Sulfuric acid made by the obsolete chamber process.

chamber process An obsolete method of manufacturing sulfuric acid in which sulfur dioxide, air, and steam are reacted in a lead chamber with oxides of nitrogen as the catalyst.

channel In percolation filtration, a portion of the clay bed where there is a preponderance of flow.

channel black *See* gas black.

channeling In chromatography, furrows or breaks in an ion-exchange bed which permit a solution to run through without having contact with active groups elsewhere in the bed.

channel process A carbon-black process in which iron channel beams are used as depositing surfaces for carbon black.

channel spin The vector sum of the spins of the particles involved in a nuclear reaction, either before or after the reaction takes place.

characteristic loss spectroscopy A branch of electron spectroscopy in which a solid surface is bombarded with monochromatic electrons, and backscattered particles which have lost an amount of energy equal to the core-level binding energy are detected. Abbreviated CLS.

characteristic radiation Radiation originating in an atom following removal of an electron, whose wavelength depends only on the element concerned and the energy levels involved.

characteristic x-rays Electromagnetic radiation emitted as a result of rearrangements of the electrons in the inner shells of atoms; the spectrum consists of lines whose wavelengths depend only on the element concerned and the energy levels involved.

characterization factor A number which expresses the variations in physical properties with change in character of the paraffinic stock; ranges from 12.5 for paraffinic stocks to 10.0 for the highly aromatic stocks. Also known as Watson factor.

charcoal test A determination of the natural gasoline content of natural gas by adsorbing the gasoline on activated charcoal and then recovering it by distillation.

charge-delocalized ion A charged species in which the charge is distributed over more than one atom.

charged species A chemical entity in which the overall total of electrons is unequal to the overall total of protons.

charge independence The principle that the nuclear (strong) force between a neutron and a proton is identical to the force between two protons or two neutrons in the same orbital and spin state.

charge-localized ion A charged species in which the charge is centered on a single atom.

charge transfer The process in which an ion takes an electron from a neutral atom, with a resultant transfer of charge.

charge-transfer complexes Compounds in which electrons move between molecules.

charging pump Pump that provides pressurized fluid flow for the input of another unit, such as to a triplex pump that requires positive pressure.

chavicol $C_3H_5C_6H_4OH$ A colorless phenol that is liquid at room temperature; boils at 230°C; soluble in alcohol and water; found in many essential oils. Also known as 1-allyl-4-hydroxybenzene; *para*-allyl phenol.

chE *See* cholinesterase.

cheesebox still One of the first types of vertical cylindrical stills designed with a vapor dome.

chelate A molecular structure in which a heterocyclic ring can be formed by the unshared electrons of neighboring atoms.

chelating agent An organic compound in which atoms form more than one coordinate bonds with metals in solution.

chelating resin Any of the ion-exchange resins with unusually high selectivity for specific cations; for example, phenol-formaldehyde resin with 8-quinolinol replacing part of the phenol, particularly selective for copper, nickel, cobalt, and iron(III).

chelation A chemical process involving formation of a heterocyclic ring compound which contains at least one metal cation or hydrogen ion in the ring.

chelerythrine $C_{21}H_{17}O_4H$ A poisonous, crystalline alkaloid, slightly soluble in alcohol; it is derived from the seeds of the herb celandine (*Chelidonium majus*) and has narcotic properties.

cheletropic reaction A chemical reaction involving the elimination of a molecule in which two sigma bonds terminating at a single atom are made or broken.

chelidonic acid $C_7H_4O_6$ A pyran isolated from the perennial herb celandine (*Chelidonium majus*).

chelometry Analytical technique involving the formation of 1:1 soluble chelates when a metal ion is titrated with aminopolycarboxylate and polyamine reagents; a form of complexiometric titration.

chemical 1. Related to the science of chemistry. 2. A substance characterized by definite molecular composition.

chemical bond *See* bond.

chemical cellulose *See* alpha cellulose.

chemical compound *See* compound.

chemical crystallography The geometric description, and study, of the internal arrangement of atoms in crystals formed from chemical compounds.

chemical dating The determination of the relative or absolute age of minerals and of ancient objects and materials by measurement of their chemical compositions.

chemical deposition Precipitation of a metal from a solution of a salt by introducing another metal.

chemical element *See* element.

chemical energy Energy of a chemical compound which, by the law of conservation of energy, must undergo a change equal and opposite to the change of heat energy in a reaction; the rearrangement of the atoms in reacting compounds to produce new compounds causes a change in chemical energy.

chemical equilibrium A condition in which a chemical reaction is occurring at equal rates in its forward and reverse directions, so that the concentrations of the reacting substances do not change with time.

chemical exchange process A method of separating isotopes of the lighter elements by the repetition of a process of chemical change which involves exchange of the isotopes.

chemical fire extinguisher Any of three types of fire extinguishers (vaporizing liquid, carbon dioxide, and dry chemical) which expel chemicals in solid, liquid, or gaseous form to blanket or smother a fire.

chemical formula A notation utilizing chemical symbols and numbers to indicate the chemical composition of a pure substance; examples are CH_4 for methane and HCl for hydrogen chloride.

chemical indicator A substance whose physical appearance is altered at or near the end point of a chemical titration.

chemical inhibitor A substance capable of stopping or retarding a chemical reaction.

chemical-ion pump A vacuum pump whose pumping action is based on evaporation of a metal whose vapor then reacts with the chemically active molecules in the gas to be evacuated.

chemical kinetics That branch of physical chemistry concerned with the mechanisms and rates of chemical reactions. Also known as reaction kinetics.

chemically pure Without impurities detectable by analysis. Abbreviated cp.

chemical microscopy Application of the microscope to the solution of chemical problems.

chemical polarity Tendency of a molecule, or compound, to be attracted or repelled by electrical charges because of an asymmetrical arrangement of atoms around the nucleus.

chemical potential In a thermodynamic system of several constituents, the rate of change of the Gibbs function of the system with respect to the change in the number of moles of a particular constituent.

chemical pulping Separation of wood fiber for paper pulp by chemical treatment of wood chips to dissolve the lignin that cements the fibers together.

chemical reaction A change in which a substance (or substances) is changed into one or more new substances; there is only a minute change, Δm, in the mass of the system, given by $\Delta E = \Delta mc^2$, where ΔE is the energy emitted or absorbed and c is the speed of light.

chemical reactivity The tendency of two or more chemicals to react to form one or more products differing from the reactants.

chemical reactor Vessel, tube, pipe, or other container within which a chemical reaction is made to take place; may be batch or continuous, open or packed, and can use thermal, catalytic, or irradiation actuation.

chemical shift *See* isomeric shift.

chemical similitude A procedure used to ensure satisfactory operation of a full-scale chemical process by comparison with pilot plant data.

chemical symbol A notation for one of the chemical elements, consisting of letters; for example Ne, O, C, and Na represent neon, oxygen, carbon, and sodium.

chemical synthesis The formation of one chemical compound from another.

chemical thermodynamics The application of thermodynamic principles to problems of chemical interest.

chemical weathering A weathering process whereby rocks and minerals are transformed into new, fairly stable chemical combinations by such chemical reactions as hydrolysis, oxidation, ion exchange, and solution. Also known as decay; decomposition.

chemiclearance The use of chemical analysis to establish the safe use of a substance.

chemiluminescence Emission of light as a result of a chemical reaction without an apparent change in temperature.

chemiosmosis A chemical reaction occurring through an intervening semipermeable membrane. Also known as chemosmosis.

chemisorption A chemical adsorption process in which weak chemical bonds are formed between gas or liquid molecules and a solid surface.

chemist A scientist specializing in chemistry.

chemosmosis *See* chemiosmosis.

chemosynthesis The synthesis of organic compounds from carbon dioxide by microorganisms using energy derived from chemical reactions.

chemurgy An obsolete term referring to a branch of chemistry concerned with the profitable utilization of organic raw materials, especially agricultural products, for nonfood purposes such as for paints and varnishes.

chenic acid *See* chenodeoxycholic acid.

chenodeoxycholic acid $C_{24}H_{40}O_4$ A constituent of bile; needlelike crystals with a melting point of 119°C; soluble in alcohol, methanol, and acetic acid; used on an experimental basis to prevent and dissolve gallstones. Also known as anthropodesoxycholic acid; chenic acid; gallodesoxycholic acid.

chiller Oil-refining apparatus in which the temperature of paraffin distillates is lowered preparatory to filtering out the solid wax components.

chinaldine *See* quinaldine.

Chinese vermilion   *See* mercuric sulfide.

Chinese white   A term used in the paint industry for zinc oxide and kaolin used as a white pigment. Also known as zinc white.

chinidine   *See* quinidine.

chinoidine   *See* quinoidine.

chinone   *See* quinone.

chirality   The handedness of an asymmetric molecule.

chiral molecules   Molecules which are not superposable with their mirror images.

chiral twinning   *See* optical twinning.

chitin   A white or colorless amorphous polysaccharide that forms a base for the hard outer integuments of crustaceans, insects, and other invertebrates.

chitinase   An externally secreted digestive enzyme produced by certain microorganisms and invertebrates that hydrolyzes chitin.

chloflurecol methyl ester $C_{15}H_{11}ClO_3$ A white, crystalline compound with a melting point of 152°C; slight solubility in water; used as a growth regulator for grass and weeds. Also known as methyl-2-chloro-9-hydroxyfluorene-9-carboxylate.

chloracetone   *See* chloroacetone.

chloracetophenone   *See* chloroacetophenone.

chloral CCl_3CHO A colorless, oily liquid soluble in water; used industrially to prepare DDT; a hypnotic. Also known as trichloroacetic aldehyde; trichloroethanal.

chloralase $C_8H_{11}Cl_3O_6$ Colorless, water-soluble crystals, melting at 185°C; made by heating chloral with dextrose; used as a hypnotic.

chloral hydrate $CCl_3CH(OH)_2$ Colorless, deliquescent needles with slightly bitter caustic taste, soluble in water; a hypnotic. Also known as crystalline chloral; hydrated chloral.

chloralkane   Chlorinated aliphatic hydrocarbon of the methane series (C_nH_{2n+2}).

chloralosane   *See* chloralose.

chloralose $C_8H_{11}O_6Cl_3$ A crystalline compound with a melting point of 178°C; used as a repellent for birds. Also known as glucochloralose.

α-chloralose $C_8H_{11}O_6Cl_3$ Needlelike crystals with a melting point of 87°C; soluble in glacial acetic acid and ether; used on seed grains as a bird repellent and as a hypnotic for animals. Also known as anhydroglucochloral; chloralosane; glucochloral; α-D-glucochloralose.

chloramben   *See* 3-amino-2,5-dichlorobenzoic acid.

chloramine T $CH_3C_6H_4SO_2NClNa\cdot3H_2O$ A white, crystalline powder that decomposes slowly in air, freeing chlorine; used as an antiseptic, a germicide, and an oxidizing agent and chlorinating agent.

chloranil $C_6Cl_4O_2$ Yellow leaflets melting at 290°C; soluble in organic solvents; made from phenol by treatment with potassium chloride and hydrochloric acid; used as an agricultural fungicide and as an oxidizing agent in the manufacture of dyes. Also known as tetrachloroquinone.

chloranilic acid $C_6H_2Cl_2O_4$ A relatively strong dibasic acid whose crystals are red and melt between 283–284°C; used in spectrophotometry.

chlorate ClO_3^- **1.** A negative ion derived from chloric acid. **2.** A salt of chloric acid.

chlorbenside $C_{13}H_{10}SCl_2$ White crystals with a melting point of 72°C; used as a miticide for spider mites on fruit trees and ornamentals. Also known as *para*-chlorobenzyl-*para*-chlorophenyl sulfide.

chlorbromuron $C_9H_{10}ONBrCl$ A white solid with a melting point of 94–96°C; used as a pre- and postemergence herbicide for annual grass and for broadleaf weeds on crops, soybeans, and Irish potatoes. Also known as 3-(4-bromo-3-chlorophenyl)-1-methoxy-1-methylurea.

chlordan *See* chlordane.

chlordane $C_{10}H_6Cl_8$ A volatile liquid insecticide; a chlorinated hexahydrome-thanoindene. Also spelled chlordan.

chlordimeform $C_{10}H_{13}ClN_2$ A tan-colored solid, melting point 35°C; used as a miticide and insecticide for fruits, vegetables, and cotton. Also known as N'-(4-chloro-*ortho*-tolyl)-N,N-dimethylformamidine.

chlorendic acid $C_9H_4Cl_6O_4$ White, fine crystals used in fire-resistant polyester resins and as an intermediate for dyes, fungicides, and insecticides.

chlorendic anhydride $C_9H_2Cl_6O_3$ White, fine crystals used in fire-resistant po-lyester resins, in hardening epoxy resins, and as a chemical intermediate.

Chlorex process A method of extracting lubricating-oil stocks; the solvent is Chlorex, a trademark for β,β-dichlorodiethyl ether.

chlorfenethol $C_{14}H_{12}Cl_2O$ A colorless, crystalline compound with a melting point of 69.5–70°C; insoluble in water; used for control of mites in ornamentals and shrub trees. Also known as 1,1-bis(4-chlorophenyl)ethanol.

chlorfenpropmethyl $C_{10}H_{10}OCl_2$ A colorless to brown liquid used as a post-emergence herbicide of wild oats, cereals, fodder beets, sugarbeets, and peas. Also known as 2-chloro-3-(4-chlorophenyl)-methylpropionate.

chlorfensulfide $C_{12}H_6Cl_4N_2S$ A yellow, crystalline compound with a melting point of 123.5–124°C; used as a miticide for citrus. Also known as 4-chlo-rophenyl-2,4,5-trichlorophenylazosulfide.

chlorfenvinphos $C_{12}H_{14}Cl_3O_4P$ An amber liquid with a boiling point of 168–170°C; used as an insecticide for ticks, flies, lice, and mites on cattle. Also known as 2-chloro-1-(2,4-dichlorophenyl)-vinyl diethylphosphate.

chlorhydrin *See* chlorohydrin.

chloric acid $HClO_3$ A compound that exists only in solution and as chlorate salts; breaks down at 40°C.

chloride **1.** A compound which is derived from hydrochloric acid and contains the chlorine atom in the −1 oxidation state. **2.** In general, any binary com-pound containing chlorine.

chloride benzilate *See* lachesne.

chloridization *See* chlorination.

chlorimide *See* dichloramine.

chlorinated acetone *See* chloroacetone.

chlorinated paraffin One of a group of chlorine derivatives of paraffin compounds.

chlorination 1. Introduction of chlorine into a compound. Also known as chloridization. 2. Water sterilization by chlorine gas.

chlorinator The apparatus used in chlorinating.

chlorine A chemical element, symbol Cl, atomic number 17, atomic weight 35.453; used in manufacture of solvents, insecticides, and many non-chlorine-containing compounds, and to bleach paper and pulp.

chlorine dioxide ClO_2 A green gas used to bleach cellulose and to treat water.

chlorine water A clear, yellowish liquid used as a deodorizer, antiseptic, and disinfectant.

chlorite A salt of chlorous acid.

chloritization The introduction of, production of, replacement by, or conversion into chlorite.

chlormephos $C_5H_{12}O_2S_2ClP$ A liquid used as an insecticide for soil. Also known as *S*-chloromethyl-*O*,*O*-diethyl phosphorodithioate.

chloro- A prefix describing an organic compound which contains chlorine atoms substituted for hydrogen.

chloroacetic acid $ClCH_2COOH$ White or colorless, deliquescent crystals that are soluble in water, ether, chloroform, benzene, and alcohol; used as an herbicide and in the manufacture of dyes and other organic molecules. Also known as chloroethanoic acid; monochloroacetic acid.

chloroacetic acid *tert*-butyl ester *See* *tert*-butyl chloroacetate.

chloroacetic anhydride $C_4H_4Cl_2O_3$ Crystals with a melting point of 46°C; soluble in chloroform and ether; used in the preparation of cellulose chloracetates and in the *N*-acetylation of amino acids in alkaline solution. Also known as *sym*-dichloroacetic anhydride; monochloroacetic acid anhydride.

chloroacetone CH_3COCH_2Cl Pungent, colorless liquid used as military tear gas and in organic synthesis. Also known as chloracetone; chlorinated acetone; monochloroacetone.

chloroacetonitrile $ClCH_2CN$ A colorless liquid with a pungent odor; soluble in hydrocarbons and alcohols; used as a fumigant. Also known as chloroethane nitrile; chloromethyl cyanide.

chloroacetophenone $C_6H_5COCH_2Cl$ Rhombic crystals melting at 59°C; an intermediate in organic synthesis. Also known as chloracetophenone; phenacyl chloride.

chloroacrolein $H_2C{:}ClCHO$ A colorless liquid with a boiling point of 29–31°C; used as a tear gas.

2-chloroallyl diethyldithiocarbamate *See* sulfallate.

chloroazotic acid *See* aqua regia.

chlorobenzal *See* benzal chloride.

chlorobenzaldehyde C_6H_4CHOCl A colorless to yellowish liquid (ortho form) or powder (para form) with a boiling range of 209–215°C; soluble in alcohol, ether, and acetone; used in dye manufacture.

chlorobenzene C_6H_5Cl A colorless, mobile, volatile liquid with an almondlike odor; used to produce phenol, DDT, and aniline. Also known as chlorobenzol; monochlorobenzene; phenyl chloride.

chlorobenzilate $C_{16}H_{14}Cl_2O_3$ A yellow-brown, viscous liquid with a melting point of 35–37°C; used as a miticide in agriculture and horticulture. Also known as ethyl 4,4'-dichlorobenzilate.

***para*-chlorobenzoic acid** ClC_6H_4COOH A white powder with a melting point of 238°C; soluble in methanol, absolute alcohol, and ether; used in the manufacture of dyes, fungicides, and pharamaceuticals.

chlorobenzoyl chloride ClC_6H_4COCl A colorless liquid with a boiling range of 227–239°C; soluble in alcohol, acetone, and water; used in dye and pharmaceuticals manufacture.

chlorobenzyl chloride $ClC_6H_4CH_2Cl$ A colorless liquid with a boiling range of 216–222°C; soluble in acetone, alcohol, and ether; used in the manufacture of organic chemicals.

***para*-chlorobenzyl-*para*-chlorophenyl sulfide** *See* chlorbenside.

***S*-(4-chlorobenzyl)-*N,N*-diethylthiocarbamate** *See* benthiocarb.

2-chloro-4,6-bis(ethylamino)-*s*-triazine *See* simazine.

2-chloro-4,6-bis(isopropylamino)-*s*-triazine *See* propazine.

chlorobium chlorophyll $C_{51}H_{67}O_4N_4Mg$ Either of two spectral forms of chlorophyll occurring as esters of farnesol in certain (*Chlorobium*) photosynthetic bacteria. Also known as bacterioviridin.

chlorobutadiene *See* chloroprene.

2-chloro-1,3-butadiene *See* chloroprene.

chlorobutanol $Cl_3CC(CH_3)_2OH$ Colorless to white crystals with a melting point of 78°C; soluble in alcohol, glycerol, ether, and chloroform; used as a plasticizer and a preservative for biological solutions. Also known as acetone chloroform; trichloro-*tert*-butyl alcohol.

4-chloro-2-butynyl-*meta*-chlorocarbanilate *See* barban.

chlorocarbon A compound of chlorine and carbon only, such as carbon tetrachloride, CCl_4.

2-chloro-3-(4-chlorophenyl)-methylpropionate *See* chlorfenpropmethyl.

chlorochromic anhydride *See* chromyl chloride.

2-chloro-1-(2,4-dichlorophenyl)-vinyl diethylphosphate *See* chlorfenvinphos.

***O*-[2-chloro-1-(2,5-dichlorophenyl)-vinyl]-*O,O*-diethyl phosphorothioate** $C_{12}H_{14}O_3P$ SCl_2 A brown liquid with a boiling point of 145°C at 0.005 mmHg (0.667 newtons per square meter); used as an insecticide for lawn and turf pests.

2-chloro-2',6'-diethyl-*N*-(butoxymethyl)-acetanilide *See* butachlor.

2-chloro-2-diethylcarbamoyl-1-methylvinyl dimethylphosphate *See* phosphamidon.

1,1,1-chlorodifluoroethane CH_3CClF_2 A colorless gas with a boiling point of $-130.8°C$; used as a refrigerant, solvent, and aerosol propellant. Also known as 1,1,1-difluorochloroethane; difluoromonochloroethane.

chlorodifluoromethane $CHClF_2$ A colorless gas with a boiling point of $-40.8°C$ and freezing point of $-160°C$; used as an aerosol propellant and refrigerant. Also known as difluoromonochloromethane; monochlorodifluoromethane.

10-chloro-5,10-dihydrophenarsazine See phenarsazine chloride.

3-chloro-1,2-dihydroxypropane See 3-chloro-1,2-propanediol.

1-chloro-2,4-dinitrobenzene $C_6H_3ClN_2O_4$ Yellow crystals with a melting point of 52–54°C; soluble in hot alcohol, ether, and benzene; used as a reagent in the determination of pyridine compounds such as nicotinic acid, and nicotinamide. Also known as 4-chloro-1,3-dinitrobenzene; 6-chloro-1,3-dinitrobenzene; 2,4-dinitro-1-chlorobenzene.

4-chloro-1,3-dinitrobenzene See 1-chloro-2,4-dinitrobenzene.

6-chloro-1,3-dinitrobenzene See 1-chloro-2,4-dinitrobenzene.

1-chloro-2,3-epoxypropane See epichlorohydrin.

chloroethane See ethyl chloride.

chloroethane nitrile See chloroacetonitrile.

chloroethanoic acid See chloroacetic acid.

chloroethene See vinyl chloride.

3-chloro-N-ethoxy-2,6-dimethoxybenzimidic acid anhydride with benzoic acid   See benzomate.

chloroethyl alcohol See ethylene chlorohydrin.

2-chloro-4-ethylamino-6-isopropylamino-s-triazine $C_8H_{14}N_5Cl$ White crystals with a melting point of 173–175°C; solubility in water is 33 parts per million at 20°C; used as an herbicide for weed control in corn, sorghum, and other crops. Also known as atrazine.

2-(4-chloro-6-ethylamino-s-triazin-2-yl-amino)-2-methyl propionitrile See cyanazine.

chloroethylene See vinyl chloride.

2-chloro-N-(2-ethyl-6-methylphenyl)-N-(2-methoxy-1-methyl) acetamide $C_{15}H_{22}O_2N$ A white liquid that boils at 100°C; used as an herbicide on corn and soybeans.

2-(chloroethyl)phosphonic acid See ethephon.

chlorofenac $C_8H_5Cl_3O_2$ A colorless solid with a melting point of 159–160°C; slightly soluble in water; used as a preemergence herbicide for noncrop areas. Also known as 2,3,6-trichlorophenylacetic acid.

chloroform $CHCl_3$ A colorless, sweet-smelling, nonflammable liquid; used at one time as an anesthetic. Also known as trichloromethane.

chloroformic acid benzyl ester See benzyl chloroformate.

chloroformyl chloride See phosgene.

chlorogenic acid $C16H_{18}O_9$ An important factor in plant metabolism; isolated from green coffee beans; the hemihydrate crystallizes in needlelike crystals from water. Also known as 3-caffeoylquinic acid; 3-(3,4-dihydroxycinnamoyl)quinic acid.

chloroguanide *See* chloroguanide hydrochloride.

chlorohydrin Any of the compounds derived from a group of glycols or polyhydroxy alcohols by chlorine substitution for part of the hydroxyl groups. Also spelled chlorhydrin.

α-chlorohydrin *See* 3-chloro-1,2-propanediol.

chlorohydrocarbon A carbon- and hydrogen-containing compound with chlorine substituted for some hydrogen in the molecule.

chlorohydroquinone $ClC_6H_3(OH)_2$ White to light tan crystals with a melting point of 100°C; soluble in water and alcohol; used as a photographic developer and bactericide and for dyestuffs.

2-chloro-5-hydroxy-1,3-dimethylbenzene *See* 4-chloro-3,5-xylenol.

5-chloro-8-hydroxyquinoline C_9H_6ClNO Crystals with a melting point of 130°C; used as a fungicide and bactericide. Also known as 5-chloro-8-oxychinolin; 5-chloro-8-quinolinol.

2-chloro-5-hydroxy-*meta*-xylene *See* 4-chloro-3,5-xylenol.

2-chloro-*N*-isopropylacetanilide *See* propachlor.

6-chloro-4-isopropyl-1-methyl-3-phenol *See* chlorothymol.

chloromethane CH_3Cl A colorless, noncorrosive, liquefiable gas which condenses to a colorless liquid; used as a refrigerant, and as a catalyst carrier in manufacture of butyl rubber. Also known as methyl chloride.

chloromethapyrilene citrate *See* chlorothen citrate.

2-chloro-4-(methylamino)-6-(ethylamino)-*s*-triazine *See* triistazine.

4-chloro-5-(methylamino)-2-(α,α,α,-trifluoro-*meta*-tolyl)-3-(2*H*)-pyridazine *See* norflurazon.

chloromethyl cyanide *See* chloroacetonitrile.

***S*-chloromethyl-*O,O*-diethyl phosphorodithioate** *See* chlormephos.

***N'*-(3-chloro-4-methylphenyl)-*N*-dimethylurea** *See* chlorotoluron.

2-chloro-*N*-(1-methyl-2-propynyl)acetanilide *See* prynachlor.

1-chloronaphthalene $C_{10}H_7Cl$ An oily liquid used as an immersion medium in the microscopic determination of refractive index of crystals and as a solvent for oils, fats, and DDT.

chloroneb $C_8H_8O_2Cl_2$ A white solid with a melting point of 133–135°C; used as a fungicide for cotton, soybean seeds, and bean seeds. Also known as 1,4-dichloro-2-5-dimethoxybenzene.

chloronitrous acid *See* aqua regia.

4-chloro-2-oxo-3-benzothiazolin-acetic acid $C_8H_6O_3NSCl$ White crystals with a melting point of 193°C; used as a preemergence herbicide for cereals and clovers.

5-chloro-8-oxychinolin *See* 5-chloro-8-hydroxyquinoline.

chlorophenol red $C_{19}H_{12}.ICl_2O_5S$ A dye that is used as an acid-base indicator; yellow in acid solution, red in basic solution. Also spelled chlrphenol red.

para-**chlorophenoxyacetic acid** $C_8H_7O_3Cl$ A solid with a melting point of 157–158°C; used as a growth regulator for tomatoes and peaches.

3-(*para*-chlorophenyl)-1,1-dimethylurea $C_9H_{11}ClN_2O$ A white, crystalline compound with a melting point of 174–175°C; limited solubility in water; used as an herbicide for grasses and broadleaf weeds in noncropland areas. Also known as monuron.

S-**{[*para*-chlorophenyl)thio]methyl}-*O,O*-diethyl phosphorodithioate** *See* carbophenothion.

chlorophenoxypropionic acid $C_9H_9O_3Cl$ A growth regulator used as a fruit thinner for plums and prunes. Also known as 2-(3-chlorophenoxy)-propionic acid.

2-(3-chlorophenoxy)-propionic acid *See* chlorophenoxypropionic acid.

para-**chlorophenylbenzenesulfonate** *See* fenson.

3-(*para*-chlorophenyl)-1,1-dimethylurea *See* telvar.

4-(2-chlorophenylhydrazono)-3-methyl-5-isoxazolone *See* drazoxolon.

3-(4-chlorophenyl)-1-methoxyl-1-methylurea $C_9H_{11}N_2O_2Cl$ A colorless solid with a melting point of 75–78°C; solubility in water is 735 parts per million; used as a pre- and postemergence herbicide for crops and ornamental plants. Also known as monolinuron.

ortho-**chlorophenyl-methylcarbamate** $C_8H_8ClNO_2$ A white, crystalline compound with a melting point of 90–91°C; used as an insecticide for rice. Abbreviated CPMC.

para-**chlorophenyl-*N*-methylcarbamate** $C_8H_8ClNO_2$ A white, crystalline compound with a melting point of 116°C; used as an herbicide in combination with other herbicides.

S-**{[(*para*-chlorophenyl)thio]methyl}-*O,O*-diethyl phosphorodithioate** *See* carbophenothion.

4-chlorophenyl-2,4,5-trichlorophenylazosulfide *See* chlorfensulfide.

S-*para*-**chlorophenyl-2,4,5-trichlorophenyl sulfone** *See* tetradifon.

S-**(2-chloro-1-phthalimidoethyl)-*O,O*-diethyl phosphorodithioate** *See* dialifor.

chlorophyll The generic name for any of several oil-soluble green tetrapyrrole plant pigments which function as photoreceptors of light energy for photosynthesis.

chlorophyll a $C_{55}H_{72}O_5N_4Mg$ A magnesium chelate of dihydroporphyrin that is esterified with phytol and has a cyclopentanone ring; occurs in all higher plants and algae.

chlorophyllase An enzyme that splits or hydrolyzes chlorophyll.

chlorophyll b $C_{55}H_{70}O_6N_4Mg$ An ester similar to chlorophyll *a* but with a —CHO substituted for a —CH$_3$; occurs in small amounts in all green plants and algae.

chloropicrin CCl_3NO_2 A colorless liquid with a sweet odor whose vapor is very irritating to the lungs and causes vomiting, coughing, and crying; used as a soil fumigant. Also known as nitrochloroform; trichloronitromethane.

chloroplatinate 1. A double salt of platinic chloride and another chloride. 2. A salt of chloroplatinic acid. Also known as platinochloride.

chloroplatinic acid H_2PtCl_6 An acid obtained as red-brown deliquescent crystals; used in chemical analysis. Also known as platinic chloride.

chloroprene C_4H_5Cl A colorless liquid which polymerizes to chloroprene resin. Also known as chlorobutadiene; 2-chloro-1,3-butadiene.

chloroprene resin A polymer of chloroprene used to form materials resembling natural rubber.

chloropropane Propane molecules with chlorine substituted in various amounts for the hydrogen atoms.

3-chloro-1,2-propanediol $ClCH_2CH(OH)CH_2OH$ A sweetish-tasting liquid that has a tendency to turn a straw color; soluble in ether, alcohol, and water; used to manufacture dye intermediates and to lower the freezing point of dynamite. Also known as 3-chloro-1,2-dihydroxypropane; α-chlorohydrin; glycerol α-monochlorohydrin; α-monochlorohydrin.

3-chloropropanonitrile *See* β-chloroproprionitrile.

chloropropene Propene molecules with chlorine substituted for some hydrogen atoms.

3-chloro-1-propene *See* allyl chloride.

β-chloropropionitrile $ClCH_2CH_2CN$ A liquid with an acrid odor; miscible with various organic solvents such as ethanol, ether, and acetone; used in polymer synthesis and in the synthesis of pharmaceuticals. Also known as 3-chloropropanonitrile.

chloropropylene oxide *See* epichlorohydrin.

3-chloro-1-propyne *See* propargyl bromide.

5-chloro-8-quinolinol *See* 5-chloro-8-hydroxyquinoline.

N-chlorosuccinimide $C_4H_4ClNO_2$ Orthorhombic crystals with the smell of chlorine; melting point is 150–151°C; soluble in water, benzene, and alcohol; used as a chlorinating agent.

chlorosulfonic acid $ClSO_2OH$ A fuming liquid that decomposes in water to sulfuric acid and hydrochloric acid; used in pharmaceuticals, pesticides, and dyes, and as a chemical intermediate. Also known as chlorosulfuric acid; sulfuric chlorohydrin.

chlorosulfuric acid *See* chlorosulfonic acid; sulfuryl chloride.

chlorothalonil $C_8Cl_4N_2$ Colorless crystals with a melting point of 250–251°C; used as a fungicide for crops, turf, and ornamental flowers. Also known as 2,4,5,6-tetrachloroisophthalonitrile.

chlorothymol $CH_3C_6H_2(OH)(C_3H_7)Cl$ White crystals melting at 59–61°C; soluble in benzene alcohol, insoluble in water; used as a bactericide. Also known as 6-chloro-4-isopropyl-1-methyl-3-phenol.

***ortho*-chlorotoluene** $CH_3C_6H_4Cl$ A liquid with a boiling point of 158.97°C; soluble in alcohol, chloroform, benzene, and ether; used in organic synthesis, as a solvent, and as an intermediate in dyestuff manufacture.

chlorotoluron $C_{10}H_{11}ClN_2O$ A colorless, crystalline compound with a melting point of 147–148°C; used as a preemergence herbicide on cereals. Also known as N'-(3-chloro-4-methylphenyl)-N-dimethylurea.

N′-(4-chloro-*ortho*-tolyl)-N,N-dimethylformamidine   *See* chlordimeform.

4-(4-chloro-*ortho*-tolyl)oxy butyric acid $C_{10}H_{13}ClO_3$ A white, crystalline compound with a melting point of 99–100°C; used as a postemergence herbicide for peas.

2-[(4-chloro-*ortho*-tolyl)oxy]propionic acid   *See* mecoprop.

2-chloro-1-(2,4,5-trichlorophenyl)vinyl dimethylphosphate   *See* tetrachlorvinphos.

chlorotrifluoroethylene polymer A colorless, noninflammable, heat-resistant resin, soluble in most organic solvents, and with a high impact strength; can be made into transparent filling and thin sheets; used for chemical piping, fittings, and insulation for wire and cables, and in electronic components. Also known as fluorothene; polytrifluorochloroethylene resin.

chlorotrifluoromethane $CClF_3$ A colorless gas having a boiling point of -81.4°C and a freezing point of -181°C; used as a dielectric and aerospace clinical, refrigerant, and aerosol propellant, and for metals hardening and pharmaceuticals manufacture. Also known as monochlorotrifluoromethane; trifluorochloromethane.

chloroxine $C_9H_5Cl_2NO$ Crystals with a melting point of 179–180°C; soluble in benzene and in sodium and potassium hydroxides; used as an analytical reagent. Also known as 5,7-dichloro-8-hydroxyquinoline.

chloroxuron $C_{15}H_{15}ClO_2N_2$ A tan solid with a melting point of 151–152°C; used as a pre- and postemergence herbicide for annual broadleaf weeds in common crops. Also known as 3-[*para*-(*para*-chlorophenoxy)phenyl]-1,1-dimethylurea.

chloroxylenol   *See* 4-chloro-3,5-xylenol.

2-chloro-*meta*-xylenol   *See* 4-chloro-3,5-xylenol.

4-chloro-3,5-xylenol $ClC_6H_2(CH_3)_2OH$ Crystals with a melting point of 115.5°C; soluble in water, 95% alcohol, benzene, terpenes, ether, and alkali hydroxides; used as an antiseptic and germicide and to stop mildew; used in humans as a topical and urinary antiseptic and as a topical antiseptic in animals. Also known as 2-chloro-5-hydroxy-1,3-dimethylbenzene; 2-chloro-5-hydroxy-*meta*-xylene; chloroxylenol; 2-chloro-*meta*-xylenol; parachlorometaxylenol.

chlorpropham $C_{10}H_{12}NO_2Cl$ A tan solid with a melting point of 38–39°C; used as a preemergence herbicide for fruit and vegetable crops. Also known as isopropyl-*meta*-chlorocarbanilate.

chlorpyrifos $C_9H_{11}Cl_3NO_3PS$ A white, granular, crystalline compound with a melting point of 41–43°C; used as an insecticide in the home and on turf and ornamental flowers. Also known as *O,O*-diethyl-*O*-(3,5,6-trichloro-2-pyridyl)phosphorothioate.

chlorthiamid $C_7H_5Cl_2NS$ An off-white, crystalline compound with a melting point of 151–152°C; used as a herbicide for selective weed control in industrial sites. Also known as 2,6-dichlorothiobenzamide.

cholaic acid   *See* taurocholic acid.

cholane $C_{24}H_{42}$ A tetracyclic hydrocarbon which may be considered as the parent substance of sterols, hormones, bile acids, and digitalis aglycons.

cholate   Any salt of cholic acid.

cholecystokinin   A hormone produced by the mucosa of the upper intestine which stimulates contraction of the gallbladder.

choleglobin   Combined native protein (globin) and open-ring iron-porphyrin, which is bile pigment hemoglobin; a precursor of biliverdin.

cholesteric material   A liquid crystal material in which the elongated molecules are parallel to each other within the plane of a layer, but the direction of orientation is twisted slightly from layer to layer to form a helix through the layers.

cholesterol $C_{27}H_{46}O$ A sterol produced by all vertebrate cells, particularly in the liver, skin, and intestine, and found most abundantly in nerve tissue.

cholic acid $C_{24}H_{40}O_5$ An unconjugated, crystalline bile acid.

choline $C_5H_{15}O_2N$ A basic hygroscopic substance constituting a vitamin of the B complex; used by most animals as a precursor of acetylcholine and a source of methyl groups.

cholinesterase   An enzyme found in blood and in various other tissues that catalyzes hydrolysis of choline esters, including acetylcholine. Abbreviated chE.

choline succinate dichloride dihydrate   *See* succinylcholine chloride.

cholytaurine   *See* taurocholic acid.

chondrin   A horny gelatinous protein substance obtainable from the collagen component of cartilage.

chondrodendrin   *See* bebeerine.

chondroitin   A nitrogenous polysaccharide occurring in cartilage in the form of condroitinsulfuric acid.

chondromucoid   A mucoid found in cartilage; a glycoprotein in which chondroitinsulfuric acid is the prosthetic group.

chondroprotein   A protein (glycoprotein) occurring normally in cartilage.

chorionic gonadotropin   *See* human chorionic gonadotropin.

Christiansen effect   Monochromatic transparency effect when finely powdered substances, such as glass or quartz, are immersed in a liquid having the same refractive index.

Christmas factor   A soluble protein blood factor involved in blood coagulation. Also known as factor IX; plasma thromboplastin component (PTC).

chromate CrO_4^{2-} 1. An ion derived from the unstable acid H_2CrO_4. 2. A salt or ester of chromic acid.

chromatin   The deoxyribonucleoprotein complex forming the major portion of the nuclear material and of the chromosomes.

chromatogram   The pattern formed by zones of separated pigments and of colorless substance in chromatographic procedures.

chromatograph   To employ chromatography to separate substances.

chromatographic adsorption   Preferential adsorption of chemical compounds (gases or liquids) in an ascending molecular-weight sequence onto a solid adsorbent material, such as activated carbon, alumina, or silica gel; used for analysis and separation of chemical mixtures.

chromatography A method of separating and analyzing mixtures of chemical substances by chromatographic adsorption.

chromatophorotrophin Any crustacean neurohormone which controls the movement of pigment granules within chromatophores.

chrome alum $KCr(SO_4)_2 \cdot 12H_2O$ An alum obtained as purple crystals and used as a mordant, in tanning, and in photography in the fixing bath. Also known as potassium chromium sulfate.

chrome dye One of a class of acid dyes used on wool with a chromium compound as mordant.

chrome green *See* chromic oxide.

chrome red 1. A pigment containing basic lead chromate. 2. Any of several mordant acid dyes.

chrome tanning Tanning treatment of animal skin with chromium salts.

chrome yellow 1. A yellow pigment composed of normal lead chromate, $PbCrO_4$, or other lead compounds. 2. Any of several mordant acid dyes.

chromic acid H_2CrO_4 The hydrate of CrO_3; exists only as salts or in solution.

chromic chloride $CrCl_3$ Crystals that are pinkish violet shimmering plates, almost insoluble in water, but easily soluble in presence of minute traces of chromous chloride; used in calico printing, as a mordant for cotton and silk.

chromic fluoride $CrF_3 \cdot 4H_2O$ Crystals that are green, soluble in water; used in dyeing cottons.

chromic hydroxide $Cr(OH)_3 \cdot 2H_2O$ Gray-green, gelatinous precipitate formed when a base is added to a chromic salt; the precipitate dries to a bluish, amorphous powder; prepared as an intermediate in the manufacture of other soluble chromium salts.

chromic nitrate $Cr(NO_3)_3 \cdot 9H_2O$ Purple, rhombic crystals that are soluble in water; used as a mordant in textile dyeing.

chromic oxide Cr_2O_3 A dark green, amorphous powder, forming hexagonal crystals on heating that are insoluble in water or acids; used as a pigment to color glass and ceramic ware and as a catalyst. Also known as chrome green.

chromium A metallic chemical element, symbol Cr, atomic number 24, atomic weight 51.996.

chromium-51 A radioactive isotope with atomic mass 51 made by neutron bombardment of chromium; radiates gamma rays.

chromium carbide Cr_3C_2 Orthorhombic crystals with a melting point of 1890°C; resistant to oxidation, acids, and alkalies; used for hot-extrusion dies, in spray-coating materials, and as a component for pumps and valves.

chromium chloride A group of compounds of chromium and chloride; chromium may be in the +2, +3, or +6 oxidation state.

chromium dioxide Cr_2O_2 Black, acicular crystals; a semiconducting material with strong magnetic properties used in recording tapes.

chromium oxide   A compound of chromium and oxygen; chromium may be in the $+2$, $+3$, or $+6$ oxidation state.

chromium oxychloride   *See* chromyl chloride.

chromium stearate $Cr(C_{18}H_{35}O_2)_3$ A dark-green powder, melting at 95–100°C; used in greases, ceramics, and plastics.

chromogen   A pigment precursor.

chromogenesis   Production of colored substances as a result of metabolic activity; characteristic of certain bacteria and fungi.

chromolipid   *See* lipochrome.

chromometer   *See* colorimeter.

chromophore   An arrangement of atoms that gives rise to color in many organic substances.

chromophyll   Any plant pigment.

chromoprotein   Any protein, such as hemoglobin, with a metal-containing pigment.

chromotropic acid $C_{10}H_8O_8S_2$ White, needlelike crystals that are soluble in water; used as an analytical reagent and azo dye intermediate. Also known as 1,8-dihyroxynaphthalene-3,6-disulfonic acid.

chromyl chloride CrO_2Cl_2 A dark-red, toxic, fuming liquid that boils at 116°C; reacts with water to form chromic acid; used to make dyes and chromium complexes. Also known as chlorochromic anhydride; chromium oxychloride.

chronoamperometry   Electroanalysis by measuring at a working electrode the rate of change of current versus time during a titration; the potential is controlled.

chronopotentiometry   Electroanalysis based on the measurement at a working electrode of the rate of change in potential versus time; the current is controlled.

chrysazin   *See* 1,8-dihydroxyanthraquinone.

chrysene $C_{18}H_{12}$ An organic, polynuclear hydrocarbon which when pure gives a bluish fluorescence; a component of short afterglow or luminescent paint.

chrysoidine $C_6H_5NNC_6H_3(NH_2)_2 \cdot HCl$ Large, black crystals or a red-brown powder that melts at 117°C; soluble in water and alcohol; used as an orange dye for silk and cotton. Also known as *meta*-diaminoazobenzene hydrochloride.

chrysophanic acid $C_{15}H_{10}O_4$ Yellow leaves that melt at 196°C; soluble in ether, chloroform, and hot alcohol; extracted from senna leaves and rhubarb root; used in medicine as a mild laxative.

Chugaev reaction   The thermal decomposition of methyl esters of xanthates to yield olefins without rearrangement.

chylomicron   One of the extremely small lipid droplets, consisting chiefly of triglycerides, found in blood after ingestion of fat.

chymosin   *See* rennin.

chymotrypsin   A proteinase in the pancreatic juice that clots milk and hydrolyzes casein and gelatin.

chymotrypsinogen An inactive proteolytic enzyme of pancreatic juice; converted to the active form, chymotrypsin, by trypsin.

cigarette burning In rocket propellants, black powder, gasless delay elements, and pyrotechnic candles, the type of burning induced in a solid grain by permitting burning on one end only, so that the burning progresses in the direction of the longitudinal axis.

ciguatoxin A toxin produced by the benthic dinoflagellate *Gambierdiscus toxicus*.

cinchonamine $C_{19}H_{24}N_2O$ A yellow, crystalline, water-insoluble alkaloid that melts at 184°C; derived from the bark of *Remijia purdieana*, a member of the madder family of shrubs.

cinchonine $C_{19}H_{22}N_2O$ A colorless, crystalline alkaloid that melts at about 245°C; extracted from cinchona bark, it is used as a substitute for quinine and as a spot reagent for bismuth.

cinnamaldehyde *See* cinnamic aldehyde.

cinnamate A salt of cinnamic acid, containing the radical $C_9H_7O_2^-$.

cinnamein *See* benzyl cinnamate.

cinnamic acid $C_6H_5CHCHCOOH$ Colorless, monoclinic acid; forms scales, slightly soluble in water; found in natural balsams. Also known as benzalacetic acid; β-phenylacrylic acid; styrylformic acid.

cinnamic alcohol $C_6H_5CH{:}CHCH_2OH$ White needles that congeal upon heating and are soluble in alcohol; used in perfumery. Also known as cinnamyl alcohol; phenylallylic alcohol; 3-phenyl-2-propen-1-ol; styryl carbinol.

cinnamic aldehyde $C_6H_5CH{:}CHCHO$ A yellow oil with a cinnamon odor, sweet taste, and a boiling point of 248°C; used in flavors and perfumes. Also known as cinnamaldehyde; cinnamyl aldehyde; 3-phenylpropenal.

cinnamoyl chloride $C_6H_5CHCHCOCl$ Yellow crystals that melt at 35°C, and decompose in water; used as a chemical intermediate. Also known as phenylacrylyl chloride.

cinnamyl alcohol *See* cinnamic alcohol.

cinnamyl aldehyde *See* cinnamic aldehyde.

cinnamyl methyl ketone *See* benzylideneacetone.

circular chromatography *See* radial chromatography.

circular deoxyribonucleic acid A single- or double-stranded ring of deoxyribonucleic acid found in certain bacteriophages and in human wart virus. Also known as ring deoxyribonucleic acid.

circular paper chromatography A paper chromatographic technique in which migration from a spot in the sheet takes place in 360° so that zones separate as a series of concentric rings.

circulating pump Pump used to circulate process liquid out of and back into a process system, as in the circulation of distillation column bottoms through an external heater, or the circulation of storage tank bottoms to mix tank contents.

circulating system Fluid system in which the process fluid is taken from and pumped back into the system, as in the circulation of distillation column bottoms through an external heater.

cis A descriptive term indicating a form of isomerism in which atoms are located on the same side of an asymmetric molecule.

cis-trans isomerism A type of geometrical isomerism found in alkenic systems in which it is possible for each of the doubly bonded carbons to carry two different atoms or groups; two similar atoms or groups may be on the same side (cis) or on opposite sides (trans) of a plane bisecting the alkenic carbons and perpendicular to the plane of the alkenic system.

citiolone *See* N-acetylhomocysteinethiolactone.

citraconic acid $C_5H_6O_4$ A dicarboxylic acid; hygroscopic crystals that melt at 91°C; derived from citric acid by heating. Also known as methyl maleic acid.

citral $C_{10}H_{16}O$ A pale-yellow liquid that in commerce is a mixture of two isomeric forms, alpha and beta; insoluble in water, soluble in glycerin or benzyl benzoate; used in perfumery and as an intermediate to form other compounds. Also known as 3,7-dimethyl-2,6-octadienal; geranial; geranialdehyde.

citramalase An enzyme that is involved in the fermentation of glutamate by *Clostridium tetanomorphum;* catalyzes the breakdown of citramalic acid to acetate and pyruvate.

citrate A salt or ester of citric acid.

citric acid $C_6H_8O_7 \cdot H_2O$ A colorless crystalline or white powdery organic, tricarboxylic acid occurring in plants, especially citrus fruits, and used as a flavoring agent, as an antioxidant in foods, and as a sequestering agent; the commercially produced form melts at 153°C.

citric acid cycle *See* Krebs cycle.

citronellal hydrate *See* hydroxycitronellal.

citronellol $C_{10}H_{19}OH$ A liquid derived from citronella oil; soluble in alcohol; used in perfumery. Also known as 3,7-dimethyl-6(or 7)-octen-1-ol.

citrulline $C_6H_{13}O_3N_3$ An amino acid formed in the synthesis of arginine from ornithine.

citrus flavanoid compound *See* bioflavanoid.

civetone $C_{17}H_{30}O$ 9-Cycloheptadecen-1-one, a macrocyclic ketone component of civet used in perfumes because of its pleasant odor and lasting quality; believed to function as a sex attractant among civet cats.

Cl *See* chlorine.

Claisen condensation 1. Condensation, in the presence of sodium ethoxide, of esters or of esters and ketones to form β-dicarbonyl compounds. 2. Condensation of arylaldehydes and acylphenons with esters or ketones in the presence of sodium ethoxide to yield unsaturated esters. Also known as Claisen reaction.

Claisen flask A glass flask with a U-shaped neck, used for distillation.

Claisen-Schmidt condensation A reaction employed for preparation of unsaturated aldehydes and ketones by condensation of aromatic aldehydes with aliphatic aldehydes or ketones in the presence of sodium hydroxide.

clarity Measure of the amount of opaque suspended solids in a liquid, determined by visual or optical methods.

Clark degree *See* English degree.

clarke A unit of the average abundance of an element in the earth's crust, expressed as a percentage. Also known as crustal abundance.

Clark process Softening of water by adding alkaline solutions of calcium hydroxide so that the acid carbonates are converted to normal carbonates.

clathrate A well-defined addition compound formed by inclusion of molecules in cavities formed by crystal lattices or present in large molecules; examples include hydroxyquinone, urea, and cyclodextrin.

Claude process A process of ammonia synthesis which uses high operating pressures and a train of converters.

Clausius-Dickel column *See* thermogravitational column.

Claus method Industrial method of obtaining sulfur by a partial oxidation of gaseous hydrogen sulfide in the air to give water and sulfur.

clay refining A treating process for vaporized gasoline or other light petroleum product; the material is passed through a bed of granular clay, and certain olefins are polymerized to gums and absorbed by the clay.

clay regeneration Cleaning coarse-grained absorbent clays for reuse in percolation processes by deoiling them with naphtha, steaming out excess naphtha, and roasting in a stream of air to remove carbonaceous matter.

cleavage Splitting, or the tendency to split, along planes determined by crystal structure and always parallel to a possible face.

cleavage crystal A crystal fragment bounded by cleavage faces giving it a regular form.

cleavage fracture 1. Manner of breaking a crystalline substance along the cleavage plane. 2. The appearance of such a broken surface.

cleavage plane Plane along which a crystalline substance may be split.

Cleveland open-cup tester A laboratory apparatus used to determine flash point and fire point of petroleum products.

clinoaxis The inclined lateral axis that makes an oblique angle with the vertical axis in the monoclinic system. Also known as clinodiagonal.

clinodiagonal *See* clinoaxis.

clinohedral class A rare class of crystals in the monoclinic system having a plane of symmetry but no axis of symmetry. Also known as domatic class.

clinopinacoid A form of monoclinic crystal whose faces are parallel to the inclined and vertical axes.

close-packed crystal A crystal structure in which the lattice points are centers of spheres of equal radius arranged so that the volume of the interstices between the spheres is minimal.

cloud The nucleons that are in the nucleus of an atom but not in closed shells.

cloud point The temperature at which paraffin wax or other solid substance begins to separate from a solution of petroleum oil; a cloudy appearance is seen in the oil at this point.

cloud test   An ASTM method for determining the cloud point of petroleum oil.

cloudy crystal-ball model   An optical analogy used in explaining scattering of nucleons by nuclei, in which the nucleus is thought of as a sphere of nuclear matter which partially refracts and partially absorbs the incident nucleon (de Broglie) wave. Also known as optical model.

CLS   *See* characteristic loss spectroscopy.

clupanodonic acid   *See* docosatrienynoic acid.

clupeine   A protamine found in salmon sperm, mainly composed of arginine (74.1%) and small percentages of threonine, serine, proline, alanine, valine, and isoleucine.

Cm   *See* curium.

CM-cellulose   *See* carboxymethyl cellulose.

Co   *See* cobalt.

CoA   *See* coenzyme A.

coacervate   An aggregate of colloidal droplets bound together by the force of electrostatic attraction.

coacervation   The separation, by addition of a third component, of an aqueous solution of a macromolecule colloid (polymer) into two liquid phases, one of which is colloid-rich (the coacervate) and the other an aqueous solution of the coacervating agent (the equilibrium liquid).

coagulant   An agent that causes coagulation.

coagulase   Any enzyme that causes coagulation of blood plasma.

coagulation   A separation or precipitation from a dispersed state of suspensoid particles resulting from their growth; may result from prolonged heating, addition of an electrolyte, or from a condensation reaction between solute and solvent; an example is the setting of a gel.

coalescent   Chemical additive used in immiscible liquid-liquid mixtures to cause small droplets of the suspended liquid to unite, preparatory to removal from the carrier liquid.

coalescent pack   High-surface-area packing to consolidate liquid droplets for gravity separation from a second phase (for example, gas or immiscible liquid); packing must be wettable by the droplet phase; Berl saddles, Raschig rings, knitted wire mesh, excelsior, and similar materials are used.

coalescer   Mechanical process vessel with wettable, high-surface area packing on which liquid droplets consolidate for gravity separation from a second phase (for example, gas or immiscible liquid).

coal gasification   The conversion of coal, char, or coke to a gaseous product by reaction with air, oxygen, steam, carbon dioxide, or mixtures of these.

coal hydrogenation   *See* Bergius process.

coal liquefaction   The process of preparing a liquid mixture of hydrocarbons by destructive distillation of coal.

coal tar dye   Dye made from a coal tar hydrocarbon or a derivative such as benzene, toluene, xylene, naphthalene, or aniline.

cobalamin *See* vitamin B_{12}.

cobalt A metallic element, symbol Co, atomic number 27, atomic weight 58.93; used chiefly in alloys.

cobalt-60 A radioisotope of cobalt, symbol ^{60}Co, having a mass number of 60; emits gamma rays and has many medical and industrial uses; the most commonly used isotope for encapsulated radiation sources.

cobalt blue A green-blue pigment formed of alumina and cobalt oxide. Also known as cobalt ultramarine; king's blue.

cobalt bromide *See* cobaltous bromide.

cobalt chloride *See* cobaltous chloride.

cobaltic fluoride *See* cobalt trifluoride.

cobalt-molybdate desulfurization A process for desulfurization of petroleum by using cobalt molybdate as a catalyst.

cobalt nitrate *See* cobaltous nitrate.

cobaltous acetate $Co(C_2H_3O_2)_2 \cdot 4H_2O$ Reddish-violet, deliquescent crystals; soluble in water, alcohol, and acids; used in paint and varnish driers, for anodizing, and as a feed additive mineral supplement. Also known as cobalt acetate.

cobaltous bromide $CoBr_2 \cdot 6H_2O$ Red-violet crystals with a melting point of 47–48°C; soluble in water, alcohol, and ether; used in hygrometers. Also known as cobalt bromide.

cobaltous chloride $CoCl_2$ or $CoCl_2 \cdot 6H_2O$ A compound whose anhydrous form consists of blue crystals and sublimes when heated, and whose hydrated form consists of red crystals and melts at 86.8°C; both forms are used as an absorbent for ammonia in dyes and as a catalyst. Also known as cobalt chloride.

cobaltous fluorosilicate $CoSiF_6 \cdot H_2O$ A water-soluble, orange-red powder, used in toothpastes.

cobaltous nitrate $Co(NO_3)_2 \cdot 6H_2O$ A red crystalline compound with a melting point of 56°C; soluble in organic solvents; used in sympathetic inks, as an additive to soils and animal feeds, and for vitamin preparations and hair dyes. Also known as cobalt nitrate.

cobalt oxide CoO A grayish brown powder that decomposes at 1935°C, insoluble in water; used as a colorant in ceramics and in manufacture of glass.

cobalt potassium nitrite $K_3Co(NO_2)_6$ A yellow powder which decomposes at the melting point of 200°C; used in medicine and as a yellow pigment. Also known as cobalt yellow; Fischer's salt; potassium cobaltinitrite.

cobalt sulfate Any compound of either divalent or trivalent cobalt and the sulfate group; anhydrous cobaltous sulfate, $CoSO_4$, contains divalent cobalt, has a melting point of 96.8°C, is soluble in methanol, and is utilized to prepare pigments and cobalt salts; cobaltic sulfate, $Co_2(SO_4)_3 \cdot 18H_2O$, contains trivalent cobalt, is soluble in sulfuric acid, and functions as an oxidizing agent.

cobalt trifluoride CoF_3 A brownish powder that reacts with water to form a precipitate of cobaltic hydroxide; used as a fluorinating agent. Also known as cobaltic fluoride.

cobalt ultramarine *See* cobalt blue.

cobalt yellow   *See* cobalt potassium nitrite.

cocarboxylase   *See* thiamine pyrophosphate.

cocculin   *See* picrotoxin.

cochineal   A red dye made of the dried bodies of the female cochineal insect (*Coccus cacti*), found in Central America and Mexico; used as a biological stain and indicator.

cochineal solution   An indicator in acid-base titration.

cochinilin   *See* carminic acid.

cocodyl oxide $(CH_3)_2AsOAs(CH_3)_2$ A liquid that has an obnoxious odor; slightly soluble in water, soluble in alcohol and ether; boils at 150°C. Also known as alkarsine; bisdimethyl arsenic oxide; dicacodyl oxide.

codecarboxylase   The prosthetic component of the enzyme carboxylase which catalyzes decarboxylation of D-amino acids. Also known as pyridoxal phosphate.

codehydrogenase I   *See* diphosphopyridine nucleotide.

codehydrogenase II   *See* triphosphopyridine nucleotide.

codimer   **1.** A copolymer formed from the polymerization of two dissimilar olefin molecules.   **2.** The product of polymerization of isobutylene with one of the two normal butylenes.

coenzyme   The nonprotein portion of an enzyme; a prosthetic group which functions as an acceptor of electrons or functional groups.

coenzyme A $C_{21}H_{36}O_{16}N_7P_3S$ A coenzyme in all living cells; required by certain condensing enzymes to act in acetyl or other acyl-group transfer and in fatty-acid metabolism. Abbreviated CoA.

coenzyme I   *See* diphosphopyridine nucleotide.

coenzyme II   *See* triphosphopyridine nucleotide.

cognac oil   *See* ethyl enanthate.

coherent precipitate   A precipitate that is a continuation of the lattice structure of the solvent and has no phase or grain boundary.

coincidence boundary   A grain boundary separating crystal lattices which are rotated with respect to each other by an angle with a special value, resulting in a periodic grain boundary structure and an extension of a sublattice of the original lattice across the boundary.

coion   Any of the small ions entering a solid ion exchanger and having the same charge as that of the fixed ions.

coke drum   A vessel in which coke is produced.

coke number   A number used to report the results of the Ramsbottom carbon residue test.

coke oven   A retort in which coal is converted to coke by carbonization.

coke-oven regenerator   Arrangement of refractory blocks in the flue system of a coke oven to recover waste heat from hot, exiting combustion gases; the blocks, in turn, release heat to warm, incoming fuel gas.

coker   The processing unit in which coking occurs.

coking   1. Destructive distillation of coal to make coke   2. A process for thermally converting the heavy residual bottoms of crude oil entirely to lower-boiling petroleum products and by-product petroleum coke.

coking still   A still in which coking is done; usually, it is a batch still.

Colburn method   Graphical method, and equations to calculate the theoretical number of plates (trays) needed to separate light and heavy liquids in a distillation column.

colchicine $C_{22}H_{25}O_6N$ An alkaloid extracted from the stem of the autumn crocus; used experimentally to inhibit spindle formation and delay centromere division, and medicinally in the treatment of gout.

colcothar   Red ferric oxide made by heating ferrous sulfate in the air; used as a pigment and as an abrasive in polishing glass.

cold cure   Vulcanization of rubber at nonelevated temperatures with a solution of a sulfur compound.

cold lime-soda process   A water-softening process in which water is treated with hydrated lime (sometimes in combination with soda ash), which reacts with dissolved calcium and magnesium compounds to form precipitates that can be removed as sludge.

cold settling   A process that removes wax from high-viscosity stocks.

cold-spot hygrometer   *See* dew-point hygrometer.

cold test   A test to determine the temperature at which clouding or coagulation is first visible in a sample of oil, as the temperature of the sample is reduced.

collagen   A fibrous protein found in all multicellular animals, especially in connective tissue.

collagenase   Any proteinase that decomposes collagen and gelatin.

collateral series   A radioactive decay series, initiated by transmutation, that eventually joins into one of the four radioactive decay series encountered in natural radioactivity.

collection trap   Cooled device to collect gas-chromatographic eluent, holding it for subsequent compound-identification analysis.

collective motion   Motion of nucleons in a nucleus correlated so that their overall space pattern is essentially constant or undergoes changes which are slow compared to the motions of individual nucleons.

collective transition   A nuclear transition from one state of collective motion to another.

2,4,6-collidine $(CH_3)_3C_5H_2N$ A liquid boiling at 170.4°C; slightly soluble in water, soluble in alcohol; used as a chemical intermediate. Also known as 2,4,6-trimethylpyridine.

colligative properties   Properties dependent on the number of molecules but not their nature.

collision broadening   *See* collision line-broadening.

collision excitation   The excitation of a gas by collisions of moving charged particles.

collision ionization The ionization of atoms or molecules of a gas or vapor by collision with other particles.

collision line-broadening Spreading of a spectral line due to interruption of the radiation process when the radiator collides with another particle. Also known as collision broadening.

collision-radiative recombination The capture of an electron by an ion in a gas, accompanied by the emission of one or more photons.

collision theory Theory of chemical reaction proposing that the rate of product formation is equal to the number of reactant-molecule collisions multiplied by a factor that corrects for low-energy-level collisions.

collodion Cellulose nitrate deposited from a solution of 60% ether and 40% alcohol, used for making fibers and film and in membranes for dialysis.

collodion cotton *See* pyroxylin.

collodion replication Production of a faithful collodion-film mold of a specimen surface (for example, powders, bones, microorganisms, crystals) which is sufficiently thin to be studied by electron microscopy.

colloid The phase of a colloidal system made up of particles having dimensions of 10–10,000 angstroms (1–1000 nanometers) and which is dispersed in a different phase.

colloidal dispersion *See* colloidal system.

colloidal suspension *See* colloidal system.

colloidal system An intimate mixture of two substances, one of which, called the dispersed phase (or colloid), is uniformly distributed in a finely divided state through the second substance, called the dispersion medium (or dispersing medium); the dispersion medium or dispersed phase may be a gas, liquid, or solid. Also known as colloidal dispersion; colloidal suspension.

colloid chemistry The scientific study of matter whose size is approximately 10 to 10,000 angstroms (1 to 1000 nanometers), and which exists as a suspension in a continuous medium, especially a liquid, solid, or gaseous substance.

color comparator A photoelectric instrument that compares an unknown color with that of a standard color sample for matching purposes. Also known as photoelectric color comparator.

colorimeter A device for measuring concentration of a known constituent in solution by comparison with colors of a few solutions of known concentration of that constituent. Also known as chromometer.

color stability Resistance of materials to change in color that can be caused by light or aging, as of petroleum or whiskey.

color standard Liquid solution of known chemical composition and concentration, hence of known and standardized color, used for optical analysis of samples of unknown strength.

color test The quantitative analysis of a substance by comparing the intensity of the color produced in a sample by a reagent with a standard color produced similarly in a solution of known strength.

color throw In an ion-exchange process, discoloration of the liquid passing through the bed.

column   *See* tower.

column bleed   The loss of carrier liquid during gas chromatography due to evaporation into the gas under analysis.

column chromatography   Chromatographic technique of two general types: packed columns usually contain either a granular adsorbent or a granular support material coated with a thin layer of high-boiling solvent (partitioning liquid); open-tubular columns contain a thin film of partitioning liquid on the column walls and have an opening so that gas can pass through the center of the column.

column development chromatography   Columnar apparatus for separating or concentrating one or more components from a physical mixture by use of adsorbent packing; as the specimen percolates along the length of the adsorbent, its various components are preferentially held at different rates, effecting a separation.

combination principle   *See* Ritz's combination principle.

combination unit   A processing unit that combines more than one process, such as straight-run distillation together with selective cracking.

combination vibration   A vibration of a polyatomic molecule involving the simultaneous excitation of two or more normal vibrations.

combined carbon   Carbon that is chemically combined within a compound, as contrasted with free or uncombined elemental carbon.

combined cyanide   The cyanide portion of a complex ion composed of cyanide and a metal.

combining-volumes principle   The principle that when gases take part in chemical reactions the volumes of the reacting gases and those of the products (if gaseous) are in the ratio of small whole numbers, provided that all measurements are made at the same temperature and pressure. Also known as Gay-Lussac law.

combining weight   The weight of an element that chemically combines with 8 grams of oxygen or its equivalent.

combustion   The burning of gas, liquid, or solid, in which the fuel is oxidized, evolving heat and often light.

combustion efficiency   The ratio of heat actually developed in a combustion process to the heat that would be released if the combustion were perfect.

combustion furnace   A heating device used in the analysis of organic compounds for elements.

combustion rate   The rate of burning of any substance.

combustion train   The arrangement of apparatus for elementary organic analysis.

combustion tube   A glass, silica, or porcelain tube, resistant to high temperatures, that is a component of a combustion train.

combustion wave   **1.** A zone of burning propagated through a combustible medium. **2.** The zoned, reacting, gaseous material formed when an explosive mixture is ignited.

common-ion effect The lowering of the degree of ionization of a compound when another ionizable compound is added to a solution; the compound added has a common ion with the other compound.

common salt *See* halite; sodium chloride.

comonomer One of the compounds used to produce a specific polymeric product.

comparator-densitometer Device that projects a labeled spectrum onto a screen adjacent to an enlarged image of the spectrum to be analyzed, allowing visual comparison.

comparison spectrum A line spectrum whose wavelengths are accurately known, and which is matched with another spectrum to determine the wavelengths of the latter.

competing equilibria condition The competition for a reactant in a complex chemical system in which several reactions are taking place at the same time.

competitive enzyme inhibition Prevention of an enzymatic process resulting from the reversible interaction of an inhibitor with a free enzyme.

competitive inhibition Enzyme inhibition in which the inhibitor competes with the natural substrate for the active site of the enzyme; may be overcome by increasing substrate concentration.

complete combustion Combustion in which the entire quantity of oxidizable constituents of a fuel is reacted.

complexation *See* complexing.

complex chemical reaction A chemical system in which a number of chemical reactions take place simultaneously, including reversible reactions, consecutive reactions, and concurrent or side reactions.

complex compound Any of a group of chemical compounds in which a part of the molecular bonding is of the coordinate type.

complexing Formation of a complex compound. Also known as complexation.

complexing agent A substance capable of forming a complex compound with another material in solution.

complex ion A complex, electrically charged group of atoms or radical, for example, $Cu(NH_3)_2^{2+}$.

complexometric titration A technique of volumetric analysis in which the formation of a colored complex is used to indicate the end point of a titration. Also known as chelatometry.

complex salt A class of salts in which there are no detectable quantities of each of the metal ions existing in solution; an example is $K_3Fe(CN)_6$, which in solution has K^+ but no Fe^{3+} because Fe is strongly bound in the complex ion, $Fe(CN)_6^{3-}$.

component 1. A part of a mixture. 2. The smallest number of chemical substances which are able to form all the constituents of a system in whatever proportion they may be present.

component distillation A distillation process in which a fraction that cannot normally be separated by distillation is removed by forming an azeotropic mixture.

component substances law   The law that each substance, singly or in mixture, composing a material exhibits specific properties that are independent of the other substances in that material.

composition   The elements or compounds making up a material or produced from it by analysis.

composition diagram   Graphical plots to show the solvent-solute concentration relationships during various stages of extraction operations (leaching, or solid-liquid extraction; and liquid-liquid extraction).

composition face   *See* composition surface.

composition plane   A planar composition surface in a crystal uniting two individuals of a contact twin.

composition surface   The surface uniting individuals of a crystal twin; may or may not be planar. Also known as composition face.

compound   A substance whose molecules consist of unlike atoms and whose constituents cannot be separated by physical means. Also known as chemical compound.

compound elastic scattering   Scattering in which the final state is the same as the initial state, but there is an intermediate state with the colliding systems amalgamating to form a compound system.

compound nucleus   An intermediate state in a nuclear reaction in which the incident particle combines with the target nucleus and its energy is shared among all the nucleons of the system.

compound sugar   *See* oligosaccharide.

compound twins   Individuals of one mineral group united in accordance with two or more different twin laws.

compressible fluid flow   Gas flow when the pressure drop due to the flow of a gas through a system is large enough, compared with the inlet pressure, to cause a 10% or greater decrease in gas density.

compression process   The recovery of natural gasoline from gas containing a high proportion of hydrocarbons.

Compton incoherent scattering   Scattering of gamma rays by individual nucleons in a nucleus or electrons in an atom when the energy of the gamma rays is large enough so that binding effects may be neglected.

Compton rule   An empirical law stating that the heat of fusion of an element times its atomic weight divided by its melting point in degrees Kelvin equals approximately 2.

computational chemistry   The use of calculations to predict molecular structure, properties, and reactions.

concave grating   A reflection grating which both collimates and focuses the light falling upon it, made by spacing straight grooves equally along the chord of a concave spherical or paraboloid mirror surface. Also known as Rowland grating.

concentrate   To increase the amount of a dissolved substance by evaporation.

concentration   In solutions, the mass, volume, or number of moles of solute present in proportion to the amount of solvent or total solution.

concentration cell   **1.** Electrochemical cell for potentiometric measurement of ionic concentrations where the electrode potential electromotive force produced is determined as the difference in emf between a known cell (concentration) and the unknown cell. **2.** An electrolytic cell in which the electromotive force is due to a difference in electrolyte concentrations at the anode and the cathode.

concentration gradient   The graded difference in the concentration of a solute throughout the solvent phase.

concentration polarization   That part of the polarization of an electrolytic cell resulting from changes in the electrolyte concentration due to the passage of current through the solution.

concentration potential   Tendency for a univalent electrolyte to concentrate in a specific region of a solution.

concentration scale   Any of several numerical systems defining the quantitative relation of the components of a mixture; for solutions, concentration is expressed as the mass, volume, or number of moles of solute present in proportion to the amount of solvent or total solution.

concentric tube column   A carefully insulated distillation apparatus which is capable of very high separating power, and in which the outer vapor-rising annulus of the column is concentric around an inner, bottom-discharging reflux return.

concerted reaction   A reaction in which there is a simultaneous occurrence of bond making and bond breaking.

conchiolin   A nitrogenous substance that is the organic basis of many molluscan shells.

condensable vapors   Gases or vapors which when subjected to appropriately altered conditions of temperature or pressure become liquids.

condensate flash   Partial evaporation (flash) of hot condensed liquid by a stepwise reduction in system pressure, the hot vapor supplying heat to a cooler evaporator step (stage).

condensation   Transformation from a gas to a liquid.

condensation polymer   A high-molecular-weight compound formed by condensation polymerization.

condensation polymerization   The formation of high-molecular-weight polymers from monomers by chemical reactions of the condensation type.

condensation reaction   One of a class of chemical reactions involving a combination between molecules or between parts of the same molecule.

condensation resin   A resin formed by polycondensation.

condensation temperature   In boiling-point determination, the temperature established on the bulb of a thermometer on which a thin moving film of liquid coexists with vapor from which the liquid has condensed, the vapor phase being replenished at the moment of measurement from a boiling-liquid phase.

condensed system   A chemical system in which the vapor pressure is negligible or in which the pressure maintained on the system is greater than the vapor pressure of any portion.

conductimetry The scientific study of conductance measurements of solutions; to avoid electrolytic complications, conductance measurements are usually taken with alternating current.

conductometric titration A titration in which electrical conductance of a solution is measured during the course of the titration.

configuration The three-dimensional spatial arrangement of atoms in a stable or isolable molecule.

configuration interaction Interaction between two different possible arrangements of the electrons in an atom (or molecule); the resulting electron distribution, energy levels, and transitions differ from what would occur in the absence of the interaction.

confining liquid A liquid seal (most often mercury or sodium sulfate brine) that is displaced during the no-loss transfer of a gas sample from one container to another.

conformation In a molecule, a specific orientation of the atoms that varies from other possible orientations by rotation or rotations about single bonds; generally in mobile equilibrium with other conformations of the same structure. Also known as conformational isomer; conformer.

conformational analysis The determination of the arrangement in space of the constituent atoms of a molecule that may rotate about a single bond.

congener A chemical substance that is related to another substance, such as a derivative of a compound or an element belonging to the same family as another element in the periodic table.

congestin A toxin produced by certain sea anemones.

congo red $C_{32}H_{22}N_6Na_2O_6S_2$ An azo dye, sodium diphenyldiazo-bis-αα-naphthylamine sulfonate, used as a biological stain and as an acid-base indicator; it is red in alkaline solution and blue in acid solution.

conjugase Any of a group of enzymes which catalyze the breakdown of pteroylglutamic acid.

conjugate acid-base pair An acid and a base related by the ability of the acid to generate the base by loss of a proton.

conjugated diene An acyclic hydrocarbon with a molecular structure containing two carbon-carbon double bonds separated by a single bond.

conjugated polyene An acyclic hydrocarbon with a molecular structure containing alternating carbon-carbon double and single bonds.

conjugated protein A protein combined with a nonprotein group, other than a salt or a simple protein.

conode *See* tie line.

Conrad-Limpach method Synthesis of 2-methyl-4-quinolone from aniline and acetoacetic ester.

Conradson carbon test *See* carbon residue test.

consolute Of or pertaining to liquids that are perfectly miscible in all proportions under certain conditions.

constant-current electrolysis Electrolysis in which a constant current flows through the cell; used in electrodeposition analysis.

constant-potential electrolysis Electrolysis in which a constant voltage is applied to the cell; used in electrodeposition analysis.

constant series *See* displacement series.

constitutional isomers Isomers which differ in the manner in which their atoms are linked.

constitutive enzyme An enzyme whose concentration in a cell is constant and is not influenced by substrate concentration.

constitutive property Any physical or chemical property that depends on the constitution or structure of the molecule.

contact acid Sulfuric acid produced by the contact process.

contact adsorption Process for removal of minor constituents from fluids by stirring in direct contact with powdered or granulated adsorbents, or by passing the fluid through fixed-position adsorbent beds (activated carbon or ion-exchange resin); used to decolorize petroleum lubricating oils and to remove solvent vapors from air.

contact catalysis Process of change in the structure of gas molecules adsorbed onto solid surfaces; the basis of many industrial processes.

contact filtration A process in which finely divided adsorbent clay is mixed with oil to remove color bodies and to improve the oil's stability.

contact process Catalytic manufacture of sulfuric acid from sulfur dioxide and oxygen.

contact twin Twinned crystals whose members are symmetrically arranged about a twin plane.

contemporary carbon The isotopic carbon content of living matter, based on the assumption of a natural proportion of carbon-14.

continuous contact coking A thermal conversion process using the mass-flow lift principle to give continuous coke circulation; oil-wetted particles of coke move downward into the reactor in which cracking, coking, and drying take place; pelleted coke, gas, gasoline, and gas oil are products of the process.

continuous countercurrent leaching Proess of leaching by the use of continuous equipment in which the solid and liquid are both moved mechanically, and by the use of a series of leach tanks and the countercurrent flow of solvent through the tanks in reverse order to the flow of solid.

continuous distillation Separation by boiling of a liquid mixture with different component boiling points; feed is introduced continuously, with continuous removal of overhead vapors and high-boiling bottoms liquids.

continuous equilibrium vaporization *See* equilibrium flash vaporization.

continuous spectrum A radiation spectrum which is continuously distributed over a frequency region without being broken up into lines or bands.

continuous titrator A titrator so equipped that a reservoir refills the buret.

control agent In process automatic-control work, material or energy within a process system of which the manipulated (controlled) variable is a condition or characteristic.

controlled medium In process automatic-control work, material within a process system in which a variable (for example, concentration) is controlled.

convergence pressure   The pressure at which the different constant-temperature K (liquid-vapor equilibrium) factors for each member of a two-component system converge to unity.

conversion   1. Change of a compound from one isomeric form to another. 2. The chemical change from reactants to products in an industrial chemical process. 3. Nuclear transformation of a fertile substance into a fissile substance.

conversion coefficient   Also known as conversion fraction; internal conversion coefficient. 1. The ratio of the number of conversion electrons emitted per unit time to the number of photons emitted per unit time in the de-excitation of a nucleus between two given states. 2. In older literature, the ratio of the number of conversion electrons emitted per unit time to the number of conversion electrons plus the number of photons emitted per unit time in the de-excitation of a nucleus between two given states.

conversion electron   An electron which receives energy directly from a nucleus in an internal conversion process and is thereby expelled from the atom.

conversion fraction   *See* conversion coefficient.

converted water   *See* product water.

cool flame   A faint, luminous phenomenon observed when, for example, a mixture of ether vapor and oxygen is slowly heated; it proceeds by diffusion of reactive molecules which initiate chemical processes as they go.

cooling pond   Outdoor depression into which hot process water is pumped for purposes of cooling by evaporation, convection, and radiation.

coordinate bond   *See* coordinate valence; dative bond.

coordinated complex   *See* coordination compound.

coordinate valence   A chemical bond between two atoms in which a shared pair of electrons forms the bond and the pair has been supplied by one of the two atoms. Also known as coordinate bond.

coordination chemistry   The chemistry of metal ions in their interactions with other molecules or ions.

coordination compound   A compound with a central atom or ion and a group of ions or molecules surrounding it. Also known as coordinated complex; Werner complex.

coordination lattice   The crystal structure of a coordination compound.

coordination polygon   The symmetrical polygonal chemical structure of simple polyatomic aggregates having coordination numbers of 4 or less.

coordination polyhedron   The symmetrical polyhedral chemical structure of relatively simple polyatomic aggregates having coordination numbers of 4 to 8.

coordination polymer   Organic addition polymer that is neither free-radical nor simply ionic; prepared by catalysts that combine an organometallic (for example, triethyl aluminum) and a transition metal compound (for example, $TiCl_4$).

copolymer   A mixed polymer, the product of polymerization of two or more substances at the same time.

copolymerization   A polymerization reaction that forms a copolymer.

copper   A chemical element, symbol Cu, atomic number 29, atomic weight 63.546.

copper-64   Radioactive isotope of copper with mass number of 64; derived from pile-irradiation of metallic copper; used as a research aid to study diffusion, corrosion, and friction wear in metals and alloys.

copper acetate   *See* cupric acetate.

copper arsenate $Cu_3(AsO_4)_2 \cdot 4H_2O$ or $Cu_5H_2(AsO_4)_4 \cdot 2H_2O$ Bluish powder, soluble in ammonium hydroxide and dilute acids, insoluble in water and alcohol; used as a fungicide and insecticide.

copper arsenite $CuHAsO_3$ A toxic, light green powder which is soluble in acids and decomposes at the melting point; used as a pigment and insecticide. Also known as copper orthoarsenite; cupric arsenite; Scheele's green.

copper blue   *See* mountain blue.

copper bromide   *See* cupric bromide; cuprous bromide.

copper carbonate $Cu_2(OH)_2CO_3$ A toxic, green powder; decomposes at 200°C and is soluble in acids; used in pigments and pyrotechnics and as a fungicide and feed additive. Also known as artificial malachite; basic copper carbonate; cupric carbonate; mineral green.

copper chloride   *See* cupric chloride; cuprous chloride.

copper chromate   *See* cupric chromate.

copper cyanide   *See* cupric cyanide.

copper dish gum   The milligrams of gum found in 100 milliliters of gasoline when evaporated under controlled conditions in a polished copper dish.

copper fluoride   *See* cupric fluoride; cuprous fluoride.

copper gluconate $[CH_2OH(CHOH)_4COO]_2Cu$ A light blue, crystalline powder; soluble in water; used in medicine and as a dietary supplement. Also known as cupric gluconate.

copper hydroxide   *See* cupric hydroxide.

copper nitrate   *See* cupric nitrate.

copper number   The number of milligrams of copper obtained by the reduction of Benedict's or Fehling's solution by 1 gram of carbohydrate.

copper oleate $Cu[OOC(CH_2)_7CH{=}CH(CH_2)_7CH_3]_2$ A green-blue liquid, used as a fungicide for fruits and vegetables.

copperon   *See* cupferron.

copper orthoarsenite   *See* copper arsenite.

copper oxide   *See* cupric oxide; cuprous oxide.

copper-8-quinolinolate $C_{18}H_{14}N_2O_2Cu$ A khaki-colored, water-insoluble solid used as a fungicide in fruit-handling equipment.

copper resinate   Poisonous green powder, soluble in oils and ether, insoluble in water; made by heating rosin oil with copper sulfate, followed by filtering and drying of the resultant solids; used as a metal-paint preservative and insecticide.

copper sulfate   *See* cupric sulfate.

copper sulfide CuS Black, monoclinic or hexagonal crystals that break down at 220°C; used in paints on ship bottoms to prevent fouling.

copper sweetening Those refining processes using cupric chloride to oxidize mercaptans in petroleum.

coprecipitation Simultaneous precipitation of more than one substance.

coriandrol *See* linalool.

Cori ester *See* glucose-1-phosphate.

Coriolis operator An operator which gives a large contribution to the energy of an axially symmetric molecule arising from the interaction between vibration and rotation when two vibrations have equal or nearly equal frequencies.

Coriolis resonance interactions Perturbation of two vibrations of a polyatomic molecule, having nearly equal frequencies, on each other, due to the energy contribution of the Coriolis operator.

corn sugar *See* dextrose.

corresponding states The condition when two or more substances are at the same reduced pressures, the same reduced temperatures, and the same reduced volumes.

corrosion Chemical erosion by motionless or moving agents.

corrosive sublimate *See* mercuric chloride.

corticoid *See* adrenal cortex hormone.

corticosteroid 1. Any steroid hormone secreted by the adrenal cortex of vertebrates. 2. Any steroid with properties of an adrenal cortex steroid.

corticosterone $C_{21}H_{30}O_4$ A steroid hormone produced by the adrenal cortex of vertebrates that stimulates carbohydrate synthesis and protein breakdown and is antagonistic to the action of insulin.

corticotropin A hormonal preparation having adrenocorticotropic activity, derived from the adenohypophysis of certain domesticated animals.

cortisol *See* hydrocortisone.

cortisone $C_{26}H_{28}O_8I$ A steroid hormone produced by the adrenal cortex of vertebrates that acts principally in carbohydrate metabolism.

cosmic radiation *See* cosmic rays.

cosmic rays Electrons and the nuclei of atoms, largely hydrogen, that impinge upon the earth from all directions of space with nearly the speed of light. Also known as cosmic radiation; primary cosmic rays.

cosmic-ray shower The simultaneous appearance of a number of downward-directed ionizing particles, with or without accompanying photons, caused by a single cosmic ray. Also known as air shower; shower.

cotectic Referring to conditions of pressure, temperature, and composition under which two or more solid phases crystallize at the same time, with no resorption, from a single liquid over a finite range of decreasing temperature.

cotectic crystallization Simultaneous crystallization of two or more solid phases from a single liquid over a finite range of falling temperature without resorption.

Cotton effect   The characteristic wavelength dependence of the optical rotatory dispersion curve or the circular dichroism curve or both in the vicinity of an absorption band.

coudé spectrograph   A stationary spectrograph that is attached to the tube of a coudé telescope.

coudé spectroscopy   The production and investigation of astronomical spectra using a coudé spectrograph.

Coulomb barrier   1. The Coulomb repulsion which tends to keep positively charged bombarding particles out of the nucleus. 2. Specifically, the Coulomb potential associated with this force.

Coulomb excitation   Inelastic scattering of a positively charged particle by a nucleus and excitation of the nucleus, caused by the interaction of the nucleus with the rapidly changing electric field of the bombarding particle.

coulometer   An electrolytic cell for the precise measurement of electrical quantities or current intensity by quantitative determination of chemical substances produced or consumed.

coulometric analysis   A technique in which the amount of a substance is determined quantitatively by measuring the total amount of electricity required to deplete a solution of the substance.

coulometric titration   The slow electrolytic generation of a soluble species which is capable of reacting quantitatively with the substance sought; some independent property must be observed to establish the equivalence point in the reaction.

coulometry   A determination of the amount of an electrolyte released during electrolysis by measuring the number of coulombs used.

coulostatic analysis   An electrochemical technique involving the application of a very short, large pulse of current to the electrode; the pulse charges the capacitive electrode-solution interface to a new potential, then the circuit is opened, and the return of the working electrode potential to its initial value is monitored; the current necessary to discharge the electrode interface comes from the electrolysis of electroactive species in solution; the change in electrode potential versus time results in a plot, the shape of which is proportional to concentration.

coumachlor $C_{19}H_{15}ClO_4$ A white, crystalline compound with a melting point of 169–171°C; insoluble in water; used as a rodenticide. Also known as 3-(α-acetonyl-4-chlorobenzyl)-4-hydroxy-coumarin.

coumafene   *See* warfarin.

coumarin $C_9H_6O_2$ The anhydride of o-coumaric acid; a toxic, white, crystalline lactone found in many plants and made synthetically; used in making perfume and soap. Also known as 1,2-benzopyrone.

coumarin glycoside   Any of several glycosidic aromatic principles in many plants; contains coumaric acid as the aglycon group.

coumarone C_8H_6O A colorless liquid, boiling point 169°C. Also known as benzofuran.

coumarone-indene resin   A synthetic resin prepared by polymerization of coumarone and indene.

coumatetralyl $C_{19}H_{16}O_3$ A yellow-white, crystalline compound with a melting point of 172–176°C; slightly soluble in water; used as a rodenticide. Also known as 4-hydroxy-3-(1,2,3,4-tetrahydro-1-naphthyl)coumarin.

count An ionizing event.

countercurrent cascade An extraction process involving the introduction of a sample, all at once, into a continuously flowing countercurrent system where both phases are moving in opposite directions and are continuously at equilibrium.

countercurrent distribution A profile of a compound's concentration in different ratios of two immiscible liquids.

countercurrent extraction A liquid-liquid extraction process in which the solvent and the process stream in contact with each other flow in opposite directions. Also known as countercurrent separation.

countercurrent leaching A process utilizing a series of leach tanks and countercurrent flow of solvent through them in reverse order to the flow of solid.

countercurrent separation *See* countercurrent extraction.

couple Joining of two molecules.

coupled reaction A reaction which involves two oxidants with a single reductant, where one reaction taken alone would be thermodynamically unfavorable.

coupons Polished metal strips of specified size and weight used to detect the corrosive action of liquid or gas products or to test the efficiency of corrosion-inhibitor additives.

covalence The number of covalent bonds which an atom can form.

covalent bond A bond in which each atom of a bound pair contributes one electron to form a pair of electrons. Also known as electron pair bond.

covalent crystal A crystal held together by covalent bonds. Also known as valence crystal.

covalent hydride A compound formed from a nonmetal and hydrogen, for example, H_2S and NH_3.

covalent radius The effective radius of an atom in a covalent bond.

Cox chart A straight-line graph of the logarithm of vapor pressure against a special nonuniform temperature scale; vapor pressure–temperature lines for many substances intersect at a common point on the Cox chart.

cp *See* centipoise.

CPMC *See* O-chlorophenyl-methyl-carbamate.

Cr *See* chromium.

CR *See* catalytic reforming.

crack To break a compound into simpler molecules.

cracked residue The residue of fuel resulting from decomposition of hydrocarbons during thermal or catalytic cracking.

cracking A process that is used to reduce the molecular weight of hydrocarbons by breaking the molecular bonds by various thermal, catalytic, or hydrocracking methods.

cracking coil   A coil used for cracking heavy petroleum products.

cracking still   The furnace, reaction chamber, and fractionator for thermal conversion of heavier charging stock to gasoline.

CRC method   A research technique formulated, approved, and published by the Coordinating Research Council, Inc.

cream of tartar   *See* potassium bitartrate.

creatine $C_4H_9O_2N_3$ α-Methylguanidine-acetic acid; a compound present in vertebrate muscle tissue, principally as phosphocreatine.

creatinine $C_4H_7ON_3$ A compound present in urine, blood, and muscle that is formed from the dehydration of creatine.

creosol $CH_3C_6H_4OH$ A combination of isomers, derived from coal tar or petroleum; a yellowish liquid with a phenolic odor; used as a disinfectant, in the manufacture of resins, and in flotation of ore. Also known as hydroxymethylbenzene; methyl phenol.

cresol $CH_3C_6H_4OH$ One of three poisonous, colorless isomeric methyl phenols: *o*-cresol, *m*-cresol, *p*-cresol; used in the production of phenolic resins, tricresyl phosphate, disinfectants, and solvents.

cresol red $C_{21}H_{18}O_5S$ A compound derived from *o*-cresol and used as an acid-base indicator; color change is yellow to red at pH 0.4 to 1.8, or 7.0 to 8.8, depending on preparation. Also known as *ortho*-cresolsulfonphthalein.

cricondenbar   Maximum pressure at which two phases (for example, liquid and vapor) can coexist.

cricondentherm   Maximum temperature at which two phases (for example, liquid and vapor) can coexist.

critical absorption wavelength   The wavelength, characteristic of a given electron energy level in an atom of a specified element, at which an absorption discontinuity occurs.

critical condensation temperature   The temperature at which the sublimand of a sublimed solid recondenses; used to analyze solid mixtures, analogous to liquid distillation. Also known as true condensing point.

critical constant   A characteristic temperature, pressure, and specific volume of a gas above which it cannot be liquefied.

critical current density   The amount of current per unit area of electrode at which an abrupt change occurs in a variable of an electrolytic process.

critical density   The density of a substance exhibited at its critical temperature and critical pressure.

critical humidity   The humidity of a system's atmosphere above which a crystal of a water-soluble salt will always become damp (absorb moisture from the atmosphere) and below which it will always stay dry (release moisture to the atmosphere).

critical locus   The line connecting the critical points of a series of liquid-gas phase-boundary loops for multicomponent mixtures plotted on a pressure versus temperature graph.

critical micelle concentration   The concentration of a micelle (oriented molecular arrangement of an electrically charged colloidal particle or ion) at which the rate of increase of electrical conductance with increase in concentration levels off or proceeds at a much slower rate.

critical moisture content The average moisture throughout a solid material being dried, its value being related to drying rate, thickness of material, and the factors that influence the movement of moisture within the solid.

critical phenomena Physical properties of liquids and gases at the critical point (conditions at which two phases are just about to become one); for example, critical pressure is that needed to condense a gas at the critical temperature, and above the critical temperature the gas cannot be liquefied at any pressure.

critical point **1.** The temperature and pressure at which two phases of a substance in equilibrium with each other become identical, forming one phase. **2.** The temperature and pressure at which two ordinarily partially miscible liquids are consolute.

critical potential The energy needed to raise an electron to a higher energy level in an atom (resonance potential) or to remove it from the atom (ionization potential).

critical properties Physical and thermodynamic properties of materials at conditions of critical temperature, pressure, and volume, that is, at the critical point.

critical solution temperature The temperature at which a mixture of two liquids, immiscible at ordinary temperatures, ceases to separate into two phases.

critical state Unique condition of pressure, temperature, and composition wherein all properties of coexisting vapor and liquid become identical.

critical temperature The temperature of the liquid-vapor critical point, that is, the temperature above which the substance has no liquid-vapor transition.

crosscurrent extraction Procedure of batchwise liquid-liquid extraction in a separatory funnel; solvent is added to the sample in the funnel, which is then shaken, and the extract phase is allowed to coalesce, then is drawn off.

cross-linking The setting up of chemical links between the molecular chains of polymers.

cross section per atom The microscopic cross section for a given nuclear reaction referred to the natural element, even though the reaction involves only one of the natural isotopes.

crotonaldehyde C_3H_5CHO A colorless liquid boiling at 104°C, soluble in water; vapors are lacrimatory; used as an intermediate in manufacture of n-butyl alcohol and quinaldine. Also known as propylene aldehyde.

crotonic acid C_3H_5COOH An unsaturated acid, with colorless, monoclinic crystals, soluble in water; used in the preparation of synthetic resins, plasticizers, and pharmaceuticals. Also known as α-butenic acid.

crown ether A macrocyclic polyether whose structure exhibits a conformation with a so-called hole capable of trapping cations by coordination with a lone pair of electrons on the oxygen atoms.

crude assay A procedure for determining the general distillation characteristics and other quality information of crude oil.

crude desalting The washing of crude oil with water in order to remove materials such as dirt, silt, and water-soluble minerals.

crude still The distillation equipment in which crude oil is separated into various products.

crufomate $C_{12}H_{19}ClNO_3P$ A white, crystalline compound, with a melting point of 61.8°C, which is insoluble in water; used both internally and externally for cattle parasites. Also known as 4-*tert*-butyl-2-chlorophenyl-*ortho*-methyl methylphosphoramidate.

crustecdysone $C_{27}H_{44}O_7$ 20-Hydroxyecdysone, the molting hormone produced by Y organs in crustaceans.

cryochem process A freeze-drying technique involving conduction heat transfer to the frozen solid held on a metallic surface.

cryohydrate A salt that contains water of crystallization at low temperatures. Also known as cryosel.

cryohydric point The eutectic point of an aqueous salt solution.

cryoprecipitate The precipitate of a cryoglobulin.

cryoscopic constant Equation constant expressed in degrees per mole of pure solvent; used to calculate the freezing-point-depression effects of a solute.

cryoscopy A phase-equilibrium technique to determine molecular weight and other properties of a solute by dissolving it in a liquid solvent and then ascertaining the solvent's freezing point.

cryptand A macropolycyclic polyazo-polyether, where the three-coordinate nitrogen atoms provide the vertices of a three-dimensional structure.

cryptoxanthin $C_{40}H_{57}O$ A xanthophyll carotenoid pigment found in plants; convertible to vitamin A by many animal livers.

crystal A homogeneous solid made up of an element, chemical compound or isomorphous mixture throughout which the atoms or molecules are arranged in a regularly repeating pattern.

crystal abundance *See* clarke.

crystal aerugo *See* cupric acetate.

crystal axis A reference axis used for the vectoral properties of a crystal.

crystal carbonate *See* metahydrate sodium carbonate.

crystal chemistry The study of the crystalline structure and properties of a mineral or other solid.

crystal class One of 32 categories of crystals according to the inversions, rotations about an axis, reflections, and combinations of these which leaves the crystal invariant. Also known as symmetry class.

crystal defect Any departure from crystal symmetry caused by free surfaces, disorder, impurities, vacancies and interstitials, dislocations, lattice vibrations, and grain boundaries. Also known as lattice defect.

crystal face One of the outward planar surfaces which define a crystal and reflect its internal structure. Also known as face.

crystal field theory The theory which assumes that the ligands of a coordination compound are the sources of negative charge which perturb the energy levels of the central metal ion and thus subject the metal ion to an electric field analogous to that within an ionic crystalline lattice.

crystal gliding Slip along a crystal plane due to plastic deformation; often produces crystal twins. Also known as translation gliding.

crystal grating A diffraction grating for gamma rays or x-rays which uses the equally spaced lattice planes of a crystal.

crystal growth The growth of a crystal, which involves diffusion of the molecules of the crystallizing substance to the surface of the crystal, diffusion of these molecules over the crystal surface to special sites on the surface, incorporation of molecules into the surface at these sites, and diffusion of heat away from the surface.

crystal habit The size and shape of the crystals in a crystalline solid. Also known as habit.

crystal indices *See* Miller indices.

crystal lattice A lattice from which the structure of a crystal may be obtained by associating with every lattice point an assembly of atoms identical in composition, arrangement, and orientation.

crystalline Of, pertaining to, resembling, or composed of crystals.

crystalline chloral *See* chloral hydrate.

crystalline polymer A polymer whose sections of adjacent chains are packed in a regular array.

crystallinity The quality or state of being crystalline.

crystallization The formation of crystalline substances from solutions or melts.

crystallizer Process vessel within which dissolved solids in a supersaturated solution are forced out of solution by cooling or evaporation, and then recovered as solid crystals.

crystallogram A photograph of the x-ray diffraction pattern of a crystal.

crystallographic axis One of three lines (sometimes four, in the case of a hexagonal crystal), passing through a common point, that are chosen to have definite relation to the symmetry properties of a crystal, and are used as a reference in describing crystal symmetry and structure.

crystal monochromator A spectrometer in which a collimated beam of slow neutrons from a reactor is incident on a single crystal of copper, lead, or other element mounted on a divided circle.

crystal pulling A method of crystal growing in which the developing crystal is gradually withdrawn from a melt.

crystals of Venus *See* cupric acetate.

crystal structure The arrangement of atoms or ions in a crystalline solid.

crystal symmetry The existence of nontrivial operations, consisting of inversions, rotations around an axis, reflections, and combinations of these, which bring a crystal into a position indistinguishable from its original position.

crystal system One of seven categories (cubic, hexagonal, tetragonal, trigonal, orthorhombic, monoclinic, and triclinic) into which a crystal may be classified according to the shape of the unit cell of its Bravais lattice, or according to the dominant symmetry elements of its crystal class.

crystal violet *See* methyl violet.

crystal whiskers Single crystals that have grown in a filamentary form.

crystogen *See* cystamine.

Cs *See* cesium.

C stage The final stage in a thermosetting resin reaction in which the material is relatively insoluble and infusible; the resin in a fully cured thermoset molding is in this stage. Also known as resite.

cubic cleavage Isometric crystal cleavage occurring parallel to the faces of a cube.

cubic crystal A crystal whose lattice has a unit cell with perpendicular axes of equal length.

cubic packing The spacing pattern of uniform solid spheres in a clastic sediment or crystal lattice in which the unit cell is a cube.

cubic plane A plane that is at right angles to any one of the three crystallographic axes of the cubic system.

cull In a plastics molding operation, material remaining in the transfer chamber after the mold has been filled.

cumene $C_6H_5CH(CH_3)_2$ A colorless, oily benzenoid hydrocarbon cooling at 152.4°C; used as an additive for high-octane motor fuel. Also known as isopropylbenzene.

cumene hydroperoxide $C_6H_5C(CH_3)_2OOH$ An isopropyl hydroperoxide of cumene; an oily liquid, used to make phenol and acetone.

cumidine $C_9H_{13}N$ A colorless, water-insoluble liquid, boiling at 225°C. Also known as *para*-isopropylaniline.

cumulated double bond Two double bonds on the same carbon atom, as in $>C{=}C{=}C<$.

cumulative excitation Process by which the atom is raised from one excited state to a higher state by collision, for example, with an electron.

cumulative ionization Ionization of an excited atom in the metastable state by means of cumulative excitation.

cumulene A compound with a molecular structure which contains two or more double bonds in succession.

cumulus The accumulation of minerals which have precipitated from a liquid without having been modified by later crystallization.

cupferron $NH_4ONONC_6H_5$ A colorless salt that forms crystals with a melting point of 164°C; its acid solution is a precipitating reagent. Also known as copperon; nitrosophenylhydroxylamine.

cupreine $C_{19}H_{22}O_2N_2{\cdot}H_2O$ Colorless, anhydrous crystals with a melting point of 198°C; soluble in chloroform and ether; used in medicine. Also known as hydroxycinchonine.

cupric The divalent ion of copper.

cupric acetate $Cu(C_2H_3O_2)_2{\cdot}H_2O$ Blue-green crystals, soluble in water; used as a raw material to make paris green. Also known as copper acetate; crystal aerugo; crystals of Venus; verdigris.

cupric arsenite *See* copper arsenite.

cupric bromide $CuBr_2$ Black prismatic crystals; used in photography as an intensifier and in organic synthesis as a brominating agent. Also known as copper bromide.

cupric carbonate *See* copper carbonate.

cupric chloride Also known as copper chloride. **1.** $CuCl_2$ Yellowish-brown, deliquescent powder soluble in water, alcohol, and ammonium chloride. **2.** $CuCl_2 \cdot H_2O$ A dihydrate of cupric chloride forming green crystals soluble in water; used as a mordant in dyeing and printing textile fabrics and in the refining of copper, gold, and silver.

cupric chromate $CuCrO_4$ A yellow liquid, used as a mordant. Also known as copper chromate.

cupric cyanide $Cu(CN)_2$ A green powder, insoluble in water; used in electroplating copper on iron. Also known as copper cyanide.

cupric fluoride CuF_2 White crystalline powder used in ceramics and in the preparation of brazing and soldering fluxes. Also known as copper fluoride.

cupric gluconate *See* copper gluconate.

cupric hydroxide $Cu(OH)_2$ Blue macro- or microscopic crystals; used as a mordant and pigment, in manufacture of many copper salts, and for staining paper. Also known as copper hydroxide.

cupric nitrate $Cu(NO_3)_2 \cdot 3H_2O$ Green powder or blue crystals soluble in water; used in electroplating copper on iron. Also known as copper nitrate.

cupric oxide CuO Black, monoclinic crystals, insoluble in water; used in making fibers and ceramics, and in organic and gas analyses. Also known as copper oxide.

cupric sulfate $CuSO_4$ A water-soluble salt used in copper-plating baths; crystallizes as hydrous copper sulfate, which is blue. Also known as copper sulfate.

cuprous bromide Cu_2Br_2 White or gray crystals slightly soluble in cold water. Also known as copper bromide.

cuprous chloride $CuCl$ or Cu_2Cl_2 Green, tetrahedral crystals, insoluble in water. Also known as copper chloride; resin of copper.

cuprous fluoride Cu_2F_2 Red, crystalline powder, melting point 908°C. Also known as copper fluoride.

cuprous oxide Cu_2O An oxide of copper found in nature as cuprite and formed on copper by heat; used chiefly as a pigment and as a fungicide. Also known as copper oxide.

cure To change the properties of a resin material by chemical polycondensation or addition reactions.

curine *See* bebeerine.

curing temperature That temperature at which a resin or adhesive is subjected to curing.

curing time The period of time in which a part is subjected to heat or pressure to cure the resin.

curium An element, symbol Cm, atomic number 96; the isotope of mass 244 is the principal source of this artificially produced element.

curium-242 An isotope of curium, mass number 242; half-life is 165.5 days for α-particle emission; 7.2×10^6 years for spontaneous fission.

curium-244 An isotope of curium, mass number 244; half-life is 16.6 years for α-particle emission; 1.4×10^7 years for spontaneous fission; potential use as compact thermoelectric power source.

current efficiency The ratio of the amount of electricity, in coulombs, theoretically required to yield a given quantity of material in an electrochemical process, to the amount actually consumed.

curtain coating A method in which the substrate to be coated with low-viscosity resins or solutions is passed through, and is perpendicular to, a freely falling liquid curtain.

Curtius reaction A laboratory method for degrading a carboxylic acid to a primary amine by converting the acid to an acyl azide to give products which can be hydrolyzed to amines.

cut 1. A fraction obtained by a separation process. 2. A section of a crystal having two parallel major surfaces; cuts are specified by their orientation with respect to the axes of the natural crystal, such as X cut, Y cut, BT cut, and AT cut.

cutback Blending of heavier oils with lighter ones to bring the heavier to desired specifications.

cutin A mixture of fatty substances characteristically found in epidermal cell walls and in the cuticle of plant leaves and stems.

cut point The boiling-temperature division between cuts of a crude oil or base stock.

cyanalcohol *See* cyanohydrin.

cyanamide NHCNH An acidic compound that forms colorless needles, melting at 46°C, soluble in water. Also known as urea anhydride.

cyanate A salt or ester of cyanic acid containing the radical CNO.

cyanazine $C_9H_{13}N_6Cl$ A white solid with a melting point of 166.5–167°C; used as a pre- and postemergence herbicide for corn, sorghum, soybeans, alfalfa, cotton, and wheat. Also known as 2-(4-chloro-6-ethylamino-s-triazin-2-yl-amino)-2-methylpropionitrile.

cyanic acid HCNO A colorless, poisonous liquid, which polymerizes to cyamelide and fulminic acid.

4-cyanic acid *See* fulminic acid.

cyanidation Joining of cyanide to an atom or molecule.

cyanide Any of a group of compounds containing the CN group and derived from hydrogen cyanide, HCN.

cyanine dye $C_{29}H_{35}N_2I$ Green metallic crystals, soluble in water; unstable to light, the dye is used in the photography industry as a chemical sensitizer for film. Also known as iodocyanin; quinoline blue.

cyano- Combining form indicating the radical CN.

cyanoacetamide $C_3H_4N_2O$ Needlelike crystals with a melting point of 119.5°C; soluble in water; used in organic synthesis. Also known as malonamide nitrile.

cyanoacetic acid $NCCH_2COOH$ Hygroscopic crystals with a melting point of 66°C; decomposes at 160°C; soluble in ether, water, and alcohol; used in the synthesis of intermediates and in the commercial preparation of barbital. Also known as malonic mononitrile.

cyanocarbon A derivative of hydrocarbon in which all of the hydrogen atoms are replaced by the CN group.

cyanocobalamin *See* vitamin B_{12}.

cyano complex A coordination compound containing the CN group.

2-cyanoethanol *See* ethylene cyanohydrin.

cyanoethylation A chemical reaction involving the addition of acrylonitrile to compounds with a reactive hydrogen.

cyanogen 1. A univalent radical, CN. 2. C_2N_2 A colorless, highly toxic gas with a pungent odor; a starting material for the production of complex thiocyanates used as insecticides. Also known as dicyanogen.

cyanogen bromide CNBr White crystals melting at 52°C, vaporizing at 61.3°C, and having toxic fumes that affect nerve centers; used in the synthesis of organic compounds and as a fumigant.

cyanogen chloride ClCN A poisonous, colorless gas or liquid, soluble in water; used in organic synthesis.

cyanogen fluoride CNF A toxic, colorless gas, used as a tear gas.

cyanogen iodide *See* iodine cyanide.

cyanohydrin A compound containing the radicals CN and OH. Also known as cyanalcohol.

cyanonitroacetamide *See* fulminuric acid.

2-cyano-2-nitroethanamide *See* fulminuric acid.

cyanophosphos $C_{15}H_{14}NO_2PS$ A white, crystalline solid with a melting point of 83°C; used as an insecticide to control larval pests on rice and vegetables. Also known as O-*para*-cyanophenyl-O-ethylphenylphosphonothioate.

cyanophycin A granular protein food reserve in the cells of blue-green algae, especially in the peripheral cytoplasm.

cyanoplatinate *See* platinocyanide.

cyanuric acid $HOC(NCOH)_2N \cdot 2H_2O$ Colorless, monoclinic crystals, slightly soluble in water; formed by polymerization of cyanic acid. Also known as pyrolithic acid; pyrouric acid; pyruric acid; s-triazinetriol; trihydroxycyanidine; trioxycyanidine.

cyanurtriamide *See* melamine.

cyclamate The calcium or sodium salt of cyclohexylsulfamate, an artificial sweetener.

cyclane *See* alicyclic.

cycle plant A plant in which the liquid hydrocarbons are removed from natural gas and then the gas is put back into the earth to maintain pressure in the oil reservoir.

cycle stock The unfinished product taken from a stage of a refinery process and recharged to the process at an earlier stage in the operation.

cyclethrin $C_{21}H_{28}O_3$ A viscous, brown liquid, soluble in organic solvents; used as an insecticide.

cyclic adenylic acid $C_{10}H_{12}N_5O_6P$ An isomer of adenylic acid; crystal platelets with a melting point of 219–220°C; a key regulator which acts to control the rate of a number of cellular processes in bacteria, most animals, and some higher plants. Abbreviated cAMP. Also known as adenosine 3′,5′-cyclic monophosphate; adenosine 3′,5′-cyclic phosphate; adenosine 3′,5′-monophosphate; 3′,5′-AMP; cyclic AMP.

cyclic amide An amide arranged in a ring of carbon atoms.

cyclic AMP *See* cyclic adenylic acid.

cyclic anhydride A ring compound formed by the removal of water from a compound; an example is phthalic anhydride.

cyclic catalytic reforming process A method for the production of low-Btu reformed gas consisting of the conversion of carbureted water-gas sets by installing a bed of nickel catalyst in the superheater and using the carburetor as a combustion chamber and process steam superheater. Abbreviated CCR process.

cyclic chronopotentiometry An analytic electrochemical method in which instantaneous current reversal is imposed at the working electrode, and its potential is monitored with time.

cyclic compound A compound that contains a ring of atoms.

cyclic twinning Repeated twinning of three or more individuals in accordance with the same twinning law but without parallel twinning axes.

cyclic voltammetry An electrochemical technique for studying variable potential at an electrode involving application of a triangular potential sweep, allowing one to sweep back through the potential region just covered.

cycling A series of operations in petroleum refining or natural-gas processing in which the steps are repeated periodically in the same sequence.

cyclization Changing an open-chain hydrocarbon to a closed ring.

cycloaddition A reaction in which unsaturated molecules combine to form a cyclic compound.

cycloaliphatic *See* alicyclic.

cycloalkane *See* alicyclic.

cycloalkene An unsaturated, monocyclic hydrocarbon having the formula C_nH_{2n-2}. Also known as cycloolefin.

cycloalkyne A cyclic compound containing one or more triple bonds between carbon atoms.

cycloate $C_{11}H_{21}NOS$ A yellow liquid with limited solubility in water; boiling point is 145–146°C; used as an herbicide to control weeds in sugarbeets, spinach, and table beets. Also known as S-ethyl-N-ethyl-N-cyclohexylthiocarbamate.

cyclobutadiene C_4H_4 A cyclic compound containing two alternate double bonds; used in organic synthesis. Also known as tetramethyldiene; $\Delta^{1,3}$-tetramethylene.

cyclobutane C_4H_8 An alicyclic hydrocarbon, boiling point 11°C; synthesized as a condensable gas; used in organic synthesis. Also known as tetramethylene.

cyclobutene C_4H_6 An asymmetrical cyclic hydrocarbon occurring in several isomeric forms. Also known as cyclobutylene.

cyclodextrin Cyclic degradation products of starch which contain six, seven, or eight glucose residues and have the shape of large-ring molecules.

cyclodiolefin A cycloalkene with two double bonds; sometimes included with alkenes, cycloalkenes, and hydrocarbons containing more than one ethylene bond as olefins in a generic sense.

cyclododecatriene $C_{12}H_{18}$ One of two cyclic hydrocarbons with three double bonds; the two forms are stereoisomeric; used to make nylon-6 and nylon-12.

1,3-cyclohexadiene C_6H_8 A partly saturated benzene compound with two double bonds; used in organic synthesis. Also known as dihydrobenzene.

cyclohexane C_6H_{12} A colorless liquid that is a cyclic hydrocarbon synthesized by hydrogenation of benzene; used in organic synthesis. Also known as hexamethylene.

cyclohexanol $C_6H_{11}OH$ An oily, colorless, hygroscopic liquid with a camphorlike odor and a boiling point of 160.9°C; used in soapmaking, insecticides, dry cleaning, plasticizers, and germicides. Also known as hexahydrophenol.

cyclohexanone $C_6H_{10}O$ An oily liquid with an odor suggesting peppermint and acetone; soluble in alcohol, ether, and other organic solvents; used as an industrial solvent, in the production of adipic acid, and in the preparation of cyclohexanone resins. Also known as ketohexamethylene; pimelic ketone.

cyclohexene C_6H_{10} A compound that occurs in coal tar; a liquid that is used as an alkylation component; used in the manufacture of hexahydrobenzoic acid, adipic acid, and maleic acid. Also known as 1,2,3,4-tetrahydrobenzene.

1-cyclohexene-1,2-dicarboximidomethyl-2,2-dimethyl-3(2-methylpropenyl)-cyclopropanecarboxylate *See* tetramethrin.

cyclohexylamine $C_6H_{11}NH_2$ A liquid with a strong, fishy, amine odor; miscible with water and common organic solvents; used in organic synthesis and in the manufacture of plasticizers, rubber chemicals, corrosion inhibitors, dyestuffs, dry-cleaning soaps, and emulsifying agents. Also known as aminocyclohexane; hexahydroaniline.

cyclohexylbenzene *See* phenylcyclohexane.

1-cyclohexyl-2-methylaminopropane *See* propylhexedrine.

3-cyclohexyl-5,6-trimethyleneuracil *See* lenacil.

cycloidal mass spectrometer Small mass spectrometer of limited mass range fitted with a special-type analyzer that generates a cycloidal-path beam of the sample mass.

cyclone A static reaction vessel in which fluids under pressure form a vortex.

cyclonite $(CH_2)_3N_3(NO_2)_3$ A white, crystalline explosive, consisting of hexahydro-trinitro-triazine, and having high sensitivity and brisance; mixed with other explosives or substances. Abbreviated RDX. Also known as cyclotrimethylenetrinitramine.

1,5-cyclooctadiene C_8H_{12} A cyclic hydrocarbon with two double bonds; prepared from butadiene and used to make cyclooctene and cyclooctane, which are intermediates for the production of plastics, fibers, and so on.

cyclooctane $(CH_2)_8$ A cyclic alkane melting at 9.5°C; used as an intermediate in production of plastics, fibers, adhesives, and coatings. Also known as octomethylene.

cyclooctatetraene C_8H_8 A cyclic olefin with alternate double bonds; highly reactive; rearranges to styrene.

cycloolefin *See* cycloalkene.

cycloparaffin *See* alicyclic.

1,3-cyclopentadiene C_5H_6 A colorless liquid boiling at 41.5°C; used to make resins.

cyclopentadienyl anion C_5H_5- A radical formed from cyclopentadiene.

cyclopentane C_5H_{10} A cyclic hydrocarbon that is a colorless liquid; present in crude petroleum, it is converted during refining to aromatics which improve antiknock and combustion properties of gasoline.

cyclopentanol C_5H_9OH A colorless liquid boiling at 139°C; used as a solvent for perfumes and pharmaceuticals. Also known as cyclopentyl alcohol.

cyclopentanone C_5H_8O A saturated monoketone; a colorless liquid boiling at 130°C; used as an intermediate in pharmaceutical preparation. Also known as adipinketone.

cyclopentene $(CH_2)_3CHCH$ A colorless liquid boiling at 45°C; used as a chemical intermediate in petroleum chemistry. Also known as Δ^1-pentamethylene.

cyclopentenylundecylic acid *See* hydnocarpic acid.

cyclopentyl alcohol *See* cyclopentanol.

α-cyclopropyl-α-(*para*-methoxyphenyl)-5-pyrimidinemethanol $C_{12}H_{11}N_2O$ White crystals with a melting point of 110–111°C; solubility in water is 650 parts per million; used as a growth regulator for greenhouse plants.

cyclopropane C_3H_6 A colorless gas, insoluble in water; used as an anesthetic.

cyclotrimethylenetrinitramine *See* cyclonite.

cyhexatin $C_{18}H_{34}OSn$ A whitish solid, insoluble in water; used as a miticide to control plant-feeding mites. Also known as tricyclohexyltin hydroxide.

cymene Any of the isomeric hydrocarbons metacymene, paracymene, and orthocymene; paracymene is a liquid that is colorless, has a pleasant odor, and is made from oil of cumin or oil of wild thyme.

***meta*-cym-*s*-yl-methylcarbamate** *See* promecarb.

cystamine $(CH_2)_6N_4$ A white, crystalline powder, melting at 280°C; used to make synthetic resins. Also known as aminiform; crystogen; cystamine methenamine; hexamethylene tetramine; urotropin.

cystamine methenamine *See* cystamine.

L-cystathionine $C_7H_{14}N_2O_4S$ An amino acid formed by condensation of homocysteine with serine, catalyzed by an enzyme transsulfurase; found in high concentration in the brain of primates.

cysteamine *See* 2-aminoethanethiol.

cysteine $C_3H_7O_2NS$ A crystalline amino acid occurring as a constituent of glutathione and cystine.

cystine $C_6H_{12}N_2S_2$ A white, crystalline amino acid formed biosynthetically from cysteine.

cytase Any of several enzymes in the seeds of cereals and other plants, which hydrolyze the cell-wall material.

cytidine $C_9H_{13}N_3O_5$ Cytosine riboside, a nucleoside composed of one molecule each of cytosine and D-ribose.

cytidine monophosphate *See* cytidilic acid.

cytidylic acid $C_9H_{14}O_8N_3P$ A nucleotide synthesized from the base cytosine and obtained by hydrolysis of nucleic acid. Also known as cytidine monophosphate.

cytochalasin One of a series of structurally related fungal metabolic products which, among other effects on biological systems, selectively and reversibly block cytokinesis while not affecting karyokinesis; the molecule with minor variations consists of a benzyl-substituted hydroaromatic isoindolone system, which in turn is fused to a small macrolide-like cyclic ring.

cytochrome Any of the complex protein respiratory pigments occurring within plant and animal cells, usually in mitochondia, that function as electron carriers in biological oxidation.

cytochrome a₃ $C_9H_{13}N_3O_5$ *See* cytochrome oxidase.

cytochrome oxidase Any of a family of respiratory pigments that react directly with oxygen in the reduced state. Also known as cytochrome a_3.

cytokinin Any of a group of plant hormones which elicit certain plant growth and development responses, especially by promoting cell division.

cytosine $C_4H_5ON_3$ A pyrimidine occurring as a fundamental unit or base of nucleic acids.

Czochralski process A method of producing large single crystals by inserting a small seed crystal of germanium, silicon, or other semiconductor material into a crucible filled with similar molten material, then slowly pulling the seed up from the melt while rotating it.

D

d *See* deuterium; deuteron.

D *See* deuterium.

2,4-D *See* 2,4-dichlorophenoxyacetic acid.

dalapon Generic name for 2,2-dichloropropionic acid; a liquid with a boiling point of 185–190°C at 760 mmHg; soluble in water, alcohol, and ether; used as an herbicide.

Dalton's atomic theory Theory forming the basis of accepted modern atomic theory, according to which matter is made of particles called atoms, reactions must take place between atoms or groups of atoms, and atoms of the same element are all alike but differ from atoms of another element.

Damköhler number V *See* Reynolds number.

Danckwerts model Theory applied to liquid flow across packing in a liquid-gas absorption tower; allows for liquid eddies that bring fresh liquid from the interior of the liquid body to the surface, thus contacting the gas in the column.

Daniell cell A primary cell with a constant electromotive force of 1.1 volts, having a copper electrode in a copper sulfate solution and a zinc electrode in dilute sulfuric acid or zinc sulfate, the solutions separated by a porous partition or by gravity.

dansyl chloride $(CH_3)_2NC_{10}H_6SO_2Cl$ A reagent for fluorescent labeling of amines, amino acids, proteins, and phenols.

DAP *See* diaminopimelate.

dapsone *See* 4,4′-sulfonyldianiline.

dark-line spectrum The absorption spectrum that results when white light passes through a substance, consisting of dark lines against a bright background.

Darzen's procedure Preparation of alkyl halides by refluxing a molecule of an alcohol with a molecule of thionyl chloride in the presence of a molecule of pyridine.

Darzen's reaction Condensation of aldehydes and ketones with α-haloesters to produce glycidic esters.

daughter The immediate product of radioactive decay of an element, such as uranium. Also known as decay product; radioactive decay product.

Dauphine law A twin law in which the twinned parts are related by a rotation of 180° around the c axis.

dazomet $C_5H_{10}N_2S_2$ A white, crystalline compound that decomposes at 100°C; used as an herbicide and nematicide for soil fungi and nematodes, weeds, and soil insects. Also known as tetrahydro-3,5-dimethyl-2H-1,3,5-thiadiazine-6-thione.

DBCP *See* dibromochloropropane.

DCB *See* 1,4-dichlorobutane.

DCC *See* dicyclohexylcarbodiimide.

DCCI *See* dicyclohexylcarbodiimide.

DCMO *See* carboxin.

DCNA *See* 2,6-dichloro-4-nitroaniline.

DCPA *See* dimethyl-2,3,5,6-tetrachloroterephthalate.

DDD *See* 2,2-bis(*para*-chlorophenyl)-1,1-dichloroethane.

DDT Common name for an insecticide; melting point 108.5°C, insoluble in water, very soluble in ethanol and acetone, colorless, and odorless; especially useful against agricultural pests, flies, lice, and mosquitoes. Also known as dichlorodiphenyltrichloroethane.

DDTA *See* derivative differential thermal analysis.

deacetylate The removal of an acetyl grouping from a molecule.

deacidify 1. Removal of acid. 2. A process of reducing acidity.

Deacon process A method of chlorine production by passing a hot mixture of gaseous hydrochloric acid with oxygen over a cuprous chloride catalyst.

deactivation 1. Rendering inactive, as of a catalyst. 2. Loss of radioactivity.

DEAE-cellulose *See* diethylaminoethyl cellulose.

deagglomeration Size-reduction process in which loosely adhered clumps (agglomerates) of powders or crystals are broken apart without further disintegration of the powder or crystal particles themselves.

dealkalize 1. The removal of alkali. 2. The reduction of alkalinity as in the neutralization process.

dealkylate To remove alkyl groups from a compound.

dealuminization Removal of aluminum.

deamidase An enzyme that catalyzes the removal of an amido group from a compound.

deamidate Removal from a molecule of the amido group.

deaminase An enzyme that catalyzes the hydrolysis of amino compounds, removing the amino group.

deaminate Removal from a molecule of the amino group.

deanol *See* 2-dimethylaminoethanol.

deashing A form of deionization in which inorganic salts are removed from solution by the adsorption of both the anions and cations by ion-exchange resins.

deasphalting The process of removing asphalt from petroleum fractions.

debenzylation Removal from a molecule of the benzyl group.

deblooming The process by which the fluorescence, or bloom, is removed from petroleum oils by exposing them in shallow tanks to the sun and atmospheric conditions or by using chemicals.

debromoaplysiatoxin A bislactone toxin related to aplysiatoxin and produced by the blue-green alga *Lyngbya majuscula*.

de Brun–van Eckstein rearrangement The isomerization of an aldose or ketose when mixed with aqueous calcium hydroxide to form a mixture of various monosaccharides and unfermented ketoses; used to prepare certain ketoses.

debutanization Removal of butane and lighter components in a natural-gasoline plant.

debutanizer The fractionating column in a natural-gasoline plant in which butane and lighter components are removed.

Debye-Falkenhagen effect The increase in the conductance of an electrolytic solution when the applied voltage has a very high frequency.

Debye force *See* induction force.

Debye-Hückel theory A theory of the behavior of strong electrolytes, according to which each ion is surrounded by an ionic atmosphere of charges of the opposite sign whose behavior retards the movement of ions when a current is passed through the medium.

Debye relaxation time According to the Debye-Hückel theory, the time required for the ionic atmosphere of a charge to reach equilibrium in a current-carrying electrolyte, during which time the motion of the charge is retarded.

decaborane (14) $B_{10}H_{14}$ A binary compound of boron and hydrogen that is relatively stable at room temperature; melting point 99.5°C, boiling point 213°C.

decahydrate A compound that has 10 water molecules.

decahydronaphthalene $C_{10}H_{18}$ A liquid hydrocarbon, used in some paints and lacquers as a solvent.

decalcification Loss or removal of calcium or calcium compounds from a calcified material such as bone or soil.

decanal *See* decyl aldehyde.

decane $C_{10}H_{22}$ Any of several saturated aliphatic hydrocarbons, especially $CH_3(CH_2)_8CH_3$.

decanedioic acid *See* sebacic acid.

***n*-decanoic acid** *See* capric acid.

decanol *See* decyl alcohol.

decarbonize To remove carbon by chemical means.

decarboxycysteine *See* 2-aminoethanethiol.

decarboxylase An enzyme that hydrolyzes the carboxyl radical, COOH.

decarboxylate To remove the carboxyl radical, especially from amino acids and protein.

decatoic acid *See* capric acid.

decatyl alcohol *See* decyl alcohol.

decay *See* chemical weathering; radioactive decay.

decay chain *See* radioactive series.

decay curve A graph showing how the activity of a radioactive sample varies with time; alternatively, it may show the amount of radioactive material remaining at any time.

decay family *See* radioactive series.

decay gammas The characteristic gamma rays emitted during the decay of most radioisotopes.

decay mode A possible type of decay of a radionuclide or elementary particle.

decay rate The time rate of disintegration of radioactive material, generally accompanied by emission of particles or gamma radiation.

decay series *See* radioactive series.

dechlorination Removal of chlorine from a substance.

decicaine *See* tetracaine hydrochloride.

decinormal Pertaining to a chemical solution that is one-tenth normality in reference to a 1 normal solution.

***n*-decoic acid** *See* capric acid.

decoking Removal of petroleum coke from equipment.

decolorize To remove the color from, as from a liquid.

decolorizer An agent used to decolorize; the removal of color may occur by a chemical reaction or a physical reaction.

decolorizing carbon Porous or finely divided carbon (activated or bone) with large surface area; used to adsorb colored impurities from liquids, such as lube oils.

decomposition 1. The more or less permanent structural breakdown of a molecule into simpler molecules or atoms. 2. *See* chemical weathering.

decomposition potential The electrode potential at which the electrolysis current begins to increase appreciably. Also known as decomposition voltage.

decomposition voltage *See* decomposition potential.

decyclic acid *See* capric acid.

decyl An isomeric grouping of univalent radicals, all with formulas $C_{10}H_{21}$, and derived from the decanes by removing one hydrogen.

decyl acetate $CH_3(CH_2)_9OOCCH_3$ Perfumery liquid with a floral orange-rose aroma.

decyl alcohol $C_{10}H_{21}OH$ A colorless oil, boiling at 231°C; used in plasticizers, synthetic lubricants, and detergents. Also known as decanol; decatyl alcohol; nonylcarbinol.

decyl aldehyde $CH_3(CH_2)_8CHO$ A liquid aldehyde, found in essential oils; used in flavorings and perfumes. Also known as capraldehyde; capric aldehyde; decanal.

decylene Any of a group of isomeric hydrocarbons with formula $C_{10}H_{20}$; the group is part of the ethylene series.

decylethylene *See* 1-dodecene.

decyltrichlorosilane n-$C_{10}H_{21}SiCl_3$ An organochlorosilane that boils at 183°C at 84 mm Hg; used in coupling agents or primers to obtain improved bonding between organic polymers and mineral surfaces.

deep inelastic transfer *See* quasifission.

DEET *See* diethyltoluamide.

deethanize To separate and remove ethane and sometimes lighter fractions from heavy substances, such as propane, by distillation.

deethanizer The equipment used to deethanize.

defecation Industrial purification, or clarification, of sugar solutions.

defect cluster A macroscopic cluster of crystal defects which can arise from attraction among defects.

defect motion Movement of a point defect from one lattice point to another.

definite composition law The law that a given chemical compound always contains the same elements in the same fixed proportions by weight. Also known as definite proportions law.

definite proportions law *See* definite composition law.

deflagrating spoon A long-handled spoon used in chemistry to demonstrate deflagration.

deflagration A chemical reaction accompanied by vigorous evolution of heat, flame, sparks, or spattering of burning particles.

deflocculant An agent that causes deflocculation; examples are sodium carbonate and other basic materials used to deflocculate clay slips.

deflocculate To break up and disperse agglomerates and form a stable colloid.

defluorination Removal of fluorine.

defoaming Reduction or elimination of foam.

deformation energy The energy which must be supplied to an initially spherical nucleus to give it a certain deformation in the Bohr-Wheeler theory.

degasser *See* getter.

degradation Conversion of an organic compound to one containing a smaller number of carbon atoms.

degrease 1. To remove grease from wool with chemicals. 2. To remove grease from hides or skins in tanning by tumbling them in solvents.

degree Any one of several units for measuring hardness of water, such as the English or Clark degree, the French degree, and the German degree.

degree of freedom Any one of the variables, including pressure, temperature, composition, and specific volume, which must be specified to define the state of a system.

degree of polymerization The number of structural units in the average polymer molecule in a particular sample. Abbreviated D.P.

degritting Removal of fine solid particles (grit) from a liquid carrier by gravity separation (settling) or centrifugation.

dehydrase An enzyme which catalyzes the removal of water from a substrate.

dehydration Removal of water from any substance.

dehydration tank A tank in which warm air is blown through oil to remove moisture.

dehydrator 1. A substance that removes water from a material; an example is sulfuric acid.   2. Vessel or process system for the removal of liquids from gases or solids by the use of heat, absorbents, or adsorbents.

dehydroacetic acid $C_8H_8O_4$ Crystals that melt at 108.5°C and are insoluble in water, soluble in acetone; used as a fungicide and bactericide. Also known as DHA; methylacetopyranone.

dehydroascorbic acid $C_6H_6O_6$ A relatively inactive acid resulting from elimination of two hydrogen atoms from ascorbic acid when the latter is oxidized by air or other agents; has potential ascorbic acid activity.

dehydrocholesterol $C_{27}H_{43}OH$ A provitamin of animal origin found in the skin of man, in milk, and elsewhere, which upon irradiation with ultraviolet rays becomes vitamin D.

dehydrocholic acid $C_{24}H_{34}O_5$ A white powder melting at 231–240°C, very slightly soluble in water; used as a pharmaceutical intermediate and in medicine.

dehydrocyclization Any process involving both dehydrogenation and cyclization, as in petroleum refining.

dehydroepiandrosterone $C_{19}H_{28}O_2$ Dimorphous crystals with a melting point of 140–141°C, or leaflet crystals with a melting point of 152–153°C; soluble in alcohol, benzene, and ether; used as an androgen. Also known as dehydroisoandrosterone; 3β-hydroxy-5-androsten-17-one.

dehydrogenase An enzyme which removes hydrogen atoms from a substrate and transfers it to an acceptor other than oxygen.

dehydrogenation Removal of hydrogen from a compound.

dehydrohalogenation Removal of hydrogen and a halogen from a compound.

dehydroisoandrosterone *See* dehydroepiandrosterone.

deionization An ion-exchange process in which all charged species or ionizable organic and inorganic salts are removed from solution.

de la Tour method Measurement of critical temperature, involving sealing the sample in a tube and heating it; the temperature at which the meniscus disappears is the critical temperature.

delayed alpha particle An alpha particle emitted by an excited nucleus that was formed an appreciable time after a beta disintegration process.

delayed coking A semicontinuous thermal process for converting heavy petroleum stock to lighter material.

delayed neutron A neutron emitted spontaneously from a nucleus as a consequence of excitation left from a preceding radioactive decay event; in particular, a delayed fission neutron.

delayed neutron fraction   The ratio of the mean number of delayed fission neutrons per fission to the mean total number of neutrons (prompt plus delayed) per fission.

delayed proton   A proton emitted spontaneously from a nucleus as a consequence of excitation left from a previous radioactive decay event.

Delbrück scattering   Elastic scattering of gamma rays by a nucleus caused by virtual electron-positron pair production.

d electron   An atomic electron that has an orbital angular momentum of 2 in the central field approximation.

Delepine reaction   Slow ammonolysis of alkyl halides in acid to primary amines in the presence of hexamethylenetetramine.

deliquescence   The absorption of atmospheric water vapor by a crystalline solid until the crystal eventually dissolves into a saturated solution.

delphidenolon   See myricetin.

delphinidin $C_{15}H_{11}O_7Cl$ A purple or brownish-red anthocyanin compound occurring widely in plants.

delta ray   An electron or proton ejected by recoil when a rapidly moving alpha particle or other primary ionizing particle passes through matter.

deltohedron   A polyhedron which has 12 quadrilateral faces, and is the form of a crystal belonging to the cubic system and having hemihedral symmetry. Also known as deltoid dodecahedron; tetragonal tristetrahedron.

deltoid dodecahedron   See deltohedron.

demal   A unit of concentration, equal to the concentration of a solution in which 1 gram-equivalent of solute is dissolved in 1 cubic decimeter of solvent.

demasking   A process by which a masked substance is made capable of undergoing its usual reactions; can be brought about by a displacement reaction involving addition of, for example, another cation that reacts more strongly with the masking ligand and liberates the masked ion.

demethanator   The apparatus in which demethanization is conducted.

demethanization   The process of distillation in which methane is separated from the heavier components. Also known as demethanation.

demethylation   Removal of the methyl group from a compound.

demeton-S-methyl $C_6H_{15}O_3PS_2$ An oily liquid with a 0.3% solubility in water; used as an insecticide and miticide to control aphids. Also known as $O,O,$-dimethyl-$S[2(ethylthio)ethyl]$phosphorothioate.

demeton-S-methyl sulfoxide $C_6H_{15}\ O_4PS_2$ A clear, amber liquid; limited solubility in water; used as an insecticide and miticide for pests of vegetable, fruit, and field crops, ornamental flowers, shrubs, and trees. Also known as S-$[2$-(ethylsulfinyl)ethyl]O,O-dimethyl phosphorodithioate; oxydemeton methyl.

demineralization   Removal of mineral constituents from water.

Demjanov rearrangement   A structural rearrangement that accompanies treatment of certain primary aliphatic amines with nitrous acid; the amine will undergo a ring contraction or expansion.

DEMO   *See* carboxin.

demulsification   Prevention or breaking of liquid-liquid emulsions by chemical, mechanical or electrical demulsifiers.

demulsifier   A chemical, mechanical, or electrical system that either breaks liquid-liquid emulsions or prevents them from forming.

denaturant   An inert, bad-tasting, or poisonous chemical substance added to a product such as ethyl alcohol to make it unfit for human consumption.

denature   1. To change a protein by heating it or treating it with alkali or acid so that the original properties such as solubility are changed as a result of the protein's molecular structure being changed in some way. 2. To add a denaturant, such as methyl alcohol, to grain alcohol to make the grain alcohol poisonous and unfit for human consumption.

denatured alcohol   Ethyl alcohol containing a poisonous substance, such as methyl alcohol or benzene, which makes it unfit for human consumption.

dendrite   A crystal having a treelike structure.

denitration   Removal of nitrates or nitrogen. Also known as denitrification.

denitrification   *See* denitration.

De Nora cell   Mercury-cathode cell used for production of chlorine and caustic soda by electrolysis of sodium chloride brine.

density gradient centrifugation   Separation of particles according to density by employing a gradient of varying densities; at equilibrium each particle settles in the gradient at a point equal to its density.

deodorizing   A process for removing odor-creating substances from oil or fat, in which the oil or fat is held at high temperatures and low pressure while steam is blown through.

deoil   To reduce the amount of liquid oil entrained in solid wax.

deoxidant   *See* deoxidizer.

deoxidation   1. The condition of a molecule's being deoxidized. 2. The process of deoxidizing.

deoxidize   1. To remove oxygen by any of several processes. 2. To reduce from the state of an oxide.

deoxidizer   Any substance which reduces the amount of oxygen in a substance, especially a metal, or reduces oxide compounds. Also known as deoxidant.

deoxycholate   A salt or ester of deoxycholic acid.

deoxycholic acid $C_{24}H_{40}O_4$ One of the unconjugated bile acids; in bile it is largely conjugated with glycine or taurine.

deoxycorticosterone $C_{21}H_{30}O_3$ A steroid hormone secreted in small amounts by the adrenal cortex.

6-deoxy-L-galactose   *See* L-fucose.

deoxygenation   Removal of oxygen from a substance, such as blood or polluted water.

1-(2-deoxy-β-D-ribofuranosyl)-5-iodouracil   *See* 5-iodo-2′-deoxyuridine.

deoxyribonuclease   An enzyme that catalyzes the hydrolysis of deoxyribonucleic acid to nucleotides.

deoxyribonucleic acid   A linear polymer made up of deoxyribonucleotide repeating units (composed of the sugar 2-deoxyribose, phosphate, and a purine or pyrimidine base) linked by the phosphate group joining the 3′ position of one sugar to the 5′ position of the next; most molecules are double-stranded and antiparallel, resulting in a right-handed helix structure kept together by hydrogen bonds between a purine on one chain and a pyrimidine on another; carrier of genetic information, which is encoded in the sequence of bases; present in chromosomes and chromosomal material of cell organelles such as mitochondria and chloroplasts, and also present in some viruses. Abbreviated DNA.

deoxyribonucleic acid–directed RNA polymerase   An enzyme which transcribes a ribonucleic acid molecule complementary to DNA; required for initiation of DNA replication as well as transcription of RNA.

deoxyribonucleic acid ligase   An enzyme which joins the ends of two DNA chains by catalyzing the synthesis of a phosphodiester bond between a 3′-hydroxyl group at the end of one chain and a 5′-phosphate at the end of the other.

deoxyribonucleic acid polymerase I   An enzyme which catalyzes the addition of deoxyribonucleotide residues to the end of a DNA strand; generally considered to function in the repair of damaged DNA.

deoxyribonucleic acid polymerase II   An enzyme similar in action to DNA polymerase I but with lower activity.

deoxyribonucleic acid polymerase III   An enzyme thought to be the primary enzyme involved in DNA replication.

deoxyribonucleoprotein   A protein containing molecules of deoxyribonucleic acid in close association with protein molecules.

deoxyribose $C_5H_{10}O_4$ A pentose sugar in which the hydrogen replaces the hydroxyl groups of ribose; a major constituent of deoxyribonucleic acid.

deoxy sugar   A substance which has the characteristics of a sugar, but which shows a deviation from the required hydrogen-to-oxygen ratio.

DEPC   See diethyl pyrocarbonate.

depentanizer   A fractionating column for removal of pentane and lighter fractions from a hydrocarbon mixture.

dephlegmation   In a distillation operation, the partial condensation of vapor to form a liquid richer in higher boiling constituents than the original vapor.

dephlegmator   An apparatus used in fractional distillation to cool the vapor mixture, thereby condensing higher-boiling fractions.

depolarizer   A substance added to the electrolyte of a primary cell to prevent excessive buildup of hydrogen bubbles by combining chemically with the hydrogen gas as it forms. Also known as battery depolarizer.

depolymerization   Decomposition of macromolecular compounds into relatively simple compounds.

deposition potential   The smallest potential which can produce electrolytic deposition when applied to an electrolytic cell.

depressor An agent that prevents or retards a chemical reaction or process.

depropanize In petroleum processing, the removal of propane and sometimes lighter fractions from the petroleum.

depropanizer A fractionating column in a gasoline plant for removal of propane and lighter components.

deproteinize To remove protein from a substance.

depside One of a class of esters that form from the joining of two or more molecules of phenolic carboxylic acid.

depsidone One of a class of compounds that consists of esters such as depsides, but are also cyclic ethers.

depth-type filtration Removal of solids by passing the carrier fluid through a mass-filter medium that provides a tortuous path with many entrapments to catch the solids.

depurination Detachment of guanine from sugar in a deoxyribonucleic acid molecule.

derichment In gravimetric analysis by coprecipitation of salts, a system with λ less than unity, when λ is the logarithmic distribution coefficient expressed by the ratio of the logarithms of the ratios of the initial and final solution concentrations of the two salts.

derivative A substance that is made from another substance.

derivative differential thermal analysis A method for precise determination in thermograms of slight temperature changes by taking the first derivative of the differential thermal analysis curve (thermogram) which plots time versus differential temperature as measured by a differential thermocouple. Also known as DDTA.

derivative polarography Polarography technique in which the rate of change of current with respect to applied potential is measured as a function of the applied potential (di/dE vs. E, where i is current and E is applied potential).

derivative thermometric titration The use of a special resistance-capacitance network to record first and second derivatives of a thermometric titration curve (temperature versus weight change upon heating) to produce a sharp endpoint peak.

derosination Removing excess resins from wood by saponification with alkaline aqueous solutions or organic solvents.

desalination Removal of salt, as from water or soil. Also known as desalting.

desalting 1. The process of extracting inorganic salts from oil. 2. *See* desalination.

descending chromatography A type of paper chromatography in which the sample-carrying solvent mixture is fed to the top of the developing chamber, being separated as it works downward.

desiccant *See* drying agent.

desiccator A closed vessel, usually made of glass and having an airtight lid, used for drying solid chemicals by means of a desiccant.

desilication Removal of silica, as from rock or a magma.

desmetryn $C_9H_{17}N_5S$ A white, crystalline compound with a melting point of 84–86°C; used as a postemergence herbicide for broadleaf and grassy weeds. Also known as 2-methylmercapto-4-methylamino-6-isopropylamino-S-triazine.

desmin The muscle protein forming the Z lines in striated muscle.

desmolase Any of a group of enzymes which catalyze rupture of atomic linkages that are not cleaved through hydrolysis, such as the bonds in the carbon chain of D-glucose.

desorption The process of removing a sorbed substance by the reverse of adsorption or absorption.

destearinate A process of removing from a fatty oil the lower melting point compounds.

destraction A high-pressure technique for separating high-boiling or nonvolatile material by dissolving it with application of supercritical gases.

destructive distillation Decomposition of organic compounds by heat without the presence of air.

desulfonate Removal from a sulfonated molecule of the sulfonate group.

desulfurization The removal of sulfur, as from molten metals or petroleum oil.

desulfurization unit A unit in petroleum refining for removal of sulfur compounds or sulfur.

desyl The radical $C_6H_5COCH(C_6H_5)$—; may be formed from desoxybenzoin. Also known as α-phenyl phenacyl.

DET *See* diethyltoluamide.

detergent alkylate *See* dodecylbenzene.

determination The finding of the value of a chemical or physical property of a compound, such as reaction-rate determination or specific-gravity determination.

detonation An exothermic chemical reaction that propagates with such rapidity that the rate of advance of the reaction zone into the unreacted material exceeds the velocity of sound in the unreacted material; that is, the advancing reaction zone is preceded by a shock wave.

detoxification The act or process of removing a poison or the toxic properties of a substance in the body.

deuteration The addition of deuterium to a chemical compound.

deuteride A hydride in which the hydrogen is deuterium.

deuterium The isotope of the element hydrogen with one neutron and one proton in the nucleus; atomic weight 2.0144. Designated D, d, H^2, or ^2H.

deuterium cycle *See* proton-proton chain.

deuterium oxide *See* heavy water.

deuteron The nucleus of a deuterium atom, consisting of a neutron and a proton. Designated d. Also known as deuton.

deuteron capture The absorption of a deuteron by a nucleus, giving rise to a compound nucleus which subsequently decays.

deuton *See* deuteron.

developed dye A direct azo dye that can be further diazotized by a developer after application to the fiber; it couples with the fiber to form colorfast shades. Also known as diazo dye.

developer An organic compound which interacts on a textile fiber to develop a dye.

devitrification The process by which the glassy texture of a material is converted into a crystalline texture.

De Vries effect A relatively short-term oscillation, on the order of 100 years, in the radiocarbon content of the atmosphere, and the resulting variation in the apparent radiocarbon age of samples.

devrinol $C_{17}H_{21}O_2N$ A brown solid with a melting point of 68.5–70.5°C; slight solubility in water; used as an herbicide for crops. Also known as 2-(α-naphthoxy)-N,N-diethylpropionamide.

Dewar structure A structural formula for benzene that contains a bond between opposite atoms.

dewaxing Removing wax from a material or object; a process used to separate solid hydrocarbons from petroleum.

dew point The temperature at which water vapor begins to condense.

dew-point boundary On a phase diagram for a gas-condensate reservoir (pressure versus temperature with constant gas-oil ratios), the area along which the gas-oil ratio approaches zero.

dew-point composition The water vapor–air composition at saturation, that is, at the temperature at which water exerts a vapor pressure equal to the partial pressure of water vapor in the air-water mixture.

dew-point curve On a PVT phase diagram, the line that separates the two-phase (gas-liquid) region from the one-phase (gas) region, and indicates the point at a given gas temperature or pressure at which the first dew or liquid phase occurs.

dew-point depression Reduction of the liquid-vapor dew point of a gas by removal of a portion of the liquid (such as water) from the gas (such as air).

dew-point hygrometer An instrument for determining the dew point by measuring the temperature at which vapor being cooled in a silver vessel begins to condense. Also known as cold-spot hygrometer.

dew-point pressure The gas pressure at which a system is at its dew point, that is, the conditions of gas temperature and pressure at which the first dew or liquid phase occurs.

dextran Any of the several polysaccharides, $(C_5H_{10}O_5)_n$, that yield glucose units on hydrolysis.

dextrin A polymer of D-glucose which is intermediate in complexity between starch and maltose.

dextrinization Any process that involves dextrinizing.

dextrinize To convert a starch into dextrins.

dextropimeric acid $C_{19}H_{29}COOH$ A compound found in particular in oleoresins of pine trees.

dextrose $C_6H_{12}O_6 \cdot H_2O$ A dextrorotatory monosaccharide obtained as a white, crystalline, odorless, sweet powder, which is soluble in about one part of water; an important intermediate in carbohydrate metabolism; used for nutritional purposes, for temporary increase of blood volume, and as a diuretic. Also known as corn sugar; grape sugar.

dezincification Removal of zinc.

DFP *See* isofluorphate.

DHA *See* dehydroacetic acid; dihydroxyacetone.

Di *See* didymium.

diacetate An ester or salt that contains two acetate groups.

diacetic acid *See* acetoacetic acid.

diacetic ether *See* ethyl acetoacetate.

diacetin $C_3H_5(OH)(CH_3COO)_2$ A colorless, hygroscopic liquid that is soluble in water, alcohol, ether, and benzene; boiling point 259°C; used as a plasticizer and softening agent and as a solvent. Also known as glyceryl diacetate.

diacetone alcohol $CH_3COCH_2C(CH_3)_2OH$ A colorless liquid used as a solvent for nitrocellulose and resins.

diacetyl **1.** $CH_3COCOCH_3$ A yellowish-green liquid with a boiling point of 88°C; has a strong odor that resembles quinone; occurs naturally in bay oil and butter and is produced from methyl ethyl ketone or by a special fermentation of glucose; used as an aroma carrier in food manufacturing. Also known as biacetyl; 2-3-butanedione. **2.** A prefix indicating two acetyl groups.

diacetylmethane *See* acetylacetone.

diacetyl peroxide *See* acetyl peroxide.

diacetylurea $C_5H_8O_3N_2$ An acyl derivative of urea containing two acetyl groups.

diacid An acid that has two acidic hydrogen atoms; an example is oxalic acid.

diadochy Replacement or ability to be replaced of one atom or ion by another in a crystal lattice.

dialdehyde A molecule that has two aldehyde groups, such as dialdehyde starch.

dialifor $C_{14}H_{17}ClNO_4S_2P$ A white, crystalline compound with a melting point of 67–69°C; insoluble in water; used to control pests in citrus fruits, grapes, and pecans. Also known as *S*-(2-chloro-1-phthalimidoethyl) *O,O*-diethyl phosphorodithioate.

dialkyl A molecule that has two alkyl groups.

dialkyl amine An amine that has two alkyl groups bonded to the amino nitrogen.

4-diallylamino-3,5-xylyl-*N*-methylcarbamate *See* allyxycarb.

***N,N*-diallyl-2-chloroacetamide** *See* allidochlor.

diallyl phthalate $C_6H_4(COOCH_2CH:CH_2)_2$ A colorless, oily liquid with a boiling range of 158–165°C; used as a plasticizer and for polymerization. Abbreviated DAP.

diallyl sulfide *See* allyl sulfide.

dial pulsing *See* loop pulsing.

dialuric acid $C_4H_4N_2O$ An acid that is derived by oxidation of uric acid or by the reduction of alloxan; may be used in organic synthesis. Also known as 5-hydroxy-barbituric acid.

dialysis A process of selective diffusion through a membrane; usually used to separate low-molecular-weight solutes which diffuse through the membrane from the colloidal and high-molecular-weight solutes which do not.

dialyzate The material that does not diffuse through the membrane during dialysis; alternatively, it may be considered the material that has diffused.

dialyzer 1. The semipermeable membrane used for dialyzing liquid. 2. The container used in dialysis; it is separated into compartments by membranes.

diamide A molecule that has two amide (—$CONH_2$) groups.

diamidine A molecule that has two amidine (—C=$NHNH_2$) groups.

diamine Any compound containing two amino groups.

diamine oxidase A flavoprotein which catalyzes the aerobic oxidation of amines to the corresponding aldehyde and ammonia.

diamino A term used in chemical nomenclature to indicate the presence in a molecule of two amino (—NH_2) groups.

6-diaminoacridinium hydrogen sulfate *See* proflavine sulfate.

diaminobenzene *See* phenylenediamine.

3,5-diaminobenzoic acid $C_7H_8N_2O_2$ Monohydrate crystals with a melting point of 228°C; soluble in organic solvents such as alcohol and benzene; used in the detection and determination of nitrites.

diaminoditolyl *See ortho*-tolidine.

diaminoethane *See* ethylenediamine.

2,7-diaminofluorene $C_{13}H_{12}N_2$ A compound crystallizing as needlelike crystals from water; melting point is 165°C; soluble in alcohol; used to detect bromide, chloride, nitrate, persulfate, cadmium, zinc, copper, and cobalt. Also known as 2,7-fluorenediamine.

1,6-diaminohexane *See* hexamethylenediamine.

2,4-diaminophenol hydrochloride *See* amidol.

diaminopimelate $C_7H_{14}O_4N_2$ A compound that serves as a component of cell wall mucopeptide in some bacteria and as a source of lysine in all bacteria. Abbreviated DAP. Also known as diaminopimelic acid.

diaminopimelic acid *See* diaminopimelate.

diamond structure A crystal structure in which each atom is the center of a tetrahedron formed by its nearest neighbors.

diamyl phenol $(C_5H_{11})_2C_6H_3OH$ A straw-colored liquid with a boiling range of 280–295°C; used in synthetic resins, lubricating oil additives, plasticizers, detergents, and fungicides.

diamyl sulfide $(C_5H_{11})_2S$ A combustible, yellow liquid with a distillation range of 170–180°C; used as a flotation agent and an odorant.

diaphorase Mitochondrial flavoprotein enzymes which catalyze the reduction of dyes, such as methylene blue, by reduced pyridine nucleotides such as reduced diphosphopyridine nucleotide.

diaphragm cell An electrolytic cell used to produce sodium hydroxide and chlorine from sodium chloride brine; porous diaphragm separates the anode and cathode compartments.

diarsine An arsenic compound containing an As-As bond with the general formula $(R_2As)_2$, where R represents a functional group such as CH_3.

diarylamine A molecule that contains an amine group and two aryl groups joined to the amino nitrogen.

diastase An enzyme that catalyzes the hydrolysis of starch to maltose. Also known as vegetable diastase.

diastereoisomer One of a pair of optical isomers which are not mirror images of each other. Also known as diastereomer.

diastereotopic ligand A ligand whose replacement or addition gives rise to diastereomers.

diatomic Consisting of two atoms.

diazine 1. A hydrocarbon consisting of an unsaturated hexatomic ring of two nitrogen atoms and four carbons. 2. Suffix indicating a ring compound with two nitrogen atoms.

diazinon $C_{12}H_{21}N_2O_3PS$ A light amber to dark brown liquid with a boiling point of 83–84°C; used as an insecticide for soil and household pests, and as an insecticide and nematicide for fruits and vegetables. Also known as O,O-diethyl O-(2-isopropyl-4-methyl-6-primidinyl)phosphorothioate.

diazoalkane A compound with the general formula $R_2C{=}N_2$ in which two hydrogen atoms of an alkane molecule have been replaced by a diazo group.

diazoamine The grouping —N=NNH—. Also known as azimino.

diazoaminobenzene $C_6H_5NNNHC_6H_5$ Golden yellow scales with a melting point of 96°C; soluble in alcohol, ether, and benzene; used for dyes and insecticides. Also known as benzeneazoanilide; diazobenzeneanilide; 1,3-diphenyltriazine.

diazoate A salt with molecular formula of the type $C_6H_5N{=}NOOM$, where M is a nonvalent metal.

diazobenzeneanilide *See* diazoaminobenzene.

diazo compound An organic compound containing the radical —N=N—.

diazo dye *See* developed dye.

diazo group A functional group with the formula $=N_2$.

diazoic acid $C_6H_5N{=}NOOH$ An isomeric form of phenylnitramine.

diazoimide *See* hydrazoic acid.

1,3-diazole *See* imidazole.

diazole A cyclic hydrocarbon with five atoms in the ring, two of which are nitrogen atoms and three are carbon.

diazomethane CH_2N_2 A poisonous gas used in organic synthesis to methylate compounds. Also known as azimethane; azimethylene.

diazonium The grouping $=N{\equiv}N$.

diazonium salts Compounds of the type $R{\cdot}X{\cdot}N{:}N$, with X being an acid radical such as chlorine.

diazo oxide   An organic molecule or a grouping of organic molecules that have a diazo group and an oxygen atom joined to ortho positions of an aromatic nucleus. Also known as diazophenol.

diazophenol   *See* diazo oxide.

diazo process   *See* diazotization.

diazosulfonate   A salt formed from diazosulfonic acid.

diazosulfonic acid $C_6H_5N\!\!=\!\!NSO_3H$ Any of a group of aromatic acids containing the diazo group bonded to the sulfonic acid group.

diazotization   Reaction between a primary aromatic amine and nitrous acid to give a diazo compound. Also known as diazo process.

dibasic   **1.** Compounds containing two hydrogens that may be replaced by a monovalent metal or radical. **2.** An alcohol that has two hydroxyl groups, for example, ethylene glycol.

dibasic acid   An acid having two hydrogen atoms capable of replacement by two basic atoms or radicals.

dibasic calcium phosphate   *See* calcium phosphate.

dibasic magnesium citrate $MgHC_6H_5O_7 \cdot 5H_2O$ A white or yellowish powder soluble in water; used as a dietary supplement or in medicine. Also known as acid magnesium citrate.

dibenzofuran   *See* diphenylene oxide.

dibenzopyrone   *See* xanthone.

dibenzothiazine   *See* phenothiazine.

dibenzoyl   *See* benzil.

dibenzoyl peroxide   *See* benzoyl peroxide.

dibenzyl   *See* bibenzyl.

dibenzyl disulfide $C_6H_5CH_2SSCH_2C_6H_5$ A compound crystallizing in leaflets with a melting point of 71–72°C; soluble in hot methanol, benzene, ether, and hot ethanol; used as an antioxidant in compounding of rubber and as an additive to silicone oils. Also known as benzyl disulfide; α-(benzyldithio)toluene; di(phenylmethyl)disulfide.

dibenzyl ether   *See* benzyl ether.

diborane B_2H_6 A colorless, volatile compound that is soluble in ether; boiling point −92.5°C, melting point −165.5°C; can be used to produce pentaborane and decaborane, proposed for use as rocket fuels; also used to synthesize organic boron compounds. Also known as boroethane; diboron hexahydride.

diborate   *See* borax.

diboron hexahydride   *See* diborane.

dibromide   Indicating the presence of two bromine atoms in a molecule.

dibromo-   A prefix indicating two bromine atoms.

2,4′-dibromoacetophenone   *See* para-bromophenacyl bromide.

dibromochloropropane $C_3H_5Br_2Cl$ A light yellow liquid with a boiling point of 195°C; used as a nematicide for crops. Abbreviated DBCP. Also known as 1,2-dibromo-3-chloropropane.

1,2-dibromo-3-chloropropane   *See* dibromochloropropane.

2,6-dibromo-N-chloroquinonimine   *See* 2,6-dibromoquinone-4-chlorimide.

1,2-dibromo-2,2-dichloroethyl dimethyl phosphate   *See* naled.

dibromodifluoromethane CF_2Br_2 A colorless, heavy liquid with a boiling point of 24.5°C; soluble in methanol and ether; used in the synthesis of dyes and pharmaceuticals and as a fire-extinguishing agent.

1,2-dibromoethane   *See* ethylene dibromide.

3,5-dibromo-4-hydroxybenzaldehyde O-(2′,4′-dinitrophenyl)oxime   *See* bromofenoxim.

3,5-dibromo-4-hydroxybenzonitrile   *See* bromoxynil.

5,7-dibromo-8-hydroxyquinoline   *See* broxyquinoline.

dibromomethane   *See* methylene bromide.

3,5-dibromo-4-octanoyloxybenzonitrile   *See* bromoxynil.

2,6-dibromoquinone-4-chlorimide $C_6H_2Br_2ClNO$ Yellow prisms, soluble in water; used as a reagent for phenol and phosphatases. Also known as 2,6-dibromo-N-chloroquinonimine.

3,5-dibromosalicylaldehyde $Br_2C_6H_2(OH)$ CHO   Pale yellow crystals with a melting point of 86°C; readily soluble in ether, chloroform, benzene, alcohol, and glacial acetic acid; used as an antibacterial agent.

dibromothymolsulfonphthalein   *See* bromthymol blue.

para-α-dibromotoluene   *See* para-bromobenzyl bromide.

dibucaine $C_{20}H_{29}O_2N_3$ A local anesthetic used both as the base and the hydrochloride salt.

dibutalin   *See* N-sec-butyl-4-tert-butyl-2,6-dinitroaniline.

dibutyl   Indicating the presence of two butyl groupings bonded through a third atom or group in a molecule.

dibutyl amine $C_8H_{19}N$ A colorless, clear liquid with amine aroma; either di-n-butylamine, $(C_4H_9)_2NH$, boiling at 160°C, insoluble in water, soluble in hydrocarbon solvents, or di-sec-butylamine, $(CH_3CHCH_2CH_3)_2NH$, boiling at 133°C, flammable; used in the manufacture of dyes.

dibutyl maleate $C_4H_9OOCCHCHCOOC_4H_9$ Oily liquid used for copolymers and plasticizers and as a chemical intermediate.

dibutyl oxalate $(COOC_4H_9)_2$ High-boiling, water-white liquid with mild odor, used as a solvent and in organic synthesis.

3,4-di-(tert-butyl)-phenyl-N-methylcarbamate $C_{16}H_{25}O_2N$ A white, crystalline compound that melts at 102–103°C; used as an insecticide for sheep blowflies.

dibutyl phthalate $C_{16}H_{22}O_4$ A colorless liquid, used as a plasticizer and insect repellent.

dibutyl succinate $C_{12}H_{22}O_4$ A colorless liquid, insoluble in water; used as a repellent for cattle flies, cockroaches, and ants around barns.

dibutyl tartrate $(COOC_4H_9)_2(CHOH)_2$ Liquid used as a solvent and plasticizer for cellulosics and as a lubricant.

2,6-di-*tert*-butyl-*para*-tolylmethylcarbamate $C_{17}H_{27}O_2N$ A colorless solid with a melting point of 200–201°C; solubility in water is 6–7 parts per million; used as a preemergence herbicide on established turf.

dicalcium A molecule containing two atoms of calcium.

dicalcium orthophosphate *See* calcium phosphate.

dicalcium phosphate *See* calcium phosphate.

dicarbamylamine *See* biuret.

dicarbocyanine **1.** A member of a group of dyes termed the cyanine dyes; structure consists of two heterocyclic rings joined to the five-carbon chain: =CH—CH=CH—CH=CH—. **2.** A particular dicarbocyanine dye containing two quinoline heterocyclic rings. Also known as pentamethine.

dicarboxylic acid A compound with two carboxyl groups.

dication A doubly charged cation with the general formula X^{2+}.

dichlobenil $C_7H_3Cl_2N$ A colorless, crystalline compound with a melting point of 139–145°C; used as an herbicide to control weeds in orchards and nurseries. Also known as 2,6-dichlorobenzonitrile.

dichlofenthion $C_{10}H_{13}Cl_2O_3PS$ A white, liquid compound, insoluble in water; used as an insecticide and nematicide for ornamentals, flowers, and lawns. Also known as *O,O*-diethyl-*O*-2,4-dichlorophenyl phosphorothioate.

dichlofluanid $C_9H_{11}Cl_2FN_2O_2S$ A white powder with a melting point of 105–105.6°C; insoluble in water; used as a fungicide for fruits, garden crops, and ornamental flowers. Also known as *N'*-dichlorofluoromethylthio-*N,N*-dimethyl-*N'*-phenylsulfamide.

dichlone $C_{10}H_4O_2Cl_2$ A yellow, crystalline compound, used as a fungicide for foliage and as an algicide. Also known as 2,3-dichloro-1,4-naphthoquinone.

dichloramine **1.** NH_2Cl_2 An unstable molecule considered to be formed from ammonia by action of chlorine. Also known as chlorimide. **2.** Any chloramine with two chlorine atoms joined to the nitrogen atom.

dichloride Any inorganic salt or organic compound that has two chloride atoms in its molecule.

dichloroacetic acid $CHCl_2COOH$ A strong liquid acid, formed by chlorinating acetic acid; used in organic synthesis.

3,6-dichloro-*ortho*-anisic acid C_6H_2 Cl_2OCH_3COOH A light tan, granular solid with a melting point of 114–116°C; used as an herbicide on roadways, crops, and rangelands.

dichlorobenzene $C_6H_4Cl_2$ Any of a group of substitution products of benzene and two atoms of chlorine; the three forms are *meta*-dichlorobenzene, colorless liquid boiling at 172°C, soluble in alcohol and ether, insoluble in water, or *ortho*-, colorless liquid boiling at 179°C, used as a solvent and chemical intermediate, or *para*-, volatile white crystals, insoluble in water, soluble in organic solvents, used as a germicide, insecticide, and chemical intermediate.

2,6-dichlorobenzonitrile *See* dichlobenil.

1,4-dichlorobutane $Cl(CH_2)_4Cl$ A colorless, flammable liquid with a pleasant odor; boiling point 155°C; soluble in organic solvents; used in organic synthesis, including adiponitrile. Abbreviated DCB. Also known as tetramethylene dichloride.

4,6-dichloro-*N*-(2-chlorophenyl)-1,3,5-triazin-2-amine   *See* anilazine.

2,2′-dichlorodiethyl ether   *See* dichloroethyl ether.

dichlorodiethylsulfide   *See* mustard gas.

dichlorodifluoromethane CCl_2F_2 A nontoxic, nonflammable, colorless gas made from carbon tetrachloride; boiling point $-30°C$; used as a refrigerant and as a propellant in aerosols.

1,4-dichloro-2,5-dimethoxybenzene   *See* chloroneb.

***p,p′*-dichlorodiphenylmethyl carbinol** $(ClC_6H_4)_2C(CH_3)OH$ Crystals with a melting point of 69–69.5°C; soluble in organic solvents; used as an insecticide. Abbreviated DMC. Also known as 1,1-bis(*para*-chlorophenyl)methyl carbinol.

dichlorodiphenyltrichloroethane   *See* DDT.

dichloroether   *See* dichloroethyl ether.

***sym*-dichloroethylene**   CHClCHCl   Colorless, toxic liquid with pleasant aroma, boiling at 59°C; decomposes in light, air, and moisture; soluble in organic solvents, insoluble in water; exists in cis and trans forms; used as solvent, in medicine, and for chemical synthesis. Also known as acetylene dichloride.

dichloroethyl ether $ClCH_2CH_2OCH_2\ CH_2Cl$ A colorless liquid insoluble in water, soluble in organic solvents; used as a solvent in paints, varnishes, lacquers, and as a soil fumigant. Also known as 2,2′-dichlorodiethyl ether; dichloroether; dichloroethyl oxide.

dichloroethyl oxide   *See* dichloroethyl ether.

dichlorofluoromethane $CHCl_2F$ A colorless, heavy gas with a boiling point of 8.9°C and a freezing point of $-135°C$; soluble in alcohol and ether; used in fire extinguishers and as a solvent, refrigerant, and aerosol propellant. Also known as fluorocarbon 21; fluorodichloromethane.

2-dichlorohydrin   *See* 1,3-dichloro-2-propanol.

α-dichlorohydrin $CH_2ClCHOHCH_2Cl$ Unstable liquid, the commercial product consisting of a mixture of two isomers; used as a solvent and a chemical intermediate. Also known as GDCH; glycerol dichlorohydrin; α-propenyl-dichlorohydrin.

di-(5-chloro-2-hydroxyphenyl)methane   *See* dichlorophen.

5,7-dichloro-8-hydroxyquinoline   *See* chloroxine.

dichloromethane   *See* methylene chloride.

2,3-dichloro-1,4-naphthoquinone   *See* dichlone.

2,6-dichloro-4-nitroaniline $C_6H_4Cl_2N_2O_2$ A yellow, crystalline compound that melts at 192–194°C; used as a fungicide for fruits, vegetables, and ornamental flowers. Abbreviated DCNA.

dichloropentane $C_5H_{10}Cl_2$ Mixed dichloro derivatives of normal pentane and isopentane; clear, light-yellow liquid used as solvent, paint and varnish remover, insecticide, and soil fumigant.

dichlorophen $C_{13}H_{10}Cl_2O_2$ A white, crystalline compound with a melting point of 177–178°C; used as an agricultural fungicide, germicide in soaps, and antihelminthic drug in humans. Also known as di-(5-chloro-2-hydroxyphenyl)methane.

2,4-dichlorophenoxyacetic acid Cl_2C_6 H_3OCH_2COOH Yellow crystals, melting at 142°C; used as an herbicide and pesticide. Also known as 2,4-D.

2-(2,4-dichlorophenoxy)propionic acid *See* dichlorprop.

3-(3,4-dichlorophenyl)-1,1-dimethylurea *See* diuron.

3-(3,4-dichlorophenyl)-1-methoxy-1-methylurea *See* linuron.

2-(3,4-dichlorophenyl)-4-methyl-1,2,4-oxadiazolidine-3,5-dione *See* methazole.

2,4-dichlorophenyl *para*-nitrophenyl ether $C_{12}H_7Cl_2NO_3$ A dark brown, crystalline compound with a melting point of 70–71°C; slightly soluble in water; used as a pre- and postemergence herbicide for vegetable crops and paddy rice.

1,2-dichloropropane *See* propylene dichloride.

1,3-dichloro-2-propanol $ClCH_2CHOH$ CH_2Cl A liquid soluble in water and miscible with alcohol and ether; used as a solvent for nitrocellulose and hard resins, as a binder for watercolors, in the production of photographic lacquer, and in the determination of vitamin A. Also known as α-dichlorohydrin; *sym*-dichloroisopropyl alcohol; *sym*-glycerol dichlorohydrin.

3,4-dichloropropionanilide *See* propanil.

2,6-dichlorothiobenzamide *See* chlorthiamid.

dichlorotoluene $C_7H_6Cl_2$ A colorless liquid, soluble in organic solvents, insoluble in water; isomers are 2,4-$CH_3C_6H_3Cl_2$, boiling at 200–202°C, and 3,4-($CH_3C_6H_3Cl_2$), boiling at 209°C; used as solvent and chemical intermediate.

2,2-dichlorovinyldimethyl phosphate *See* dichlorvos.

dichlorprop $C_9H_8Cl_2O_3$ A colorless, crystalline solid with a melting point of 117–118°C; used as an herbicide and fumigant for brush control on rangeland and rights-of-way. Also known as 2-(2,4-dichlorophenoxy)-propionic acid; 2,4-DP.

dichlorvos $C_4H_7O_4Cl_2P$ An amber liquid, used as an insecticide and miticide on public health pests, stored products, and flies on cattle. Also known as DDVP; 2,2-dichlorovinyldimethyl phosphate.

dichromate A salt of dichromic acid, usually orange or red.

dichromatic dye Dye or indicator in which different colors are seen, depending upon the thickness of the solution.

dichromic Pertaining to a molecule with two atoms of chromium.

dichromic acid $H_2Cr_2O_7$ An acid known only in solution, especially in the form of dichromates.

dicofol $C_{14}H_9Cl_5O$ A white solid with a melting point of 77–78°C; used as a miticide on crops. Also known as 1,1-bis(*para*-chlorophenyl)-2,2,2-trichloroethanol.

dicovalent carbon *See* divalent carbon.

dicrotophos $C_8H_{16}O_2P$ The dimethyl phosphate of 3-hydroxy-N,N-dimethyl-*cis*-crotonamide; a brown liquid with a boiling point of 400°C; miscible with water; used as an insecticide and miticide for cotton, soybeans, seeds, and ornamental flowers.

dicyandiamide $NH_2C(NH)(NHCN)$ White crystals with a melting range of 207–209°C; soluble in water and alcohol; used in fertilizers, explosives, oil well drilling muds, pharmaceuticals, and dyestuffs. Also known as cyanoguanidine.

dicyanide   A salt that has two cyanide groups.

dicyanoargentates I   *See* argentocyanides.

dicyclohexylamine $(C_6H_{11})_2NH$ A clear, colorless liquid with a boiling point of 256°C; used for insecticides, corrosion inhibitors, antioxidants, and detergents, and as a plasticizer and catalyst.

dicyclohexylcarbodiimide $C_{13}H_{22}N_2$ Crystals with a melting point of 35–36°C; used in peptide synthesis. Abbreviated DCC; DCCI. Also known as carbodicyclohexylimide.

dicyclopentadienyl iron   *See* ferrocene.

dicyclopentadienyl nickel   *See* nickelocene.

DIDA   *See* diisodecyl adipate.

didodecyl ether   *See* dilauryl ether.

DIDP   *See* diisodecyl phthalate.

didymium   A mixture of the rare-earth elements praeseodymium and neodymium. Abbreviated Di.

Dieckman condensation   Any condensation of esters of dicarboxylic acids which produce cyclic β-ketoesters.

dieldrin $C_{12}H_8Cl_6O$ A white, crystalline contact insecticide obtained by oxidation of aldrin; used in mothproofing carpets and other furnishings.

dielectric vapor detector   Apparatus to measure the change in the dielectric constant of gases or gas mixtures; used as a detector in gas chromatographs to sense changes in carrier gas.

dielectronic recombination   The combination of an electron with a positive-ion in a gas, so that the energy released is taken up by two electrons of the resulting atom.

dielectrophoresis   The ability of an uncharged material to move when subjected to an electric field.

Diels-Alder reaction   The 1,4 addition of a conjugated diolefin to a compound, known as a dienophile, containing a double or triple bond; the dienophile may be activated by conjugation with a second double bond or with an electron acceptor.

diene   One of a class of organic compounds containing two ethylenic linkages (carbon-to-carbon double bonds) in the molecules. Also known as diolefin.

diene resin   Material containing the diene group of double bonds that may polymerize.

diene value A number that represents the amount of conjugated bonds in a fatty acid or fat.

dienophile The alkene component of a reaction between an alkene and a diene.

diesel index An empirical expression for the correlation between the aniline number of a diesel fuel and its ignitability.

diester A compound containing two ester groupings.

diesterase An enzyme such as a nuclease which splits the linkages binding individual nucleotides of a nucleic acid.

diethanolamine $(HOCH_2CH_2)_2NH$ Colorless, water-soluble, deliquescent crystals, or liquid boiling at 217°C; soluble in alcohol and acetone, insoluble in ether and benzene; used in detergents, as an absorbent of acid gases, and as a chemical intermediate. Also known as DEA.

diether A molecule that has two oxygen atoms with ether bonds.

1,1-diethoxyethane *See* acetal.

diethyl Pertaining to a molecule with two ethyl groups.

diethyl adipate $C_2H_5OCO(CH_2)_4OCOC_2H_5$ Water-insoluble, colorless liquid, boiling at 245°C; used as a plasticizer.

diethylamine $(C_2H_5)_2NH$ Water-soluble, colorless liquid with ammonia aroma, boiling at 56°C; used in rubber chemicals and pharmaceuticals and as a solvent and flotation agent.

diethylaminoethyl cellulose A positively charged resin used in ion-exchange chromatography; an anion exchanger. Also known as DEAE-cellulose.

2-diethylamino-6-methylpyrimidin-4-yl diethyl phosphorothioate *See* pirimiphosethyl.

2-diethylamino-6-methylpyrimidin-4-yl dimethyl phosphorothionate $C_{11}H_{20}O_3N_3PS$ A straw-colored liquid which decomposes below 100°C; solubility in water is 5 parts per million at 30°C; used as a pesticide. Also known as pirimiphosmethyl.

diethylbarbituric acid *See* barbital.

5,5-diethylbarbituric acid *See* barbital.

diethylbenzene $C_6H_4(C_2H_5)_2$ Colorless liquid, boiling at 180–185°C; soluble in organic solvents, insoluble in water; usually a mixture of three isomers, which are 1,2- (or *ortho*-diethylbenzene), boiling at 183°C, and 1,3- (or *meta*-), boiling at 181°C, and 1,4- (or *para*-), boiling at 184°C; used as a solvent.

diethylcarbamazine $C_{16}H_{29}O_8N_3$ White, water-soluble, hygroscopic crystals, melting at 136°C; used as an anthelminthic.

diethyl carbinol $(CH_3CH_2)_2CHOH$ Colorless, alcohol-soluble liquid, boiling at 116°C; slightly soluble in water; used in pharmaceuticals and as a solvent and flotation agent. Also known as sec-*n*-amyl alcohol.

diethyl carbonate $(C_2H_5)_2CO_3$ Stable, colorless liquid with mild aroma, boiling at 126°C; soluble with most organic solvents; used as a solvent and for chemical synthesis. Also known as ethyl carbonate.

diethylenediamine *See* piperazine.

diethylene dioxide *See* 1,4-dioxane.

diethylene glycol $CH_2OHCH_2OCH_2$ CH_2OH Clear, hygroscopic, water-soluble liquid, boiling at 245°C; soluble in many organic solvents; used as a softener, conditioner, lubricant, and solvent, and in antifreezes and cosmetics.

diethylene glycol distearate *See* diglycol stearate.

diethylene glycol monoethyl ether $C_6H_{14}O_3$ A hygroscopic liquid used as a solvent for cellulose esters and in lacquers, varnishes, and enamels. Also known as 2-(2-ethoxyethoxy)-ethanol.

diethylene glycol monolaurate *See* diglycol laurate.

diethylenetriamine $(NH_2C_2H_4)_2NH$ A yellow, hygroscopic liquid with a boiling point of 206.7°C; soluble in water and hydrocarbons; used as a solvent, saponification agent, and fuel component.

diethyl ether $C_4H_{10}O$ A colorless liquid, slightly soluble in water; used as a reagent and solvent. Also known as ethyl ether; ethyl oxide; ethylic ether.

diethyl ketone *See* pentanone.

diethyl maleate $(HCCOOC_2H_5)_2$ Clear, colorless liquid, boiling at 225°C; slightly soluble in water, soluble in most organic solvents; used as a chemical intermediate.

diethylmalonylurea *See* barbital.

diethylmethylmethane *See* 3-methylpentane.

diethyl 4,4′-O-phenylene-bis(3-thioallophanate) *See* thiophanate.

diethyl phosphite $(C_2H_5O)_2HPO$ A colorless liquid with a boiling point of 138°C; soluble in water and common organic solvents; used as a paint solvent, antioxidant, and reducing agent.

diethyl phthalate $C_6H_4(CO_2C_2H_5)_2$ Clear, colorless, odorless liquid with bitter taste, boiling at 298°C; soluble in alcohols, ketones, esters, and aromatic hydrocarbons, partly soluble in aliphatic solvents; used as a cellulosic solvent, wetting agent, alcohol denaturant, mosquito repellent, and in perfumes.

diethyl pyrocarbonate $C_6H_{10}O_5$ A viscous liquid, soluble in alcohols, esters, and ketones; used as a gentle esterifying agent, as a preservative for fruit juices, soft drinks, and wines, and as an inhibitor for ribonuclease. Abbreviated DEPC. Also known as diethyl oxydiformate; pyrocarbonic acid diethyl ester.

diethylstilbesterol $C_{18}H_{20}O_2$ A white, crystalline, nonsteroid estrogen that is used therapeutically as a substitute for natural estrogenic hormones. Also known as stilbestrol.

diethyl succinate $(CH_2COOC_2H_5)_2$ Water-white liquid with pleasant aroma, boiling at 216°C; soluble in alcohol and ether, slightly soluble in water; used as a chemical intermediate and plasticizer.

diethyl sulfate $(C_2H_5)_2SO_4$ A colorless oil with a peppermint odor, and boiling at 208°C; used as an intermediate in organic synthesis. Also known as ethyl sulfate.

diethyl sulfide *See* ethyl sulfide.

diethylsulfide 2,2′-dicarboxylic acid *See* 3,3′-thiodiproprionic acid.

2 - (*O,O* - diethyl thionophosphoryl) - 5 - methyl - 6 - carbethoxypyrazolo - (1,5a)pyrimidine $C_{13}H_{20}N_3O_4P$ A colorless, crystalline solid with a melting point of 50–51°C; insoluble in water; used as a fungicide for powdery mildew on fruits. Also known as pyrazophos.

diethyltoluamide $C_{12}H_{17}ON$ A liquid whose color ranges from off-white to light yellow; used as an insect repellent for people and clothing. Also known as DEET; DET; *N,N*-diethyl-*meta*-toluamide.

difference number *See* neutron excess.

differential chemical reactor A flow reactor operated at constant temperature and very low concentrations (resulting from very short residence times), with product and reactant concentrations essentially constant at the levels in the feed.

differential ebuliometer Apparatus for precise and simultaneous measurement of both the boiling temperature of a liquid and the condensation temperature of the vapors of the boiling liquid.

differential extraction Theoretical limiting case of crosscurrent extraction in a single vessel where feed is continuously extracted with infinitesimal amounts of fresh solvent; true differential extraction cannot be achieved.

differential heat of dilution *See* heat of dilution.

differential polarography Technique of polarographic analysis which measures the difference in current flowing between two identical dropping-mercury electrodes at the same potential but in different solutions.

differential process A process in which a system is caused to move through a bubble point and as a result to form two phases, the minor phase being removed from further contact with the major phase; thus the system continuously changes in quantity and composition.

differential reaction rate The order of a chemical reaction expressed as a differential equation with respect to time; for example, $dx/dt = k(a - x)$ for first order, $dx/dt = k(a - x)(b - x)$ for second order, and so on, where k is the specific rate constant, a is the concentration of reactant A, b is the concentration of reactant B, and dx/dt is the rate of change in concentration for time t.

differential separation Release of gas (vapor) from liquids by a reduction in pressure that allows the vapor to come out of the solution, so that the vapor can be removed from the system; differs from flash separation, in which the vapor and liquid are kept in contact following pressure reduction.

differential spectrophotometry Spectrophotometric analysis of a sample when a solution of the major component of the sample is placed in the reference cell; the recorded spectrum represents the difference between the sample and the reference cell.

differential thermometric titration Thermometric titration in which titrant is added simultaneously to the reaction mixture and to a blank in identically equipped cells.

diffraction grating An optical device consisting of an assembly of narrow slits or grooves which produce a large number of beams that can interfere to produce spectra. Also known as grating.

diffraction spectrum Parallel light and dark or colored bands of light produced by diffraction.

diffraction symmetry Any symmetry in a crystal lattice which causes the systematic annihilation of certain beams in x-ray diffraction.

diffractometry The science of determining crystal structures by studying the diffraction of beams of x-rays or other waves.

diffuse series A series occurring in the spectra of many atoms having one, two, or three electrons in the outer shell, in which the total orbital angular momentum quantum number changes from 2 to 1.

diffuse spectrum Any spectrum having lines which are very broad even when there is no possibility of line broadening by collisions.

diffusion barrier Porous barrier through which gaseous mixtures are passed for enrichment of the lighter-molecular-weight constituent of the diffusate; used as a many-stage cascade system for the recovery of $^{235}UF_6$ isotopes from a $^{238}UF_6$ stream.

diffusion current In polarography with a dropping-mercury electrode, the flow that is controlled by the rate of diffusion of the active solution species across the concentration gradient produced by the removal of ions or molecules at the electrode surface.

diffusion flame A long gas flame that radiates uniformly over its length and precipitates free carbon uniformly.

diffusiophoresis A process in a scrubber whereby water vapor moving toward the cold water surface carries particulates with it.

diffusivity analysis Analysis of difficult-to-separate materials in solution by diffusion effects, using, for example, dialysis, electrodialysis, interferometry, amperometric titration, polarography, or voltammetry.

1,1,1-difluorochloroethane *See* 1,1-chlorodifluoroethane.

difluoromonochloroethane *See* 1,1-chlorodifluoroethane.

difluoromonochloromethane *See* chlorodifluoromethane.

difluron $C_{14}H_9N_2F_2ClO_2$ A solid with a melting point of 239°C; slightly soluble in water; used as an insecticide for larvae of mosquitoes and of leaf-sucking insects.

digallic acid *See* tannic acid.

digester A vessel used to produce cellulose pulp from wood chips by cooking under pressure.

digestion 1. Preferential dissolving of mineral constituents in concentrations of ore. 2. Liquefaction of organic waste materials by action of microbes. 3. Separation of fabric from tires by the use of hot sodium hydroxide. 4. Removing lignin from wood in manufacture of chemical cellulose paper pulp.

digestive enzyme Any enzyme that causes or aids in digestion.

digicitrin $C_{21}H_{21}O_{10}$ A flavone compound that is found in foxglove leaves.

digitoxigenin $C_{23}H_{34}O_4$ The steroid aglycone formed by removal of three molecules of the sugar digitoxose from digitoxin.

digitoxin $C_{41}H_{64}O_{13}$ A poisonous steroid glycoside found as the most active principle of digitalis, from the foxglove leaf.

diglucoside A compound containing two glucose molecules.

diglycerol   A compound that is a diester of glycerol.

diglycine   *See* iminodiacetic acid.

diglycolic acid $O(CH_2COOH)_2$ A white powder that forms a monohydrate; used in the manufacture of plasticizers and in organic synthesis, and to break emulsions.

diglycol laurate $C_{11}H_{23}COOC_2H_4OC_2H_4OH$ A light, straw-colored, oily liquid; soluble in methanol, ethanol, toluene, and mineral oil; used in emulsions and as an antifoaming agent. Also known as diethylene glycol monolaurate.

diglycol stearate $(C_{17}H_{35}COOC_2H_4)_2O$ A white, waxy solid with a melting point of 54–55°C; used as an emulsifying agent, suspending medium for powders in the manufacture of polishes, and thickening agent, and in pharmaceuticals. Also known as diethylene glycol distearate.

digoxin $C_{41}H_{64}O_{14}$ A crystalline steroid obtained from a foxglove leaf (*Digitalis lanata*); similar to digitalis in pharmacological effects.

dihalide   A molecule containing two atoms of halogen combined with a radical or element.

dihexagonal   Of crystals, having a symmetrical form with 12 sides.

dihexagonal-dipyramidal   Characterized by the class of crystals in the hexagonal system in which any section perpendicular to the sixfold axis is dihexagonal.

dihexahedron   A type of crystal that has 12 faces, such as a double six-sided pyramid.

dihydrate   A compound with two molecules of water of hydration.

dihydrazone   A molecule containing two hydrazone radicals.

dihydro-   A prefix indicating combination with two atoms of hydrogen.

dihydrocamphane   *See* camphane.

dihydrochloride   A compound containing two molecules of hydrochloric acid.

dihydrocoumarin   *See* benzodihydropyrone.

dihydrodiketoanthracene   *See* anthraquinone.

dihydroxy   A molecule containing two hydroxyl groups.

dihydroxyacetone $(HOCH_2)_2CO$ A colorless, crystalline solid with a melting point of 80°C; soluble in water and alcohol; used in medicine, fungicides, plasticizers, and cosmetics. Abbreviated DHA. Also known as dihydroxypropanone.

dihydroxyacetonephosphoric acid $C_3H_7O_6P$ A phosphoric acid ester of dehydroxyacetone, produced as an intermediate substance in the conversion of glycogen to lactic acid during muscular contraction. Also known as dihydroxyacetone phosphate.

2,4'-dihydroxyacetophenone $(HO)_2C_6H_3COCH_3$ Needlelike or leafletlike crystals with a melting point of 145–147°C; soluble in pyridine, warm alcohol, and glacial acetic acid; used as a reagent for the determination of iron. Also known as resacetophenone.

1,2-dihydroxyanthraquinone   *See* alizarin.

1,8-dihydroxyanthraquinone $C_{14}H_8O_4$ Orange, needlelike crystals that dissolve in glacial acetic acid; used as an intermediate in the commercial preparation of indanthrene and alizarin dyestuffs. Also known as chrysazin.

3-(3,4-dihydroxycinnamoyl)quinic acid *See* chlorogenic acid.

6,7-dihydroxycoumarin-6-glucoside *See* esculin.

2,2′-dihydroxy-4,4′-dimethoxybenzophenone $[CH_3OC_6H_3(OH)]_2CO$ Crystals with a melting point of 139–140°C; used in paint and plastics as a light absorber.

dihydroxymaleic acid $C_4H_4O_6$ Crystals soluble in alcohol; used in the detection of titanium and fluorides. Also known as 1,2-dihydroxyethylenedicarboxylic acid.

dihydroxyphenylalanine $C_9H_{11}NO_4$ An amino acid that can be formed by oxidation of tyrosine; it is converted by a series of biochemical transformations, utilizing the enzyme dopa oxidase, to melanins. Also known as dopa.

3,5-diiodosalicylic acid $C_7H_4I_2O_3$ Crystals with a sweetish, bitter taste and a melting point of 235–236°C; soluble in most organic solvents; used as a source of iodine in foods and a growth promoter in poultry, hog, and cattle feeds.

di-iron enneacarbonyl *See* iron nonacarbonyl.

diisobutylene C_8H_{16} Any one of a number of isomers, but most often 2,4,4-trimethylpentene-1 and 2,4,4-trimethylpentene-2; used in alkylation and as a chemical intermediate.

diisobutyl ketone $(CH_3)_2CHCH_2COCH_2CH(CH_3)_2$ Stable liquid, boiling at 168°C; soluble in most organic liquids; toxic and flammable; used as a solvent, in lacquers and coatings, and as a chemical intermediate.

diisocyanate A compound that contains two NCO (isocyanate) groups; used to produce polyurethane foams, resins, and rubber.

diisodecyl adipate $(C_{10}H_{21}OOC)_2(CH_2)_4$ A light-colored, oily liquid with a boiling range of 239–246°C; used as a primary plasticizer for polymers. Abbreviated DIDA.

diisodecyl phthalate $C_6H_4(COOC_{10}H_{21})_2$ A clear liquid with a boiling point of 250–257°C; used as a plasticizer. Abbreviated DIDP.

diisopropanolamine $(CH_3CHOHCH_2)_2NH$ A white, crystalline solid with a boiling point of 248.7°C; used as an emulsifying agent for polishes, insecticides, and water paints. Abbreviated DIPA.

diisopropyl 1. A molecule containing two isopropyl groups. 2. *See* 2,3-dimethylbutane.

diketene $CH_3COCHCO$ A colorless, readily polymerized liquid with pungent aroma; insoluble in water, soluble in organic solvents; used as a chemical intermediate. Also known as acetyl ketene.

diketone A molecule containing two ketone carbonyl groups.

diketopiperazine 1. $C_4H_6N_2O_2$ A compound formed by dehydration of two molecules of glycine. Also known as 2,5-piperazine-dione. 2. Any of the cyclic molecules formed from α-amino acids other than glycine or by partial hydrolysis of protein.

dilactone A molecule that contains two lactone groups.

dilatancy The property of a viscous suspension which sets solid under the influence of pressure.

dilatant A material with the ability to increase in volume when its shape is changed.

dilauryl ester *See* dilauryl thiodipropionate.

dilauryl ether $(C_{12}H_{25})_2NH$ A liquid with a boiling point of 190–195°C; used for electrical insulators, water repellents, and antistatic agents. Also known as didodecyl ether.

dilauryl thiodipropionate $(C_{12}H_{25}OOCCH_2CH_2)_2S$ White flakes with a melting point of 40°C; soluble in most organic solvents; used as an antioxidant, plasticizer, and preservative, and in food wraps and edible fats and oils. Also known as dilauryl ester; thiodipropionic acid.

dilinoleic acid $C_34H_{62}(COOH)_2$ A light yellow, viscous liquid used as an emulsifying agent and shellac substitute.

diluent An inert substance added to some other substance or solution so that the volume of the latter substance is increased and its concentration per unit volume is decreased.

dilute To make less concentrated.

dilute phase In liquid-liquid extraction, the liquid phase that is dilute with respect to the material being extracted.

dilution Increasing the proportion of solvent to solute in any solution and thereby decreasing the concentration of the solute per unit volume.

dimer A condensation product consisting of two molecules.

dimeric water Water in which pairs of molecules are joined by hydrogen bonds.

dimerization Formation of dimers.

dimetan The generic name for 5,5-dimethyldehydroresorcinol dimethylcarbamate, a synthetic carbamate insecticide.

dimethachlon $C_{10}H_7Cl_2NO_2$ A yellowish, crystalline solid with a melting point of 136.5–138°C; insoluble in water; used as a fungicide. Also known as *N*-(3,5-dichlorophenyl)succinimide.

dimethoate $C_5H_{12}NO_3PS_2$ A crystalline compound, soluble in most organic solvents; used as an insecticide.

dimethoxy neotetrazolium *See* blue tetrazolium.

dimethoxy strychnine *See* brucine.

dimethrin $C_{19}H_{28}O_2$ An amber liquid with a boiling point of 175°C; soluble in petroleum hydrocarbons, alcohols, and methylene chloride; used as an insecticide for mosquitoes, body lice, stable flies, and cattle flies. Also known as 2,4-dimethylbenzyl 2,2-dimethyl-3-(2-methylpropenyl)cyclopropane carboxylate.

dimethyl A compound that has two methyl groups.

dimethylamine $(CH_3)_2NH$ Flammable gas with ammonia aroma, boiling at 7°C; soluble in water, ether, and alcohol; used as an acid-gas absorbent, solvent, and flotation agent, in pharmaceuticals and electroplating, and in dehairing hides.

para-dimethylaminobenzalrhodanine $C_{12}H_{12}N_2OS_2$ Deep red, needlelike crystals that decompose at 270°C; soluble in strong acids; used in acetone solution for the detection of ions such as silver, mercury, copper, gold, palladium, and platinum.

2-dimethylaminoethanol $(CH_3)_2NCH_2CH_2OH$ A colorless liquid with a boiling point of 134.6°C; used for the synthesis of dyestuffs, pharmaceuticals, and corrosion inhibitors, in medicine, and as an emulsifier. Also known as deanol; dimethylethanolamine.

4-(dimethylamino)-meta-tolyl-methylcarbamate *See* aminocarb.

N,N-dimethylaniline $C_6H_5N(CH_3)_2$ A yellowish liquid slightly soluble in water; used in dyes and solvent and in the manufacture of vanillin. Also known as aniline N,N-dimethyl.

dimethylarsinic acid *See* cacodylic acid.

2,2-dimethyl-1,3-benzodioxol-4-yl-N-methylcarbamate $C_{10}H_{13}O_4N$ A white solid with a melting point of 129–130°C; solubility in water is 40 parts per million at 25°C; used as an insecticide for cockroaches. Also known as bendiocarb; ficam.

2,3-dimethylbutane $(CH_3)_2CHCH(CH_3)_2$ A colorless liquid with a boiling point of 57.9°C; used as a high-octane fuel. Also known as diisopropyl.

dimethyl carbate $C_{11}H_{14}O_4$ A colorless liquid with a boiling point of 114–115°C; used as an insect repellent. Also known as dimethyl cis-bicyclo(2,2,1)-5-heptane-2,3-decarboxylate.

5,5-dimethyl-1,3-cyclohexanedione $C_8H_{12}O_2$ Crystals that decompose at 148–150°C; soluble in water and inorganic solvents such as methanol and ethanol; used as a reagent for the identification of aldehydes. Also known as dimedone; 1,1-dimethyl-3,5-cyclohexanedione; dimethyldihydroresorcinol; 1,1-dimethyl-3,5-diketocyclohexane.

2,2-dimethyl-1,3-dioxolane-4-methanol $C_6H_{12}O_3$ The acetone ketal of glycerin; a liquid miscible with water and many organic solvents; used as a plasticizer and a solvent. Also known as acetone glycerol; glycerol dimethylketal; isopropylidene glycerol.

dimethyl ether CH_3OCH_3 A flammable, colorless liquid, boiling at -25°C; soluble in water and alcohol; used as a solvent, extractant, reaction medium, and refrigerant. Also known as methyl ether; wood ether.

N,N-dimethylformamide $HCON(CH_3)_2$ A liquid that boils at 152.8°C; extensively used as a solvent for organic compounds. Abbreviated DMF.

dimethylglyoxime $(CH_3)_2C_2(NOH)_2$ White, crystalline or powdered solid, used in analytical chemistry as a reagent for nickel.

uns-dimethylhydrazine $(CH_3)_2NNH_2$ A flammable, highly toxic, colorless liquid; used as a component of rocket and jet fuels and as a stabilizer for organic peroxide fuel additives. Also known as 1,1-dimethylhydrazine; UDMH.

dimethylisopropanolamine $(CH_3)_2NCN_2$ $CH(OH)CH_3$ A colorless liquid with a boiling point of 125.8°C; soluble in water; used in methadone synthesis.

1,6-dimethyl-4-isopropyl naphthalene *See* cadalene.

dimethylketol *See* acetoin.

3,3-dimethyl-2-methylenenorcamphane   *See* camphene.

dimethylolurea $CO(NHCH_2OH)_2$ Colorless crystals melting at 126°C, soluble in water; used to increase fire resistance and hardness of wood, and in textiles to prevent wrinkles. Also known as 1,3-bis-hydroxymethylurea; DMU.

3,4-dimethylphenyl-*N*-methyl carbamate $C_{10}H_{13}NO_2$ A white, crystalline compound with a melting point of 79–80°C; insoluble in water; used as an insecticide for hoppers on rice and scales on fruit.

dimethyl phthalate $C_6H_4(COOCH_3)_2$ Odorless, colorless liquid, boiling at 282°C; soluble in organic solvents, slightly soluble in water; used as a plasticizer, in resins, lacquers, and perfumes, and as an insect repellent.

dimethyl sebacate $[(CH_2)_4COOCH_3]_2$ Clear, colorless liquid, boiling at 294°C; used as a vinyl resin, nitrocellulose solvent, or plasticizer.

dimethyl sulfate $(CH_3)_2SO_4$ Poisonous, corrosive, colorless liquid, boiling at 188°C; slightly soluble in water, soluble in ether and alcohol; used to methylate amines and phenols. Also known as methyl sulfate.

2-4-dimethylsulfolane $C_6H_{12}O_2S$ A yellow to colorless liquid miscible with lower aromatic hydrocarbons; used as a solvent in liquid-liquid and vapor-liquid extraction processes. Also known as 2,4-dimethylcyclotetramethylene sulfone; 2,4-dimethyltetramethylene sulfone; 2,4-dimethylthiacyclopentane-1,1-dioxide.

dimethyl sulfoxide $(CH_3)_2SO$ A colorless liquid used as a local analgesic and anti-inflammatory agent, as a solvent in industry, and in laboratories as a medium for carrying out chemical reactions. Abbreviated DMSO.

dimethyl terephthalate $C_6H_4(COOCH_3)_2$ Colorless crystals, melting at 140°C and subliming above 300°; slightly soluble in water, soluble in hot alcohol and ether; used to make polyester fibers and film. Abbreviated DMT.

dimethyl-2,3,5,6-tetrachloroterephthalate $C_{10}H_6Cl_4O_4$ A colorless, crystalline compound with a melting point of 156°C; used as an herbicide for turf, ornamental flowers, and certain vegetables and berries. Also known as DCPA.

dimetilan   A generic name for 1-(dimethylcarbamoyl)-5-methyl-3-pyrazolyl dimethylcarbamate, a synthetic carbamate insecticide.

dimorphism   Having crystallization in two forms with the same chemical composition.

dineutron   **1.** A hypothetical bound state of two neutrons, which probably does not exist. **2.** A combination of two neutrons which has a transitory existence in certain nuclear reactions.

dinitramine $C_{11}H_{13}N_3O_4F_3$ A yellow solid with a melting point of 98–99°C; used as a preemergence herbicide for annual grass and broadleaf weeds in cotton and soybeans. Also known as N^3,N^3-diethyl-2,4-dinitro-6-trifluoromethyl-*meta*-phenylenediamine.

dinitrate   A molecule that contains two nitrate groups.

dinitrite   A molecule that has two nitrite groups.

2,4-dinitroaniline $(NO_2)_2C_6H_3NH_2$ A compound which crystallizes as yellow needles or greenish-yellow plates, melting at 187.5–188°C; soluble in alcohol; used in the manufacture of azo dyes.

2,4-dinitrobenzaldehyde $(NO_2)_2C_6H_3CHO$ Yellow to light brown crystals with a melting point of 72°C; soluble in alcohol, ether, and benzene; used to make Schiff bases.

dinitrobenzene Any one of three isomeric substitution products of benzene having the empirical formula $C_6H_4(NO_2)_2$.

2,4-dinitrobenzenesulfenyl chloride $(NO_2)_2$ C_6H_3SCl Crystals soluble in glacial acetic acid, with a melting point of 96°C; used as a reagent for separation and identification of naturally occurring indoles.

3,4-dinitrobenzoic acid $C_7H_4N_2O_6$ Crystals with a bitter taste and a melting point of 166°C; used in quantitative sugar analysis.

4,6-dinitro-2-*sec*-butylphenylacetate $C_{12}H_{13}O_6N_2$ A brownish liquid with a melting point of 170°C; used as an herbicide for annual broadleaf weeds in food crops such as peas, potatoes, and alfalfa.

3′,5′-dinitro-4′-(di-*N*-propyl)aminoacetophenone *See* bulab-37.

dinitrogen fixation *See* nitrogen fixation.

dinitrogen tetroxide *See* nitrogen dioxide.

dinitrophenol Any one of six isomeric substituent products of benzene having the empirical formula $(NO_2)_2C_6H_3OH$.

2,4-dinitrophenylhydrazine $(NO_2)_2C_6H_3NHNH_2$ A red, crystalline powder with a melting point of approximately 200°C; soluble in dilute inorganic acids; used as a reagent for determination of ketones and aldehydes.

dinitrotoluene Any one of six isomeric substitution products of benzene having the empirical formula $CH_3C_6H_3(NO_2)_2$; they are high explosives formed by nitration of toluene. Abbreviated DNT.

dinobuton $C_{14}H_{18}N_2O_7$ A yellow solid with a melting point of 57–60°C; used as a fungicide and miticide for fruit and vegetable crops and for cotton. Also known as 2-*sec*-butyl-4,6-dinitrophenylisopropyl carbonate.

dinoseb $C_{10}H_{12}O_5N_2$ A reddish-brown liquid with a melting point of 32°C; used as an insecticide and herbicide for numerous crops and in fruit and nut orchards. Also known as 2-(sec-butyl)-4,6-dinitrophenol.

dinoterb acetate $C_{12}H_{14}N_2O_6$ A yellow, crystalline compound with a melting point of 133–134°C; used as a preemergence herbicide for sugarbeets, legumes, and cereals, and as a postemergence herbicide for maize, sorghum, and alfalfa.

dioctahedral Pertaining to a crystal structure in which only two of the three available octahedrally coordinated positions are occupied.

dioctyl A compound that has two octyl groups.

dioctyl phthalate $(C_8H_{17}OOC)_2C_6H_4$ Pale, viscous liquid, boiling at 384°C; insoluble in water; used as a plasticizer for acrylate, vinyl, and cellulosic resins, and as a miticide in orchards. Abbreviated DOP. Also known as di-(2-ethylhexyl)phthalate.

dioctyl sebacate $(CH_2)_8(COOC_8H_{17})_2$ Water-insoluble, straw-colored liquid, boiling at 248°C; used as a plasticizer for vinyl, cellulosic, and styrene resins.

diodide A molecule that contains two iodine atoms bonded to an element or radical.

diolefin *See* diene.

diolefin hydrogenation A fixed-bed catalytic process used to hydrogenate diolefins in C_4 and C_5 fractions to mono-olefin in alkylation feedstocks.

-dione Suffix indicating the presence of two keto groups.

dioxacarb $C_{11}H_{13}NO_4$ A white, crystalline compound with a melting point of 114–115°C; used as an insecticide for potato and cocoa crops and in restaurants. Also known as 2-(1,3-dioxolan-2-yl)phenyl-*N*-methylcarbamate.

1,4-dioxane $C_4H_8O_2$ The cyclic ether of ethylene glycol; it is soluble in water in all proportions and is used as a solvent. Also known as diethylene dioxide; *para*-dioxane; glycol ethylene ether.

dioxathion $C_{12}H_{26}O_6P_2S_4$ A brown liquid, used as a miticide on cotton and fruit and ornamental crops, and as an insecticide on ticks, lice, and flies on cattle. Also known as 2,3-*para*-dioxanedithiol-*S,S*-bis(*O,O*-diethyl phosphorodithioate).

dioxide A compound containing two atoms of oxygen.

2,4-dioxo-5-fluoropyrimidine *See* fluorouracil.

dioxolane $C_3H_6O_2$ A cyclic acetal that is a liquid; used as a solvent and extractant.

2-(1,3-dioxolan-2-yl)phenyl-*N*-methylcarbamate *See* dioxacarb.

dioxopurine *See* xanthine.

dioxygenase Any of a group of enzymes which catalyze the insertion of both atoms of an oxygen molecule into an organic substrate according to the generalized formula $AH_2 + O_2 \rightarrow A(OH)_2$.

dipentene The racemic mixture of dextro and levo isomers of limonene.

dipeptidase An enzyme that hydrolyzes a dipeptide.

diphacinone $C_{23}H_{16}O_3$ A yellow powder with a melting point of 145–147°C; used to control rats, mice, and other rodents; acts as an anticoagulant. Also known as 2-diphenylacetyl-1,3-indandione.

diphenamid $C_{16}H_{17}ON$ An off-white, crystalline compound with a melting point of 134–135°C; used as a preemergence herbicide for food crops, fruits, and ornamentals. Also known as *N,N*-dimethyl-2,2-diphenylacetamide.

diphenatrile $C_{14}H_{11}N$ A yellow, crystalline compound with a melting point of 73–73.5°C; used as a preemergence herbicide for turf. Also known as diphenylacetonitrile.

diphenol A compound that has two phenol groups, for example, resorcinol.

diphenyl *See* biphenyl.

diphenylamine $(C_6H_5)_2NH$ Colorless leaflets, sparingly soluble in water; melting point 54°C; used as an additive in propellants to increase the storage life by neutralizing the acid products formed upon decomposition of the nitrocellulose. Also known as phenylaniline.

diphenylcarbazide $CO(NHNHC_6H_5)_2$ White powder, melting point 170°C; used as an indicator, pink for alkalies, colorless for acids.

diphenylcarbinol *See* benzhydrol.

diphenyl carbonate $(C_6H_5O)_2CO$ Easily hydrolyzed, white crystals, melting at 78°C; soluble in organic solvents, insoluble in water; used as a solvent, plasticizer, and chemical intermediate.

diphenylchloroarsine $(C_6H_5)_2AsCl$ Colorless crystals used during World War I as an antipersonnel device to generate a smoke causing sneezing and vomiting.

diphenyldiazene *See* azobenzene.

diphenylene oxide $C_{12}H_8O$ A crystalline solid derived from coal tar; melting point is 87°C; used as an insecticide. Also known as dibenzofuran.

diphenylglyoxal *See* benzil.

diphenylguanidine $HNC(NHC_6H_5)_2$ A white powder, melting at 147°C; used as a rubber accelerator. Also known as DPG; melaniline.

diphenyl ketone *See* benzophenone.

diphenylmethane $(C_6H_5)_2CH_2$ Combustible, colorless crystals melting at 26.5°C; used in perfumery, dyes, and organic synthesis. Also known as benzylbenzene.

diphenyl oxide $(C_6H_5)_2O$ A colorless liquid or crystals with a melting point of 27°C and a boiling point of 259°C; soluble in alcohol and ether; used in perfumery, soaps, and resins for laminated electrical insulation. Also known as diphenyl ether; phenyl ether.

4,7-diphenyl-1,10-phenanthroline *See* phenanthroline indicator.

diphenyl phthalate $C_6H_4(COOC_6H_5)_2$ White powder, melting at 80°C; soluble in chlorinated hydrocarbons, esters, and ketones, insoluble in water; used as a plasticizer for cellulosic and other resins.

diphosphate A salt that has two phosphate groups.

diphosphatidyl glycerol *See* cardiolipin.

diphosphoglyceric acid $C_3H_8O_9P_2$ An ester of glyceric acid, with two molecules of phosphoric acid, characterized by a high-energy phosphate bond.

diphosphopyridine nucleotide $C_{21}H_{27}O_{14}N_7P_2$ An organic coenzyme that functions in enzymatic systems concerned with oxidation-reduction reactions. Abbreviated DPN. Also known as codehydrogenase 1; coenzyme 1; nicotinamide adenine dinucleotide (NAD).

dipicolinic acid $C_7H_5O_4N \cdot 1\frac{1}{2}H_2O$ A chelating agent composing 5–15% of the dry weight of bacterial spores.

dipicrylamine $[(NO_2)_3C_6H_2]_2NH$ Yellow, prismlike crystals used in the gravimetric determination of potassium.

diploid A crystal form in the isometric system having 24 similar quadrilateral faces arranged in pairs.

dipnone $C_{16}H_{14}O$ A liquid ketone, formed by condensation of two acetophenone molecules; used as a plasticizer. Also known as 1,3-diphenyl-3-butene-1-one; β-methylchalcone.

dipolar ion An ion carrying both a positive and a negative charge. Also known as zwitterion.

dipole-dipole force *See* orientation force.

dipole transition A transition of an atom or nucleus from one energy state to another in which dipole radiation is emitted or absorbed.

dipping acid *See* sulfuric acid.

dipropetryn $C_{11}H_{21}N_5S$ A colorless solid with a melting point of 104–106°C; slight solubility in water; used to control weeds in cotton, corn, carrots, lima beans, okra, peas, and potatoes. Also known as 2-ethylthio-4,6-bis(isopropylamino)-S-triazine.

dipropyl A compound containing two propyl groups.

dipropylene glycol $(CH_3CHOHCH_2)_2O$ A colorless, slightly viscous liquid with a boiling point of 233°C; soluble in toluene and in water; used as a solvent and for lacquers and printing inks.

dipropyl ketone *See* 4-heptanone.

diproton A hypothetical bound state of two protons, which probably does not exist.

dipyramid A crystal having the form of two pyramids that melt at a plane of symmetry.

dipyre *See* mizzonite.

2,2′-dipyridyl $C_{10}H_8N_2$ A crystalline substance soluble in organic solvents; melting point is 69.7°C; used as a reagent for the determination of iron. Also known as 2,2′-bipyridine.

dipyrine *See* aminopyrine.

direct effect A chemical effect caused by the direct transfer of energy from ionizing radiation to an atom or molecule in a medium.

direct-fired evaporator An evaporator in which the flame and combustion gases are separated from the boiling liquid by a metal wall, or other heating surface.

direct nuclear reaction A nuclear reaction which is completed in the time required for the incident particle to transverse the target nucleus, so that it does not combine with the nucleus as a whole but interacts only with the surface or with some individual constituent.

direct-vision spectroscope A spectroscope that allows the observer to look in the direction of the light source by means of an Amici prism.

disaccharide Any of the class of compound sugars which yield two monosaccharide units upon hydrolysis.

discharged solids *See* residue.

discharge liquor Liquid that has passed through a processing operation. Also known as effluent; product.

discontinuous phase *See* disperse phase.

discrete spectrum A spectrum in which the component wavelengths constitute a discrete sequence of values rather than a continuum of values.

disilane Si_2H_6 A spontaneously flammable compound of silicon and hydrogen; it exists as a liquid at room temperature.

disilicate A silicate compound that has two silicon atoms in the molecule.

disilicide A compound that has two silicon atoms joined to a radical or another element.

disilver salt *See* silver methylarsonate.

disinclination A type of crystal imperfection in which one part of the crystal is rotated and therefore displaced relative to the rest of the crystal; observed in liquid crystals and protein coats of viruses.

disintegration Any transformation of a nucleus, whether spontaneous or induced by irradiation, in which particles or photons are emitted.

disintegration chain *See* radioactive series.

disintegration energy The energy released, or the negative of the energy absorbed, during a nuclear or particle reaction. Designated Q. Also known as Q value; reaction energy.

disintegration family *See* radioactive series.

disintegration series *See* radioactive series.

disk-and-doughnut A type of fractionating tower construction of alternating disks and plates that are doughnut-shaped, to provide mixing.

disk colorimeter A device for comparing standard and sample colors by means of rotating color disks.

dislocation A defect occurring along certain lines in the crystal structure and present as a closed ring or a line anchored at its ends to other dislocations, grain boundaries, the surface, or other structural feature. Also known as line defect.

disodium aurothiomalate *See* gold sodium thiomalate.

disodium ethylene-bis-dithiocarbamate *See* nabam.

disodium hydrogen phosphate *See* disodium phosphate.

disodium 5,5'-indigotin disulfonate *See* indigo carmine.

disodium methanearsonate *See* disodium methylarsonate.

disodium methylarsonate $CH_3AsO(ONa)_2$ A colorless, hygroscopic, crystalline solid; soluble in water and methanol; used in pharmaceuticals and as an herbicide. Abbreviated DMA. Also known as disodium methanearsonate.

disodium phenoltetrabromophthalein sulfonate *See* sulfobromophthalein sodium.

disodium phosphate Na_2HPO_4 Transparent crystals, soluble in water; used in the textile processing and other industries to control pH in the range 4–9, as an additive in processed cheese to maintain spreadability, and as a laxative and antacid. Also known as disodium hydrogen phosphate.

disodium tartrate *See* sodium tartrate.

disorder Departures from regularity in the occupation of lattice sites in a crystal containing more than one element.

dispersal pattern Distribution pattern of metals in soil, rock, water, or vegetation.

dispersed elements Elements which form few or no independent minerals but are present as minor ingredients in minerals of abundant elements.

disperse phase The phase of a disperse system consisting of particles or droplets of one substance distributed through another system. Also known as discontinuous phase; internal phase.

disperse system A two-phase system consisting of a dispersion medium and a disperse phase.

dispersion A distribution of finely divided particles in a medium.

dispersion force The force of attraction that exists between molecules that have no permanent dipole.

dispersion relation A relation between the cross section for a given effect and the de Broglie wavelength of the incident particle, which is similar to a classical dispersion formula.

dispersoid Matter in a form produced by a disperse system.

disphenoid 1. A crystal form with four similar triangular faces combined in a wedge shape; can be tetragonal or orthorhombic. 2. A crystal form with eight scalene triangles combined in pairs.

displacement A chemical reaction in which an atom, radical, or molecule displaces and sets free an element of a compound.

displacement chromatography Variation of column-development or elution chromatography in which the solvent is sorbed more strongly than the sample components; the freed sample migrates down the column, pushed by the solvent.

displacement law *See* radioactive displacement law.

displacement series The elements in decreasing order of their negative potentials. Also known as constant series; electromotive series; Volta series.

disproportionation The changing of a substance, usually by simultaneous oxidation and reduction, into two or more dissimilar substances.

dissociation Separation of a molecule into two or more fragments (atoms, ions, radicals) by collision with a second body or by the absorption of electromagnetic radiation.

dissociation constant A constant whose numerical value depends on the equilibrium between the undissociated and dissociated forms of a molecule; a higher value indicates greater dissociation.

dissociation energy The energy required for complete separation of the atoms of a molecule.

dissociation pressure The pressure, for a given temperature, at which a chemical compound dissociates.

dissociative recombination The combination of an electron with a positive molecular ion in a gas followed by dissociation of the molecule in which the resulting atoms carry off the excess energy.

dissolution Dissolving of a material.

dissolve 1. To cause to disperse. 2. To cause to pass into solution.

dissolved air flotation A liquid-solid separation process wherein the main mechanism of suspended-solids removal is the change of apparent specific gravity of those suspended solids in relation to that of the suspending liquid by the attachment of small gas bubbles formed by the release of dissolved gas to the solids.

dissymmetry coefficient Ratio of the intensities of scattered light at 45 and 135°, used to correct for destructive interference encountered in light-scattering–photometric analyses of liquid samples.

distannoxane *See* fenbutatin oxide.

distillate The products of distillation formed by condensing vapors.

distillation The process of producing a gas or vapor from a liquid by heating the liquid in a vessel and collecting and condensing the vapors into liquids.

distillation column A still for fractional distillation.

distillation curve The graphical plot of temperature versus overhead product (distillate) volume or weight for a distillation operation.

distillation loss In a laboratory distillation, the difference between the volume of liquid introduced into the distilling flask and the sum of the residue and condensate received.

distillation range The difference between the temperature at the initial boiling point and at the end point of a distillation test.

distillation test A standardized procedure for finding the initial, intermediate, and final boiling points in the boiling range of petroleum products.

distilled mustard gas A delayed-action casualty gas (mustard gas) that has been distilled, or purified, to greatly reduce the odor and thereby increase its difficulty of detection.

distilled water Water that has been freed of dissolved or suspended solids and organisms by distillation.

distilling flask A round-bottomed glass flask that is capable of holding a liquid to be distilled.

distribution coefficient The ratio of the amounts of solute dissolved in two immiscible liquids at equilibrium.

distribution law The law stating that if a substance is dissolved in two immiscible liquids, the ratio of its concentration in each is constant.

disulfate A compound that has two sulfate radicals.

disulfide 1. A compound that has two sulfur atoms bonded to a radical or element. 2. One of a group of organosulfur compounds RSSR' that may be symmetrical (R = R') or unsymmetrical (R and R', different).

disulfonate A molecule that has two sulfonate groups.

disulfonic acid A molecule that has two sulfonic acid groups.

disulfoton $C_8H_{19}O_2PS_3$ A pale yellow liquid, used as an insecticide and miticide on foliage, cotton, beets, potatoes, and ornamental flowers. Also known as O,O-diethyl-S[(2-ethylthio)-ethyl]phosphorodithioate.

disulfram *See* tetraethylthiuram disulfide.

diterpene $C_{20}H_{32}$ 1. A group of terpenes that have twice as many atoms in the molecule as monoterpenes. 2. Any derivative of diterpene.

ditetrazolium chloride *See* blue tetrazolium chloride.

dithiocarbamate 1. A salt of dithiocarbamic acid. 2. Any other derivative of dithiocarbamic acid.

dithiocarbamic acid NH_2CS_2H A colorless, unstable powder; various metal salts are readily obtained, and used as strong accelerators for rubber. Also known as aminodithioformic acid.

dithioethyleneglycol *See* 1,2-ethanedithiol.

dithioic acid An organic acid in which sulfur atoms have replaced both oxygen atoms of the carboxy group.

dithiol *See* dimercaprol.

dithionate Any salt formed from dithionic acid.

dithionic acid $H_2S_2O_6$ A strong acid formed by the oxidation of sulfurous acid, and known only by its salts and in solution.

dithiooxamide $NH_2CSCSNH_2$ Red crystals soluble in alcohol; used as a reagent for copper, cobalt, and nickel, and for the determination of osmium. Also known as ethanedithioamide; rubeanic acid.

1,4-dithiothreitol $C_4H_{10}O_2S_2$ Needlelike crystals soluble in water, ethanol acetone, ethylacetate; used as a protective agent for thiol (SH) groups.

ditungsten carbide W_2C A gray powder having hardness approaching that of diamond; forms hexagonal crystals with specific gravity 17.2; melting point 2850°C.

diuretic hormone A neurohormone that promotes water loss in insects by increasing the volume of fluid secreted into the Malpighian tubules.

diuron $C_9H_{10}Cl_2N_2O$ A white solid with a melting point of 158–159°C; used as an herbicide for weeds and grass in numerous crops, including cotton, sugarcane, and pineapple. Also known as 3-(3,4-dichlorophenyl)-1,1-dimethylurea.

divalent carbon A charged or uncharged carbon atom that has formed only two covalent bonds. Also known as dicovalent carbon.

divalent metal A metal whose atoms are each capable of chemically combining with two atoms of hydrogen.

divanadyl tetrachloride *See* vanadyl chloride.

diver method Measure of the size of suspended solid particles; small glass divers of known density sink to the level where the liquid-suspension density is equal to that of the diver, allowing calculation of particle size. Also known as Berg's diver method.

divinyl 1. A molecule that has two vinyl groups. 2. *See* 1,3-butadiene.

divinyl acetylene C_6H_6 A linear trimer of acetylene, made by passing acetylene into a hydrochloric acid solution that has metallic catalysts; used as an intermediate in neoprene manufacture.

divinylbenzene $C_6H_4(CHCH_2)_2$ Polymerizable, water-white liquid used to make rubbers, drying oils, and ion-exchange resins and other polymers; forms include ortho, meta, and para isomers. Also known as vinylstyrene.

divinyl ether *See* vinyl ether.

divinyl oxide *See* vinyl ether.

D line The yellow line that is the first line of the major series of the sodium spectrum; the doublet in the Fraunhofer lines whose almost equal components have wavelengths of 5895.93 and 5889.96 angstroms respectively.

DMA *See* disodium methylarsonate.

DMB *See* hydroquinone dimethyl ether.

DMC *See* p,p'-dichlorodiphenylmethyl carbinol; penicillamine.

DMDT *See* methoxychlor.

DMF *See* N,N-dimethylformamide.

DMSO *See* dimethyl sulfoxide.

DMT *See* dimethyl terephthalate.

DMU *See* dimethylolurea.

DNA *See* deoxyribonucleic acid.

Dobbin's reagent A mercuric chloride–potassium iodide reagent used to test for caustic alkalies in soap.

Dobson spectrophotometer A photoelectric spectrophotometer used in the determination of the ozone content of the atmosphere; compares the solar energy at two wavelengths in the absorption band of ozone by permitting the radiation of each to fall alternately upon a photocell.

docosane $C_{22}H_{46}$ A paraffin hydrocarbon, especially the normal isomer $CH_3(CH_2)_{20}CH_3$.

docosanoic acid $CH_3(CH_2)_{20}CO_2H$ A crystalline fatty acid, melting at 80°C, slightly soluble in water and alcohol, and found in the fats and oils of some seeds such as peanuts. Also known as behenic acid.

1-docosanol *See* behenyl alcohol.

docosapentanoic acid $C_{21}H_{33}CO_2H$ A pale-yellow liquid, boils at 236°C (5 mm Hg), insoluble in water, soluble in ether, and found in fish blubber. Also known as clupanodonic acid.

cis-13-docosenoic acid *See* erucic acid.

doctor roll Roller device used to remove accumulated filter cake from rotary filter drums.

doctor solution Sodium plumbite solution used to remove mercaptan sulfur from gasoline and other light petroleum distillates; used in doctor treatment.

doctor test A procedure using doctor solution (sodium plumbite) to detect sulfur compounds in light petroleum distillates which react with the sodium plumbite.

doctor treatment Refining process to sweeten (reduce the odor) of gasoline, solvents, and kerosine; sodium plumbite and sulfur convert the odoriferous mercaptans into disulfides.

dodecahydrate A hydrated compound that has a total of 12 water molecules associated with it.

dodecane $CH_3(CH_2)_{10}CH_3C_{12}H_{26}$ An oily paraffin compound, a colorless liquid, boiling at 214.5°C, insoluble in water; used as a solvent and in jet fuel research. Also known as dihexyl; propylene tetramer; tetrapropylene.

dodecanoic acid *See* lauric acid.

1-dodecene $CH_2CH(CH_2)_9CH_3$ A colorless liquid, boiling at 213°C, insoluble in water; used in flavors, dyes, perfumes, and medicines. Also known as decylethylene; dodecylene.

dodecyl $C_{12}H_{25}$ A radical derived from dodecane by removing one hydrogen atom; in particular, the normal radical, $CH_3(CH_2)_{10}CH_2$—.

dodecyl alcohol *See* lauryl alcohol.

dodecylbenzene Blend of isomeric (mostly monoalkyl) benzenes with saturated side chains averaging 12 carbon atoms; used in the alkyl amyl sulfonate type of detergents. Also known as detergent alkylate.

dodecylene *See* 1-dodecene.

dodecyl sodium sulfate *See* sodium lauryl sulfate.

Doebner-Miller synthesis Synthesis of methylquinoline by heating aniline with paraldehyde in the presence of hydrochloric acid.

dolichol Any of a group of long-chain unsaturated isoprenoid alcohols containing up to 84 carbon atoms; found free or phosphorylated in membranes of the endoplasmic reticulum and Golgi apparatus.

dolomol *See* magnesium stearate.

dome An open crystal form consisting of two faces astride a symmetry plane.

Donnan equilibrium The particular equilibrium set up when two coexisting phases are subject to the restriction that one or more of the ionic components cannot pass from one phase into the other; commonly, this restriction is caused by a membrane which is permeable to the solvent and small ions but impermeable to colloidal ions or charged particles of colloidal size. Also known as Gibbs-Donnan equilibrium.

DOP *See* dioctyl phthalate.

dopa *See* dihydroxyphenylalanine.

dopamine $C_8H_{11}O_2N$ An intermediate in epinephrine and norepinephrine biosynthesis; the decarboxylation product of dopa.

dopa oxidase An enzyme that catalyzes the oxidation of dihydroxyphenylalanine to melanin; occurs in the skin.

Doppler broadening Frequency spreading that occurs in single-frequency radiation when the radiating atoms, molecules, or nuclei do not all have the same velocity and may each give rise to a different Doppler shift.

Doppler-free spectroscopy Any of several techniques which make use of the intensity and monochromatic nature of a laser beam to overcome the Doppler broadening of spectral lines and measure their wavelengths with extremely high accuracy.

Doppler-free two-photon spectroscopy A version of Doppler free spectroscopy in which the wavelength of a transition induced by the simultaneous absorption of two photons is measured by placing a sample in the path of a laser beam reflected on itself, so that the Doppler shifts of the incident and reflected beams cancel.

Dorn effect A difference in a potential resulting from the motions of particles through water; the potential exists between the particles and the water.

dotriacontane $C_{32}H_{66}$ A paraffin hydrocarbon, in particular, the normal isomer $CH_3(CH_2)_{30}CH_3$, which is crystalline.

double-beam spectrophotometer An instrument that uses a photoelectric circuit to measure the difference in absorption when two closely related wavelengths of light are passed through the same medium.

double beta decay A nuclear transformation in which the atomic number changes by 2 and the mass number does not change; either two electrons are emitted or two orbital electrons are captured.

double bond A type of linkage between atoms in which two pair of electrons are shared equally.

double-bond isomerism Isomerism in which two or more substances possess the same elementary composition but differ in having double bonds in different positions.

double-bond shift In an organic molecular structure, the occurrence when a pair of valence bonds that join a pair of carbons (or other atoms) shifts, via chemical reaction, to a new position, for example, $H_2C\!=\!C\!-\!C\!-\!CH_2$ (butene-1) to $H_2C\!-\!C\!=\!C\!-\!CH_2$ (butene-2).

double decomposition The simple exchange of elements of two substances to form two new substances; for example, $CaSO_4 + 2NaCl \rightarrow CaCl_2 + Na_2SO_4$.

double distribution The product distribution resulting from counter double-current extraction, a scheme in which each of the two liquid phases is transferred simultaneously and continuously in opposite directions through an interconnected train of contact vessels.

double-hump fission barrier Two separated maxima in a plot of potential energy against nuclear deformation of an actinide nucleus, which inhibit spontaneous fission of the nucleus and give rise to isomeric states in the valley between the two maxima.

double layer *See* electric double-layer.

double nickel salt *See* nickel ammonium sulfate.

double-pipe exchanger Fluid-fluid heat exchanger made of two concentric pipe sections; one fluid (such as a coolant) flows in the annular space between pipes, and the other fluid (such as hot process stream) flows through the inner pipe.

double salt 1. A salt that upon hydrolysis forms two different anions and cations. 2. A salt that is a molecular combination of two other salts.

double-solvent refining Petroleum-refining process using two solvents to simultaneously deasphalt and solvent-treat lubricating-oil stocks.

doublet 1. Two stationary states which have the same orbital and spin angular momentum but which have different total angular momenta, and therefore have slightly different energies due to spin-orbit coupling. 2. Two electrons which are shared between two atoms and give rise to a nonpolar valence bond. 3. Two closely separated spectral lines arising from a transition between a single state and a pair of states forming a doublet as described in the atomic physics definition.

downcomer A method of conveying liquid from one tray to the one below in a bubble-tray column.

downflow   In an ion-exchange system, the direction of the flow of the solution being processed.

Downs cell   A brick-lined steel vessel with four graphite anodes projecting upward from the bottom, with cathodes in the form of steel cylinders concentric with the anodes, containing an electrolyte which is 40% NaCl and 60% $CaCl_2$ at 590°C; used to make sodium.

Down's process   A method for producing sodium and chlorine from sodium chloride; potassium chloride and fluoride are added to the sodium chloride to reduce the melting point; the fused mixture is electrolyzed, with sodium forming at the cathode and chlorine at the anode.

downstream   Portion of a product stream that has already passed through the system; that portion located after a specific process unit.

Dow process   A process to produce magnesium by electrolysis of its fused chloride.

2,4-DP   *See* dichlorprop.

DPN   *See* diphosphopyridine nucleotide.

drag factor   Ratio of hindered diffusion rate to unhindered rate through a swollen dialysis membrane. Also known as Faxen drag factor; hindrance factor.

drain valve   A valve used to drain off material that has separated from a fluid or gas stream, or one used to empty a process line, vessel, or storage tank.

Draper effect   The increase in volume at constant pressure at the start of the reaction of hydrogen and chlorine to form hydrogen chloride; the volume increase is caused by an increase in temperature of the reactants, due to heat released in the reaction.

drawing   Removing ceramic ware from a kiln after it has been fired.

drazoxolon $C_{10}H_8ClN_3O_2$ A yellow, crystalline compound with a melting point of 167°C; used as a fungicide for diseases of coffee and tea. Also known as 4-(2-chlorophenylhydrazono)-3-methyl-5-isoxazolone.

Drew number   A dimensionless group used in the study of diffusion of a solid material A into a stream of vapor initially composed of substance B, equal to

$$\frac{Z_A(M_A - M_B) + M_B}{(Z_A - Y_{AW})(M_B - M_A)} \cdot \ln\frac{M_V}{M_W}$$

where M_A and M_B are the molecular weights of components A and B, M_V and M_W are the molecular weights of the mixture in the vapor and at the wall, and Y_{AW} and Z_A are the mole fractions of A at the wall and in the diffusing stream, respectively. Symbolized N_D.

drop model of nucleus   *See* liquid-drop model of nucleus.

dropping-mercury electrode   An electrode consisting of a fine-bore capillary tube above which a constant head of mercury is maintained; the mercury emerges from the tip of the capillary at the rate of a few milligrams per second and forms a spherical drop which falls into the solution at the rate of one every 2–10 seconds.

dropping point   The temperature at which grease changes from a semisolid to a liquid state under standardized conditions.

drum Tower or vessel in a refinery into which heated products are conducted so that volatile portions can separate.

dry acid Nonaqueous acetic acid used for oil-well reservoir acidizing treatment.

dry ashing The conversion of an organic compound into ash (decomposition) by a burner or in a muffle furnace.

dry box A container or chamber filled with argon, or sometimes dry air or air with no CO_2, to provide an inert atmosphere in which manipulation of very reactive chemicals is carried out in the laboratory.

dry-box process The passing of coke-oven or other industrial gases through boxes containing trays of iron oxide coated on wood shavings or other supporting material in order to remove hydrogen sulfide.

dry-chemical fire extinguisher A dry powder, consisting principally of sodium bicarbonate, which is used for extinguishing small fires, especially electrical fires.

dry coloring A plastics coloring method in which uncolored particles of the plastic material are tumble-blended with selected dyes and pigments.

dry-desiccant dehydration Use of silica gel or other solid absorbent to remove liquids from gases, such as water from air, or liquid hydrocarbons from natural gas.

dry distillation Distillation of materials that are dry.

dry ice Carbon dioxide in the solid form, usually made in blocks to be used as a coolant; changes directly to a gas at $-78.5°C$ as heat is absorbed.

drying 1. An operation in which a liquid, usually water, is removed from a wet solid in equipment termed a dryer. 2. A process of oxidation whereby a liquid such as linseed oil changes into a solid film.

drying agent Soluble or insoluble chemical substance that has such a great affinity for water that it will abstract water from a great many fluid materials; soluble chemicals are calcium chloride and glycerol, and insoluble chemicals are bauxite and silica gel. Also known as desiccant.

dry point The temperature at which the last drop of liquid evaporates from the bottom of the flask.

dry spot 1. An open area of an incomplete surface film on laminated plastic. 2. A section of laminated glass where the interlayer and glass are not bonded.

Dualayer distillate process A process for the removal of mercaptan and oxygenated compounds from distillate fuel oils; treatment is with concentrated caustic Dualayer solution and electrical precipitation of the impurities.

Dualayer solution A concentrated potassium or sodium hydroxide solution containing a solubilizer; used in the Dualayer distillate process.

dual-gravity valve A float-operated valve that operates on the interface between two immiscible liquids of different specific gravities.

dual radioactive decay Property exhibited by a nucleus which has two or more independent and alternative modes of decay.

Dubbs cracking A continuous, liquid-phase, thermal cracking process.

Duhem's equation *See* Gibbs-Duhem equation.

Dühring's rule   The rule that a plot of the temperature at which a liquid exerts a particular vapor pressure against the temperature at which a similar reference liquid exerts the same vapor pressure produces a straight or nearly straight line.

Dukler theory   Relationship of velocity and temperature distribution in thin films on vertical walls; used to calculate eddy viscosity and thermal conductivity near the solid boundary.

dulcin   *See* dulcitol.

dulcite   *See* dulcitol.

dulcitol $C_6H_8(OH)_6$ A sugar with a slightly sweet taste; white, crystalline powder with a melting point of 188.5°C; soluble in hot water; used in medicine and bacteriology. Also known as dulcin; dulcite; dulcose; euonymit; galactitol; melampyrit.

dulcose   *See* dulcitol.

Dumas method   A procedure for the determination of nitrogen in organic substances by combustion of the substance.

dunnite   *See* ammonium picrate.

durene $C_6H_2(CH_3)_4$ Colorless crystals with camphor aroma; boiling point 190°C; soluble in organic solvents, insoluble in water; used as a chemical intermediate. Also known as durol.

durol   *See* durene.

Dutch liquid   *See* ethylene chloride.

Dutch process   A process for making white lead; metallic lead is placed in vessels containing a dilute acetic acid, and the vessels are stacked in bark or manure.

Dy   *See* dysprosium.

dye   A colored substance which imparts more or less permanent color to other materials. Also known as dyestuff.

dyeing   The application of color-producing agents to material, usually fibrous or film, in order to impart a degree of color permanence demanded by the projected end use.

dyeing assistant   Material such as sodium sulfate added to a dye bath to control or promote the action of a textile dye.

dyestuff   *See* dye.

dynamic holdup   Liquid held by a tank or process vessel, with constant introduction of fresh material and counteracting withdrawal of held material to maintain a constant liquid level.

dypnone $C_6H_5COCHC(CH_3)C_6H_5$ A light-colored liquid with a boiling point of 246°C at 50 mmHg; used as a plasticizer and perfume base and in light-stable coatings.

Dyson notation   A notation system for representing organic chemicals developed by G. Malcolm Dyson; the compound is described on a single line, symbols are used for the chemical elements involved as well as for the functional groups and various ring systems; for example, methyl alcohol is C.Q and phenol is B6.Q.

dysprosium   A metallic rare-earth element, symbol Dy, atomic number 66, atomic weight 162.50.

dystetic mixture   A mixture of two or more substances that has the highest possible melting point of all mixtures of these substances.

easin $C_{20}H_6O_5I_4Na_2$ The sodium salt of tetraiodofluorescein; a brown powder, insoluble in water; used as a dye and a pH indicator (hydrogen ion) at pH 2.0. Also known as iodoeasin; sodium tetrafluorescein.

ebullating-bed reactor A type of fluidized bed in which catalyst particles are held in suspension by the upward movement of the liquid reactant and gas flow. Also known as slurry-bed reactor.

ebulliometer The instrument used for ebulliometry. Also known as ebullioscope.

ebulliometry The precise measurement of the absolute or differential boiling points of solutions.

ecdysone The molting hormone of insects.

ecgonine $C_9H_{15}NO_3$ An alkaloid obtained in crystalline form by the hydrolysis of cocaine.

echelette grating A diffraction grating with coarse groove spacing, designed for the infrared region; has grooves with comparatively flat sides and concentrates most of the radiation by reflection into a small angular coverage.

echelle grating A diffraction grating designed for use in high orders and at angles of illumination greater than 45° to obtain high dispersion and resolving power by the use of high orders of interference.

echelon grating A diffraction grating which consists of about 20 plane-parallel plates about 1 centimeter thick, cut from one sheet, each plate extending beyond the next by about 1 millimeter, and which has a resolving power on the order of 10^6.

echinopsine $C_{10}H_9O$ An alkaloid obtained from *Echinops* species; crystallizes as needles from benzene solution, melts at 152°C; physiological action is similar to that of brucine and strychnine.

eclipsed conformation A particular arrangement of constituent atoms that may rotate about a single bond in a molecule; for ethane it is such that when viewed along the axis of the carbon-carbon bond the hydrogen atoms of one methyl group are exactly in line with those of the other methyl group.

ectoenzyme An enzyme which is located on the external surface of a cell.

Edeleanu process A process for removal of compounds of sulfur from petroleum fractions by an extraction procedure utilizing liquid sulfur dioxide, or liquid sulfur dioxide and benzene.

edge-bridging ligand A ligand that forms a bridge over one edge of the polyhedron of a metal cluster structure.

edge dislocation A dislocation which may be regarded as the result of inserting an extra plane of atoms, terminating along the line of the dislocation. Also known as Taylor-Orowan dislocation.

edifenphos $C_{14}H_{15}O_2PS_2$ A clear, yellow liquid, insoluble in water; used as a fungicide on rice. Also known as O-ethyl-S,S-diphenyldithiophosphate.

Edman degradation technique In protein analysis, an approach to amino-end-group determination involving the use of a reagent, phenylisothiocyanate, that can be applied to the liberation of a derivative of the amino-terminal residue without hydrolysis of the remainder of the peptide chain.

effective area Absolute or cross-sectional area of process media involved in the process, such as the actual area of filter media through which a fluid passes, or the available surface area of absorbent contacted by a gas or liquid.

effective molecular diameter The general extent of the electron cloud surrounding a gas molecule as calculated in any of several ways.

effective permeability The observed permeability exhibited by a porous medium to one fluid phase when there is physical interaction between this phase and other fluid phases present.

effector An activator of an allosteric enzyme.

effervescence The bubbling of a solution of an element or chemical compound as the result of the emission of gas without the application of heat; for example, the escape of carbon dioxide from carbonated water.

efficiency In an ion-exchange system, a measurement of the effectiveness of a system expressed as the amount of regenerant required to remove a given unit of adsorbed material.

efflorescence The property of hydrated crystals to lose water of hydration and crumble when exposed to air.

effluent *See* discharge liquor.

effusion The movement of a gas through an opening which is small as compared with the average distance which the gas molecules travel between collisions.

Ehrlich's reagent $(CH_3)_2NC_6H_4CHO$ Granular or leafletlike crystals that are soluble in many organic solvents; melting point is 74°C; used in the preparation of dyes, as a reagent for arsphenamine, anthranilic acid, antipyrine, indole, and skatole, and as a differentiating agent between true scarlet fever and serum eruptions. Also known as *para*-dimethylaminobenzaldehyde; 4-dimethylaminobenzenecarbonal.

eicosanoic acid $CH_3(CH_2)_{18}COOH$ A white, crystalline, saturated fatty acid, melting at 75.4°C; a constituent of butter. Also known as arachic acid; arachidic acid.

einschluss thermometer All-glass, liquid-filled thermometer, temperature range -201 to $+360$°C, used for laboratory test work.

einsteinium Synthetic radioactive element, symbol Es, atomic number 99; discovered in debris of 1952 hydrogen bomb explosion; now made in cyclotrons.

Einstein photochemical equivalence law The law that each molecule taking part in a chemical reaction caused by electromagnetic radiation absorbs one photon of the radiation. Also known as Stark-Einstein law.

Einstein's absorption coefficient The proportionality constant governing the absorption of electromagnetic radiation by atoms, equal to the number of quanta absorbed per second divided by the product of the energy of radiation per unit volume per unit wave number and the number of atoms in the ground state.

Einstein viscosity equation An equation which gives the viscosity of a sol in terms of the volume of dissolved particles divided by the total volume.

elaidic acid $CH_3(CH_2)_7CH:CH(CH_2)_7COOH$ A transisomer of an unsaturated fatty acid, oleic acid; crystallizes as colorless leaflets, melts at 44°C, boils at 288°C (100 mm Hg), insoluble in water, soluble in alcohol and ether; used in chromatography as a reference standard. Also known as *trans*-9-octadecenoic acid.

elaidinization The process of changing the geometric cis form of an unsaturated fatty acid or a compound related to it into the trans form, resulting in an acid that is more resistant to oxidation.

elaidin reaction A test that differentiates nondrying oils such as olein from semidrying oils and drying oils; nitrous acid converts olein into its solid isomer, while semidrying oils in contact with nitrous acid thicken slowly, and drying oils such as tung oil become hard and resinous.

elastase An enzyme which acts on elastin to change it chemically and render it soluble.

elastin An elastic protein composing the principal component of elastic fibers.

Elbs reaction The formation of anthracene derivatives by dehydration and cyclization of diaryl ketone compounds which have a methyl group or methylene group; heating to an elevated temperature is usually required.

electrical calorimeter Device to measure heat evolved (from fusion or vaporization, for example); measured quantities of heat are added electrically to the sample, and the temperature rise is noted.

electrical equivalent In conductometric analyses of electrolyte solutions, an outside, calibrated current source as compared to (equivalent to) the current passing through the sample under analysis; for example, a Wheatstone-bridge balanced reading.

electric desalting A process to remove impurities such as inorganic salts from crude oil by settling out in an electrostatic field.

electric dipole transition A transition of an atom or nucleus from one energy state to another, in which electric dipole radiation is emitted or absorbed.

electric double layer A phenomenon found at a solid-liquid interface; it is made up of ions of one charge type which are fixed to the surface of the solid and an equal number of mobile ions of the opposite charge which are distributed through the neighboring region of the liquid; in such a system the movement of liquid causes a displacement of the mobile counterions with respect to the fixed charges on the solid surface. Also known as double layer.

electric precipitation A process that utilizes an electric field to improve the separation of hydrocarbon reagent dispersions.

electric quadrupole transition A transition of an atom or molecule from one energy state to another, in which electric quadrupole radiation is emitted or absorbed.

electride A member of a class of ionic compounds in which the anion is believed to be an electron.

electrobalance Analytical microbalance utilizing electromagnetic weighing; the sample weight is balanced by the torque produced by current in a coil in a magnetic field, with torque proportional to the current.

electrocatalysis Any one of the mechanisms which produce a speeding up of half-cell reactions at electrode surfaces.

electrochemical cell A combination of two electrodes arranged so that an overall oxidation-reduction reaction produces an electromotive force; includes dry cells, wet cells, standard cells, fuel cells, solid-electrolyte cells, and reserve cells.

electrochemical effect Conversion of chemical to electric energy, as in electrochemical cells; or the reverse process, used to produce elemental aluminum, magnesium, and bromine from compounds of these elements.

electrochemical emf Electrical force generated by means of chemical action, in manufactured cells (such as dry batteries) or by natural means (galvanic reaction).

electrochemical equivalent The weight in grams of a substance produced or consumed by electrolysis with 100% current efficiency during the flow of a quantity of electricity equal to 1 faraday ($96,487.0 \pm 1.6$ coulombs).

electrochemical potential The difference in potential that exists when two dissimilar electrodes are connected through an external conducting circuit and the two electrodes are placed in a conducting solution so that electrochemical reactions occur.

electrochemical process 1. A chemical change accompanying the passage of an electric current, especially as used in the preparation of commercially important quantities of certain chemical substances. 2. The reverse change, in which a chemical reaction is used as the source of energy to produce an electric current, as in a battery.

electrochemical reduction cell The cathode component of an electrochemical cell, at which chemical reduction occurs (while at the anode, chemical oxidation occurs).

electrochemical series A series in which the metals and other substances are listed in the order of their chemical reactivity or electrode potentials, the most reactive at the top and the less reactive at the bottom. Also known as electromotive series.

electrochemical techniques The experimental methods developed to study the physical and chemical phenomena associated with electron transfer at the interface of an electrode and solution.

electrochemistry A branch of chemistry dealing with chemical changes accompanying the passage of an electric current; or with the reverse process, in which a chemical reaction is used to produce an electric current.

electrochromatography Type of chromatography that utilizes application of an electric potential to produce an electric differential. Also known as electropherography.

electrocratic Referring to the repulsion exhibited by soap films and other colloids in solutions; such repulsion involves a strong osmotic contribution but is largely controlled by electrical forces.

electrocyclic reaction The interconversion of a linear π-system containing n π-electrons and a cyclic molecule containing $n - 2$ π-electrons which is formed by joining the ends of the linear molecule.

electrodecantation A modification of electrodialysis in which a cell is divided into three sections by two membranes and electrodes are placed in the end sections; colloidal matter is concentrated at the sides and bottom of the middle section, and the liquid that floats to the top is drawn off.

electrodeposition analysis An electroanalytical technique in which an element is quantitatively deposited on an electrode.

electrode potential Also known as electrode voltage. The voltage existing between an electrode and the solution or electrolyte in which it is immersed; usually, electrode potentials are referred to a standard electrode, such as the hydrogen electrode.

electrode voltage *See* electrode potential.

electrodialysis Dialysis that is conducted with the aid of an electromotive force applied to electrodes adjacent to both sides of the membrane.

electrodialyzer An instrument used to conduct electrodialysis.

electrodisintegration The breakup of a nucleus into two or more fragments as a result of bombardment by electrons.

electrofocusing *See* isoelectric focusing.

electrogravimetry Electrodeposition analysis in which the quantities of metals deposited may be determined by weighing a suitable electrode before and after deposition.

electrohydrodynamic ionization mass spectroscopy A technique for analysis of nonvolatile molecules in which the nonvolatile material is dissolved in a volatile solvent with a high dielectric constant such as glycerol, and high electric-field gradients at the surface of droplets of the liquid solution induce ion emission.

electrokinetic phenomena The phenomena associated with movement of charged particles through a continuous medium or with the movement of a continuous medium over a charged surface.

electrolysis A method by which chemical reactions are carried out by passage of electric current through a solution of an electrolyte or through a molten salt.

electrolyte A chemical compound which when molten or dissolved in certain solvents, usually water, will conduct an electric current.

electrolytic analysis Basic electrochemical technique for quantitative analysis of conducting solutions containing oxidizable or reducible material; measurement is based on the weight of material plated out onto the electrode.

electrolytic cell A cell consisting of electrodes immersed in an electrolyte solution, for carrying out electrolysis.

electrolytic conductance The transport of electric charges, under electric potential differences, by charged particles (called ions) of atomic or larger size.

electrolytic conductivity The conductivity of a medium in which the transport of electric charges, under electric potential differences, is by particles of atomic or larger size.

electrolytic dissociation   The ionization of a compound in a solution.

electrolytic mercaptan process   A process in which an aqueous caustic solution is used to extract mercaptans from refinery streams.

electrolytic potential   Difference in potential between an electrode and the immediately adjacent electrolyte, expressed in terms of some standard electrode difference.

electrolytic process   An electrochemical process involving the principles of electrolysis, especially as relating to the separation and deposition of metals.

electrolytic separation   Separation of isotopes by electrolysis, based on differing rates of discharge at the electrode of ions of different isotopes.

electrolytic solution   A solution made up of a solvent and an ionically dissociated solute; it will conduct electricity, and ions can be separated from the solution by deposition on an electrically charged electrode.

electromigration   A process used to separate isotopes or ionic species by the differences in their ionic mobilities in an electric field.

electromodulation   Modulation spectroscopy in which changes in transmission or reflection spectra induced by a perturbing electric field are measured.

electromotance   *See* electromotive force.

electromotive force   **1.** The difference in electric potential that exists between two dissimilar electrodes immersed in the same electrolyte or otherwise connected by ionic conductors. **2.** The resultant of the relative electrode potential of the two dissimilar electrodes at which electrochemical reactions occur. Abbreviated emf. Also known as electromotance.

electromotive series   *See* displacement series; electrochemical series.

electron acceptor   An atom or part of a molecule joined by a covalent bond to an electron donor.

electron affinity   The work needed in removing an electron from a negative ion, thus restoring the neutrality of an atom or molecule.

electron attachment   *See* electron capture.

electron capture   **1.** The process in which an atom or ion passing through a material medium either loses or gains one or more orbital electrons. **2.** A radioactive transformation of nuclide in which a bound electron merges with its nucleus. Also known as electron attachment.

electron capture detector   Extremely sensitive gas chromatography detector that is a modification of the argon ionization detector, with conditions adjusted to favor the formation of negative ions.

electron cloud   Picture of an electron state in which the charge is thought of as being smeared out, with the resulting charge density distribution corresponding to the probability distribution function associated with the Schrödinger wave function.

electron configuration   The orbital and spin arrangement of an atom's electrons, specifying the quantum numbers of the atom's electrons in a given state.

electron distribution curve   A curve indicating the electron distribution among the different available energy levels of a solid substance.

electron donor   An atom or part of a molecule which supplies both electrons of a duplet forming a covalent bond.

electronegative   Pertaining to an atom or group of atoms that has a relatively great tendency to attract electrons to itself.

electronegative potential   Potential of an electrode expressed as negative with respect to the hydrogen electrode.

electron energy level   A quantum-mechanical concept for energy levels of electrons about the nucleus; electron energies are functions of each particular atomic species.

electron energy loss spectroscopy   A technique for studying atoms, molecules, or solids in which a substance is bombarded with monochromatic electrons, and the energies of scattered electrons are measured to determine the distribution of energy loss.

electroneutrality principle   The principle that in an electrolytic solution the concentrations of all the ionic species are such that the solution as a whole is neutral.

electronic absorption spectrum   Spectrum resulting from absorption of electromagnetic radiation by atoms, ions, and molecules due to excitations of their electrons.

electronic angular momentum   The total angular momentum associated with the orbital motion of the spins of all the electrons of an atom.

electronic band spectrum   Bands of spectral lines associated with a change of electronic state of a molecule; each band corresponds to certain vibrational energies in the initial and final states and consists of numerous rotational lines.

electronic emission spectrum   Spectrum resulting from emission of electromagnetic radiation by atoms, ions, and molecules following excitations of their electrons.

electronic energy curve   A graph of the energy of a diatomic molecule in a given electronic state as a function of the distance between the nuclei of the atoms.

electronic magnetic moment   The total magnetic dipole moment associated with the orbital motion of all the electrons of an atom and the electron spins; opposed to nuclear magnetic moment.

electronic spectrum   Spectrum resulting from emission or absorption of electromagnetic radiation during changes in the electron configuration of atoms, ions, or molecules, as opposed to vibrational, rotational, fine-structure, or hyperfine spectra.

electron magnetic moment   The magnetic dipole moment which an electron possesses by virtue of its spin. Also known as electron dipole moment.

electron multiplicity   In an atom with Russell-Saunders coupling, the quantity $2S + 1$, where S is the total spin quantum number.

electron nuclear double resonance   A type of electron paramagnetic resonance (EPR) spectroscopy permitting greatly enhanced resolution, in which a material is simultaneously irradiated at one of its EPR frequencies and by a second oscillatory field whose frequency is swept over the range of nuclear frequencies. Abbreviated ENDOR.

electron number The number of electrons in an ion or atom.

electron pair A pair of valence electrons which form a nonpolar bond between two neighboring atoms.

electron probe microanalysis A technique in analytical chemistry in which a finely focused beam of electrons is used to excite an x-ray spectrum characteristic of the elements in the sample; can be used with samples as small as 10^{-11} cubic centimeter.

electron shell 1. The collection of all the electron states in an atom which have a given principal quantum number. 2. The collection of all the electron states in an atom which have a given principal quantum number and a given orbital angular momentum quantum number.

electron spectroscopy The study of the energy spectra of photoelectrons or Auger electrons emitted from a substance upon bombardment by electromagnetic radiation, electrons, or ions; used to investigate atomic, molecular, or solid-state structure, and in chemical analysis.

electron spectroscopy for chemical analysis *See* x-ray photoelectron spectroscopy.

electron spectrum Visual display, photograph, or graphical plot of the intensity of electrons emitted from a substance bombarded by x-rays or other radiation as a function of the kinetic energy of the electrons.

electron transport system The components of the final sequence of reactions in biological oxidations; composed of a series of oxidizing agents arranged in order of increasing strength and terminating in oxygen.

electroosmosis The movement in an electric field of liquid with respect to colloidal particles immobilized in a porous diaphragm or a single capillary tube.

electropherography *See* electrochromatography.

electrophilic Any chemical process in which electrons are acquired from or shared with other molecules or ions.

electrophilic reagent A reactant which accepts an electron pair from a molecule, with which it forms a covalent bond.

electrophoresis An electrochemical process in which colloidal particles or macromolecules with a net electric charge migrate in a solution under the influence of an electric current. Also known as cataphoresis.

electrophoretic effect Retarding effect on the characteristic motion of an ion in an electrolytic solution subjected to a potential gradient, which results from motion in the opposite direction by the ion atmosphere.

electrophoretic mobility A characteristic of living cells in suspension and biological compounds (proteins) in solution to travel in an electric field to the positive or negative electrode, because of the charge on these substances.

electrophoretic variants Phenotypically different proteins that are separable into distinct electrophoretic components due to differences in mobilities; an example is erythrocyte acid phosphatase.

electropositive Pertaining to elements, ions, or radicals that tend to give up or lose electrons.

electropositive potential Potential of an electrode expressed as positive with respect to the hydrogen electrode.

electrorefining Petroleum refinery process for light hydrocarbon streams in which an electrostatic field is used to assist in separation of chemical treating agents (acid, caustic, doctor) from the hydrocarbon phase.

electroreflectance Electromodulation in which reflection spectra are studied. Abbreviated ER.

electrostatic bond A valence bond in which two atoms are kept together by electrostatic forces caused by transferring one or more electrons from one atom to the other.

electrostatic valence rule The postulate that in a stable ionic structure the valence of each anion, with changed sign, equals the sum of the strengths of its electrostatic bonds to the adjacent cations.

electrosynthesis A reaction in which synthesis occurs as the result of an electric current.

electrovalence The valence of an atom that has formed an ionic bond.

electrovalent bond *See* ionic bond.

Elektrion process A process of condensation and polymerization in which a mixture of a relatively light mineral oil and a fatty oil is subjected to an electric discharge in an atmosphere of hydrogen; the product is a very viscous oil used for blending with lighter lubricating oils.

element A substance made up of atoms with the same atomic number; common examples are hydrogen, gold, and iron. Also known as chemical element.

element 104 The first element beyond the actinide series, and the twelfth transuranium element; the atoms of element 104, of mass number 260, were first produced by irradiating plutonium-242 with neon-22 ions in a heavy-ion cyclotron.

element 105 An artificial element whose isotope of mass number 260 was discovered by bombarding californium-249 with nitrogen-15 ions in a heavy-ion linear accelerator.

element 106 An artificial element whose isotope of mass number 263 was discovered by bombarding californium-249 with oxygen-18 ions in a heavy-ion linear accelerator, and whose isotope of mass number 259 was discovered by bombarding lead-207 and lead-208 with chromium-54 ions in a heavy-ion cyclotron.

element 107 An artificial element whose isotope of mass number 261 has been tentatively identified as a reaction product in the bombardment of bismuth-209 with chromium-54 ions and lead-208 with manganese-55 ions in a heavy-ion cyclotron.

elementary process In chemical kinetics, the particular events at the atomic or molecular level which make up an overall reaction.

elementary reaction A reaction which involves only a single transition state with no intermediates. Also known as step.

eleostearic acid $CH_3(CH_2)_7(CH:CH)_3(CH_2)_3COOH$ A colorless, water-insoluble, crystalline, unsaturated fatty acid; the glycerol ester is a chief component of tung oil. Also known as octadeca-9,11,13-trienoic acid.

Elgin extractor Spray-tower, multistage, counterflow extractor in which the diameter of the base section is expanded to eliminate flow restriction at the light-liquid distribution location.

elimination reaction A chemical reaction involving elimination of some portion of a reactant compound, with the production of a second compound.

ellagic acid $C_{14}H_6O_8$ A compound isolated from tannins as yellow crystals that are minimally soluble in hot water. Also known as gallogen.

elongation factor Any of several proteins required for elongation of growing polypeptide chains during protein synthesis.

eluant A liquid used to extract one material from another, as in chromatography.

eluate The solution that results from the elution process.

elution The removal of adsorbed species from a porous bed or chromatographic column by means of a stream of liquid or gas.

elutriation The process of removing substances from a mixture through washing and decanting.

emanation *See* radioactive emanation.

Embden-Meyerhof pathway *See* glycolytic pathway.

emf *See* electromotive force.

emission flame photometry A form of flame photometry in which a sample solution to be analyzed is aspirated into a hydrogen-oxygen or acetylene-oxygen flame; the line emission spectrum is formed, and the line or band of interest is isolated with a monochromator and its intensity measured photoelectrically.

emission lines Spectral lines resulting from emission of electromagnetic radiation by atoms, ions, or molecules during changes from excited states to states of lower energy.

emission spectrometer A spectrometer that measures percent concentrations of preselected elements in samples of metals and other materials; when the sample is vaporized by an electric spark or arc, the characteristic wavelengths of light emitted by each element are measured with a diffraction grating and an array of photodetectors.

emission spectrum Electromagnetic spectrum produced when radiations from any emitting source, excited by any of various forms of energy, are dispersed.

emodin $C_{14}H_4O_2(OH)_3CH_3$ Orange needles crystallizing from alcohol solution, melting point 256–257°C, practically insoluble in water, soluble in alcohol and aqueous alkali hydroxide solutions, occurs as the rhamnoside in plants such as rhubarb root and alder buckthorn; used as a laxative. Also known as archen; frangula emodin; frangulic acid; rheum emodin.

empirical formula A chemical formula indicating the variety and relative proportions of the atoms in a molecule but not showing the manner in which they are linked together.

emulsification The process of dispersing one liquid in a second immiscible liquid; the largest group of emulsifying agents are soaps, detergents, and other compounds, whose basic structure is a paraffin chain terminating in a polar group.

emulsification test Standard laboratory procedure for evaluating the resistance of insulating oils, turbine oils, and other lubricating oils to emulsification.

emulsion A stable dispersion of one liquid in a second immiscible liquid, such as milk (oil dispersed in water).

emulsion breaking In an emulsion, the combined sedimentation and coalescence of emulsified drops of the dispersed phase so that they will settle out of the carrier liquid; can be accomplished mechanically (in settlers, cyclones, or centrifuges) with or without the aid of chemical additives to increase the surface tension of the droplets.

emulsion cleaner A cleaner composed of organic solvents dispersed in an aqueous solution with the aid of an emulsifying agent.

emulsion polymerization A polymerization reaction that occurs in one phase of an emulsion.

enallachrome *See* esculin.

enantiomer *See* enantiomorph.

enantiomorph One of an isomeric pair of crystalline forms or compounds whose molecules are nonsuperimposable mirror images. Also known as enantiomer; optical antipode; optical isomer.

enantiomorphism A phenomenon of mirror-image relationship exhibited by right-handed and left-handed crystals or by the molecular structures of two stereoisomers.

enantiotopic ligand A ligand whose replacement or addition gives rise to enantiomers.

enantiotropy The relation of crystal forms of the same substance in which one form is stable above the transition-point temperature and the other stable below it, so that the forms can change reversibly one into the other.

endergonic Of or pertaining to a biochemical reaction in which the final products possess more free energy than the starting materials; usually associated with anabolism.

endo- Prefix that denotes inward-directed valence bonds of a six-membered ring in its boat form.

endocyclic double bond In a molecular structure, a double bond that is part of the ring system.

endoenzyme An intracellular enzyme, retained and utilized by the secreting cell.

endoergic *See* endothermic.

endogenous Relating to the metabolism of nitrogenous tissue elements.

endogenous pyrogen A fever-inducing substance (protein) produced by cells of the host body, such as leukocytes and macrophages.

endonuclease Any of a group of enzymes which degrade deoxyribonucleic acid or ribonucleic acid molecules by attaching nucleotide linkages within the polynucleotide chain.

endopeptidase An enzyme that acts upon the centrally located peptide bonds of a protein molecule.

ENDOR *See* electron nuclear double resonance.

β-endorphin   A 31–amino acid peptide fragment of pituitary β-lipotropic hormone having morphinelike activity.

endosulfan $C_9H_6Cl_6O_3S$ A tan solid that melts between -10 and $100°C$; used as an insecticide and miticide on vegetable and forage crops, on ornamental flowers, and in controlling termites and tsetse flies. Also known as 6,7,8,9,10,10-hexachloro-1,5,5a,6,9,9a-hexahydro-6,9-methano-2,4,3-benzo(e)-dioxathiepin-3-oxide.

endotherm   In differential thermal analysis, a graph of the temperature difference between a sample compound and a thermally inert reference compound (commonly aluminum oxide) as the substances are simultaneously heated to elevated temperatures at a predetermined rate, and the sample compound undergoes endothermal or exothermal processes.

endothermic   Pertaining to a reaction which absorbs heat. Also known as endoergic.

end point   1. In the distillation analysis of crude petroleum and its products, the highest reading of a thermometer when a specified proportion of the liquid has boiled off. Also known as final boiling point. 2. That stage in the titration at which an effect, such as a color change, occurs, indicating that a desired point in the titration has been reached.

endrin $C_{12}H_8OCl_6$ Poisonous, white crystals that are insoluble in water; it is used as a pesticide and is a stereoisomer of dieldrin, another pesticide.

energy metabolism   The chemical reactions involved in energy transformations within cells.

enfleurage   Removal of the odoriferous components from flowers by placing them near an odorless mixture of lard and tallow; this mixture absorbs the perfume, which is subsequently extracted.

Engler distillation test   A standard test for determination of the volatility characteristics of a gasoline by the measurement of the percent of gasoline distilled at various specific temperatures.

Engler flask   A standardized flask of 100-milliliter volume used in the Engler distillation test.

English degree   A unit of water hardness, equal to 1 part calcium carbonate to 70,000 parts water; equivalent to 1 grain of calcium carbonate per gallon of water. Also known as Clark degree.

English vermilion   Bright vermilion pigment of precipitated mercury sulfide; in paints, it tends to darken when exposed to light.

enhanced line   *See* enhanced spectral line.

enhanced spectral line   A spectral line of a very hot source, such as a spark, whose intensity is much greater than that of a line in a flame or arc spectrum. Also known as enhanced line.

enium ion   A cationic portion of an ionic species in which the valence shell of a positively charged nonmetallic atom has two electrons less than normal, and the charged entity has one covalent bond less than the corresponding uncharged speices; used as a suffix with the root name. Also known as ylium ion.

enkephalin   A mixture of two polypeptides isolated from the brain; central mode of action is an inhibition of neurotransmitter release.

enol An organic compound with a hydroxide group adjacent to a double bond; varies with a ketone form in the effect known as enol-keto tautomerism; an example is the compound $CH_3COH{=}CHCO_2C_2H_5$.

enolase An enzyme that catalyzes the reversible dehydration of phosphoglyceric acid to phosphopyruvic acid.

enolate anion The delocalized anion which is left after the removal of a proton from an enol, or of the carbonyl compound in equilibrium with the enol.

enol-keto tautomerism The tautomeric migration of a hydrogen atom from an adjacent carbon atom to a carbonyl group of a keto compound to produce the enol form of the compound; the reverse process of hydrogen atom migration also occurs.

enriched material Material in which the amount of one or more isotopes has been increased above that occurring in nature, such as uranium in which the abundance of ^{235}U is increased.

enriching column The portion of a countercurrent contractor (liquid-liquid extraction or vapor-liquid distillation) above the feed point in which an upward-moving, product-rich stream from the stripping column is further purified by countercurrent contact with a downward-flowing reflux stream from the overhead product-recovery vessel.

enterokinase An enzyme which catalyzes the conversion of trypsinogen to trypsin.

enthalpimetric analysis Generic designation for a group of modern thermochemical methodologies such as thermometric enthalpy titrations which rely on monitoring the temperature changes produced in adiabatic calorimeters by heats of reaction occurring in solution; in contradistinction, classical methods of thermoanalysis such as thermogravimetry focus primarily on changes occurring in solid samples in response to externally imposed programmed alterations in temperature.

enthalpy of reaction The change in enthalpy accompanying a chemical reaction.

enthalpy of transition The change of enthalpy accompanying a phase transition.

enthalpy titration *See* thermometric titration.

Entner-Doudoroff pathway A sequence of reactions for glucose degradation, with the liberation of energy; the distinguishing feature is the formation of 2-keto-3-deoxy-6-phosphogluconate from 6-phosphogluconate and the cleaving of this compound to yield pyruvate and glyceraldehyde-3-phosphate.

entrainer An additive that forms an azeotrope with one component of a liquid mixture to aid in otherwise difficult separations by distillation, as in azeotropic distillation.

entrainment A process in which the liquid boils so violently that suspended droplets of liquid are carried in the escaping vapor.

entrance slit Narrow slit through which passes the light entering a spectrometer.

entropy of activation The difference in entropy between the activated complex in a chemical reaction and the reactants.

entropy of mixing After mixing substances, the difference between the entropy of the mixture and the sum of the entropies of the components of the mixture.

entropy of transition   The heat absorbed or liberated in a phase change divided by the absolute temperature at which the change occurs.

enzyme   Any of a group of catalytic proteins that are produced by living cells and that mediate and promote the chemical processes of life without themselves being altered or destroyed.

enzyme inhibition   Prevention of an enzymic process as a result of the interaction of some substance with the enzyme so as to decrease the rate of reaction.

enzyme repression   The process by which the rate of synthesis of an enzyme is reduced in the presence of a metabolite, often the end product of a chain of reactions in which the enzyme in question operates near the beginning.

enzyme unit   The amount of an enzyme that will catalyze the transformation of 10^{-6} mole of substrate per minute or, when more than one bond of each substrate is attacked, 10^{-6} of 1 gram equivalent of the group concerned, under specified conditions of temperature, substrate concentration, and pH number.

enzymology   A branch of science dealing with the chemical nature, biological activity, and biological significance of enzymes.

eosin $C_{20}H_8O_5Br_4$ 1. A red fluorescent dye in the form of triclinic crystals that are insoluble in water; used chiefly in cosmetics and as a toner. Also known as bromeosin; bromo acid; eosine; tetrabromofluorescein. 2. The red to brown crystalline sodium or potassium salt of this dye; used in organic pigments, as a biological stain, and in pharmaceuticals. Also known as eosine; eosine G; eosine Y; eosine yellowish.

eosine   *See* eosin.

eosine G   *See* eosin.

eosine Y   *See* eosin.

eosine yellowish   *See* eosin.

ephedrine $C_{10}H_{15}NO$ A white, crystalline, water-soluble alkaloid present in several *Ephedra* species and also produced synthetically; a sympathomimetic amine, it is used for its action on the bronchi, blood pressure, blood vessels, and central nervous system.

epi-   A prefix used in naming compounds to indicate the presence of a bridge or intramolecular connection.

epichlorohydrin C_3H_5OCl A colorless, unstable liquid, insoluble in water; used as a solvent for resins. Also known as 1-chloro-2,3-epoxypropane; chloropropylene oxide.

epimer   A type of isomer in which the difference between the two compounds is the relative position of the H (hydrogen) group and OH (hydroxyl) group on the last asymmetric C (carbon) atom of the chain, as in the sugars D-glucose and D-mannose.

epimerase   A type of enzyme that catalyzes the rearrangement of hydroxyl groups on a substrate.

epimerization   A process in which in an optically active compound that contains two or more asymmetric centers, only one of these is altered by some reaction to form an epimer.

epinephrine $C_9H_{13}O_3N$ A hormone secreted by the adrenal medulla that acts to increase blood pressure due to stimulation of heart action and constriction of peripheral blood vessels. Also known as adrenaline.

epitaxy Growth of one crystal on the surface of another crystal in which the growth of the deposited crystal is oriented by the lattice structure of the substrate.

EPN *See* O-ethyl-O-*para*-nitrophenyl phenylphosphonothioate.

epoxidation Reaction yielding an epoxy compound, such as the conversion of ethylene to ethylene oxide.

epoxide A reactive group in which an oxygen atom is joined to each of two carbon atoms which are already bonded.

epoxy- A prefix indicating presence of an epoxide group in a molecule.

1,2-epoxybutane *See* 1,2-butylene oxide.

1,2-epoxyethane *See* ethylene oxide.

2,3-epoxy-1-propanol *See* glycidol.

epoxy resin A polyether resin formed originally by the polymerization of bisphenol A and epichlorohydrin, having high strength, and low shrinkage during curing; used as a coating, adhesive, casting, or foam.

epsilcapromin *See* ε-aminocaproic acid.

equalizing line A pipe or tubing interconnection between two closed vessels, containers, or process systems to allow pressure equalization.

equation A symbolic expression that represents in an abbreviated form the laboratory observations of a chemical change; an equation (such as $2H_2 + O_2 \rightarrow 2H_2O$) indicates what reactants are consumed (H_2 and O_2) and what products are formed (H_2O), the correct formula of each reactant and product, and satisfies the law of conservation of atoms in that the symbols for the number of atoms reacting equals the number of atoms in the products.

equation of state A mathematical expression which defines the physical state of a homogeneous substance (gas, liquid, or solid) by relating volume to pressure and absolute temperature for a given mass of the material.

equidensity technique Interference microscopy technique utilizing the Sabattier effect in photographic emulsions; the equidensities (lines of equal density in a photographic emulsion) are produced by exactly superimposing a positive and a negative of the same interferogram, and making a copy; used to measure photographic film emulsion density.

equilibrium constant A constant at a given temperature such that when a reversible chemical reaction $cC + bB = gG + hH$ has reached equilibrium, the value of this constant K^0 is equal to

$$\frac{a_G^g a_H^h}{a_C^c a_B^b}$$

where a_G, a_H, a_C, and a_B represent chemical activities of the species G, H, C, and B at equilibrium.

equilibrium diagram A phase diagram of the equilibrium relationship between temperature, pressure, and composition in any system.

equilibrium dialysis A technique used to determine the degree of ion bonding by protein; the protein solution, placed in a bag impermeable to protein but permeable to small ions, is immersed in a solution containing the diffusible ion whose binding is being studied; after equilibration of the ion across the membrane, the concentration of ion in the protein-free solution is determined; the concentration of ion in the protein solution is determined by subtraction; if binding has occurred, the concentration of ion in the protein solution must be greater.

equilibrium flash vaporization Process in which a continuous liquid-mixture feed stream is partly vaporized in a column or vessel, with continuous withdrawal of vapor and liquid portions, the vapor and liquid in equilibrium. Also known as continuous equilibrium vaporization; simple continuous distillation.

equilibrium moisture content The moisture content in a hydroscopic material that is being dried by contact with air at constant temperature and humidity when a definite, fixed (equilibrium) moisture content in the solid is reached.

equilibrium potential A point in which forward and reverse reaction rates are equal in an electrolytic solution, thereby establishing the potential of an electrode.

equilibrium prism Three-dimensional (solid) diagram for multicomponent mixtures to show the effects of composition changes on some key property, such as freezing point.

equilibrium ratio 1. In any system, relation of the proportions of the various components (gas, liquid) at equilibrium conditions. 2. *See* equilibrium vaporization ratio.

equilibrium solubility The maximum solubility of one material in another (for example, water in hydrocarbons) for specified conditions of temperature and pressure.

equilibrium still Recirculating distillation apparatus (no product withdrawal) used to determine vapor-liquid equilibria data.

equilibrium vaporization ratio In a liquid-vapor equilibrium mixture, the ratio of the mole fraction of a component in the vapor phase (y) to the mole fraction of the same component in the liquid phase (x), or $y/x = K$ (the K factor). Also known as equilibrium ratio.

equipartition 1. The condition in a gas where under equal pressure the molecules of the gas maintain the same average distance between each other. 2. The equal distribution of a compound between two solvents. 3. The distribution of the atoms in an orderly fashion, such as in a crystal.

equivalence point The point in a titration where the amounts of titrant and material being titrated are equivalent chemically.

equivalent conductance Property of an electrolyte, equal to the specific conductance divided by the number of gram equivalents of solute per cubic centimeter of solvent.

equivalent electrons Electrons in an atom which have the same principal and orbital quantum numbers, but not necessarily the same magnetic orbital and magnetic spin quantum numbers.

equivalent nuclei A set of nuclei in a molecule which are transformed into each other by rotations, reflections, or combinations of these operations, leaving the molecule invariant.

equivalent weight The number of parts by weight of an element or compound which will combine with or replace, directly or indirectly, 1.008 parts by weight of hydrogen, 8.00 parts of oxygen, or the equivalent weight of any other element or compound.

equiviscous temperature A measure of viscosity used in the tar industry, equal to the temperature in degrees Celsius at which the viscosity of tar is 50 seconds as measured in a standard tar efflux viscometer. Abbreviated EVT.

Er *See* erbium.

ER *See* electroreflectance.

erbia *See* erbium oxide.

erbium A trivalent metallic rare-earth element, symbol Er, of the yttrium subgroup, found in euxenite, gadolinite, fergusonite, and xenotine; atomic number 68, atomic weight 167.26, specific gravity 9.051; insoluble in water, soluble in acids; melts at 1400–1500°C.

erbium halide A compound of erbium and one of the halide elements.

erbium nitrate $Er(NO_3)_3 \cdot 5H_2O$ Pink crystals that are soluble in water, alcohol, and acetone; may explode if it is heated or shocked.

erbium oxalate $Er_2(C_2O_4)_3 \cdot 10H_2O$ A red powder that decomposes at 575°C; used to separate erbium from common metals.

erbium oxide Er_2O_3 Pink powder that is insoluble in water; used as an actuator for phosphors and in manufacture of glass that absorbs in the infrared. Also known as erbia.

erbium sulfate $Er_2(SO_4)_3 \cdot 8H_2O$ Red crystals that are soluble in water.

erbon $C_{11}H_9Cl_5O_3$ A white solid with a melting point of 49–50°C; insoluble in water; used as an herbicide for perennial broadleaf weeds. Also known as 2-(2,4,5-trichlorophenoxy)ethyl 2,2-dichloropropionate.

ergosterin *See* ergosterol.

ergosterol $C_{28}H_{44}O$ A crystalline, water-insoluble, unsaturated sterol found in ergot, yeast, and other fungi, and which may be converted to vitamin D_2 on irradiation with ultraviolet light or activation with electrons. Also known as ergosterin.

ergotinine An alkaloid and an isomer of ergotoxine that is a 1:1:1 mixture of ergocornine, ergocristine, and ergocryptine; crystallizes in long needles from acetone solutions, melting point 229°C, and soluble in chloroform, alcohol, and absolute ether.

ergotoxine An alkaloid and an isomer of ergotinine that is a 1:1:1 mixture of ergocornine, ergocristine, and ergocryptine; crystallizes in orthorhombic crystals, melts at 190°C, and is soluble in methyl alcohol, ethyl alcohol, acetone, and chloroform.

eriodictyol $C_{15}H_{22}O_6$ A compound isolated from *Eriodictyon californicum* as needlelike crystals from a dilute alcohol solution, sparingly soluble in boiling water, hot alcohol, and glacial acetic acid; used in medicine as an expectorant. Also known as 3′,4′,5,7-tetrahydroxyflavanone.

Erlenmeyer flask A conical glass laboratory flask, with a broad bottom and a narrow neck.

Erlenmeyer synthesis   Preparation of cyclic ethers by the condensation of an aldehyde with an α-acylamino acid in the presence of acetic anhydride and sodium acetate.

erucic acid $C_{22}H_{42}O_2$ A monoethenoid acid that is the cis isomer of brassidic acid and makes up 40 to 50% of the total fatty acid in rapeseed, wallflower seed, and mustard seed; crystallizes as needles from alcohol solution, insoluble in water, soluble in ethanol and methanol. Also known as *cis*-13-docosenoic acid.

erythrite   *See* erythritol.

erythritol $H(CHOH)_4H$ A tetrahydric alcohol; occurs as tetragonal prisms, melting at 121°C, soluble in water; used in medicine as a vasodilator. Also known as antierythrite; erythrite; ethroglucin; erythrol; phycite; tetrahydroxy butane.

erythrocruorin   Any of the iron-porphyrin protein respiratory pigments found in the blood and tissue fluids of certain invertebrates; corresponds to hemoglobin in vertebrates.

erythroidine $C_{16}H_{19}NO_3$ An alkaloid existing in two forms: α-erythroidine and β-erythroidine, isolated from *Erythrina* species; β-erythroidine has an action similar to that of curare as a skeletal muscle relaxant.

D-erythropentose   *See* ribulose.

erythrophleine $C_{24}H_{39}NO_5$ An alkaloid isolated from the bark of *Erythrophleum guineense;* used in medicine experimentally for its digitalislike action.

erythropoietin   A hormone, thought to be produced by the kidneys, that regulates erythropoiesis, at least in higher vertebrates.

erythrose $HOCH_2(CHOH)_2CHO$ A tetrose sugar obtained from erythrol; a syrupy liquid at room temperature.

erythrosin $C_{13}H_{18}O_6N_2$ A red compound obtained by reacting tyrosine with nitric acid.

erythrulose $C_4H_8O_4$ A ketose sugar occurring as an oxidation product of erythritol due to the action of certain bacteria.

Es   *See* einsteinium.

escaping tendency   The tendency of a solute species to escape from solution; related to the chemical potential of the solute.

Eschka mixture   A mixture of two parts magnesium oxide and one part anhydrous sodium carbonate; used as a fusion mixture for determining sulfur in coal.

Eschweiler-Clarke modification   A modification of the Leuckart reaction, involving reductive alkylation of ammonia or amines (except tertiary amines) by formaldehyde and formic acid.

esdragol   *See* estragole.

eserine   *See* physostigmine.

essential amino acid   Any of eight of the 20 naturally occurring amino acids that are indispensable for optimum animal growth but cannot be formed in the body and must be supplied in the diet.

essential fatty acid Any of the polyunsaturated fatty acids which are required in the diet of mammals; they are probably precursors of prostaglandins.

ester The compound formed by the elimination of water and the bonding of an alcohol and an organic acid.

esterase Any of a group of enzymes that catalyze the synthesis and hydrolysis of esters.

ester gum A compound obtained by forming an ester of a natural resin with a polyhydric alcohol; used in varnishes, paints, and cellulosic lacquers. Also known as rosin ester.

esterification A chemical reaction whereby esters are formed.

estersil Hydrophobic silica powder, an ester of —SiOH with a monohydric alcohol; used as a filler in silicone rubbers, plastics, and printing inks.

estradiol $C_{18}H_{24}O_2$ An estrogenic hormone produced by follicle cells of the vertebrate ovary; provokes estrus and proliferation of the human endometrium, and stimulates ICSH (interstitial-cell-stimulating hormone) secretion.

estragole $C_6H_4(C_3H_5)(OCH_3)$ A colorless liquid with the odor of anise, found in basil oil, estragon oil, and anise bark oil; used in perfumes and flavorings. Also known as *para*-allylanisole; chavicol methyl ether; esdragol; methyl chavicol.

estriol $C_{18}H_{24}O_3$ A crystalline estrogenic hormone obtained from human pregnancy urine.

estrogen Any of various natural or synthetic substances possessing the biologic activity of estrus-producing hormones.

estrogenic hormone A hormone, found principally in ovaries and also in the placenta, which stimulates the accessory sex structures and the secondary sex characteristics in the female.

estrone $C_{18}H_{22}O_2$ An estrogenic hormone produced by follicle cells of the vertebrate ovary; functions the same as estradiol.

Etard reaction Direct oxidation of an aromatic or heterocyclic bound methyl group to an aldehyde by utilizing chromyl chloride or certain metallic oxides.

ethal *See* cetyl alcohol.

ethamine *See* ethyl amine.

ethanal *See* acetaldehyde.

ethanamide *See* acetamide.

ethanamidine hydrochloride *See* acetamidine hydrochloride.

ethane CH_3CH_3 A colorless, odorless gas belonging to the alkane series of hydrocarbons, with freezing point of $-183.3°C$ and boiling point of $-88.6°C$; used as a fuel and refrigerant and for organic synthesis.

1,2-ethanediamine *See* ethylenediamine.

ethanedioyl chloride *See* oxalyl chloride.

ethanedithioamide *See* dithiooxamide.

1,2-ethanedithiol $HSCH_2CH_2SH$ A liquid, freely soluble in alcohol and in alkalies; used as a metal complexing agent. Also known as dithioethyleneglycol; ethylenedimercaptan.

ethane nitrile *See* acetonitrile.

ethanethiol *See* ethyl mercaptan.

ethanethiolic acid *See* thioacetic acid.

ethanoic acid *See* acetic acid.

ethanol *See* alcohol.

ethanol acid *See* glyoxalic acid.

ethanolamine $NH_2(CH_2)_2OH$ A colorless liquid, miscible in water; used in scrubbing H_2S and CO_2 from petroleum gas streams, for dry cleaning, in paints, and in pharmaceuticals. Also known as 2-aminoethanol; 2-hydroxyethylamine; monoethanolamine.

ethanolurea $NH_2CONHCH_2CH_2OH$ A white solid; its formaldehyde condensation products are thermoplastic and water-soluble.

ethanoylaminoethanoic acid *See* aceturic acid.

ethanoyl chloride *See* acetyl chloride.

ethbenzamide *See* salicylamide.

ethene *See* ethylene.

ethenol *See* vinyl alcohol.

ethenylamidine hydrochloride *See* acetamidine hydrochloride.

ethephon $C_2H_6ClO_3P$ A white solid with a melting point of 74.75°C; very soluble in water; used as a growth regulator for tomatoes, apples, cherries, and walnuts. Also known as CEPHA; 2-(chloroethyl)phosphonic acid.

ether **1.** One of a class of organic compounds characterized by the structural feature of an oxygen linking two hydrocarbon groups (such as R—O—R). **2.** $(C_2H_5)_2O$ A colorless liquid, slightly soluble in water; used as a reagent, intermediate, anesthetic, and solvent. Also known as ethyl ether.

etherification The process of making an ether from an alcohol.

ethidine *See* ethylidene.

ethidium bromide $C_{21}H_{20}BrN_3$ Dark red crystals with a melting point of 238–240°C; used in treating trypanosomiasis in animals and as an inhibitor of deoxyribonucleic and ribonucleic acid synthesis. Also known as homidium bromide.

ethinyl The CH_3:C— radical from acetylene. Also known as acetenyl; acetylenyl; ethynyl.

ethiofencarb $C_{11}H_{15}O_2NS$ A yellow liquid, used as an insecticide in fruits, vegetables, and ornamentals. Also known as 2-ethylmercaptomethylphenyl-*N*-methylcarbamate.

ethiolate $C_7H_{15}ONS$ A yellow liquid with a boiling point of 206°C; used as a preemergence herbicide for corn. Also known as *S*-ethyldiethylthiocarbamate.

ethion $C_9H_{22}O_4P_2S_3$ A clear liquid, slightly soluble in water; used as an insecticide and miticide in the control of aphids, mites, scales, thrips, and foliage-eating larvae on food, fiber, and ornamental crops. Also known as O,O,O',O'-tetraethyl S,S'-methylene bisphosphorodithioate.

ethionic acid $HO \cdot SO_2 \cdot CH_2 \cdot CH_2 \cdot SO_2OH$ An unstable diacid, known only in solution. Also known as ethylene sulfonic acid.

ethionine $C_5H_{13}O_2N$ An amino acid that is the ethyl analog of and the biological antagonist of methionine.

ethiops mineral *See* mercuric sulfide.

ethirimol $C_{12}H_{21}N_3O$ A white, crystalline compound with a melting point of 159–160°C; slight solubility in water; used as a fungicide for powdery mildew of barley. Also known as 5-n-butyl-2-ethylamino-4-hydroxy-6-methylpyrimidine.

ethohexadiol $C_8H_{18}O_2$ A slightly oily liquid, used as an insect repellent. Also known as 2-ethyl-1,3-hexanediol; 2-ethyl-3-propyl-1,3-propanediol; octylene glycol.

ethoprop $C_8H_{19}O_2PS_2$ A pale yellow liquid compound, insoluble in water; used as an insecticide for soil insects and as a nematicide for plant parasitic nematodes. Also known as O-ethyl-S,S-dipropylphosphorodithioate.

ethoxide A compound formed from ethanol by replacing the hydrogen of the hydroxy group by a monovalent metal. Also known as ethylate.

ethoxy The C_2H_5O — radical from ethyl alcohol. Also known as ethyoxyl.

***para*-ethoxyacetanilide** *See* acetophenetidin.

2-ethoxybenzenecarbonamide *See* salicylamide.

S-(N-ethoxycarbonyl-N-methylcarbamoylmethyl)-O,O-diethylphosphorodithioate *See* mecarbam.

6-ethoxy-1,2-dihydro-2,2,4-trimethylquinoline *See* ethoxyquin.

2-ethoxyethanol *See* cellosolve.

2-(2-ethoxyethoxy)ethanol *See* diethylene glycol monoethyl ether.

1-ethoxy-2-hydroxy-4-propenylbenzene *See* propenyl guaethol.

ethoxylation A catalytic process which involves the direct addition of ethylene oxide to an alkyl phenol or to an aliphatic alcohol.

ethoxyquin $C_{14}H_{19}NO$ A dark liquid, used as a growth regulator to protect apples and pears in storage. Also known as 6-ethoxy-1,2-dihydro-2,2,4-trimethylquinoline.

ethroglucin *See* erythritol.

ethyl **1.** The hydrocarbon radical C_2H_5. **2.** Trade name for the tetraethyllead antiknock compound in gasoline.

ethyl acetate $CH_3COOC_2H_5$ A colorless liquid, slightly soluble in water; boils at 77°C; a medicine, reagent, and solvent. Also known as acetic ester; acetic ether; acetidin.

ethyl acetic acid *See* butyric acid.

ethyl acetoacetate $CH_3COCH_2COOC_2H_5$ A colorless liquid, boiling at 181°C; used as a reagent, intermediate, and solvent. Also known as acetoacetic ester; diacetic ether.

ethyl acetone *See* pentanone.

ethyl acetylene Compound with boiling point 8.1°C; insoluble in water, soluble in alcohol; used in organic synthesis. Also known as 1-butine; 1-butyne.

ethyl acrylate $C_5H_8O_2$ A colorless liquid, boiling at 99°C; used to manufacture chemicals and resins.

ethyl alcohol *See* alcohol.

ethylallene *See* pentadiene.

ethyl amine A colorless liquid, boiling at 15°C, water-soluble; used as a solvent, as a dye intermediate, and in organic synthesis. Also known as aminoethane; ethamine.

ethyl-*para*-aminobenzoate $C_6H_4NH_2CO_2$ C_2H_5 A white powder, melting point 88–92°C, slightly soluble in ethanol and ether, very slightly soluble in water; used as a local anesthetic. Also known as benzocaine.

2-(ethylamino)-4-(isopropylamino)-6-(methylthio)-*s*-triazine $C_9H_{17}N_5S$ White crystals with a melting point of 84–85°C; solubility in water is 185 parts per million at 20°C; used as an herbicide for pineapple, sugarcane, and bananas. Also known as ametryn.

ethyl amyl ketone $C_8H_{16}O$ A colorless liquid, almost insoluble in water; used in perfumery. Also known as octanone-3.

ethylate *See* ethoxide.

ethylation Formation of a new compound by introducing the ethyl radical $(C_2H_5{}^-)$.

ethyl benzene $C_6H_5C_2H_5$ A colorless liquid that boils at 136°C, insoluble in water; used in organic synthesis, as a solvent, and in making styrene.

ethyl benzoate $C_6H_5COOCH_2CH_3$ Colorless, aromatic liquid, boiling at 213°C, insoluble in water; used as a solvent, in flavoring extracts, and in perfumery.

ethyl-*N*-benzoyl-*N*-(3,4-dichlorophenyl)-2-aminopropionate *See* benzoylpropethyl.

α-ethylbenzylalcohol *See* 1-phenyl-1-propanol.

ethyl borate $B(OC_2H_5)_3$ A salt of ethanol and boric acid; colorless, flammable liquid; used in antiseptics, disinfectants, and fireproofing. Also known as boron triethoxide; triethylic borate.

ethyl bromide C_2H_5Br A colorless liquid, boiling at 39°C; used as a refrigerant and in organic synthesis. Also known as bromic ether; bromoethane.

2-ethylbutene $CH_3CH_2(C_2H_5)CCH_2$ Colorless liquid, soluble in alcohol and organic solvents, insoluble in water; used in organic synthesis.

2-ethylbutyl acetate $C_2H_5CH(C_2H_5)CH_2O_2CCH_3$ Colorless liquid with mild odor; used as a solvent for resins, lacquers, and nitrocellulose.

2-ethylbutyl alcohol $(C_2H_5)_2CHCH_2OH$ Stable, colorless liquid, miscible in most organic solvents, slightly water-soluble; used as a solvent for resins, waxes, and dyes, and in the synthesis of perfumes, drugs, and flavorings.

ethyl butyl ketone $C_2H_5COC_4H_9$ A colorless liquid, boiling at 147°C; used in solvent mixtures. Also known as 3-heptanone.

ethyl butyrate $C_3H_7COOC_2H_5$ A colorless liquid, boiling at 121°C; used in flavoring extracts and perfumery. Also known as butyric ether.

ethyl caprate $CH_3(CH_2)_8COOC_2H_5$ A colorless liquid, used in the manufacture of wine bouquets and cognac essence. Also known as ethyl decanoate.

ethyl caproate $C_5H_{11}COOC_2H_5$ A colorless to yellow liquid, boiling at 167°C, soluble in ether and alcohol, and having a pleasant odor; used as a chemical intermediate and in the food industry as an artificial fruit essence. Also known as ethyl hexanoate; ethyl hexoate.

ethyl caprylate $CH_3(CH_2)_6COOC_2H_5$ A clear, colorless liquid with a pineapple odor; used to make fruit ethers. Also known as ethyl octanoate.

ethyl carbamate   *See* urethane.

ethyl carbinol   *See* propyl alcohol.

ethyl carbonate   *See* diethyl carbonate.

ethyl cellulose   The ethyl ester of cellulose; it has film-forming properties and is inert to alkalies and dilute acids; used in adhesives, lacquers, and coatings.

ethyl chloride C_2H_5Cl A colorless gas, liquefying at 12.2°C, slightly soluble in water; used as a solvent, in medicine, and as an intermediate. Also known as chloroethane; hydrochloric ether; monochloroethane; muriatic ether.

ethyl chloroacetate $CH_2ClCOOC_2H_5$ A colorless liquid, boiling at 145°C; used as a poison gas, solvent, and chemical intermediate.

ethyl chlorophyllide   A compound formed by replacing the phytyl ($C_{20}H_{39}$) tail of the chlorophyll molecule with a short ethyl (C_2H_5) tail; crystallizes easily and has an absorption spectrum and electrochemical properties similar to those of chlorophyll.

ethyl cinnamate $C_6H_5CH=CHCOOC_2H_5$ An oily liquid with a faint cinnamon odor; used as a fixative for perfumes. Also known as ethyl phenylacrylate.

ethyl crotonate $CH_3CHCHCO_2C_2H_5$ A compound with a pungent aroma; boiling point of 143–147°C, soluble in water, soluble in ether; one of two isomeric forms used as an organic intermediate, a solvent for cellulose esters, and as a plasticizer for acrylic resins. Also known as *trans*-2-butenoic acid.

ethyl crotonic acid CH_3CHCCR_5COOH Colorless monoclinic crystals, subliming at 40°C; used as a peppermint flavoring.

ethyl cyanide C_2H_5CN A colorless liquid that boils at 97.1°C; poisonous. Also known as hydrocyanic ether; propionitrile.

ethyl decanoate   *See* ethyl caprate.

ethyl 4,4′-dichlorobenzilate   *See* chlorobenzilate.

***S*-ethyldiethylthiocarbamate**   *See* ethiolate.

***S*-ethyl diisobutylthiocarbamate**   *See* butylate.

***O*-ethyl-*S,S*-diphenyldithiophosphate**   *See* edifenphos.

***O*-ethyl-*S,S*-dipropylphosphorodithioate**   *See* ethoprop.

***S*-ethyl-*N,N*-dipropylthiocarbamate** $C_9H_{19}NOS$ An amber liquid soluble in water at 370 parts per million; used as a pre- and postemergence herbicide on vegetable crops. Also known as EDTC.

ethyl enanthate $CH_3(CH_2)_5COOC_2H_5$ A clear, colorless oil with a boiling point of 187°C; soluble in alcohol, chloroform, and ether; taste and odor are fruity; used as a flavor for liqueurs and soft drinks. Also known as cognac oil; ethyl heptanoate; ethyl oenanthate.

ethylene C_2H_4 A colorless, flammable gas, boiling at $-102.7°C$; used as an agricultural chemical, in medicine, and for the manufacture of organic chemicals and polyethylene. Also known as ethene; olefiant gas.

ethylene alkylation A catalytic petroleum-refining process in which dry isobutane and ethylene react to form ethylene alkylate.

ethyleneamine *See* piperazine.

3,3′-ethylenebis(tetrahydro-4,6-dimethyl-2H-1,3,5-thiadiazine-2-thione) *See* milneb.

ethylene bromide *See* ethylene dibromide.

ethylene carbonate $(CH_2O)_2CO$ Odorless, colorless solid with low melting point; soluble in water and organic solvents; used as a polymer and resin solvent, in solvent extraction, and in organic syntheses.

ethylene chloride $ClCH_2CH_2Cl$ A colorless, oily liquid, boiling at 83.7°C; used as a solvent and fumigant, for organic synthesis, and for ore flotation. Also known as Dutch liquid; ethylene dichloride.

ethylene chlorobromide CH_2BrCH_2Cl Volatile, colorless liquid with chloroform-like odor; soluble in ether and alcohol but not in water; general-purpose solvent for cellulosics; used in organic synthesis.

ethylene chlorohydrin $ClCH_2CH_2OH$ A colorless, poisonous liquid, boiling at 129°C; used as a solvent and in organic synthesis. Also known as chloroethyl alcohol.

ethylene cyanide $C_2H_4(CN)_2$ Colorless crystals, melting at 57°C; used in organic synthesis. Also known as succinonitrile.

ethylene cyanohydrin C_3H_5ON A colorless liquid that is miscible with water and boils at 221°C. Also known as 2-cyanoethanol; 1-hydroxy-2-cyanoethane.

ethylene diacetate *See* ethylene glycol diacetate.

ethylenediamine $NH_2CH_2CH_2NH_2$ Colorless liquid, melting at 8.5°C, soluble in water; used as a solvent, corrosion inhibitor, and resin and in adhesive manufacture. Also known as diaminoethane; 1,2-ethanediamine.

ethylenediaminetetraacetic acid $(HOOCCH_2)_2NCH_2CH_2N(CH_2COOH)$ White crystals, slightly soluble in water and decomposing above 160°C; the sodium salt is a strong chelating agent, reacting with many metallic ions to form soluble nonionic chelate. Abbreviated EDTA.

ethylene dibromide $BrCH_2CH_2Br$ A colorless, poisonous liquid, boiling at 131°C; insoluble in water; used in medicine, as a solvent in organic synthesis, and in antiknock gasoline. Also known as 1,2-dibromoethane; ethylene bromide.

ethylene dichloride *See* ethylene chloride.

ethylenedimercaptan *See* 1,2-ethanedithiol.

ethylene glycol *See* glycol.

ethylene glycol bis(trichloroacetate) $C_4H_4Cl_6O_4$ A white solid with a melting point of 40.3°C; used as an herbicide for cotton and soybeans. Also known as EGT.

ethylene glycol diacetate CH_3COOCH_2 CH_2OOCCH_3 A liquid used as a solvent for oils, cellulose esters, and explosives. Also known as ethylene diacetate; glycol diacetate.

ethylene glycol monobutyl ether   *See* 2-butoxyethanol.

ethylene glycol monomethyl ether   *See* 2-methoxyethanol.

ethylene glycol monophenyl ether   *See* 2-phenoxyethanol.

ethylenehydrinsulfonic acid   *See* isethionic acid.

ethyleneimine C_2H_4NH Highly corrosive liquid, colorless and clear; miscible with organic solvents and water; used as an intermediate in fuel oil production, refining lubricants, textiles, and pharmaceuticals. Also known as aziridine.

ethylene nitrate $(CH_2NO_3)_2$ An explosive yellow liquid, insoluble in water. Also known as glycoldinitrate.

ethylene oxide $(CH_2)_2O$ A colorless gas, soluble in organic solvents and miscible in water, boiling point 11°C; used in organic synthesis, for sterilizing, and for fumigating. Also known as 1,2-epoxyethane.

ethylene resin A thermoplastic material composed of polymers of ethylene; the resin is synthesized by polymerization of ethylene at elevated temperatures and pressures in the presence of catalysts. Also known as polyethylene; polyethylene resin.

ethylene sulfonic acid   *See* ethionic acid.

ethylethanolamine $C_2H_5NHCH_2CH_2OH$ Water-white liquid with amine odor; soluble in alcohol, ether, and water; used in dyes, insecticides, fungicides, and surface-active agents.

ethyl ether   *See* ether.

***ortho*-ethyl ether**   *See* salicylamide.

ethyl ether–boron trifluoride complex   *See* boron trifluoride etherate.

***S*-ethyl-*N*-ethyl-*N*-cyclohexylthiocarbamate**   *See* cycloate.

ethylethylene   *See* butene-1.

ethyl formate $HCOOC_2H_5$ A colorless liquid, boiling at 54.4°C; used as a solvent, fumigant, and larvicide and in flavors, resins, and medicines.

ethyl heptanoate   *See* ethyl enanthate.

***S*-ethyl hexahydro-1*H*-azepine-1 carbothioate**   *See* molinate.

2-ethyl-1,3-hexanediol   *See* ethohexadiol.

ethyl hexanoate   *See* ethyl caproate.

2-ethyl-1-hexanol   *See* 2-ethyl-hexyl alcohol.

ethyl hexoate   *See* ethyl caproate.

2-ethyl hexoic acid $C_4H_9CH(C_2H_5)COOH$ A liquid that is slightly soluble in water, boils at 226.9°C, and has a mild odor; used as an intermediate to make metallic salts for paint and varnish driers, esters for plasticizers, and light metal salts for conversion of some oils to grease. Also known as butyl ethylacetic acid.

2-ethylhexyl acetate $CH_3COOCH_2CHC_2H_5C_4H_9$ Water-white, stable liquid; used as a solvent for nitrocellulose, resins, and lacquers. Also known as octyl acetate.

2-ethylhexyl acrylate $CH_2CHCOOCH_2CH(C_2H_5)C_4H_9$ Pleasant-smelling liquid; used as monomer for plastics, protective coatings, and paper finishes.

2-ethylhexyl alcohol $C_4H_9CH(C_2H_5)CH_2OH$ Colorless, slightly viscous liquid; used as a defoaming or wetting agent, as a solvent for protective coatings, waxes, and oils, and as a raw material for plasticizers. Also known as 2-ethyl-1-hexanol; octyl alcohol.

2-ethylhexylamine $C_4H_9CH(C_2H_5)CH_2NH_2$ Water-white liquid with slight ammonia odor; slightly water-soluble; used to synthesize detergents, rubber chemicals, and oil additives.

2-ethylhexyl bromide $C_4H_9CH(C_2H_5)CH_2Br$ Water-white, water-insoluble liquid; used to prepare pharmaceuticals and disinfectants.

2-ethylhexyl chloride $C_4H_9CH(C_2H_5)CH_2Cl$ Colorless liquid; used to synthesize cellulose derivatives, pharmaceuticals, resins, insecticides, and dyestuffs.

ethyl-*para*-hydroxybenzoate $HOC_6H_4COOC_2H_5$ Crystals with a melting point of 116°C that are soluble in water, alcohol, and ether; used as a preservative for pharmaceuticals. Also known as ethylparaben.

ethyl-*para*-hydroxyphenyl ketone *See* *para*-hydroxypropiophenone.

ethyl-2-hydroxypropionate *See* ethyl lactate.

ethylic compound Generic term for ethyl compounds.

ethylic ether *See* diethyl ether.

ethylidine The $CH_3 \cdot CH\!\!=$ radical from ethane, C_2H_5. Also known as ethidine.

ethyl iodide C_2H_5I A colorless liquid, boiling at 72.3°C; used in medicine and in organic synthesis. Also known as hydroiodic ether; iodoethane.

ethyl isobutylmethane *See* 2-methylhexane.

ethyl isovalerate $(CH_3)_2CHCH_2COOC_2H_5$ A colorless, oily liquid with an apple odor, soluble in water and miscible with alcohol, benzene, and ether; used for flavoring beverages and confectioneries.

ethyl lactate $CH_3CHOHCOOC_2H_5$ A colorless liquid that boils at 154°C, has a mild odor, and is miscible with water and organic solvents such as alcohols, ketones, esters, and hydrocarbons; used as a flavoring and as a solvent for cellulose compounds such as nitrocellulose, cellulose acetate, and cellulose ethers. Also known as ethyl-2-hydroxypropionate.

ethyl malonate $CH_2(COOC_2H_5)_2$ A colorless liquid, boiling at 198°C; used as an intermediate and a plasticizer. Also known as malonic ester.

ethyl mercaptan C_2H_5SH A colorless liquid, boiling at 36°C. Also known as ethanethiol; ethyl sulfhydrate; thioethyl alcohol.

2-ethylmercaptomethylphenyl-*N*-methylcarbamate *See* ethiofencarb.

ethyl methacrylate $CH_2CCH_3COOC_2H_5$ Colorless, easily polymerized liquid, water-insoluble; used to produce polymers and chemical intermediates.

ethyl (2*E*,4*E*)-11-methoxy-3,7,11-trimethyl-2,4-dodecadienethiolate *See* triprene.

ethylmethylacetylene *See* pentyne.

ethyl methyl ketone *See* methyl ethyl ketone.

ethyl 4-(methylthio)-*meta*-tolyl isopropylphosphoramidate $C_{13}H_{22}NO_2PS$ A brown, crystalline compound with a melting point of 49°C; limited solubility in water; used as a nematicide and fumigant to treat seeds and turf grasses.

ethyl nitrate $C_2H_5NO_3$ A colorless, flammable liquid, boiling at 87.6°C; used in perfumes, drugs, and dyes and in organic synthesis.

ethyl nitrite $C_2H_5NO_2$ A colorless liquid, boiling at 16.4°C; used in medicine and in organic synthesis. Also known as sweet spirits of niter.

O-ethyl-O-*para*-nitrophenyl phenylphosphonothioate $C_2H_5O_4NPS$ A yellow, crystalline compound with a melting point of 36°C; used as an insecticide and miticide on fruit crops.

ethyl octanoate *See* ethyl caprylate.

ethyl oenanthate *See* ethyl enanthate.

ethyl oleate $C_{20}H_{38}O_2$ A yellow oil, insoluble in water; used as a solvent, plasticizer, and lubricant.

ethyl orthosilicate *See* ethyl silicate.

ethyl oxalate $(COOC_2H_5)_2$ Oily, unstable, colorless liquid that is combustible; miscible with organic solvents, very slightly soluble in water; used as a solvent for cellulosics and resins, and as an intermediate for dyes and pharmaceuticals.

ethyl oxide *See* diethyl ether.

ethylparaben *See* ethyl-*para*-hydroxybenzoate.

ethyl phenylacrylate *See* ethyl cinnamate.

ethyl phenyl carbinol *See* 1-phenyl-1-propanol.

N-ethyl-5-phenylisoxazolium-3'-sulfonate $C_{11}H_{11}NO_4S$ Crystals that decompose at 207–208°C; used to form peptide bonds. Also known as Woodward's Reagent K.

1-ethyl-3-piperidinol *See* 1-ethyl-3-hydroxypiperidine.

ethyl propionate $C_2H_5COOC_2H_5$ A colorless liquid, slightly soluble in water, boiling at 99°C; used as solvent and pyroxylin cutting agent. Also known as propionic ether.

2-ethyl-3-propyl-1,3-propanediol *See* ethohexadiol.

ethyl salicylate $(HO)C_6H_4COOC_2H_5$ A clear liquid with a pleasant odor; used in commercial preparation of artificial perfumes. Also known as sal ethyl; salicylic acid ethyl ether; salicylic ether.

ethyl silicate $(C_2H_5)_4SiO_4$ A colorless, flammable liquid, hydrolyzed by water; used as a preservative for stone, brick, and masonry, in lacquers, and as a bonding agent. Also known as ethyl orthosilicate.

ethyl sulfate *See* diethyl sulfate.

ethyl sulfhydrate *See* ethyl mercaptan.

ethyl sulfide $(C_2H_5)_2S$ A colorless, oily liquid, boiling at 92°C; used as a solvent and in organic synthesis. Also known as diethyl sulfide; ethylthioethane.

2-ethylthio-4,6-bis(isopropylamino)-S-triazine *See* dipropetryn.

ethylthioethane *See* ethyl sulfide.

ortho-**ethyl(*O*-2,4,5-trichlorophenyl)ethylphosphonothioate** $C_{10}H_{12}OPSCl_2$ An amber liquid with a boiling point of 108°C at 0.01 mmHg; solubility in water is 50 parts per million; used as an insecticide for vegetable crops and soil pests on meadows. Also known as trichloronate.

ethyl urethane *See* urethane.

ethyl vanillin $C_2H_5O(OH)C_6H_3CHO$ A compound, crystallizing in fine white crystals that melt at 76.5°C, has a strong vanilla odor and four times the flavor of vanilla, soluble in organic solvents such as alcohol, chloroform, and ether; used in the food industry as a flavoring agent to replace or fortify vanilla.

ethyne *See* acetylene.

ethynyl *See* ethinyl.

ethynylation Production of an acetylenic derivative by the condensation of acetylene with a compound such as an aldehyde; for example, production of butynediol from the union of formaldehyde with acetylene.

ethyoxyl *See* ethoxy.

etioallocholane *See* androstane.

etioporphyrin $C_{31}H_{34}N_4$ A synthetic porphyrin that has four ethyl and four methyl groups in a red-pigmented compound whose crystals melt at 280°C.

eucalyptol $C_{10}H_{18}O$ A colorless oil with a camphorlike odor; boiling point is 174–177°C; used in pharmaceuticals, perfumery, and flavoring. Also known as cajeputol; cineol.

eugenol $CH_2CHCH_2C_6H_3(OCH_3)OH$ A colorless or yellowish aromatic liquid with spicy odor and taste, soluble in organic solvents, and extracted from clove oil; used in flavors, perfumes, medicines, and the manufacture of vanilla. Also known as 4-allyl-2-methoxyphenol.

euglobulin True globulin; a simple protein that is soluble in distilled water and dilute salt solutions.

euonymit *See* dulcitol.

europium A member of the rare-earth elements in the cerium subgroup, symbol Eu, atomic number 63, atomic weight 151.96, steel gray and malleable, melting at 1100–1200°C.

europium halide Any of the compounds of the element europium and the halogen elements; for example, europium chloride, $EuCl_3 \cdot xH_2O$.

europium oxide Eu_2O_3 A white powder, insoluble in water; used in red- and infrared-sensitive phosphors.

eutectic An alloy or solution that has the lowest possible constant melting point.

eutectic point The point in the constitutional diagram indicating the composition and temperature of the lowest melting point of a eutectic.

eutectic system The particular composition and temperature of materials at the eutectic point.

eutectic temperature The temperature at the lowest melting point of a eutectic.

eutectogenic system A multicomponent liquid-solid mixture in which pure solid phases of each component are in equilibrium with the remaining liquid mixture at a specific (usually minimum) temperature for a given composition, that is, the eutectic point.

eutectoid The point in an equilibrium diagram for a solid solution at which the solution on cooling is converted to a mixture of solids.

evaporation loss The loss of a stored volatile liquid component or mixture by evaporation; controlled by temperature, pressure, and the presence or absence of vapor-recovery systems.

evaporator A device used to vaporize part or all of the solvent from a solution; the valuable product is usually either a solid or concentrated solution of the solute.

even-even nucleus A nucleus which has an even number of neutrons and an even number of protons.

even-odd nucleus A nucleus which has an even number of protons and an odd number of neutrons.

EVT *See* equiviscous temperature.

exchange adsorption Ion exchange process in which the fluid phase contains (or consists of) two adsorbable components which together entirely saturate the surfaces of the adsorbent.

exchange broadening The broadening of a spectral line by some type of chemical or spin exchange process which limits the lifetime of the absorbing or emitting species and produces the broadening via the Heisenberg uncertainty principle.

exchange narrowing The phenomenon in which, when a spectral line is split and thereby broadened by some variable perturbation, the broadening may be narrowed by a dynamic process that exchanges different values of the perturbation.

exchange reaction Reaction in which two atoms or ions exchange places either in two different molecules or in the same molecule.

exchange velocity In an ion-exchange process, the speed with which one ion is displaced from an exchanger in favor of another ion.

excimer An excited diatomic molecule where both atoms are of the same species and are dissociated in the ground state.

exciplex An excited electron donor-acceptor complex which is dissociated in the ground state.

excision enzyme A bacterial enzyme that removes damaged dimers from the deoxyribonucleic acid molecule of a bacterial cell following light or ultraviolet radiation or nitrogen mustard damage.

excitation A process in which an atom or molecule gains energy from electromagnetic radiation or by collision, raising it to an excited state.

excitation curve A curve showing the relative yield of a specified nuclear reaction as a function of the energy of the incident particles or photons. Also known as excitation function.

excitation function 1. The cross section for an incident electron to excite an atom to a particular excited state expressed as a function of the electron energy. 2. *See* excitation curve.

excitation index In emission spectroscopy, the ratio of intensities of a pair of extremely nonhomologous spectra lines; used to provide a sensitive indication of variation in excitation conditions.

excitation purity The ratio of the departure of the chromaticity of a specified color to that of the reference source, measured on a chromaticity diagram; used as a guide of the wavelength of spectrum color needed to be mixed with a reference color to give the specified color.

excitation spectrum The graph of luminous efficiency per unit energy of the exciting light absorbed by a photoluminescent body versus the frequency of the exciting light.

exciting line The frequency of electromagnetic radiation, that is, the spectral line from a noncontinuous source, which is absorbed by a system in connection with some particular process.

exergonic Of or pertaining to a biochemical reaction in which the end products possess less free energy than the starting materials; usually associated with catabolism.

exhaustion point In an ion-exchange process, the state of an adsorbent at which it no longer can produce a useful ion exchange.

exo- A conformation of carbon bonds in a six-membered ring such that the molecule is boat-shaped with one or more substituents directed outward from the ring.

exocyclic double bond A double bond that is connected to and external to a ring structure.

exoenzyme An enzyme that functions outside the cell in which it was synthesized.

exonuclease Any of a group of enzymes which catalyze hydrolysis of single nucleotide residues from the end of a deoxyribonucleic acid chain.

exopeptidase An enzyme that acts on the terminal peptide bonds of a protein chain.

exotherm The graphical plotting of heat rise and fall versus time for an exothermic reaction or process system.

explosion A chemical reaction or change of state which is effected in an exceedingly short space of time with the generation of a high temperature and generally a large quantity of gas.

explosive D *See* ammonium picrate.

explosive limits The upper and lower limits of percentage composition of a combustible gas mixed with other gases or air within which the mixture explodes when ignited.

expression Separation of liquid from a two-phase solid-liquid system by compression under conditions that permit liquid to escape while the solid is retained between the compressing surfaces. Also known as mechanical expression.

extended dislocation A dislocation in a close-packed structure consisting of a strip of stacking fault edged by two lines across which slip through a fraction of a lattice constant, into one of the alternative stacking positions, has occurred.

extender A material used to dilute or extend or change the properties of resins, ceramics, paints, rubber, and so on.

extensive property A noninherent property of a system, such as volume or internal energy, that changes with the quantity of material in the system; the quantitative value equals the sum of the values of the property for the individual constituents.

external circuit All connecting wires, devices, and current sources which achieve desired conditions within an electrolytic cell.

extinction *See* absorbance.

extinction coefficient *See* absorptivity.

extract Material separated from liquid or solid mixture by a solvent.

extractant The liquid used to remove a solute from another liquid.

extraction A method of separation in which a solid or solution is contacted with a liquid solvent (the two being essential mutually insoluble) to transfer one or more components into the solvent.

extraction column Vertical-process vessel in which a desired product is separated from a liquid by countercurrent contact with a solvent in which the desired product is preferentially soluble.

extractive distillation A distillation process to separate components from eutectic mixtures; a solution of the mixture is cooled, causing one component to crystallize out and the other to remain in solution; used to separate *p*-xylene and *m*-xylene, using *n*-pentane as the solvent.

extractor An apparatus for solvent-contact with liquids or solids for removal of specified components.

extreme narrowing approximation A mathematical approximation in the theory of spectral-line shapes to the effect that the exchange narrowing of a perturbation is complete.

extrinsic factor *See* vitamin B_{12}.

extrinsic sol A colloid whose stability is attributed to electric charge on the surface of the colloidal particles.

exudation *See* sweating.

Eyring equation An equation, based on statistical mechanics, which gives the specific reaction rate for a chemical reaction in terms of the heat of activation, entropy of activation, the temperature, and various constants.

F

F *See* fluorine.

face *See* crystal face.

face-bridging ligand A ligand that forms a bridge over one triangular face of the polyhedron of a metal cluster structure.

face-centered cubic lattice A lattice whose unit cells are cubes, with lattice points at the center of each face of the cube, as well as at the vertices. Abbreviated fcc lattice.

face-centered orthorhombic lattice An orthorhombic lattice which has lattice points at the center of each face of a unit cell, as well as at the vertices.

factor I *See* fibrinogen.

factor III *See* thromboplastin.

factor IV Calcium ions involved in the mechanism of blood coagulation.

factor V *See* proaccelerin.

factor VII A procoagulant, related to prothrombin, that is involved in the formation of a prothrombin-converting principle which transforms prothrombin to thrombin. Also known as stable factor.

factor VIII *See* antihemophilic factor.

factor IX *See* Christmas factor.

factor X *See* Stuart factor.

factor XI A procoagulant present in normal blood but deficient in hemophiliacs. Also known as plasma thromboplastin antecedent (PTA).

factor XII A blood clotting factor effective experimentally only in vitro; deficient in hemophiliacs. Also known as Hageman factor.

FAD *See* flavin adenine dinucleotide.

falling-film still Special molecular distillation apparatus designed for high evaporative and separation efficiency. Also known as falling-film molecular still.

false body The property of certain colloidal substances, such as paints and printing inks, of solidifying when left standing.

family A group of elements whose chemical properties, such as valence, solubility of salts, and behavior toward reagents, are similar.

famphur $C_{10}H_{16}NO_5PS_2$ A crystalline compound with a melting point of 55°C; slightly soluble in water; used as an insecticide for lice and grubs of reindeer and cattle. Also known as O,O-dimethyl-O-[(*para*-dimethylsulfamoyl) phenyl]phosphorothioate.

Faraday's laws of electrolysis 1. The amount of any substance dissolved or deposited in electrolysis is proportional to the total electric charge passed. 2. The amounts of different substances dissolved or desposited by the passage of the same electric charge are proportional to their equivalent weights.

farnesol $C_1H_{25}OH$ A colorless liquid extracted from oils of plants such as citronella, neroli, cyclamen, and tuberose; it has a delicate floral odor, and is an intermediate step in the biological synthesis of cholesterol from mevalonic acid in vertebrates; used in perfumery.

fast chemical reaction A reaction with a half-life of milliseconds or less; such reactions occur so rapidly that special experimental techniques are required to observe their rate.

fast fission Fission caused by fast neutrons.

fast-neutron spectrometry Neutron spectrometry in which nuclear reactions are produced by or yield fast neutrons; such reactions are more varied than in the slow-neutron case.

fat Any of the glyceryl esters of fatty acids which form a class of neutral organic compounds.

fat-metabolizing hormone *See* ketogenic hormone.

fatty acid An organic monobasic acid of the general formula $C_nH_{2n+1}COOH$ derived from the saturated series of aliphatic hydrocarbons; examples are palmitic acid, stearic acid, and oleic acid; used as a lubricant in cosmetics and nutrition, and for soaps and detergents.

fatty acid peroxidase An enzyme present in germinating plant seeds which catalyzes the oxidation of the carboxyl carbon of fatty acids to carbon dioxide.

fatty acyl carnitine Transport form of fatty acids which allows them to cross the mitochondrial membrane; formed by reaction of fatty acyl-coenzyme A with carnitine by employing the enzyme carnitine acyltransferase. Also known as acyl carnitine.

fatty acyl-coenzyme A Activated form of fatty acids formed by the enzyme acyl-coenzyme A synthetase at the expense of adenosinetriphosphate. Also known as acyl-coenzyme A.

fatty alcohol A high-molecular-weight, straight-chain primary alcohol derived from natural fats and oils; includes lauryl, stearyl, oleyl, and linoleyl alcohols; used in pharmaceuticals, cosmetics, detergents, plastics, and lube oils and in textile manufacture.

fatty amine RCH_2NH_2 A normal aliphatic amine from oils and fats; used as a plasticizer, in medicine, as a chemical intermediate, and in rubber manufacture.

fatty ester $RCOOR'$ A fatty acid in which the alkyl group (R') of a monohydric alcohol replaces the active hydrogen; for example, $RCOOCH_3$ from reaction of $RCOOH$ with methane.

fatty nitrile RCN An ester of hydrogen cyanide derived from fatty acid; used in lube oil additives and plasticizers, and as a chemical intermediate.

Favorskii rearrangement A reaction in which α-halogenated ketones undergo rearrangement in the presence of bases, with loss of the halogen and formation of carboxylic acids or their derivatives with the same number of carbon atoms.

Faxen drag factor *See* drag factor.

fcc lattice *See* face-centered cubic lattice.

Fe *See* iron.

^{55}Fe *See* iron-55.

^{59}Fe *See* iron-59.

feedback inhibition A cellular control mechanism by which the end product of a series of metabolic reactions inhibits the activity of the first enzyme in the sequence.

feed preparation unit A processing unit (such as distillation or desulfurization units) providing feedstock for subsequent processing.

feed tray For a tray-type distillation column, that tray on which fresh feedstock is introduced into the system.

Fehling's reagent A solution of cupric sulfate, sodium potassium tartrate, and sodium hydroxide, used to test for the presence of reducing compounds such as sugars.

fenaminosulf $C_8H_{10}N_3SO_3Na$ A yellow-brown powder, decomposing at 200°C; used as a fungicide for seeds and seedlings in crops. Also known as *para*-dimethylaminobenzenediazo sodium sulfonate.

fenazaflor $C_{15}H_7Cl_2F_3N_2O_2$ A greenish-yellow, crystalline compound with a melting point of 103°C; used as an insecticide and miticide for spider mites and eggs. Also known as phenyl 5,6-dichloro-2-trifluoromethyl-1-benzimidazole carboxylate.

fenbutatin oxide $C_{60}H_{78}OSn_2$ A white, crystalline compound, insoluble in water; used to control mites in deciduous and citrus fruits. Also known as distannoxane; hexakis(β,β-dimethylphenethyl).

fenchol *See* fenchyl alcohol.

fenchone $C_{10}H_{16}O$ An isomer of camphor; a colorless oil that boils at 193°C and is soluble in ether; a constituent of fennel oil; used as a flavoring.

fenchyl alcohol $C_{10}H_{18}O$ A colorless solid or oily liquid, boiling at 198–204°C, isolated from pine oil and turpentine and also made synthetically; used as a solvent, an intermediate in organic synthesis, and as a flavoring. Also known as fenchol; 1-hydroxy fenchane.

fenitrothion $C_9H_{12}NO_5PS$ A yellow-brown liquid, insoluble in water; used as a miticide and insecticide for rice, orchards, vegetables, cereals, and cotton, and for fly and mosquito control. Also known as *O,O*-dimethyl-*O*-(4-nitro-*meta*-tolyl)phosphorothioate.

fenolactine *See* *para*-lactophenetide.

Fenske equation *See* Fenske-Underwood equation.

Fenske-Underwood equation Equation in plate-to-plate distillation-column calculations relating the number of theoretical plates needed at total reflux to overall relative volatility and the liquid-vapor composition ratios on upper and lower plates. Also known as Fenske equation.

fenson $C_{12}H_9ClO_2S$ A colorless, crystalline compound with a melting point of 61–62°C; insoluble in water; used as a miticide for pears, peaches, and apples. Also known as *para*-chlorophenylbenzenesulfonate; CPBS: PCPBS.

fensulfothion $C_{11}H_{17}S_2O_2P$ A brown liquid with a boiling point of 138–141°C; used as an insecticide and nematicide in soils. Also known as *O,O*-diethyl-*O-para*-(methylsulfinylphenyl)phosphorothioate.

fenthion *See O,O*-dimethyl *O*-[3-methyl-4(methylthio)phenyl]phosphorothioate.

fentiazon $(C_6H_5)_2C_3HN_2S_2$ A yellow, crystalline compound with a melting point of 296.4°C; used as a fungicide for rice. Also known as 3-benzylideneamino-4-phenylthiazoline-2-thione.

fentinacetate $C_{20}H_{18}O_2Sn$ A yellow to brown, crystalline solid that melts at 124–125°C; used as a fungicide, molluscicide, and algicide for early and late blight on potatoes, sugarbeets, peanuts, and coffee. Also known as triphenyltinacetate.

fenuron $C_9H_{12}N_2O$ A white, crystalline compound with a melting point of 133–134°C; soluble in water; used as an herbicide to kill weeds and bushes. Also known as 3-phenyl-1,1-dimethylurea.

fenuron-TCA $C_{11}H_{13}Cl_3N_2O_3$ A white, crystalline compound with a melting point of 65–68°C; moderately soluble in water; used as an herbicide for non-crop areas. Also known as 3-phenyl-1,1-dimethylurea trichloroacetate.

FEP resin *See* fluorinated ethylene propylene resin.

ferbam $C_9H_{18}FeN_3S_6$ A black powder that decomposes at 180°C; slight solubility in water; used as a fungicide for fruit, tobacco, and other crops. Also known as ferric dimethyl dithiocarbamate.

ferment An agent that can initiate fermentation and other metabolic processes.

Fermi beta-decay theory Theory in which a nucleon source current interacts with an electron-neutrino field to produce beta decay, in a manner analogous to the interaction of an electric current with an electromagnetic field during the emission of a photon of electromagnetic radiation.

Fermi constant A universal constant, introduced in beta-disintegration theory, that expresses the strength of the interaction between the transforming nucleon and the electron-neutrino field.

Fermi plot *See* Kurie plot.

Fermi resonance In a polyatomic molecule, the relationship of two vibrational levels that have in zero approximation nearly the same energy; they repel each other, and the eigenfunctions of the two states mix.

Fermi selection rules Selection rules for beta decay in a Fermi transition; that is, there is no change in total angular momentum or parity of the nucleus in an allowed transition.

Fermi transition Beta decay subject to Fermi selection rules.

fermium A synthetic radioactive element, symbol Fm, with atomic number 100; discovered in debris of the 1952 hydrogen bomb explosion, and now made in nuclear reactors.

ferrate A multiple iron oxide with another oxide, for example, Na_2FeO_4.

ferredoxins Iron-containing proteins that transfer electrons, usually at a low potential, to flavoproteins; the iron is not present as a heme.

ferric   The term for a compound of trivalent iron, for example, ferric bromide, $FeBr_3$.

ferric acetate $Fe_2(C_2H_3O_2)_3$ A brown compound, soluble in water; used as a tonic and dye mordant.

ferric ammonium alum   *See* ferric ammonium sulfate.

ferric ammonium citrate $Fe(NH_4)_3(C_6H_5O_7)_2$ Red, deliquescent scales or granules; odorless, water soluble, and affected by light; used in medicine and blueprint photography.

ferric ammonium oxalate $(NH_4)_3Fe(C_2O_4)_3 \cdot 3H_2O$ Green, crystalline material, soluble in water and alcohol, sensitive to light; used in blueprint photography.

ferric ammonium sulfate $FeNH_4(SO_4)_2 \cdot 12H_2O$ Efflorescent, water-soluble crystals; used in medicine, in analytical chemistry, and as a mordant in textile dyeing. Also known as ferric ammonium alum; iron ammonium sulfate.

ferric arsenate $FeAsO_4 \cdot 2H_2O$ A green or brown powder, insoluble in water, soluble in dilute mineral acids; used as an insecticide.

ferric bromide $FeBr_3$ Red, deliquescent crystals that decompose upon heating; soluble in water, ether, and alcohol; used in medicine and analytical chemistry. Also known as ferric sesquibromide; ferric tribromide; iron bromide.

ferric chloride $FeCl_3$ Brown crystals, melting at 300°C, that are soluble in water, alcohol, and glycerol; used as a coagulant for sewage and industrial wastes, as an oxidizing and chlorinating agent, as a disinfectant, in copper etching, and as a mordant. Also known as anhydrous ferric chloride; ferric trichloride; flores martis; iron chloride.

ferric citrate $FeC_6H_5O_7 \cdot 3H_2O$ Red scales that react to light; soluble in water, insoluble in alcohol; used as a medicine for certain blood disorders, and for blueprint paper. Also known as iron citrate.

ferric dichromate $Fe_2(CrO_4)_3$ A red-brown, granular powder, miscible in water; used as a mordant.

ferric dimethyl dithiocarbamate   *See* ferbam.

ferric ferrocyanide $Fe_4[Fe(CN)_6]_3$ Dark-blue crystals, used as a pigment, and with oxalic acid in blue ink. Also known as iron ferrocyanide.

ferric fluoride FeF_3 Green, rhombohedral crystals, soluble in water and acids; used in porcelain and pottery manufacture. Also known as iron fluoride.

ferric hydrate   *See* ferric hydroxide.

ferric hydroxide $Fe(OH)_3$ A brown powder, insoluble in water; used as arsenic poisoning antidote, in pigments, and in pharmaceutical preparations. Also known as ferric hydrate; iron hydroxide.

ferric nitrate $Fe(NO_3)_3 \cdot 9H_2O$ Colorless crystals, soluble in water and decomposed by heat; used as a dyeing mordant, in tanning, and in analytical chemistry. Also known as iron nitrate.

ferric oxalate $Fe_2(COO)_3$ Yellow scales, soluble in water, decomposing when heated at about 100°C; used as a catalyst and in photographic printing papers.

ferric oxide Fe_2O_3 Red, hexagonal crystals or powder, insoluble in water and soluble in acids, melting at 1565°C; used as a catalyst and pigment for metal polishing, in metallurgy, and in medicine. Also known as ferric oxide red; jeweler's rouge; red ocher.

ferric oxide red   *See* ferric oxide.

ferric phosphate $FePO_4 \cdot 2H_2O$ Yellow, rhombohedral crystals, insoluble in water, soluble in acids; used in medicines and fertilizers. Also known as iron phosphate.

ferric resinate   Reddish-brown, water-insoluble powder; used as a drier for paints and varnishes. Also known as iron resinate.

ferric sesquibromide   *See* ferric bromide.

ferric stearate $Fe(C_{18}H_{35}O_2)_3$ A light-brown, water-insoluble powder; used as a varnish drier. Also known as iron stearate.

ferric sulfate $Fe_2(SO_4)_3 \cdot 9H_2O$ Yellow, water-soluble, rhombohedral crystals, decomposing when heated; used as a chemical intermediate, disinfectant, soil conditioner, pigment, and analytical reagent, and in medicine. Also known as iron sulfate.

ferric tribromide   *See* ferric bromide.

ferric trichloride   *See* ferric chloride.

ferric vanadate $Fe(VO_3)_3$ Grayish-brown powder, insoluble in water and alcohol; used in metallurgy. Also known as iron metavanadate.

ferricyanic acid $H_3Fe(CN)_6$ A red-brown unstable solid.

ferricyanide   A salt containing the radical $Fe(CN)_6^{3-}$.

ferrihemoglobin   Hemoglobin in the oxidized state. Also known as methemoglobin.

ferriporphyrin   A red-brown to black complex of iron and porphyrin in which the iron is in the 3+ oxidation state.

ferrisulphas   *See* ferrous sulfate.

ferrite   An unstable compound of a strong base and ferric oxide which exists in alkaline solution, such as $NaFeO_2$.

ferritin   An iron-protein complex occurring in tissues, probably as a storage form of iron.

ferrocene $(CH_2)_5Fe(CH_2)_5$ Orange crystals that are soluble in ether, melting point 174°C; used as a combustion control additive in fuels, and for heat stabilization in greases and plastics. Also known as dicyclopentadienyl iron.

ferrochelatase   A mitochondrial enzyme which catalyzes the incorporation of iron into the protoporphyrin molecule.

ferrocyanic acid $H_4Fe(CN)_6$ A white solid obtained by treating ferrocyanides with acid.

ferrocyanide   A salt containing the radical $Fe(CN)_6^{4-}$.

ferrocyanide process   A regenerative chemical treatment for removal of mercaptans from petroleum fuels; uses caustic–sodium ferrocyanide reagent.

ferroporphyrin   A red complex of porphyrin and iron in which the iron is in the 2+ oxidation state.

ferrous The term or prefix used to denote compounds of iron in which iron is in the divalent (2 +) state.

ferrous acetate $Fe(CH_3COO)_2 \cdot 4H_2O$ Soluble green crystals, soluble in water and alcohol, that are combustible and that oxidize to basic ferric acetate in air; used as textile dyeing mordant, as wood preservative, and in medicine. Also known as iron acetate.

ferrous ammonium sulfate $Fe(SO_4) \cdot (NH_2)_2SO_4 \cdot 6H_2O$ Light-green, water-soluble crystals; used in medicine, analytical chemistry, and metallurgy. Also known as iron ammonium sulfate; Mohr's salt.

ferrous arsenate $Fe_3(AsO_4)_2 \cdot 6H_2O$ Water-insoluble, toxic green amorphous powder, soluble in acids; used in medicine and as an insecticide. Also known as iron arsenate.

ferrous carbonate $FeCO_3$ Green rhombohedral crystals that are soluble in carbonated water and decompose when heated; used in medicine.

ferrous chloride $FeCl_2 \cdot 4H_2O$ Green, monoclinic crystals, soluble in water; used as a mordant in dyeing, for sewage treatment, in metallurgy, and in pharmaceutical preparations. Also known as iron chloride; iron dichloride.

ferrous hydroxide $Fe(OH)_2$ A white, water-insoluble, gelatinous solid that turns reddish-brown as it oxidizes to ferric hydroxide.

ferrous oxalate $Fe(COO)_2$ A water-soluble, yellow powder; used in photography and medicine. Also known as iron oxalate.

ferrous oxide FeO A black powder, soluble in water, melting at 1419°C. Also known as black iron oxide; iron monoxide.

ferrous sulfate $FeSO_4 \cdot 7H_2O$ Blue-green, water-soluble, monoclinic crystals; used as a mordant in dyeing wool, in the manufacture of ink, and as a disinfectant. Also known as ferrisulphas; green copperas; green vitriol; iron sulfate.

ferrous sulfide FeS Black crystals, insoluble in water, soluble in acids, melting point 1195°C; used to generate hydrogen sulfide in ceramics manufacture. Also known as iron sulfide.

ferrum Latin term for iron; derivation of the symbol Fe.

fertilizin A mucopolysaccharide, derived from the jelly coat of an egg, that plays a role in sperm recognition and the stimulation of sperm motility and metabolic activity.

ferulic acid $C_{10}H_{10}O_4$ A compound widely distributed in small amounts in plants, having two isomers: the cis form is a yellow oil, and the trans form is obtained from water solutions as orthorhombic crystals.

fetal hemoglobin A normal embryonic hemoglobin having alpha chains identical to those of normal adult human hemoglobin, and gamma chains similar to adult beta chains.

Feulgen reaction An aldehyde specific reaction based on the formation of a purple-colored compound when aldehydes react with fuchsin-sulfuric acid; deoxyribonucleic acid gives this reaction after removal of its purine bases by acid hydrolysis; used as a nuclear stain.

fibrin The fibrous, insoluble protein that forms the structure of a blood clot; formed by the action of thrombin.

fibrinase An enzyme that catalyzes the formation of covalent bonds between fibrin molecules. Also known as fibrin-stabilizing factor.

fibrinogen A plasma protein synthesized by the parenchymal cells of the liver; the precursor of fibrin. Also known as factor I.

fibrinoid A homogeneous, refractile, oxyphilic substance occurring in degenerating connective tissue, as in term placentas, rheumatoid nodules, and Aschoff bodies, and in pulmonary alveoli in some prolonged pneumonitides.

fibrinolysin *See* plasmin.

fibrin-stabilizing factor *See* fibrinase.

fibroin A protein secreted by spiders and silkworms which rapidly solidifies into strong, insoluble thread that is used to form webs or cocoons.

fibrous protein Any of a class of highly insoluble proteins representing the principal structural elements of many animal tissues.

ficam *See* 2,2-dimethyl-1,3-benzodioxol-4-yl-*N*-methylcarbomate.

ficin A proteolytic enzyme obtained from fig latex or sap; hydrolyzes casein, meat, fibrin, and other proteinlike materials; used in the food industry and as a diagnostic aid in medicine.

field desorption mass spectroscopy A technique for analysis of nonvolatile molecules in which a sample is deposited on a thin tungsten wire containing sharp microneedles of carbon on the surface; a voltage is applied to the wire, thus producing high electric-field gradients at the points of the needles, and moderate heating then causes desorption from the surface or molecular ions, which are focused into a mass spectrometer.

field shift The portion of the mass shift produced by the change in the size and shape of the nuclear charge distribution when neutrons are added to the nucleus. Also known as volume shift.

Fierz interference Interference between the axial vector and tensor parts of the weak interaction of nucleon and lepton (electron-neutrino) fields in beta decay; measurements of the beta-particle energy spectrum indicate that it vanishes.

film-development chromatography Liquid-analysis chromatographic technique in which the stationary phase (adsorbent) is a strip or layer, as in paper or thin-layer chromatography.

filter cake washing An operation performed at the end of a filtration, in which residual liquid impurities are washed out of the cake by the flow of another liquid through the cake.

filter flask A flask with a side arm to which a vacuum can be applied; usually filter flasks have heavy side walls to withstand high vacuum.

filter leaf The frame or structure in a filter press that holds the filter cloth or other filter medium; a number of leaves in series usually comprises a filter press.

filter photometry 1. Colorimetric analysis of solution colors with a filter applied to the eyepiece of a conventional colorimeter. 2. Inspection of a pair of Nessler tubes through a filter.

filter-press cell An electrolytic cell consisting of several units in series, as in a filter press, in which each electrode, except the two end ones, acts as an anode on one side and a cathode on the other, and the space between electrodes is divided by porous asbestos diaphragms.

filter spectrophotometer Spectrophotographic analyzer of spectral radiations in which a filter is used to isolate narrow portions of the spectrum.

final boiling point *See* end point.

fine structure The splitting of spectral lines in atomic and molecular spectra caused by the spin angular momentum of the electrons and the coupling of the spin to the orbital angular momentum.

fire The manifestation of rapid combustion, or combination of materials with oxygen.

firedamp reforming process A process in which methane (firedamp) is mixed with steam and passed over a nickel catalyst for conversion to a mixture of hydrogen and carbon monoxide; this mixture is blended with pure methane, and the result is a fuel of high calorific value.

fire point The lowest temperature at which a volatile combustible substance vaporizes rapidly enough to form above its surface an air-vapor mixture which burns continuously when ignited by a small flame.

first-order reaction A chemical reaction in which the rate of decrease of concentration of component A with time is proportional to the concentration of A.

first-order spectrum A spectrum, produced by a diffraction grating, in which the difference in path length of light from adjacent slits is one wavelength.

fir wood oil *See* pine needle oil.

Fischer-Hepp rearrangement The rearrangement of a nitroso derivative of a secondary aromatic amine to a p-nitrosoarylamine; the reaction is brought about by an alcoholic solution of hydrogen chloride.

Fischer indole synthesis A reaction to form indole derivatives by means of a ring closure of aromatic hydrazones.

Fischer polypeptide synthesis A synthesis of peptides in which α-amino acids or those peptides with a free amino group react with acid halides of α-haloacids, followed by amination with ammonia.

Fischer-Tropsch process A catalytic process to synthesize hydrocarbons and their oxygen derivatives by the controlled reaction of hydrogen and carbon monoxide.

fissiochemistry The process of producing chemical change by means of nuclear energy.

fission The division of an atomic nucleus into parts of comparable mass; usually restricted to heavier nuclei such as isotopes of uranium, plutonium, and thorium. Also known as atomic fission; nuclear fission.

fission barrier One or more maxima in the plot of potential energy against nuclear deformation of a heavy nucleus, which inhibits spontaneous fission of the nucleus.

fission cross section The cross section for a bombarding neutron, gamma ray, or other particle to induce fission of a nucleus.

fission isomer A highly deformed nuclear state lying in the second well of a double-hump fission barrier.

fission neutron   A neutron emitted as a result of nuclear fission.

fission product   Any radioactive or stable nuclide resulting from fission, including both primary fission fragments and their radioactive decay products.

fission spectrum   The energy distribution of neutrons arising from fission.

fission threshold   The minimum kinetic energy of a bombarding neutron required to induce fission of a nucleus.

fission yield   The percent of fissions that gives a particular nuclide or group of isobars.

Fittig's synthesis   The synthesis of aromatic hydrocarbons by the condensation of aryl halides with alkyl halides, using sodium as a catalyst.

fixed-bed hydroforming   A cyclic petroleum process that utilizes a fixed bed of molybdenum oxide catalyst deposited on activated alumina.

fixed-bed operation   An operation in which the additive material (catalyst, absorbent, filter media, ion-exchange resin) remains stationary in the chemical reactor.

fixed carbon   Solid, combustible residue remaining after removal of moisture, ash, and volatile materials from coal, coke, and bituminous materials; expressed as a percentage.

fixed ion   An ion in the lattice of a solid ion exchanger.

flaking   Continuous process operation to remove heat from material in the liquid state to cause its solidification.

flame   A hot, luminous reaction front (or wave) in a gaseous medium into which the reactants flow and out of which the products flow.

flame emission spectroscopy   A flame photometry technique in which the solution containing the sample to be analyzed is optically excited in an oxyhydrogen or oxyacetylene flame.

flame excitation   Use of a high-temperature flame (such as oxyacetylene) to excite spectra emission lines from alkali and alkaline-earth elements and metals.

flame ionization detector   A device in which the measured change in conductivity of a standard flame (usually hydrogen) due to the insertion of another gas or vapor is used to detect the gas or vapor.

flame photometer   One of several types of instruments used in flame photometry, such as the emission flame photometer and the atomic absorption spectrophotometer, in each of which a solution of the chemical being analyzed is vaporized; the spectral lines resulting from the light source going through the vapors enters a monochromator that selects the band or bands of interest.

flame photometry   A branch of spectrochemical analysis in which samples in solution are excited to produce line emission spectra by introduction into a flame.

flame propagation   The spread of a flame in a combustible environment outward from the point at which the combustion started.

flame spectrometry   A procedure used to measure the spectra or to determine wavelengths emitted by flame-excited substances.

flame spectrophotometry A method used to determine the intensity of radiations of various wavelengths in a spectrum emitted by a chemical inserted into a flame.

flame spectrum An emission spectrum obtained by evaporating substances in a nonluminous flame.

flame speed The rate at which combustion moves through an explosive mixture.

flammability A measure of the extent to which a material will support combustion. Also known as inflammability.

flammability limits The stoichiometric composition limits (maximum and minimum) of an ignited oxidizer-fuel mixture what will burn indefinitely at given conditions of temperature and pressure without further ignition.

flare A device for disposing of combustible gases from refining or chemical processes by burning in the open, in contrast to combustion in a furnace or closed vessel or chamber.

flash chamber A conventional oil-and-gas separator operated at low pressure, with the liquid from a higher-pressure vessel being flashed into it. Also known as flash trap; flash vessel.

flash drum A facility, such as a tower, which receives the products of a preheater or heat exchanger to release pressure; volatile components are vaporized and separated for further fractionation.

flash dry The rapid evaporation of moisture from a porous or granular solid by a sudden reduction in pressure or by placing the material in an updraft of warm air.

flashing Vaporization of volatile liquids by either heat or vacuum.

flashing flow The condition when a liquid at its boiling point flows through a heated conduit and is further heated to cause partial vaporization (flashing), with a resultant two-phase (vapor-liquid) flow.

flash photolysis A method of studying fast photochemical reactions in gas molecules; a powerful lamp is discharged in microsecond flashes near a reaction vessel holding the gas, and the products formed by the flash are observed spectroscopically.

flash point The lowest temperature at which vapors from a volatile liquid will ignite momentarily upon the application of a small flame under specified conditions; test conditions can be either open- or closed-cup.

flash process Liquid-vapor system in which the composition remains constant, but the proportion of gas and liquid phases changes as pressure or temperature change.

flash separation Process for separation of gas (vapor) from liquid components under reduced pressure; the liquid and gas remain in contact as the gas evolves from the liquid.

flash spectroscopy The study of the electronic states of molecules after they absorb energy from an intense, brief light flash.

flash trap *See* flash chamber.

flash vaporization Rapid vaporization achieved by passing a volatile liquid through continuously heated coils.

flash vessel   *See* flash chamber.

flask   A long-necked vessel, frequently of glass, used for holding liquids.

flavan $C_{15}H_{14}O$ 2-Phenylbenzopyran, an aromatic heterocyclic compound from which all flavonoids are derived.

flavanol   Yellow needles with a melting point of 169°C, derived from flavanone; a flavanoid pigment used as a dye. Also known as 3-hydroxyflavone.

flavanone $C_{15}H_{12}O_2$ A colorless crystalline derivative of flavone.

flavin   1. A yellow dye obtained from the bark of quercitron trees. 2. Any of several water-soluble yellow pigments occurring as coenzymes of flavoproteins.

flavin adenine dinucleotide $C_{27}H_{33}N_9O_{15}P_2$ A coenzyme that functions as a hydrogen acceptor in aerobic dehydrogenases (flavoproteins). Abbreviated FAD.

flavin mononucleotide   *See* riboflavin 5'-phosphate.

flavin phosphate   *See* riboflavin 5'-phosphate.

flavone   1. Any of a number of ketones composing a class of flavonoid compounds. 2. $C_{15}H_{10}O_2$ A colorless crystalline compound occurring as dust on the surface of many primrose plants.

flavonoid   Any of a series of widely distributed plant constituents related to the aromatic heterocyclic skeleton of flavan.

flavonol   1. Any of a class of flavonoid compounds that are hydroxy derivatives of flavone. 2. $C_{16}H_{10}O_2$ A colorless, crystalline compound from which many yellow plant pigments are derived.

flavoprotein   Any of a number of conjugated protein dehydrogenases containing flavin that play a role in biological oxidations in both plants and animals; a yellow enzyme.

Fleming cracking process   An obsolete liquid-phase thermal cracking process for heavy petroleum fractions; the charge was heated under pressure in a vertical shell still.

Flesh-Demag process   A gas-making process in which a cyclic water-gas apparatus is used for feeding and charring the coal charge and for gas generation, with periodic automatic removal of the resultant ash.

F line   A green-blue line in the spectrum of hydrogen, at a wavelength of 486.133 nanometers.

floc   Small masses formed in a fluid through coagulation, agglomeration, or biochemical reaction of fine suspended particles.

flocculant   *See* flocculating agent.

flocculate   To cause to aggregate or coalesce into a flocculent mass.

flocculating agent   A reagent added to a dispersion of solids in a liquid to bring together the fine particles to form flocs. Also known as flocculant.

flocculent   Pertaining to a material that is cloudlike and noncrystalline.

floc point   The temperature at which wax or solids separate from kerosene and other illuminating oils as a definite floc.

floc test A quantitative test applied to kerosine and other illuminating oils to detect substances rendered insoluble by heat.

flooding Condition in a liquid-vapor counterflow device (such as a distillation column) in which the rate of vapor rise is such as to prevent liquid downflow, causing a buildup of the liquid (flooding) within the device.

Flood's equation A relation used to determine the liquidus temperature in a binary fused salt system.

florentium *See* promethium-147.

flores A form of a chemical compound made by the process of sublimation.

flores martis *See* ferric chloride.

florigen A plant hormone that stimulates buds to flower.

flotation agent A chemical which alters the surface tension of water or which makes it froth easily.

flow birefringence Orientation of long, thin asymmetric molecules in the direction of flow of a solution forced to flow through a capillary tube.

flowers of tin *See* stannic oxide.

flow reactor A dynamic reactor system in which reactants flow continuously into the vessel and products are continuously removed, in contrast to a batch reactor.

fluid-bed process A type of process based on the tendency of finely divided powders to behave in a fluidlike manner when supported and moved by a rising gas or vapor stream; used mainly for catalytic cracking of petroleum distillates.

fluid catalyst Finely divided solid particles utilized as a catalyst in a fluid-bed process.

fluid catalytic cracking An oil refining process in which the gas-oil is cracked by a catalyst bed fluidized by using oil vapors.

fluid coking A thermal process utilizing the fluidized solids technique for continuous conversion of heavy, low-grade petroleum oils into petroleum coke and lighter hydrocarbon products.

fluid hydroforming A type of fluid catalytic cracking process used by petroleum refineries to upgrade low-octane-number stocks.

fluidization A roasting process in which finely divided solids are suspended in a rising current of air (or other fluid), producing a fluidized bed; used in the calcination of various minerals, in Fischer-Tropsch synthesis, and in the coal industry.

fluidized adsorption Method of vapor- or gas-fractionation (separation via adsorption-desorption cycles) in a fluidized bed of adsorbent material.

fluoborate **1.** Any of a group of compounds related to the borates in which one or more oxygens have been replaced by fluorine atoms. **2.** The BF_4^- ion, which is derived from fluoboric acid, HBF_4.

fluoboric acid HBF_4 Colorless, clear, water-miscible acid; used for electrolytic brightening of aluminum and for forming stabilized diazo salts.

fluometuron $C_{10}H_{11}F_3N_2O$ A white, crystalline solid with a melting point of 163–164.5°C; used as an herbicide for cotton and sugarcane. Also known as 1,1-dimethyl-3-(α,α,α-trifluoro-*meta*-tolyl)urea.

fluoranthene $C_{10}H_{10}$ A tetracyclic hydrocarbon found in coal tar fractions and petroleum, forming needlelike crystals, boiling point 250°C, and soluble in organic solvents such as ether and benzene.

fluorene $C_{13}H_{10}$ A hydrocarbon chemical present in the middle oil fraction of coal tar; insoluble in water, soluble in ether and acetone, melting point 116–117°C; used as the basis for a group of dyes. Also known as 2,3-benzindene; diphenylenemethane.

fluorescein $C_{20}H_{12}O_5$ A yellowish to red powder, melts and decomposes at 290°C, insoluble in water, benzene, and chloroform, soluble in glacial acetic acid, boiling alcohol, ether, dilute acids, and dilute alkali; used in medicine, in oceanography as a marker in sea water, and in textiles to dye silk and wool.

fluorescence **1.** Emission of electromagnetic radiation that is caused by the flow of some form of energy into the emitting body and which ceases abruptly when the excitation ceases. **2.** Emission of electromagnetic radiation that is caused by the flow of some form of energy into the emitting body and whose decay, when the excitation ceases, is temperature-independent. **3.** Gamma radiation scattered by nuclei which are excited to and radiate from an excited state.

fluorescence spectra Emission spectra of fluorescence in which an atom or molecule is excited by absorbing light and then emits light of characteristic frequencies.

fluorescent dye A highly reflective dye that serves to intensify color and add to the brilliance of a fabric.

fluorescent pigment A pigment capable of absorbing both visible and nonvisible electromagnetic radiations and releasing them quickly as energy of desired wavelength; examples are zinc sulfide or cadmium sulfide.

fluoridation Formation in rocks of fluorine-containing minerals such as fluorite or topaz.

fluoride A salt of hydrofluoric acid, HF, in which the fluorine atom is in the minus-one oxidation state.

fluorinated ethylene propylene resin Copolymers of tetrafluoroethylene and hexafluoropropylene. Abbreviated FEP resin.

fluorination A chemical reaction in which fluorine is introduced into a chemical compound.

fluorine A gaseous or liquid chemical element, symbol F, atomic number 9, atomic weight 18.998; a member of the halide family, it is the most electronegative element and the most chemically energetic of the nonmetallic elements; highly toxic, corrosive, and flammable; used in rocket fuels and as a chemical intermediate.

fluoroacetate Acetate in which carbon-connected hydrogen atoms are replaced by fluorine atoms.

fluoroacetic acid CH_2FCOOH A poisonous, crystalline compound obtained from plants, such as those of the Dichapetalaceae family, South Africa, soluble in water and alcohol, and burns with a green flame; the sodium salt is used as a water-soluble rodent poison. Also known as fluoroethanoic acid; gifblaar poison.

fluoroalkane   Straight-chain, saturated hydrocarbon compound (or analog thereof) in which some of the hydrogen atoms are replaced by fluorine atoms.

***para*-fluoroaniline** $FC_6H_4NH_2$ A liquid that is an intermediate in the manufacture of herbicides and plant growth regulators. Also known as 4-fluorobenzenamine.

4-fluorobenzenamine   *See para*-fluoroaniline.

fluorobenzene C_6H_5F A colorless liquid with a boiling point of 84.9°C; used as an insecticide intermediate. Also known as phenyl fluoride.

4-fluorobenzeneacetic acid   *See para*-fluorophenylacetic acid.

fluorocarbon   A hydrocarbon such as Freon in which part or all hydrogen atoms have been replaced by fluorine atoms; can be liquid or gas and is nonflammable and heat-stable; used as refrigerant, aerosol propellant, and solvent. Also known as fluorohydrocarbon.

fluorocarbon-11   *See* trichlorofluoromethane.

fluorocarbon-21   *See* dichlorofluoromethane.

fluorocarbon fiber   Fiber made from a fluorocarbon resin, such as polytetrafluoroethylene resin.

fluorocarbon resin   Polymeric material made up of carbon and fluorine with or without other halogens (such as chlorine) or hydrogen; the resin is extremely inert and more dense than corresponding fluorocarbons such as Teflon.

fluorochemical   Any chemical compound containing fluorine; usually refers to the fluorocarbons.

fluorodichloromethane   *See* dichlorofluoromethane.

fluorodifen $C_{13}H_7F_3N_2O_4$ A yellow, crystalline compound with a melting point of 93°C; used as a pre- and postemergence herbicide for food crops. Also known as *para*-nitrophenyl α,α,α,-trifluoro-2-nitro-*para*-tolyl ether.

1-fluoro-2,4-dinitrobenzene $(NO_2)_2C_6H_3F$ Crystals that are soluble in benzene, propylene glycol, and ether; used as a reagent for labeling terminal amino acid groups and in the detection of phenols. Also known as 2,4-dinitro-1-fluorobenzene; Sanger's reagent.

fluoroethanoic acid   *See* fluoroacetic acid.

fluoroform CHF_3 A colorless, nonflammable gas, boiling point 84°C at 1 atmosphere (101, 325 newtons per square meter), freezing point 160°C at 1 atmosphere; used in refrigeration and as an intermediate in organic synthesis. Also known as propellant 23; refrigerant 23; trifluoromethane.

fluorogenic substrate   A nonfluorescent material that is acted upon by an enzyme to produce a fluorescent compound.

fluorohydrocarbon   *See* fluorocarbon.

fluorometric analysis   A method of chemical analysis in which a sample, exposed to radiation of one wavelength, absorbs this radiation and reemits radiation of the same or longer wavelength in about 10^{-9} second; the intensity of reemited radiation is almost directly proportional to the concentration of the fluorescing material. Also known as fluorescence analysis; fluorometry.

fluorometry   *See* fluorometric analysis.

***para*-fluorophenylacetic acid** $FC_6H_4CH_2\ COOH$ Crystals with a melting point of 86°C; used as an intermediate in the manufacture of fluorinated anesthetics. Also known as 4-fluorobenzeneacetic acid.

fluorophosphoric acid H_2PO_3F A colorless, viscous liquid that is miscible with water; used in metal cleaners and as a catalyst.

fluorothene   *See* chlorotrifluoroethylene polymer.

fluorotrichloromethane   *See* trichlorofluoromethane.

fluosilicate   A salt derived from fluosilicic acid, H_2SiF_6, and containing the $SiF_6{}^{2-}$ ion.

fluosilicic acid H_2SiF_6 A colorless acid, soluble in water, which attacks glass and stoneware; highly corrosive and toxic; used in water fluoridation and electroplating. Also known as hydrofluorosilicic acid; hydrofluosilicic acid.

fluosulfonic acid HSO_3F Colorless, corrosive, fuming liquid; soluble in water with partial decomposition; used as organic synthesis catalyst and in electroplating.

flurecol-*n*-butyl-ester   *See* flurenol.

flurenol $C_{18}H_{18}O_3$ A solid, crystalline compound with a melting point of 70–71°C; used as an herbicide for vegetables, cereals, and ornamental flowers. Also known as *n*-butyl-9-hydroxy fluorene-9-carboxylate; flurecol-*n*-butyl-ester.

fluxional compound   1. Any of a group of molecules which undergo rapid intramolecular rearrangements in which the component atoms are interchanged among equivalent structures. 2. Molecules in which bonds are broken and reformed in the rearrangement process.

Fm   *See* fermium.

FMN   *See* riboflavin 5'-phosphate.

foam   An emulsionlike two-phase system where the dispersed phase is gas or air.

folic acid $C_{19}H_{19}N_7O_6$ A yellow, crystalline vitamin of the B complex; it is slightly soluble in water, usually occurs in conjugates containing glutamic acid residues, and is found especially in plant leaves and vertebrate livers. Also known as pteroylglutamic acid (PGA).

folic acid sodium salt   *See* sodium folate.

folimat $C_5H_{12}NO_4PS$ An oily liquid that decomposes at 135°C; soluble in water; used as an insecticide and miticide on fruit and vegetable crops and on ornamental flowers. Also known as dimethyl *S*-(*N*-methylcarbamoyl-methyl)phosphorothioate; omethioate.

Folin solution   An aqueous solution of 500 grams of ammonium sulfate, 5 grams of uranium acetate, and 6 grams of acetic acid in a volume of 1 liter; used to test for uric acid.

follicle-stimulating hormone   A protein hormone released by the anterior pituitary of vertebrates which stimulates growth and secretion of the Graafian follicle and also promotes spermatogenesis. Abbreviated FSH.

folpet $C_9H_4Cl_3NO_2S$ A buff or white, crystalline compound with a melting point of 177–178°C; insoluble in water; used as a fungicide on fruits, vegetables, and ornamental flowers. Also known as N-(trichloromethylthio)-phthalimide.

foot's oil The oil sweated out of slack wax; it takes its name from the fact that it goes to the bottom, or foot, of the pan when sweated.

forbidden line A spectral line associated with a transition forbidden by selection rules; optically this might be a magnetic dipole or electric quadrupole transition.

force constant An expression for the force acting to restrain the relative displacement of the nuclei in a molecule.

forensic chemistry The application of chemistry to the study of materials or problems in cases where the findings may be presented as technical evidence in a court of law.

formal charge The apparent charge of an element in a compound; for example, magnesium has a formal charge of $+2$ in MgO and oxygen has a charge of -2.

formaldehyde HCHO The simplest aldehyde; a gas at room temperature, and a poisonous, clear, colorless liquid solution with pungent odor; used to make synthetic resins by reaction with phenols, urea, and melamine, as a chemical intermediate, as an embalming fluid, and as a disinfectant. Also known as formol; methylene oxide.

formaldehyde sodium bisulfite $CH_3 NaO_4S$ A compound used as a fixing agent for fibers containing keratin, in metallurgy for flotation of lead-zinc ores, and in photography. Also known as methylolsulfonic acid sodium salt; sodium formaldehyde-bisulfite.

formamidase An enzyme involved in tryptophane catabolism; catalyzes the conversion of N-formylkynurenine to kynurenine and formate.

formamide 1. A compound containing the radical HCONH. 2. $HCONH_2$ A clear, colorless hygroscopic liquid, boiling at 200–212°C; soluble in water and alcohol; used as a solvent, softener, and chemical intermediate. Also known as formylamine; methanamide.

formamidinesulfinic acid $H_2NC(NH) SO_2H$ A reagent for the reduction of ketones to secondary alcohols. Also known as aminoiminomethanesulfinic acid.

formate A compound containing the HCOO — radical.

formation factor The ratio between the conductivity of an electrolyte and that of a rock saturated with the same electrolyte. Also known as resistivity factor.

formic acid HCOOH A colorless, pungent, toxic, corrosive liquid melting at 8.4°C; soluble in water, ether, and alcohol; used as a chemical intermediate and solvent, in dyeing and electroplating processes, and in fumigants. Also known as methanoic acid.

formic ether *See* ethyl formate.

formol *See* formaldehyde.

formonitrile *See* hydrocyanic acid.

formula 1. A combination of chemical symbols that expresses a molecule's composition. 2. A reaction formula showing the interrelationship between reactants and products.

formulation The particular mixture of base chemicals and additives required for a product.

formula weight 1. The gram-molecular weight of a substance. 2. In the case of a substance of uncertain molecular weight such as certain proteins, the molecular weight calculated from the composition, assuming that the element present in the smallest proportion is represented by only one atom.

formyl The formic acid radical, HCO —; it is characteristic of aldehydes.

2-formyl-3,4-dihydro-2H-pyran *See* acrolein dimer.

formyl methionine Formylated methionine; initiates peptide chain synthesis in bacteria.

Fortrat parabola Graph of wave numbers of lines in a molecular spectral band versus the serial number of the successive lines.

Foulger's test A test for fructose in which urea, sulfuric acid, and stannous chloride are added to the solution to be tested, the solution is boiled, and in the presence of fructose a blue coloration forms.

fouling factor In heat transfer, the lowering of clear-film transfer rates resulting from corrosion, dirt, or roughness of the surface of tube walls of heat exchangers.

four-degree calorie The heat needed to change the temperature of 1 gram of water from 3.5 to 4.5°C.

fp *See* freezing point.

Fr *See* francium.

fraction One of the portions of a volatile liquid within certain boiling point ranges, such as petroleum naphtha fractions or gas-oil fractions.

fractional condensation Separation of components of vaporized liquid mixtures by condensing the vapors in stages (partial condensation); highest-boiling-point components condense in the first condenser stage, allowing the remainder of the vapor to pass on to subsequent condenser stages.

fractional distillation A method to separate a mixture of several volatile components of different boiling points; the mixture is distilled at the lowest boiling point, and the distillate is collected as one fraction until the temperature of the vapor rises, showing that the next higher boiling component of the mixture is beginning to distill; this component is then collected as a separate fraction.

fractional precipitation Method for separating elements or compounds with similar solubilities by a series of analytical precipitations, each one improving the purity of the desired element.

fractionating column An apparatus used widely for separation of fluid (gaseous or liquid) components by vapor-liquid fractionation or liquid-liquid extraction or liquid-solid adsorption.

fractionation Separation of a mixture in successive stages, each stage removing from the mixture some proportion of one of the substances, as by differential solubility in water-solvent mixtures.

francium A radioactive alkali-metal element, symbol Fr, atomic number 87, atomic weight distinguished by nuclear instability; exists in short-lived radioactive forms, the chief isotope being francium-223.

Franck-Condon principle The principle that in any molecular system the transition from one energy state to another is so rapid that the nuclei of the atoms involved can be considered to be stationary during the transition.

frangula emodin *See* emodin.

frangulic acid *See* emodin.

Frankland's method Reaction of dialkyl zinc compounds with alkyl halides to form hydrocarbons; may be used to form paraffins containing a quaternary carbon atom.

Frank partial dislocation A partial dislocation whose Burger's vector is not parallel to the fault plane, so that it can only diffuse and not glide, in contrast to a Schockley partial dislocation.

Fraude's reagent *See* perchloric acid.

fraunhofer A unit for measurement of the reduced width of a spectrum line such that a spectrum line's reduced width in fraunhofers equals 10^6 times its equivalent width divided by its wavelength.

Fraunhofer lines The dark lines constituting the Fraunhofer spectrum.

Fraunhofer spectrum The absorption lines in sunlight, due to the cooler outer layers of the sun's atmosphere.

Frazer-Brace extraction method A method used to extract oil from citrus fruit; utilizes a machine which has abrasive carborundum rolls to rasp the peel from the fruit under a water spray; the water-and-peel mixture is screened and settled to allow oil separation.

free atom An atom, as in a gas, whose properties, such as spectrum and magnetic moment, are not significantly affected by other atoms, ions, or molecules nearby.

freeboard The space provided above the resin bed in an ion-exchange column to allow for expansion of the bed during backwashing.

free cyanide Cyanide not combined as part of an ionic complex.

free ion An ion, such as found in an ionized gas, whose properties, such as spectrum and magnetic moment, are not significantly affected by other atoms, ions, or molecules nearby.

free molecule A molecule, as in a gas, whose properties, such as spectrum and magnetic moment, are not affected by other atoms, ions, and molecules nearby.

free radical **1.** An atom or a diatomic or polyatomic molecule which possesses at least one unpaired electron. **2.** A species which is uncharged and possesses one or more unpaired electrons. Also known as a radical.

freeze To solidify a liquid by removal of heat.

freezing mixture A mixture of substances whose freezing point is lower than that of its constituents.

freezing point The temperature at which a liquid and a solid may be in equilibrium. Abbreviated fp.

freezing-point depression   The lowering of the freezing point of a solution compared to the pure solvent; the depression is proportional to the active mass of the solute in a given amount of solvent.

Fremy's salt   *See* potassium bifluoride.

frequency factor   The constant A (or ν) in the Arrhenius equation, which is the relation between reaction rate and absolute temperature T; the equation is $k = Ae - \Delta H_{act}/RT$, where k is the specific rate constant, ΔH_{act} is the heat of activation, and R is the gas constant.

Freund method   A method for preparation of cycloparaffins in which dihalo derivatives of the paraffins are treated with zinc to produce the cycloparaffin.

Friedel-Crafts reaction   A substitution reaction, catalyzed by aluminum chloride in which an alkyl (R—) or an acyl (RCO —) group replaces a hydrogen atom of an aromatic nucleus to produce hydrocarbon or a ketone.

Friedel's law   The law that x-ray or electron diffraction measurements cannot determine whether or not a crystal has a center of symmetry.

Friedlander synthesis   A synthesis of quinolines; the method is usually catalyzed by bases and consists of condensation of an aromatic *o*-amino-carbonyl derivative with a compound containing a methylene group in the alpha position to the carbonyl.

Fries rearrangement   The conversion of a phenolic ester into the corresponding *o*- and *p*-hydroxyketone by treatment with catalysts of the type of aluminum chloride.

Fries' rule   The rule that the most stable form of the bonds of a polynuclear compound is that arrangement which has the maximum number of rings in the benzenoid form, that is, three double bonds in each ring.

front-end volatility   The volatility of the lower-boiling fractions of gasoline, such as butanes.

front pinacoid   The {100} pinacoid in an orthorhombic, monoclinic, or triclinic crystal. Also known as macropinacoid; orthopinacoid.

frother   Substance used in flotation processes to make air bubbles sufficiently permanent, principally by reducing surface tension.

froth promoter   A chemical compound used with a frothing agent.

D-fructopyranose   *See* fructose.

fructose $C_6H_{12}O_5$ The commonest of ketoses and the sweetest of sugars, found in the free state in fruit juices, honey, and nectar of plant glands. Also known as D-fructopyranose.

frustrated internal reflectance   *See* attenuated total reflectance.

FSH   *See* follicle-stimulating hormone.

f-sum rule   The rule that the sum of the f values (or oscillator strengths) of absorption transitions of an atom in a given state, minus the sum of the f values of the emission transitions in that state, equals the number of electrons which take part in these transitions. Also known as Thomas-Reiche-Kuhn sum rule.

fuchsin $C_{20}H_{19}N_3$ Brownish-red crystals, used as a dye or in the commercial preparation of other dyes, and as an antifungal drug. Also known as magenta; rosaniline.

fucoidin A gum composed of L-fucose and sulfate acid ester groups obtained from *Fucus* species and other brown algae.

L-fucopyranose *See* L-fucose.

L-fucose $C_6H_{12}O_5$ A methyl pentose present in some algae and a number of gums and identified in the polysaccharides of blood groups and certain bacteria. Also known as 6-deoxy-L-galactose; L-fucopyranose; L-galactomethylose; L-rhodeose.

fucoxanthin $C_{40}H_{60}O_6$ A carotenoid pigment; a partial xanthophyll ester found in diatoms and brown algae.

fuel-cell catalyst A substance, such as platinum, silver, or nickel, from which the electrodes of a fuel cell are made, and which speeds the reaction of the cell; it is especially important in a fuel cell which does not operate at high temperatures.

fuel-cell electrolyte The substance which conducts electricity between the electrodes of a fuel cell.

fuel-cell fuel A substance, such as hydrogen, carbon monoxide, sodium, alcohol, or a hydrocarbon, which reacts with oxygen to generate energy in a fuel cell.

fugitive dye A dye that is unstable, that is, not fast; used in the textile processing for purposes of identity.

fulminate 1. A salt of fulminic acid. 2. $HgC_2N_2O_2$ An explosive mercury compound derived from the fulminic acid; used for the caps or exploders by means of which charges of gunpowder, dynamite, and other explosives are fired. Also known as mercury fulminate.

fulminic acid CNOH An unstable isomer of cyanic acid, whose salts are known for their explosive characteristics. Also known as 4-cyanic acid; paracyanic acid.

fulminuric acid $CN \cdot CH(NO_2) \cdot CONH_2$ A trimer of cyanuric acid; a water-soluble compound, crystallizing in colorless needles, melting at 138°C, and exploding at 145°C. Also known as cyanonitroacetamide; 2-cyano-2-nitroethanamide; isocyanuric acid.

fulvene C_6H_6 A yellow oil, an isomer of benzene. Also known as methylenecyclopentadiene.

fumarase An enzyme that catalyzes the hydration of fumaric acid to malic acid, and the reverse dehydration.

fumaric acid $C_4H_4O_4$ A dicarboxylic organic acid produced commercially by synthesis and fermentation; the trans isomer of maleic acid; colorless crystals, melting point 287°C; used to make resins, paints, varnishes, and inks, in foods, as a mordant, and as a chemical intermediate. Also known as boletic acid.

fume hood A fume-collection device over an enclosed shelf or table, so that experiments involving poisonous or unpleasant fumes or gases may be conducted away from the experimental area.

fumes Particulate matter consisting of the solid particles generated by condensation from the gaseous state, generally after volatilization from melted substances, and often accompanied by a chemical reaction, such as oxidation.

fumigant A chemical compound which acts in the gaseous state to destroy insects and their larvae and other pests; examples are dichlorethyl ether, *p*-dichlorobenzene, and ethylene oxide.

fuming nitric acid Concentrated nitric acid containing dissolved nitrogen dioxide; may be prepared by adding formaldehyde to concentrated nitric acid.

fuming sulfuric acid Concentrated sulfuric acid containing dissolved sulfur trioxide. Also known as oleum.

functional group An atom or group of atoms, acting as a unit, that has replaced a hydrogen atom in a hydrocarbon molecule and whose presence imparts characteristic properties to this molecule; frequently represented as R—.

functionality Ability of a compound to form covalent bonds; compounds may be mono-, di-, tri-, or polyfunctional, that is, one, two, three, or many functional groups may participate in a reaction.

fundamental series A series occurring in the line spectra of many atoms and ions having one, two, or three electrons in the outer shell, in which the total orbital angular momentum quantum number changes from 3 to 2.

funicular distribution The distribution of a two-phase, immiscible liquid mixture (such as oil and water, one a wetting phase, the other nonwetting) in a porous system when the wetting phase is continuous over the surface of the solids.

2-furaldehyde *See* furfural.

furan 1. One of a group of organic heterocyclic compounds containing a diunsaturated ring of four carbon atoms and one oxygen atom. 2. $C_4H_4O_4$ The simplest furan type of molecule; a colorless, mildly toxic liquid, boiling at 32°C, insoluble in water, soluble in alcohol and ether; used as a chemical intermediate. Also known as furfuran; tetrol.

2-furancarbinol *See* furfuryl alcohol.

furancarboxylic acid *See* furoic acid.

2,5-furandione *See* maleic anhydride.

furanose A sugar whose cyclic or ring structure resembles that of furan.

furanoside A glycoside whose cyclic sugar component resembles that of furan.

furan resin A liquid, thermosetting resin in which the furan ring is an integral part of the polymer chain, made by the condensation of furfuryl alcohol; used as a cement and adhesive, casting resin, coating, and impregnant.

furfural C_4H_3OCHO When pure, a colorless liquid, soluble in organic solvents, slightly soluble in water; used as a lube oil–refining solvent, in cellulosic formulations, in making resins, as a weed killer, as a fungicide, and as a chemical intermediate. Also known as 2-furaldehyde; furfuraldehyde; furfurol; furol.

furfur alcohol *See* furfuryl alcohol.

furfuraldehyde *See* furfural.

furfural extraction Process for the refining of lubricating oils and other organic materials by contact with furfural.

furfuran *See* furan.

furfurol *See* furfural.

furfuryl The univalent radical C_5H_5O — from furfural.

furfuryl alcohol $C_5H_6O_2$ A liquid with a faint burning odor and bitter taste, soluble in alcohol and ether, usually prepared from furfural; used as a solvent in the manufacturing of wetting agents and resins. Also known as 2-furancarbinol; furfur alcohol; furylcarbinol; 2-furylcarbinol; 2-hydroxy methyl furan.

furnace black A carbon black formed by partial combustion of liquid and gaseous hydrocarbons in a closed furnace with a deficiency of oxygen; used as a reinforcing filler for synthetic rubber.

furoic acid $C_5H_4O_3$ Long monoclinic prisms crystallized from the water solution, soluble in ether and alcohol; used as a preservative and bactericide. Also known as furancarboxylic acid; pyromucic acid.

furol *See* furfural.

furylcarbinol *See* furfuryl alcohol.

2-furylcarbinol *See* furfuryl alcohol.

fused aromatic ring A molecular structure in which two or more aromatic rings have two carbon atoms in common.

fused-salt electrolysis Electrolysis with use of purified fused salts as raw material and as an electrolyte.

fusion 1. Combination of two light nuclei to form a heavier nucleus (and perhaps other reaction products) with release of some binding energy. Also known as atomic fusion; nuclear fusion. 2. A change of the state of a substance from the solid phase to the liquid phase. Also known as melting.

fusion tube Device used for the analysis of the elements in a compound by fusing them with another compound; for example, analysis of nitrogen in organic compounds by fusing the compound with sodium and analyzing for sodium cyanide.

f value *See* oscillator strength.

G

Ga *See* gallium.

GABA *See* α-aminobutyric acid.

GABOB *See* 4-amino-3-hydroxybutyric acid.

Gabriel's synthesis A synthesis of primary amines by the hydrolysis of N-alkylphthalimides; the latter are obtained from potassium phthalimide and alkyl halides.

gadoleic acid $C_{20}H_{38}O_2$ A fatty acid derived from cod liver oil, and melting at 20°C.

gadolinium A rare-earth element, symbol Gd, atomic number 64, atomic weight 157.25; highly magnetic, especially at low temperatures.

galactan Any of a number of polysaccharides composed of galactose units. Also known as galactosan.

galactaric acid *See* mucic acid.

galactitol *See* dulcitol.

galactogen A polysaccharide, in snails, that yields galactose on hydrolysis.

galactoglucomannan Any of a group of polysaccharides which are prominent components of coniferous woods; they are soluble in alkali and consist of D-glucopyranose and D-mannopyranose units.

galactokinase An enzyme which reacts D-galactose with adenosinetriphosphate to give D-galactose-1-phosphate and adenosinediphosphate.

galactolipid *See* cerebroside.

galactomannan Any of a group of polysaccharides which are composed of D-galactose and D-mannose units, are soluble in water, and form highly viscous solutions; they are plant mucilages existing as reserve carbohydrates in the endosperm of leguminous seeds.

L-galactomethylose *See* L-fucose.

galactonic acid $C_6H_{12}O_7$ A monobasic acid derived from galactose, occurring in three optically different forms, and melting at 97°C. Also known as pentahydroxyhexoic acid.

galactosamine $C_6H_{14}O_5N$ A crystalline amino acid derivative of galactose; found in bacterial cell walls.

galactosan *See* galactan.

galactose $C_6H_{12}O_6$ A monosaccharide occurring in both levo and dextro forms as a constituent of plant and animal oligosaccharides (lactose and raffinose) and polysaccharides (agar and pectin). Also known as cerebrose.

galactosidase An enzyme that hydrolyzes galactosides.

galactoside A glycoside formed by the reaction of galactose with an alcohol; yields galactose on hydrolysis.

galacturonic acid The monobasic acid resulting from oxidation of the primary alcohol group of D-galactose to carboxyl; it is widely distributed as a constituent of pectins and many plant gums and mucilages.

galipol $C_{15}H_{26}O$ A terpene alcohol derived from the oil of the angostura bark; colorless crystals that melt at 89°C.

gallacetophenone $C_8H_8O_4$ A white to brownish-gray, crystalline powder, melting at 173°C, soluble in water, alcohol, and ether; used as an antiseptic. Also known as alizarin yellow C.

gallein $C_{20}H_{10}O_7$ A brown powder or green scales, broken down by heat; used as a pH indicator in the analysis of phosphates in urine and as an intermediate in the manufacture of dyes. Also known as anthracene violet; gallin; pyrogallolphthalein.

gallic acid $C_7H_6O_5$ A crystalline compound that forms needles from solutions of absolute methanol or chloroform, dissolves in water, alcohol, ether, and glycerol; obtained from nutgall tannins or from *Penicillium notatum* fermentation; used to make antioxidants and ink dyes and in photography.

gallin *See* gallein.

gallium A chemical element, symbol Ga, atomic number 31, atomic weight 69.72.

gallium arsenide GaAs A crystalline material, melting point 1238°C; frequently alloys of this material are formed with gallium phosphide or indium arsenide.

gallium halide A compound formed by bonding of gallium to either chlorine, bromine, iodine, fluorine, or astatine.

gallium phosphide GaP Transparent crystals made by reacting phosphorus and gallium suboxide at low temperature.

gallocyanine $C_{15}H_{13}ClN_2O_5$ Green crystals soluble in alcohol, glacial acetic acid, alkali carbonates, and concentrated hydrochloric acid; used as a dye and as a reagent for the determination of lead.

gallodesoxycholic acid *See* chenodeoxycholic acid.

gallogen *See* ellagic acid.

gallotannic acid *See* tannic acid.

gallotannin *See* tannic acid.

galvanic series The relative hierarchy of metals arranged in order from magnesium (least noble) at the anodic, corroded end through platinum (most noble) at the cathodic, protected end.

gamma The gamma position (the third carbon atom in an aliphatic carbon chain) on a chemical compound.

gamma acid $C_{10}H_5NH_2OHSO_3H$ White crystals, slightly soluble in water; an intermediate in dyestuff manufacture. Also known as 2-amino-8-naphthol-6-sulfonic acid; 7-amino-1-naphthol-3-sulfonic acid; 2,5-naphthylamine sulfonic acid; 3-sulfonic acid; 6-sulfonic acid.

gamma cross section The cross section for absorption or scattering of gamma rays by a nucleus or atom.

gamma decay *See* gamma emission.

gamma emission A quantum transition between two energy levels of a nucleus in which a gamma ray is emitted. Also known as gamma decay.

gamma flux density The number of gamma rays passing through a unit area in a unit time.

gamma ray A high-energy photon, especially as emitted by a nucleus in a transition between two energy levels.

gamma transition *See* glass transition.

gammil A unit of concentration, equal to a concentration of 1 milligram of solute in 1 liter of solvent. Also known as micril; microgammil.

Gamow barrier The potential barrier which retards the escape of alpha particles from the nucleus according to the Gamow-Condon-Gurney theory.

Gamow-Condon-Gurney theory An early quantum-mechanical theory of alpha-particle decay according to which the alpha particle penetrates a potential barrier near the surface of the nucleus by a tunneling process.

Gamow-Teller interaction Interaction between a nucleon source current and a lepton field which has an axial vector or tensor form.

Gamow-Teller selection rules Selection rules for beta decay caused by the Gamow-Teller interaction; that is, in an allowed transition there is no parity change of the nuclear state, and the spin of the nucleus can either remain unchanged or change by $\pm$ 1; transitions from spin 0 to spin 0 are excluded, however.

ganglioside One of a group of glycosphingolipids found in neuronal surface membranes and spleen; they contain an N-acyl fatty acid derivative of sphingosine linked to a carbohydrate (glucose or galactose); they also contain N-acetylglucosamine or N-acetylgalactosamine, and N-acetylneuraminic acid.

gas absorption operation The recovery of solute gases present in gaseous mixtures of noncondensables; this recovery is generally achieved by contacting the gas stream with a liquid that offers specific or selective solubility for the solute gas to be recovered, or with an adsorbent (for example, synthetic or natural zeolite) that accepts only specific molecule sizes or shapes.

gas adsorption The concentration of a gas upon the surface of a solid substance by attractive forces between the surface and the gas molecules.

gas analysis Analysis of the constituents or properties of a gas (either pure or mixed); composition can be measured by chemical adsorption, combustion, electrochemical cells, indicator papers, chromatography, mass spectroscopy, and so on; properties analyzed for include heating value, molecular weight, density, and viscosity.

gas black Fine particles of carbon formed by partial combustion or thermal decomposition of natural gas; used to reinforce rubber products such as tires. Also known as carbon black; channel black.

gas-cell frequency standard   An atomic frequency standard in which the frequency-determining element is a gas cell containing rubidium, cesium, or sodium vapor.

gas chromatograph   The instrument used in gas chromatography to detect volatile compounds present; also used to determine certain physical properties such as distribution or partition coefficients and adsorption isotherms, and as a preparative technique for isolating pure components or certain fractions from complex mixtures.

gas chromatography   A separation technique involving passage of a gaseous moving phase through a column containing a fixed adsorbent phase; it is used principally as a quantitative analytical technique for volatile compounds.

gas-condensate liquid   A hydrocarbon, such as propane, butane, and pentane, obtained as condensate when wet natural gas is compressed or refrigerated.

gas dehydrator   A device or system to remove moisture vapor from a gas stream, usually incorporates desiccant-type packed towers.

gaseous diffusion   1. Pressure-induced free-molecular transfer of gas through microporous barriers as in the process of making fissionable fuel. 2. Selective solubility diffusion of gas through nonporous polymers by absorption and solution of the gas in the polymer matrix.

gas filter   A device used to remove liquid or solid particles from a flowing gas stream.

gas generator   1. A device used to generate gases in the laboratory. 2. A chemical plant for producing gas from coal, for example, water gas.

gasification   Any chemical or heat process used to convert a substance to a gas; coal is converted by the Hygas process to a gaseous fuel.

gasifier   A unit for producing gas, particularly synthesis gas from coal.

gas lift   Solids movement operation in which an upward-flowing gas stream in a closed conduit or vessel is used to lift and move powdered or granular solid material.

gas-liquid chromatography   A form of gas chromatography in which the fixed phase (column packing) is a liquid solvent distributed on an inert solid support. Abbreviated GLC. Also known as gas-liquid partition chromatography.

gas making   Making water gas or air gas by the action of steam and air upon hot coke.

gasometric method   An analytical technique for gases; the gas may be measured by instrumental methods or through chemical reactions with specific reagents.

gas producer   A device for complete gasification of coal by utilizing simultaneously the air and water-gas reactions.

gas reversion   A process which combines thermal cracking or reforming of naphtha with thermal polymerization or alkylation of hydrocarbon gases carried out in the same reaction zone.

gas scrubbing   Removal of gaseous or liquid impurities from a gas by the action of a liquid; the gas is contacted with the liquid which removes the impurities by dissolving or by chemical combination.

gas-solid chromatography A form of gas chromatography in which the moving phase is a gas and the stationary phase is a surface-active sorbent (charcoal, silica gel, or activated alumina). Abbreviated GSC.

gas solubility The extent that a gas dissolves in a liquid to produce a homogeneous system.

gas-treating system A process system to remove nonhydrocarbon impurities (such as water vapor, hydrogen sulfide, or carbon dioxide) from wellhead gas.

gastric acid Hydrochloric acid secreted by parietal cells in the fundus of the stomach.

gastric enzyme Any digestive enzyme secreted by cells lining the stomach.

gastrin A polypeptide hormone secreted by the pyloric mucosa which stimulates the pancreas to release pancreatic fluid and the stomach to release gastric acid.

gastrointestinal hormone Any hormone secreted by the gastrointestinal system.

gas-tube boiler *See* waste-heat boiler.

gatinon *See* benzthiazuron.

Gatterman-Koch synthesis A synthesis of aldehydes; aldehydes form when an aromatic hydrocarbon is heated in the presence of hydrogen chloride, certain metallic chloride catalysts, and either carbon monoxide or hydrogen cyanide.

Gatterman reaction 1. Reaction of a phenol or phenol ester, and hydrogen chloride or hydrogen cyanide, in the presence of a metallic chloride such as aluminum chloride to form, after hydrolysis, an aldehyde. 2. Reaction of an aqueous ethanolic solution of diazonium salts with precipitated copper powder or other reducing agent to form diaryl compounds.

gaultheria oil *See* methyl salicylate.

Gay-Lussac law *See* combining-volumes principle.

Gay-Lussac tower A component part in the chamber process for sulfuric acid production that absorbs nitrogen oxides to form nitrous vitriol.

Gd *See* gadolinium.

Ge *See* germanium.

Geiger-Nutall rule The rule that the logarithm of the decay constant of an alpha emitter is linearly related to the logarithm of the range of the alpha particles emitted by it.

gel A two-phase colloidal system consisting of a solid and a liquid in more solid form than a sol.

gelatin A protein derived from the skin, white connective tissue, and bones of animals; used as a food and in photography, the plastics industry, metallurgy, and pharmaceuticals.

gelatinase An enzyme, found in some yeasts and molds, that liquefies gelatin.

gelation 1. The act or process of freezing. 2. Formation of a gel from a sol.

gel electrophoresis Electrophoresis performed in silica gel, a porous, inert medium.

gel filtration A type of column chromatography which separates molecules on the basis of size; higher-molecular-weight substances pass through the column first. Also known as molecular exclusion chromatography; molecular sieve chromatography.

gel permeation chromatography   Analysis by chromatography in which the stationary phase consists of beads of porous polymeric material such as a crosslinked dextran carbohydrate derivative sold under the trade name Sephadex; the moving phase is a liquid.

gel point   Stage at which a liquid begins to exhibit elastic properties and increased viscosity.

geminal   Referring to like atoms or groups attached to the same atom in a molecule.

general formula   A formula that can apply not only to one specific compound but to a series of related compounds; for example, the general formula for an aldehyde RCHO, where R is hydrogen in formaldehyde (the simplest aldehyde) and is a hydrocarbon radical for other aldehydes in the series such as CH_3 for acetaldehyde and C_2H_5 for proprionaldehyde.

Geneva system   An international system of nomenclature for organic compounds based on hydrocarbon derivatives; names correspond to the longest straight carbon chain in the molecule.

genicide $C_{13}H_8O_2$ A compound with needlelike crystals and a melting point of 174°C; insoluble in water; used as an insecticide, miticide, and ovicide. Also known as oxoxanthone; 9-xanthenone; xanthone.

genistin $C_{21}H_{20}O_{10}$ A pale-yellow glucoside derived from soybean meal, crystallizes from 80% methanol solution, melting point 256°C, soluble in hot 80% ethanol, hot 80% methanol, and hot acetone. Also known as 7-D-glucoside.

gentianic acid   *See* gentisic acid.

gentian violet   *See* methyl violet.

gentisic acid $C_7H_6O_4$ A crystalline compound that forms monoclinic prisms from a water solution, sublimes at 200°C, melts at 250°C, and is soluble in water, alcohol, ether, sodium, and salt; used in medicine. Also known as gentianic acid; hydroquinone carboxylic acid; 5-hydroxysalicylic acid.

geochemical anomaly   Above-average concentration of a chemical element in a sample of rock, soil, vegetation, stream, or sediment; indicative of nearby mineral deposit.

geochemical balance   The proportional distribution, and the migration rate, in the global fractionation of elements, minerals, or compounds; for example, the distribution of quartz in igneous rocks, its liberation by weathering, and its redistribution into sediments and, in solution, into lakes, rivers, and oceans.

geochemical cycle   During geologic changes, the sequence of stages in the migration of elements between the lithosphere, hydrosphere, and atmosphere.

geochemical evolution   **1.** A change in any constituent of a rock beyond that amount present in the parent rock. **2.** A change in chemical composition of a major segment of the earth during geologic time, as the oceans.

geometrical isomerism   The phenomenon in which isomers contain atoms attached to each other in the same order and with the same bonds but with different spatial, or geometrical, relationships; the explicit geometry imposed upon a molecule by, say, a double bond between carbon atoms makes possible the existence of these isomers.

geranial *See* citral.

geranialdehyde *See* citral.

geraniol $(CH_3)_2CCH(CH_2)_2C(CH_3)CH\ CH_2OH$ A colorless to pale-yellow liquid, an alcohol and a terpene, boiling point 230°C; soluble in alcohol and ether, insoluble in water; used in perfumery and flavoring.

geranyl $C_{10}H_{17}$ The radical from geraniol, $(CH_3)_2{:}CHCH_2CH_2{\cdot}CHCH_3{:}$ $CH{\cdot}CH_2OH$.

Gerard reagent The quaternary ammonium compounds, acethydrazide-pyridinium chloride and trimethylacethydrazide ammonium chloride; used to separate aldehydes and ketones from oily or fatty natural materials and to extract sex hormones from urine.

germane **1.** A hydride of germanium whose general formula is Ge_nH_{2n+2}. **2.** The compound GeH_4, a hydride of germanium, a colorless gas that is combustible in air and burns with a blue flame.

germanide A compound of an alkaline earth or alkali metal with germanium; an example is magnesium germanide, Mg_2Ge; the germanides are reactive with water.

germanium A brittle, water-insoluble, silvery-gray metallic element in the carbon family, symbol Ge, atomic number 32, atomic weight 72.59, melting at 959°C.

germanium halide A dihalide or tetrahalide of fluorine, chlorine, bromine, or iodine with germanium.

germanium oxide The monoxide GeO or dioxide GeO_2; a study of GeO indicates it exists in polymeric form; GeO_2 is a white powder, soluble in alkalies; used in special glass and in medicine.

getter **1.** A substance, such as thallium, that binds gases on its surface and is used to maintain a high vacuum in a vacuum tube. **2.** A special metal alloy that is placed in a vacuum tube during manufacture and vaporized after the tube has been evacuated; when the vaporized metal condenses, it absorbs residual gases. Also known as degasser.

g factor *See* Landé g factor.

ghost crystal *See* phantom crystal.

gibberellic acid $C_{18}H_{22}O_6$ A crystalline acid occurring in plants that is similar to the gibberellins in its growth-promoting effects.

gibberellin Any member of a family of naturally derived compounds which have a gibbane skeleton and a broad spectrum of biological activity but are noted as plant growth regulators.

Gibbs adsorption equation A formula for a system involving a solvent and a solute, according to which there is an excess surface concentration of solute if the solute decreases the surface tension, and a deficient surface concentration of solute if the solute increases the surface tension.

Gibbs adsorption isotherm An equation for the surface pressure of surface monolayers,

$$\phi = RT \int_0^p \Gamma d(\ln p)$$

where ϕ is surface pressure, T is absolute temperature, R is the gas constant, Γ is the number of molecules adsorbed per gram per unit surface area, and p is the pressure of the gas.

Gibbs diaphragm cell   A type of electrolytic diaphragm cell for chlorine production, with graphite electrodes and a cylindrical shape.

Gibbs-Donnan equilibrium   *See* Donnan equilibrium.

Gibbs-Duhem equation   A relation that imposes a condition on the composition variation of the set of chemical potentials of a system of two or more components,

$$ SdT - VdP + \sum_{i=1}^{r} n_i d\mu_i = 0 $$

where S is entropy, T absolute temperature, P pressure, n_i the number of moles of the ith component, and μ_i the chemical potential of the ith component. Also known as Duhem's equation.

Gibbs-Helmholtz equation   An expression for the influence of temperature upon the equilibrium constant of a chemical reaction, $(d \ln K°/dT)_P = \Delta H°/RT^2$, where $K°$ is the equilibrium constant, $\Delta H°$ the standard heat of the reaction at the absolute temperature T, and R the gas constant.

Gibbs phase rule   A relation describing the nature of a heterogeneous chemical system at equilibrium, $F = C + 2 - P$, where F is the degrees of freedom, P the number of phases, and C the number of components. Also known as Gibbs rule.

Gibbs-Poynting equation   An expression relating the effect of the total applied pressure P upon the vapor pressure p of a liquid, $(dp/dP)_T = V_l/V_g$, where V_l and V_g are molar volumes of the liquid and vapor.

Gibbs rule   *See* Gibbs phase rule.

Giegy-Hardisty process   The production of sebacic acid from castor oil or its acids by reaction of the acid at a high temperature with caustic alkali.

Giemsa stain   A stain for hemopoietic tissue and hemoprotozoa consisting of a stock glycerol methanol solution of eosinates of Azure B and methylene blue with some excess of the basic dyes.

gifblaar poison   *See* fluoroacetic acid.

Gillespie equilibrium still   A recirculating equilibrium distillation apparatus used to establish azeotropic properties of liquid mixtures.

Gilliland correlation   Approximation method for distillation-column calculations; correlates reflux ratio and number of plates for the column as functions of minimum reflux and minimum plates.

Girbotal process   A regenerative absorption process to remove carbon dioxide, hydrogen sulfide, and other acid impurities from natural gas, using mono-, di-, or triethanolamine as the reagent.

gitonin   The gitogenin tetraglycoside in *Digitalis purpurea* seed; resembles digitonin.

glacial acetic acid CH_3COOH Pure acetic acid (containing less than 1% water); a clear, colorless, caustic hygroscopic liquid, boiling at 118°C, soluble in water, alcohol, and ether, and crystallizing readily; used as a solvent for oils and resins.

glassed steel Process piping or vessels lined with glass; a glass-steel composite has structural strength of steel and corrosion resistance of glass.

glass electrode An electrode or half cell in which potential measurements are made through a glass membrane, which acts as a cation-exchange membrane; thus, the potential arises from phase-boundary and diffusion potentials which, depending on the composition of the glass, are logarithmic functions of the activity of the cations such as H^+, Na^+, or K^+ of the solutions in which the electrode is immersed.

glass transition The change in an amorphous region of a partially crystalline polymer from a viscous or rubbery condition to a hard and relatively brittle one; usually brought about by changing the temperature. Also known as gamma transition; glassy transition.

glassy transition *See* glass transition.

Glauber's salt $Na_2SO_4 \cdot 10H_2O$ Crystalline hydrated sodium sulfate; loses water when exposed to air; water soluble, alcohol insoluble; used in textile dyeing and medicine.

glaze stain Colorant for ceramic glazes; made of a finely ground calcined oxide, such as of cobalt, copper, manganese, or iron.

glide *See* slip.

glide plane A lattice plane in a crystal on which translation or twin gliding occurs. Also known as slip plane.

globin Any of a class of soluble histone proteins obtained from animal hemoglobins.

globoside A glycoside of ceramide containing several sugar residues, but not neuraminic acid; obtained from human, sheep, and hog erythrocytes.

globular protein Any protein that is readily soluble in aqueous solvents.

globulin A heat-labile serum protein precipitated by 50% saturated ammonium sulfate and soluble in dilute salt solutions.

glost firing The process of glazing and firing ceramic ware which has previously been fired at a higher temperature.

Glover tower A tower in the lead chamber process for manufacturing sulfuric acid; in this tower the nitrogen oxide, sulfur dioxide, and air mixture is passed upward and sprayed with a sulfuric acid–nitrosyl sulfuric acid mixture.

glucagon The protein hormone secreted by α-cells of the pancreas which plays a role in carbohydrate metabolism. Also known as hyperglycemic factor; hyperglycemic glycogenoiytic factor.

glucamine $C_6H_{15}O_4N$ An amine formed by reduction of glucosylamine or of glucose oxime.

glucan A polysaccharide composed of the hexose sugar D-glucose.

glucinium The former name for the element beryllium, coined because the salts of beryllium are sweet-tasting.

glucocerebroside A glycoside of ceramide that contains glucose.

glucochloral *See* chloralose.

glucochloralose *See* chloralose.

α-D-glucochloralose *See* chloralose.

glucocorticoid A corticoid that affects glucose metabolism; secreted principally by the adrenal cortex.

glucogenesis Formation of glucose within the animal body from products of glycolysis.

glucokinase An enzyme that catalyzes the phosphorylation of D-glucose to glucose-6-phosphate.

glucolipid A glycolipid that yields glucose on hydrolysis.

glucomannan A polysaccharide composed of D-glucose and D-mannose; a prominent component of coniferous trees.

gluconate A salt of gluconic acid.

gluconeogenesis Formation of glucose within the animal body from substances other than carbohydrates, particularly proteins and fats.

gluconic acid $C_6H_{12}O_7$ A crystalline acid obtained from glucose by oxidation; used in cleaning metals.

gluconic acid sodium salt *See* sodium gluconate.

D-glucopyranose *See* glucose.

glucopyranoside Any glucoside that contains a six-membered ring.

glucosamine $C_6H_{13}O_5$ An amino sugar; the most abundant in nature, occurring in glycoproteins and chitin.

glucose $C_6H_{12}O_6$ A monosaccharide; occurs free or combined and is the most common sugar. Also known as cerelose; D-glucopyranose.

glucose-6-phosphatase An enzyme found in liver which catalyzes the hydrolysis of glucose-6-phosphate to free glucose and inorganic phosphate.

glucose phosphate A phosphoric derivative of glucose, as glucose-1-phosphate.

glucose-1-phosphate $C_6H_{12}O_8P$ An ester of glucopyranose in which a phosphate group is attached to carbon atom 1; there are two types: α-D- and β-D-glucose-1-phosphates. Also known as Cori ester.

glucose-6-phosphate $C_6H_{13}O_9P$ An ester of glucose with phosphate attached to carbon atom 6. Also known as Robisonester.

glucose-6-phosphate dehydrogenase The mammalian enzyme that catalyzes the oxidation of glucose-6-phosphate by TPN$^+$ (triphosphopyridine nucleotide).

glucosidase An enzyme that hydrolyzes glucosides.

glucoside One of a group of compounds containing the cyclic forms of glucose, in which the hydrogen of the hemiacetal hydroxyl has been replaced by an alkyl or aryl group.

glucosulfone sodium *See* sodium glucosulfone.

glucosyltransferase An enzyme that catalyzes the glucosylation of hydroxymethyl cytosine; a constituent of bacteriophage deoxyribonucleic acid.

glucuronic acid $C_6H_{10}O_7$ An acid resulting from oxidation of the CH_2OH radical of D-glucose to COOH; a component of many polysaccharides and certain vegetable gums. Also known as glycuronic acid.

glucuronidase An enzyme that catalyzes hydrolysis of glucuronides. Also known as glycuronidase.

glucuronide A compound resulting from the interaction of glucuronic acid with a phenol, an alcohol, or an acid containing a carboxyl group. Also known as glycuronide.

D-glucuronolactone $C_6H_8O_6$ A water-soluble crystalline compound found in plant gums in polymers with other carbohydrates, and an important structural component of almost all fibrous and connective tissues in animals; used in medicine as an antiarthritic.

glutamate A salt or ester of glutamic acid.

glutamic acid $C_5H_9O_4N$ A dicarboxylic amino acid of the α-ketoglutaric acid family occurring widely in proteins.

glutaminase The enzyme which catalyzes the conversion of glutamine to glutamic acid and ammonia.

glutamine $C_5H_{10}O_3N_2$ An amino acid; the monamide of glutamic acid; found in the juice of many plants and essential to the development of certain bacteria.

glutamine synthetase An enzyme which catalyzes the formation of glutamine from glutamic acid and ammonia, using adenosinetriphosphate as a source of energy.

glutaraldehyde $OHC(CH_2)_3CHO$ A liquid with a boiling point of 188°C; soluble in water and alcohol; used as a biological solution (50%) and for leather tanning.

glutarate The salt or ester of glutaric acid.

glutaric acid $C_5H_5O_4$ A water-soluble, crystalline acid that occurs in green sugarbeets and in water extracts of crude wool.

glutathione $C_{10}H_{17}O_6N_3S$ A widely distributed tripeptide that is important in plant and animal tissue oxidation reactions.

glutelin A class of simple, heat-labile proteins occurring in seeds of cereals; soluble in dilute acids and alkalies.

gluten 1. A mixture of proteins found in the seeds of cereals; gives dough elasticity and cohesiveness. 2. An albuminous element of animal tissue.

glutenin A glutelin of wheat.

glyceraldehyde $CH_2OHCOHCHO$ A colorless solid, isomeric with dehydroxyacetone; soluble in water and insoluble in organic solvents; an important intermediate in carbohydrate metabolism; used as a chemical intermediate in biochemical research and nutrition. Also known as 2,3-dihydroxypropanal; glyceric aldehyde.

glycerate A salt or ester of glyceric acid.

glyceric acid $C_3H_6O_4$ A hydroxy acid obtained by oxidation of glycerin.

glyceric aldehyde *See* glyceraldehyde.

glyceride An ester of glycerin and an organic acid radical; fats are glycerides of certain long-chain fatty acids.

glycerin *See* glycerol.

glycerinated vaccine virus *See* smallpox vaccine.

glycerokinase An enzyme that catalyzes the phosphorylation of glycerol to glycerophosphate during microbial fermentation of propionic acid.

glycerol $CH_2OHCHOHCH_2OH$ The simplest trihedric alcohol; when pure, it is a colorless, odorless, viscous liquid with a sweet taste; it is completely soluble in water and alcohol but only partially soluble in common solvents such as ether and ethyl acetate; used in manufacture of alkyd resins, explosives, antifreezes, medicines, inks, perfumes, cosmetics, soaps, and finishes. Also known as glycerin; glycyl alcohol.

glycerol dichlorohydrin *See* α-dichlorohydrin.

sym-glycerol dichlorohydrin *See* 1,3-dichloro-2-propanol.

glycerol dimethyl ketal *See* 2,2-dimethyl-1,3-dioxolane-4-methanol.

glycerol α-monochlorohydrin *See* 3-chloro-1,2-propanediol.

glycerophosphate Any salt of glycerophosphoric acid.

glycerophosphoric acid $C_3H_5(OH)_2OPO_3H_2$ Either of two pale-yellow, water-soluble, isomeric dibasic acids occurring in nature in combined form as cephalin and lecithin.

glyceryl $OCH_2OCHOCH_2 \equiv$ The radical group from glycerol, $(CH_2OH)_2CHOH$.

glyceryl diacetate *See* diacetin.

glyceryl monoacetate *See* acetin.

glyceryl triacetate *See* triacetin.

glyceryl tripalmitate *See* tripalmitin.

glyceryl tristearate *See* stearin.

glycide *See* glycidol.

glycidic acid $C_2H_3O \cdot CO_2H$ A volatile liquid. Also known as epoxy-propionic acid.

glycidol $C_3H_6O_2$ A colorless, liquid epoxide that boils at 162°C and is miscible with water; used in organic synthesis. Also known as epihydrin alcohol; 2,3-epoxy-1-propanol; glycide.

glycin $C_8H_9NO_3$ A crystalline compound that forms shiny leaflets from water solution, melts at 245–247°C, and is soluble in alkalies and mineral acids; used as a photographic developer and in the analytical determination of iron, phosphorus, and silicon. Also known as *para*-hydroxyanilinoacetic acid; *para*-hydroxyphenylaminoacetic acid; photoglycine.

glycine $C_2H_5O_2N$ A white, crystalline amino acid found as a constituent of many proteins. Also known as aminoacetic acid.

glyco- Chemical prefix indicating sweetness, or relating to sugar or glycine.

glycocholic acid $C_{26}H_{43}NO_6$ A bile obtained by the conjugation of cholic acid with glycine.

glycocoll-*para*-phenetidine hydrochloride *See* phenocoll hydrochloride.

glycocyamine $C_3H_7N_3O_2$ A product of interaction of aminocetic acid and arginine, which on transmethylation with methionine is converted to creatine. Also known as guanidine-acetic acid.

glycogen A nonreducing, white, amorphous polysaccharide found as a reserve carbohydrate stored in muscle and liver cells of all higher animals, as well as in cells of lower animals.

glycogenesis The metabolic formation of glycogen from glucose.

glycogenolysis The metabolic breakdown of glycogen.

glycogen synthetase An enzyme that catalyzes the synthesis of the amylose chain of glycogen.

glycol 1. $C_nH_{2n}(OH)_2$ An organic chemical with two hydroxyl groups on an essentially aliphatic carbon chain. Also known as dihydroxy alcohol. 2. $HOCH_2CH_2OH$ A colorless dihydroxy alcohol used as an antifreeze, in hydraulic fluids, and in the manufacture of dynamites and resins. Also known as ethlene glycol.

glycol diacetate See ethylene glycol diacetate.

glycoldinitrate See ethylene nitrate.

glycol ester Chemical compound composed of the reaction products of a glycol, $C_nH_{2n}(OH)_2$, and an organic acid; an example is ethylene glycol diacetate, the product of ethylene glycol and acetic acid.

glycol ether A colorless liquid used as a solvent, in detergents, and as a diluent; a typical example is ethylene glycol diethyl ether, $C_2H_5OCH_2CH_2OC_2H_5$.

glycol ethylene ether See 1,4-dioxane.

glycolic acid $CH_2OHCOOH$ Colorless, deliquescent leaflets, decomposing about 78°C; soluble in water, alcohol, and ether; used as a chemical intermediate in fabric dyeing. Also known as glycollic acid; hydroxyacetic acid.

glycolipid Any of a class of complex lipids which contain carbohydrate residues.

glycollic acid See glycolic acid.

glycolysis The enzymatic breakdown of glucose or other carbohydrate, with the formation of lactic acid or pyruvic acid and the release of energy in the form of adenosinetriphosphate.

glycolythiourea See 2-thiohydantoin.

glycolytic pathway The principal series of phosphorylative reactions involved in pyruvic acid production in phosphorylative fermentations. Also known as Embden-Meyerhof pathway; hexose diphosphate pathway.

glycolyurea See hydantoin.

glyconeogenesis The metabolic process of glycogen formation from noncarbohydrate precursors.

glycopeptide See glycoprotein.

glycoprotein Any of a class of conjugated proteins containing both carbohydrate and protein units. Also known as glycopeptide.

glycosidase An enzyme that hydrolyzes a glycoside.

glycoside A compound that yields on hydrolysis a sugar (glucose, galactose) and an aglycon; many of the glycosides are therapeutically valuable.

glycotropic Acting to antagonize the action of insulin.

glycuronic acid   *See* glucuronic acid.

glycuronidase   *See* glucuronidase.

glycuronide   *See* glucuronide.

glycyl NH_2CH_2COO- or $NHCH_2COO=$ The radical from glycine, NH_2CH_2COOH; found in peptides.

glycyl alcohol   *See* glycerol.

glyoxal $(CHO)_2$ Colorless, deliquescent powder or liquid with mild odor, melting point 15°C, boiling point 51°C; used to insolubilize starches, cellulosic materials, and proteins, in embalming fluids, for leather tanning, and for rayon shrinkproofing.

glyoxalase   An enzyme present in various body tissues which catalyzes the conversion of methylglyoxal into lactic acid.

glyoxalic acid $CHOCOOH$ Colorless crystals that are soluble in water, forming glyoxylic acid. Also known as ethanol acid; oxaldehydic acid; oxoethanoic acid.

glyoxaline   *See* imidazole.

glyoxylate cycle   A sequence of biochemical reactions related to respiration in germinating fatty seeds by which acetyl coenzyme A is converted to succinic acid and then to hexose.

glyoxylic acid $CH(OH)_2COOH$ An aldehyde acid found in many plant and animal tissues, especially unripe fruit.

glyphosate $C_3H_8NO_5P$ A white solid with a melting point of 200°C; slight solubility in water; used as an herbicide in postharvest treatment of crops. Also known as N-(phosphonomethyl)glycine.

glyphosine $C_4H_{11}NO_8P_2$ A white solid with a melting point of 203°C; quite soluble in water; used as a growth regulator in sugarcane. Also known as N,N-bis(phosphonomethyl)glycine.

glyptal resin   A phthalic anhydride glycerol made from an emulsion of an alkyd resin; used in lacquers and insulation.

GMP   *See* guanylic acid.

gnomonic projection   A projection for displaying the poles of a crystal in which the poles are projected radially from the center of a reference sphere onto a plane tangent to the sphere.

gold   A chemical element, symbol Au, atomic number 79, atomic weight 196.967; soluble in aqua regia; melts at 1065°C.

gold-198   The radioisotope of gold, atomic mass number 198 and half-life 2.7 days; used in medical treatment of tumors by injecting it in colloidal form directly into tumor tissue.

gold chloride $AuCl_3$ A red, soluble compound made by reaction of gold and chlorine or by reaction of $HAuCl_4$ with chlorine; decomposes by heat; soluble in water, alcohol, and ether; used in photography, plating, inks, medicine, and ceramics.

golden antimony sulfide   *See* antimony pentasulfide.

gold hydroxide $Au(OH)_3$ A yellow-brown, light-sensitive, water-insoluble powder; dissolves in most acids; easily reduced to metallic gold; used in medicine, porcelain, gold plating, and daguerreotypes.

gold number A measure of the amount of protective colloid which must be added to a standard red gold sol mixed with sodium chloride solution to prevent the solution from causing the sol to coagulate, as manifested by a change in color from red to blue.

gold oxide Au_2O_3 Water-insoluble, heat-decomposable, brownish-black powder; soluble in hydrochloric acid; used to gild, in medicine and porcelain, and for daguerreotypes. Also known as auric oxide; gold trioxide.

gold potassium chloride See potassium gold chloride.

gold potassium cyanide See potassium gold cyanide.

gold salt See sodium gold chloride.

gold size A solution of white and red lead and yellow ocher in linseed oil; used to seal permanently microscopical preparations.

gold sodium chloride See sodium gold chloride.

gold sodium cyanide See sodium gold cyanide.

gold tin precipitate See gold tin purple.

gold tin purple A brown powder which is a mixture of gold chloride and brown tin oxide, soluble in ammonia; used in coloring enamels, manufacturing ruby glass, and painting porcelain. Also known as gold tin precipitate; purple of Cassius.

gold trioxide See gold oxide.

Gomberg-Bachmann-Hey reaction Production of diaryl compounds by adding alkali to a mixture of a diazonium salt and a liquid aromatic hydrocarbon or a derivative.

Gomberg reaction The production of free radicals by reaction of metals with triarylmethyl halides.

gonadotropic hormone Either of two adenohypophyseal hormones, FSH (follicle-stimulating hormone) or ICSH (interstitial-cell-stimulating hormone), that act to stimulate the gonads.

gonadotropin A substance that acts to stimulate the gonads.

gonyautoxin One of a group of saxitoxin-related compounds that are produced by the dinoflagellates *Gonyaulax catenella* and *tamarensis*.

Gooch crucible A ceramic crucible with a perforated base; in analysis it is used for filtration through asbestos or glass.

good geometry An arrangement of source and detecting equipment such that little error is produced by the finite sizes of the source and the detector aperture.

good oil See raffinate.

gorgonin The protein, frequently containing iodine and bromine, composing the horny skeleton of members of the Gorgonacea; contains iodine and bromine.

gorlic acid $C_5H_7(C_{12}H_{22})COOH$ An unsaturated acid derived from sapucainha oil, obtained from the seeds of a tree in the Amazon Valley.

gossypose *See* raffinose.

gouy An electrokinetic unit equal to the product of the electrokinetic potential and the electric displacement divided by 4π times the polarization of the electrolyte.

Gouy balance Device for measurement of diamagnetic and paramagnetic susceptibilities of samples (solid, liquid, solution).

gradient elution analysis A form of gas-liquid chromatography in which the eluting solvent is changed with time, either by gradually mixing a second solvent of greater eluting power with the first, less powerful solvent, or by a gradual change in pH or other property.

graduate A cylindrical vessel that is calibrated in fluid ounces or milliliters or both; used to measure the volume of liquids.

Graebe-Ullman reaction 1. Production of fluorenone by boiling 2-benzoylbenzenediazonium salts in dilute acid solution. 2. Reaction of 2-aminodiphenylamines with nitrous acid to form a benzotriazole which on heating loses nitrogen to form a carbazole.

graft copolymer Any high polymer composed of two or more different polymeric entities chemically united.

grain alcohol *See* alcohol.

grainer process A salt production method in which salt is produced by surface evaporation of brine in open-air flat pans.

gram-atomic weight The atomic weight of an element expressed in grams, that is, the atomic weight on a scale on which the atomic weight of carbon-12 isotope is taken as 12 exactly.

gram-equivalent weight The equivalent weight of an element or compound expressed in grams on a scale in which carbon-12 has an equivalent weight of 3 grams in those compounds in which its formal valence is 4.

gram-molecular volume The volume occupied by a gram-molecular weight of a chemical in the gaseous state at 0°C and 760 millimeters of pressure (101,325 newtons per square meter).

gram-molecular weight The molecular weight of compound expressed in grams, that is, the molecular weight on a scale on which the atomic weight of carbon-12 isotope is taken as 12 exactly.

granulate To form or crystallize into grains, granules, or small masses.

graphite anode One of the electrodes of graphite used in a mercury cell to produce chlorine by electrolysis.

graphitization The formation of graphitelike material from organic compounds.

Grassmann's laws Seven laws of color identification and mixing that form the basis of modern analytical colorimetry.

grass-roots plant A complete plant erected on a virgin site.

grating *See* diffraction grating.

grating constant The distance between consecutive grooves of a diffraction grating.

grating spectrograph A grating spectroscope provided with a photographic camera or other device for recording the spectrum.

grating spectroscope A spectroscope which employs a transmission or reflection grating to disperse light, and usually also has a slit, a mirror or lenses to collimate the light sent through the slit and to focus the light dispersed by the grating into spectrum lines, and an eyepiece for viewing the spectrum.

gravimetric absorption method A method of measuring the moisture content of a gas in which a known volume of gas is passed through a suitable desiccant, such as phosphorus pentoxide or silica gel, and the change in weight of the desiccant is observed.

gravimetric analysis That branch of quantitative analytical chemistry in which a desired constituent is converted, usually by precipitation or combustion, to a pure compound or element, of definite known composition, and is weighed; in a few cases a compound or element is formed which does not contain the constituent but bears a definite mathematical relationship to it.

gravity cell An electrolytic cell in which two ionic solutions are separated by means of gravity.

Gray clay treating A fixed-bed, vapor-phase treating process used to polymerize selectively unsaturated gum-forming constituents (diolefins); a fixed bed is used of 30- to 60-mesh fuller's earth.

green copperas *See* ferrous sulfate.

green nickel oxide *See* nickel oxide.

green salt *See* uranium tetrafluoride.

green vitriol *See* ferrous sulfate.

grid spectrometer A grating spectrometer in which a large increase in light flux without loss of resolution is achieved by replacing entrance and exit slits with grids consisting of opaque and transparent areas, patterned to have large transmittance only when the entrance grid image coincides with that of the exit grid.

Griess reagent A reagent used to test for nitrous acid; it is a solution of sulfanilic acid, α-naphthylamine and acetic acid in water.

Grignard reaction A reaction between an alkyl or aryl halide and magnesium metal in a suitable solvent, usually absolute ether, to form an organometallic halide.

Grignard reagent RMgX The organometallic halide formed in the Grignard reaction; an example is C_2H_5MgCl; it is useful in organic synthesis.

Grignard synthesis Use of the Grignard reagent in any one of a vast number of reactions, usually condensations; typical syntheses involve formation of a hydrocarbon, acid, ketone, or secondary or tertiary alcohol.

grinding-type resin Vinyl or other resin that requires grinding before dispersal into plastisols or organosols.

groove Any of a group of depressions in the double helix of deoxyribonucleic acid that are believed to be sites occupied by nuclear proteins.

Grotthus' chain theory An early theory used to explain the conductivity of an electrolyte, in which it was assumed that the cathode and anode attract hydrogen and oxygen respectively, and the molecules of the electrolyte are stretched out in chains between the electrodes, with decomposition occurring in molecules closest to the electrodes.

group 1. A family of elements with similar chemical properties. 2. A combination of bonded atoms that behave as a single unit under certain conditions.

Grove's synthesis Production of alkyl chlorides by passing hydrochloric acid into an alcohol in the presence of anhydrous zinc chloride.

growth hormone 1. A polypeptide hormone secreted by the anterior pituitary which promotes an increase in body size. Abbreviated GH. 2. Any hormone that regulates growth in plants and animals.

growth regulator A synthetic substance that produces the effect of a naturally occurring hormone in stimulating plant growth; an example is dichlorophenoxyacetic acid.

growth spiral A structure on a crystal surface, observed after growth, consisting of a growth step winding downward and outward in an Archimedean spiral which may be distorted by the crystal structure.

growth step A ledge on a crystal surface, one or more lattice spacings high, where crystal growth can take place.

GR-S rubber Former designation for general-purpose synthetic rubbers formed by copolymerization of emulsions of styrene and butadiene; used in tires and other rubber products; previously also known as Buna-S, currently known as SBR (styrene-butadiene rubber).

GSC *See* gas-solid chromatography.

guaiacol $C_6H_4(OH)OCH_3$ A colorless, crystalline compound, soluble in water; used as a reagent to determine the presence of such substances as lignin, narceine, and nitrous acid. Also known as *ortho*-hydroxyanisole; 1-hydroxy-2-methoxybenzene; methylpyrocatechin.

guanidine CH_5N_2 Aminomethanamidine, a product of protein metabolism found in urine. Also known as carbamidine; iminourea.

guanine $C_5H_5ON_5$ A purine base; occurs naturally as a fundamental component of nucleic acids.

guanosine $C_{10}H_{13}O_5N_5$ Guanine riboside, a nucleoside composed of guanine and ribose. Also known as vernine.

guanosine monophosphate *See* guanylic acid.

guanosine phosphoric acid *See* guanylic acid.

guanosine tetraphosphate A nucleotide which participates in the regulation of gene transcription in bacteria by turning off the synthesis of ribosomal ribonucleic acid.

guanylic acid A nucleotide composed of guanine, a pentose sugar, and phosphoric acid and formed during the hydrolysis of nucleic acid. Abbreviated GMP. Also known as guanosine monophosphate; guanosine phosphoric acid.

Guerbet reaction A condensation of alcohols at high temperatures through the action of sodium alkoxides.

guest element *See* trace element.

Guldberg and Waage law   *See* mass action law.

Guldberg-Waage group   A dimensionless number used in studying chemical reactions in blast furnaces; it is given by an equation relating volumes of reacting gases and reacting products. Symbolized N_{GW}.

Gulfining   A catalytic hydrogen-treating process used to remove sulfur from cracked and straight-run distillates and fuel oils, also improving carbon residue, color, and general stability and effecting a slight increase in gravity.

gum accroides   *See* acaroid resin.

gum test   A standard ASTM test to determine the amount of gums in gasolines.

gunbarrel   An atmospheric vessel used for treatment of waterflood waste water.

Günzberg reagent   A solution of 2 grams of vanillin and 4 grams of phloroglucinol in 80 milliliters of 95% alcohol; used as a test reagent for determining free hydrochloric acid in gastric juice.

Gurney-Mott theory   A theory of the photographic process that proposes a two-stage mechanism; in the first stage, a light quantum is absorbed at a point within the silver halide gelatin, releasing a mobile electron and a positive hole; these mobile defects diffuse to trapping sites (sensitivity centers) within the volume or on the surface of the grain; in the second stage, trapped (negatively charged) electron is neutralized by an interstitial (positively charged) silver ion, which combines with the electron to form a silver atom; the silver atom is capable of trapping a second electron, after which the process repeats itself, causing the silver speck to grow.

Gutzeit test   A test for arsenic; zinc and dilute sulfuric acid are added to the substance, which is then covered with a filter paper moistened with mercuric chloride solution; a yellow spot forms on the paper if arsenic is in the sample.

H *See* hydrogen.

²H *See* deuterium.

³H *See* tritium.

Haber-Bosch process Early nitrogen-fixation process for production of ammonia from hydrogen and nitrogen, catalyzed by iron; now replaced by more efficient ammonia synthesis processes. Also known as Haber process.

Haber process *See* Haber-Bosch process.

habit *See* crystal habit.

habit plane The crystallographic plane or system of planes along which certain phenomena such as twinning occur.

H acid $H_2NC_{10}H_4(OH)(SO_3H)_2$ A gray powder or crystalline substance that is soluble in water, ether, and alcohol; used as a dye intermediate. Also known as 8-amino-1-naphthol-3,6-disulfonic acid.

hadronic atom An atom consisting of a negatively charged, strongly interacting particle orbiting around an ordinary nucleus.

hafnium A metallic element, symbol Hf, atomic number 72, atomic weight 178.49; melting point 2000°C, boiling point above 5400°C.

hafnium carbide HfC Gray powder, melting at 3887°C; used in the control rods of nuclear reactors.

Hageman factor *See* factor XII.

Haggenmacher equation Equation to calculate latent heats of vaporizations of pure compounds by using critical conditions with Antoine constants.

halazone $COOHC_6H_4SO_2NCl_2$ White crystals, with strong chlorine aroma; slightly soluble in water and chloroform; used as water disinfectant.

half-cell A single electrode immersed in an electrolyte.

half-cell potential In electrochemical cells, the electrical potential developed by the overall cell reaction; can be considered, for calculation purposes, as the sum of the potential developed at the anode and the potential developed at the cathode, each being a half-cell.

half-life The time required for one-half of a given material to undergo chemical reactions.

halide A compound of the type MX, where X is fluorine, chlorine, iodine, bromine, or astatine, and M is another element or organic radical.

Hall cyclic thermal reforming   A gas-making process that uses component parts of carbureted–water gas apparatus to generate high-Btu gas from feedstocks ranging from naphtha to Bunker C.

halmyrolysis   Postdepositional chemical changes that occur while sediment is on the sea floor.

haloalkane   Halogenated aliphatic hydrocarbon.

halocarbon   A compound of carbon and a halogen, sometimes with hydrogen.

halocarbon plastic   Plastic made from halocarbon resins.

halocarbon resin   Resin produced by the polymerization of monomers made of halogenated hydrocarbons, such as tetrafluoroethylene, C_2F_4, and trifluorochloroethylene, C_2F_3Cl.

haloform CHX_3 A compound made by reaction of acetaldehyde or methyl ketones with NaOX, where X is a halogen; an example is iodoform, HCl_3, or bromoform, $HCBr_3$ or chloroform, $HCCl_3$.

haloform reaction   Halogenation of acetaldehyde or a methyl ketone in aqueous basic solution; the reaction is characteristic of compounds containing a CH_3CO group linked to a hydrogen or to another carbon.

halogen   Any of the elements of the halogen family, consisting of fluorine, chlorine, bromine, iodine, and astatine.

halogenated hydrocarbon   One of a group of halogen derivatives of organic hydrogen- and carbon-containing compounds; the group includes monohalogen compounds (alkyl or aryl halides) and polyhalogen compounds that contain the same or different halogen atoms.

halogenation   A chemical process or reaction in which a halogen element is introduced into a substance, generally by the use of the element itself.

halohydrin   A compound with the general formula X—R—OH where X is a halide such as Cl^-; an example is chlorohydrin.

halomorphic   Referring to an intrazonal soil whose features have been strongly affected by either neutral or alkali salts, or both.

Hammett acidity function   An expression for the acidity of a medium, defined as $h_0 = K_{BH^+}[BH^+]/[B]$, where K_{BH^+} is the dissociation constant of the acid form of the indicator, and $[BH^+]$ and $[B]$ are the concentrations of the protonated base and the unprotonated base respectively.

hand sugar refractometer   Portable device to read refractive indices of sugar solutions. Also known as proteinometer.

Hansa yellow   Group of organic azo pigments with strong tinting power, but poor opacity in paints; used where nontoxicity is important.

Hantzsch synthesis   The reaction whereby a pyrrole compound is formed when a β-ketoester, chloroacetone, and a primary amine condense.

Hanus solution   Iodine monobromide in glacial acetic acid; used to determine iodine values in oils containing unsaturated organic compounds.

haptoglobin   An alpha globulin that constitutes 1–2% of normal blood serum; contains about 5% carbohydrate.

hard acid   A Lewis acid of low polarizability, small size, and high positive oxidation state; it does not have easily excitable outer electrons; some examples are H^+, Li^+, and Al^+.

hard base   A Lewis base (electron donor) that has high polarizability and low electronegativity, is easily oxidized, or possesses lowlying empty orbitals; some examples are H_2O, HO^-, OCH_3^-, and F^-.

hard detergent   A nonbiodegradable detergent.

hardener   Compound reacted with a resin polymer to harden it, such as the amines or anhydrides that react with epoxides to cure or harden them into plastic materials. Also known as curing agent.

hardness   The amount of calcium carbonate dissolved in water, usually expressed as parts of calcium carbonate per million parts of water.

hardness test   Test to determine the calcium and magnesium content of water.

hard-sphere collision theory   A theory for calculating reaction rate constants for biomolecular gas-phase reactions in which the molecules are considered to be colliding, hard spheres.

hard water   Water that contains certain salts, such as those of calcium or magnesium, which form insoluble deposits in boilers and form precipitates with soap.

Hardy-Schulz rule   An increase in the charge of ions results in a large increase in their flocculating power.

Hargreaves process   A process for the manufacture of salt cake (sodium sulfate) by passing a mixture of sulfur dioxide and air through sodium chloride brine in a countercurrent manner.

Haring cell   An electrolytic cell with four electrodes used to measure electrolyte resistance and polarization of electrodes.

harman $C_{12}H_{10}N_2$ Crystals that melt at 237-238°C; inhibits growth of molds and certain bacteria. Also known as arabine; loturine; 2-methyl-β-carboline; 3-methyl-4-carboline; passiflorin.

harmonic vibration-rotation band   A vibration-rotation band of a molecule in which the harmonic oscillator approximation holds for the vibrational levels, so that the vibrational levels are equally spaced.

Hartmann diaphragm   Comparison device for positive-element-identification readings from emission spectra.

Hartmann test   A test for spectrometers in which light is passed through different parts of the entrance slit; any resulting changes of the spectrum indicate a fault in the instrument.

Hartman's solution   Solution of thymol, ethyl alcohol, and sulfuric ether; used for selective dentin analysis.

hartree   A unit of energy used in studies of atomic spectra and structure, equal (in centimeter-gram-second units) to $4\pi^2me^4/h^2$, where e and m are the charge and mass of the electron, and h is Planck's constant; equal to approximately 27.21 electronvolts or 4.360×10^{-18} joule.

Hartree units   A system of units in which the unit of angular momentum is Planck's constant divided by 2π, the unit of mass is the mass of the electron, and the unit of charge is the charge of the electron. Also known as atomic units.

Hasche process   A thermal reforming process for hydrocarbon fuels; it is a non-catalytic regenerative method in which a mixture of hydrocarbon gas or vapor and air is passed through a regenerative mass that is progressively hotter in the direction of the gas flow; partial combustion occurs, liberating heat to crack the remaining hydrocarbons in a combustion zone.

Hatch-Slack pathway   A metabolic cycle involved in the non-light-requiring phase of photosynthesis in certain plants having specific metabolic and anatomical modifications in their mesophyll and bundle sheath cells which facilitate the temporary fixation of carbon dioxide (CO_2) into four-carbon organic acid; these acids are next broken down to three-carbon organic acids plus CO_2 in bundle sheath cells, where this freed CO_2 is then fixed into carbohydrates in a normal Calvin cycle pathway.

Haüy law   The law that for a given crystal there is a set of ratios such that the ratios of the intercepts of any crystal plane on the crystal axes are rational fractions of these ratios.

HCB   *See* hexachlorobenzene.

HCG   *See* human chorionic gonadotropin.

hcp structure   *See* hexagonal close-packed structure.

He   *See* helium.

heat exchange   A unit operation based on heat transfer which functions in the heating and cooling of fluids with or without phase change.

heating value   *See* heat of combustion.

heat of activation   The increase in enthalpy when a substance is transformed from a less active to a more reactive form at constant pressure.

heat of association   Increase in enthalpy accompanying the formation of 1 mole of a coordination compound from its constituent molecules or other particles at constant pressure.

heat of combustion   The amount of heat released in the oxidation of 1 mole of a substance at constant pressure, or constant volume. Also known as heat value; heating value.

heat of decomposition   The change in enthalpy accompanying the decomposition of 1 mole of a compound into its elements at constant pressure.

heat of dilution   **1.** The increase in enthalpy accompanying the addition of a specified amount of solvent to a solution of constant pressure. Also known as integral heat of dilution; total heat of dilution. **2.** The increase in enthalpy when an infinitesimal amount of solvent is added to a solution at constant pressure. Also known as differential heat of dilution.

heat of dissociation   The increase in enthalpy at constant pressure, when molecules break apart or valence linkages rupture.

heat of formation   The increase in enthalpy resulting from the formation of 1 mole of a substance from its elements at constant pressure.

heat of hydration   The increase in enthalpy accompanying the formation of 1 mole of a hydrate from the anhydrous form of the compound and from water at constant pressure.

heat of ionization   The increase in enthalpy when 1 mole of a substance is completely ionized at constant pressure.

heat of linkage   The bond energy of a particular type of valence linkage between atoms in a molecule, as determined by the energy required to dissociate all bonds of the type in 1 mole of the compound divided by the number of such bonds in a compound.

heat of reaction   1. The negative of the change in enthalpy accompanying a chemical reaction at constant pressure. 2. The negative of the change in internal energy accompanying a chemical reaction at constant volume.

heat of solution   The enthalpy of a solution minus the sum of the enthalpies of its components. Also known as integral heat of solution; total heat of solution.

heat value   *See* heat of combustion.

heavy acid   *See* phosphotungstic acid.

heavy hydrogen   Hydrogen consisting of isotopes whose mass number is greater than one, namely deuterium or tritium.

heavy meromyosin   The larger of two fragments obtained from the muscle protein myosin following limited proteolysis by trypsin or chymotrypsin.

heavy oxygen   *See* oxygen-18.

heavy water   A compound of hydrogen and oxygen containing a higher proportion of the hydrogen isotope deuterium than does naturally occurring water. Also known as deuterium oxide.

Hefner lamp   A flame lamp that burns amyl acetate.

Hehner number   Weight percent of water-insoluble fatty acids in fats and oils.

height equivalent of theoretical plate   In a packed fractionating column, a height of packing that makes a separation equivalent to that of a theoretical plate; used in sorption and distillation calculations. Abbreviated HETP.

height of transfer unit   A dimensionless parameter used to calculate countercurrent sorption tower operations; it is proportional to the apparent resident time of the fluid. Abbreviated HTU.

Heisenberg force   A force between two nucleons derivable from a potential with an operator which exchanges both the positions and the spins of the particles.

Heitler-London covalence theory   A calculation of the binding energy and the distance between the atoms of a diatomic hydrogen molecule, which assumes that the two electrons are in atomic orbitals about each of the nuclei, and then combines these orbitals into a symmetric or antisymmetric function.

helcosol   *See* bismuth pyrogallate.

helical-fin section   Helical-shaped, extended-surface addition for the external surfaces of process-fluid tubes to increase heat-exchange efficiency; used for gas heating and cooling and in fuel oil residuum exchangers.

helicin   *See* salicylaldehyde.

heliotropin   *See* piperonal.

helium   A gaseous chemical element, symbol He, atomic number 2, and atomic weight 4.0026; one of the noble gases in group 0 of the periodic table.

helium-3   The isotope of helium with mass number 3, constituting approximately 1.3 parts per million of naturally occurring helium.

helium-4   The isotope of helium with mass number 4, constituting nearly all naturally occurring helium.

helium burning   The synthesis of elements in stars through the fusion of three alpha particles to form a carbon-12 nucleus, followed by further captures of alpha particles.

helium spectrometer   A small mass spectrometer used to detect the presence of helium in a vacuum system; for leak detection, a jet of helium is applied to suspected leaks in the outer surface of the system.

Hell-Volhard-Zelinsky reaction   Preparation of an ester or α-halo substituted acid (chloro or bromo) by reacting the halogen on the acid in the presence of phosphorus or phosphorus halide, and then followed by hydrolysis or alcoholysis of the haloacyl halide resulting.

Helmholtz equation   The relationship stating that the emf (electromotive force) of a reversible electrolytic cell equals the work equivalent of the chemical reaction when charge passes through the cell plus the product of the temperature and the derivative of the emf with respect to temperature.

helminthosporin $C_{15}H_{10}O_5$ A maroon, crystalline pigment formed by certain fungi growing on a sugar substrate.

hematein $C_{16}H_{12}O_6$ A brownish stain and chemical indicator obtained by oxidation of hematoxylin.

hematin $C_{34}H_{33}O_5N_4Fe$ The hydroxide of ferriheme derived from oxidized heme.

hematochrome   A red pigment occurring in green algae, especially when plants are exposed to intense light on subaerial habitats.

hematopoietin   A substance which is produced by the juxtaglomerular apparatus in the kidney and controls the rate of red cell production. Also known as hemopoietin.

hematoporphyrin $C_{34}H_{38}O_6N$ Iron-free heme, a porphyrin obtained by treating hemoglobin with sulfuric acid in vitro. Also known as hemoporphyrin.

hematoxylin $C_{16}H_{14}O_6$ A colorless, crystalline compound occurring in hematoxylon; upon oxidation, it is converted to hematein which forms deeply colored lakes with various metals; used as a stain in microscopy.

heme $C_{34}H_{32}O_4N_4Fe$ An iron-protoporphyrin complex associated with each polypeptide unit of hemoglobin.

hemerythrin   See hemoerythrin.

heme synthetase   An enzyme which combines protoporphyrin IX, ferrous iron, and globin to form the intact hemoglobin molecule.

hemiacetal   A class of compounds that have the grouping $>$C(OH)—(OR) and that result from the reaction of an aldehyde and alcohol.

hemicellulose $(C_6H_{10}O_5)_n$ A type of polysaccharide found in plant cell walls in association with cellulose and lignin; it is soluble in and extractable by dilute alkaline solutions. Also known as hexosan.

hemihedral symmetry   The possession by a crystal of only half of the elements of symmetry which are possible in the crystal system to which it belongs.

hemiholohedral   Of hemihedral form but with half the octants having the full number of planes.

hemiketal   A carbonyl compound that results from the addition of an alcohol to the carbonyl group of a ketone, with the general formula (R)(R')C(OH)(OR).

hemimellitic acid $C_6H_3(COOH)_3$ A compound crystallizing in colorless needles; melting point 196°C; slightly soluble in water. Also known as benzene-1,2,3-tricarboxylic acid.

hemimorphic crystal A crystal with no transverse plane of symmetry and no center of symmetry; composed of forms belonging to only one end of the axis of symmetry.

hemin $C_{34}H_{32}O_4N_4FeCl$ The crystalline salt of ferriheme, containing iron in the ferric state.

hemiprism A pinacoid that cuts two crystallographic axes.

hemitropic Pertaining to a twinned structure in which, if one part were rotated 180°, the two parts would be parallel.

hemobilirubin Bilirubin in normal blood serum before passage through the liver.

hemoconia Round or dumbbell-shaped, refractile, colorless particles found in blood plasma.

hemocyanin A blue respiratory pigment found only in mollusks and in arthropods other than insects.

hemoerythrin A red respiratory pigment found in a few annelid and sipunculid worms and in the brachiopod *Lingula*. Also known as hemerythrin.

hemoglobin The iron-containing, oxygen-carrying molecule of the red blood cells of vertebrates comprising four polypeptide subunits in a heme group.

hemoglobin A The type of hemoglobin found in normal adults, which moves as a single component in an electrophoretic field, is rapidly denatured by highly alkaline solutions, and contains two titratable sulfhydryl groups per molecule.

hemopexin A heme-binding protein in human plasma that may be a regulator of heme and drug metabolism, and a distributor of heme.

hemopoietin *See* hematopoietin.

hemoporphyrin *See* hematoporphyrin.

hemosiderin An iron-containing glycoprotein found in most tissues and especially in liver.

hemotrophe The nutritive substance supplied via the placenta to embryos of viviparous animals.

hendecanal *See* undecanal.

hendecane *See* undecane.

hendecanoic acid *See* undecanoic acid.

10-hendecenyl acetate *See* undecylenyl acetate.

hendecyl *See* undecyl.

Henderson equation for pH An equation for the pH of an acid during its neutralization: $pH = pK_a + \log [salt]/[acid]$ where pK_a is the logarithm to base 10 of the reciprocal of the dissociation constant of the acid; the equation is found to be useful for the pH range 4–10, providing the solutions are not too dilute.

heneicosane $C_{21}H_{44}$ Saturated hydrocarbon of the methane series; the crystals melt at 40°C and boil at 215°C (at 15 mm Hg).

Hengstebeck approximation A method of calculation to estimate the distribution of non-key components in distillation column products.

Henry's law The law that at sufficiently high dilution in a liquid solution, the fugacity of a nondissociating solute becomes proportional to its concentration.

hentriacontane $C_{31}H_{64}$ A hydrocarbon; a crystalline material melting at 68°C and boiling at 302°C (at 15 mm Hg); derived from roots of *Oenanthe crocata* and found in beeswax.

hepar calcies *See* calcium sulfide.

heparin An acid mucopolysaccharide acting as an antithrombin, antithromboplastin, and antiplatelet factor to prolong the clotting time of whole blood; occurs in a variety of tissues, most abundantly in liver.

heptachlor $C_{10}H_7Cl_7$ An insecticide; a white to tan, waxy solid; insoluble in water, soluble in alcohol and xylene; melts at 95–96°C.

heptacosane $C_{27}H_{56}$ A hydrocarbon; water-insoluble crystals melting at 60°C and boiling at 270°C (at 15 mm Hg); soluble in alcohol; found in beeswax.

heptadecane $C_{17}H_{36}$ A hydrocarbon; water-insoluble, alcohol-soluble solid melting at 23°C and boiling at 303°C; used as a chemical intermediate. Also known as dioctylmethane.

heptadecanol $C_{17}H_{35}OH$ An alcohol; colorless liquid boiling at 309°C; slightly soluble in water; used as a chemical intermediate, as a perfume fixative, in cosmetics and soaps, and to manufacture surfactants.

heptadione-2,3 *See* acetyl valeryl.

heptaldehyde $C_6H_{13}CHO$ An aldehyde; ether-soluble, colorless oil with fruity aroma; slightly soluble in water; boils at 153°C; used as a chemical intermediate and for perfumes and pharmaceuticals. Also known as heptanal.

heptanal *See* heptaldehyde.

heptane $CH_3(CH_2)_5CH_3$ A hydrocarbon; water-insoluble, flammable, colorless liquid boiling at 98°C; soluble in alcohol, chloroform, and ether; used as an anesthetic, solvent, and chemical intermediate, and in standard octane-rating tests.

1,7-heptanedicarboxylic acid *See* azelaic acid.

heptanedioic acid *See* pimelic acid.

heptanoic acid $CH_3(CH_2)_5COOH$ Clear oil boiling at 223°C; soluble in alcohol and ether, insoluble in water; used as a chemical intermediate. Also known as enanthic acid; oenanthic acid.

1-heptanol $C_7H_{15}OH$ An alcohol; a fragrant, colorless liquid boiling at 174°C; soluble in water, ether, or alcohol; used as a chemical intermediate, as a solvent, and in cosmetics. Also known as heptyl alcohol.

2-heptanol *See* methyl amyl carbinol.

3-heptanol $CH_3CH_2CH(OH)C_4H_9$ An alcohol; a liquid boiling at 156°C; used as a coating, solvent, and diluent, as a chemical intermediate, and as a flotation frother.

2-heptanone *See* methyl *n*-amyl ketone.

3-heptanone *See* ethyl butyl ketone.

4-heptanone $(CH_3CH_2CH_2)_2CO$ A colorless liquid that is stable and has a pleasant odor; boils at approximately 98°C; used to put nitrocellulose and raw and blown oils into solution, and used in lacquers and as a flavoring in foods. Also known as butyrone; dipropyl ketone.

heptene $C_{17}H_{14}$ A liquid that is a mixture of isomers; boils at 189.5°C; used as an additive in lubricants, as a catalyst, and as a surface active agent. Also known as heptylene.

heptose Any member of the group of monosaccharides containing seven carbon atoms.

heptoxide An oxide whose molecule contains seven atoms of oxygen.

heptulose The generic term for a ketose formed from a seven-carbon monosaccharide.

heptyl $CH_3(CH_2)_6-$ The radical from heptane, $CH_3(CH_2)_5CH_3$.

n-heptylacetylene *See* nonyne.

heptyl alcohol *See* 1-heptanol.

heptylene *See* heptene.

Hercules trap Water-measuring liquid trap used in aquametry when the material collected is heavier than water.

Hermann-Mauguin symbols Symbols representing the 32 symmetry classes, consisting of series of numbers giving the multiplicity of symmetry axes in descending order, with other symbols indicating inversion axes and mirror planes.

Hess's law The law that the evolved or absorbed heat in a chemical reaction is the same whether the reaction takes one step or several steps. Also known as the law of constant heat summation.

hetero- Chemical prefix meaning different; for example, a heterocyclic compound is one in which the ring is made of more than one kind of atom.

heteroauxin $C_{10}H_9O_2N$ A plant growth hormone with an indole skeleton.

heteroazeotrope Liquid mixture that is not completely miscible in all proportions in the liquid phase, yet does not form an azeotrope. Also known as heterogeneous zeotrope.

heterocyclic compound Compound in which the ring structure is a combination of more than one kind of atom; for example, pyridine, C_5H_5N.

heterodesmic Pertaining to those atoms bonded in more than one way in crystals.

heterogeneous Pertaining to a mixture of phases such as liquid-vapor, or liquid-vapor-solid.

heterogeneous catalysis Catalysis occurring at a phase boundary, usually a solid-fluid interface.

heterogeneous chemical reaction Chemical reaction system in which the reactants are of different phases; for example, gas with liquid, liquid with solid, or a solid catalyst with liquid or gaseous reactants.

heterogeneous zeotrope   *See* heteroazeotrope.

heterolytic cleavage   Breaking of a covalent bond involving carbon to produce two oppositely charged fragments.

heteropoly acid   Complex acids of metals, whose specific gravity is greater than 4, with phosphoric acid; an example is phosphomolybdic acid.

heteropoly compound   Polymeric compounds of molybdates with anhydrides of other elements such as phosphorus; the yellow precipitate $(NH_4)_3[P(Mo_3O_{10})_4]$ is such a compound.

heteropolysaccharide   A polysaccharide which is a polymer consisting of two or more different monosaccharides.

heterotopic faces   On molecules, faces of double bonds where addition gives rise to isomeric structures.

heterotopic ligands   Constitutionally identical ligands whose separate replacement by a different ligand gives rise to isomeric structures.

heterotrophic effect   The interaction between different ligands, such as the effect of an inhibitor or activator on the binding of a substrate by an enzyme.

heterotropic enzyme   A type of allosteric enzyme in which a small molecule other than the substrate serves as the allosteric reflector.

hexachlorobenzene C_6Cl_6 Colorless, needlelike crystals with a melting point of 231°C; used in organic synthesis and as a fungicide. Abbreviated HCB. Also known as perchlorobenzene.

hexachlorobutadiene $Cl_2C:CClCCl:CCl_2$ A colorless liquid with mild aroma, boiling at 210–220°C; soluble in alcohol and ether, insoluble in water; used as solvent, heat-transfer liquid, and hydraulic fluid.

1,2,3,4,5,6-hexachlorocyclohexane $C_6H_6Cl_6$ A white or yellow powder or flakes with a musty odor; a systemic insecticide toxic to flies, cockroaches, aphids, and boll weevils. Also known as benzene hexachloride (BHC); TBH.

hexachloroethane Cl_3CCCl_3 Colorless crystals with a camphorlike odor, melting point 185°C, toxic; used in organic synthesis, as a retarding agent in fermentation, and as a rubber accelerator. Also known as carbon trichloride; perchloroethane.

hexachlorophene $(C_6HCl_3OH)_2CH_2$ A white powder melting at 161°C; soluble in alcohol, ether, acetone, and chloroform, insoluble in water; bacteriostat used in antiseptic soaps, cosmetics, and dermatologicals.

hexachloropropylene $CCl_3CCl:CCl_2$ Water-white liquid boiling at 210°C, soluble in alcohol, ether, and chlorinated solvents, insoluble in water; used as a solvent, plasticizer, and hydraulic fluid.

hexacontane $C_{60}H_{122}$ Solid, saturated hydrocarbon of the methane series; melts at 101°C.

hexacosane $C_{26}H_{54}$ Saturated hydrocarbon of the methane series; colorless crystals melting at 57°C.

1-hexadecene $CH_3(CH_2)_{13}CH:CH_2$ A colorless liquid made by treating cetyl alcohol with phosphorous pentoxide; boils at 274°C; soluble in organic solvents such as alcohol, ether, and petroleum; used as an intermediate in organic synthesis. Also known as cetene; cetylene; α-hexadecylene.

α-hexadecylene *See* 1-hexadecene.

hexadentate ligand A chelating agent having six groups capable of attachment to a metal ion. Also known as sexadentate ligand.

1,6-hexadiamine *See* hexamethylene diamine.

hexadiene C_6H_{10} A group of unsaturated hydrocarbons with two double bonds; some members of the group are 1,4-hexadiene, 1,5-hexadiene, and 2,4-hexadiene.

2,4-hexadienoic acid *See* sorbic acid.

hexaflurate $K(ASF_6)$ A white, crystalline compound with a melting point of 440°C; used as an herbicide for rangeland. Also known as potassium hexafluoroarsenate.

hexagonal close-packed structure Close-packed crystal structure characterized by the regular alternation of two layers; the atoms in each layer lie at the vertices of a series of equilateral triangles, and the atoms in one layer lie directly above the centers of the triangles in neighboring layers. Abbreviated hcp structure.

hexagonal lattice A Bravais lattice whose unit cells are right prisms with hexagonal bases and whose lattice points are located at the vertices of the unit cell and at the centers of the bases.

hexagonal system A crystal system that has three equal axes intersecting at 120° and lying in one plane; a fourth, unequal axis is perpendicular to the other three.

hexahydric alcohol A member of the mannitol-sorbitol-dulcitol sugar group; isomer of $C_6H_8(OH)_6$.

hexahydrocresol *See* methyl cyclohexanol.

hexahydrocymene *See* menthane.

hexahydromethyl phenol *See* methyl cyclohexanol.

hexahydrothymol *See* menthol.

hexahydrotoluene *See* methyl cyclohexane.

hexametapol $C_6H_{18}N_3OP$ A liquid used as a solvent in organic synthesis, as a deicing additive for jet engine fuel, and as an insect pest chemosterilant and chemical mutagen. Also known as hexamethylphosphoramide (HMPA).

hexamethyldiaminoisopropanol diiodide *See* propiodal.

hexamethylenediamine $H_2N(CH_2)_6NH_2$ Colorless solid boiling at 205°C; slightly soluble in water, alcohol, and ether; used to make nylon and other high polymers. Also known as 1,6-diamino-hexane; 1,6-hexanediamine.

hexamethylene glycol *See* 1,6-hexanediol.

hexamethylphosphoramide *See* hexametapol.

hexamethylphosphorictriamide *See* bempa.

hexane C_6H_{14} Water-insoluble, toxic, flammable, colorless liquid with faint aroma; forms include: *n*-hexane, a straight-chain compound boiling at 68.7°C and used as a solvent, paint diluent, alcohol denaturant, and polymerization-reaction medium; isohexane, a mixture of hexane isomers boiling at 54–61°C and used as a solvent and freezing-point depressant; and neohexane.

hexanedinitrile *See* adiponitrile.

hexanedioic acid *See* adipic acid.

1,6-hexanediol $HO(CH_2)_6OH$ A crystalline substance, soluble in water and alcohol; used in gasoline refining, as an intermediate in nylon manufacturing, and in making polyesters and polyurethanes. Also known as 1,6-dihydroxy-hexane; hexamethylene glycol.

hexanitrodiphenyl amine $(NO_2)_3C_6H_2NHC_6H_2(NO_2)_3$ Explosive, yellow solid melting at 238–244°C; insoluble in water, ether, alcohol, or benzene; soluble in alkalies and acetic and nitric acids; used as an explosive and in potassium analysis.

hexanitromannite *See* mannitol hexanitrate.

1-hexanol *See* hexyl alcohol.

hexaphenylethane $(C_6H_5)_3CC(C_6H_5)_3$ The dimer of triphenylmethyl radical.

1-hexene $CH_3(CH_2)_3HC{:}CH_2$ Colorless, olefinic hydrocarbon boiling at 64°C; soluble in alcohol, acetone, ether, and hydrocarbons, insoluble in water; used as a chemical intermediate and for resins, drugs, and insecticides. Also known as hexylene.

1,6-hexenediamine *See* hexamethylenediamine.

hexoctahedron A cubic crystal form that has 48 equal triangular faces, each of which cuts the three crystallographic axes at different distances.

hexokinase Any enzyme that catalyzes the phosphorylation of hexoses.

hexone *See* methyl isobutyl ketone.

hexosamine A primary amine derived from a hexose by replacing the hydroxyl with an amine group.

hexosaminidase A An enzyme which catalyzes the hydrolysis of the N-acetyl-galactosamine residue from certain gangliosides.

hexosan *See* hemicellulose.

hexose Any monosaccharide that contains six carbon atoms in the molecule.

hexose diphosphate pathway *See* glycolytic pathway.

hexose monophosphate cycle A pathway for carbohydrate metabolism in which one molecule of hexose monophosphate is completely oxidized.

hexose phosphate Any one of the phosphoric acid esters of a hexose, notably glucose, formed during the metabolism of carbohydrates by living organisms.

hextetrahedron A 24-faced form of crystal in the tetrahedral group of the isometric system.

hexulose A ketose made from a six-carbon-chain monosaccharide.

hexylacetylene *See* octyne.

hexyl alcohol $CH_3(CH_2)_4CH_2OH$ Colorless liquid boiling at 156°C; soluble in alcohol and ether, slightly soluble in water; used as a chemical intermediate for pharmaceuticals, perfume esters, and antiseptics. Also known as amyl carbinol; 1-hexanol.

hexylamine $CH_3(CH_2)_5NH_2$ Poisonous, water-white liquid with amine aroma; boils at 129°C; a ptomaine base from the autolysis of protoplasm. Also known as aminohexane.

hexylene   *See* 1-hexene.

hexylene glycol $C_6H_{14}O_2$ Water-miscible, colorless liquid boiling at 198°C; used in hydraulic brake fluids, in printing inks, and in textile processing.

hexylresorcinol $C_6H_{13}C_6H_3(OH)_2$ Sharp-tasting, white to yellowish crystals melting at 64°C; slightly soluble in water, soluble in glycerin, vegetable oils, and organic solvents; used in medicine.

1-hexyne C_4H_9CCH A colorless, water-white liquid, either n-butylacetylene, boiling at 71.5°C, or methylpropylacetylene, boiling at 84°C.

Hf   *See* hafnium.

HF akylation   Petroleum refinery alkylation process in which olefins (C_3, C_4, C_5) are reacted with isobutane in the presence of hydrofluoric acid catalyst.

hfs   *See* hyperfine structure.

Hg   *See* mercury.

Higbie model   Mass-transfer theory for packed absorption towers, stating that liquid flows across each packing piece in laminar flow and is mixed with other liquids meeting it at the points of discontinuity between packing elements.

high-energy bond   Any chemical bond yielding a decrease in free energy of at least 5 kilocalories per mole.

high-frequency titration   A conductimetric titration in which two electrodes are mounted on the outside of the beaker or vessel containing the solution to be analyzed and an alternating current source in the megahertz range is used to measure the course of a titration.

high polymer   A large molecule (of molecular weight greater than 10,000) usually composed of repeat units of low-molecular-weight species; for example, ethylene or propylene.

high-pressure chemistry   The study of chemical reactions and phenomena that occur at pressures exceeding 10,000 bars (a bar is nearly equivalent to a kilogram per square centimeter), mainly concerned with the properties of the solid state.

high-pressure process   A chemical process operating at elevated pressure; for example, phenol manufacture at 330 atmospheres (1 atmosphere = 101,325 newtons per square meter), ethylene polymerization at 2000 atm, ammonia synthesis at 100–1000 atm, and synthetic-diamond manufacture up to 100,000 atm.

high-temperature chemistry   The study of chemical phenomena occurring above about 500 K.

high-vacuum insulation   High vacuum between the walls of double-wall vessels to serve as thermal insulation at ultralow (cryogenic) temperatures, such as in Dewar vessels.

Hill plot   A graphic representation of the Hill reaction.

Hill reaction   1. The release of molecular oxygen by isolated chloroplasts in the presence of a suitable electron receptor, such as ferricyanide. 2. Production of substituted phenylacetic acids by the oxidation of the corresponding alkylbenzene by potassium permanganate in the presence of acetic acid.

hindrance factor   See drag factor.

Hinsberg test   A test to distinguish between primary and secondary amines; it involves reaction of an amine with benzene disulforyl chloride in alkaline solution; primary amines give sulfonamides that are soluble in basic solution; secondary amines give insoluble derivatives; tertiary amines do not react with the reagent.

hippuric acid $C_6H_5CONHCH_2 \cdot COOH$ Colorless crystals melting at 188°C; soluble in hot water, alcohol, and ether; used in medicine and as a chemical intermediate.

histamine $C_5H_9N_3$ An amine derivative of histadine which is widely distributed in human tissues.

histidase   An enzyme found in the liver of higher vertebrates that catalyzes the deamination of histidine to urocanic acid.

histidine $C_6H_9O_2N_3$ A crystalline basic amino acid present in large amounts in hemoglobin and resulting from the hydrolysis of most proteins.

histochemistry   A science that deals with the distribution and activities of chemical components in tissues.

histone   Any of the strong, soluble basic proteins of cell nuclei that are precipitated by ammonium hydroxide.

Hittorf method   A procedure for determining transference numbers in which one measures changes in the composition of the solution near the cathode and near the anode of an electrolytic cell, due to passage of a known amount of electricity.

HMF   See 5-(hydroxymethyl)-2-furaldehyde.

HMPA   See hexametapol.

Ho   See holmium.

Hodgson number   Method of predicting the metering error during pulsating gas flow when a surge tank is located between the pulsation source (pump or compressor) and the meter (orifice, nozzle, or venturi).

Hofmann amine separation   A technique to separate a mixture of primary, secondary, and tertiary amines; they are heated with ethyl oxalate; there is no reaction with tertiary amines, primary amines form a diamide, and the secondary amines form a monoamide; when the reaction mixture is distilled, the mixture is separated into components.

Hofmann degradation   The action of bromine and an alkali on an amide so that it is converted into a primary amine with one less carbon atom.

Hofmann exhaustive methylation reaction   The thermal decomposition of quaternary ammonium hydroxide compounds to yield an olefin and water; an exception is tetramethylammonium hydroxide, which decomposes to give an alcohol.

Hofmann mustard-oil reaction Preparation of alkylisothiocyanates by heating together a primary amine, mercuric chloride, and carbon disulfide.

Hofmann reaction A reaction in which amides are degraded by treatment with bromine and alkali (caustic soda) to amines containing one less carbon; used commercially in the production of nylon.

Hofmann rearrangement A chemical rearrangement of the hydrohalides of N-alkylanilines upon heating to give aminoalkyl benzenes.

Hofmeister series An arrangement of anions or cations in order of decreasing ability to produce coagulation when their salts are added to lyophilic sols. Also known as lyotopic series.

holdup 1. Volume of material held or contained in a process vessel or line. 2. Liquid held up (suspended) in a vertical process vessel or line by rising gas or vapor streams.

hole-burning spectroscopy A method of observing extremely narrow line widths in certain ions and molecules embedded in crystalline solids, in which broadening produced by crystal-site-dependent statistical field variations is overcome by having a monochromatic laser temporarily remove ions or molecules at selected crystal sites from their absorption levels, and observing the resulting dip in the absorption profile with a second laser beam.

holmium A rare-earth element belonging to the yttrium subgroup, symbol Ho, atomic number 67, atomic weight 164.93, melting point 1400–1525°C.

holoaxial Having all possible axes of symmetry.

holocellulose The total polysaccharide fraction of wood, straw, and so on, that is composed of cellulose and all of the hemicelluloses and that is obtained when the extractives and the lignin are removed from the natural material.

holoenzyme A complex, fully active enzyme, containing an apoenzyme and a coenzyme.

holohedral Pertaining to a crystal structure having the highest symmetry in each crystal class. Also known as holosymmetric; holosystemic.

holohedron A crystal form of the holohedral class, having all the faces needed for complete symmetry.

holopulping process A process for making paper pulp by alkaline oxidation of extremely thin wood chips at low temperature and pressure and then solubilization of the lignin fraction.

holosymmetric *See* holohedral.

holosystemic *See* holohedral.

homatropine $C_{16}H_{21}O_3N$ An alkaloid that causes pupil dilation and paralysis of accommodation.

homeomorph A crystal that displays a form similar to that of a crystal with a different chemical composition.

homidium bromide *See* ethidium bromide.

homo- 1. Indicating the homolog of a compound differing in formula from the latter by an increase of one CH_2 group. 2. Indicating a homopolymer made up of a single type of monomer, such as polyethylene from ethylene. 3. Indicating that a skeletal atom has been added to a well-known structure.

homocyclic compound A ring compound that has one type of atom in its structure; an example is benzene.

homocysteine $C_4H_9O_2NS$ An amino acid formed in animals by demethylation of methionine.

homodesmic Of a crystal, having atoms bonded in a single way.

homogeneous Pertaining to a substance having uniform composition or structure.

homogeneous catalysis Catalysis occurring within a single phase, usually a gas or liquid.

homogeneous chemical reaction Chemical reaction system in which all constituents (reactants and catalyst) are of the same phase.

homogentisase The enzyme that catalyzes the conversion of homogentisic acid to fumaryl acetoacetic acid.

homogentisic acid $C_8H_8O_4$ An intermediate product in the metabolism of phenylalanine and tyrosine; found in excess in persons with phenylketonuria and alkaptonuria.

homologation A type of hydroformylation in which carbon monoxide reacts with certain saturated alcohols to yield either aldehydes or alcohols (or a mixture of both) containing one more carbon atom than the parent.

homology 1. The relation among elements of the same group, or family, in the periodic table. 2. That state, in a series of organic compounds that differ from each other by a CH_2 such as the methane series C_nH_{2n+2}, in which there is a similarity between the compounds in the series and a graded change of their properties.

homolysis Symmetrical breaking of a covalent electron bond; for example, $A:B = A· + B·$.

homometric pair A pair of crystal structures whose x-ray diffraction patterns are identical.

homomorphs Chemical molecules that are similar in size and shape, but not necessarily having any other characteristics in common.

homopolar bond A covalent bond whose total dipole moment is zero.

homopolymer A polymer formed from a single monomer; an example is polyethylene, formed by polymerization of ethylene.

homopolysaccharide A polysaccharide which is a polymer of one kind of monosaccharide.

ortho-**homosalicylic acid** See 3-methylsalicylic acid.

homoserine $C_4H_9O_3N$ An amino acid formed as an intermediate product in animals in the metabolic breakdown of cystathionine to cysteine.

4-homosulfanilamide See *para*-(aminomethyl)benzenesulfonamide.

homotropic enzyme A type of allosteric enzyme in which the substrate serves as the allosteric reflector.

homovanillic acid $HOC_6H_3(OCH_3)CH_2CO_2H$ A major metabolite of 3-*O*-methyl dopa; used in enzyme determination. Abbreviated HVA. Also known as 4-hydroxy-3-methoxyphenylacetic acid.

homozeotrope Mixture in which the liquid components are miscible in all proportions in the liquid phase, and may be separated by ordinary distillation.

Hooker diaphragm cell A device used in industry for the electrolysis of brine (sodium chloride) to make chlorine and caustic soda (sodium hydroxide) or caustic potash (potassium hydroxide); saturated brine fed around the anode passes through the diaphragm to the cathode; chlorine is formed at the anode and hydrogen released at the cathode, leaving sodium hydroxide and residual sodium chloride in the cell liquor; the diaphragm prevents the products from mixing.

Hopkins-Cole reaction The appearance of a violet ring when concentrated sulfuric acid is added to a mixture that includes a protein and glyoxylic acid; however, gelatin and zein do not show the reaction.

horizontal chromatography Paper chromatography in which the chromatogram is horizontal instead of vertical.

hormone 1. A chemical messenger produced by endocrine glands and secreted directly into the bloodstream to exert a specific effect on a distant part of the body. 2. An organic compound that is synthesized in minute quantities in one part of a plant and translocated to another part, where it influences a specific physiological process.

Hortvet sublimator Device for the determination of the condensation temperature (sublimation point) of sublimed solids.

hot spot An area or point within a reaction system at which the temperature is appreciably higher than in the bulk of the reactor; usually locates the reaction front.

Houben-Hoesch synthesis Condensation of cyanides with polyhydric phenols in the presence of hydrogen chloride and zinc chloride to yield phenolic ketones.

Houdry butane dehydrogenation A catalytic process for dehydrogenating light hydrocarbons from crude oil to their corresponding mono- or diolefins; chromia-alumina catalysts with inert material are used in pellet form.

Houdry fixed-bed catalytic cracking A cyclic, regenerable process for cracking of petroleum distillates to produce high-octane gasoline from higher-boiling petroleum fractions; synthetic or natural bead catalysts of activated hydrosilicate of alumina may be used. Also known as Houdry process.

Houdry hydrocracking A catalytic process combining cracking and desulfurization of crude petroleum oil in the presence of hydrogen; catalysts may be nickel oxide or nickel sulfide on silica alumina, and cobalt molybdate on alumina.

Houdry process *See* Houdry fixed-bed catalytic cracking.

H substance An agent similar to histamine and believed to play a role in local blood vessel response in tissue damage.

HTU *See* height of transfer unit.

Huber's reagent Aqueous solution of ammonium molybdate and potassium ferrocyanide used as a reagent to detect free mineral acid.

Hubl's reagent Solution of iodine and mercuric chloride in alcohol used to determine the iodine content of oils and fats.

Hull cell An electrodeposition cell that operates within a simultaneous range of known current densities.

human chorionic gonadotropin   A gonadotropic and luteotropic hormone secreted by the chorionic vesicle. Abbreviated HCG. Also known as chorionic gonadotropin.

humectant   A substance which absorbs or retains moisture; examples are glycerol, propylene glycol, and sorbitol; used in preparing confectioneries and dried fruit.

humic acid   Any of various complex organic acids obtained from humus; insoluble in acids and organic solvents.

humidity indicator   Cobalt salt (for example, cobaltous chloride) that changes color as the surrounding humidity changes; changes from pink when hydrated, to greenish-blue when anhydrous.

humin   An insoluble pigment formed in the acid hydrolysis of a protein that contains tryptophan.

Humphreys series   A series of lines in the infrared spectrum of atomic hydrogen whose wave numbers are given by $R_H[(1/36) - (1/n^2)]$, where R_H is the Rydberg constant for hydrogen, and n is any number greater than 6.

Hund rules   Two rules giving the order in energy of atomic states formed by equivalent electrons: of the terms given by equivalent electrons, the ones with greatest multiplicity have the least energy, and of these the one with greatest orbital angular momentum is lowest; the state of a multiplet with lowest energy is that in which the total angular momentum is the least possible, if the shell is less than half-filled, and the greatest possible, if more than half filled.

Hundsdieke reaction   Production of an alkyl halide by boiling a silver carboxylate with an equivalent weight of bromine in carbon tetrachloride.

HVA   *See* homovanillic acid.

hyaline   A clear, homogeneous, structureless material found in the matrix of cartilage, vitreous body, mucin, and glycogen.

hyaluronate   A salt or ester of hyaluronic acid.

hyaluronate lyase   *See* hyaluronidase.

hyaluronic acid   A polysaccharide found as an integral part of the gellike substance of animal connective tissue.

hyaluronidase   Any one of a family of enzymes which catalyze the breakdown of hyaluronic acid. Also known as hyaluronate lyase; spreading factor.

hybrid enzyme   A form of polymeric enzyme occurring in heterozygous individuals that shows a hybrid molecular form made up of subunits differing in one or more amino acids.

hybridized orbital   A molecular orbital which is a linear combination of two or more orbitals of comparable energy (such as $2s$ and $2p$ orbitals), is concentrated along a certain direction in space, and participates in formation of a directed valence bond.

hybrid molecule   A single molecule, usually protein, peculiar to heterozygotes and containing two structurally different polypeptide chains determined by two different alleles.

hydantoin $C_3N_2O_2H$ A white, crystalline compound, melting point 220°C; used as an intermediate in certain pharmaceutical manufacturing and as a textile softener and lubricant. Also known as glycolyurea.

hydnocarpic acid $C_{16}H_{28}O_2$ A nonedible fat and oil isolated from chaulmoogra oil, forming white crystals that melt at 60°C; used to treat Hansen's disease. Also known as cyclopentenylundecylic acid.

hydracrylic acid $CH_2OH \cdot CH_2COOH$ An oily liquid that is an isomer of lactic acid and that breaks down on heating to acrylic acid. Also known as 3-hydroxypropanoic acid; β-hydroxypropionic acid.

hydrase An enzyme that catalyzes removal or addition of water to a substrate without hydrolyzing it.

hydrastine $C_{21}H_{21}NO_6$ An alkaloid isolated from species of the family Ranunculaceae and from *Hydrastis canadensis;* orthorhombic prisms crystallize from alcohol solution, melting point 132°C; highly soluble in acetone and benzene, soluble in chloroform, less soluble in ether and alcohol.

hydrastinine $C_{11}H_{13}O_3N$ A compound formed by the decomposition of hydrastine; crystallizes as needles from petroleum-ether solution, soluble in organic solvents such as alcohol, chloroform, and ether; used in medicine as a stimulant in coronary disease and as a hemostatic in uterine hemorrhage.

hydrate 1. A form of a solid compound which has water in the form of H_2O molecules associated with it; for example, anhydrous copper sulfate is a white solid with the formula $CuSO_4$, but when crystallized from water a blue crystalline solid with formula $CuSO_4 \cdot 5H_2O$ results, and the water molecules are an integral part of the crystal. 2. A crystalline compound resulting from the combination of water and a gas; frequently a constituent of natural gas that is under pressure.

hydrated alumina *See* alumina trihydrate.

hydrated aluminum oxide *See* alumina trihydrate.

hydrated chloral *See* chloral hydrate.

hydrated electron An electron released during ionization of a water molecule by water and surrounded by water molecules oriented so that the electron cannot escape. Also known as aqueous electron.

hydrated halloysite *See* endellite.

hydrated lime *See* calcium hydroxide.

hydrated manganic hydroxide *See* manganic hydroxide.

hydrated mercurous nitrate $Hg_2(NO_3)_2 \cdot 2H_2O$ Poisonous, light-sensitive crystals, soluble in warm water, decomposes at 70°C; used as an analytical reagent and in cosmetics and medicine.

hydrated silica *See* silicic acid.

hydrate inhibitor A material (such as alcohol or glycol) added to a gas stream to prevent the formation and freezing of gas hydrates in low-temperature systems.

hydration The incorporation of molecular water into a complex molecule with the molecules or units of another species; the complex may be held together by relatively weak forces or may exist as a definite compound.

hydrazide An acyl hydrazine; a compound of the formula

$$R—C—NH—NH_2$$

where R may be an alkyl group.

hydrazine H_2NNH_2 A colorless, hygroscopic liquid, boiling point 114°C, with an ammonialike odor; it is reducing, decomposable, basic, and bifunctional; used as a rocket fuel, in corrosion inhibition in boilers, and in the synthesis of biologically active materials, explosives, antioxidants, and photographic chemicals.

hydrazine hydrate $H_2NNH_2OH_2O$ A colorless, fuming liquid that boils at 119.4°C; used as a component in jet fuels and as an intermediate in organic synthesis.

hydrazinobenzene *See* phenylhydrazine.

2-hydrazinoethanol *See* 2-hydroxyethylhydrazine.

hydrazobenzene $C_{12}H_{12}N_2$ A colorless, crystalline compound, melts at 132°C, slightly soluble in water, soluble in alcohol; used as an intermediate in the synthesis of benzidine. Also known as N,N'-diphenylhydrazine.

hydrazoic acid NHN:N Explosive liquid, a strong protoplasmic poison boiling at 37°C. Also known as azoimide; diazoimide; hydronitric acid.

hydrazone A compound containing the grouping —NH·N:C—, and obtained from a condensation reaction involving hydrazines with aldehydes or ketones; has been used as an exotic fuel.

hydride A compound containing hydrogen and another element; examples are H_2S, which is a hydride although it may be properly called hydrogen sulfide, and lithium hydride, LiH.

hydrindantin $C_{18}H_{10}O_6$ A compound used as a reagent for the photometric determination of amino acids.

hydriodic acid A yellow liquid that is a water solution of the gas hydrogen iodide; a solution of 59% hydrogen iodide produces a liquid that is constant-boiling; it is a strong acid used in organic synthesis and as a reagent in analytical chemistry.

hydriodic acid gas *See* hydrogen iodide.

hydrobenzoin $C_{14}H_{14}O_2$ A colorless, crystalline compound formed by action of sodium amalgam on benzaldehyde, melts at 136°C, and is slightly soluble in water. Also known as benzyleneglycol; diphenyldihydroxyethane.

hydroboration The process of producing organoboranes by the addition of a compound with a B-H bond to an unsaturated hydrocarbon; for example, the reaction of diborane ion with a carbonyl compound.

hydrobromic acid HBr A solution of hydrogen bromide in water, usually 40%; a clear, colorless liquid; used in medicine, analytical chemistry, and synthesis of organic compounds.

hydrocarbon One of a very large group of chemical compounds composed only of carbon and hydrogen; the largest source of hydrocarbons is from petroleum crude oil.

hydrocarbon resins Brittle or gummy materials prepared by the polymerization of several unsaturated constituents of coal tar, rosin, or petroleum; they are inexpensive and find uses in rubber and asphalt formulations and in coating and caulking compositions.

hydrochinone *See* hydroquinone.

hydrochloric acid HCl A solution of hydrogen chloride gas in water; a poisonous, pungent liquid forming a constant-boiling mixture at 20% concentration in water; widely used as a reagent, in organic synthesis, in acidizing oil wells, ore reduction, food processing, and metal cleaning and pickling. Also known as muriatic acid.

hydrochloric ether *See* ethyl chloride.

hydrocinnamic acid $C_6H_5CH_2CH_2COOH$ A compound whose crystals have a floral odor (hyacinth-rose) and melt at 46°C; used in perfumes and flavoring. Also known as 3-phenylpropionic acid.

hydrocinnamic alcohol *See* phenylpropyl alcohol.

hydrocinnamic aldehyde *See* phenylpropyl aldehyde.

hydroclone A device for separating a solid-liquid mixture during an industrial process by using a conical vortex and centrifugal force.

hydrocortisone $C_{21}H_{30}O_5$ The generic name for 17-hydroxycorticosterone; an adrenocortical steroid occurring naturally and prepared synthetically; its effects are similar to cortisone, but it is more active. Also known as cortisol.

hydrocrackate The product of a hydrocracker.

hydrocracker A high-pressure processing unit that cracks long hydrocarbon molecules under a high-hydrogen-content atmosphere.

hydrocracking A catalytic, high-pressure petroleum refinery process that is flexible enough to produce either high-octane gasoline or aviation jet fuel; the two main reactions are the adding of hydrogen to petroleum-derived molecules too massive and complex for gasoline and then the cracking of them to the required fuels; the catalyst is an acidic solid and a hydrogenating metal component.

hydrocyanic acid HCN A highly toxic liquid that has the odor of bitter almonds and boils at 25.6°C; used to manufacture cyanide salts, acrylonitrile, and dyes, and as a fumigant in agriculture. Also known as formonitrile; hydrogen cyanide; prussic acid.

hydrocyanic ether *See* ethyl cyanide.

hydrodealkylation A petroleum refining operation in which heat and pressure are used to remove methyl groups or larger alkyl groups from hydrocarbons, or to change positions of these groups on the molecule; used to upgrade low-value products.

hydrodesulfurization A catalytic process in which the petroleum feedstock is reacted with hydrogen to reduce the sulfur content in the oil.

hydrofining A fixed-bed catalytic process to desulfurize and hydrogenate a wide range of charge stocks, from gases through waxes; the catalyst comprises cobalt oxide and molybdenum oxide on an extruded alumina support and may be regenerated in place by air and steam or flue gas.

hydrofluoric acid   An aqueous solution of hydrogen fluoride, HF; colorless, fuming, poisonous liquid; extremely corrosive, it is a weak acid as compared to hydrochloric acid, but will attack glass and other silica materials; used to polish, frost, and etch glass, to pickle copper, brass, and alloy steels, to clean stone and brick, to acidize oil wells, and to dissolve ores.

hydrofluorosilicic acid   *See* fluorosilicic acid.

hydrofluosilicic acid   *See* fluosilicic acid.

hydroforming   A petroleum-refinery process in which naphthas are passed over a catalyst at elevated temperatures and moderate pressures in the presence of added hydrogen or hydrogen-containing gases, to form high-octane BTX aromatics for motor fuels or chemical manufacture.

hydroformylation   The reaction of adding hydrogen and the —CHO group to the carbon atoms across a double bond to yield oxygenated derivatives; an example is in the oxo process where the term hydroformylation applies to those reactions brought about by treating olefins with a mixture of hydrogen and carbon monoxide in the presence of a cobalt catalyst.

hydrogasification   A technique to manufacture synthetic pipeline gas from coal; pulverized coal is reacted with hot, raw, hydrogen-rich gas containing a substantial amount of steam at 1000 psig (6.9×10^6 newtons per square meter, gage) to form methane.

hydrogel   The formation of a colloid in which the disperse phase (colloid) has combined with the continuous phase (water) to produce a viscous jellylike product; for example, coagulated silicic acid.

hydrogen   The first chemical element, symbol H, in the periodic table, atomic number 1, atomic weight 1.00797; under ordinary conditions it is a colorless, odorless, tasteless gas composed of diatomic molecules, H_2; used in manufacture of ammonia and methanol, for hydrofining, for desulfurization of petroleum products, and to reduce metallic oxide ores.

hydrogenase   Enzyme that catalyzes the oxidation of hydrogen.

hydrogenated oil   Unsaturated liquid vegetable oil that has had hydrogen catalytically added so as to convert the oil to a hydrogen-saturated solid.

hydrogenation   1. Saturation of diolefin impurities in gasolines to form a stable product. 2. Catalytic reaction of hydrogen with other compounds, usually unsaturated; for example, unsaturated cottonseed oil is hydrogenated to form solid fats.

hydrogen bond   A type of bond formed when a hydrogen atom bonded to atom A in one molecule makes an additional bond to atom B either in the same or another molecule; the strongest hydrogen bonds are formed when A and B are highly electronegative atoms, such as fluorine, oxygen, or nitrogen.

hydrogen bromide   HBr A hazardous, toxic gas used as a chemical intermediate and as an alkylation catalyst; forms hydrobromic acid in aqueous solution.

hydrogen chloride   HCl A fuming, highly toxic, colorless gas soluble in water, alcohol, and ether; used in the production of vinyl chloride and alkyl chloride, and in polymerization, isomerization, and other reactions.

hydrogen cyanide   *See* hydrocyanic acid.

hydrogen cycle   The complete process of a cation-exchange operation in which the adsorbent is used in the hydrogen or free acid form.

hydrogen disulfide *See* hydrogen sulfide.

hydrogen electrode A noble metal (such as platinum) of large surface area covered with hydrogen gas in a solution of hydrogen ion saturated with hydrogen gas; metal is used in a foil form and is welded to a wire sealed in the bottom of a hollow glass tube, which is partially filled with mercury; used as a standard electrode with a potential of zero to measure hydrogen ion activity.

hydrogen equivalent The number of replaceable hydrogen atoms or hydroxyl groups in a molecule of an acid or a base.

hydrogen fluoride HF The hydride of fluoride; anhydrous HF is a mobile, colorless, liquid that fumes in air, melts at $-83°C$, boils at $19.8°C$; used to make fluorine-containing refrigerants (such as Freon) and organic fluorocarbon compounds, as a catalyst in alkylate gasoline manufacture, as a fluorinating agent, and in preparation of hydrofluoric acid.

hydrogen iodide HI A water-soluble, colorless gas that may be used in organic synthesis and as a reagent. Also known as hydriodic acid gas.

hydrogen ion *See* hydronium ion.

hydrogen ion concentration The normality of a solution with respect to hydrogen ions, H^+; it is related to acidity measurements in most cases by pH $=$ log $1/2$ $[1/(H^+)]$, where (H^+) is the hydrogen ion concentration in gram equivalents per liter of solution.

hydrogen ion exponent An expression of pH as $-\log c_H$, where c_H $=$ hydrogen ion concentration.

hydrogen line A spectral line emitted by neutral hydrogen having a frequency of 1420 megahertz and a wavelength of 21 centimeters; radiation from this line is used in radio astronomy to study the amount and velocity of hydrogen in the Galaxy.

hydrogenolysis A reaction in which hydrogen gas causes a chemical change that is similar to the role of water in hydrolysis.

hydrogenous Of, pertaining to, or containing hydrogen.

hydrogen oxide *See* water.

hydrogen peroxide H_2O_2 Unstable, colorless, heavy liquid boiling at $158°C$; soluble in water and alcohol; used as a bleach, chemical intermediate, rocket fuel, and antiseptic. Also known as peroxide.

hydrogen phosphide *See* phosphine.

hydrogen selenide H_2Se A toxic, colorless gas, soluble in water, carbon disulfide, and phosgene; used to make metallic selenides and organoselenium compounds and in the preparation of semiconductor materials.

hydrogen sulfide H_2S Flammable, toxic, colorless gas with offensive odor, boiling at $-60°C$; soluble in water and alcohol; used as an analytical reagent, as a sulfur source, and for purification of hydrochloric and sulfuric acids. Also known as hydrogen disulfide.

hydrogen tellurate *See* telluric acid.

hydrogeochemistry The study of the chemical characteristics of ground and surface waters as related to areal and regional geology.

hydroiodic ether *See* ethyl iodide.

hydrolase Any of a class of enzymes which catalyze the hydrolysis of proteins, nucleic acids, starch, fats, phosphate esters, and other macromolecular substances.

hydrolysis 1. Decomposition or alteration of a chemical substance by water. 2. In aqueous solutions of electrolytes, the reactions of cations with water to produce a weak base or of anions to produce a weak acid.

hydrolytic enzyme A catalyst that acts like a hydrolase.

hydrolytic process A reaction of both organic and inorganic chemistry wherein water effects a double decomposition with another compound, hydrogen going to one compound and hydroxyl to another.

hydronitric acid *See* hydrazoic acid.

hydronium ion H_3O^+ A proton combined with a molecule of water; found in pure water and in all aqueous solutions. Also known as hydrogen ion; oxonium ion.

hydrophile-lipophile balance The relative simultaneous attraction of an emulsifier for two phases of an emulsion system; for example, water and oil.

hydrophilic Having an affinity for, attracting, adsorbing, or absorbing water.

hydrophobic Lacking an affinity for, repelling, or failing to adsorb or absorb water.

hydroquinol *See* hydroquinone.

hydroquinone $C_6H_4(OH)_2$ White crystals melting at 170°C and boiling at 285°C; soluble in alcohol, ether, and water; used in photographic dye chemicals, in medicine, as an antioxidant and inhibitor, and in paints, varnishes, and motor fuels and oils. Also known as hydrochinone; hydroquinol; quinol.

hydroquinone carboxylic acid *See* gentisic acid.

hydroquinone dimethyl ether $C_6H_4(OCH_3)_2$ White flakes with a melting point of 56°C; used as a weathering agent in paint, as a flavoring, and in dyes and cosmetics. Also known as 1,4-dimethoxybenzene (DMB); dimethyl hydroquinone.

hydroquinone monomethyl ether $CH_3OC_6H_4OH$ A white, waxy solid with a melting point of 52.5°C; soluble in benzene, acetone, and alcohol; used for antioxidants, pharmaceuticals, and dyestuffs. Also known as *para*-hydroxyanisole; 4-methoxyphenol.

hydroquinone monopentyl ether *See para*-pentyloxyphenol.

hydrosilylation The addition of a Si-H bond to a C-C double bond of an olefin.

hydrosol A colloidal system in which the dispersion medium is water, and the dispersed phase may be a solid, a gas, or another liquid.

hydrosulfide A compound that has the SH— radical; for example, sulfhydrates, sulfhydryls, thioalcohols, thiols, sulfur alcohols, and mercaptans.

hydrothermal crystal growth Formation of simple crystals of quartz at elevated temperatures and pressures in an autoclave with an alkaline solution.

hydrotreating Oil refinery catalytic process in which hydrogen is contacted with petroleum intermediate or product streams to remove impurities, such as oxygen, sulfur, nitrogen, or unsaturated hydrocarbons.

hydrotrope Compound with the ability to increase the solubilities of certain slightly soluble organic compounds.

hydrous Indicating the presence of an indefinite amount of water.

hydroxamic acid An organic compound that contains the group —C(=O)NHOH.

hydroxide Compound containing the OH⁻ group; the hydroxides of metals are usually bases and those of nonmetals are usually acids; a hydroxide can be organic or inorganic.

hydroximino *See* nitroso.

hydroxisoxazole $C_4H_5NO_2$ A colorless, crystalline compound with a melting point of 86–87°C; used as a fungicide in soil and as a growth regulator for seeds. Also known as 3-hydroxy-5-methylisoxazole; hymexazol.

hydroxy- Chemical prefix indicating the OH⁻ group in an organic compound, such as hydroxybenzene for phenol, C_6H_5OH; the use of just oxy- for the prefix is incorrect. Also spelled hydroxyl-.

para-**hydroxyacetanilide** *See* acetaminophen.

hydroxyacetic acid *See* glycolic acid.

hydroxy acid Any organic acid, with an OH⁻ group, such as hydroxyacetic acid.

2-hydroxy-3-amylene-α-naphthoquinone *See* lapachol.

para-**hydroxyanilinoacetic acid** *See* glycin.

ortho-**hydroxyanisole** *See* guaiacol.

para-**hydroxyanisole** *See* hydroquinone monomethyl ether.

5-hydroxybarbituric acid *See* dialuric acid.

ortho-**hydroxybenzaldehyde** *See* salicylaldehyde.

ortho-**hydroxybenzamide** *See* salicylamide.

ortho-**hydroxybenzoic acid** *See* salicylic acid.

para-**hydroxybenzoic acid** $C_6H_4(OH)$ COOH·2H₂O Colorless crystals melting at 210°C; soluble in alcohol, water, and ether; used as a chemical intermediate and for synthetic drugs.

ortho-**hydroxybenzyl alcohol** *See* salicyl alcohol.

hydroxybenzoic acid $C_7H_6O_3$ Any one of three crystalline derivatives of benzoic acid: ortho, meta, and para forms; the ester of the para compound is used as a bacteriostatic agent.

2-hydroxybenzoic-5-sulfonic acid *See* sulfosalicylic acid.

2-hydroxybiphenyl *See* phenylphenol.

3-hydroxybutanal *See* aldol.

3-hydroxy-2-butanone *See* acetoin.

β-**hydroxybutyric dehydrogenase** The enzyme that catalyzes the conversion of L-β-hydroxybutyric acid to acetoacetic acid by dehydrogenation.

2-hydroxycamphane *See* borneol.

hydroxycarbamide *See* hydroxyurea.

meta-hydroxycarbanilate *See* phenmediphan.

hydroxycarbonyl compound A compound possessing one or more hydroxy (— OH) groups and one or more carbonyl (=C=O) groups.

2-hydroxy-5-chlorobenzoxazole *See* chlorzoxazone.

hydroxycholine *See* muscarine.

hydroxycitronellal $C_{10}H_{20}O_2$ A colorless or light yellow, viscous liquid with a boiling range of 94–96°C; soluble in 50% alcohol and fixed oils; used in perfumery and flavoring. Also known as citronellal hydrate; 3,7-dimethyl-7-hydroxyoctanal.

1-hydroxy-2-cyanoethane *See* ethylene cyanohydrin.

4-hydroxy-3,5-diiodobenzonitrile(4-cyano-2,6-diiodo- phenol) *See* ioxynil.

hydroxydimethylbenzene *See* xylenol.

para-hydroxydiphenyl *See* phenylphenol.

2-(hydroxydiphenyl)methane $C_6H_5CH_2C_6H_4OH$ A crystalline substance with a melting point of 20.2–20.9°C, or a liquid; used as a germicide, preservative, and antiseptic. Also known as *ortho*-benzylphenol.

hydroxyethylacetamide *See* N-acetylethanolamine.

2-hydroxyethylamine *See* ethanolamine.

2-hydroxyethylhydrazine $HOCH_2CH_2NHNH_2$ A colorless, slightly viscous liquid with a melting point of $-70°C$; soluble in lower alcohols; used as an abscission agent in fruit. Also known as 2-hydrazinoethanol.

1-hydroxy fenchane *See* fenchyl alcohol.

3-hydroxyflavone *See* flavanol.

hydroxyhydroquinone *See* 1,2,4-benzenetriol.

hydroxyl- *See* hydroxy-.

hydroxylamine NH_2OH A colorless, crystalline compound produced commercially by acid hydrolysis of nitroparaffins, decomposes on heating, melts at 33°C; used in organic synthesis and as a reducing agent.

hydroxylamine hydrochloride $(NH_2OH)Cl$ A crystalline substance with a melting point of 151°C; soluble in glycerol and propylene glycol; used as a reducing agent in photography and in synthetic and analytic chemistry; as an antioxidant in fatty acids and soaps; and as a reagent for enzyme reactivation. Also known as oxammonium hydrochloride.

ortho-hydroxylaniline $C_6H_4NH_2OH$ White crystals that turn brownish upon standing for some time; melts at 172-173°C, and will sublime upon more heating; soluble in cold water and benzene; used as a dye for hair and furs, and as a dye intermediate. Also known as *ortho*-aminophenol; oxammonium.

hydroxylase Any of several enzymes that catalyze certain hydroxylation reactions involving atomic oxygen.

hydroxylation reaction One of several types of reactions used to introduce one or more hydroxyl groups into organic compounds; an oxidation reaction as opposed to hydrolysis.

3-hydroxymenthane *See* menthol.

4-hydroxy-3-methoxyacetophenone   *See* acetovanillon.

1-hydroxy-2-methoxybenzene   *See* guaiacol.

4-hydroxy-3-methoxyphenylacetic acid   *See* homovanillic acid.

2-hydroxy methyl furan   *See* furfuryl alcohol.

3-hydroxy-5-methylisoxazole   *See* hydroxisoxazole.

7-hydroxy-4-methyl-2-oxo-3-chromene   *See* hymecromone.

3-hydroxy-2-methyl-6-γ-pyrone   *See* maltol.

1-hydroxynaphthalene   *See* α-naphthol.

2-hydroxynaphthalene   *See* β-naphthol.

3-hydroxy-2-naphthoic acid   *See* β-hydroxynaphthoic acid.

β-hydroxynaphthoic acid $C_{10}H_6OH$ COOH   A yellow solid that is soluble in ether and alcohol and melts at about 218°C; used as a dye and a pigment. Also known as 3-hydroxy-2-naphthoic acid; 3-naphthol-2-carboxylic acid; β-oxynaphthoic acid.

4-hydroxy-3-nitrobenzenearsonic acid $HOC_6H_3(NO_2)AsO(OH)_2$ Crystals used as a reagent for zirconium; also used to control enteric infections and to improve growth and feed efficiency in animals. Also known as 2-nitro-1-hydroxybenzene-4-arsonic acid; nitrophenolarsonic acid; roxarsone.

12-hydroxyoleic acid   *See* ricinoleic acid.

15-hydroxypentadecanoic acid lactone   *See* pentadecanolide.

1-hydroxy-2-phenoxyethane   *See* 2-phenoxyethanol.

2-hydroxy-2-phenyl acetophenone   *See* benzoin.

para-**hydroxyphenylaminoacetic acid**   *See* glycin.

α-(4-hydroxyphenyl)-β-aminoethane   *See* tyramine.

para-**hydroxyphenylethanolamine**   *See* octopamine.

2-*para*-hydroxyphenylethylamine   *See* tyramine.

hydroxyproline $C_5H_9O_3N$ An amino acid that is essentially limited to structural proteins of the collagen type.

2-hydroxy-2 propanesulfonic acid sodium salt   *See* acetone sodium bisulfite.

2-hydroxy-1,2,3-propane-tricarboxylic acid   *See* isocitric acid.

2-hydroxypropanoic acid   *See* lactic acid.

3-hydroxypropanoic acid   *See* hydracrylic acid.

α-hydroxypropanoic acid   *See* lactic acid.

1-hydroxy-2-propanone   *See* acetol.

β-hydroxypropionic acid   *See* hydracrylic acid.

α-hydroxypropionitrile   *See* lactonitrile.

hydroxypropyl alginate   *See* propylene glycol alginate.

2-hydroxypropylamine   *See* isopropanolamine.

α-hydroxypropylbenzene   *See* 1-phenyl-1-propanol.

(2-hydroxypropyl)trimethyammonium chloride acetate   *See* methacholine chloride.

hydroxyquinol   *See* 1,2,4-benzenetriol.

8-hydroxyquinoline C_9H_6NOH White crystals or powder that darken on exposure to light, slightly soluble in water, soluble in benzene, melting at 73–75°C; used in preparing fungicides and in the separation of metals by acting as a precipitating agent. Also known as oxine; oxyquinoline; 8-quinolinol.

8-hydroxyquinoline sulfuric acid salt   *See* 8-hydroxyquinoline sulfate.

3-hydroxyquinuclidine   *See* 3-quinuclidinol.

4-hydroxysalicylic acid   *See* β-resorcylic acid.

5-hydroxysalicylic acid   *See* gentisic acid.

4-hydroxy-3-(1,2,3,4-tetrahydro-1-naphthyl)coumarin   *See* coumatetralyl.

5-hydroxytryptamine   *See* serotonin.

5-hydroxytryptophan $C_{11}H_{12}N_2O_3$ Minute rods or needlelike crystals; the biological precursor of serotonin.

3-hydroxytyramine hydrobromide $(HO)_2 C_6H_3CH_2CH_2NH_2$·HBr   A source of dopamine for the synthesis of catecholamine analogs.

8-hydroxyxanthine   *See* uric acid.

2-hydroxy-*para*-cymene   *See* carvacrol.

2-hydroxy-*meta*-toluic acid   *See* 3-methylsalicylic acid.

hygroscopic   **1.** Possessing a marked ability to accelerate the condensation of water vapor; applied to condensation nuclei composed of salts which yield aqueous solutions of a very low equilibrium vapor pressure compared with that of pure water at the same temperature. **2.** Pertaining to a substance whose physical characteristics are appreciably altered by effects of water vapor. **3.** Pertaining to water absorbed by dry soil minerals from the atmosphere; the amounts depend on the physicochemical character of the surfaces, and increase with rising relative humidity.

hygroscopic depression   The measure of a desiccant's capacity to take on water.

Hylleraas coordinates   Coordinates for two particles used in studying the helium atom; they comprise the distance between the two particles, the sum of the distances of the particles from the origin, and the difference of the distances of the particles from the origin.

hymecromone $C_{10}H_8O_3$ A crystalline substance with a melting point of 194–195°C; soluble in methanol and glacial acetic acid; used as choleretic and antispasmodic drugs and as a standard for the fluorometric determination of enzyme activity. Also known as 7-hydroxy-4-methyl-2-oxo-3-chromene; imecromone; β-methylumbelliferone.

hymexazol   *See* hydroxisoxazole.

hyoscyamine $C_{17}H_{23}O_3N$ A white, crystalline alkaloid isolated from henbane, belladonna, and other plants of the family Solanaceae, which is freely soluble in alcohol and dilute acids; used in medicine as an anticholinergic.

hyperchromicity   An increase in the absorption of ultraviolet light by polynucleotide solutions due to a loss of the ordered secondary structure.

hyperconjugation An arrangement of bonds in a molecule that is similar to conjugation in its formulation and manifestations, but the effects are weaker; it occurs when a CH_2 or CH_3 group (or in general, an AR_2 or AR_3 group where A may be any polyvalent atom and R any atom or radical) is adjacent to a multiple bond or to a group containing an atom with a lone π-electron, π-electron pair or quartet, or π-electron vacancy; it can be sacrificial (relatively weak) or isovalent (stronger).

hyperfine structure A splitting of spectral lines due to the spin of the atomic nucleus or to the occurrence of a mixture of isotopes in the element. Abbreviated hfs.

hyperforming A catalytic, petroleum-refinery hydrogenation process to improve naphtha octane number by removal of sulfur and nitrogen compounds; the catalyst is cobalt molybdate on a silica-alumina base.

hyperglycemic factor *See* glucagon.

hyperglycemic glycogenolytic factor *See* glucagon.

hypergolic Capable of igniting spontaneously upon contact.

hypersorption Process with recirculating bed of activated-carbon adsorbent for continuous recovery of ethylene from methane and other low-molecular-weight gases.

hypertensin *See* angiotensin.

hypervalent atom A central atom in a single-bonded structure that imparts more than eight valence electrons in forming covalent bonds.

hypnotoxin A supposed hormone produced by brain tissue and inducing sleep.

hypo *See* sodium thiosulfate.

hypochlorite $ClO_3{}^-$ A negative ion derived from hypochlorous acid, HClO; the ion is an oxidizing agent and a constituent of bleaching agents.

hypochlorite sweetening A petroleum refinery process to oxidize gasoline mercaptans by agitation with an aqueous, alkaline hypochlorite solution.

hypochlorous acid HOCl Weak, unstable acid existing in solution only; its salts (such as calcium hypochlorite) are used as bleaching agents.

hypochromicity A decrease in the absorption of ultraviolet light by polynucleotide solutions due to the formation of an ordered secondary structure.

hypoiodous acid HIO A very weak unstable acid that occurs as the result of the weak hydrolysis of iodine in water.

hypothalamic releasing factor Any of the hormones secreted by the hypothalamus which travel by way of nerve fibers to the anterior pituitary, where they cause selective release of specific pituitary hormones.

hypoxanthine $C_5H_4ON_4$ An intermediate product derived from adenine in the hydrolysis of nucleic acid.

***n*-heptadecanoic acid** $CH_3(CH_2)_{15}COOH$ A fatty acid that is saturated; soluble in ether and alcohol, insoluble in water; colorless crystals melt at 61°C. Also known as margaric acid.

***n*-hexadecane** $C_{16}H_{34}$ A colorless, solid hydrocarbon, melting point 20°C; a standard reference fuel in determining the ignition quality (cetane number) of diesel fuels. Also known as cetane.

n-hexaldehyde $CH_3(CH_2)_4CHO$ Colorless liquid with sharp aroma, boiling at 128.6°C; used as an intermediate for plasticizers, dyes, insecticides, resins, and rubber chemicals. Also known as caproaldehyde; caproic aldehyde.

n-hexyl acetate $CH_3COOC_6H_{13}$ Colorless liquid boiling at 169°C; soluble in alcohol and ether, insoluble in water; used as a solvent for resins and cellulosic esters.

n-hexyl ether $C_6H_{13}OC_6H_{13}$ Faintly colored liquid with a characteristic odor, only slightly water-soluble; used in solvent extraction and in the manufacture of collodion and various cellulosic products.

I

I *See* iodine.

IAA *See* indoleacetic acid.

IBA *See* indolbutyric acid.

IBIB *See* isobutyl isobutyrate.

ibogaine $C_{26}H_{32}O_2N_2$ An alkaloid isolated from the stems and leaves of the shrub *Tabernanthe iboga*, crystallizing from absolute ethanol as prismatic needles, melting at 152–153°C, soluble in ethanol, ether, and chloroform; used in medicine.

ice **1.** The dense substance formed by the freezing of water to the solid state; has a melting point of 32°F (0°C) and commonly occurs in the form of hexagonal crystals. **2.** A layer or mass of frozen water.

ice color *See* azoic dye.

ice crystal Any one of a number of macroscopic crystalline forms in which ice appears, including hexagonal columns, hexagonal platelets, dendritic crystals, ice needles, and combinations of these forms; although the crystal lattice of ice is hexagonal in its symmetry, varying conditions of temperature and vapor pressure can lead to growth of crystalline forms in which the simple hexagonal pattern is almost undiscernible.

ice needle A long, thin ice crystal whose cross section perpendicular to its long dimension is typically hexagonal. Also called ice spicule.

ice point The true freezing point of water; the temperature at which a mixture of air-saturated pure water and pure ice may exist in equilibrium at a pressure of 1 standard atmosphere (101,325 newtons per square meter).

ice spicule *See* ice needle.

ice splinters Minute, electrically charged fragments of ice which have been observed under laboratory conditions to be torn away from dendritic crystals or spatial aggregates exposed to moving air.

icositetrahedron *See* trapezohedron.

ICP-AES *See* inductively coupled plasma-atomic emission spectroscopy.

IDA *See* iminodiacetic acid.

ideal solution A solution that conforms to Raoult's law over all ranges of temperature and concentration and shows no internal energy change on mixing and no attractive force between components.

IDU　*See* 5-iodo-2′-deoxyuridine.

IDUR　*See* 5-iodo-2′-deoxyuridine.

ignite　To start a fuel burning.

ignition　The process of starting a fuel mixture burning, or the means for such a process.

ignition point　*See* ignition temperature.

ignition temperature　The lowest temperature at which combustion begins and continues in a substance when it is heated in air. Also known as autogenous ignition temperature; ignition point.

Ilkovič equation　Mathematical relationship between diffusion current, diffusion coefficient, and active-substance concentration; used for polarographic analysis calculations.

illinium　*See* promethium-147.

imbibition　Absorption of liquid by a solid or a semisolid material.

imecromone　*See* hymecromone.

imidazole　$C_3H_4N_2$　One of a group of organic heterocyclic compounds containing a five-membered diunsaturated ring with two nonadjacent nitrogen atoms as part of the ring; the particular compound imidazole is a member of the group. Also known as 1,3-diazole; glyoxaline; iminazole.

imidazoletrione　*See* parabanic acid.

imidazolyl　$C_3H_3N_2$　The radical from imidazole.

imide　1. A compound derived from acid anhydrides by replacing the oxygen (O) with the =NH group.　2. A compound that has either the =NH group or a secondary amine in which R is an acyl radical, as R_2NH.

iminazole　*See* imidazole.

imine　A class of compounds that are the product of condensation reactions of aldehydes or ketones with ammonia or amines; they have the NH radical attached to the carbon with the double bond, as R—HC=NH; an example is benzaldimine.

imino acid　Organic acid in which the =NH group is attached to one or two carbons; for example, acetic acid, $NH(CH_2COOH)_2$.

imino compound　A compound that has the =NH radical attached to one or two carbon atoms.

iminodiacetic acid　$C_4H_7NO_4$　A crystalline substance used as an intermediate in the manufacture of chelating agents, surface-active agents, and complex salts. Abbreviated IDA. Also known as diglycine; iminodiethanoic acid.

iminodiethanoic acid　*See* iminodiacetic acid.

imino nitrogen　Nitrogen combined with hydrogen in the imino group. Also known as ammonia nitrogen.

iminourea　*See* guanidine.

immersion sampling　Collection of a liquid sample for laboratory or other analysis by immersing a container in the liquid and filling it.

immiscible Pertaining to liquids that will not mix with each other.

impact parameter In a nuclear collision, the perpendicular distance from the target nucleus to the initial line of motion of the incident particle.

imperial red Any of the red varieties of ferric oxide used as pigment.

implosion The sudden reduction of pressure by chemical reaction or change of state which causes an inrushing of the surrounding medium.

In *See* indium.

inactive tartaric acid *See* racemic acid.

incineration The process of burning a material so that only ashes remain.

inclusion 1. A crystal or fragment of a crystal found in another crystal. 2. A small cavity filled with gas or liquid in a crystal.

inclusion complex Crystalline mixture in which the molecules of one component are contained within the crystal lattice of another component. Also known as inclusion compound.

inclusion compound *See* inclusion complex.

incomplete combustion Combustion in which oxidation of the fuel is incomplete.

incomplete fusion *See* quasi-fission.

incubation Maintenance of chemical mixtures at specified temperatures for varying time periods to study chemical reactions, such as enzyme activity.

indamine $HN:C_6H_4:N\cdot C_6H_4NH_2$ An unstable dye obtained by the reaction of *para*-phenylenediamine and aniline. Also known as phenylene blue.

indan $C_6H_4(CH_2)_3$ Colorless liquid boiling at 177°C; soluble in alcohol and ether, insoluble in water; derived from coal tar.

indanthrene *See* indanthrone.

indanthrone $C_{28}H_{14}N_2O_4$ A blue pigment or vat dye soluble in dilute base solutions; used in cotton dyeing and as a pigment in paints and enamels. Also known as 6,15-dihydro-5,9,14,18-anthrazinetetrone; indanthrene.

indene C_9H_8 A colorless, liquid, polynuclear hydrocarbon; boils at 181°C and freezes at -2°C; derived from coal tar distillates; copolymers with benzofuran have been manufactured on a small scale for use in coatings and floor coverings.

independent migration law The law that each ion in a conductiometric titration contributes a definite amount to the total conductance, irrespective of the nature of the other ions in the electrolyte.

index of unsaturation A numerical value that represents the number of rings or double bonds in a molecule; a triple bond is considered to have the numerical value of 2.

indican 1. C_8H_6NOSOK The potassium salt of indoxylsulfate found in urine as a result of bacterial action on tryptophan in the bowel. 2. $C_{14}H_{17}O_6N$ A glucoside of indoxyl occurring in the indigo plant; on hydrolysis indican gives rise to indoxyl, which is oxidized to indigo by air.

indigo 1. A blue dye extracted from species of the *Indigofera* bush. 2. *See* indigo blue.

indigo blue $C_{16}H_{10}O_2N_2$ A component of the dye indigo, crystallizing as dark-blue rhomboids that break down at 30°C, that are soluble in hot aniline and hot chloroform, and that are also made synthetically; used as a reagent and a dye. Also known as indigo.

indigo carmine $C_{16}H_8N_2Na_2O_8S_2$ A dark blue powder with coppery luster; used as a dye in testing kidney function and as a reagent in detecting chlorate and nitrate. Also known as disodium 5,5'-indigotin disulfonate; sodium indigotin disulfonate; soluble indigo blue.

indigoid dye Any of the vat dyes with $C_{16}H_{10}O_2N_2$ (indigo) or $C_{16}H_8S_2O_2$ (thioindigo) groupings; used to dye cotton and rayon, sometimes silk.

indigo red $C_{16}H_{10}O_2N_2$ A red isomer of indigo obtained in the manufacture of indigo. Also known as indirubin.

indirect effect A chemical effect of ionizing radiation on a dilute solution caused by the interaction of solute molecules with highly reactive transient molecules or ions formed by reaction of the radiation with the solvent.

indirubin *See* indigo red.

indium A metallic element, symbol In, atomic number 49, atomic weight 114.82; soluble in acids; melts at 156°C, boils at 1450°C.

indium antimonide InSb Crystals that melt at 535°C; an intermetallic compound having semiconductor properties and the highest room-temperature electron mobility of any known material; used in Hall-effect and magneto-resistive devices and as an infrared detector.

indium arsenide InAs Metallic crystals that melt at 943°C; an intermetallic compound having semiconductor properties; used in Hall-effect devices.

indium chloride $InCl_3$ Hygroscopic white powder, soluble in water and alcohol.

indium phosphide InP A metallic mass that is brittle and melts at 1070°C; an intermetallic compound having semiconductor properties.

indium sulfate $In_2(SO_4)_3$ Deliquescent, water-soluble, grayish powder; decomposes when heated.

indogen The radical $C_6H_4(NH)COC\!\!=\!\!$; it occurs, for example, in the molecule indigo.

indogenide A compound containing the radical $C_6H_4(NH)\cdot CO\cdot C\!\!=\!\!$ from indogen.

indoleacetic acid $C_{10}H_9O_2N$ A decomposition product of tryptophan produced by bacteria and occurring in urine and feces; used as a hormone to promote plant growth. Abbreviated IAA.

indolebutyric acid $C_{12}H_{13}O_2N$ A crystalline acid similar to indoleacetic acid in auxin activity. Abbreviated IBA.

indole Also known as 2,3-benzopyrrole. **1.** C_6H_4 (CHNH)CH A decomposition product of tryptophan formed in the intestine during putrefaction and by certain cultures of bacteria. **2.** Carcinogenic, white to yellowish scales with unpleasant aroma; reagent and in perfumery and medicine.

indoxyl $(C_8H_6N)OH$ A yellow crystalline glycoside, used as an intermediate in the manufacture of indigo.

induced fission Fission which takes place only when a nucleus is bombarded with neutrons, gamma rays, or other carriers of energy.

inducible enzyme An enzyme which is present in trace quantities within a cell but whose concentration increases dramatically in the presence of substrate molecules.

induction force A type of van der Waals force resulting from the interaction of the dipole moment of a polar molecule and the induced dipole moment of a nonpolar molecule. Also known as Debye force.

induction period A time of acceleration of a chemical reaction from zero to a maximum rate.

inductively coupled plasma-atomic emission spectroscopy A type of atomic spectroscopy in which the light emitted by atoms and ions in an inductively coupled plasma is observed. Abbreviated ICP-AES.

industrial alcohol Ethyl alcohol that has been denatured by acetates, ketones, gasoline, or other additives to make it unfit for beverage purposes.

inert atmosphere A nonreactive gas atmosphere, such as nitrogen, carbon dioxide, or helium; used to blanket reactive liquids in storage, to purge process lines and vessels of reactive gases and liquids, and to cover a reaction mix in a partially filled vessel.

inert gas *See* noble gas.

inflammability *See* flammability.

infrared spectrometer Device used to identify and measure the concentrations of heteroatomic compounds in gases, in many nonaqueous liquids, and in some solids by arc or spark excitation and subsequent measurement of the electromagnetic emissions in the wavelength range of 0.78 to 300 micrometers.

infrared spectrophotometry Spectrophotometry in the infrared region, usually for the purpose of chemical analysis through measurement of absorption spectra associated with rotational and vibrational energy levels of molecules.

infrared spectroscopy The study of the properties of material systems by means of their interaction with infrared radiation; ordinarily the radiation is dispersed into a spectrum after passing through the material.

infusion The aqueous solution of a soluble constituent of a substance as the result of the substance's steeping in the solvent for a period of time.

ingrain color *See* azoic dye.

inhibition index The amount of antimetabolite that will overcome the biological effect of a unit weight of metabolite.

inhibitor A substance which is capable of stopping or retarding a chemical reaction; to be technically useful, it must be effective in low concentration.

inhibitor sweetening Petroleum-refinery treating process to sweeten gasoline (convert mercaptans to disulfides) of low mercaptan content; uses a phenylenediamine inhibitor, air, and caustic.

initial boiling point According to ASTM petroleum-analysis distillation procedures, the recorded temperature when the first drop of distilled vapor is liquefied and falls from the end of the condenser.

initiator The substance or molecule (other than reactant) that initiates a chain reaction, as in polymerization; an example is acetyl peroxide.

inner quantum number A quantum number J which gives an atom's total angular momentum, excluding the nuclear spin.

inorganic Pertaining to or composed of chemical compounds that do not contain carbon as the principal element (excepting carbonates, cyanides, and cyanates), that is, matter other than plant or animal.

inorganic acid A compound composed of hydrogen and a nonmetal element or radical; examples are hydrochloric acid, HCl, sulfuric acid, H_2SO_4, and carbonic acid, H_2CO_3.

inorganic chemistry The study of chemical reactions and properties of all the elements and their compounds, with the exception of hydrocarbons, and usually including carbides, oxides of carbon, metallic carbonates, carbon-sulfur compounds, and carbon-nitrogen compounds.

inorganic peroxide An inorganic compound containing an element at its highest state of oxidation (such as perchloric acid, $HClO_4$), or having the peroxy group, —O—O— (such as perchromic acid, $H_3CrO_8\cdot 2H_2O$).

inorganic pigment A natural or synthetic metal oxide, sulfide, or other salt used as a coloring agent for paints, plastics, and inks.

inorganic polymer Large molecules, usually linear or branched chains with atoms other than carbon in their backbone; an example is glass, an inorganic polymer made up of rings and chains of repeating silicate units.

inosine $C_{10}H_{12}N_4O_5$ A compound occurring in muscle; a hydrolysis product of inosinic acid.

inosinic acid $C_{10}H_{13}N_4O_8P$ A nucleotide constituent of muscle, formed by deamination of adenylic acid; on hydrolysis it yields hypoxanthine and D-ribose-5-phosphoric acid. Also known as hypoxanthine riboside-5-phosphoric acid; hypoxanthylic acid; inosine monophosphate (IMP).

inositol $C_6H_6(OH)_6\cdot 2H_2O$ A water-soluble alcohol often grouped with the vitamins; there are nine stereoisomers of hexahydroxycyclohexane, and the only one of biological importance is optically inactive *meso*-inositol, comprising white crystals, widely distributed in animals and plants; it serves as a growth factor for animals and microorganisms.

insol *See* insoluble.

insoluble Incapable of being dissolved in another material; usually refers to solid-liquid or liquid-liquid systems. Abbreviated insol.

insoluble anode An anode that resists dissolution during electrolysis.

inspissation 1. The process of thickening a liquid by evaporation. 2. Thickening of an oil deposit by evaporation or oxidation, resulting, for example, after long exposure in pitch or gum formation.

insulin A protein hormone produced by the beta cells of the islets of Langerhans which participates in carbohydrate and fat metabolism.

insulinase An enzyme produced by the liver which is able to inactivate insulin.

integral heat of dilution *See* heat of dilution.

integral heat of solution *See* heat of solution.

integral procedure decomposition temperature Decomposition temperatures derived from graphical integration of the thermogravimetric analysis of a polymer.

intensive properties Properties independent of the quantity or shape of the substance under consideration; for example, temperature, pressure, or composition.

intercept One of the distances cut off a crystal's reference axis by planes.

interdiffusion The self-mixing of two fluids, initially separated by a diaphragm.

interface The boundary between any two phases: among the three phases (gas, liquid, and solid), there are five types of interfaces: gas-liquid, gas-solid, liquid-liquid, liquid-solid, and solid-solid.

interface mixing The mixing of two immiscible or partially miscible liquids at the plane of contact (interface).

interfacial angle The angle between two crystal faces.

interference spectrum A spectrum that results from interference of light, as in a very thin film.

interferon A protein produced by intact animal cells when infected with viruses; acts to inhibit viral reproduction and to induce resistance in host cells.

interhalogen Any of the compounds formed from the elements of the halogen family that react with each other to form a series of binary compounds; for example, iodine monofluoride.

interionic attraction The coulomb attraction between ions of opposite sign in a solution.

intermediary metabolism Intermediate steps in the chemical synthesis and breakdown of foodstuffs within body cells.

intermediate A precursor to a desired product; ethylene is an intermediate for polyethylene, and ethane is an intermediate for ethylene.

intermedin A hormonal substance produced by the intermediate portion of the hypophysis of certain animal species which influences pigmentation; similar to melanocyte-stimulating hormone in humans.

intermolecular force The force between two molecules; it is that negative gradient of the potential energy between the interacting molecules, if energy is a function of the distance between the centers of the molecules.

internal conversion A nuclear de-excitation process in which energy is transmitted directly from an excited nucleus to an orbital electron, causing ejection of that electron from the atom.

internal conversion coefficient *See* conversion coefficient.

internal diffusion The diffusion of liquid or gaseous reactants to the innermost pore depths of an adsorbent-base catalyst, necessary for full catalytic effect.

internal phase *See* disperse phase.

internal reflectance spectroscopy *See* attenuated total reflectance.

internal standard The principal line in spectrum analysis by the logarithmic sector method, a quantitative spectroscopy procedure.

internuclear distance The distance between two nuclei in a molecule.

interpenetration twin Two or more individual crystals so twinned that they appear to have grown through one another. Also known as penetration twin.

interpolymer A mixed polymer made from two or more starting materials.

interstitial A crystal defect in which an atom occupies a position between the regular lattice positions of a crystal.

interstitial atom A displaced atom which is forced into a nonequilibrium site within a crystal lattice.

interstitial-cell–stimulating hormone *See* luteinizing hormone.

intestinal hormone Either of two hormones, secretin and cholecystokinin, secreted by the intestine.

intracavity absorption spectroscopy A highly sensitive technique in which an absorbing sample is placed inside the resonator of a broad-band dye laser, and absorption lines are detected as dips in the laser emission spectrum.

intracellular enzyme An enzyme that remains active only within the cell in which it is formed. Also known as organized ferment.

intrastratal solution A chemical attrition of the constituents of a rock after deposition.

intrinsic factor A substance, produced by the stomach, which combines with the extrinsic factor (vitamin B_{12}) in food to yield an antianemic principle; lack of the intrinsic factor is believed to be a cause of pernicious anemia. Also known as Castle's intrinsic factor.

intrinsic tracer An isotope that is present naturally in a form suitable for tracing a given element through chemical and physical processes.

intrinsic viscosity The ratio of a solution's specific viscosity to the concentration of the solute, extrapolated to zero concentration. Also known as limiting viscosity number.

introfaction Change in fluidity and specific wetting properties (for impregnation acceleration) of an impregnating compound, caused by an introfier (impregnation accelerator).

inulase An enzyme produced by certain molds that catalyzes the conversion of inulin to levulose.

inulin A polysaccharide made up of polymerized fructofuranose units.

inverse beta decay A reaction providing evidence for the existence of the neutrino, in which an antineutrino (or neutrino) collides with a proton (or neutron) to produce a neutron (or proton) and a positron (or electron).

inverse Stark effect The Stark effect as observed with absorption lines, in contrast to emission lines.

inverse Zeeman effect A splitting of the absorption lines of atoms or molecules in a static magnetic field; it is the Zeeman effect observed with absorption lines.

inversion 1. Change of a compound into an isomeric form. 2. A change from one crystal polymorph to another. Also known as transformation.

inversion spectrum Lines in the microwave spectra of certain molecules (such as ammonia) which result from the quantum-mechanical analog of an oscillation of the molecule between two configurations which are mirror images of each other.

invertase *See* saccharase.

invertin *See* saccharase.

Io *See* ionium.

iodate A salt of iodic acid containing the IO_3^- radical; sodium and potassium iodates are the most important salts and are used in medicine.

iodic acid HIO_3 Water-soluble, moderately strong acid; colorless or white powder or crystals; decomposes at 110°C; used in analytical chemistry and medicine.

iodic acid anhydride *See* iodine pentoxide.

iodide **1.** A compound which contains the iodine atom in the −1 oxidation state and which may be considered to be derived from hydriodic acid (HI); examples are KI and NaI. **2.** A compound of iodine, such as CH_3CH_2I, in which the iodine has combined with a more electropositive group.

iodine A nonmetallic halogen element, symbol I, atomic number 53, atomic weight 126.9044; melts at 114°C, boils at 184°C; the poisonous, corrosive, dark plates or granules are readily sublimed; insoluble in water, soluble in common solvents; used as germicide and antiseptic, in dyes, tinctures, and pharmaceuticals, in engraving lithography, and as a catalyst and analytical reagent.

iodine-131 A radioactive, artificial isotope of iodine, mass number 131; its half-life is 8 days with beta and gamma radiation; used in medical and industrial radioactive tracer work; moderately radiotoxic.

iodine bisulfide *See* sulfur iodine.

iodine cyanide ICN Poisonous, colorless needles with pungent aroma and acrid taste; melts at 147°C; soluble in water, alcohol, and ether; used in taxidermy as a preservative. Also known as cyanogen iodide.

iodine disulfide *See* sulfur iodine.

iodine number A measure of the iodine absorbed in a given time by a chemically unsaturated material, such as a vegetable oil or a rubber; used to measure the unsaturation of a compound or mixture. Also known as iodine value.

iodine pentoxide I_2O_5 White crystals, decomposing at 275°C, very soluble in water, insoluble in absolute alcohol, ether, and chloroform; used as an oxidizing agent to oxidize carbon monoxide to dioxide at ordinary temperatures, and in organic synthesis. Also known as iodic acid anhydride.

iodine test Placing a few drops of potassium iodide solution on a sample to detect the presence of starch; test is positive if sample turns blue.

iodine value *See* iodine number.

iodisan *See* propiodal.

iodoacetic acid CH_2ICOOH White or colorless crystals that are soluble in water and alcohol, and melt at 82–83°C; used in biological research for its inhibitive effect on enzymes.

iodoalkane An alkane hydrocarbon in which an iodine atom replaces one or more hydrogen atoms in the molecule; an example is iodomethane, CH_3I, better known as methyl iodide.

iodocyanin *See* cyanine dye.

iodoeasin *See* easin.

iodoethane *See* ethyl iodide.

iodoethylene *See* tetraiodoethylene.

iodofenphos *See* jodfenphos.

iodoform CHI_3 A yellow, hexagonal solid; melting point 119°C; soluble in chloroform, ether, and water; has weak bactericidal qualities and is used in ointments for minor skin diseases. Also known as triiodomethane.

iodohydrocarbon A hydrocarbon in which an iodine atom replaces one or more hydrogen atoms in the molecule, as in an alkane, aromatic, or olefin.

iodomethane *See* methyl iodide.

iodometry An application of iodine chemistry to oxidation-reduction titrations for the quantitative analysis in certain chemical compounds, in which iodine is used as a reductant and the iodine freed in the associated reaction is titrated, usually in neutral or slightly acid mediums with a standard solution of a reductant such as sodium thiosulfate or sodium arsenite; examples of chemicals analyzed are copper(III), gold(VI), arsenic(V), antimony(V), chlorine, and bromine.

iodophor Any compound that is a carrier of iodine.

iodopsin The visual pigment found in the retinal cones, consisting of retinene, combined with photopsin.

iodosobenzene C_6H_5IO A yellowish-white amorphous solid that explodes at 200°C, soluble in hot water and alcohol; a strong oxidizing agent.

iodoxybenzene $C_6H_5IO_2$ Clear white crystals that explode at 227–228°C, slightly soluble in water, insoluble in chloroform, acetone, and benzene; a strong oxidizing agent.

iodyrite *See* iodargyrite.

ion An isolated electron or positron or an atom or molecule which by loss or gain of one or more electrons has acquired a net electric charge.

ion atmosphere *See* ion cloud.

ion cloud A slight preponderance of negative ions around a positive ion in an electrolyte, and vice versa, according to the Debye-Hückel theory. Also known as ion atmosphere.

ion detector Device for detection of presence or concentration of liquid solution ions, such as with a pH meter or by conductimetric techniques.

ion exchange A chemical reaction in which mobile hydrated ions of a solid are exchanged, equivalent for equivalent, for ions of like charge in solution; the solid has an open, fishnetlike structure, and the mobile ions neutralize the charged, or potentially charged, groups attached to the solid matrix; the solid matrix is termed the ion exchanger.

ion-exchange chromatography A chromatographic procedure in which the stationary phase consists of ion-exchange resins which may be acidic or basic.

ion exclusion Ion-exchange resin system in which the mobile ions in the resin-gel phase electrically neutralize the immobilized charged functional groups attached to the resin, thus preventing penetration of solvent electrolyte into the resin-gel phase; used in separations where electrolyte is to be excluded from the resin, but not nonpolar materials, as the separation of salt from nonpolar glycerin.

ion-exclusion chromatography Chromatography in which the adsorbent material is saturated with the same mobile ions (cationic or anionic) as are present in the sample-carrying eluent (solvent), thus repelling the similar sample ions.

ion fractionation Separation of cations or anions from an ionic solution by use of a membrane permeable to the desired ion; equipment includes electro-dialyzers and ion-fractionation stills.

ionic bond A type of chemical bonding in which one or more electrons are transferred completely from one atom to another, thus converting the neutral atoms into electrically charged ions; these ions are approximately spherical and attract one another because of their opposite charge. Also known as electrovalent bond.

ionic conductance The contribution of a given type of ion to the total equivalent conductance in the limit of infinite dilution.

ionic crystal A crystal in which the lattice-site occupants are charged ions held together primarily by their electrostatic interaction.

ionic equilibrium The condition in which the rate of dissociation of nonionized molecules is equal to the rate of combination of the ions.

ionic equivalent conductance The contribution made by each ion species of a salt toward an electrolyte's equiviconductance.

ionic gel A gel with ionic groups attached to the structure of the gel; the groups can not diffuse out into the surrounding solution.

ionicity The ionic character of a solid.

ionic lattice The lattice of an ionic crystal.

ionic membrane Semipermeable membrane that conducts electricity; the application of an electric field to the membrane achieves an electrophoretic movement of ions through the membrane; used in electrodialysis.

ionic polymerization Polymerization that proceeds via ionic intermediates (carbonium ions or carbanions) than through neutral species (olefins or acetylenes).

ionic radii Radii which can be assigned to ions because the rapid variation of their repulsive interaction with distance makes them repel like hard spheres; these radii determine the dimensions of ionic crystals.

ionic strength A measure of the average electrostatic interactions among ions in an electrolyte; it is equal to one-half the sum of the terms obtained by multiplying the molality of each ion by its valence squared.

ionium A naturally occurring radioisotope, symbol Io, of thorium, atomic weight 230.

ionization A process by which a neutral atom or molecule loses or gains electrons, thereby acquiring a net charge and becoming an ion; occurs as the result of the dissociation of the atoms of a molecule in solution ($NaCl \rightarrow Na^+ + Cl^-$) or of a gas in an electric field ($H_2 \rightarrow 2H^+$).

ionization constant Analog of the dissociation constant, where $k = [H^+][A^-]/[HA]$; used for the application of the law of mass action to ionization; in the equation HA represents the acid, such as acetic acid.

ionization degree The proportion of potential ionization that has taken place for an ionizable material in a solution or reaction mixture.

ionization energy The amount of energy needed to remove an electron from a given kind of atom or molecule to an infinite distance; usually expressed in electron volts, and numerically equal to the ionization potential in volts.

ionization potential The energy per unit charge needed to remove an electron from a given kind of atom or molecule to an infinite distance; usually expressed in volts. Also known as ion potential.

ionized atom An atom with an excess or deficiency of electrons, so that it has a net charge.

ion kinetic energy spectrometry A spectrometric technique that uses a beam of ions of high kinetic energy passing through a field-free reaction chamber from which ionic products are collected and energy analyzed; it is a generalization of metastable ion studies in which both unimolecular and bimolecular reactions are considered.

ion mean life The average time between the ionization of an atom or molecule and its recombination with one or more electrons, or its loss of excess electrons.

ionography A type of electrochromatography involving migration of ions.

ionomer Polymer with covalent bonds between the elements of the chain, and ionic bonds between the chains.

ionomer resin A polymer which has ethylene as the major component, but which contains both covalent and ionic bonds.

ionone $C_{13}H_{20}O$ A colorless to light yellow liquid with a boiling point of 126–128°C at 12 mmHg (1600 newtons per square meter); soluble in alcohol, ether, and mineral oil; used in perfumery, flavoring, and vitamin A production. Also known as irisone.

ionophore Any of a class of compounds, generally cyclic, having the ability to carry ions across lipid barriers due to the property of cation selectivity; examples are valinomycin and nonactin.

ion potential *See* ionization potential.

ion retardation Sorbent extraction of strong electrolytes with an anion-exchange resin in which a cationic monomer has been polymerized, or vice versa.

ion scattering spectroscopy A spectroscopic technique in which a low-energy (about 1000 electronvolts) beam of inert-gas ions in directed at a surface, and the energies and scattering angles of the scattered ions are used to identify surface atoms. Abbreviated ISS.

ioxynil $C_7H_3I_2NO$ A colorless solid with a melting point of 212–213°C; used for postemergence control of seedling weeds in cereals and sports turf. Also known as 4-hydroxy-3,5-diiodobenzonitrile(4-cyano-2,6-diiodophenol).

ioxynil octanoate $C_{15}H_{17}I_2NO_2$ A waxy solid with a melting point of 59–60°C; insoluble in water; used as an insecticide for cereals and sugarcane.

IPC *See* propham.

Ir *See* iridium.

iridic chloride $IrCl_4$ A hygroscopic brownish-black mass, soluble in water and alcohol; used to analyze for nitric acid, HNO_3, and in analytical microscopic work. Also known as iridium chloride; iridium tetrachloride.

iridium A metallic element, symbol Ir, atomic number 77, atomic weight 192.2, in the platinum group; insoluble in acids, melting at 2454°C.

iridium-192 Radioactive isotope of iridium with a 75-day half-life; β and γ radiation; used in cancer treatment and for radiography of light metal castings.

iridium chloride *See* iridic chloride.

iridium tetrachloride *See* iridic chloride.

irisone *See* ionone.

iron A silvery-white metallic element, symbol Fe, atomic number 26, atomic weight 55.847, melting at 1530°C.

iron-55 Radioactive isotope of iron, symbol ^{55}Fe, with a 2.91-year half-life; highly toxic.

iron-59 Radioactive isotope of iron, symbol ^{59}Fe, 46.3-day half-life; β and γ radiation; highly toxic; used to study metallic welds, corrosion mechanisms, engine wear, and bodily functions.

iron acetate *See* ferrous acetate.

iron ammonium sulfate *See* ferric ammonium sulfate; ferrous ammonium sulfate.

iron arsenate *See* ferrous arsenate.

iron-binding protein A serum protein, such as hemoglobin, for the transport of iron ions.

iron black Fine black antimony powder used to give a polished-steel look to papier-maché and plaster of paris; made by reaction of zinc with acid solution of an antimony salt and precipitation of black antimony powder.

iron blue Ferric ferrocyanide used as blue pigment by the paint industry for permanent body and trim paints; also used in blue ink, in paper dyeing, and as a fertilizer ingredient.

iron bromide *See* ferric bromide.

iron carbonyl *See* iron pentacarbonyl.

iron chloride *See* ferric chloride; ferrous chloride.

iron citrate *See* ferric citrate.

iron dichloride *See* ferrous chloride.

irone $C_{14}H_{22}O$ A colorless liquid terpene; a component of essential oil from the orrisroot; used in perfumes.

iron ferrocyanide *See* ferric ferrocyanide.

iron fluoride *See* ferric fluoride.

iron hydroxide *See* ferric hydroxide.

iron metabolism   The chemical and physiological processes involved in absorption of iron from the intestine and in its role in erythrocytes.

iron metavanadate   *See* ferric vanadate.

iron monoxide   *See* ferrous oxide.

iron nitrate   *See* ferric nitrate.

iron nonacarbonyl $Fe_2(CO)_9$ Orange-yellow crystals that break down at 100°C to yield tetracarbonyl, slightly soluble in alcohol and acetone, almost insoluble in water, ether, and benzene. Also known as di-iron enneacarbonyl.

iron oxalate   *See* ferrous oxalate.

iron oxide   Any of the hydrated, synthetic, or natural oxides of iron: ferrous oxide, ferric oxide, ferriferous oxide.

iron oxide process   A process by which a gas is passed through iron oxide and wood shavings to remove sulfides.

iron pentacarbonyl $Fe(CO)_5$ An oily liquid that decomposes upon exposure to light, soluble in most organic solvents; used as a source of a pure iron catalyst and for magnet cores. Also known as iron carbonyl.

iron phosphate   *See* ferric phosphate.

iron-porphyrin protein   Any protein containing iron and porphyrin; examples are hemoglobin, the cytochromes, and catalase.

iron resinate   *See* ferric resinate.

iron stearate   *See* ferric stearate.

iron sulfate   *See* ferric sulfate; ferrous sulfate.

iron sulfide   *See* ferrous sulfide.

iron tetracarbonyl $Fe_3(CO)_{12}$ Dark-green lustrous crystals that break down at 140–150°C; soluble in organic solvents. Also known as tri-iron dodecacarbonyl.

isatin $C_6H_5NO_2$ An indole substituted with oxygen at carbon position 2 and 3; crystallizes as red needles that are soluble in hot water; used in dye manufacture.

isethionic acid $CH_2OH \cdot CH \cdot SO_2OH$ A water-soluble liquid, boiling at 100°C; used in the manufacture of detergents. Also known as ethylenehydrinsulfonic acid; oxyethylsulfonic acid.

iso-   1. A prefix indicating an isomer of an element in which there is a difference in the nucleus when compared to the most prevalent form of the element. 2. A prefix indicating a single branching at the end of the carbon chain.

isoactyl thioglycolate $HSCH_2COOCH_2C_7H_{15}$ A colorless liquid with a slight fruity odor and a boiling point of 125°C; used in antioxidants, insecticides, oil additives, and plasticizers.

isoalkane   An alkane with a branched chain whose next-to-last carbon atom is bonded to a single methyl group.

isoalkyl group   A group of atoms resulting from the removal of a hydrogen atom from a methyl group situated at the end of the straight-chain segment of an isoalkane.

isoalloxazine mononucleotide *See* riboflavin 5'-phosphate.

isoamyl acetate *See* amyl acetate.

isoamyl alcohol *See* isobutyl carbinol.

isoamyl benzoate $C_6H_5COOC_5H_{11}$ Colorless liquid with fruity aroma; boils at 260°C: soluble in alcohol, insoluble in water; used in flavors and perfumes. Also known as amyl benzoate.

isoamyl bromide $(CH_3)_2CHCH_2CH_2Br$ A colorless liquid with a boiling point of 120–121°C; miscible with alcohol and with ether; used in organic synthesis. Also known as 1-bromo-3-methylbutane.

isoamyl butyrate $C_5H_{11}COOC_3H_7$ A water-white liquid boiling at 150–180°C; soluble in alcohol and ether; used as a solvent and plasticizer for cellulose acetate and in flavor extracts.

isoamyl chloride $C_5H_{11}Cl$ Water-insoluble, colorless liquid boiling at 100°C; it can be any one of several compounds, such as 1-chloro-3-methylbutane, $(CH_3)_2CH(CH_2)_2Cl$, or mixtures thereof; used as a solvent, in inks, for soil fumigation, and as a chemical intermediate.

isoamyl isovalerate *See* isoamyl valerate.

isoamyl nitrite *See* amyl nitrite.

isoamyl salicylate *See* amyl salicylate.

isoamyl valerate $C_4H_9CO_2C_5H_{11}$ Clear liquid with apple aroma; boils at 204°C; soluble in alcohol and ether, insoluble in water; used in medicine and fruit flavors. Also known as apple essence; apple oil; isoamyl isovalerate.

isobar One of two or more nuclides having the same number of nucleons in their nuclei but differing in their atomic numbers and chemical properties.

isobaric spin *See* isotopic spin.

isobornyl acetate $C_{10}H_{17}OOCCH_3$ A colorless liquid with an odor of pine needles and a boiling point of 220–224°C; soluble in fixed oils and mineral oil; used in toiletries and soaps and antiseptics, and as a flavoring agent.

isobornyl thiocyanoacetate $C_{10}H_{17}OOCCH_2SCN$ An oily, yellow liquid; soluble in alcohol, benzene, chloroform, and ether; used in medicine and as an insecticide.

isobutane $(CH_3)_2CHCH_3$ A colorless, stable gas, noncorrosive to metals, nonreactive with water; boils at $-11.7°C$; used as a chemical intermediate, refrigerant, and fuel. Also known as 2-methyl propane.

isobutanol *See* isobutyl alcohol.

isobutene *See* isobutylene.

isobutyl The radical $(CH_3)_2CHCH_2-$, occurring, for example, in isobutanol (isobutyl alcohol), $(CH_3)_2CHCH_2OH$.

isobutyl acetate $C_4H_9OOCCH_3$ Colorless liquid with fruitlike aroma; soluble in alcohols, ether, and hydrocarbons, insoluble in water; boils at 116°C; used as a solvent for lacquer and nitrocellulose.

isobutyl alcohol $(CH_3)_2CHCH_2OH$ A colorless liquid that is a by-product of the synthetic production of methanol, boils at 107°C; soluble in water, ether, and alcohol; used as a solvent in paints and lacquers, in organic synthesis, and in resin coatings. Also known as isobutanol; isopropylcarbinol; 2-methyl-1-propanol.

isobutyl aldehyde $(CH_3)_2CHCHO$ Colorless, transparent liquid with pungent aroma; soluble in alcohol, insoluble in water; boils at 64°C; used as a chemical intermediate. Also known as isobutyraldehyde.

isobutyl carbinol $(CH_3)_2CH(CH_2)_2OH$ Colorless liquid with pungent taste and disagreeable aroma; soluble in alcohol and ether, slightly soluble in water; boils at 132°C; used as a chemical intermediate and solvent, and in pharmaceutical products and medicines. Also known as isoamyl alcohol; 3-methyl-1-butanol; primary isoamyl alcohol.

isobutylene $(CH_3)_2CCH_2$ Flammable, colorless, volatile liquid boiling at −7°C; easily polymerized; used in gasolines, as a chemical intermediate, and to make butyl rubber. Also known as isobutene; 2-methylpropene.

isobutyl isobutyrate $(CH_3)_2CHCOOCH_2CH(CH_3)_2$ A colorless liquid with a fruity odor and a boiling point of 148.7°C; soluble in alcohol and ether; used for flavoring and as an insect repellent. Abbreviated IBIB.

isobutyraldehyde   *See* isobutyl aldehyde.

isobutyric acid $(CH_3)_2CHCOOH$ Colorless liquid boiling at 154°C; soluble in water, alcohol, and ether; used as a chemical intermediate and disinfectant, in flavor and perfume bases, and for leather treating.

isobutyryl $(CH_3)_2C\cdot CHO$ The radical group from isobutyric acid, $(CH_3)_2CHCOOH$.

isocarb   A line on a map that connects points of equal content of fixed carbon in coal.

isocetyl laurate $C_{11}H_{23}COOC_{16}H_{33}$ An oily, combustible liquid, soluble in most organic solvents; used in cosmetics and pharmaceuticals and as a plasticizer and textile softener.

isocitric acid $HOOCCH_2CH(COOH)CH(OH)COOH$ An isomer of citric acid that is involved in the Krebs tricarboxylic acid cycle in bacteria and plants. Also known as 2-hydroxy-1,2,3-propanetricarboxylic acid.

isoconcentration   Constant concentration values.

isoconcentration map   Map or diagram or a liquid or gas system's concentration with respect to a single component of the system, shown by constant-concentration contour lines.

isocracking   A hydrocracking process for conversion of hydrocarbons into more valuable, lower-boiling products; operates at relatively low temperatures and pressures in the presence of hydrogen and a catalyst.

isocyanate   **1.** One of a group of neutral derivatives of primary amines; its formula is R—N=C=O, where R may be an alkyl or aryl group; an example is 2,4-toluene diisocyanate. **2.** Any compound containing the isocyanato functional group.

isocyanate resin   A linear alkyd resin lengthened by reaction with isocyanates, then treated with a glycol or diamine to cross-link the molecular chain; the product has good abrasion resistance.

isocyanato group   A functional group (—N=C—O) which forms isocyanates by replacing the hydrogen atom of a hydrocarbon.

isocyanic acid $HN{=}C{=}O$ One of two forms of cyanic acid; a gas used as an intermediate in the preparation of polyurethane and other resins.

isocyanide   A compound with the general formula $RN{\equiv}C$ in which the hydrogen of a hydrocarbon has been replaced by the $-N{\equiv}C$ group.

isocyanine   Any one of a series of dyes whose structure has two heterocyclic or quinoline rings connected by an odd number chain of carbon atoms containing conjugated double bonds; for example, cyanine blue.

isocyanuric acid   See fulminuric acid.

isocyclic compound   A compound in which the ring structure is made up of one kind of atom.

isodecyl chloride $C_{10}H_{21}Cl$ A colorless liquid with a boiling point of 210.6°C; used as a solvent and in extractants, cleaning compounds, pharmaceuticals, insecticides, and plasticizers.

isodiaspheres   Nuclides which have the same difference in the number of neutrons and protons.

isodulcitol   See rhamnose.

isoelectric focusing   Protein separation technique in which a mixture of protein molecules is resolved into its components by subjecting the mixture to an electric field in a supporting gel having a previously established pH gradient. Also known as electrofocusing.

isoelectric point   The pH value of the dispersion medium of a colloidal suspension at which the colloidal particles do not move in an electric field.

isoelectric precipitation   Precipitation of materials at the isoelectric point (the pH at which the net charge on a molecule in solution is zero); proteins coagulate best at this point.

isoelectronic   Pertaining to atoms having the same number of electrons outside the nucleus of the atom.

isoelectronic sequence   A set of spectra produced by different chemical elements ionized so that their atoms or ions contain the same number of electrons.

isoenzyme   Any of the electrophoretically distinct forms of an enzyme, representing different polymeric states but having the same function. Also known as isozyme.

isoeugenol $C_{10}H_{12}O_2$ An oily liquid prepared from eugenol by heating, slightly soluble in water; used in the manufacture of vanillin.

isoeugenol acetate   See acetylisoeugenol.

isoflavone $C_{15}H_{10}O_2$ A colorless, crystalline ketone, occurring in many plants, generally in the form of a hydroxy derivative.

isoforming   A petroleum refinery process in which olefinic naphtha is contacted with an alumina catalyst at high temperature and low pressure to produce isomers of higher octane number.

isohexane C_6H_{14} A liquid mixture of isomeric hydrocarbons, flammable and explosive, insoluble in water, soluble in most organic solvents, boils at 54–61°C; used as a solvent, freezing-point depressant, and chemical intermediate.

isohydric   Referring to a set of solutions with the same hydrogen ion concentration and not affecting the conductivity of each of the various solutions on mixing.

isolate   To separate two portions of a process system by means of valving or line blanks; used as safety measure during maintenance or repair, or to redirect process flows.

isolation   Separation of a pure chemical substance from a compound or mixture; as in distillation, precipitation, or absorption.

isoleucine $C_6H_{13}O_2$ An essential monocarboxylic amino acid occurring in most dietary proteins.

isomer   **1.** One of two or more chemical substances having the same elementary percentage composition and molecular weight but differing in structure, and therefore in properties; there are many ways in which such structural differences occur; one example is provided by the compounds n-butane, $CH_3(CH_2)_2CH_3$, and isobutane, $CH_3CH(CH_3)_2$. **2.** One of two or more nuclides having the same mass number and atomic number, but existing for measurable times in different quantum states with different energies and radioactive properties.

isomerase   An enzyme that catalyzes isomerization reactions.

isomeric shift   Shift in the Mössbauer resonance caused by the effect of the valence of the atom on the interaction of the electron density at the nucleus with the nuclear charge. Also known as chemical shift.

isomeric transition   A radioactive transition from one nuclear isomer to another of lower energy.

isomerism   **1.** The phenomenon whereby certain chemical compounds have structures that are different although the compounds possess the same elemental composition. **2.** The occurrence of nuclear isomers.

isomerization   A process whereby a compound is changed into an isomer; for example, conversion of butane into isobutane.

isometric system   The crystal system in which the forms are referred to three equal, mutually perpendicular axes. Also known as cubic system.

isomorphism   A condition present when an ion at high dilution is incorporated by mixed crystal formation into a precipitate, even though such formation would not be predicted on the basis of crystallographic and ionic radii; an example is coprecipitation of lead with potassium chloride.

isonicotinic acid $C_6H_5NO_2$ White platelets or powder, slightly soluble in water, sublimes at 260°C; used in the manufacture of isonicotinic acid hydrazide, an antitubercular agent. Also known as pyridine-4-carboxylic acid.

isonicotinic acid hydrazide   *See* isoniazid.

isonitrosoacetophenone $C_8H_7NO_2$ Platelike crystals with a melting point of 126–128°C; soluble in alkalies and alkali carbonates; used to detect ferrous ions and palladium. Also known as benzoylformaldoxime.

isooctane $(CH_3)_2CHCH_2C(CH_3)_3$ Flammable, colorless liquid boiling at 99°C; slightly soluble in alcohol and ether, insoluble in water; used in motor fuels and as a chemical intermediate. Also known as 2,2,4-trimethylpentane.

isooctyl alcohol $C_7H_{15}CH_2OH$ Mixture of isomers from oxo-process synthesis; boils at 182–195°C; used as a chemical intermediate, resin solvent, emulsifier, and antifoaming agent.

isoparaffin A branched-chain version of a straight-chain (normal) saturated hydrocarbon; for example, isooctane, or 2,2,4-trimethyl pentane, $(CH_3)_3C_5H_9$, is the branched-chain version of n-octane, $CH_3(CH_2)_6CH_3$.

isopentane $CH_3CHCH_3CH_2CH_3$ Flammable, colorless liquid with pleasant aroma; boils at 28°C; soluble in oils, ether, and hydrocarbons, insoluble in water; used as a solvent and chemical intermediate. Also known as 2-methylbutane.

isopentanoic acid C_4H_9COOH A colorless, combustible liquid with a boiling point of 183.2°C; used for manufacture of plasticizers, pharmaceuticals, and synthetic lubricants.

isophorone $COCHC(CH_3)CH_2C(CH_3)_2CH_2$ A water-white liquid boiling at 215°C; used as a solvent for lacquers and polyvinyl and nitrocellulose resins.

isophthalic acid $C_6H_4(COOH)_2$ Colorless crystals subliming at 345°C; slightly soluble in water, soluble in alcohol and acetic acid, and insoluble in benzene; used as an intermediate for polyester and polyurethane resins, and as a plasticizer. Also known as *meta*-phthalic acid.

isopolymolybdate A class of compounds formed by the acidification of a molybdate solution, or in some cases by heating normal molybdates.

isopolytungstate A compound formed by the condensation of tungstate compounds, usually classified into metatungstates, such as $Na_6W_{12}O_{40}\cdot xH_2O$, and paratungstates, such as $Na_{10}W_{12}O_{41}\cdot xH_2O$.

isoprene C_5H_8 A conjugated diolefin; a mobile, colorless liquid having a boiling point of 34.1°C; insoluble in water, soluble in alcohol and ether; polymerizes readily to form dimers and high-molecular-weight elastomer resins.

isopropaline $C_{15}H_{23}N_3O_4$ An orange liquid with limited solubility in water; used as a preemergence herbicide for control of grass and broadleaf weeds on tobacco. Also known as 2,6-dinitro-N,N-dipropylcumidine.

isopropanol *See* isopropyl alcohol.

isopropanolamine $CH_3CH(OH)CH_2NH_2$ A combustible liquid with a faint ammonia odor and a boiling point of 159.9°C; soluble in water; used as an emulsifying agent and for dry-cleaning soaps, wax removers, cosmetics, plasticizers, and insecticides. Also known as 1-amino-2-propanol; 2-hydroxypropylamine.

isopropenyl acetate $CH_3CO_2C(CH_3){=}CH_2$ A liquid with a boiling point of 97°C; used for acylation of potential enols. Also known as 1-propen-2-yl acetate.

2-isopropoxyphenyl *N*-methylcarbamate $C_{11}H_{15}O_3N$ A colorless solid with a melting point of 91°C; used as an insecticide for cockroaches, flies, mosquitoes, and lawn insects.

isopropyl The radical $(CH_3)_2CH$, from isopropane; an example of its occurrence is in isopropyl alcohol, $(CH_3)_2CHOH$.

isopropyl acetate $CH_3COOCH(CH_3)_2$ A colorless, aromatic liquid with a boiling point of 89.4°C; used as a solvent and for paints and printing inks.

isopropyl alcohol $(CH_3)_2CHOH$ A colorless liquid that boils at 82.4°C; soluble in water, ether, and ethanol; used in manufacturing of acetone and its derivatives, of glycerol, and as a solvent. Also known as isopropanol; 2-propanol; *sec*-propyl alcohol.

isopropylamine $(CH_3)_2CHNH_2$ A volatile, colorless liquid with a boiling point of 32.4°C; used as a solvent and in the manufacture of pharmaceuticals, dyes, insecticides, and bactericides. Also known as 2-aminopropane.

isopropyl arterenol   *See* isoproterenol.

isopropylbenzene   *See* cumene.

isopropyl-2-(*N*-benzoyl-3-chloro-4-fluoroanilino)propionate $C_{19}H_{19}O_3NClF$ Off-white crystals with a melting point of 56–57°C; used as a postemergence herbicide for wild oats and barley.

isopropylcarbinol   *See* isobutyl alcohol.

isopropyl *meta*-chlorocarbanilate   *See* chlorpropham.

isopropyl-*meta*-cresol   *See* thymol.

isopropyl-*ortho*-cresol   *See* carvacrol.

isopropyl 4,4′-dibromobenzilate $C_{17}H_{16}\,O_3Br_2$ A brownish solid with a melting point of 77°C; solubility in water is less than 0.5 part per million at 20°C; used as a miticide for deciduous fruit and citrus.

isopropyl 4,4′-dichlorobenzilate $C_{17}H_{16}\,O_3Cl_2$ A white powder with a melting point of 70–72°C; solubility in water is less than 10 parts per million at 20°C; used as a miticide for spider mites on apple and pear trees.

isopropyl ether $(CH_3)_2CHOCH(CH_3)_2$ Water-soluble, flammable, colorless liquid with etherlike aroma; boils at 68°C; used as a solvent and extractant, in paint and varnish removers, and in spotting formulas. Also known as diisopropyl ether.

isopropylideneacetone   *See* mesityl oxide.

isopropylidene glycerol   *See* 2,2-dimethyl-1,3-dioxolane-4-methanol.

2-isopropyl-5-methylbenzoquinone   *See para*-thymoquinone.

4-isopropyl-1-methylcyclohexane   *See* menthane.

7-isopropyl-1-methyl phenanthrene   *See* retene.

isopropyl-*N*-phenylcarbamate   *See* propham.

***N*-4-isopropylphenyl-*N*′,*N*′-dimethylurea** $(CH_3)_2CHC_6H_4NHCON(CH_3)_2$ A crystalline solid with a melting point of 151–153°C; solubility in water is 170 parts per million; used as an herbicide for wheat, barley, and rye.

***ortho*-isopropylphenyl-methylcarbamate** $C_{11}H_{15}O_2N$ A white, crystalline compound with a melting point of 88–89°C; used as an insecticide for rice and cacao crops. Also known as MIPC.

isopulegol $C_{10}H_{17}OH$ An alcohol derived from terpene as a water-white liquid that has a mintlike odor; used in making perfumes.

isopurpurin   *See* anthrapurpurin.

isoquinoline $C_6H_4CHNCHCH$ Colorless liquid boiling at 243°C; soluble in most organic solvents and dilute mineral acids, insoluble in water; derived from coal tar or made synthetically; used to make dyes, insecticides, pharmaceuticals, and rubber accelerators, and as a chemical intermediate.

isosafrole $C_{10}H_{10}O_2$ A liquid with the odor of anise that is obtained from safrole, and that boils at 253°C; used to make perfumes and flavors. Also known as 1,2-methylenedioxy-4-propenylbenzene.

isospin *See* isotopic spin.

isosteric Referring to similar electronic arrangements in chemical compounds.

isosterism A similarity in the physical properties of ions, compounds, or elements, as a result of electron arrangements that are identical or similar.

isostructural Pertaining to crystalline materials that have corresponding atomic positions, and have a considerable tendency for ionic substitution.

isosynthesis A process in which mixtures of hydrogen and carbon monoxide are reacted over a thorium oxide catalyst (sometimes mixed with additional substances) to produce branched hydrocarbons.

isotachophoresis A variant of electrophoresis in which ionic species move with equal velocity in the presence of an electric field.

isotactic Designating crystalline polymers in which substituents in the asymmetric carbon atoms have the same (rather than random) configuration in relation to the main chain.

isothiocyanate A compound of the type $R-N=C=S$, where R may be an alkyl or aryl group; an example is mustard oil. Also known as sulfocarbimide.

isotone One of several nuclides having the same number of neutrons in their nuclei but differing in the number of protons.

isotope One of two or more atoms having the same atomic number but different mass number.

isotope abundance The ratio of the number of atoms of a particular isotope in a sample of an element to the number of atoms of a specified isotope, or to the total number of atoms of the element.

isotope-dilution analysis Variation on paper-chromatography analysis; a labeled radioisotope of the same type as the one being quantitated is added to the solution, then quantitatively analyzed afterward via radioactivity measurement.

isotope effect The effect of difference of mass between isotopes of the same element on nonnuclear physical and chemical properties, such as the rate of reaction or position of equilibrium, of chemical reactions involving the isotopes.

isotope exchange reaction A chemical reaction in which interchange of the atoms of a given element between two or more chemical forms of the element occurs, the atoms in one form being isotopically labeled so as to distinguish them from atoms in the other form.

isotope shift A displacement in the spectral lines due to the different isotopes of an element.

isotopic element An element which has more than one naturally occurring isotope.

isotopic indicator *See* isotopic tracer.

isotopic label *See* isotopic tracer.

isotopic number *See* neutron excess.

isotopic spin A quantum-mechanical variable, resembling the angular momentum vector in algebraic structure whose third component distinguished between members of groups of elementary particles, such as the nucleons, which apparently behave in the same way with respect to strong nuclear forces, but have different charges. Also known as isobaric spin; isospin; i-spin.

isotopic tracer   An isotope of an element, either radioactive or stable, a small amount of which may be incorporated into a sample material (the carrier) in order to follow the course of that element through a chemical, biological, or physical process, and also follow the larger sample. Also known as isotopic indicator; isotopic label; label; tag.

isotypic   Pertaining to a crystalline substance whose chemical formula is analogous to, and whose structure is like, that of another specified compound.

isovalent conjugation   An arrangement of bonds in a conjugated molecule such that alternative structures with an equal number of bonds can be written; an example occurs in benzene.

isovalent hyperconjugation   An arrangement of bonds in a hyperconjugated molecule such that the number of bonds is the same in the two resonance structures but the second structure is energetically less favorable than the first structure; examples are $H_3 \equiv C—C^+ H_2$ and $H_3 \equiv C—CH_2$.

isovaleral   *See* isovaleraldehyde.

isovaleraldehyde $(CH_3)_2CHCH_2CHO$ A colorless liquid with an applelike odor and a boiling point of 92°C; soluble in alcohol and ether; used in perfumes and pharmaceuticals and for flavoring. Also known as isovaleral; isovaleric aldehyde; 3-methylbutyraldehyde.

isovaleric acid $(CH_3)_2CHCH_2COOH$ Colorless liquid with disagreeable taste and aroma; boils at 176°C; soluble in alcohol and ether; found in valeriana, hop, tobacco, and other plants; used in flavors, perfumes, and medicines.

isovaleric aldehyde   *See* isovaleraldehyde.

2-isovaleryl-1,3-indandione $C_{14}H_{14}O_3$ A yellow, crystalline compound with a melting point of 67–68°C; insoluble in water; used as a rodenticide.

isoxathion $C_{13}H_{16}NO_4PS$ A yellow liquid, insoluble in water; used as an insecticide for crops such as rice, citrus, vegetables, and tobacco. Also known as O,O-diethyl-O-(5-phenyl-3-isoxazolyl)phosphorothioate.

isozyme   *See* isoenzyme.

i-spin   *See* isotopic spin.

ISS   *See* ion scattering spectroscopy.

itaconic acid $CH_2{:}C(COOH)CH_2COOH$ A colorless crystalline compound that decomposes at 165°C, prepared by fermentation with *Aspergillus terreus;* used as an intermediate in organic synthesis and in resins and plasticizers. Also known as methylene succinic acid.

itatartaric acid $C_5H_8O_6$ A compound produced experimentally by fermentation; formed as a minor product, 5.8% of total acidity produced, of an itaconic–acid producing strain of *Aspergillus niger.*

ium ion   A positively charged group of atoms in which a charged nonmetallic ion other than carbon or silicon possesses a closed-shell electron configuration; often joined to a root word, as in carbonium ion.

Ivanov reagent   A reagent that is similar to a Grignard reagent, and that is formed by reacting an arylacetic acid or its sodium salt with isopropyl magnesium halide.

J

J acid *See* 2-amino-5-naphthol-7-sulfonic acid.

Jacquemart's reagent Analytical reagent used to test for ethyl alcohol; consists of an aqueous solution of mercuric nitrate and nitric acid.

Jahn-Teller effect The effect whereby, except for linear molecules, degenerate orbital states in molecules are unstable.

Janecke coordinates Use of a rectangular or Ponchon-type diagram to plot the solvent content of liquid-liquid equilibrium phases; used for solvent-extraction design calculations.

jasmone $C_{11}H_{16}O$ A liquid ketone found in jasmine oil and other essential oils from plants.

Jeremiassen crystallizer Device used to grow solid crystals in a supersaturated liquid solution and to separate them from it.

jeweler's rouge *See* ferric oxide.

j-j coupling A process for building up many-electron wave functions; the spin and orbital functions of each particle are combined to form eigenfunctions of the particle's total angular momentum, and then the wave functions of all the particles are combined to form eigenfunctions of the total angular momentum of the system; this coupling is used when the spin-orbit interaction is strong compared to the electrostatic interaction.

jodfenphos $C_8H_8O_3Cl_2IPS$ A crystalline compound with a melting point of 76°C; slight solubility in water; used as an insecticide in homes, farm buildings, and industrial sites. Also known as *O,O*-dimethyl-*O*-(2,5-dichloro-4-iodophenyl)thiophosphate; iodofenphos.

Johann crystal geometry The focusing shape of a diffracting crystal for x-ray dispersion used in electron-probe microanalysis; less stringent than Johannson crystal geometry.

Johannson crystal geometry The full-focusing shape of a diffracting crystal for x-ray dispersion used in electron-probe microanalyzers; more stringent than Johann crystal geometry.

Jones reductor A device used to chemically reduce solutions, such as ferric salt solutions, consisting of a vertical tube containing granular zinc into which the solution is poured.

juniperic acid $C_{16}H_{32}O_3$ A crystalline hydroxy acid that melts at 95°C, obtained from waxy exudations from conifers.

K *See* potassium.

K acid $C_{10}H_4NH_2OH(SO_3H)_2$ An acid derived from naphthylamine trisulfonic acid; used in dye manufacture. Also known as 1-amino-8-naphthol-4,6-disulfonic acid; 8-amino-1-naphthol-3,5-disulfonic acid.

K-A decay Radioactive decay of potassium-40 (^{40}K) to argon-40 (^{40}A), as the nucleus of potassium captures an orbital electron and then decays to argon-40; the ratio of ^{40}K to ^{40}A is used to determine the age of rock (K-A age).

kaempferol A flavonoid with a structure similar to that of quercetin but with only one hydroxyl in the B ring; acts as an enzyme cofactor and causes growth inhibition in plants.

kalium *See* potassium.

kallidin I *See* bradykinin.

kaonic atom An atom consisting of a negatively charged kaon orbiting around an ordinary nucleus.

Kapitza expander Reciprocating-piston gas expander used for helium liquefaction; relies on close fit rather than packing or rings on the pistons.

karbutilate $C_{14}H_{21}N_3O_3$ An off-white solid with a melting point of 176–177°C; used as an herbicide on noncroplands, railroad rights-of-way, and plant sites. Also known as *meta*-(3,3-dimethylureido)phenyl-*tert*-butylcarbamate.

Karl Fischer reagent A solution of 8 moles pyridine to 2 moles sulfur dioxide, with the addition of about 15 moles methanol and then 1 mole iodine; used to determine trace quantities of water by titration.

Karl Fischer technique A method of determining trace quantities of water by titration; the Karl Fischer reagent is added in small increments to a glass flask containing the sample until the color changes from yellow to brown or a change in potential is observed at the end point.

Karrer method An industrial method for the chemical synthesis of riboflavin.

kauri-butanol value The measure of milliliters of paint or varnish petroleum thinner needed to cause cloudiness in a solution of kauri gum in butyl alcohol.

kayser A unit of reciprocal length, especially wave number, equal to the reciprocal of 1 centimeter. Also known as rydberg.

K capture A type of beta interaction in which a nucleus captures an electron from the K shell of atomic electrons (the shell nearest the nucleus) and emits a neutrino.

Keesom force *See* orientation force.

Keesom relationship An equation for the potential energy associated with the interaction of the dipole moments of two polar molecules.

Kekulé structure A molecular structure of a cyclic conjugated system that is depicted with alternating single and double bonds.

Keldysh theory A theory of multiphoton ionization, in which an atom is ionized by rapid absorption of a sufficient number of photons; it predicts that the ionization rate depends primarily upon the ratio of the mean binding electric field to the peak strength of the incident electromagnetic field, and upon the ratio of the binding energy to the energy of photons in the field.

K electron An electron in the K shell.

keratin Any of various albuminoids characteristic of epidermal derivatives, such as nails and feathers, which are insoluble in protein solvents, have a high sulfur content, and generally contain cystine and arginine as the predominating amino acids.

kernel An atom that has been stripped of its valence electrons, or a positively charged nucleus lacking the outermost orbital electrons.

ketal 1. Former term for the $=CO$ group, as in dimethyl ketal (acetone). 2. Any of the ketone acetates from condensation of alkyl orthoformates with ketones in the presence of alcohols.

ketene C_2H_2O A colorless, toxic, highly reactive gas, with disagreeable taste; boils at $-56°C$; soluble in ether and acetone, and decomposes in water and alcohol; used as an acetylating agent in organic synthesis.

ketene lamp An electrically heated Chromel filament by the means of which acetone is hydrolyzed to produce ketene.

ketimide A compound that is represented by $R_2:C:NX$, where X is an acyl radical.

ketimine An organic compound that contains the divalent group $>C=NH$; a Schiff base is an example.

keto- Organic chemical prefix for the keto or carbonyl group, $C:O$, as in a ketone.

keto acid A compound that is both an acid and a ketone; an example is β-acetoacetic acid.

ketoadipic acid $C_6H_8O_5$ An intermediate product in the metabolism of lysine to glutaric acid.

β-ketobutyranilide *See* acetoacetic acid.

ketogenesis Production of ketone bodies.

ketogenic hormone A factor originally derived from crude anterior hypophysis extract which stimulated fatty-acid metabolism; now known as a combination of adrenocorticotropin and the growth hormone. Also known as fat-metabolizing hormone.

ketogenic substance Any foodstuff which provides a source of ketone bodies.

ketoglutaric acid $C_5H_6O_5$ A dibasic keto acid occurring as an intermediate product in carbohydrate and protein metabolism.

ketohexamethylene *See* cyclohexanone.

ketohexose Any monosaccharide composed of a six-carbon chain and containing one ketone group.

ketolase A type of enzyme that catalyzes cleavage of carbohydrates at the carbonyl carbon position.

ketolysis Dissolution of ketone bodies.

ketone One of a class of chemical compounds of the general formula RR'CO, where R and R' are alkyl, aryl, or heterocyclic radicals; the groups R and R' may be the same or different, or incorporated into a ring; the ketones, acetone, and methyl ethyl ketone are used as solvents, and ketones in general are important intermediates in the synthesis of organic compounds.

ketone body Any of various ketones which increase in blood and urine in certain conditions, such as diabetic acidosis, starvation, and pregnancy. Also known as acetone body.

α-ketopropionic acid *See* pyruvic acid.

ketose A carbohydrate that has a ketone group.

ketosteroid One of a group of neutral steroids possessing keto substitution, which produces a characteristic red color with m-dinitrobenzene in an alkaline solution; these compounds are principally metabolites of adrenal cortical and gonadal steroids.

γ-ketovaleric acid *See* levulinic acid.

kettle reboiler Tube-and-shell heat exchange device in which liquid is vaporized on the shell side from heat transferred from hot liquid flowing through the tubes; dome space allows liquid-vapor separation above the tube bundle.

Keyes process A distillation process used to obtain absolute alcohol; benzene is added to a constant-boiling 95% alcohol–water solution, and on distillation anhydrous alcohol leaves the bottom of the column.

Kikuchi lines A pattern consisting of pairs of white and dark parallel lines, obtained when an electron beam is scattered (diffracted) by a crystalline solid; the pattern gives information on the structure of the crystal.

Kiliani reaction A method of synthesizing a higher aldose from a lower aldose; monosaccharides, such as aldehydes and ketones, react with hydrogen cyanide to form cyanohydrins, which are hydrolyzed to hydroxy acids, converted to lactones, and reduced to aldoses with sodium amalgams.

kilogram-equivalent weight A unit of mass 1000 times the gram-equivalent weight.

kinase Any enzyme that catalyzes phosphorylation reactions.

kinetin $C_{10}H_9ON_5$ A cytokinin formed in many plants which has a stimulating effect on cell division.

king's blue *See* cobalt blue.

Kistiakowsky-Fishtine equation An equation to calculate latent heats of vaporization of pure compounds; useful when vapor pressure and critical data are not available.

kitol $C_{40}H_{60}O_2$ One of the provitamins of vitamin A derived from whale liver oil; crystallizes from methanol solution.

Kjeldahl method Quantitative analysis of organic compounds to determine nitrogen content by interaction with concentrated sulfuric acid; ammonia is distilled from the NH_4SO_4 formed.

Klein-Rydberg method A method for determining the potential energy function of the distance between the nuclei of a diatomic molecule from the molecule's vibrational and rotational levels.

Klein's reagent Saturated solution of borotungstate; used to separate minerals by specific gravity.

K/L ratio The ratio of the number of internal conversion electrons emitted from the K shell of an atom during de-excitation of a nucleus to the number of such electrons emitted from the L shell.

knockout vessel A vessel, drum, or trap used to remove fluid droplets from flowing gases.

Knoevenagel reaction The condensation of aldehydes with compounds containing an activated methylene ($=CH_2$) group.

Knorr synthesis A condensation reaction carried out in either glacial acetic acid or an aqueous alkali in which an α-aminoketone combines with an α-carbonyl compound to form a pyrrole; possibly the most versatile pyrrole synthesis.

knot A chiral structure in which rings containing 50 or more members have a knotlike configuration.

Knudsen cell A vessel used to measure very low vapor pressures by measuring the mass of vapor which escapes when the vessel contains a liquid in equilibrium with its vapor.

Knudsen-Langmuir equation Relationship of molecular distillation rate to vapor saturation pressure, solution temperature, and molecular weight during evaporation and no-recycle condensation.

Kohlrausch method A method of measuring the electrolytic conductance of a solution using a Wheatstone bridge.

Kojic acid $C_6H_6O_4$ A crystalline antibiotic with a melting point of 152–154°C; soluble in water, acetone, and alcohol; used in insecticides and as an antifungal and antimicrobial agent.

Kolbe hydrocarbon synthesis The production of an alkane by the electrolysis of a water-soluble salt of a carboxylic acid.

Kolbe-Schmitt synthesis The reaction of carbon dioxide with sodium phenoxide at 125°C to give salicyclic acid.

Konowaloff rule An empirical rule which states that in the vapor over a liquid mixture there is a higher proportion of that component which, when added to the liquid, raises its vapor pressure, than of other components.

Kopp's law The law that for solids the molal heat capacity of a compound at room temperature and pressure approximately equals the sum of heat capacities of the elements in the compound.

Korner's method A method for determining the absolute position of substituents for positional isomers in benzene by the experimental production of positional isomers from a given disubstituted benzene.

Korshun method   Microdetermination of carbon and hydrogen in organic compounds; the sample is prepyrolyzed (cracked) in a shortage of oxygen, then oxidized in an excess of oxygen.

Kossel-Sommerfeld law   The law that the arc spectra of the atom and ions belonging to an isoelectronic sequence resemble each other, especially in their multiplet structure.

Kovat's retention indexes   Procedure to identify compounds in gas chromatography; the behavior of a compound is indicated by its position on a scale of normal alkane values (for example, methane = 100, ethane = 200).

Kr   *See* krypton.

kraft process   *See* sulfate pulping.

kraft pulping   *See* sulfate pulping.

Krebs cycle   A sequence of enzymatic reactions involving oxidation of a two-carbon acetyl unit to carbon dioxide and water to provide energy for storage in the form of high-energy phosphate bonds. Also known as citric acid cycle; tricarboxylic acid cycle.

Krebs-Henseleit cycle   A cyclic reaction pathway involving the breakdown of arginine to urea in the presence of arginase.

Kremser formula   Equation for calculating distillation-column material balances and equilibrium, assuming the ideal distribution law, that is, the concentrations in the two phases (vapor and liquid) are proportional to each other.

krypton   A colorless, inert gaseous element, symbol Kr, atomic number 36, atomic weight 83.80; it is odorless and tasteless; used to fill luminescent electric tubes.

krypton-86   An isotope of krypton, atomic mass 86; used in measurement of the standard meter.

K shell   The innermost shell of electrons surrounding the atomic nucleus, having electrons characterized by the principal quantum number 1.

Kundt rule   The rule that the optical absorption bands of a solution are displaced toward the red when its refractive index increases because of changes in composition or other causes.

Kurie plot   Graph used in studying beta decay, in which the square root of the number of beta particles whose momenta (or energy) lie within a certain narrow range, divided by a function worked out by Fermi, is plotted against beta-particle energy; it is a straight line for allowed transitions and some forbidden transitions, in accord with the Fermi beta-decay theory. Also known as Fermi plot.

Kurrol's salt $NaPO_3(IV)$ A crystalline high-temperature form of sodium phosphate made by seeding a melt at 550°C.

kyanize   To saturate wood with mercuric chloride as a decay preventive.

kynurenic acid $C_{10}H_7O_3N$ A product of tryptophan metabolism found in the urine of mammals.

kynurenine $C_{10}H_{12}O_3N_2$ An intermediate product of tryptophan metabolism occurring in the urine of mammals.

L

La *See* lanthanum.

labile factor *See* proaccelerin.

laccal $C_{17}H_{31}C_6H_3(OH)_2$ A phenol compound which is found in the sap of lacquer trees, and which can be isolated in crystalline form.

laccase Any of a class of plant oxidases which catalyze the oxidation of phenols.

lachesne $C_{20}H_{26}ClNO_3$ A compound that crystallizes from a solution of ethanol and acetone, and whose melting point is 213°C; used in ophthalmology. Also known as chloride benzilate.

lactalbumin A simple protein contained in milk which resembles serum albumin and is of high nutritional quality.

lactam An internal (cyclic) amide formed by heating gamma (γ) and delta (δ) amino acids; thus γ-aminobutyric acid readily forms γ-butyrolactam lactam (pyrrolidone); many lactams have physiological activity.

lactase An enzyme that catalyzes the hydrolysis of lactose to dextrose and galactose.

lactate A salt or ester of lactic acid in which the acidic hydrogen of the carboxyl group has been replaced by a metal or an organic radical.

lactate dehydrogenase A zinc-containing enzyme which catalyzes the oxidation of several α-hydroxy acids to corresponding α-keto acids.

lactic acid $C_3H_6O_3$ A hygroscopic α-hydroxy acid, occurring in three optically isomeric forms: L form, in blood and muscle tissue as a product of glucose and glycogen metabolism; D form, obtained by fermentation of sucrose; and DL form, a racemic mixture present in foods prepared by bacterial fermentation, and also made synthetically. Also known as 2-hydroxypropanoic acid; α-hydroxypropanoic acid.

lactic dehydrogenase An enzyme that catalyzes the dehydrogenation of L-lactic acid to pyruvic acid. Abbreviated LDH.

lactide A cyclic, intermolecular, double ester formed from α-hydroxy acids; most lactides are relatively low melting solids and are easily hydrolyzed by base to form salts of the parent acid, such as sodium lactate.

lactim A tautomeric enol form of a lactam with which it forms an equilibrium whenever the lactam nitrogen carries a free hydrogen.

lactin *See* lactose.

lactoferrin An iron-binding protein found in milk, saliva, tears, and intestinal and respiratory secretions that interferes with the iron metabolism of bacteria; in conjunction with antibodies, it plays an important role in resistance to certain infectious diseases.

lactoflavin   *See* riboflavin.

lactogenic hormone   *See* prolactin.

lactoglobulin   A crystalline protein fraction of milk, which is soluble in half-saturated ammonium sulfate solution and insoluble in pure water.

lactonase   The enzyme that catalyzes the hydrolysis of 6-phosphoglucono-Δ-lactone to 6-phosphogluconic acid in the pentose phosphate pathway.

lactone   An internal cyclic mono ester formed by gamma (γ) or delta (δ) hydroxy acids spontaneously; thus γ-hydroxybutyric acid forms γ-butyrolactone.

lactonitrile $CH_3CHOHCN$ A straw-colored liquid boiling at 183°C; soluble in water, insoluble in carbon disulfide and petroleum ether; used as a solvent, and as a chemical intermediate in making esters of lactic acid. Also known as acetaldehyde cyanohydrin; α-hydroxypropionitrile.

lactophenene   *See para*-lactophenetide.

lactose $C_{12}H_{22}O_{11}$ A disaccharide composed of D-glucose and D-galactose which occurs in milk. Also known as lactin; milk sugar.

lactyl-*para*-phenetiden   *See para*-lactophenetide.

Ladenburg f value   *See* oscillator strength.

lambda sulfur   One of the two components of plastic (or gamma) sulfur; soluble in carbon disulfide.

Lambert-Beer law   *See* Bouguer-Lambert-Beer law.

Lambert's law   *See* Bouguer-Lambert law.

Lamb shift   A small shift in the energy levels of a hydrogen atom, and of hydrogenlike ions, from those predicted by the Dirac electron theory, in accord with principles of quantum electrodynamics.

lamellar crystal   A polycrystalline substance whose grains are in the form of thin sheets.

lanatoside   Any of three natural glycosides from the leaves of *Digitalis lanata*; on hydrolysis with acid, it yields one molecule of D-glucose, three molecules of digitoxose, and one molecule of acetic acid; all three glycosides are cardioactive.

Landé g factor   1. The negative ratio of the magnetic moment of an electron or atom, in units of the Bohr magneton, to its angular momentum, in units of Planck's constant divided by 2π. 2. The ratio of the difference in energy between two energy levels which differ only in magnetic quantum number to the product of the Bohr magneton, the applied magnetic field, and the difference between the magnetic quantum numbers of the levels; identical to the first definition for free atoms. Also known as Landé splitting factor; spectroscopic splitting factor. 3. The ratio of the magnetic moment of a nucleon, in units of the nuclear magneton, to its angular momentum in units of Planck's constant divided by 2π.

Landé interval rule   The rule that when the spin-orbit interaction is weak enough to be treated as a perturbation, an energy level having definite spin angular momentum and orbital angular momentum is split into levels of differing total angular momentum, so that the interval between successive levels is proportional to the larger of their total angular momentum values.

Landé Γ-permanence rule The rule that the sum of the shifts of energy levels produced by the spin-orbit interaction, over a series of states having the same spin and orbital angular momentum quantum numbers (or the same total angular momentum quantum numbers for individual electrons) but different total angular momenta, and having the same total magnetic quantum number, is independent of the strength of an applied magnetic field.

Landé splitting factor *See* Landé g factor.

Langmuir isotherm equation An equation, useful chiefly for gaseous systems, for the amount of material adsorbed on a surface as a function of pressure, while the temperature is held constant, assuming that a single layer of molecules is adsorbed; it is $f = ap/(1 + ap)$, where f is the fraction of surface covered, p is the pressure, and a is a constant.

lanosterol $C_{30}H_{50}O$ An unsaturated sterol occurring in wool fat and yeast.

lanthana *See* lanthanum oxide.

lanthanide contraction A phenomenon encountered in the rare-earth elements; the radii of the atoms of the members of the series decrease slightly as the atomic numbers increase; starting with element 58 in the periodic table, the balancing electron fills in an inner incomplete $4f$ shell as the charge on the nucleus increases.

lanthanide series Rare-earth elements of atomic numbers 57 through 71; their chemical properties are similar to those of lanthanum, atomic number 57.

lanthanum A chemical element, symbol La, atomic number 57, atomic weight 138.91; it is the second most abundant element in the rare-earth group.

lanthanum nitrate $La(NO_3)_3 \cdot 6H_2O$ Hygroscopic white crystals melting at 40°C; soluble in alcohol and water; used as an antiseptic and in gas mantles.

lanthanum oxide La_2O_3 A white powder melting at about 2000°C; soluble in acid, insoluble in water; used to replace lime in calcium lights and in optical glass. Also known as lanthana; lanthanum sesquioxide; lanthanum trioxide.

lanthanum sesquioxide *See* lanthanum oxide.

lanthanum sulfate $La_2(SO_4)_3 \cdot 9H_2O$ White crystals; slightly soluble in water, soluble in alcohol; used for atomic weight determinations for lanthanum.

lanthanum trioxide *See* lanthanum oxide.

lapachoic acid *See* lapachol.

lapachol $C_{15}H_{14}O_3$ A yellow crystalline compound obtained from lapacho, a hardwood in Argentina and Paraguay. Also known as 2-hydroxy-3-amylene-α-naphtho-quinone; lapachoic acid; targusic acid.

Laporte selection rule The rule that an electric dipole transition can occur only between states of opposite parity.

larixinic acid *See* maltol.

laser heterodyne spectroscopy A high-resolution spectroscopic technique, used in astronomical and atmospheric observations, in which the signal to be measured is mixed with a laser signal in a solid-state diode, producing a difference-frequency signal in the radio-frequency range.

laser-induced nuclear polarization A technique for making the spin vectors of an ensemble of nuclei point preferentially in one direction by means of an optical pumping process using either circularly or linearly polarized laser light. Abbreviated LINUP.

laser spectroscopy A branch of spectroscopy in which a laser is used as an intense, monochromatic light source; in particular, it includes saturation spectroscopy, as well as the application of laser sources to Raman spectroscopy and other techniques.

lattice A regular periodic arrangement of points in three-dimensional space; it consists of all those points P for which the vector from a given fixed point to P has the form $n_1 \mathbf{a} + n_2 \mathbf{b} + n_3 \mathbf{c}$, where n_1, n_2, and n_3 are integers, and $\mathbf{a}$, $\mathbf{b}$, and $\mathbf{c}$ are fixed, linearly independent vectors. Also known as periodic lattice; space lattice.

lattice constant A parameter defining the unit cell of a crystal lattice, that is, the length of one of the edges of the cell or an angle between edges. Also known as lattice parameter.

lattice defect *See* crystal defect.

lattice parameter *See* lattice constant.

laudanidine $C_{20}H_{25}NO_4$ An optically active alkaloid found in opium that crystallizes as prisms from an alcohol solution, and melts at 185°C. Also known as *l*-laudanine; tritopine.

laudanine $C_{20}H_{25}NO_4$ An optically inactive alkaloid derived from alkaline mother liquors from morphine extraction; it crystallizes in orthorhombic prisms from alcohol and chloroform; the prisms melt at 167°C, and are soluble in hot alcohol, benzene, and chloroform. Also known as *dl*-laudanine.

laudanosine $C_{21}H_{27}NO_4$ An alkaloid that is the methyl ether of laudanine; the optically inactive form crystallizes from dilute alcohol and melts at about 115°C; the levorotatory active form crystallizes from light petroleum solution and melts at 89°C.

Laue camera The apparatus used in the Laue method; the x-ray beam usually enters through a hole in the x-ray film, which records beams bent through an angle of nearly 180° by the crystal; less commonly, the film is placed beyond the crystal.

Laue condition 1. The condition for a vector to lie in a Laue plane: its scalar product with a specified vector in the reciprocal lattice must be one-half of the scalar product of the latter vector with itself. 2. *See* Laue equations.

Laue equations Three equations which must be satisfied for an x-ray beam of specified wavelength to be diffracted through a specified angle by a crystal; they state that the scaler products of each of the crystallographic axial vectors with the difference between unit vectors in the directions of the incident and scattered beams, are integral multiples of the wavelength. Also known as Laue condition.

Laue method A method of studying crystalline structures by x-ray diffraction, in which a finely collimated beam of polychromatic x-rays falls on a single crystal whose orientation can be set as desired, and diffracted beams are recorded on a photographic film.

Laue pattern The characteristic photographic record obtained in the Laue method.

Laue plane   A plane which is the perpendicular bisector of a vector in the reciprocal lattice; such planes form the boundaries of Brillouin zones.

Laue theory   A theory of diffraction of x-rays by crystals, based on the Laue equations.

laughing gas   *See* nitrous oxide.

lauric acid $CH_3(CH_2)_{10}COOH$ A fatty acid melting at 44°C, boiling at 225°C (100 mmHg; 13,332 newtons per square meter); colorless needles soluble in alcohol and ether, insoluble in water; found as the glyceride in vegetable fats, such as coconut and laurel oils; used for wetting agents, in cosmetics, soaps, resins, and insecticides, and as a chemical intermediate. Also known as dodecanoic acid.

lauric aldehyde   *See* lauryl aldehyde.

lauryl alcohol $CH_3(CH_2)_{11}OH$ A colorless solid which is obtained from coconut oil fatty acids, has a floral odor, and boils at 259°C; used in detergents, lubricating oils, and pharmaceuticals. Also known as alcohol C-12; n-dodecanol; dodecyl alcohol.

lauryl aldehyde $CH_3(CH_2)_{10}CHO$ A constituent of an essential oil from the silver fir; a colorless solid or a liquid, with a floral odor, that is soluble in 90% alcohol; used in perfumes. Also known as aldehyde C-12 lauric; lauraldehyde; lauric aldehyde.

lauryl mercaptan $C_{12}H_{25}SH$ Pale-yellow or water-white liquid with mild odor; insoluble in water, soluble in organic solvents; used to manufacture plastics, pharmaceuticals, insecticides, fungicides, and elastomers.

law of constant angles   The law that the angles between the faces of a crystal remain constant as the crystal grows.

law of constant heat summation   *See* Hess's law.

law of corresponding states   The law that when, for two substances, any two ratios of pressure, temperature, or volume to their respective critical properties are equal, the third ratio must equal the other two.

law of definite composition   *See* law of definite proportion.

law of definite proportion   The law that a given chemical compound always contains the same elements in the same fixed proportion by weight. Also known as law of definite composition.

law of mass action   The law stating that the rate at which a chemical reaction proceeds is directly proportional to the molecular concentrations of the reacting compounds.

law of rational intercepts   *See* Miller law.

lawrencium   A chemical element, symbol Lr, atomic number 103; two isotopes have been discovered, mass number 257 or 258 and mass number 256.

layer lattice   *See* layer structure.

layer structure   A crystalline structure found in substances such as graphites and clays, in which the atoms are largely concentrated in a set of parallel planes, with the regions between the planes comparatively vacant. Also known as layer lattice.

L capture   A type of generalized beta interaction in which a nucleus captures an electron from the L shell of atomic electrons (the shell second closest to the nucleus).

LDH   *See* lactic dehydrogenase.

leachate   A solution formed by leaching.

leaching   **1.** The dissolving, by a liquid solvent, of soluble material from its mixture with an insoluble solid; leaching is an industrial separation operation based on mass transfer; examples are the washing of a soluble salt from the surface of an insoluble precipitate, and the extraction of sugar from sugarbeets. **2.** The separation or dissolving out of soluble constituents from a rock or ore body by percolation of water.

lead   A chemical element, symbol Pb, atomic number 82, atomic weight 207.19.

lead-208   Lead isotope, atomic mass number of 208, formed by the radioactive decay of thorium.

lead acetate $Pb(C_2H_3O_2)_2 \cdot 3H_2O$ Poisonous, water-soluble white crystals decomposing at 280°C; loses water at 75°C; used in hair dyes, medicines, and textile mordants, for waterproofing, for manufacture of varnishes and pigments, and as an analytical reagent. Also known as sugar of lead.

lead antimonite $Pb_3(SbO_4)_2$ Poisonous, water-insoluble orange-yellow powder; used as a paint pigment and to stain glass and ceramics. Also known as antimony yellow; Naples yellow.

lead arsenate $Pb_3(AsO_4)_2$ Poisonous, water-insoluble white crystals; soluble in nitric acid; used as an insecticide.

lead azide $Pb(N_3)_2$ Unstable, colorless needles that explode at 350°C; lead azide is shipped submerged in water to reduce sensitivity; used as a detonator for high explosives.

lead borate $Pb(BO_2)_2 \cdot H_2O$ Poisonous, water-insoluble white powder; soluble in dilute nitric acid; used as varnish and paint drier, for galvanoplastic work, in lead glass, and in waterproofing paints.

lead bromide $PbBr_2$ An alcohol-insoluble white powder melting at 373°C, boiling at 916°C; slightly soluble in hot water.

lead carbonate $PbCO_3$ Poisonous, acid-soluble white crystals decomposing at 315°C; insoluble in alcohol and water; used as a paint pigment.

lead-chamber process   A process for the preparation of impure or dilute (60–78%) sulfuric acid; sulfur dioxide is oxidized by moist air with nitrogen oxide catalysts in a series of lead-lined chambers, the Gay-Lussac tower and the Glover tower; used primarily in the manufacture of fertilizer.

lead chloride $PbCl_2$ Poisonous white crystals melting at 498°C, boiling at 950°C; slightly soluble in hot water, insoluble in alcohol and cold water; used to make lead salts and lead chromate pigments and as an analytical reagent.

lead chromate $PbCrO_4$ Poisonous, water-insoluble yellow crystals melting at 844°C; soluble in acids; used as a paint pigment.

lead cyanide $Pb(CN)_2$ Poisonous white to yellow powder; slightly soluble in water, decomposed by acids; used in metallurgy.

lead dioxide PbO_2 Poisonous brown crystals that decompose when heated; insoluble in water and alcohol, soluble in glacial acetic acid; used as an oxidizing agent, in electrodes, batteries, matches, and explosives, as a textile mordant, in dye manufacture, and as an analytical reagent. Also known as anhydrous plumbic acid; brown lead oxide; lead peroxide.

lead fluoride PbF_2 A crystalline solid with a melting point of 824°C; used for laser crystals and electronic and optical applications.

lead fluosilicate   *See* lead hexafluorosilicate.

lead formate $Pb(CHO_2)_2$ Poisonous, water-soluble brownish-white crystals that decompose at 190°C; used as an analytical reagent.

lead halide PbX_2, where X is a halogen (such as F, Br, Cl, or I).

lead hexafluorosilicate $PbSiF_6 \cdot 2H_2O$ Poisonous, colorless, water-soluble crystals; used in the electrolytic method for refining lead. Also known as lead fluosilicate; lead silicofluoride.

lead iodide PbI_2 Poisonous, water- and alcohol-insoluble golden-yellow crystals melting at 402°C, boiling at 954°C; used in photography, medicine, printing, mosaic gold, and bronzing.

lead metasilicate   *See* lead silicate.

lead molybdate $PbMoO_4$ Poisonous, acid-soluble yellow powder; insoluble in water and alcohol; used in pigments and as an analytical reagent.

lead monoxide   PbO   Yellow, tetragonal crystals that melt at 888°C and are soluble in alkalies and acids; used in storage batteries, ceramics, pigments, and paints. Also known as litharge; plumbous oxide; yellow lead oxide.

lead nitrate $Pb(NO_3)_2$ Strongly oxidizing, poisonous, water- and alcohol-soluble white crystals that decompose at 205–223°C; used as a textile mordant, paint pigment, and photographic sensitizer and in medicines, matches, explosives, tanning, and engraving.

lead oleate $Pb(C_{18}H_{33}O_2)_2$ Poisonous, water-insoluble, white, ointmentlike material; soluble in alcohol, benzene, and ether; used in varnishes, lacquers, and high-pressure lubricants, and as a paint drier.

lead orthoplumbate   *See* lead tetroxide.

lead oxide red   *See* lead tetroxide.

lead peroxide   *See* lead dioxide.

lead phosphate Pb_3PO_4 A poisonous, white powder that melts at 1014°C; soluble in nitric acid and in fixed alkali hydroxide; used as a stabilizer in plastics.

lead pigments   Chemical compounds of lead used in paints to give color; examples are white lead; basic lead carbonate; lead carbonate; lead thiosulfate; lead sulfide; basic lead sulfate (sublimed white lead); silicate white lead; basic lead silicate; lead chromate; basic lead chromate; lead oxychloride; and lead oxide (monoxide and dioxide).

lead resinate $Pb(C_{20}H_{29}O_2)_2$ Poisonous, insoluble, brown, lustrous, translucent lumps; used as a paint and varnish drier and for textile waterproofing.

lead silicate $PbSiO_3$ Toxic, insoluble white crystals; used in ceramics, paints, and enamels, and to fireproof fabrics. Also known as lead metasilicate.

lead silicofluoride   *See* lead hexafluorosilicate.

lead sodium thiosulfate $Na_4Pb(S_2O_3)_3$ Poisonous, small, white, heavy crystals that are soluble in thiosulfate solutions; used in the manufacture of matches. Also known as lead sodium hyposulfate; sodium lead hyposulfate; sodium lead thiosulfate.

lead stearate $Pb(C_{18}H_{35}O_2)_2$ Poisonous white powder; soluble in alcohol and ether, insoluble in water; used as a lacquer and varnish drier and in high-pressure lubricants.

lead sulfate $PbSO_4$ Poisonous white crystals melting at 1170°C; slightly soluble in hot water, insoluble in alcohol; used in storage batteries and as a paint pigment.

lead sulfide PbS Blue, metallic, cubic crystals that melt at 1120°C, derived from the mineral galena or by reacting hydrogen sulfide gas with a solution of lead nitrate; used in semiconductors and ceramics. Also known as plumbous sulfide.

lead susceptibility   The increase in octane number of gasoline imparted by the addition of a specified amount of TEL (tetraethyllead).

lead telluride $PbTe$ A crystalline solid that is very toxic if inhaled or ingested; melts at 902°C; used as a semiconductor and photoconductor in the form of single crystals.

lead tetraacetate $Pb(CH_3COO)_4$ Crystals that are faintly pink or colorless; melts at 175°C; used as an oxidizing agent in organic chemistry, cleaving 1,2-diols to form aldehydes or ketones.

lead tetroxide Pb_3O_4 A poisonous, bright-red powder, soluble in excess glacial acetic acid and dilute hydrochloric acid; used in medicine, in cement for special applications, in manufacture of colorless glass, and in ship paint. Also known as lead orthoplumbate; lead oxide red; red lead.

lead thiocyanate $Pb(SCN)_2$ Yellow, monoclinic crystals, soluble in potassium thiocyanate and slightly soluble in water; used in the powder mixture that primes small arm cartridges, in dyes, and in safety matches.

lead titanate $PbTiO_3$ A water-insoluble, pale-yellow solid; used as coloring matter in paints.

lead tungstate $PbWO_4$ A yellowish powder, melting at 1130°C; insoluble in water, soluble in acid; used as a pigment. Also known as lead wolframate.

lead vanadate $Pb(VO_3)_2$ A water-insoluble, yellow powder; used as a pigment and for the preparation of other vanadium compounds.

lead wolframate   *See* lead tungstate.

leakage   A phenomenon occurring in an ion-exchange process in which some influent ions are not adsorbed by the ion-exchange bed and appear in the effluent.

leakage halo   The dispersion of elements along channels and paths followed by mineralizing solutions leading into and away from the central focus of mineralization.

leaving group   The group of charged or uncharged atoms that departs during a substitution or displacement reaction. Also known as nucleofuge.

lecithin   Any of a group of phospholipids having the general composition $CH_2OR_1 \cdot CHOR_2 \cdot CH_2OPO_2OHR_3$, in which R_1 and R_2 are fatty acids and R_3 is choline, and with emulsifying, wetting, and antioxidant properties.

lecithinase An enzyme that catalyzes the breakdown of a lecithin into its constituents.

lecithinase A An enzyme that catalyzes the removal of only one fatty acid from lecithin, yielding lipolecithin.

lecithinase C An enzyme that catalyzes the removal of the nitrogenous base of lecithin to produce the base and a phosphatidic acid.

lecithinase D An enzyme that catalyzes the removal of the phosphorylated base from lecithins, producing $\alpha\beta$-diglyceride.

lectin Any of various proteins that agglutinate erythrocytes and other types of cells and also have other properties, including mitogenesis, agglutination of tumor cells, and toxicity toward animals; found widely in plants, predominantly in legumes, and also occurring in bacteria, fish, and invertebrates.

left-handed Having a crystal structure with a mirror-image relationship to a right-handed structure.

L electron An electron in the L shell.

lenacil $C_{13}H_{18}N_2O_2$ A colorless, crystalline compound with a melting point of 316–317°C; slight solubility in water; used as an herbicide to control weeds in sugarbeets, cereal grains, and strawberries. Also known as 3-cyclohexyl-5,6-trimethyleneuracil.

Lennard-Jones potential A semiempirical approximation to the potential of the force between two molecules, given by $V = (A/R^{12}) - (B/R^6)$, where R is the distance between the centers of the molecules, and A and B are constants.

lepidine $C_9H_6NCH_3$ An alkaloid derived as an oily liquid from cinchona bark; boils at 266°C; soluble in ether, benzene, and alcohol; used in organic synthesis. Also known as cincholepidine; γ-methylquinoline.

leptophos $C_{13}H_{10}BrCl_2O_2PS$ A white solid with a melting point of 70.2–70.6°C; slight solubility in water; used as an insecticide on vegetables, fruit, turf, and ornamentals. Also known as O-(4-bromo-2,5-dichlorophenyl) O-methyl phenylphosphorothioate.

leucaenine *See* mimosine.

leucaenol *See* mimosine.

leucenine *See* mimosine.

leucenol *See* mimosine.

leucine $C_6H_{13}O_2N$ A monocarboxylic essential amino acid obtained by hydrolysis of protein-containing substances such as milk.

leucine amino peptidase An enzyme that acts on peptides to catalyze the release of terminal amino acids, especially leucine residues, having free α-amino groups.

leucitohedron *See* trapezohedron.

leucoline *See* quinoline.

leucosin A simple protein of the albumin type found in wheat and other cereals.

leukocidin A toxic substance released by certain bacteria which destroys leukocytes.

leukocytolysin   A leukocyte-disintegrating lysin.

leukotriene   Any of a family of oxidized metabolites of certain polyunsaturated fatty acids, predominantly arachidonic acid, that mediate responses in allergic reactions and inflammations, produced in specific cells upon stimulation.

leurocristine   *See* vincristine.

levan $(C_6H_{10}O_5)_n$ A polysaccharide consisting of repeating units of D-fructose and produced by a range of microorganisms, such as *Bacillus mesentericus*.

Levenstein process   A process for the manufacture of mustard gas from ethene, $CH_2=CH_2$, and sulfur chloride, S_2Cl_2.

levigate   **1.** To separate a finely divided powder from a coarser material by suspending in a liquid in which both substances are insoluble. Also known as elutriation. **2.** To grind a moist solid to a fine powder.

levo form   An optical isomer which induces levorotation in a beam of plane polarized light.

levulinic acid $CH_3COCH_2CH_2COOH$ Crystalline compound forming plates or leaflets that melt at 37°C; freely soluble in alcohol, ether, and chloroform; used in the manufacture of pharmaceuticals, plastics, rubber, and synthetic fibers. Also known as acetylpropionic acid; γ-ketovaleric acid; 4-oxypentanoic acid.

levulose   Levorotatory D-fructose.

Lewis acid   A substance that can accept an electron pair from a base; thus, $AlCl_3$, BF_3, and SO_3 are acids.

Lewis base   A substance that can donate an electron pair; examples are the hydroxide ion, OH^-, and ammonia, NH_3.

lewisite $C_2H_2AsCl_3$ An oily liquid, colorless to brown or violet; forms a toxic gas, used in World War I.

Lewis-Matheson method   Trial-and-error calculation method for the design of multicomponent distillation columns, or for the determination of the separating ability of an existing column.

Li   *See* lithium.

Libby effect   The increase, since about 1950, in the carbon-14 content of the atmosphere, produced by the detonation of thermonuclear devices.

lidocaine $C_{14}H_{22}N_2O$ A crystalline compound, used as a local anesthetic. Also known as lignocaine.

life test   In petroleum testing, an ASTM oxidation test made on inhibited steam-turbine oils to determine their stability under oxidizing conditions.

ligand   The molecule, ion, or group bound to the central atom in a chelate or a coordination compound; an example is the ammonia molecules in $[Co(NH_3)_6]^{3+}$.

ligand membrane   A solvent immiscible with water and a reagent and acting as an extractant and complexing agent for an ion.

ligase   An enzyme that catalyzes the union of two molecules, involving the participation of a nucleoside triphosphate which is converted to a nucleoside diphosphate or monophosphate. Also known as synthetase.

light hydrogen *See* protium.

light meromyosin The smaller of two fragments obtained from the muscle protein myosin following limited proteolysis by trypsin or chymotrypsin.

lignin A substance that together with cellulose forms the woody cell walls of plants and cements them together; a colorless to brown substance removed from paper-pulp sulfite liquor.

lignin plastic A plastic based on resins derived from lignin; used as a binder or extender.

lignocaine *See* lidocaine.

lignocellulose Any of a group of substances in woody plant cells consisting of cellulose and lignin.

lignosulfonate Any of several substances manufactured from waste liquor of the sulfate pulping process of soft wood; used in the petroleum industry to reduce the viscosity of oil well muds and slurries, and as extenders in glues, synthetic resins, and cements.

limed rosin *See* calcium resinate.

lime kiln Furnace-type apparatus, usually a long, tilted cylinder that is slowly rotated, used to heat calcium carbonate, $CaCO_3$, above 900°C to produce lime.

liming Soaking hides and skins in milk of lime and causing them to swell, to facilitate the removal of hair.

limiting current density The maximum current density to achieve a desired electrode reaction before hydrogen or other extraneous ions are discharged simultaneously.

limiting viscosity number *See* intrinsic viscosity.

limonene $C_{10}H_{16}$ A terpene with a lemon odor that is optically active and is found in oils from citrus fruits and in oils from peppermint and spearmint; a colorless, water-insoluble liquid that boils at 176°C.

linalool $(CH_3)_2C:CH(CH_2)_2CCH_3OHCH:CH_2$ A terpene that is a colorless liquid, has a bergamot odor, boils at 195–196°C, and is found in many essential oils, particularly bergamot and rosewood; used as a flavoring agent and in perfumes. Also known as coriandrol; 3,7-dimethyl-1,6-octadiene-3-ol.

linalyl acetate $(CH_3)_2C:CH(CH_2)_2CCH_3(OCOCH_3)CH:CH_2$ The acetic acid ester of linalool, a colorless oily liquid with a bergamot odor that boils at 108–110°C; used in perfumes and as a flavoring agent.

Linde copper sweetening A petroleum-refinery process to treat gasolines and distillates with a slurry of clay and cupric chloride to remove mercaptans.

lineage structure An imperfection structure characterizing a crystal, parts of which have slight differences in orientation.

linear molecule A molecule whose atoms are arranged so that the bond angle between each is 180°; an example is carbon dioxide, CO_2.

linear polymer A polymer whose molecule is arranged in a chainlike fashion with few branches or bridges between the chains.

linear Stark effect A splitting of spectral lines of hydrogenlike atoms placed in an electric field; each energy level of principal quantum number n is split into $2n - 1$ equidistant levels of separation proportional to the field strength.

line defect *See* dislocation.

line-formula method A system of notation for hydrocarbons showing the chemical elements, functional groups, and ring systems in linear form; an example is acetone, CH_3COCH_3.

line pair In spectrographic analysis, a particular spectral line and the internal standard line with which it is compared to determine the concentration of a substance.

line spectrum 1. A spectrum of radiation in which the quantity being studied, such as frequency or energy, takes on discrete values. 2. Conventionally, the spectra of atoms, ions, and certain molecules in the gaseous phase at low pressures; distinguished from band spectra of molecules, which consist of a pattern of closely spaced spectral lines which could not be resolved by early spectroscopes.

line strength The intensity of a spectrum line.

linoleic acid $C_{17}H_{31}COOH$ A yellow unsaturated fatty acid, boiling at 229°C (14 mm Hg), occurring as a glyceride in drying oils; obtained from linseed, safflower, and tall oils; a principal fatty acid in plants, and considered essential in animal nutrition; used in medicine, feeds, paints, and margarine. Also known as linolic acid; 9,12-octadecadienoic acid.

linolenate A salt or ester of linolenic acid.

linolenic acid $C_{17}H_{29}COOH$ One of the principal unsaturated fatty acids in plants and an essential fatty acid in animal nutrition; a colorless liquid that boils at 230°C (17 mmHg or 2266 newtons per square meter), soluble in many organic solvents; used in medicine and drying oils. Also known as 9,12,15-octadecatrienoic acid.

linolenyl alcohol $C_{18}H_{32}O$ A colorless, combustible solid used for paints, paper, leather, and flotation processes. Also known as octadecatrienol.

linolic acid *See* linoleic acid.

LINUP *See* laser-induced nuclear polarization.

linuron $C_9H_{10}Cl_2N_2O_2$ A white, crystalline compound that melts at 93–94°C; slight solubility in water; used as an herbicide for roadsides and fence rows, and weeds in crops. Also known as 3-(3,4-dichlorophenyl)-1-methoxy-1-methylurea.

lipase An enzyme that catalyzes the hydrolysis of fats or the breakdown of lipoproteins.

lipid One of a class of compounds which contain long-chain aliphatic hydrocarbons and their derivatives, such as fatty acids, alcohols, amines, amino alcohols, and aldehydes; includes waxes, fats, and derived compounds. Also known as lipin; lipoid.

lipid metabolism The physiologic and metabolic processes involved in the assimilation of dietary lipids and the synthesis and degradation of lipids.

lipin 1. A compound lipid, such as a cerebroside. 2. *See* lipid.

lipochrome Any of various fat-soluble pigments, such as carotenoid, occurring in natural fats. Also known as chromolipid.

lipofuscin Any of a group of lipid pigments found in cardiac and smooth muscle cells, in macrophages, and in parenchyma and interstitial cells; differential reactions include sudanophilia, Nile blue staining, fatty acid, glycol, and ethylene.

lipoic acid $C_8H_{14}O_2S_2$ A compound which participates in the enzymatic oxidative decarboxylation of α-keto acids in a stage between thiamine pyrophosphate and coenzyme A.

lipoid 1. A fatlike substance. 2. *See* lipid.

lipophilic 1. Having a strong affinity for fats. 2. Promoting the solubilization of lipids.

lipopolysaccharide Any of a class of conjugated polysaccharides consisting of a polysaccharide combined with a lipid.

lipoprotein Any of a class of conjugated proteins consisting of a protein combined with a lipid.

lipotropic Having an affinity for lipid compounds.

lipotropic hormone Any hormone having lipolytic activity on adipose tissue.

lipoxidase An enzyme catalyzing the oxidation of the double bonds of an unsaturated fatty acid.

liquid chromatography A form of chromatography employing a liquid as the moving phase and a solid or a liquid on a solid support as the stationary phase; techniques include column chromatography, gel permeation chromatography, and partition chromatography.

liquid crystal A liquid which is not isotropic; it is birefringent and exhibits interference patterns in polarized light; this behavior results from the orientation of molecules parallel to each other in large clusters.

liquid dioxide *See* nitrogen dioxide.

liquid-drop model of nucleus A model of the nucleus in which it is compared to a drop of incompressible liquid, and the nucleons are analogous to molecules in the liquid; used to study binding energies, fission, collective motion, decay, and reactions. Also known as drop model of nucleus.

liquid filter A device for the removal of solids or coalesced droplets out of a liquid stream by use of a filter medium, such as a screen, cartridge, or granular bed.

liquid glass *See* sodium silicate.

liquid hydrocarbon A hydrocarbon that has been converted from a gas to a liquid by pressure or by reduction in temperature; usually limited to butanes, propane, ethane, and methane.

liquid junction emf The emf (electromotive force) generated at the area of contact between the salt bridge and the test solution in a pH cell electrode.

liquid-liquid chemical reaction Chemical reaction in which the reactants, two or more, are liquids.

liquid-liquid extraction The removal of a soluble component from a liquid mixture by contact with a second liquid, immiscible with the carrier liquid in which the component is preferentially soluble.

liquid-phase hydrogenation Hydrogen reaction with liquid-phase hydrogenatable material, such as unsaturated aliphatic or aromatic hydrocarbons.

liquid seal 1. The depth of liquid above an opening from which gas or vapor issues, as for a riser in a distillation-column tray. 2. Product drawoff in which a depth of liquid prevents the outflow of gas or vapor.

liquid-solid chemical reaction Chemical reaction in which at least one of the reactants is a liquid, and another of the reactants is a solid.

liquid-solid equilibrium *See* solid-liquid equilibrium.

liquid sulfur dioxide–benzene process A petroleum-refinery process using a mixed solvent (SO_2 and benzene) to dewax lubricating oils or improve their viscosity indices.

liquid-vapor chemical reaction Chemical reaction in which at least one of the reactants is a liquid, and another of the reactants is a vapor.

liquid-vapor equilibrium The equilibrium relationship between the liquid and its vapor phase for a partially vaporized compound or mixture at specified conditions of pressure and temperature; for mixtures, it is expressed by $K = x/y$, where K is the equilibrium constant, x the mole fraction of a key component in the vapor, and y the mole fraction of the same key component in the liquid. Also known as vapor-liquid equilibrium.

liquor 1. Supernatant liquid decanted from a liquid-solids mixture in which the solids have settled. 2. Liquid overflow from a liquid-liquid extraction unit.

lithamide *See* lithium amide.

litharge *See* lead monoxide.

lithium A chemical element, symbol Li, atomic number 3, atomic weight 6.939; an alkali metal.

lithium aluminum hydride $LiAlH_2$ A compound made by the reaction of lithium hydride and aluminum chloride; a powerful reducing agent for specific linkages in complex molecules; used in organic synthesis.

lithium amide $LiNH_2$ A compound crystallizing in the cubic form, and melting at 380–400°C; used in organic synthesis. Also known as lithamide.

lithium bromide $LiBr \cdot H_2O$ A white, deliquescent, granular powder with a bitter taste, melting at 547°C; soluble in alcohol and glycol; used to add moisture to air-conditioning systems and as a sedative and hypnotic in medicine.

lithium carbonate Li_2CO_3 A colorless, crystalline compound that melts at 700°C and has slight solubility in water; used in ceramic industries in the manufacture of powdered glass for porcelain enamel formulation.

lithium cell An electrolytic cell for the production of metallic lithium.

lithium chloride $LiCl \cdot 2H_2O$ A colorless, water-soluble compound, forming octahedral crystals and melting at 614°C; used to form concentrated brine in commercial air-conditioning systems and as a pyrotechnic in welding and brazing fluxes.

lithium citrate $Li_3C_6H_5O_7 \cdot 4H_2O$ White powder that decomposes when heated; slightly soluble in alcohol; soluble in water; used in beverages and pharmaceuticals.

lithium fluoride LiF Poisonous, white powder melting at 870°C, boiling at 1670°C; insoluble in alcohol, slightly soluble in water, and soluble in acids; used as a heat-exchange medium, as a welding and soldering flux, in ceramics, and as crystals in infrared instruments.

lithium halide A binary compound of lithium, LiX, where X is a halide; examples are lithium chloride, LiCl, and lithium fluoride, LiF.

lithium hydride LiH Flammable, brittle, white, translucent crystals; decomposes in water; insoluble in ether, benzene, and toluene; used as a hydrogen source and desiccant, and to prepare lithium amide and double hydrides.

lithium hydroxide LiOH; LiOH·H_2O Colorless crystals; used as a storage-battery electrolyte, as a carbon dioxide absorbent, and in lubricating greases and ceramics.

lithium iodide LiI; LiI·$3H_2O$ White, water- and alcohol-soluble crystals; LiI melts at 446°C; LiI·$3H_2O$ loses water at 72°C; used in medicine, photography, and mineral waters.

lithium molybdate Li_2MoO_4 Water-soluble white crystals melting at 705°C; used as a catalytic cracking (petroleum) catalyst and as a mill additive for steel.

lithium nitrate $LiNO_3$ Water- and alcohol-soluble colorless powder melting at 261°C; used as a heat-exchange medium and in ceramics, pyrotechnics, salt baths, and refrigeration systems.

lithium perchlorate $LiClO_4·3H_2O$ A compound with high oxygen content (60% available oxygen), used as a source of oxygen in rockets and missiles.

lithium stearate $LiC_{18}H_{35}O_2$ A white, crystalline compound with a melting point of 220°C; used in cosmetics, plastics, and greases, and as a corrosion inhibitor in petroleum.

lithium tetraborate $Li_2B_4O_7·5H_2O$ White crystals that lose water at 200°C; insoluble in alcohol, soluble in water; used in ceramics.

lithium titanate Li_2TiO_3 A water-insoluble white powder with strong fluxing ability when used in titanium-containing enamels; also used as a mill additive in vitreous and semivitreous glazes.

lithocholic acid $C_{24}H_{40}O_3$ A crystalline substance with a melting point of 184–186°C; soluble in hot alcohol; found in ox, human, and rabbit bile.

lithogeochemical survey A geochemical survey that involves the sampling of rocks.

lithophile 1. Pertaining to elements that have become concentrated in the silicate phase of meteorites or the slag crust of the earth. 2. Pertaining to elements that have a greater free energy of oxidation per gram of oxygen than iron. Also known as oxyphile.

Littrow grating spectrograph A spectrograph having a plane grating at an angle to the axis of the instrument, and a lens in front of the grating which both collimates and focuses the light.

Littrow mounting The arrangement of the grating and other components of a Littrow grating spectrograph, which is analogous to that of a Littrow quartz spectrograph.

Littrow quartz spectrograph A spectrograph in which dispersion is accomplished by a Littrow quartz prism with a rear reflecting surface that reverses the light; a lens in front of the prism acts as both collimator and focusing lens.

liver phosphorylase An enzyme that catalyzes the breakdown of liver glycogen to glucose-1-phosphate.

lixiviate   To extract a soluble component from a solid mixture by washing or percolation processes.

L/M   The ratio of the number of internal conversion electrons emitted from the L shell in the de-excitation of a nucleus to the number of such electrons emitted from the M shell.

loading   Condition of vapor overcapacity in a liquid-vapor-contact tower, in which rising vapor lifts or holds falling liquid.

loaming   In geochemical prospecting, a method in which samples of material from the surface are tested for traces of a sought-after metal; its presence on the surface presumably indicates a near-surface ore body.

Lobry de Bruyn–Ekenstein transformation   The change in which an aldose sugar treated with dilute alkali results in a mixture of an epimeric pair and 2-keto-hexose due to the production of enolic forms in the presence of hydroxyl ions, followed by a rearrangement.

London dispersion force   *See* van der Waals force.

longitudinal baffle   Baffle sheets or plates within a process vessel (such as a heat exchanger) that are parallel to the long dimension of the vessel; used to direct fluid flow in the desired flow pattern.

longitudinal flow reactor   Theoretical reactor system in which there is no longitudinal mixing (back mixing) of reactants and products as they flow through the reactor, but in which there is complete radial (side-to-side) mixing.

long-tube vertical evaporator   A liquid evaporator in which the material is force-fed into the bottom of a bundle of long, vertical tubes; hot liquid on the outsides of the tubes transfers heat to the rising liquid feed, causing partial evaporation.

look box   Box with glass windows built into distillation-column rundown lines (or other flow lines) so that the stream of condensate from the condenser can be watched.

Loomis-Wood diagram   A graph used to assign lines in a molecular spectrum to the various branches of rotational bands when these branches overlap, in which the difference between observed wave numbers and wave numbers extrapolated from a few lines that apparently belong to one branch are plotted against arbitrary running numbers for that branch.

loop seal   Antivapor seal for liquid drawoffs from process or storage vessels; liquid drawoff is made to flow through an immersed loop or beneath an obstruction, thus sealing off vapor flow.

lophine $C_{21}H_{16}O_2$ A colorless, crystalline, water-insoluble compound that melts at 275°C; used as an indicator in fluorescent neutralization tests.

Lorentz line-splitting theory   A theory predicting that when a light source is placed in a strong magnetic field, its spectral lines are each split into three components, one of them retaining the zero-field frequency, and the other two shifted upward and downward in frequency by the Larmor frequency (the normal Zeeman effect).

Lorentz theory of light sources   A theory according to which light is emitted by vibrations of electrons, which are damped harmonic oscillators attached to atoms.

Lorentz unit A unit of reciprocal length used to measure the difference, in wave numbers, between a (zero field) spectrum line and its Zeeman components; equal to $eH/4\pi mc^2$, where H is the magnetic field strength, c is the speed of light, and e and m are the charge and mass of the electron respectively (gaussian units).

low-boiling butene-2 *See* butene-2.

low-energy np scattering An elastic collision of a neutron, having an energy from less than 1 electronvolt to 10,000,000 electronvolts, with a proton (usually the nucleus of a hydrogen atom).

low-energy pp scattering An elastic collision of a proton, having an energy of less than 10,000,000 electronvolts, with another proton (usually the nucleus of a hydrogen atom).

low-frequency spectrum Spectrum of atoms and molecules in the microwave region, arising from such causes as the coupling of electronic and nuclear angular momenta, and the Lamb shift.

low-temperature carbonization Low-temperature destructive distillation of coal to produce liquid products.

low-temperature separation Liquid condensate recovery from wet gases at temperatures of 20 to $-20°F$ (-6.7 to $-28.9°C$), the temperature range at which the gas-oil separator operates.

l-process The synthesis of certain light nuclides through the breakup of heavier nuclides, probably by cosmic-ray bombardment of the interstellar medium.

Lr *See* lawrencium.

LSD *See* lysergic acid diethylamide.

LSD-25 *See* lysergic acid diethylamide.

L shell The second shell of electrons surrounding the nucleus of an atom, having electrons whose principal quantum number is 2.

Lu *See* lutetium.

luciferase An enzyme that catalyzes the oxidation of luciferin.

luciferin A species-specific pigment in many luminous organisms that emits heatless light when combined with oxygen.

Luggin probe A device which transmits a significant current density on the surface of an electrode to measure its potential.

Lugol solution A solution of 5 grams of iodine and 10 grams of potassium iodide per 100 milliliters of water; used in medicine.

luminol $NH_2 \cdot C_6H_3 \cdot (CO \cdot NH)_2$ A white, water-soluble, crystalline compound that melts at 320°C; used in an alkaline solution for analytical testing in chemistry. Also known as 3-aminophthalic acid cyclic hydrazide.

Lundegardh vaporizer A device used for emission flame photometry in which a compressed air aspirator vaporizes the solution within a chamber; smaller droplets are carried into the fuel-gas stream and to the burner orifice where the solvent is evaporated, dissociated, and optically excited.

lutein 1. A dried, powdered preparation of corpus luteum. 2. *See* xanthophyll.

luteinizing hormone A glycoprotein hormone secreted by the adenohypophysis of vertebrates which stimulates hormone production by interstitial cells of gonads. Also known as interstitial-cell-stimulating hormone (ICSH).

luteinizing-hormone releasing factor A small peptide hormone released from the hypothalamus which acts on the pituitary gland to cause release of luteinizing hormone.

luteotropic hormone *See* prolactin.

lutetium A chemical element, symbol Lu, atomic number 71, atomic weight 174.97; a very rare metal and the heaviest member of the rare-earth group.

lyase An enzyme that catalyzes the nonhydrolytic cleavage of its substrate with the formation of a double bond; examples are decarboxylases.

lycine *See* betaine.

lycopene $C_{40}H_{50}$ A red, crystalline hydrocarbon that is the coloring matter of certain fruits, as tomatoes; it is isomeric with carotene.

lye **1.** A solution of potassium hydroxide or sodium hydroxide used as a strong alkaline solution in industry. **2.** The alkaline solution that is obtained from the leaching of wood ashes.

Lyman-alpha radiation Radiation emitted by hydrogen associated with the spectral line in the Lyman series whose wavelength is 1215 angstrom units.

Lyman band A band in the ultraviolet spectrum of molecular hydrogen, extending from 1250 to 1610 angstrom units.

Lyman continuum A continuous range of wavelengths (or wave numbers or frequencies) in the spectrum of hydrogen at wavelengths less than the Lyman limit, resulting from transitions between the ground state of hydrogen and states in which the single electron is freed from the atom.

Lyman ghost A false line observed in a spectroscope as a result of a combination of periodicities in the ruling.

Lyman limit The lower limit of wavelengths of spectral lines in the Lyman series (912 angstrom units), or the corresponding upper limit in frequency, energy of quanta, or wave number (equal to the Rydberg constant for hydrogen).

Lyman series A group of lines in the ultraviolet spectrum of hydrogen covering the wavelengths of 1215–912 angstrom units.

lyngbyatoxin A An indole alkaloid toxin produced by *Lyngbya majuscula*.

lyophilic Referring to a substance which will readily go into colloidal suspension in a liquid.

lyophilization Rapid freezing of a material, especially biological specimens for preservation, at a very low temperature followed by rapid dehydration by sublimation in a high vacuum.

lyophobic Referring to a substance in a colloidal state that has a tendency to repel liquids.

lyotopic series *See* Hofmeister series.

lyotropic liquid crystal A liquid crystal prepared by mixing two or more components, one of which is polar in character (for example, water).

lysergic acid $C_{16}H_{16}N_2O_2$ A compound that crystallizes in the form of hexagonal plates that melt and decompose at 240°C; derived from ergot alkaloids; used as a psychotomimetic agent.

lysergic acid diethylamide $C_{15}H_{15}N_2CON(C_2H_5)_2$ A psychotomimetic drug synthesized from compounds derived from ergot. Abbreviated LSD; LSD-25.

lysine $C_6H_{14}O_2N_2$ An essential, basic amino acid obtained from many proteins by hydrolysis.

lysozyme An enzyme present in certain body secretions, principally tears, which acts to hydrolyze certain bacterial cell walls.

lysyl oxidase Any of a group of enzymes found in bone and connective tissue which oxidize terminal amino groups of lysine residues in tropocollagen molecules to aldehyde residues.

M

M acid $NH_2C_{10}H_5(OH)SO_3H$ A sulfonic acid formed by alkaline fusion of a disulfonic acid of α-naphthylamine; used as a dye intermediate. Also known as 5-amino-1-naphthol-3-sulfonic acid.

macle A twinned crystal.

Macquer's salt *See* potassium arsenate.

macroanalysis Qualitative or quantitative analysis of chemicals that are in quantities of the order of grams.

macrocyclic An organic molecule with a large ring structure usually containing over 15 atoms.

macrodome Dome of a crystal in which planes are parallel to the longer lateral axis.

macroglobulin Any gamma globulin with a sedimentation constant of 195.

macrolide A large ring molecule with many functional groups bonded to it.

macromolecular Composed of or characterized by large molecules.

macromolecule A large molecule in which there is a large number of one or several relatively simple structural units, each consisting of several atoms bonded together.

macronutrient An element, such as potassium and nitrogen, essential in large quantities for plant growth.

macropinacoid *See* front pinacoid.

macroporous resin A member of a class of very small, highly cross-linked polymer particles penetrated by channels through which solutions can flow; used as ion exchanger. Also known as macroreticular resin.

Maerz and Paul color system A dictionary of 7056 different colors in the form of two-impression screen-plate printing on semiglossy paper.

mafenide *See para*-(aminomethyl)benzenesulfonamide.

magic acid A superacid consisting of equal molar quantities of fluorosulfonic acid (FSO_2OH) and antimony pentafluoride (SbF_5).

magic numbers The integers 8, 20, 28, 50, 82, 126; nuclei in which the number of protons, neutrons, or both is magic have a stability and binding energy which is greater than average, and have other special properties.

magister of sulfur Amorphous sulfur produced by acid precipitation from solutions of hyposulfites or polysulfides.

magnesia Magnesium oxide that is processed for a particular purpose.

magnesia mixture Reagent used to analyze for phosphorus; consists of the filtered liquor from an aqueous mixture of ammonium chloride, magnesium sulfate, and ammonia.

magnesium A metallic element, symbol Mg, atomic number 12, atomic weight 24.312.

magnesium acetate $Mg(OOCCH_3)_2 \cdot 4H_2O$ or $Mg(OOCCH_3)_2$ A compound forming colorless crystals that are soluble in water and melt at 80°C; used in textile printing, in medicine as an antiseptic, and as a deodorant.

magnesium arsenate $Mg_3(AsO_4)_2 \cdot xH_2O$ A white, poisonous, water-insoluble powder used as an insecticide.

magnesium benzoate $Mg(C_7H_5O_2) \cdot 3H_2O$ A crystalline white powder melting at 200°C; soluble in alcohol and hot water; used in medicine.

magnesium borate $3MgO \cdot B_2O_3$ Crystals that are white or colorless and transparent; soluble in alcohol and acids, slightly soluble in water; used as a fungicide, antiseptic, and preservative.

magnesium bromate $Mg(BrO_3)_2 \cdot 6H_2O$ A white crystalline compound, insoluble in alcohol, soluble in water; a fire hazard; used as an analytical reagent.

magnesium bromide $MgBr_2 \cdot 6H_2O$ Deliquescent, colorless, bitter-tasting crystals, melting at 172°C; soluble in water, slightly soluble in alcohol; used in medicine and in the synthesis of organic chemicals.

magnesium carbonate $MgCO_3$ A water-insoluble, white powder, decomposing at about 350°C; used as a refractory material.

magnesium chlorate $Mg(ClO_3)_2 \cdot 6H_2O$ A white powder, bitter-tasting and hygroscopic; slightly soluble in alcohols, soluble in water; used in medicine.

magnesium chloride $MgCl_2 \cdot 6H_2O$ Deliquescent white crystals; soluble in water and alcohol; used in disinfectants and fire extinguishers, and in ceramics, textiles, and paper manufacture.

magnesium fluoride MgF_2 White, fluorescent crystals; insoluble in water and alcohol, soluble in nitric acid; melts at 1263°C; used in ceramics and glass. Also known as magnesium flux.

magnesium fluosilicate $MgSiF_6 \cdot 6H_2O$ Water-soluble, efflorescent white crystals; used in ceramics, in mothproofing and waterproofing, and as a concrete hardener. Also known as magnesium silicofluoride.

magnesium flux *See* magnesium fluoride.

magnesium formate $Mg(CHO_2)_2 \cdot 2H_2O$ Colorless, water-soluble crystals; insoluble in alcohol and ether; used in analytical chemistry and medicine.

magnesium gluconate $Mg(C_6H_{11}O_7)_2 \cdot 2H_2O$ An odorless, tasteless, water-soluble powder; used in medicine.

magnesium halide A compound formed from the metal magnesium and any of the halide elements; an example is magnesium bromide.

magnesium hydrate *See* magnesium hydroxide.

magnesium hydride MgH_2 A hydride compound formed from the metal magnesium; it decomposes violently in water, and in a vacuum at about 280°C.

magnesium hydroxide $Mg(OH)_2$ A white powder, very slightly soluble in water, decomposing at 350°C; used as an intermediate in extraction of magnesium metal, and as a reagent in the sulfite wood pulp process. Also known as magnesium hydrate.

magnesium hyposulfite *See* magnesium thiosulfate.

magnesium iodide $MgI_2 \cdot 8H_2O$ Crystalline powder, white and deliquescent, discoloring in air; soluble in water, alcohol, and ether; used in medicine.

magnesium lactate $Mg(C_3H_5O_3)_2 \cdot 3H_2O$ Bitter-tasting, water-soluble white crystals; slightly soluble in alcohol; used in medicine.

magnesium methoxide $(CH_3O)_2Mg$ Colorless crystals that decompose when heated; used as a catalyst, dielectric coating, and cross-linking agent, and to form gels. Also known as magnesium methylate.

magnesium methylate *See* magnesium methoxide.

magnesium nitrate $Mg(NO_3)_2 \cdot 6H_2O$ Deliquescent white crystals; soluble in alcohol and water; a fire hazard; used as an oxidizing material in pyrotechnics.

magnesium oleate $Mg(C_{18}H_{33}O_2)_2$ Water-insoluble, yellowish mass; soluble in hydrocarbons, alcohol, and ether; used as a plasticizer lubricant and emulsifying agent, and in varnish driers and dry-cleaning solutions.

magnesium oxide MgO A white powder that (depending on the method of preparation) may be light and fluffy, or dense; melting point 2800°C; insoluble in acids, slightly soluble in water; used in making refractories, and in cosmetics, pharmaceuticals, insulation, and medicine.

magnesium perchlorate $Mg(ClO_4)_2 \cdot 6H_2O$ White, deliquescent crystals; soluble in water and alcohol; explosive when in contact with reducing materials; used as a drying agent for gases.

magnesium peroxide MgO_2 A tasteless, odorless white powder; soluble in dilute acids, insoluble in water; a fire hazard; used as a bleaching and oxidizing agent, and in medicine.

magnesium phosphate A compound with three forms: monobasic, $MgH_4(PO_4)_2 \cdot 2H_2O$, used in medicine and wood fireproofing; dibasic, $MgHPO_4 \cdot 3H_2O$, used in medicine and as a plastics stabilizer; tribasic, $Mg_3(PO_4)_2 \cdot 8H_2O$, used in dentifrices, as an adsorbent, and in pharmaceuticals.

magnesium salicylate $Mg(C_7H_5O_3)_3 \cdot 4H_2O$ Efflorescent colorless crystals; soluble in water and alcohol; used in medicine.

magnesium silicate $3MgSiO_3 \cdot 5H_2O$ White, water-insoluble powder, containing variable proportions of water of hydration; used as a filler for rubber and in medicine.

magnesium silicofluoride *See* magnesium fluosilicate.

magnesium stearate $Mg(C_{18}H_{35}O_2)_2$ Tasteless, odorless white powder; soluble in hot alcohol, insoluble in water; melts at 89°C; used in paints and medicine, and as a plastics stabilizer and lubricant. Also known as dolomol.

magnesium sulfate $MgSO_4$ Colorless crystals with a bitter, saline taste; soluble in glycerol; used in fireproofing, textile processes, ceramics, cosmetics, and fertilizers.

magnesium sulfite $MgSO_3 \cdot 6H_2O$ A white, crystalline powder; insoluble in alcohol, slightly soluble in water; used in medicine and paper pulp.

magnesium thiosulfate $MgS_2O_3 \cdot 6H_2O$ Colorless crystals that lose water at 170°C; used in medicine. Also known as magnesium hyposulfite.

magnesium trisilicate $Mg_2Si_3O_8 \cdot 5H_2O$ A white, odorless, tasteless powder; insoluble in water and alcohol; used as an industrial odor absorbent and in medicine.

magnesium tungstate $MgWoO_4$ White crystals, insoluble in alcohol and water, soluble in acid; used in luminescent paint and for fluorescent x-ray screens.

magneson $C_{12}H_9N_3O_4$ A brownish-red powder, soluble in dilute aqueous sodium hydroxide; used in the detection of magnesium and molybdenum. Also known as 2,4-dihydroxy-4'-nitroazobenzene.

magnetic filter Filtration device in which the filter screen is magnetized to trap and remove fine iron from liquids or liquid suspensions being filtered.

magnetic quantum number The eigenvalue of the component of an angular momentum operator in a specified direction, such as that of an applied magnetic field, in units of Planck's constant divided by 2π.

magnetic scanning The magnetic field sorting of ions into their respective spectrums for analysis by mass spectroscopy; accomplished by varying the magnetic field strength while the electrostatic field is held constant.

magnetochemistry A branch of chemistry which studies the interrelationship between the bulk magnetic properties of a substance and its atomic and molecular structure.

magnetofluid A Newtonian or shear-thinning fluid whose flow properties become viscoplastic when it is modulated by a magnetic field.

Maillard reaction A reaction in which the amino group in an amino acid tends to form condensation products with aldehydes; believed to cause the Browning reaction when an amino acid and a sugar coexist, evolving a characteristic flavor useful in food preparations.

Majorana force A force between two nucleons postulated to explain various phenomena, which can be derived from a potential containing an operator which exchanges the nucleons' positions but not their spins.

makeup water Water feed needed to replace that which is lost by evaporation or leakage in a closed-circuit, recycle operation.

malate dehydrogenase *See* malic enzyme.

malathion $C_{10}H_{19}O_6PS_2$ A yellow liquid, slightly soluble in water; malathion is the generic name for S-[1,2-bis(ethoxycarbonyl)ethyl] O,O-dimethylphosphorodithioate; used as an insecticide.

maleate An ester or salt of maleic acid.

maleic acid HOOCCH:CHCOOH A colorless, crystalline dibasic acid; soluble in water, acetone, and alcohol; melting point 130–131°C; used in textile processing, and as an oil and fat preservative.

maleic anhydride $C_4H_2O_3$ Colorless crystals, soluble in acetone, hydrolyzing in water; used to form polyester resins. Also known as 2,5-furandione.

maleic hydrazide $C_4N_2H_4O_2$ Solid material, decomposing at 260°C; slightly soluble in alcohol and water; used as a weed killer and growth inhibitor.

malic acid $COOH \cdot CH_2 \cdot CHOH \cdot COOH$ Hydroxysuccinic acid: a dibasic hydroxy acid existing in two optically active isomers and a racemic form; found in apples and many other fruits.

malic dehydrogenase An enzyme in the Krebs cycle that catalyzes the conversion of L-malic acid to oxaloacetic acid.

malic enzyme An enzyme which utilizes nicotinamide-adenine dinucleotide phosphate (NADP) to catalyze the oxidative decarboxylation of malic acid to pyruvic acid and carbon dioxide. Also known as malate dehydrogenase.

malonic acid $CH_2(COOH)_2$ A white, crystalline dicarboxylic acid, melting at 132–134°C; used to manufacture pharmaceuticals. Also known as methanedicarbonic acid; methanedicarboxylic acid.

malonic ester See ethyl malonate.

malonic methyl ester nitrile See methyl cyanoacetate.

malonic mononitrile See cyanacetic acid.

malonyl $CH_2(COO)_2$ A bivalent radical formed from malonic acid.

malonyl urea See barbituric acid.

maltase An enzyme that catalyzes the conversion of maltose to dextrose.

maltobiose See maltose.

maltol $C_6H_6O_3$ Crystalline substance with a melting point of 161–162°C and a fragrant caramellike odor; used as a flavoring agent in bread and cakes. Also known as 3-hydroxy-2-methyl-γ-pyrone; larixinic acid.

maltose $C_{12}H_{22}O_{11}$ A crystalline disaccharide that is a product of the enzymatic hydrolysis of starch, dextrin, and glycogen; does not appear to exist free in nature. Also known as maltobiose; malt sugar.

maltose phosphorylase An enzyme which reacts maltose with inorganic phosphate to yield glucose and glucose-1-phosphate.

malt sugar See maltose.

mammary-stimulating hormone 1. Estrogen and progesterone considered together as the hormones which induce proliferation of the mammary ductile and acinous elements respectively. 2. See prolactin.

mammogen See prolactin.

mammogenic hormone 1. Any hormone that stimulates or induces development of the mammary gland. 2. See prolactin.

mammotropin See prolactin.

mandelic acid $C_6H_5CHOHCOOH$ A white, crystalline compound, melting at 117–119°C, darkening upon exposure to light; used in organic synthesis. Also known as amygdalic acid; benzoglycolic acid; phenylglycolic acid; α-phenylhydroxyacetic acid.

mandelic acid nitrile See mandelonitrile.

mandelonitrile $C_6H_5CH(OH)CN$ A liquid used to prepare bitter almond water. Also known as benzaldehyde cyanohydrin; mandelic acid nitrile.

maneb $Mn[SSCH(CH_2)_2NHCSS]$ A generic term for manganese ethylene-1,2-bisdithiocarbamate; irritating to eyes, nose, skin, and throat; used as a fungicide.

manganate 1. Salts that have manganese in the anion. 2. In particular, a salt of manganic acid formed by fusion of manganese dioxide with an alkali.

manganese A metallic element, symbol Mn, atomic weight 54.938, atomic number 25; a transition element whose properties fall between those of chromium and iron.

manganese acetate $Mn(C_2H_3O_2)_2 \cdot 4H_2O$ A pale-red crystalline compound melting at 80°C; soluble in water and alcohol; used in textile dyeing, as a catalyst, and for leather tanning.

manganese binoxide *See* manganese dioxide.

manganese black *See* manganese dioxide.

manganese borate MnB_4O_7 Water-insoluble, reddish-white powder; used as a varnish and oil drier.

manganese bromide *See* manganous bromide.

manganese carbonate $MnCO_3$ Rose-colored crystals found in nature as rhodocrosite; soluble in dilute acids, insoluble in water; used in medicine, in fertilizer, and as a paint pigment.

manganese citrate $Mn_3(C_6H_5O_7)_2$ A white powder, water-insoluble in the presence of sodium citrate; used in medicine.

manganese dioxide MnO_2 A black, crystalline, water-insoluble compound, decomposing to manganese sesquioxide, Mn_2O_3, and oxygen when heated to 535°C; used as a depolarizer in certain dry-cell batteries, as a catalyst, and in dyeing of textiles. Also known as battery manganese; manganese binoxide; manganese black; manganese peroxide.

manganese fluoride *See* manganous fluoride.

manganese gluconate $Mn(C_6H_{11}O_7)_2 \cdot 2H_2O$ A pinkish powder, insoluble in benzene and alcohol, soluble in water; used in medicine, in vitamin tablets, and as a feed additive and dietary supplement.

manganese green *See* barium manganate.

manganese halide Compound of manganese with a halide, such as chlorine, bromine, fluorine, or iodine.

manganese heptoxide Mn_2O_7 A compound formed as an explosive dark-green oil by the action of concentrated sulfuric acid on permanganate compounds.

manganese hydrogen phosphate *See* acid manganous phosphate.

manganese hydroxide *See* manganous hydroxide.

manganese hypophosphite $Mn(H_2PO_2)_2 \cdot H_2O$ Odorless, tasteless pink crystals which explode if heated with oxidants; used in medicine.

manganese iodide *See* manganous iodide.

manganese lactate $Mn(C_3H_5O_3)_2 \cdot 3H_2O$ Pale-red crystals; insoluble in water and alcohol; used in medicine.

manganese linoleate $Mn(C_{18}H_{31}O_2)_2$ A dark-brown mass, soluble in linseed oil; used in pharmaceutical preparations and as a varnish and paint drier.

manganese monoxide *See* manganese oxide.

manganese naphthenate Hard brown resinous mass, soluble in mineral spirits; melts at 135°C; contains 6% manganese in commercial solutions; used as a paint and varnish drier.

manganese oleate $Mn(C_{18}H_{33}O_2)_2$ Granular brown mass, soluble in oleic acid and ether, insoluble in water; used in medicine and as a varnish drier.

manganese oxalate $MnC_2O_4 \cdot 2H_2O$ A white crystalline compound, soluble in dilute acids, only slightly soluble in water; used as a paint and varnish drier.

manganese oxide MnO Green powder, soluble in acids, insoluble in water; melts at 1650°C; used in medicine, in textile printing, as a catalyst, in ceramics, and in dry batteries. Also known as manganese monoxide; manganous oxide.

manganese peroxide See manganese dioxide.

manganese resinate $Mn(C_{20}H_{29}O_2)_2$ Water-insoluble mass, flesh-colored or brownish black; used as a varnish and oil drier.

manganese silicate See manganous silicate.

manganese sulfate See manganous sulfate.

manganese sulfide See manganous sulfide.

manganic fluoride MnF_3 Poisonous red crystals, decomposed by heat and water; used as a fluorinating agent.

manganic hydroxide $Mn(OH)_3$ A brown powder that rapidly loses water to form MnO(OH); used in ceramics and as a fabric pigment. Also known as hydrated manganic hydroxide.

manganic oxide Mn_2O_3 Hard black powder, insoluble in water, soluble in cold hydrochloric acid, hot nitric acid, and sulfuric acid; occurs in nature as manganite. Also known as manganese sesquioxide.

manganous bromide $MnBr_2 \cdot 4H_2O$ Water-soluble, deliquescent red crystals. Also known as manganese bromide.

manganous chloride $MnCl_2 \cdot 4H_2O$ Water-soluble, deliquescent rose-colored crystals melting at 88°C; used as a catalyst and in paints, dyeing, and pharmaceutical preparations.

manganous fluoride MnF_2 Reddish powder, insoluble in water, soluble in acid. Also known as manganese fluoride.

manganous hydroxide $Mn(OH)_2$ Heat-decomposable white-pink crystals; insoluble in water and alkali, soluble in acids; occurs in nature as pyrochroite. Also known as manganese hydroxide.

manganous iodide $MnI_2 \cdot 4H_2O$ Water-soluble, deliquescent yellowish-brown crystals. Also known as manganese iodide.

manganous oxide See manganese oxide.

manganous silicate $MnSiO_3$ Water-insoluble red crystals or yellowish-red powder; occurs in nature as rhodonite. Also known as manganese silicate.

manganous sulfate $MnSO_4 \cdot 4H_2O$ Water-soluble, translucent, efflorescent rose-red prisms; melts at 30°C; used in medicine, textile printing, and ceramics, as a fungicide and fertilizer, and in paint manufacture. Also known as manganese sulfate.

manganous sulfide   MnS   An almost water-insoluble powder that decomposes on heating; used as a pigment and as an additive in making steel. Also known as manganese sulfide.

manganous sulfite $MnSO_3$ Grayish-black or brownish-red powder, soluble in sulfur dioxide, insoluble in water.

manifold of states   A set of states sufficient to form a representation of an operator or a Lie group of operators.

mannan   Any of a group of polysaccharides composed chiefly or entirely of D-mannose units.

manna sugar   *See* mannitol.

Mannich reaction   Condensation of a primary or secondary amine or ammonia (usually as the hydrochloride) with formaldehyde and a compound containing at least one reactive hydrogen atom, for example, acetophenone. Also known as Mannich condensation reaction.

mannite   *See* mannitol.

mannitol $C_6H_8(OH)_6$ A straight-chain alcohol with six hydroxyl groups; a white, water-soluble, crystalline powder; used in medicine and as a dietary supplement. Also known as manna sugar; mannite.

mannitol hexanitrate $C_6H_8(ONO_2)_6$ Explosive colorless crystals; soluble in alcohol, acetone, and ether, insoluble in water; melts at 112°C; used in explosives and medicine. Also known as hexanitromannite (HNM); nitromannite; nitromannitol.

mannose $C_6H_{12}O_6$ A fermentable monosaccharide obtained from manna.

manure salts   Potash salts that have a high proportion of chloride and 20–30% potash; used in fertilizers.

maphenide   *See* para-(aminomethyl)benzenesulfonamide.

Marangoni effect   The effect that a disturbance of the liquid-liquid interface (due to interfacial tension) has on mass transfer in a liquid-liquid extraction system.

Mark-Houwink equation   The relationship between intrinsic viscosity and molecular weight for homogeneous linear polymers.

Markovnikoff's rule   In an addition reaction, the additive molecule RH adds as H and R, with the R going to the carbon atom with the lesser number of hydrogen atoms bonded to it.

marsh gas   Combustible gas, consisting chiefly of methane, produced as a result of decay of vegetation in stagnant water.

Marsh test   A test for the presence of arsenic in a compound; the substance to be tested is mixed with granular zinc, and dilute hydrochloric acid is added to the mixture; gaseous arsine forms, which decomposes to a black deposit of arsenic, when the gas is passed through a heated glass tube. Also known as Marsh-Berzelius test.

Mars pigments   A group of five pigments produced when milk of lime is added to a ferrous sulfate solution, and the precipitate is calcined; color is controlled by calcination temperature to give yellow, orange, brown, red, or violet.

mass action law   The law that the rate of a chemical reaction for a uniform system at constant temperature is proportional to the concentrations of the substances reacting. Also known as Guldberg and Waage law.

mass-analyzed ion kinetic energy spectrometry   A type of ion kinetic energy spectrometry in which the ionic products undergo mass analysis followed by energy analysis. Abbreviated MIKES.

mass defect   The difference between the mass of an atom and the sum of the masses of its individual components in the free (unbound) state.

Massenfilter   *See* quadrupole spectrometer.

Massey formula   A formula for the probability that an excited atom approaching the surface of a metal will emit secondary electrons.

mass formula   An equation giving the atomic mass of a nuclide as a function of its atomic number and mass number.

mass number   The sum of the numbers of protons and neutrons in the nucleus of an atom or nuclide. Also known as nuclear number; nucleon number.

mass shift   The portion of the isotope shift which results from the difference between the nuclear masses of different isotopes.

mass spectrometry   An analytical technique for identification of chemical structures, determination of mixtures, and quantitative elemental analysis, based on application of the mass spectrometer.

mass susceptibility   Magnetic susceptibility of a compound per gram. Also known as specific susceptibility.

mass-to-charge ratio   In analysis by mass spectroscopy, the measurement of the sample mass as a ratio to its ionic charge.

master equation   An equation which determines the rate of change of the population of an energy level in terms of the populations of other levels and transition probabilities.

masticate   To process rubber on a machine to make it softer and more pliable before mixing with other substances.

material balance   A calculation to inventory material inputs versus outputs in a process system.

material well   In a plastics process, the space provided in a compression or transfer mold to allow for the bulk factor.

matrix effects   The enhancement or suppression of minor element spectral lines from metallic oxides during emission spectroscopy by the matrix element (such as graphite) used to hold the sample.

matrix isolation   A spectroscopic technique in which reactive species can be characterized by maintaining them in a very cold, inert environment while they are examined by an absorption, electron-spin resonance, or laser excitation spectroscope.

matrix spectrophotometry   Spectrophotometric analysis in which the specimen is irradiated in sequence at more than one wavelength, with the visible spectrum evaluated for the energy leaving for each wavelength of irradiation.

matromycin   *See* oleandomycin.

mb   *See* millibarn.

McCabe-Thiele diagram   Graphical method for calculation of the number of theoretical plates or contacting stages required for a given binary distillation operation.

Md   *See* mendelevium.

MEA   *See* 2-aminoethanethiol.

mean-average boiling point   Pseudo boiling point for a hydrocarbon mixture; calculated from the ASTM distillation curve's volumetric average boiling point.

mean temperature difference   In heat exchange calculations, a pseudo average temperature difference between the warmer and colder fluids at inlet and outlet conditions.

mecarbam $C_{10}H_{20}O_5PS_2$ An oily liquid, used as an insecticide and miticide for pests of rice, cabbage, and onions. Also known as S-(N-ethoxycarbonyl-N-methylcarbamoylmethyl)-O,O-diethylphosphorodithioate.

mecarphon $C_7H_{14}O_4PS_2$ A solid with a melting point of 36°C; insoluble in water; used as an insecticide and miticide for soil insects and for mite and aphid control. Also known as S-(N-methoxycarbonyl-N-methylcarbamoylmethyl)methyl-O-methyl phosphorodithioate.

mechanochemical effect   Changes in the dimensions of certain polymers, particularly photoelectrolytic gels and crystalline polymers, in response to changes in their chemical environment.

mechanochemistry   The study of the conversion of mechanical energy into chemical energy in polymers.

mechanophotochemistry   The study of changes in the dimensions of certain photoresponsive polymers upon exposure to light.

meconin $C_{10}H_{10}O_4$ A neutral principle of opium; white crystals, soluble in hot water and alcohol and melting at 102–103°C. Also known as opianyl.

mecoprop $C_{10}H_{11}ClO_3$ A colorless, crystalline compound with a melting point of 94–95°C; slightly soluble in water; used as an herbicide to control broadleaf weeds in cereals, turf, and lawns. Also known as 2-[(4-chloro-*ortho*-tolyl)oxy]propionic acid.

media migration   Carryover of fibers or other filter material by liquid effluent from a filter unit.

medinoterb acetate $C_{13}H_{16}N_2O_6$ A yellow, crystalline compound with a melting point of 86–87°C; used as a preemergence herbicide for broadleaf weeds in cotton, sugarbeets, and leguminous crops, and as a postemergence herbicide for cereal crops. Also known as 2-*tert*-butyl-5-methyl-4,6-dinitrophenyl acetate.

medium   1. The carrier in which a chemical reaction takes place. 2. Material of controlled pore size used to remove foreign particles or liquid droplets from fluid carriers.

meglumine   *See* N-methyl glucamine.

MEK   *See* methyl ethyl ketone.

melamine $C_3H_6N_4$ A white crystalline compound that is slightly soluble in water, melts at 354°C and is a cyclic trimer of cyanamide; used to make melamine resins and in tanning of leather. Also known as cyanurtriamide; 2,4,6-triamino-s-triazine.

melampyrit *See* dulcitol.

melaniline *See* diphenylguanidine.

melanin Any of a group of brown or black pigments occurring in plants and animals.

melanocyte-stimulating hormone A protein substance secreted by the intermediate lobe of the pituitary of man which causes dispersion of pigment granules in the skin; similar to intermedins in other vertebrates. Abbreviated MSH. Also known as melanophore-dilating principle; melanophore hormone.

melanogen A colorless precursor of melanin.

melanogenesis The formation of melanin.

melanophore-dilating principle *See* melanocyte-stimulating hormone.

melanophore hormone *See* melanocyte-stimulating hormone.

melanoprotein A conjugated protein in which melanin is the associated chromagen.

M electron An electron whose principal quantum number is 3.

melissic acid $CH_3(CH_2)_{29}COOH$ Fatty acid found in beeswax; soluble in benzene and hot alcohol; melts at 90°C; used in biochemical research.

melitose *See* raffinose.

melitriose *See* raffinose.

mellitate An ester or salt of mellitic acid.

mellitic acid $C_6(COOH)_6$ A water-soluble compound forming colorless needles that melt at 287°C.

melt 1. To change a solid to a liquid by the application of heat. 2. A melted material.

melting *See* fusion.

membrane 1. The medium through which the fluid stream is passed for purposes of filtration. 2. The ion-exchange medium used in dialysis, diffusion, osmosis and reverse osmosis, and electrophoresis.

MEMC *See* methoxyethylmercury chloride.

menadione $C_{11}H_8O_2$ Yellow crystals, soluble in alcohol, benzene, and vegetable oils, insoluble in water; affected by sunlight; used in medicine and fungicides.

menazon $C_6H_8N_5O_2PS_2$ A colorless, crystalline compound that decomposes at 160–162°C; slightly soluble in water; used as an insecticide for the control of aphids. Also known as S-(4,6-diamino-S-triazin-2-yl-methyl) O,O-dimethylphosphorodithioate.

mendelevium Synthetic radioactive element, symbol Md, with atomic number 101; made by bombarding lighter elements with light nuclei accelerated in cyclotrons.

menthacamphor *See* menthol.

menthane $C_{10}H_{20}$ A colorless, water-insoluble liquid hydrocarbon; used in organic synthesis. Also known as hexahydrocymene; 4-isopropyl-1-methylcyclohexane; menthonaphthene; terpane.

menthene $C_{10}H_{18}$ A colorless, water-insoluble, liquid hydrocarbon; used in organic synthesis.

para-**menthan-3-ol** *See* menthol.

menthol $CH_3C_6H_9(C_3H_7)OH$ An alcohol-soluble, white crystalline compound that may exist in levo form or a mixture of dextro and levo isomers; used in medicines and perfumes, and as a flavoring agent. Also known as hexahydrothymol; 3-hydroxymenthane; menthacamphor; *para*-menthan-3-ol; methylhydroxyisopropylcyclohexane; peppermint camphor.

menthonaphthene *See* menthane.

menthone $C_{10}H_{18}O$ Oily, colorless ketonic liquid with slight peppermint odor; slightly soluble in water, soluble in organic solvents.

menthyl $C_{10}H_{19}$ A univalent radical that is derived from menthol by removal of the hydroxyl group.

meperidine hydrochloride $C_{15}H_{21}O_2N \cdot HCl$ A white, odorless crystalline compound, melting at 186–189°C; soluble in water and alcohol; used in medicine.

mephentermine sulfate $(C_{11}H_{17}N)_2 \cdot H_2SO_4 \cdot 2H_2O$ White odorless crystals; slightly soluble in alcohol, soluble in water; used in medicine.

mephosfolan $C_8H_{16}O_3PNS_2$ A yellow to amber liquid, used as an insecticide and miticide for agricultural crops.

mepyrapone *See* metyrapone.

-mer A combining form denoting the repeating structure unit of any high polymer.

merbromin $C_{20}H_8O_6Na_2Br_2Hg$ A green crystalline powder that gives a deep-red solution in water; used as an antiseptic.

mercamine *See* 2-aminoethanethiol.

mercaptal A group of organosulfur compounds that contain the group $=C(SR)_2$.

mercaptamine *See* 2-aminoethanethiol.

mercaptan A group of organosulfur compounds that are derivatives of hydrogen sulfide in the same way that alcohols are derivatives of water; have a characteristically disagreeable odor, and are found with other sulfur compounds in crude petroleum; an example is methyl mercaptan. Also known as thiol.

mercaptide A compound consisting of a metal and a mercaptan.

mercapt-, mercapto- A combining form denoting the presence of the thiol (SH) group.

mercaptoacetic acid *See* thioglycollic acid.

2-mercaptobenzoic acid *See* thiosalicylic acid.

mercaptobenzothiazole C_7H_5NS A yellow powder, melting at 164–174°C; used in rubber as a vulcanization accelerator with stearic acid. Also known as MBT.

mercapto compound *See* sulfhydryl compound.

mercaptodimethur $C_{11}H_{15}NO_2S$ A white, crystalline compound with a melting point of 121°C; insoluble in water; used as an insecticide, molluscide, and bird repellent for vegetable, fruit crops, and ornamentals. Also known as 4-(methylthio)-3,5-xylylmethylcarbamate.

mercaptoethanol $HSCH_2CH_2OH$ Mobile liquid, water-white; soluble in water, benzene, ether, and most organic solvents; boils at 157°C; used as a solvent, chemical intermediate, and reducing agent.

β-mercaptoethylamine   *See* 2-aminoethanethiol.

mercaptol   A compound formed by combining a mercaptal and a ketone.

N-(mercaptomethyl)-phthalimide S-(0,0-dimethylphosphorodithioate)   *See* phosmet.

2-mercaptoproprionic acid   *See* thiolactic acid.

mercaptosuccinic acid   *See* thiomalic acid.

β-mercaptovaline   *See* penicillamine.

mercuric   The mercury ion with a 2+ oxidation state, for example $Hg(NO_3)_2$.

mercuric acetate $Hg(C_2H_3O_2)_2$ Poisonous, light-sensitive white crystals; soluble in alcohol and water; used in medicine and as a catalyst in organic synthesis. Also known as mercury acetate.

mercuric arsenate $HgHAsO_4$ A poisonous yellow powder; soluble in hydrochloric acid, insoluble in water; used in antifouling and waterproof paints and in medicine. Also known as mercury arsenate; mercury arseniate.

mercuric barium iodide $HgI_2 \cdot BaI_2 \cdot 5H_2O$ Crystals that are yellow or reddish and deliquescent; soluble in alcohol and water; used in aqueous solution as Rohrbach's solution for mineral separation on the basis of density. Also known as barium mercury iodide; mercury barium iodide.

mercuric benzoate $Hg(C_7H_5O_2)_2 \cdot H_2O$ Poisonous white crystals, sensitive to light, melting at 165°C; slightly soluble in alcohol and water; used in medicine. Also known as mercury benzoate.

mercuric bromide $HgBr_2$ Poisonous white crystals, sensitive to light, melting at 235°C; soluble in alcohol and ether; used in medicine. Also known as mercury bromide.

mercuric chloride $HgCl_2$ An extremely toxic compound that forms white, rhombic crystals which sublime at 300°C and are soluble in alcohol or benzene; used for the manufacture of other mercuric compounds, as a fungicide, and in medicine and photography. Also known as bichloride of mercury; corrosive sublimate.

mercuric cyanate   *See* mercury fulminate.

mercuric cyanide $Hg(CN)_2$ Poisonous, colorless, transparent crystals that darken in light, decompose when heated; soluble in water and alcohol; used in photography, medicine, and germicidal soaps. Also known as mercury cyanide.

mercuric fluoride HgF_2 Poisonous, transparent crystals that decompose when heated; moderately soluble in alcohol and water; used to synthesize organic fluorides.

mercuric iodide HgI_2 Poisonous red crystals that turn yellow when heated to 150°C; soluble in boiling alcohol; used in medicine and in Nessler's and Mayer's reagents.

mercuric lactate $Hg(C_3H_5O_3)_2$ A poisonous white powder that decomposes when heated; soluble in water; used in medicine.

mercuric nitrate $Hg(NO_3)_2 \cdot H_2O$ Poisonous, colorless crystals that decompose when heated; soluble in water and nitric acid, insoluble in alcohol; a fire hazard; used in medicine, in nitrating organic aromatics, and in felt manufacture. Also known as mercury nitrate; mercury pernitrate.

mercuric oleate $Hg(C_{18}H_{33}O_2)_2$ A poisonous yellowish-to-red liquid or solid mass; insoluble in water; used in medicine and antifouling paints, and as an antiseptic. Also known as mercury oleate.

mercuric oxide   HgO   A compound of mercury that exists in two forms, red mercuric oxide and yellow mercuric oxide; the red form decomposes upon heating, is insoluble in water, and is used in pigments and paints, and in ceramics; the yellow form is insoluble in water, decomposes upon heating, and is used in medicine. Also known as mercury oxide; red precipitate; yellow precipitate.

mercuric phosphate $Hg_3(PO_4)_2$ Poisonous yellowish or white powder; insoluble in alcohol and water, soluble in acids; used in medicine. Also known as mercury phosphate; neutral, normal, or tertiary mercuric phosphate; trimercuric orthophosphate.

mercuric salicylate $Hg(C_7H_5O_3)_2$ Poisonous, white powder; odorless and tasteless; almost insoluble in water and alcohol; variable composition; used in medicine. Also known as salicylated mercury.

mercuric stearate $Hg(C_{17}H_{35}CO_2)_2$ Poisonous yellow powder; soluble in fatty acids, slightly soluble in alcohol; used as a germicide and in medicine. Also known as mercury stearate.

mercuric sulfate $HgSO_4$ A toxic, white, crystalline powder, soluble in acids; used in medicine, as a catalyst, and for galvanic batteries. Also known as mercury persulfate; mercury sulfate.

mercuric sulfide   HgS   **1.** The black variety is a poisonous powder; insoluble in water, alcohol, and nitric acid, soluble in sodium sulfide solution; sublimes at 583°C; used as a pigment. Also known as black mercury sulfide; ethiops mineral. **2.** The red variety is a poisonous powder; insoluble in water and alcohol; sublimes at 446°C; used as a medicine and pigment. Also known as Chinese vermilion; quicksilver vermilion; red mercury sulfide; vermilion.

mercuric sulfocyanate   *See* mercuric thiocyanate.

mercuric sulfocyanide   *See* mercuric thiocyanate.

mercuric thiocyanate $Hg(SCN)_2$ Poisonous white powder; soluble in alcohol, slightly soluble in water; decomposes when heated; used in photography. Also known as mercuric sulfocyanide; mercury sulfocyanate; mercury thiocyanate.

mercurous   Referring to mercury with a $1+$ valence; for example, mercurous chloride, Hg_2Cl_2, where the mercury is covalently bonded, as Cl—Hg—Hg—Cl.

mercurous acetate $HgC_2H_3O_2$ Poisonous colorless plates or scales; decomposed by boiling water and by light; soluble in dilute nitric acids, slightly soluble in water. Also known as mercury acetate; mercury protoacetate.

mercurous bromide   HgBr   Poisonous white powder, crystals, or fibrous mass; odorless and tasteless; darkens in light; soluble in hot sulfuric acid and fuming nitric acid, insoluble in alcohol and ether; used in medicine. Also known as mercury bromide.

mercurous chlorate $Hg_2(ClO_3)_2$ Poisonous white crystals that decompose at 250°C; soluble in alcohol and water; explodes in contact with combustible substances. Also known as mercury chlorate.

mercurous chloride Hg_2Cl_2 Odorless, nonpoisonous white crystals that darken in light; insoluble in water, alcohol, and ether; melts at 302°C; used in medicine and pyrotechnics. Also known as mercury monochloride; mercury protochloride; mild mercury chloride.

mercurous chromate Hg_2CrO_4 Red powder with variable composition; decomposes when heated; soluble in nitric acid, insoluble in water and alcohol; used to color ceramics green. Also known as mercury chromate.

mercurous iodide Hg_2I_2 Odorless, tasteless, poisonous yellow powder; darkens when heated; insoluble in water, alcohol, and ether; sublimes at 140°C; used as external medicine. Also known as mercury protoiodide.

mercurous oxide Hg_2O A poisonous black powder; insoluble in water, soluble in acids; decomposes at 100°C.

mercurous phosphate Hg_3PO_4 Light-sensitive white powder with variable composition; insoluble in alcohol and water, soluble in nitric acids; used in medicine. Also known as mercury phosphate; neutral, normal, or tertiary mercurous phosphate; trimercurous orthophosphate.

mercurous sulfate Hg_2SO_4 Poisonous yellow-to-white powder; soluble in hot sulfuric acid or dilute nitric acid, insoluble in water; used as a catalyst and in laboratory batteries.

mercury A metallic element, symbol Hg, atomic number 80, atomic weight 200.59, existing at room temperature as a silvery, heavy liquid. Also known as quicksilver.

mercury acetate See mercuric acetate; mercurous acetate.

mercury arsenate See mercuric arsenate.

mercury arseniate See mercuric arsenate.

mercury barium iodide See mercuric barium iodide.

mercury benzoate See mercuric benzoate.

mercury bromide See mercuric bromide; mercurous bromide.

mercury-cathode cell Electrolytic cell used to manufacture chlorine and caustic soda from sodium chloride brine; includes Castner and DeNora cells.

mercury chlorate See mercurous chlorate.

mercury chromate See mercurous chromate.

mercury cosmetic See ammoniated mercury.

mercury cyanide See mercuric cyanide.

mercury fulminate $Hg(CNO)_2$ A gray, crystalline powder; explodes at the melting point; soluble in alcohol, ammonium hydroxide, and hot water; used for explosive caps and detonators. Also known as mercuric cyanate.

mercury monochloride See mercurous chloride.

mercury naphthenate Poisonous dark-amber liquid; soluble in mineral oils; used in gasoline antiknock compounds and as a paint antimildew promoter.

mercury nitrate *See* mercuric nitrate.

mercury oleate *See* mercuric oleate.

mercury oxide *See* mercuric oxide.

mercury pernitrate *See* mercuric nitrate.

mercury persulfate *See* mercuric sulfate.

mercury phosphate *See* mercuric phosphate; mercurous phosphate.

mercury protoacetate *See* mercurous acetate.

mercury protochloride *See* mercurous chloride.

mercury protoiodide *See* mercurous iodide.

mercury sulfocyanate *See* mercuric thiocyanate.

mercury stearate *See* mercuric stearate.

mercury sulfate *See* mercuric sulfate.

mercury sulfucyanate *See* mercuric thiocyanate.

mercury thiocyanate *See* mercuric thiocyanate.

merohedral Of a crystal class in a system, having a general form with only one-half, one-fourth, or one-eighth the number of equivalent faces of the corresponding form in the holohedral class of the same system. Also known as merosymmetric.

meromyosin Protein fragments of a myosin molecule, produced by limited proteolysis.

merosymmetric *See* merohedral.

mesaconic acid $C_5H_6O_4$ An unsaturated dibasic acid, an isomer of citraconic acid, that melts at 202°C. Also known as methyl fumaric acid.

mescaline $C_{11}H_{17}NO_3$ The alkaloid 3,4,5-trimethoxyphenethylamine, found in mescal buttons; produces unusual psychic effects and visual hallucinations.

mesityl oxide $(CH_3)_2C{=}CHCOCH_3$ A colorless, oily liquid with a honeylike odor; solidifies at -41.5°C; used as a solvent for resins, particularly vinyl resins, many gums, and nitrocellulose; also used in lacquers, paints, and varnishes. Also known as isopropylideneacetone.

meso- A prefix meaning intermediate or middle, as in denoting inactive optical isomers, the form of intermediate inorganic acid, the middle position in cyclic organic compounds, or a ring system with middle ring positions.

mesobilirubin $C_{33}H_{40}O_6N_4$ Yellow, crystalline by-product of bilirubin reduction.

mesobilirubinogen $C_{33}H_{44}O_6N_4$ Colorless, crystalline by-product of bilirubin reduction; may be converted to urobilin, stercobilinogen, or stercobilin.

mesobiliverdin $C_{28}H_{38}O_6N_4$ A structural isomer of phycoerythrin and phycocyanobilin released by certain biliprotein by treatment with alkali.

mesoinositol *See* myoinositol.

mesomorphism   A state of matter intermediate between a crystalline solid and a normal isotropic liquid, in which long rod-shaped organic molecules contain dipolar and polarizable groups.

mesoxalyurea   *See* alloxan.

messenger ribonucleic acid   A linear sequence of nucleotides which is transcribed from and complementary to a single strand of deoxyribonucleic acid and which carries the information for protein synthesis to the ribosomes. Abbreviated mRNA.

meta-   A prefix for benzene-ring compounds when two side chains are connected to carbon atoms with an unsubstituted carbon atom between them.

metabolite   A product of intermediary metabolism.

metachromasia   **1.** The property exhibited by certain pure dyestuffs, chiefly basic dyes, of coloring certain tissue elements in a different color, usually of a shorter wavelength absorption maximum, than most other tissue elements. **2.** The assumption of different colors or shades by different substances when stained by the same dye. Also known as metachromatism.

metachromatism   *See* metachromasia.

metaformaldehyde   *See* *sym*-trioxane.

metahydrate sodium carbonate $Na_2CO_3 \cdot H_2O$ Water-soluble, white crystals with an alkaline taste, loses water at 109°C, melts at 851°C; used in medicine, photography, and water pH control, and as a food additive. Also known as crystal carbonate; soda crystals.

metal alkyl   One of the family of organometallic compounds, a combination of an alkyl organic radical with a metal atom or atoms.

metal cluster compound   A compound in which two or more metal atoms aggregate so as to be within bonding distance of one another and each metal atom is bonded to at least two other metal atoms; some nonmetal atoms may be associated with the cluster.

metaldehyde $(CH_3CHO)_n$ White acetaldehyde-polymer prisms; soluble in organic solvents, insoluble in water; used as a pesticide or fuel.

metallic bond   The type of chemical bond that is present in all metals, and may be thought of as resulting from a sea of valence electrons which are free to move throughout the metal lattice.

metallic element   An element generally distinguished (from a nonmetallic one) by its luster, electrical conductivity, malleability, and ability to form positive ions.

metallic soap   A salt of stearic, oleic, palmitic, lauric, or erucic acid with a heavy metal such as cobalt or copper; used as a drier in paints and inks, in fungicides, decolorizing varnish, and waterproofing.

metallocene   Organometallic coordination compound which is obtained as a cyclopentadienyl derivative of a transition metal or a metal halide.

metallocycle   A compound whose structure consists of a cyclic array of atoms of which one is a metal atom; frequently the ring contains three or four carbon atoms and one transition-metal atom.

metalloid   A nonmetallic element, such as carbon or nitrogen, which can combine with a metal to form an alloy.

metalloporphyrin A compound, such as heme, consisting of a porphyrin combined with a metal such as iron, copper, silver, zinc, or magnesium.

metalloprotein A protein enzyme containing a metallic atom as an inherent portion of its molecule.

metallothionein A group of vertebrate and invertebrate proteins that bind heavy metals; it may be involved in zinc homeostasis and resistance to heavy-metal toxicity.

metamer One of two or more chemical compounds that exhibits isomerism with the others.

metanephrine An inactive metabolite of epinephrine (3-O-methylepinephrine) that is excreted in the urine; it is recovered and measured as a test for pheochromocytoma.

metanilic acid $C_6H_4(NH_2)SO_3H$ A water-soluble, crystalline compound, isomeric with sulfanilic acid; used in medicines and dyes. Also known as *meta*-aminobenzenesulfonic acid; *meta*-sulfanilic acid.

metastable equilibrium A state of pseudoequilibrium having higher free energy than the true equilibrium state.

metastable ion In mass spectroscopy, an ion formed by a secondary dissociation process in the analyzer tube (formed after the parent or initial ion has passed through the accelerating field).

metastable phase Existence of a substance as either a liquid, solid, or vapor under conditions in which it is normally unstable in that state.

metathesis A reaction involving the exchange of elements or groups as in the general equation $AX + BY \rightarrow AY + BX$.

metathetical salts Salts that form a four-component, ternary equilibrium system in which there are four possible binary systems, resulting in two quadruple points.

metatitanic acid *See* titanic acid.

metering pump Plunger-type pump designed to control accurately small-scale fluid-flow rates; used to inject small quantities of materials into continuous-flow liquid streams. Also known as proportioning pump.

methabenzthiazuron $C_{10}H_{11}N_3OS$ A white, crystalline compound with a melting point of 119–120°C; used to control weeds in wheat, peas, garlic, broad beans, and onions. Also known as 1,3-dimethyl-3-(2-benzothiazolyl)-urea.

methacetin $CH_3OC_6H_4NHCOCH_3$ A water-insoluble, white powder, melting at 127.1°C. Also known as acetanisidine; *para*-methoxyacetanilide.

methacrolein $CH_2C(CH_3)CHO$ Liquid with 68°C boiling point; slightly soluble in water; used to make resins and copolymers.

methacrylate ester $CH_2{:}C(CH_3)COOR$ Methacrylic acid ester in which R can be methyl, ethyl, isobutyl, or 50-50 n-butyl-isobutyl groups; used to make thermoplastic polymers or copolymers.

methacrylic acid $CH_2C(CH_3)COOH$ Easily polymerized, colorless liquid melting at 15–16°C; soluble in water and most organic solvents; used to make water-soluble polymers and as a chemical intermediate.

methacrylonitrile $CH_2{:}C(CH_3)CN$ Clear, colorless liquid boiling at 90°C; used to make solvent-resistant thermoplastic polymers and copolymers.

methallyl alcohol $H_2C{:}C(CH_3)CH_2OH$ Flammable, toxic, water-soluble, colorless liquid boiling at 115°C; has pungent aroma; soluble in most organic solvents; used as a chemical intermediate. Also spelled methyl allyl alcohol.

methamidophos *See* acephatemet.

methanamide *See* formamide.

methane CH_4 A colorless, odorless, and tasteless gas, lighter than air and reacting violently with chlorine and bromine in sunlight, a chief component of natural gas; used as a source of methanol, acetylene, and carbon monoxide. Also known as methyl hydride.

methanearsonic acid $CH_3AsO(OH)_2$ A white solid with a melting point of 161°C; very soluble in water; used as an herbicide for cotton crops and for noncrop areas. Abbreviated MAA.

methanedicarbonic acid *See* malonic acid.

methanedicarboxylic acid *See* malonic acid.

methane-disulfonic acid *See* methionic acid.

methanesulfonic acid CH_3SO_2OH A solid with a melting point of 20°C; used as a catalyst in polymerization, esterification, and alkylation reactions, and as a solvent. Also known as methysulfonic acid.

methanesulfonic acid tetramethylene ester *See* busulfan.

methanethiomethane *See* methyl sulfide.

methanogenesis The biosynthesis of the hydrocarbon methane; common in certain bacteria. Also known as bacterial methanogenesis.

methanoic acid *See* formic acid.

methanol *See* methyl alcohol.

methazole $C_9H_6Cl_2N_2O_3$ A tan solid with a melting point of 123–124°C; slight solubility in water; used as an herbicide for pre- and postemergence control of weeds in crops. Also known as 2-(3,4-dichlorophenyl)-4-methyl-1,2,4-oxadiazolidine-3,5-dione.

methemoglobin *See* ferrihemoglobin.

methenyl *See* methine group.

methidathion $C_4H_{11}O_4N_2PS_3$ A colorless, crystalline compound with a melting point of 39–40°C; used as an insecticide and miticide for pests on alfalfa, citrus, and cotton.

methide A binary compound consisting of methyl and, most commonly, a metal, such as sodium (sodium methide, $NaCH_3$).

methine group $HC{\equiv}$ A radical consisting of a single carbon and a single hydrogen. Also known as methenyl; methylidyne.

methionic acid $CH_2(SO_3H)_2$ An acid that exists as hygroscopic crystals; used in organic synthesis. Also known as methane-disulfonic acid.

methionine $C_5H_{11}O_2NS$ An essential amino acid; furnishes both labile methyl groups and sulfur necessary for normal metabolism.

methomyl $C_5H_{10}N_2O_2S$ A white, crystalline compound with a melting point of 78–79°C; limited solubility in water; used as an insecticide, nematicide, and fumigant on crops. Also known as S-methyl-N-[(methylcarbamoyl)-oxy]thioacetimidate.

methone *See* 5,5-dimethyl-1,3-cyclohexanedione.

methopyrapone *See* metyrapone.

methotrexate *See* amethopterin.

methoxide A compound formed from a metal and the methoxy radical; an example is sodium methoxide. Also known as methylate.

methoxy- OCH_3— A combining form indicating the oxygen-containing methane radical, found in many organic solvents, insecticides, and plasticizer intermediates.

para-**methoxyacetanilide** *See* methacetin.

4-methoxybenzaldehyde *See* anisaldehyde.

para-**methoxybenzaldehyde** *See* anisaldehyde.

methoxybenzene *See* anisole.

2-methoxy-4H-1,3,2-benzodioxaphosphorin-2-sulfide $C_8H_9O_3PS$ A light yellow, crystalline compound with a melting point of 52–55°C; used to control pests of fruits, rice, and vegetable and fiber crops.

para-**methoxybenzoic acid** *See* anisic acid.

para-**methoxybenzyl alcohol** *See* anisic alcohol.

2-(methoxycarbonylamino)-benzimidazole $C_9H_9N_3O_2$ A gray powder that decomposes at 307–312°C; used as a fungicide for cereals, vegetables, fruits, cotton, and ornamental flowers. Also known as carbendazim.

methoxychlor $Cl_3CCH(C_6H_4OCH_3)_2$ White, water-insoluble crystals melting at 89°C; used as an insecticide. Also known as DMDT; methoxy DDT.

methoxy DDT *See* methoxychlor.

2-methoxyethanol $CH_3OCH_2CH_2OH$ A poisonous liquid, used as a solvent for low-viscosity cellulose acetate, natural and some synthetic resins, and alcohol-soluble dyes, and also used in dyeing leather. Also known as ethylene glycol monomethyl ether.

methoxyethylmercury chloride $CH_3OCH_2CH_2HgCl$ A white, crystalline compound with a melting point of 65°C; used as a fungicide in diseases of sugar-cane, pineapples, seed potatoes, and flower bulbs, and as seed dressings for cereals, legumes, and root crops. Abbreviated MEMC.

4-methoxy-2-hydroxybenzophenone *See* oxybenzone.

methoxyl CH_3O— A radical which is univalent.

2-methoxynaphthalene *See* β-naphthyl methyl ether.

4-methoxyphenol *See* hydroquinone monomethyl ether.

4-(*para*-methoxyphenyl)-2-butanone $C_{11}H_{14}O_2$ A yellow liquid with a melting point of 8°C; used as an insect attractant. Also known as anisylacetone.

1-methoxy-4-propenyl benzene *See* anethole.

para-**methoxypropenylbenzene** *See* anethole.

6-methoxy-α-tetralone $C_{11}H_{12}O_2$ A crystalline substance with a melting point of 80°C; used in the synthesis of derivatives of estrane and 19-norsteroids.

methyl The alkyl group derived from methane and usually written CH_3—. Also known as carbinyl.

methyl abietate $C_{19}H_{29}COOCH_3$ Colorless to yellow liquid boiling at 365°C; miscible with most organic solvents; used as a solvent and plasticizer for lacquers, varnishes, and coatings.

methyl acetate $CH_3CO_2CH_3I$ Flammable, colorless liquid with fragrant odor; boils at 54°C; partially soluble in water, miscible with hydrocarbon solvents; used as a solvent and extractant.

methylacetic acid *See* propionic acid.

methyl acetoacetate $CH_3COCH_2CO_2CH_3$ Alcohol-soluble, colorless liquid boiling at 172°C; used as a chemical intermediate and as a solvent for cellulosics.

methyl acetophenone $CH_3C_6H_4COCH_3$ Fragrant (coumarin aroma), colorless or pale-yellow liquid, soluble in alcohol; used in perfumery. Also known as methyl toly ketone.

methylacetopyranone *See* dehydroacetic acid.

methyl acetylene *See* allylene.

methyl acid phosphate *See* methylphosphoric acid.

methyl acrylate $CH_2{:}CHCOOCH_3$ A readily polymerized, volatile, colorless liquid boiling at 80°C; slightly soluble in water; used as a chemical intermediate and in making polymers.

methylal $CH_3OCH_2OCH_3$ Flammable, volatile, colorless liquid boiling at 42°C; soluble in ether, hydrocarbons, and alcohol, partially soluble in water; used as a solvent and chemical intermediate, and in perfumes, adhesives, coatings. Also known as dimethoxymethane; formal.

methyl alcohol CH_3OH A colorless, toxic, flammable liquid, boiling at 64.5°C, miscible with water, ether, alcohol; used in manufacture of formaldehyde, chemical synthesis, antifreeze for autos, and as a solvent. Also known as methanol; wood alcohol.

methyl allyl alcohol *See* methallyl alcohol.

methyl allyl chloride $CH_2{:}C(CH_3)CH_2Cl$ Volatile, flammable, colorless liquid boiling at 72°C; has disagreeable odor; used as an insecticide and fumigant, and for chemical synthesis.

methylamine CH_3NH_2 A colorless gas that is highly toxic and flammable; used to prepare dyes, and as a chemical intermediate. Also known as aminomethane; monomethylamine.

methyl aminoacetic acid *See* sarcosine.

methyl-4-aminobenzene sulfonyl-carbamate $H_2NC_6H_4SO_2NHCO_2CH_3$ White crystals with a melting point of 143–145°C; used as a pre- and postemergence herbicide. Also known as asulam.

methyl *ortho*-aminobenzoate *See* methyl anthranilate.

N-methyl-*para*-aminophenol $CH_3NHC_6H_4OH$ Colorless, combustible needles with a melting point of 87°C; soluble in water, alcohol, and ether; used as a photographic developer.

386   methyl amyl acetate

methyl amyl acetate $CH_3COOCH(CH_3)CH_2CH(CH_3)_2$ Toxic, flammable, colorless liquid with mild, agreeable odor; boils at 146°C; used as nitrocellulose lacquer solvent. Also known as methyl isobutyl carbinol acetate.

methyl amyl alcohol $(CH_3)_2CHCH_2CHOHCH_3$ Toxic, flammable, colorless liquid; boils at 132°C; miscible with water and most organic solvents; used as a solvent and as a chemical intermediate. Also known as methyl isobutyl carbinol (MIBC).

methyl amyl carbinol $CH_3(CH_2)_4CHOHCH_3$ Colorless liquid with mild aroma; boils at 160°C; miscible with most organic liquids; used as an ore-flotation frothing agent and as a synthetic-resin solvent. Also known as 2-heptanol.

methyl-*n*-amyl ketone $CH_3(CH_2)_4COCH_3$ Stable, water-white liquid; miscible with organic lacquer solvents, slightly soluble in water; used as an inert reaction medium and as a solvent for nitrocellulose lacquers. Also known as 2-heptanone.

***N*-methylaniline** $C_6H_5NH(CH_3)$ Oily liquid, colorless to reddish-brown; soluble in water and organic solvents; boils at 190°C; used as an acid acceptor, solvent, and chemical intermediate.

α-methylanisalacetone $CH_3OC_6H_4CH{:}CHCOCH_2CH_3$ A white to pale yellow, combustible solid with a melting point of 60°C; used as a flavoring.

methyl anisole *See* methyl *para*-cresol.

methyl anthranilate $H_2NC_6H_4CO_2CH_3$ A yellowish to colorless liquid, slightly soluble in water; used in flavoring and in perfumery. Also known as artificial neroli oil; methyl *ortho*-aminobenzoate.

2-methyl anthraquinone *See* tectoquinone.

methyl arachidate $CH_3(CH_2)_{18}COOCH_3$ A waxlike solid with a melting point of 45.8°C; soluble in alcohol and ether; used in medical research and as a reference standard for gas chromatography. Also known as methyl eicosanoate.

methylarsinic sulfide CH_3AsS A colorless compound whose flakes melt at 110°C; insoluble in water; used as a fungicide in treating cotton seeds. Also known as rhizoctol.

methylate *See* methoxide.

2-methylaziridine *See* propyleneimine.

methyl behenate $CH_3(CH_2)_{20}COOCH_3$ A combustible, waxlike solid with a melting point of 53.2°C; soluble in alcohol and ether; used in medical and biochemical research and as a reference standard for gas chromatography. Also known as methyl docosanoate.

methylbenzene *See* toluene.

methylbenzethonium chloride $C_{27}H_{44}O_2Cl{\cdot}H_2O$ Colorless crystals with a melting point of 161–163°C; soluble in alcohol, hot benzene, chloroform, and water; used as a bactericide.

methyl benzoate $C_6H2_5CO_2CH_3$ Colorless, fragrant liquid boiling at 199°C; slightly soluble in alcohol and water, soluble in ether; used in perfumery and as a solvent. Also known as niobe oil.

2-methylbenzoic acid *See* *ortho*-toluic acid.

3-methylbenzoic acid   *See meta*-toluic acid.

4-methylbenzoic acid   *See para*-toluic acid.

2-methylbenzophenone   *See ortho*-phenyl tolyl ketone.

1-methyl-3-(2-benzothiazolyl)-urea   *See* benzthiazuron.

methyl *ortho*-benzoylbenzoate $C_6H_5COC_6H_4COOCH_3$ A colorless, combustible liquid with a boiling point of 351°C; slightly soluble in water; used as a plasticizer.

α-methylbenzyl acetate $C_6H_5CH(CH_3)OOCCH_3$ A colorless, combustible liquid with a strong floral odor; soluble in glycerin, mineral oil, and 70% alcohol; used in perfumes and as a flavoring. Also known as methylphenylcarbinyl acetate; phenylmethylcarbinyl acetate; styralyl acetate.

α-methylbenzyl alcohol $C_6H_5CH(CH_3)OH$ A colorless, combustible liquid with a mild floral odor and a boiling point of 204°C; soluble in water; used in perfumes and dyes and as a flavoring agent. Also known as methylphenylcarbinol; *sec*-phenethyl alcohol; phenylmethylcarbinol; styralyl alcohol.

α-methylbenzylamine $C_6H_5CH(CH_3)NH_2$ A colorless, combustible liquid with a boiling point of 188.5°C; soluble in most organic solvents; used as an emulsifying agent.

***meta*-methylbenzyl bromide**   *See* α-bromo-*meta*-xylene.

α-methylbenzyl ether $C_6H_5CH(CH_3)OCH(CH_3)C_6H_5$ A straw-colored, combustible liquid with a boiling point of 286.3°C; at 760 mmHg (101,325 newtons per square meter); slightly soluble in water; used as a solvent and as a synthetic rubber softener.

α-methyl bivinyl   *See* pentadiene.

methyl blue   Dark-blue powder or dye; sodium triphenyl *para*-rosaniline sulfonate; used as a biological and bacteriological stain and as an antiseptic.

methyl borate   *See* trimethyl borate.

methyl bromide CH_3Br A toxic, colorless gas that forms a crystalline hydrate with cold water; used in synthesis of organic compounds, and as a fumigant. Also known as bromomethane.

2-methyl-1,3-butadiene   *See* isoprene.

2-methylbutanal   *See* 2-methylbutyraldehyde.

2-methylbutane   *See* isopentane.

2-methyl-1-butanol $C_5H_{12}O$ A liquid with a boiling point of 128°C, miscible with alcohol and with ether, slightly soluble in water; used as a solvent, in organic synthesis, and as an additive in oils and paints. Also known as active amyl alcohol.

3-methyl-1-butanol   *See* isobutyl carbinol.

methyl butene C_5H_{10} Either of two colorless, flammable, volatile liquid isomers; soluble in alcohol, insoluble in water: 3-methyl-1-butene boils at 20°C, is used as a chemical intermediate and in the manufacture of high-octane fuel, and is also known as isopropylethylene; 3-methyl-2-butene boils at 38°C, is used as an anesthetic and high-octane fuel and as a chemical intermediate, and is also known as trimethylethylene.

2-methyl-2-butene　*See* amylene.

methyl butyl ketone　$CH_3COC_4H_9$　A liquid boiling at 127°C; soluble in water, alcohol, and ether; used as a solvent. Also known as propylacetone.

methylbutynol　$HC:CCOH(CH_3)_2$　Water-miscible, colorless liquid boiling at 104°C; soluble in most organic solvents; used as a stabilizer for chlorinated organic compounds, as a solvent, and as a chemical intermediate.

2-methylbutyraldehyde　$CH_3CH_2CH(CH_3)CHO$　A combustible liquid with a boiling point of 92.93°C; soluble in alcohol and ether; used as a brightener in electroplating. Also known as 2-methylbutanal.

3-methylbutyraldehyde　*See* isovaleraldehyde.

methyl butyrate　$CH_3CH_2CH_2COOCH_3$　Liquid boiling at 102°C; used as a solvent for cellulosic materials.

methyl caprate　$CH_3(CH_2)_8COOCH_3$　A colorless, combustible liquid with a boiling point of 244°C; soluble in alcohol and ether; used in the manufacture of detergents, stabilizers, plasticizers, textiles, and lubricants. Also known as methyl decanoate.

methyl caproate　$CH_3(CH_2)_4COOCH_3$　Colorless liquid boiling at 150°C; soluble in alcohol and ether, insoluble in water; used as an intermediate to make caproic acid. Also known as methyl hexanoate.

methyl caprylate　$CH_3(CH_2)_6COOCH_3$　Colorless liquid boiling at 193°C; soluble in ether and alcohol, insoluble in water; used as an intermediate to make caprylic acid.

***meta*-methylcarbanilate**　*See* phenmedipham.

2-methyl-6-carboline　*See* harman.

3-methyl-4-carboline　*See* harman.

methyl carbonate　$CO(OCH_3)_2$　Water-insoluble, colorless liquid boiling at 91°C; has pleasant odor; miscible with acids and alkalies; used as a chemical intermediate.

methylcellulose　A grayish-white powder derived from cellulose; swells in water to a colloidal solution; soluble in glacial acetic acid; used in water-based paints and ceramic glazes, for leather tanning, and as a thickening and sizing agent, adhesive, and food additive. Also known as cellulose methyl ether.

β-methylchalcone　*See* dipnone.

methyl chavicol　*See* estragole.

methyl chloride　*See* chloromethane.

methyl chloroacetate　$ClHC_2COOCH_3$　Colorless liquid boiling at 131°C; miscible with ether and alcohol, slightly soluble in water; used as a solvent.

methyl chlorocarbonate　*See* methyl chloroformate.

methyl chloroform　*See* trichloroethane.

methyl chloroformate　$ClCOOCH_3$　A toxic, corrosive, colorless liquid with a boiling point of 71.4°C; soluble in benzene, ether, and methanol; used as a lacrimator in military poison gas and for insecticides. Also known as methyl chlorocarbonate.

methyl-2-chloro-9-hydroxyfluorene-9-carboxylate *See* chloflurecol methyl ester.

2-methyl-4-chlorophenoxyacetic acid *See* bromate.

methyl cinnamate $C_6H_5CH:CHCO_2CH_3$ A white crystalline compound with strawberry aroma; soluble in ether and alcohol, insoluble in water; boils at 260°C; used to flavor confectioneries and in perfumes.

methyl para-cresol $CH_3C_6H_4OCH_3$ Colorless liquid with floral aroma; used in perfumery. Also known as para-cresyl methyl ether; methyl anisole.

methyl cyanide *See* acetonitrile.

methyl cyanoacetate $CNCH_2COOCH_3$ A toxic, combustible, colorless liquid with a boiling point of 203°C; soluble in water, ether, and alcohol; used in pharmaceuticals and dyes. Also known as malonic methyl ester nitrile.

methyl cyclohexane C_7H_{14} Colorless liquid boiling at 101°C; used as a cellulosic solvent and as a chemical intermediate. Also known as hexahydrotoluene.

methyl cyclohexanol $CH_3C_6H_{10}OH$ A toxic, colorless liquid with menthol aroma; a mixture of three isomers; used as a solvent for lacquer and cellulosics, as a lubricant antioxidant, and in detergents and textile soaps. Also known as hexahydrocresol; hexahydromethyl phenol.

methyl cyclohexanone $CH_3C_5H_9CO$ A toxic, clear to pale-yellow liquid with acetonelike aroma; a mixture of cyclic ketones; used as a solvent and in lacquers.

1-(2-methylcyclohexyl)-3-phenylurea *See* siduron.

methyl-1,3-cyclopentadiene *See* methylcyclopentadiene dimer.

methylcyclopentadiene dimer $C_{12}H_{16}$ A flammable, colorless liquid with a boiling range of 78–183°C; soluble in alcohol, benzene, and ether; used in high-energy fuels, plasticizers, dyes, and pharmaceuticals. Also known as methyl-1,3-cyclopentadiene.

methyl cyclopentane $C_5H_9CH_3$ Flammable, colorless liquid boiling at 72°C; used as a chemical intermediate.

methyl decanoate *See* methyl caprate.

methyl 5-(2′,4′-dichlorophenoxy)-2-nitrobenzoate *See* bifenox.

methyl-N-(3,4-dichlorophenyl)carbamate *See* swep.

methyl diethanolamine $CH_3N(C_2H_4OH)_2$ A colorless liquid miscible with water and benzene; has amine aroma; boils at 247°C; used as a chemical intermediate and as an acid-gas absorbent.

6-methyldihydromorphinone hydrochloride *See* metopon hydrochloride.

2-methyl-5,6-dihydro-4-H-pyran-3-carboxylic acid anilide *See* pyracarbolid.

methyl N′,N′-dimethyl N[(methylcarbamoyl)oxy]-1-thiooxamimidate *See* oxamyl.

methyl dioxolane $C_4H_7O_2$ Water-soluble, clear liquid boiling at 81°C; used as a solvent and extractant. Also known as 2-methyl-1,3-dioxolane.

2-methyl-1,3-dioxolane *See* methyl dioxolane.

methyl dipropylmethane *See* methyl heptane.

methyl docosanoate *See* methyl behenate.

methyl eicosanoate *See* methyl arachidate.

methylene $—CH_2—$ A radical that contains a bivalent carbon.

methylene blue Dark green crystals or powder; soluble in water (deep blue solution), alcohol, and chloroform; $C_{16}H_{18}N_3SCl\cdot3H_2O$ used in medicine; $(C_{16}H_{18}N_3SCl)_2\cdot ZnCl_2\cdot H_2O$ used as a textile dye, biological stain, and indicator. Also known as methylthionine chloride.

methylene bromide $CH_2Br_2\cdot$ Colorless, clear liquid boiling at 97°C; miscible with organic solvents, slightly soluble in water; used as a solvent and chemical intermediate. Also known as dibromomethane.

methylene chlorabromide *See* bromochloromethane.

methylene chloride CH_2Cl_2 A colorless liquid, practically nonflammable and nonexplosive; used as a refrigerant in centrifugal compressors, a solvent for organic materials, and a component in nonflammable paint-remover mixtures. Also known as carrene; dichloromethane.

methylenecyclopentadiene *See* fulvene.

3,4-methylenedioxy-1-allyl benzene *See* safrole.

3,4-methylenedioxybenzaldehyde *See* piperonal.

1,2-methylenedioxy-4-propenylbenzene *See* isosafrole.

methylene iodide CH_2I_2 Yellow liquid boiling at 180°C; soluble in ether and alcohol, insoluble in water; used as a chemical intermediate and to separate mineral mixtures. Also known as diiodomethane.

methylene oxide *See* formaldehyde.

methylene succinic acid *See* itaconic acid.

methyl ester An ester that forms methanol when hydrolyzed.

methyl ether *See* dimethyl ether.

methylethylcellulose A combustible, white to cream-colored, fibrous solid or powder; disperses in cold water, forming solutions which undergo reversible transformation from sol to gel; used as an emulsifier and foaming agent.

methyl ethyl diketone *See* acetyl propionyl.

methyl ethylene *See* propylene.

methyl ethylene glycol *See* propylene glycol.

methyl ethyl glyoxal *See* acetyl propionyl.

methyl ethyl ketone $CH_3COC_2H_5$ A water-soluble, colorless liquid that is miscible in oil; used as a solvent in vinyl films and nitrocellulose coatings, and as a reagent in organic synthesis. Also known as 2-butanone; ethyl methyl ketone; MEK.

methyleugenol $C_{11}H_{14}O_2$ A crystalline compound, soluble in water; melting point is -4°C; used as an attractant for the oriental fruit fly. Also known as 1-allyl-3,4-dimethoxybenzene.

methyl formate $HCOOCH_3$ A flammable, colorless liquid with a boiling point of 31.8°C; soluble in ether, water, and alcohol; used in military poison gases and larvicides, and as a fumigant.

methyl fumaric acid *See* mesaconic acid.

2-methylfuran $C_4H_3OCH_3$ A colorless liquid with ether aroma; boils at 64°C; used as a chemical intermediate.

methyl furoate $C_4H_3OCO_2CH_3$ Colorless liquid that turns yellow in light; soluble in ether and alcohol, insoluble in water; used as a solvent and chemical intermediate.

methyl GAG *See* methylglyoxal bis(guanylhydrazone).

methyl glucoside $C_7H_{14}O_6$ Odorless, water-soluble white crystals; used to make resins, drying oils, plasticizers, and surfactants.

methyl glycocoll *See* sarcosine.

methyl glycol *See* propylene glycol.

methyl heptane C_8H_{18} Either of two colorless, water-insoluble liquids, soluble in alcohol and ether, used as chemical intermediates: 2-methylheptane boils at 118°C, is flammable, and is also known as isooctane; 4-methylheptane boils at 122°C and is known as methyl dipropylmethane.

methylheptenone $(CH_3)_2C$:$CH(CH_2)_2COCH_3$ A combustible, colorless liquid with a boiling point of 173–174°C; a constituent of many essential oils; used in perfumes and for flavoring. Also known as 6-methyl-5-hepten-2-one.

6-methyl-5-hepten-2-one *See* methylheptenone.

2-(1-methylheptyl)-4,6-dinitrophenyl crotonate $C_{18}H_{24}O_6N_2$ A brown liquid, insoluble in water; used as a fungicide and miticide for fruit and vegetable crops.

methyl hexadecanoate *See* methyl palmitate.

2-methylhexane C_7H_{16} Colorless liquid boiling at 90°C; insoluble in alcohol and water; used as a chemical intermediate. Also known as ethyl isobutylmethane.

methyl hexanoate *See* methyl caproate.

methyl hexyl ketone $CH_3COC_6H_{13}$ A combustible, colorless liquid with a boiling point of 173.5°C; soluble in alcohol, hydrocarbons, ether, and esters; used in perfumes and as a flavoring and odorant. Also known as 2-octanone.

methyl hydride *See* methane.

methyl-*meta*-hydroxycarbanilate *See* phenmedipham.

methylhydroxyisopropylcyclohexane *See* menthol.

α-methyl-3-(*para*-hydroxyphenyl)alanine *See* α-methyl-*para*-tyrosine.

methyl hydroxystearate $C_{19}H_{38}O_3$ A white, waxy material; slightly soluble in organic solvents, insoluble in water; used in cosmetics, inks, and adhesives.

methylidyne *See* methine group.

3-methylindole *See* skatole.

methyl iodide CH_3I Flammable colorless liquid that turns brown in light; boils at 42°C; soluble in ether and alcohol, insoluble in water; used as a chemical intermediate, in medicine, and in analytical chemistry. Also known as iodomethane.

methyl isobutyl carbinol   *See* methyl amyl alcohol.

methyl isobutyl carbinol acetate   *See* methyl amyl acetate.

methyl isobutyl ketone $(CH_3)_2CHCH_2COCH_3$ Flammable colorless liquid with pleasant aroma; boils at 116°C, miscible with most organic solvents; used as a solvent, extractant, and chemical intermediate. Also known as hexone.

methylisopropylphenanthrene   *See* retene.

2-methyl-5-isopropyl phenol   *See* carvacrol.

methylisothiocyanate C_2H_3NS A crystalline compound, with a melting point of 35–36°C; soluble in alcohol and ether; used as a pesticide and in amino acid sequence analysis. Also known as methyl mustard oil.

methyl lactate $CH_3CHCHCOOCH_3$ Liquid boiling at 145°C; miscible with water and most organic liquids; used as a solvent for lacquers, stains, and cellulosic materials.

methyl laurate $CH_3(CH_2)_{10}COOCH_3$ Water-insoluble, clear, colorless liquid boiling at 262°C; used as a chemical intermediate to make rust removers, and for leather treatment.

methyl linoleate $C_{19}H_{34}O_2$ A combustible, colorless liquid with a boiling point of 212°C; soluble in alcohol and ether; used in the manufacture of detergents, emulsifiers, lubricants, and textiles, and in medical research.

methyl maleic acid   *See* citraconic acid.

methyl mercaptan CH_3SH Colorless, toxic, flammable gas with unpleasant odor; boils at 6.2°C; insoluble in water, soluble in organic solvents; used as a chemical intermediate.

2-methylmercapto-4-methylamino-6-isopropylamino-S- triazine   *See* desmetryn.

methylmercury cyanide   *See* methylmercury nitrile.

methylmercury nitrile CH_3HgCN A crystalline solid with a melting point of 95°C; soluble in water; used as a fungicide to treat seeds of cereals, flax, and cotton. Also known as methylmercury cyanide.

methyl methacrylate $CH_2C(CH_3)COOCH_3$ A flammable, colorless liquid, soluble in most organic solvents but insoluble in water; used as a monomer for polymethacrylate resins.

2-methyl-2-(methylthio)propionaldehyde-O-(methylcarbamoyl)oxime   *See* aldicarb.

methyl mustard oil   *See* methylisothiocyanate.

methyl myristate $CH_3(CH_2)_{12}COOCH_3$ A colorless liquid with a boiling point of 186.8°C; used in the manufacture of detergents, plasticizers, resins, textiles, and animal feeds, and as a flavoring. Also known as methyl tetradecanoate.

methylnaphthalene $C_{10}H_7CH_3$ A solid melting at 34°C; used in insecticides and organic synthesis.

methyl naphthyl ether   *See* β-naphthyl methyl ether.

methyl nitrate CH_3NO_3 Explosive liquid boiling at 60°C; slightly soluble in water, soluble in ether and alcohol; used as a rocket propellant.

meta-**methylnitrobenzene**   *See* *meta*-nitrotoluene.

ortho-methylnitrobenzene   *See ortho*-nitrotoluene.

3-methyl-4-nitro-1-(para-nitrophenyl)-2-pyrazoline-5-one   *See* picolinic acid.

N-methyl-N-nitroso-para-toluenesulfonamide   *See   para*-tolylsulfonylmethylnitrosamide.

methyl nonanoate $CH_3(CH_2)_7COOCH_3$ A colorless liquid with a fruity odor and a boiling point of 213.5°C; soluble in alcohol and ether; used in perfumes and flavors, and for medical research. Also known as methyl pelargonate.

methyl nonyl ketone $CH_3COC_9H_{19}$ An oily liquid with a boiling point of 225°C; soluble in two parts of 70% alcohol; used in perfumes and flavoring. Also known as 2-undecanone.

methyl oleate $C_{17}H_{33}COOCH_3$ Amber liquid with faint fatty odor; soluble in organic liquids, mineral spirits, and vegetable oil, insoluble in water; used as a plasticizer and softener.

methylol riboflavin   An orange to yellow powder, soluble in water; used as a nutrient and in medicine.

methylol urea $H_2NCONHCH_2OH$ Water-soluble, colorless crystals melting at 111°C; used to treat textiles and wood, and in the manufacture of resins and adhesives.

methyl orthophosphoric acid   *See* methylphosphoric acid.

methyl palmitate $CH_3(CH_2)_{14}COOCH_3$ A colorless liquid with a boiling point of 211.5°C; soluble in alcohol and ether; used in the manufacture of detergents, resins, plasticizers, lubricants, and animal feed. Also known as methyl hexadecanoate.

methyl parathion $C_8H_{10}NO_5PS$ An amber to dark brown liquid with slight solubility in water; used as an insecticide to control boll weevils, leafhoppers, cutworms, and rice bugs. Also known as *O,O*-dimethyl-*ortho-para*-nitrophenyl phosphorothioate.

methyl pelargonate   *See* methyl nonanoate.

3-methylpentane C_6H_{14} Flammable, colorless liquid; insoluble in water, soluble in alcohol; boils at 64°C; used as a chemical intermediate. Also known as diethylmethylmethane.

2-methylpentanoic acid $(CH_3)_2CH(CH_2)_2COOH$ A colorless liquid with a boiling point of 197°C; soluble in alcohol, benzene, and acetone; used for plasticizers, vinyl stabilizers, and metallic salts.

methylpentene polymer   Thermoplastic material based on 4-methylpentene-1; has low gravity, excellent electrical properties, and 90% optical transmission.

methyl pentose   1. Any compound that is a methyl derivative of a five carbon sugar. 2. In particular, the compound $CH_3(CHOH)_4CHO$.

methyl phenyl acetate $C_6H_5CH_2COOCH_3$ A colorless liquid with honey odor; used to flavor tobacco and in perfumery.

methylphenylcarbinol   *See* α-methylbenzyl alcohol.

methylphenylcarbinyl acetate   *See* α-methylbenzyl acetate.

methylphenyl ether   *See* anisole.

methylphosphoric acid $CH_3H_2PO_4$ A straw-colored liquid used for textile- and paper-processing compounds, as a rust remover, and in soldering flux. Also known as methyl acid phosphate; methyl orthophosphoric acid.

2-methyl-3-phytyl-1,4-napthoquinone *See* phytonadione.

methylpicrylnitramine *See* tetryl.

N-methylpiperazinyl-N′-propylphenothiazine *See* perazine.

3-(2-methylpiperidino)propyl 3,4-dichlorobenzoate *See* piperalin.

3-(2-methyl-1-piperidyl)propyl benzoate hydrochloride *See* piperocaine hydrochloride.

2-methyl-1-propanal *See* isobutyl alcohol.

2-methyl propane *See* isobutane.

2-methylpropene *See* isobutylene.

methyl propionate $CH_3CH_2COOCH_3$ A flammable, colorless liquid with a boiling range of 78.0–79.5°C; soluble in most organic solvents; used as a solvent for cellulose nitrate, in lacquers, varnishes, and paints, and for flavoring.

methyl propyl carbinol $CH_3CHOHC_3H_7$ Colorless liquid boiling at 119°C; miscible with ether and alcohol, slightly soluble in water; used as a pharmaceuticals intermediate and as a paint and lacquer solvent. Also known as *sec-n*-amyl alcohol; 2-pentanol.

methyl propyl ketone *See* pentanone.

2-methyl pyridine *See* picoline.

3-methyl pyridine *See* picoline.

methylpyrocatechin *See* guaiacol.

N-methyl-2-pyrrolidone C_5H_9NO A liquid boiling at 202°C; miscible with water, castor oil, and organic solvents; used as a chemical intermediate and as a solvent for petroleum and resins, and in PVC spinning.

3-(1-methyl-2-pyrrolidyl)pyridine *See* nicotine.

α-methylquinoline *See* quinaldine.

γ-methylquinoline *See* lepidine.

6-methyl-2,3-quinoxalinedithiol cyclic-S,S-dithiocarbonate *See* oxythioquinox.

methyl red $(CH_3)_2NC_6H_4NNC_6H_4COOH$ A dark red powder or violet crystals with a melting point of 180°C; soluble in alcohol, ether, and glacial acetic acid; used as an acid-base indicator (pH 4.2–6.2).

methyl ricinoleate $C_{19}H_{36}O_3$ Clear, low-viscosity fluid used as a wetting agent, cutting oil additive, lubricant, and plasticizer.

methylrosaniline chloride *See* methyl violet.

methyl salicylate $C_6H_4OHCOOCH_3$ A colorless, yellow, or reddish liquid, slightly soluble in water, boiling at 222.2°C, with an odor of wintergreen; used in medicine and perfumery, and as a solvent for cellulose derivatives. Also known as betula oil; gaultheria oil; wintergreen oil.

3-methylsalicylic acid $C_8H_8O_3$ A white to reddish, crystalline compound with a melting point of 165–166°C; soluble in chloroform, alcohol, ether, and alkali hydroxides; used to make dyes. Also known as *ortho*-cresotic acid; *ortho*-cresotinic acid; *ortho*-homosalicylic acid; 2-hydroxy-*meta*-toluic acid.

methyl silicone $[(CH_3)_2SiO]_x$, $[C(CH_3)_2Si_2O_3]_y$, etc. The common varieties of silicones with properties of oil, resin, or rubber, depending on molecular size and arrangement.

methyl stearate $C_{17}H_{35}COOCH_3$ Colorless crystals melting at 39°C; soluble in alcohol and ether, insoluble in water; used as an intermediate for stearic acid manufacture.

methyl styrene *See* vinyltoluene.

α-methyl styrene $C_6H_5C(CH_3):CH_2$ Colorless, toxic, polymerizable liquid boiling at 165°C; used to produce polystyrene resins.

methyl styryl ketone *See* benzylideneacetone.

methyl sulfate *See* dimethyl sulfate.

methyl sulfide $(CH_3)_2S$ Flammable, colorless liquid with disagreeable aroma; soluble in ether and alcohol, insoluble in water; boils at 38°C; used as a chemical intermediate. Also known as dimethyl sulfide; methanethiomethane.

methylsulfonic acid *See* methanesulfonic acid.

4-(methylsulfonyl)-2,6-dinitro-*N*,*N*-dipropylaniline *See* nitralin.

methyl tetradecanoate *See* methyl myristate.

methyl-1,2,5,6-tetrahydro-1-methylnicotinate *See* arecoline.

N-methyl-*N*,2,4,6-tetranitroaniline *See* tetryl.

4-methyl-5-thiazoleethanol C_6H_9NOS A viscous, oily liquid; soluble in alcohol, ether, benzene, chloroform, and water; used as an intermediate in the synthesis of vitamin B_1 and as a sedative and hypnotic.

methylthionine chloride *See* methylene blue.

6-methyl-2-thiouracil *See* methylthiouracil.

4-(methylthio)-3,5-xylylmethylcarbamate *See* mercaptodimethur.

methyl tolyl ketone *See* methyl acetophenone.

methyl transferase Any of a group of enzymes which catalyze the reaction of *S*-adenosyl methionine with a suitable acceptor to yield the methylated acceptor molecule and *S*-adenosyl homocysteine.

methyltrimethylolmethane *See* trimethylolethane.

methyltrinitrobenzene *See* 2,4,6-trinitrotoluene.

α-methyl-*para*-tyrosine $C_{10}H_{13}NO_3$ A crystalline compound which acts as the inhibitor of the first and rate-limiting reaction in the biosynthesis of catecholamine; used as an inhibitor of tyrosine hydroxylase. Also known as α-methyl-3-(*para*-hydroxyphenyl)alanine.

β-methylumbelliferone *See* hymecromone.

methyl violet A derivative of pararosaniline, used as an antiallergen and bactericide, acid-base indicator, biological stain, and textile dye. Also known as crystal violet; gentian violet; methylrosaniline chloride.

metobromuron $C_9H_{11}BrN_2O_2$ A colorless, crystalline compound with a melting point of 95.5–96°C; limited solubility in water; used as a preemergence herbicide to control weeds in potatoes. Also known as 3-(para-bromophenyl)-1-methoxy-1-methylurea.

metribuzin $C_8H_{14}N_4OS$ A white, crystalline solid with a melting point of 125–126.5°C; used as a preemergence herbicide for soybeans and pre- and postemergence treatment for potatoes. Also known as 4-amino-6-tert-butyl-3-(methylthio)-1,2,4-triazin-5-(4H)-one.

mevalonic acid $HO_2C_5H_9COOH$ A dihydroxy acid used in organic synthesis. Also known as 3,5-dihydroxy-3-methyl valeric acid.

mexacarbate $C_{12}H_{18}N_2O_2$ A tan solid with a melting point of 85°C; used to control insect pests of trees, flowers, and shrubs. Also known as 4-dimethylamino 3,5-xylyl-N-methylcarbamate.

Meyer atomic volume curve A graph of the atomic volumes of the elements versus their atomic numbers; it reveals a periodicity, with peaks at the alkali elements and valleys at the transition elements.

Mg See magnesium.

MIBC See methyl amyl alcohol.

micelle A colloidal aggregate of a unique number (between 50 and 100) of amphipathic molecules, which occurs at a well-defined concentration known as the critical micelle concentration.

Michaelis constant A constant K_m such that the initial rate of reaction V, produced by an enzyme when the substrate concentration is high enough to saturate the enzyme, is related to the rate of reaction v at a lower substrate concentration c by the formula $V = v (1 + K_m/c)$.

Michler's ketone See tetramethyldiaminobenzophenone.

micril See gammil.

microanalysis Identification and chemical analysis of material on a small scale so that specialized instruments such as the microscope are needed; the material analyzed may be on the scale of 1 microgram.

microcapsule A capsule with a plastic or waxlike coating having a diameter anywhere from well below 1 micrometer to over 2000 micrometers.

microchemistry 1. The chemistry of individual cells and minute organisms. 2. The study of chemical reactions, using small quantities of materials, frequently less than 1 milligram or 1 milliliter, and often requiring special small apparatus and microscopical observation.

microcrystalline Composed of or containing crystals that are visible only under the microscope.

microdensitometer A high-sensitivity densitometer used in spectroscopy to detect spectrum lines too faint on a negative to be seen by the human eye.

microelectrolysis Electrolysis of small quantities of material.

microelectrophoresis Direct microscopic observation and measurement of the velocity of migration of ions or other charged bodies through a solution toward oppositely charged electrodes. Also known as optical cytopherometry.

microencapsulation Enclosing of materials in capsules from well below 1 micrometer to over 2000 micrometers in diameter.

microgammil *See* gammil.

microincineration Reduction of small quantities of organic substances to ash by application of heat.

microlite A microscopic crystal which polarizes light. Also known as microlith.

microlith *See* microlite.

micronutrient Trace elements and compounds required by living systems only in minute amounts.

microporous barrier A metallic or plastic membrane with micrometer-sized pores used for dialysis and other membrane-separation processes.

microprobe An instrument for chemical microanalysis of a sample, in which a beam of electrons is focused on an area less than a micrometer in diameter, and the characteristic x-rays emitted as a result are dispersed and analyzed in a crystal spectrometer to provide a qualitative and quantitative evaluation of chemical composition.

microprobe spectrometry Microanalysis of a sample, using a microprobe.

microradiography Technique for the study of surfaces of solids by monochromatic-radiation (such as x-ray) contrast effects shown via projection or enlargement of a contact radiograph.

microspectrograph A microspectroscope provided with a photographic camera or other device for recording the spectrum.

microspectrophotometer A split-beam or double-beam spectrophotometer including a microscope for the localization of the object under study, and capable of carrying out spectral analyses within the dimensions of a single cell.

microspectroscope An instrument for analyzing the spectra of microscopic objects, such as living cells, in which light passing through the sample is focused by a compound microscope system, and both this light and the light which has passed through a reference sample are dispersed by a prism spectroscope, so that the spectra of both can be viewed simultaneously.

microthrowing power Relative ability of an electroplating solution to deposit metal in a small, shallow aperture or crevice not exceeding a few thousandths of an inch in dimensions.

microwave spectrometer An instrument which makes a graphical record of the intensity of microwave radiation emitted or absorbed by a substance as a function of frequency, wavelength, or some related variable.

microwave spectroscope An instrument used to observe the intensity of microwave radiation emitted or absorbed by a substance as a function of frequency, wavelength, or some related variable.

microwave spectroscopy The methods and techniques of observing and the theory for interpreting the selective absorption and emission of microwaves at various frequencies by solids, liquids, and gases.

microwave spectrum A display, photograph, or plot of the intensity of microwave radiation emitted or absorbed by a substance as a function of frequency, wavelength, or some related variable.

migration current Additional current produced by electrostatic attraction of cations to the surface of a dropping electrode; an unpredictable and undesirable effect to be avoided during analytical voltammetry.

MIKES *See* mass-analyzed ion kinetic energy spectrometry.

mild mercury chloride *See* mercurous chloride.

milk A suspension of certain metallic oxides, as milk of magnesia, iron, or bismuth.

milk factor A filtrable, noncellular agent in the milk and tissues of certain strains of inbred mice; transmitted from the mother to the offspring by nursing. Also known as Bittner milk factor.

milk sugar *See* lactose.

Miller indices Three integers identifying a type of crystal plane; the intercepts of a plane on the three crystallographic axes are expressed as fractions of the crystal parameters; the reciprocals of these fractions, reduced to integral proportions, are the Miller indices. Also known as crystal indices.

Miller law If the edges formed by the intersections of three faces of a crystal are taken as the three reference axes, then the three quantities formed by dividing the intercept of a fourth face with one of these axes by the intercept of a fifth face with the same axis are proportional to small whole numbers, rarely exceeding 6. Also known as law of rational intercepts.

millibarn A unit of cross section equal to one-thousandth of a barn. Abbreviated mb.

milliequivalent One-thousandth of a compound's or an element's equivalent weight.

Millikan oil-drop experiment A method of determining the charge on an electron, in which one measures the terminal velocities of rise and fall of oil droplets in an electric field after the droplets have picked up charge from ionization in the surrounding gas produced by an x-ray beam.

Millon's reagent Reagent used to test for proteins; made by dissolving mercury in nitric acid, diluting, then decanting the liquid from the precipitate.

milneb $C_{12}H_{22}N_4S_4$ A white, crystalline compound with a melting point of 140–141°C; slight solubility in water; used as a fungicide for fruit, tobacco, and potato crops. Also known as 3,3′-ethylenebis(tetrahydro-4,6-dimethyl-2*H*-1,3,5-thiadiazine-2-thione); thiadiazin.

mimetic Pertaining to a crystal that is twinned or malformed but whose crystal symmetry appears to be of a higher grade than it actually is.

mimosine $C_8H_{10}N_2O_4$ A crystalline compound with a melting point of 235–236°C; soluble in dilute acids or bases; used as a depilatory agent. Also known as leucaenine; leucaenol; leucenine; leucenol.

mineral green *See* copper carbonate.

mineralocorticoid A steroid hormone secreted by the adrenal cortex that regulates mineral metabolism and, secondarily, fluid balance.

mineralogy The science which concerns the study of natural inorganic substances called minerals.

minimum ionizing speed The smallest speed at which a charged particle passing through a gas can ionize an atom or molecule.

minimum reflux ratio The smallest reflux ratio in a two-component liquid distillation system that will produce the desired overhead and bottom compositions.

minimum wetting rate The smallest liquid-flow rate through a packed column that will thoroughly wet the column packing.

MIPC *See* O-isopropylphenyl-methylcarbamate.

mirror nuclei A pair of atomic nuclei, each of which would be transformed into the other by changing all its neutrons into protons, and vice versa.

mirror plane of symmetry *See* plane of mirror symmetry.

miscibility The tendency or capacity of two or more liquids to form a uniform blend, that is, to dissolve in each other; degrees are total miscibility, partial miscibility, and immiscibility.

misfire Failure of fuel or an explosive charge to ignite properly.

Mitscherlich law of isomorphism Substances which have similar chemical properties and crystalline forms usually have similar chemical formulas.

mix crystal *See* mixed crystal.

mixed acid *See* nitrating acid.

mixed aniline point The minimum temperature at which a mixture of aniline, heptane, and hydrocarbon will form a solution; related to the aromatic character of the hydrocarbon.

mixed cryoglobulin A cryoglobulin with a monoclonal component made of immunoglobulin belonging to two different classes, one of which is monoclonal.

mixed crystal A crystal whose lattice sites are occupied at random by different ions or molecules of two different compounds. Also known as mix crystal.

mixed flow Flow stream existing in two or more phases, such as gas, hydrocarbon, and water. Also known as mixed-phase flow.

mixed indicator Color-change indicator for acid-base titration end points in which a mixture of two indicator substances is used to give sharper end-point color changes.

mixed potential The electrode potential of a material while more than one electrochemical reaction is occurring simultaneously.

mixer-settler Solvent-extraction system with alternating or combined arrangement of mixers and settlers; used for chemicals extraction, lubricating-oil refining, and uranium oxide recovery. Also known as mixer-settler extractor.

mixer-settler extractor *See* mixer-settler.

mixing The intermingling of different materials (liquid, gas, solid) to produce a homogeneous mixture.

MMSCFD Abbreviation for million standard cubic feet per day; usually refers to gas flow.

MMSCFH Abbreviation for million standard cubic feet per hour; usually refers to gas flow.

MMSCFM Abbreviation for million standard cubic feet per minute; usually refers to gas flow.

Mn *See* manganese.

Mo *See* molybdenum.

modified Lewis acid   An acid that is a halide ion acceptor.

modulated Raman scattering   Application of modulation spectroscopy to the study of Raman scattering; in particular, use of external perturbations to lower the symmetry of certain crystals and permit symmetry-forbidden modes, and the use of wavelength modulation to analyze second-order Raman spectra.

modulation spectroscopy   A branch of spectroscopy concerned with the measurement and interpretation of changes in transmission or reflection spectra induced (usually) by externally applied perturbation, such as temperature or pressure change, or an electric or magnetic field.

Mohr cubic centimeter   A unit of volume used in saccharimetry, equal to the volume of 1 gram of water at a specified temperature, usually 17.5°C, in which case, it is equal to 1.00238 cubic centimeters.

Mohr's salt   *See* ferrous ammonium sulfate.

Mohr titration   Titration with silver nitrate to determine the concentration of chlorides in a solution; silver chromate precipitation is the end-point indicator.

moiety   A part or portion of a molecule, generally complex, having a characteristic chemical or pharmacological property.

moisture   Water that is dispersed through a gas in the form of water vapor or small droplets, dispersed through a solid, or condensed on the surface of a solid.

mol   *See* mole.

molal average boiling point   A pseudo boiling point for a mixture calculated as the summation of individual mole fraction–boiling point (in degrees Rankine) products.

molal heat capacity   *See* molar heat capacity.

molality   Concentration given as moles per 1000 grams of solvent.

molal quantity   The number of moles (gram-molecular weights) present, expressed with weight in pounds, grams, or such units, numerically equal to the molecular weight; for example, pound-mole, gram-mole.

molal solution   Concentration of a solution expressed in moles of solute divided by 1000 grams of solvent.

molal volume   *See* molar volume.

molar conductivity   The ratio of the conductivity of an electrolytic solution to the concentration of electrolyte in moles per unit volume.

molar heat capacity   The amount of heat required to raise 1 mole of a substance 1° in temperature. Also known as molal heat capacity; molecular heat capacity.

molarity   Measure of the number of gram-molecular weights of a compound present (dissolved) in 1 liter of solution; it is indicated by M, preceded by a number to show solute concentration.

molar solution   Aqueous solution that contains 1 mole (gram-molecular weight) of solute in 1 liter of the solution.

molar specific heat   The ratio of the amount of heat required to raise the temperature of 1 mole of a compound 1°, to the amount of heat required to raise the temperature of 1 mole of a reference substance, such as water, 1° at a specified temperature. Also known as molal specific heat; molecular specific heat.

molar susceptibility   Magnetic susceptibility of a compound per gram-mole of that compound.

molar volume   The volume occupied by one mole of a substance in the form of a solid, liquid, or gas. Also known as molal volume; mole volume.

mole   An amount of substance of a system which contains as many elementary units as there are atoms of carbon in 0.012 kilogram of the pure nuclide carbon-12; the elementary unit must be specified and may be an atom, molecule, ion, electron, photon, or even a specified group of such units. Symbolized mol.

molecular adhesion   A particular manifestation of intermolecular forces which causes solids or liquids to adhere to each other; usually used with reference to adhesion of two different materials, in contrast to cohesion.

molecular amplitude   The difference between the molecular rotation at the extreme (peak or trough) value caused by the longer light wavelength and the molecular rotation at the extreme value caused by the shorter wavelength.

molecular association   The formation of double molecules or polymolecules from a single species as a result of specific and moderately strong intermolecular forces.

molecular asymmetry   *See* asymmetry.

molecular attraction   A force which pulls molecules toward each other.

molecular conductivity   The conductivity of a volume of electrolyte containing 1 mole of dissolved substance.

molecular crystal   A solid consisting of a lattice array of molecules such as hydrogen, methane, or more complex organic compounds, bound by weak van der Waals forces, and therefore retaining much of their individuality.

molecular diamagnetism   Diamagnetism of compounds, especially organic compounds whose susceptibilities can often be calculated from the atoms and chemical bonds of which they are composed.

molecular diameter   The diameter of a molecule, assuming it to be spherical; has a numerical value of 10^{-8} centimeter multiplied by a factor dependent on the compound or element.

molecular dipole   A molecule having an electric dipole moment, whether it is permanent or produced by an external field.

molecular distillation   A process by which substances are distilled in high vacuum at the lowest possible temperature and with least damage to their composition.

molecular energy level   One of the states of motion of nuclei and electrons in a molecule, having a definite energy, which is allowed by quantum mechanics.

molecular exclusion chromatography   *See* gel filtration.

molecular gas   A gas composed of a single species, such as oxygen, chlorine, or neon.

molecular heat capacity   *See* molar heat capacity.

molecularity   In a chemical reaction, the number of molecules which come together and form the activated complex.

molecular magnet   A molecule having a nonvanishing magnetic dipole moment, whether it is permanent or produced by an external field.

molecular orbital   A wave function describing an electron in a molecule.

molecular paramagnetism   Paramagnetism of molecules, such as oxygen, some other molecules, and a large number of organic compounds.

molecular polarizability   The electric dipole moment induced in a molecule by an external electric field, divided by the magnitude of the field.

molecular relaxation   Transition of a molecule from an excited energy level to another excited level of lower energy or to the ground state.

molecular sieve chromatography   *See* gel filtration.

molecular specific heat   *See* molar specific heat.

molecular spectroscopy   The production, measurement, and interpretation of molecular spectra.

molecular spectrum   The intensity of electromagnetic radiation emitted or absorbed by a collection of molecules as a function of frequency, wave number, or some related quantity.

molecular still   An apparatus used to conduct molecular distillation.

molecular structure   The manner in which electrons and nuclei interact to form a molecule, as elucidated by quantum mechanics and a study of molecular spectra.

molecular vibration   The theory that all atoms within a molecule are in continuous motion, vibrating at definite frequencies specific to the molecular structure as a whole as well as to groups of atoms within the molecule; the basis of spectroscopic analysis.

molecular volume   The volume that is occupied by 1 mole (gram-molecular weight) of an element or compound; equals the molecular weight divided by the density.

molecular weight   The sum of the atomic weights of all the atoms in a molecule.

molecular-weight distribution   Frequency of occurrence of the different molecular-weight chains in a homologous polymeric system.

molecule   A group of atoms held together by chemical forces; the atoms in the molecule may be identical as in H_2, S_2, and S_8, or different as in H_2O and CO_2; a molecule is the smallest unit of matter which can exist by itself and retain all its chemical properties.

mole fraction   The ratio of the number of moles of a substance in a mixture or solution to the total number of moles of all the components in the mixture or solution.

mole percent   Percentage calculation expressed in terms of moles rather than weight.

mole volume   *See* molar volume.

molinate $C_9H_{17}NOS$ A light yellow liquid with limited solubility in water; used as an herbicide to control watergrass in rice. Also known as S-ethyl hexahydro-1H-azepine-1-carbothioate.

molting hormone Any of several hormones which activate molting in arthropods.

molybdate A salt derived from a molybdic acid.

molybdenum A chemical element, symbol Mo, atomic number 42, and atomic weight 95.95.

molybdenum dioxide MoO_2 Lead-gray powder; insoluble in hydrochloric and hydrofluoric acids; used in pigment for textiles.

molybdenum disilicide $MoSi_2$ A dark gray, crystalline powder with a melting range of 1870–2030°C; soluble in hydrofluoric and nitric acids; used in electrical resistors and for protective coatings for high-temperature conditions.

molybdenum disulfide MoS_2 A black lustrous powder, melting at 1185°C, insoluble in water, soluble in aqua regia and concentrated sulfuric acid; used as a dry lubricant and an additive for greases and oils. Also known as molybdenum sulfide; molybdic sulfide.

molybdenum pentachloride $MoCl_5$ Hygroscopic gray-black needles melting at 194°C; reacts with water and air; soluble in anhydrous organic solvents; used as a catalyst and as raw material to make molybdenum hexacarbonyl.

molybdenum sesquioxide MoO_3 Water-insoluble, gray-black powder with slight solubility in acids; used as a catalyst and as a coating for metal articles.

molybdenum sulfide *See* molybdenum disulfide.

molybdenum trioxide MoO_3 A white solid at room temperature, with a melting point of 795°C; soluble in concentrated mixtures of nitric and sulfuric acids and nitric and hydrochloric acids; used as a corrosion inhibitor, in enamels and ceramic glazes, in medicine and agriculture, and as a catalyst in the petroleum industry.

molybdic acid Any acid derived from molybdenum trioxide, especially the simplest acid H_2MoO_4, obtained as white crystals.

molybdic sulfide *See* molybdenum disulfide.

monatomic Composed of one atom.

monatomic gas A gas whose molecules have only one atom; the inert gases are examples.

mono- A prefix for chemical compounds to show a single radical; for example, monoglyceride, a glycol ester on which a single acid group is attached to the glycerol group.

monoacetate A compound such as a salt or ester that contains one acetate group.

monoacetin *See* acetin.

monoacid Compound with a single acid group, such as hydrochloric acid, HCl, or 2-naphthol-7-sulfonic acid, $C_{10}H_6(OH)(SO_3H)$.

monoamine An amine compound that has only one amino group.

monoamine oxidase An enzyme that catalyzes oxidative deamination of monoamines.

monoammonium tartrate *See* ammonium bitartrate.

monobasic Pertaining to an acid with one displaceable hydrogen atom, such as hydrochloric acid, HCl.

monobasic calcium phosphate *See* calcium phosphate.

monobasic sodium phosphate NaH_2PO_4 White crystals that are slightly hygroscopic, soluble in water, insoluble in alcohol; used in baking powders and acid cleansers, and as a cattle-food supplement.

monoblastic leukemia *See* acute monocytic leukemia.

monobromoacetanilide *See* para-bromoacetanilide.

monocalcium phosphate *See* calcium phosphate.

monochloroacetic acid *See* chloroacetic acid.

monochloroacetone *See* chloroacetone.

monochlorodifluoromethane *See* chlorodifluoromethane.

monochloroethane *See* ethyl chloride.

monochlorotrifluoromethane *See* chlorotrifluoromethane.

monochromator A spectrograph in which a detector is replaced by a second slit, placed in the focal plane, to isolate a particular narrow band of wavelengths for refocusing on a detector or experimental object.

monoclinic system One of the six crystal systems characterized by a single, twofold symmetry axis or a single symmetry plane.

monoclonal cryoglobulin A cryoglobulin composed of immunoglobin with only one class or subclass of heavy and light chain.

monodispersity Polymer system that is homogeneous in molecular weight, that is, it does not have a distribution of different molecular-weight chains within the total mass.

monoester An ester that has only one ester group.

monoethanolamine *See* ethanolamine.

monoglyceride Any of the fatty-acid glycerol esters where only one acid group is attached to the glycerol group, for example, $RCOOCH_2CHOHCH_2OH$; examples are glycerol monostearate and monolaurate; used as emulsifiers in cosmetics and lubricants.

monolayer *See* monomolecular film.

monolinuron *See* 3-(4-chlorophenyl)-1-methoxyl-1-methyl-urea.

monomer A simple molecule which is capable of combining with a number of like or unlike molecules to form a polymer; it is a repeating structure unit within a polymer.

monomethylamine *See* methylamine.

monomolecular film A film one molecule thick. Also known as monolayer.

monopotassium L-glutamate *See* potassium glutamate.

monosaccharide A carbohydrate which cannot be hydrolyzed to a simpler carbohydrate; a polyhedric alcohol having reducing properties associated with an actual or potential aldehyde or ketone group; classified on the basis of the number of carbon atoms, as triose (3C), tetrose (4C), pentose (5C), and so on.

monosodium acid methanearsonate CH_4AsNaO_3 A white, crystalline solid; melting point is 132–139°C; soluble in water; used as an herbicide for grassy weeds on rights-of-way, storage areas, and noncrop areas, and as preplant treatment for cotton, citrus trees, and turf. Abbreviated MSMA.

monosodium glutamate   *See* sodium glutamate.

monosodium methanearsonate   *See* sodium methanearsonate.

monoterpene   1. A class of terpenes with molecular formula $C_{10}H_{16}$; the members of the class contain two isoprene units. 2. A derivative of a member of such a class.

monotrophic   Of crystal pairs, having one of the pair always metastable with respect to the other.

monovalent   A radical or atom whose valency is 1.

monoxide   A compound that contains a single oxygen atom, such as carbon monoxide, CO.

monuron   *See* 3-(*para*-chlorophenyl)-1,1-dimethylurea.

Mooney unit   An arbitrary unit used to measure the plasticity of raw, or unvulcanized rubber; the plasticity in Mooney units is equal to the torque, measured on an arbitrary scale, on a disk in a vessel that contains rubber at a temperature of 100°C and rotates at two revolutions per minute.

mordant   An agent, such as alum, phenol, or aniline, that fixes dyes to tissues, cells, textiles, and other materials by combining with the dye to form an insoluble compound. Also known as dye mordant.

morin $C_{15}H_{10}O_7·2H_2O$ Colorless needles soluble in boiling alcohol, slightly soluble in water; used as a mordant dye and analytical reagent.

morphine benzyl ether hydrochloride   *See* peronine.

morphine methyl bromide   *See* morphosan.

morpholine C_4H_8ONH A hygroscopic liquid, soluble in water; used as a solvent and rubber accelerator. Also known as tetrahydro-1,4-oxazine.

morphosan $C_{17}H_{19}NO_3·CH_3Br$ A solid morphine derivative without morphine's disagreeable aftereffects; used in medicine. Also known as morphine methyl bromide.

morphotropism   Similarity of structure, axial ratios, and angles between faces of one or more zones in crystalline substances whose formulas can be derived one from another by substitution.

Morse equation   An equation according to which the potential energy of a diatomic molecule in a given electronic state is given by a Morse potential.

Morse potential   An approximate potential associated with the distance r between the nuclei of a diatomic molecule in a given electronic state; it is $V(r) = D\{1 - \exp[-a(r-r_e)]\}^2$, where r_e is the equilibrium distance, D is the dissociation energy, and a is a constant.

mosaic gold   *See* stannic sulfide.

mosaic structure   In crystals, a substructure in which neighboring regions are oriented slightly differently.

Moseley's law The law that the square-root of the frequency of an x-ray spectral line belonging to a particular series is proportional to the difference between the atomic number and a constant which depends only on the series.

Mössbauer effect The emission and absorption of gamma rays by certain nuclei, bound in crystals, without loss of energy through nuclear recoil, with the result that radiation emitted by one such nucleus can be absorbed by another.

Mössbauer spectroscopy The study of Mössbauer spectra, for example, for nuclear hyperfine structure, chemical shifts, and chemical analysis.

Mössbauer spectrum A plot of the absorption, by nuclei bound in a crystal lattice, of gamma rays emitted by similar nuclei in a second crystal, as a function of the relative velocity of the two crystals.

mountain blue $2CuCO_3 \cdot Cu(OH)_2$ Ground azurite used as a paint pigment. Also known as copper blue.

moving bed Granulated solids in a process vessel that are circulated (moved) either mechanically or by gravity flow; used in catalytic and absorption processes.

moving-bed catalytic cracking Petroleum refining process for cracking (breaking) of long hydrocarbon molecules by use of heat, pressure, and a granular cracking catalyst that is continuously cycled between the reactor vessel and the catalyst regenerator.

moving-boundary electrophoresis A U-tube variation of electrophoresis analysis that uses buffered solution so that all ions of a given species move at the same rate to maintain a sharp, moving front (boundary).

MPK *See* pentanone.

mRNA *See* messenger ribonucleic acid.

MSCFD Abbreviation for thousand standard cubic feet per day; usually refers to gas flow.

MSCFH Abbreviation for thousand standard cubic feet per hour; usually refers to gas flow.

MSCFM Abbreviation for thousand standard cubic feet per minute; usually refers to gas flow.

MSG *See* sodium glutamate.

MSH *See* melanocyte-stimulating hormone.

M shell The third layer of electrons about the nucleus of an atom, having electrons characterized by the principal quantum number 3.

MSMA *See* monosodium acid methanearsonate.

MTMC *See* *meta*-tolyl-*N*-methylcarbamate.

mucic acid $HOOC(CHOH)_4COOH$ A white, crystalline powder with a melting point of 210°C; soluble in water; used as a metal ion sequestrant and to retard concrete hardening. Also known as galactaric acid; saccharolactic acid; tetrahydroxyadipic acid.

mucigen A substance from which mucin is derived; contained in mucus-secreting epithelial cells.

mucin   A glycoprotein constituent of mucus and various other secretions of humans and lower animals.

mucoid   1. Any of various glycoproteins, similar to mucins but differing in solubilities and precipitation properties and found in cartilage, in the crystalline lens, and in white of egg. 2. Resembling mucus.

mucolytic   Effecting the dissolution, liquefaction, or dispersion of mucus and mucopolysaccharides.

mucopolysaccharide   Any of a group of polysaccharides containing an amino sugar and uronic acid; a constituent of mucoproteins, glycoproteins, and blood-group substances.

mucoprotein   Any of a group of glycoproteins containing a sugar, usually chondroitinsulfuric or mucoitinsulfuric acid, combined with amino acids or polypeptides.

mud sump   Upstream area in a process vessel where, because of a velocity drop, entrained solids drop out and are collected in a sump.

mull technique   Method for obtaining infrared spectra of materials in the solid state; material to be scanned is first pulverized, then mulled with mineral oil.

multicomponent distillation   The distillation separation of a single liquid feed stream containing three or more components into a single overhead product and a single bottoms product.

multiphase flow   Mixture of two or more distinct phases (such as oil, water, and gas) flowing through a closed conduit.

multiphoton absorption   The excitation of an atom or other microscopic system to a higher quantum state by simultaneous absorption of two or more photons which together provide the necessary energy.

multiphoton ionization   The removal of one or more electrons from an atom or other microscopic system as the result of simultaneous absorption of two or more photons.

multiple cartridges   Filter medium made up of two or more filter cartridges, either fastened end to end or arranged side by side (in series or parallel flow respectively).

multiple decay   *See* branching.

multiple disintegration   *See* branching.

multiple-effect evaporation   Series-operation energy economizer system in which heat from the steam generated (evaporated liquid) in the first stage is used to evaporate additional liquid in the second stage (by reducing system pressure), and so on, up to 10 or more effects; commonly used in the pulp and paper industry.

multiplet   A collection of relatively closely spaced spectral lines resulting from transitions to or from the members of a multiplet (as in the quantum-mechanics definition).

multiplet intensity rules   Rules for the relative intensities of spectral lines in a spin-orbit multiplet, stating that the sum of the intensities of all lines which start from a common initial level, or end on a common final level, is proportional to $2J + 1$, where J is the total angular momentum of the initial level or final level respectively.

muramic acid   An organic acid found in the mucopeptide (murein) in the cell walls of bacteria and blue-green algae.

muramidase   Lysozyme when acting as an enzyme on the hydrolysis of the muramic acid-containing mucopeptide in the cell walls of some bacteria.

murein   The peptidoglycan of bacterial cell walls.

muriatic acid   *See* hydrochloric acid.

muriatic ether   *See* ethyl chloride.

Murphree efficiency   In a plate-distillation column, the ratio of the actual change in vapor composition when the vapor passes through the liquid on a tray (plate) to the composition change of the vapor if it were in vapor-liquid equilibrium with the tray liquid.

muscarine $C_8H_{19}NO_3$ A quaternary ammonium compound, the toxic ingredient of certain mushrooms, as *Amanita muscaria*. Also known as hydroxycholine.

muscle hemoglobin   *See* myoglobin.

musk ambrette $C_{12}H_{16}N_2O_5$ White to yellow powder with heavy musky aroma; soluble in various oils and phthalates, insoluble in water; congeals at 83°C; used as a perfume fixative. Also known as 2,6-dinitro-3-methoxy-4-tert-butyltoluene.

musk ketone $C_{14}H_{18}N_2O_5$ White to yellow crystals with sweet musk aroma; soluble in various oils and phthalates, insoluble in water; used as a perfume fixative. Also known as 3,5-dinitro-2,6-dimethyl-4-tert-butylacetophenone.

musk xylene   *See* musk xylol.

musk xylol $(NO_2)_3C_6(CH_3)_2C(CH_3)_3$ White to yellow crystals with powerful musk aroma; soluble in various oils and phthalates, insoluble in water; congeals at 105°C; used as a perfume fixative. Also known as musk xylene; 2,4,6-trinitro-1,3-dimethyl-5-tert-butylbenzene.

mustard gas $HS(CH_2ClCH_2)_2S$ An oil, density 1.28, boiling point 215°C; used in chemical warfare. Also known as dichlorodiethylsulfide.

mustard oil   *See* allyl isothiocyanate.

mutarotation   A change in the optical rotation of light that takes place in the solutions of freshly prepared sugars.

mutase   An enzyme able to catalyze a dismutation or a molecular rearrangement.

mutual exclusion rule   The rule that if a molecule has a center of symmetry, then no transition is allowed in both its Raman scattering and infrared emission (and absorption), but only in one or the other.

mutuality of phases   The rule that if two phases, with respect to a reaction, are in equilibrium with a third phase at a certain temperature, then they are in equilibrium with respect to each other at that temperature.

mycobactin   Any compound produced by some strains, and required for growth by other strains, of *Mycobacteria*.

myelin   A soft, white fatty substance that forms a sheath around certain nerve fibers.

myoglobin   A hemoglobinlike iron-containing protein pigment occurring in muscle fibers. Also known as muscle hemoglobin; myohemoglobin.

myohematin   A cytochrome respiratory enzyme allied to hematin.

myohemoglobin   *See* myoglobin.

myoinositol   The commonest isomer of inositol. Also known as mesoinositol.

myokinase   An enzyme that catalyzes the reversible transfer of phosphate groups in adenosinediphosphate; occurs in muscle and other tissues.

myosin   A muscle protein, comprising up to 50% of the total muscle proteins; combines with actin to form actomycin.

β-myrcene $C_{10}H_{16}$ An oily liquid with a pleasant odor; soluble in alcohol, chloroform, ether, and glacial acetic acid; used as an intermediate in the preparation of perfume chemicals. Also known as 2-methyl-6-methylene-2,7-octadiene.

myricetin $C_{15}H_{10}O_8$ A yellow, crystalline compound with a melting point of 357°C; soluble in alcohol; used as an inhibitor of adenosinetriphosphatase. Also known as cannabiscetin; delphidenolon.

myristic acid $CH_3(CH_2)_{12}COOH$ Oily white crystals melting at 58°C; soluble in ether and alcohol, insoluble in water; used to synthesize flavor and perfume esters, and in soaps and cosmetics.

myristyl alcohol $C_{14}H_{29}OH$ Liquid boiling at 264°C; soluble in ether and alcohol, insoluble in water; used as a chemical intermediate, plasticizer, and perfume fixative. Also known as 1-tetradecanol.

N

n- Chemical prefix for "normal" (straight-carbon-chain) hydrocarbon compounds.

N *See* newton; nitrogen; normality.

Na *See* sodium.

NAA *See* naphthaleneacetic acid.

nabam $NaSSCNHCH_2CH_2NHCSSNa$ Water-soluble, colorless crystals that will irritate skin and eyes; used as a pesticide and pesticides intermediate. Also known as disodium ethylene-bis-dithiocarbamate.

NAD *See* diphosphopyridine nucleotide.

nadir point *See* photograph nadir.

naled $C_4H_7Br_2Cl_2O_4$ A white solid with a melting point of 27°C; slight solubility in water; used as an insecticide and miticide for crops, farm buildings, and kennels, and for mosquito control. Also known as 1,2-dibromo-2,2-dichloroethyl dimethyl phosphate.

nantokite *See* cuprous chloride.

naphthacene $C_{18}H_{12}$ A hydrocarbon molecule that may be considered to be four benzene rings fused together; it is explosive when shocked; used in organic synthesis. Also known as rubene; tetracene.

naphthalene $C_{10}H_8$ White, volatile crystals with coal tar aroma; insoluble in water, soluble in organic solvents; structurally it is represented as two benzenoid rings fused together; boiling point 218°C, melting point 80.1°C; used for moth repellents, fungicides, lubricants, and resins, and as a solvent. Also known as naphthalin; tar camphor.

naphthaleneacetamide $C_{12}H_{11}NO$ A colorless solid with a melting point of 183°C; used as a growth regulator for root cuttings and for thinning of apples and pears. Also known as 1-naphthaleneacetamide.

1-naphthaleneacetamide *See* naphthaleneacetamide.

naphthaleneacetic acid $C_{10}H_7CH_2COOH$ White, odorless crystals, melting at 132–135°C; soluble in organic solvents, slightly soluble in water; used as an agricultural spray. Abbreviated NAA. Also known as 1-naphthylacetic acid.

naphthalene-1,5-disulfonic acid $C_{10}H_6(SO_3H)_2$ White crystals, decomposing when heated; used to make dyes. Also known as Armstrong's acid.

1-naphthalenesulfonic acid $C_{10}H_8O_3S$ A crystalline compound with a melting point of 90°C (dihydrate); soluble in water or alcohol; used to make α-naphthol.

naphthalic acid *See* phthalic acid.

naphthalin *See* naphthalene.

naphthene Any of the cycloparaffin derivatives of cyclopentane (C_5H_{10}) or cyclohexane (C_6H_{12}) found in crude petroleum.

naphthenic acid Any of the derivatives of cyclopentane, cyclohexane, cycloheptane, or other naphthenic homologs derived from petroleum; molecular weights 180 to 350; soluble in organic solvents and hydrocarbons, slightly soluble in water; used as a paint drier and wood preservative, and in metals production.

naphthionic acid $C_{10}H_6(NH_2)SO_3H$ White powder or crystals that decompose when heated; used to manufacture dyes. Also known as 1-aminonaphthalene-4-sulfonic acid; 4-amino-1-naphthalene sulfonic acid; 1-naphthylamine-4-sulfonic acid.

1-naphthol *See* α-naphthol.

2-naphthol *See* β-naphthol.

α-naphthol $C_{10}H_7OH$ Colorless to yellow powder, melting at 96°C; used to make dyes and perfumes, and in synthesis of organic molecules. Also known as 1-hydroxynaphthalene; 1-naphthol.

β-naphthol $C_{10}H_7OH$ White crystals that melt at 121.6°C; insoluble in water; used to make pigments, dyes, and antioxidants. Also known as 2-hydroxynaphthalene; 2-naphthol.

3-naphthol-2-carboxylic acid *See* β-hydroxynaphthoic acid.

β-naphthol methyl ether *See* β-naphthyl methyl ether.

naphthopyrinidine *See* 5,6-benzoquinoline.

β-naphthoquinoline *See* 5,6-benzoquinoline.

1,2-naphthoquinone $C_{10}H_6O_2$ A golden yellow, crystalline compound that decomposes at 145–147°C; soluble in benzene and ether; used as a reagent for resorcinol and thalline. Also known as β-naphthoquinone.

1,4-naphthoquinone $C_{10}H_6O_2$ Greenish-yellow powder soluble in organic solvents, slightly soluble in water; melts at 123–126°C; used as an antimycotic agent, in synthesis, and as a rubber polymerization regulator. Also known as α-naphthoquinone.

α-naphthoquinone *See* 1,4-naphthoquinone.

β-naphthoquinone *See* 1,2-naphthoquinone.

1,2-naphthoquinone-4-sulfonate *See* sodium β-naphthoquinone-4-sulfonate.

1,2-naphthoquinone-4-sulfonic acid sodium salt *See* sodium β-naphthoquinone-4-sulfonate.

β-naphthoquinone-4-sulfonic acid sodium salt *See* sodium β-naphthoquinone-4-sulfonate.

naphthoresorcinol $C_{10}H_6(OH)_2$ Crystals with a melting point of 124–125°C; soluble in ether, alcohol, and water; used as a reagent for sugars and oils, and to determine glucuronic acid in urine. Also known as 1,3-dihydroxynaphthalene.

2-naphthoxyacetic acid *See* β-naphthoxyacetic acid.

β-naphthoxyacetic acid $C_{10}H_9O_2$ A crystalline compound soluble in water, with a melting point of 156°C; used as a growth regulator to set blossoms and regulate growth for pineapples, strawberries, and tomatoes. Also known as BNOA; 2-naphthoxyacetic acid.

2-(α-naphthoxy)-N,N-diethylpropionamide   *See* devrinol.

1-naphthylacetic acid   *See* naphthaleneacetic acid.

naphthylamine $C_{10}H_7NH_2$ White, toxic crystals, soluble in alcohol and ether; used in dyes; the two forms are α-naphthylamine, boiling at 301°C, and β-naphthylamine, boiling at 306°C.

1-naphthylamine-4-sulfonic acid   *See* naphthionic acid.

2-naphthylamine-6-sulfonic acid   *See* Brönner's acid.

2,5-naphthylamine sulfonic acid   *See* gamma acid.

β-naphthylmethyl ether $C_{10}H_7OCH_3$ White, crystalline scales with a melting point of 72°C; soluble in alcohol and ether; used for soap perfumes. Also known as 2-methoxynaphthalene; methyl naphthyl ether; β-naphthol methyl ether.

N-1-naphthylphthalamic acid $C_{10}H_7NHCOC_6H_4COOH$ A crystalline solid with a melting point of 185°C; used as a preemergence herbicide.

α-naphthylthiocarbamide   *See* 1-(1-naphthyl)-2-thiourea.

1-(1-naphthyl)-2-thiourea $C_{10}H_7NHCSNH_2$ A crystalline compound with a melting point of 198°C; soluble in water, acetone, triethylene glycol, and hot alcohol; used as a poison to control the adult Norway rat. Also known as α-naphthylthiocarbamide; α-naphthylthiourea (ANTU); N-1-naphthyl-thiourea.

α-naphthylthiourea   *See* 1-(1-naphthyl)-2-thiourea.

N-1-naphthylthiourea   *See* 1-(1-naphthyl)-2-thiourea.

Naples yellow   *See* lead antimonite.

naptalam $C_{18}H_{12}O_3NNa$ A light purple solid with a melting point of 185°C; limited solubility in water; used as an herbicide and growth regulator to control weeds and to thin peaches. Also known as N-naphthylphthalamic acid, sodium salt (NPA).

narceine $C_{23}H_{27}O_8N \cdot 3H_2O$ White, odorless crystals with bitter taste; soluble in alcohol and water, insoluble in ether; melts at 170°C; used in medicine.

naringenin-7-rhamnoglucoside   *See* naringin.

naringenin-7-rutinoside   *See* naringin.

naringin $C_{27}H_{32}O_{14}$ A crystalline bioflavonoid with a melting point of 171°C; soluble in acetone and alcohol; used as a food supplement. Also known as aurantiin; naringenin-7-rhamnoglucoside; naringenin-7-rutinoside.

nascent   Pertaining to an atom or simple compound at the moment of its liberation from chemical combination, when it may have greater activity than in its usual state.

native   Pertaining to an element found in nature in a nongaseous state.

native metal   A metallic native element; includes silver, gold, copper, iron, mercury, iridium, lead, palladium, and platinum.

native uranium Uranium as found in nature; a mixture of the fertile uranium-238 isotope (99.3%), the fissionable uranium-235 isotope (0.7%), and a minute percentage of other uranium isotopes. Also known as natural uranium; normal uranium.

natrium Latin name for sodium; source of the symbol Na.

natural-gasoline plant Compression, distillation, and absorption process facility used to remove natural gasoline (mostly butanes and heavier components) from natural gas.

natural red *See* purpurin.

natural species *See* unchanged species.

natural uranium *See* native uranium.

Nb *See* niobium.

Nd *See* neodymium.

NDGA *See* nordihydroguaiaretic acid.

Ne *See* neon.

nearest neighbors Any pair of atoms in a crystal lattice which are as close to each other, or closer to each other, than any other pair.

near-infrared spectrophotometry Spectrophotometry at wavelengths in the near-infrared region, generally using instruments with quartz prisms in the monochromators and lead sulfide photoconductor cells as detectors to observe absorption bands which are harmonics of bands at longer wavelengths.

neburon $C_{12}H_{16}Cl_2N_2O$ A white, crystalline compound with a melting point of 102–103°C; used as an herbicide to control weeds in nursery ornamentals, dichondras, and wheat. Also known as 1-*n*-butyl-3-(3,4-dichlorophenyl)-1-methylurea.

negative catalysis A catalytic reaction such that the reaction is slowed down by the presence of the catalyst.

negative crystal A crystal containing a cavity, where the form of the cavity is one of the characteristic crystal forms of the mineral in question.

negative elongation In a section of an anisotropic crystal, a sign of elongation that is parallel to the faster of the two plane-polarized rays.

negative ion An atom or group of atoms which by gain of one or more electrons has acquired a negative electric charge.

negative-ion vacancy A point defect in an ionic crystal in which a negative ion is missing from its lattice site.

neighbor One of a pair of atoms or ions in a crystal which are close enough to each other for their interaction to be of significance in the physical problem being studied.

N electron An electron in the fourth (*N*) shell of electrons surrounding the atomic nucleus, having the principal quantum number 4.

Nelson diaphragm cell Obsolete carbon-electrode type of electrolytic diaphragm cell once widely used to produce chlorine and caustic soda from brine.

nematic phase A phase of a liquid crystal in the mesomorphic state, in which the liquid has a single optical axis in the direction of the applied magnetic field, appears to be turbid and to have mobile threadlike structures, can flow readily, has low viscosity, and lacks a diffraction pattern.

nematogenic solid A solid which will form a nematic liquid crystal when heated.

neo-, ne- Prefix indicating hydrocarbons where a carbon is bonded directly to at least four other carbon atoms, such as neopentane.

neodymium A metallic element, symbol Nd, with atomic weight 144.24, atomic number 60; a member of the rare-earth group of elements.

neodymium chloride $NdCl_3 \cdot xH_2O$ Water- and acid-soluble, pink lumps; used to prepare metallic neodymium.

neodymium oxide Nd_2O_3 A hygroscopic, blue-gray powder; insoluble in water, soluble in acids; used to color glass and in ceramic capacitors.

neohexane C_6H_{14} Volatile, flammable, colorless liquid boiling at 50°C; used as high-octane component of motor and aviation gasolines.

neohexane alkylation A noncatalytic petroleum-refinery alkylation process that forms neohexane from a feed of ethylene and isobutane.

neomineralization Chemical interchange within a rock whereby its mineral constituents are converted into entirely new mineral species.

neon A gaseous element, symbol Ne, atomic number 10, atomic weight 20.183; a member of the family of noble gases in the zero group of the periodic table.

neonicotine *See* anabasine.

neopentane C_5H_{12} Colorless liquid boiling at 10°C; soluble in alcohol, insoluble in water; a hydrocarbon found as a minor component of natural gasoline. Also known as 2,2-dimethylpropane.

neotenin A hormone secreted by cells of the corpus allatum in arthropod larvae and nymphs; inhibits the development of adult characters. Also known as juvenile hormone.

nephrolysin A toxic substance capable of disintegrating kidney cells.

N-nephthylphthalamic acid, sodium salt *See* naptalam.

neptunium A chemical element, symbol Np, atomic number 93, atomic weight 237.0482; a member of the actinide series of elements.

neptunium decay series Little-known radioactive elements with short lives; produced as successive series of decreasing atomic weight when uranium-237 and plutonium-241 decay radioactively through neptunium-237 to bismuth-209.

Nernst equation The relationship showing that the electromotive force developed by a dry cell is determined by the activities of the reacting species, the temperature of the reaction, and the standard free-energy change of the overall reaction.

Nernst-Thomson rule The rule that in a solvent having a high dielectric constant the attraction between anions and cations is small so that dissociation is favored, while the reverse is true in solvents with a low dielectric constant.

Nernst zero of potential An electrode potential corresponding to the reversible equilibrium between hydrogen gas at a pressure of 1 standard atmosphere and hydrogen ions at unit activity.

nerol $C_{10}H_{17}OH$ Colorless liquid with rose-neroli odor; derived from geraniol (a trans isomer); used in perfumery.

nerolidol $C_{15}H_{26}O$ A straw-colored sesquiterpene alcohol; liquid with rose and apple aroma derived from cabreuva oil, oils of orange flower, and ylang ylang; soluble in alcohol; used in perfumery.

Nessler's reagent Mercuric iodide– potassium iodide solution, used to analyze for small amounts of ammonia.

Nessler tubes Standardized glass tubes for filling with standard solution colors for visual color comparison with similar tubes filled with solution samples.

Neumann's principle The principle that the symmetry elements of the point group of a crystal are included among the symmetry elements of any property of the crystal.

neuraminic acid $C_9H_{17}NO_8$ An amino acid, the aldol condensation product of pyruvic acid and N-acetyl-D-mannosamine, regarded as the parent acid of a family of widely distributed acyl derivatives known as sialic acids.

neurine $CH_2=CHN(CH_3)_3OH$ A very poisonous, syrupy liquid with fishy aroma; soluble in water and alcohol; a product of putrefaction of choline in brain tissue and bile, and in cadavers. Also known as trimethylvinylammonium hydroxide.

neurochemistry Chemistry of the nervous system.

neurohormone A hormone produced by nervous tissue.

neurohumor A hormonal transmitter substance, such as acetylcholine, released by nerve endings in the transmission of impulses.

neurotoxin A poisonous substance in snake venom that acts as a nervous system depressant by blocking neuromuscular transmission by binding acetylcholine receptors on motor end plates, or on the innervated face of an electroplax.

neutral atom An atom in which the number of electrons that surround the nucleus is equal to the number of protons in the nucleus, so that there is no net electric charge.

neutral flame Gas flame produced by a mixture of fuel and oxygen so as to be neither oxidizing nor reducing.

neutral granulation Propellant granulation in which the surface area of a grain remains constant during burning.

neutralization equivalent For an acid or base, the same as equivalent weight; multiplication of the neutralization equivalent by the number of acidic or basic groups in the molecule gives the molecular weight.

neutralization number Petroleum product test; it is the milligrams of potassium hydroxide required to neutralize the acid in 1 gram of oil; used as an indication of oil acidity.

neutralize To make a solution neutral (neither acidic nor basic, pH of 7) by adding a base to an acidic solution, or an acid to a basic solution.

neutral mercuric phosphate *See* mercuric phosphate.

neutral mercurous phosphate *See* mercurous phosphate.

neutral molecule A molecule in which the number of electrons surrounding the nuclei is the same as the total number of protons in the nuclei, so that there is no net electric charge.

neutral potassium phosphate *See* potassium phosphate.

neutral red $(CH_3)_2NC_6H_3N_2C_6H_2CH_3NH_2 \cdot ClH$ Water- and alcohol-soluble green powder; used as pH 6.8–8.0 acid-base indicator, and as a dye to test stomach function. Also known as dimethyl diaminophenazine chloride; toluylene red.

neutron absorption *See* neutron capture.

neutron binding energy The energy required to remove a single neutron from a nucleus.

neutron capture A process in which the collision of a neutron with a nucleus results in the absorption of the neutron into the nucleus with the emission of one or more prompt gamma rays; in certain cases, beta decay or fission of the nucleus results. Also known as neutron absorption; neutron radiative capture.

neutron-capture cross section The cross section for neutron capture by nuclei in a material; it is a measure of the probability that this reaction will occur.

neutron cross section A measure of the probability that an interaction of a given kind will take place between a nucleus and an incident neutron; it is an area such that the number of interactions which occur in a sample exposed to a beam of neutrons is equal to the product of the number of nuclei in the sample and the number of neutrons in the beam that would pass through this area if their velocities were perpendicular to it.

neutron excess The number of neutrons in a nucleus in excess of the number of protons. Also known as difference number; isotopic number.

neutron magnetic moment A vector whose scalar product with the magnetic flux density gives the negative of the energy of interaction of a neutron with a magnetic field.

neutron number The number of neutrons in the nucleus of an atom.

neutron radiative capture *See* neutron capture.

neutron spectrometry A method of observing excited states of nuclei in which neutrons are used to bombard a target, causing nuclei to be transmuted into excited states by various nuclear reactions; the resultant excited states are determined by observing resonances in the reaction cross sections or by observing spectra of emitted particles or gamma rays. Also known as neutron spectroscopy.

neutron spectroscopy *See* neutron spectrometry.

neutron spectrum A plot or display of the number of neutrons at various energies, such as the neutrons emitted in a nuclear reaction, or the neutrons in a nuclear reactor.

Newland's law of octaves An arrangement of the elements that predated Mendeleeff's periodic table; Newland's arrangement was a grouping of the elements in increasing atomic weights (starting with lithium) in horizontal rows of eight elements, with each new row directly beneath the previous one.

Ni *See* nickel.

niacin *See* nicotinic acid.

niacinamide *See* nicotinamide.

nickel A chemical element, symbol Ni, atomic number 28, atomic weight 58.71.

nickel-63 Radioactive nickel with beta radiation and 92-year half-life; derived by pile-irradiation of nickel; used in radioactive composition studies and tracer studies.

nickel acetate $Ni(OOCCH_3)_2 \cdot 4H_2O$ Efflorescent green crystals that decompose upon heating; soluble in alcohol and water; used as textile dyeing mordant.

nickel ammonium sulfate $NiSO_4 \cdot (NH_4)_2SO_4 \cdot 6H_2O$ A green, crystalline compound, soluble in water; used as a nickel electrolyte for electroplating. Also known as ammonium nickel sulfate; double nickel salt.

nickel arsenate $Ni_3(AsO_4)_2 \cdot H_2O$ Poisonous yellow-green powder; soluble in acids, insoluble in water; used as a fat-hardening catalyst in soapmaking. Also known as nickelous arsenate.

nickel carbonate $NiCO_3$ Light-green crystals that decompose upon heating; soluble in acid, insoluble in water; used in electroplating.

nickel carbonyl $Ni(CO)_4$ Colorless, flammable, poisonous liquid boiling at 43°C; soluble in alcohol and concentrated nitric acid, insoluble in water; used in gas plating (vapor decomposes at 60°C) and to produce metallic nickel. Also known as nickel tetracarbonyl.

nickel cyanide $Ni(CN)_2 \cdot 4H_2O$ Poisonous, water-insoluble apple-green powder; melts and loses water at 200°C, decomposes at higher temperatures; used for electroplating and metallurgy.

nickel formate $Ni(HCOO)_2 \cdot 2H_2O$ Water-soluble green crystals; used in hydrogenation catalysts.

nickel iodide NiI_2 or $NiI_2 \cdot 6H_2O$ Hygroscopic black or blue-green solid; soluble in water and alcohol; sublimes when heated. Also known as nickelous iodide.

nickel nitrate $Ni(NO_3)_2 \cdot 6H_2O$ Fire-hazardous oxidant; deliquescent, green, water- and alcohol-soluble crystals; used for nickel plating and brown ceramic colors, and in nickel catalysts.

nickelocene $(C_5H_5)_2Ni$ Dark green crystals with a melting point of 171–173°C; soluble in most organic solvents; used as an antiknock agent. Also known as dicyclopentadienyl nickel.

nickelous arsenate *See* nickel arsenate.

nickelous iodide *See* nickel iodide.

nickelous oxide *See* nickel oxide.

nickelous phosphate *See* nickel phosphate.

nickel oxide NiO Green powder; soluble in acids and ammonium hydroxide; insoluble in water; used to make nickel salts and for porcelain paints. Also known as green nickel oxide; nickelous oxide.

nickel phosphate $Ni_3(PO_4)_2 \cdot 7H_2O$ A light-green powder; soluble in acids and ammonium hydroxide, insoluble in water; used for electroplating and production of yellow nickel. Also known as nickelous phosphate; trinickelous orthophosphate.

nickel tetracarbonyl *See* nickel carbonyl.

niclosamide $C_{13}H_8Cl_2N_2O_4$ A white powder with a melting point of 227–232°C; insoluble in water; used as an antihelminthic drug for dogs and cats. Also known as 2′,5-dichloro-4′-nitrosalcylanilide.

nicotinamide $C_6H_6ON_2$ Crystalline basic amide of the vitamin B complex that is interconvertible with nicotinic acid in the living organism; the amide of nicotinic acid. Also known as niacinamide.

nicotinamide adenine dinucleotide *See* diphosphopyridine nucleotide.

nicotine $C_{10}H_{14}N_2$ A colorless liquid with a boiling point of 247.3°C; miscible with water; used as a contact insecticide fumigant in closed spaces. Also known as 3-(1-methyl-2-pyrrolidyl)pyridine; nicotine sulfate.

nicotinic acid $C_6H_5NO_2$ A component of the vitamin B complex; a white, water-soluble powder stable to heat, acid, and alkali; used for the treatment of pellagra. Also known as niacin.

nigre Dark-colored layer formed between neat soap and lye during soap manufacture; contains more soap than lye, and a high concentration of salts and colored impurities.

ninhydrin $C_9H_4O_3 \cdot H_2O$ White crystals or powder with a melting point of 240–245°C; soluble in water and alcohol; used for the detection and assay of peptides, amines, amino acids, and amino sugars. Also known as triketo-hydrindene hydrate.

niobe oil *See* methyl benzoate.

niobic acid $Nb_2O_5 \cdot nH_2O$ Family of hydrates; white precipitate, soluble in inorganic acids and bases, insoluble in water; its formation is part of the analytical determination of niobium.

niobium A chemical element, symbol Nb, atomic number 41, atomic weight 92.906.

niobium carbide NbC A lavender gray powder with a melting point of 3500°C; used for carbide-tipped tools and special steels.

niter *See* potassium nitrate.

niter cake *See* sodium bisulfate.

nitralin $C_{13}H_{19}N_3O_6S$ A light yellow to orange, crystalline compound with a melting point of 151–152°C; slight solubility in water; used as a preemergence herbicide for weed control for cotton, food crops, and ornamentals. Also known as 4-(methylsulfonyl)-2,6-dinitro-*N,N*-dipropylaniline.

meta-**nitraniline** *See meta*-nitroaniline.

ortho-**nitraniline** *See ortho*-nitroaniline.

para-**nitraniline** *See para*-nitroaniline.

nitrate 1. A salt or ester of nitric acid. 2. Any compound containing the NO_3^- radical.

nitrating acid Sulfuric-nitric acid mix used to nitrate cellulosics and aromatic chemicals. Also known as mixed acid.

nitration Introduction of an NO_2^- group into an organic compound.

nitrene A molecular fragment that is an uncharged, electron-deficient species containing a monocovalent nitrogen.

nitric acid HNO_3 Strong oxidant that is fire-hazardous; colorless or yellowish liquid, miscible with water; boils at 86°C; used for chemical synthesis, explosives, and fertilizer manufacture, and in metallurgy, etching, engraving, and ore flotation. Also known as aqua fortis.

nitric oxide NO A colorless gas that, at room temperature, reacts with oxygen to form nitrogen dioxide (NO_2, a reddish-brown gas); may be used to form other compounds.

nitride Compound of nitrogen and a metal, such as Mg_3N_2.

nitrile RC≡N Cyanide derived by removal of water from an acid amide.

nitrilotriacetic acid $N(CH_2COOH)_3$ A white powder, melting point 240°C, with some decomposition; soluble in water; it is toxic, and birth abnormalities may result from ingestion; may be used as a chelating agent in the laboratory. Also known as NTA; TGA; triglycine; triglycolamic acid.

nitrite A compound containing the radical NO_2^-; can be organic or inorganic.

nitro- Chemical prefix showing the presence of the NO_2^- radical.

nitroalkane A compound, with the general formula RNO_2, in which the hydrogen atom of an alkane molecule has been replaced by a nitro group.

***meta*-nitroaniline** $NO_2C_6H_4NH_2$ Yellow crystals that melt at 112.5°C; a toxic material; used as a dye intermediate. Also known as *meta*-nitraniline.

***ortho*-nitroaniline** $NO_2C_6H_4NH_2$ Orange-red crystals that melt at 69.7°C, soluble in ethanol; a toxic material; used to manufacture dyes. Also known as *ortho*-nitraniline.

***para*-nitroaniline** $NO_2C_6H_4NH_2$ Yellow crystals that melt at 148°C; insoluble in water, soluble in ethanol; a toxic material; used to make dyes, and as a corrosion inhibitor. Also known as *para*-nitraniline.

nitroaromatic A nitrated benzene or benzene derivative, such as nitrobenzene, $C_6H_5NO_2$, or nitrobenzoic acid, $NO_2 \cdot C_6H_4 \cdot COOH$.

nitrobarite *See* barium nitrate.

nitrobenzene $C_6H_5NO_2$ Greenish crystals or a yellowish liquid, melting point 5.70°C; a toxic material; used in aniline manufacture. Also known as oil of mirbane.

***ortho*-nitrobiphenyl** $C_{12}H_9NO_2$ A crystalline compound with a sweetish odor; melting point is 36.7°C; used as a plasticizer for resins, cellulose acetate and nitrate, and polystyrenes, and as a fungicide for textiles. Abbreviated ONB.

nitrobromoform *See* bromopicrin.

nitrocalcite *See* calcium nitrate.

nitrocellulose *See* cellulose nitrate.

nitrochloroform *See* chloropicrin.

nitrocotton *See* cellulose nitrate.

nitro dye A dye with the NO_2 chromophore group in the molecules.

nitroethane $CH_3CH_2NO_2$ A colorless liquid, slightly soluble in water; boils at 114°C; used as a solvent for cellulosics, resins, waxes, fats, and dyestuffs, and as a chemical intermediate.

nitro explosive Explosive compound containing one or more NO_2^- groups, such as nitroglycerine, $C_3H_5(ONO_2)_3$, or trinitrotoluene, $C_6H_2(CH_3)(NO_2)_3$.

nitrogen   A chemical element, symbol N, atomic number 1, atomic weight 14.0067; it is a gas, diatomic (N_2) under normal conditions; about 78% of the atmosphere is N_2; in the combined form the element is a constituent of all proteins.

nitrogen acid anhydride   *See* nitrogen pentoxide.

nitrogenase   An enzyme that catalyzes a six-electron reduction of N_2 in the process of nitrogen fixation.

nitrogen balance   The net loss or gain of nitrogen in a soil.

nitrogen cycle   *See* carbon-nitrogen cycle.

nitrogen dioxide NO_2 A reddish-brown gas; it exists in varying degrees of concentration in equilibrium with other nitrogen oxides; used to produce nitric acid. Also known as dinitrogen tetroxide; liquid dioxide; nitrogen peroxide; nitrogen tetroxide.

nitrogen fixation   Conversion of atmospheric nitrogen into compounds such as ammonia, calcium cyanamide, or nitrogen oxides by chemical or electric-arc processes.

nitrogen monoxide   *See* nitrous oxide.

nitrogen mustard   Any of the substituted mustard gases in which the sulfur is replaced by an amino nitrogen, such as for methyl bis(2-chlorethyl)amine, $(CH_2ClCH_2)_2NCH_3$; useful in cancer research.

nitrogen pentoxide N_2O_5 Colorless crystals, soluble in water (forms HNO_3); decomposes at 46°C. Also known as nitrogen acid anhydride.

nitrogen peroxide   *See* nitrogen dioxide.

nitrogen solution   Mixture used to neutralize super-phosphate in fertilizer manufacture; consists of 60% ammonium nitrate, and the balance a 50% aqua ammonia solution.

nitrogen tetroxide   *See* nitrogen dioxide.

nitrogen trifluoride NF_3 A colorless gas that has a melting point of -206.6°C and a boiling point of -128.8°C; used as an oxidizer for high-energy fuels.

nitrogen trioxide N_2O_3 Green, water-soluble liquid; boils at 3.5°C.

nitroglycerin $CH_2NO_3CHNO_3CH_2NO_3$ Highly unstable, explosive, flammable pale-yellow liquid; soluble in alcohol; freezes at 13°C and explodes at 260°C; used as an explosive, to make dynamite, and in medicine.

nitroguanidine $H_2NC(NH)NHNO_2$ Explosive yellow solid, soluble in alcohol; melts at 246°C; used in explosives and smokeless powder.

nitrohydrochloric acid   *See* aqua regia.

2-nitro-1-hydroxybenzene-4-arsonic acid   *See* 4-hydroxy-3-nitrobenzenearsonic acid.

nitromannite   *See* mannitol hexanitrate.

nitromannitol   *See* mannitol hexanitrate.

nitrometer   Glass apparatus used to collect and measure nitrogen and other gases evolved by a chemical reaction. Also known as azotometer.

nitromethane CH_3NO_2 A liquid nitroparaffin compound; oily and colorless; boils at 101°C; used as a monopropellant for rockets, in chemical synthesis, and as an industrial solvent for cellulosics, resins, waxes, fats, and dyestuffs.

nitromuriatic acid *See* aqua regia.

nitron $CN_4(C_6H_5)_3CH$ Yellow crystals, soluble in chloroform and acetone; used as reagent to detect NO_3 ion in dilute solutions. Also known as diphenylenedianilinohydrotriazole.

nitronium Positively charged NO_2 ion, believed to be formed from HNO_3.

nitroparaffin Any organic compound in which one or more hydrogens of a paraffinic hydrocarbon are replaced by a nitro, or NO_2^-, group, such as nitromethane, CH_3NO_2, or nitroethane, $C_2H_5NO_2$; used as a solvent and a chemical intermediate.

***ortho*-nitrophenol** $C_6H_5NO_3$ A yellow, crystalline compound; melting point is 44–45°C; soluble in hot water, alcohol, benzene, ether, carbon disulfide, and alkali hydroxides; used in the commercial preparation of many compounds.

nitrophenolarsonic acid *See* 4-hydroxy-3-nitrobenzenearsonic acid.

***para*-nitrophenylazosalicylate sodium** *See* alizarin yellow.

***para*-nitrophenylhydrazine** $C_6H_7N_3O_2$ An orange-red, crystalline compound with a melting point of about 157°C; soluble in hot water or hot benzene; used as a reagent for aliphatic aldehydes and ketones.

***para*-nitrophenyl α,α,α-trifluoro-2-nitro-*para*-tolyl ether** *See* fluorodifen.

1-nitropropane $CH_3CH_2CH_2NO_2$ A colorless liquid with a boiling point of 132°C; used as a rocket propellant and gasoline additive.

2-nitropropane $CH_3CHNO_2CH_3$ A colorless liquid with a boiling point of 120°C; used as a solvent for vinyl coatings, as a rocket propellant, and as a gasoline additive.

nitroso The radical NO^- with trivalent nitrogen. Also known as hydroximino; oximido.

nitrosophenylhydroxylamine *See* cupferron.

nitrostarch $C_{12}H_{12}(NO_2)_8O_{10}$ Orange powder, soluble in ethyl alcohol; used in explosives. Also known as starch nitrate.

***meta*-nitrotoluene** $NO_2C_6H_4CH_3$ Yellow powder that melts at 15°C; insoluble in water; used in organic synthesis. Also known as *meta*-methylnitrobenzene.

***ortho*-nitrotoluene** $NO_2C_6H_4CH_3$ A yellow liquid boiling at 220.4°C; insoluble in water; used to produce toluidine and dyes. Also known as *ortho*-methylnitrobenzene.

***para*-nitrotoluene** $NO_2C_6H_4CH_3$ Yellow crystals that melt at 51.7°C; insoluble in water, soluble in ethanol; used to produce toluidine and to manufacture dyes. Also known as *para*-methylnitrobenzene.

nitrourea $NH_2CONHNO_2$ Highly explosive white crystals, melting at 159°C; soluble in ether and alcohol, slightly soluble in water; used as a chemical intermediate.

nitrous acid HNO_2 Aqueous solution of nitrogen trioxide, N_2O_3.

nitrous oxide N_2O. Colorless, sweet-tasting gas, boiling at -90°C; slightly soluble in water, soluble in alcohol; used as a food aerosol, and as an anesthetic in dentistry and surgery. Also known as laughing gas; nitrogen monoxide.

nitroxanthic acid   *See* picric acid.

nitroxylene $C_6H_3(CH_3)_2NO_2$ Any of three isomers occurring either as a yellow liquid or as crystalline needles with a melting point of 2°C and boiling point of 246°C; soluble in alcohol and ether; used in gelatinizing accelerators for pyroxylin. Also known as dimethylnitrobenzene.

nitryl halide NO_2X Compound containing a halide (X) and a nitro group (NO_2).

N line   One of the characteristic lines in an atom's x-ray spectrum, produced by excitation of an N electron.

No   *See* nobelium.

nobelium   A chemical element, symbol No, atomic number 102, atomic weight 254 when the element is produced in the laboratory; a synthetic element, in the actinium series.

noble gas   A gas in group 0 of the periodic table of the elements; it is monatomic and, with limited exceptions, chemically inert. Also known as inert gas.

noble potential   A potential equaling or approaching that of the noble elements, such as gold, silver, or copper, of the electromotive series.

NODA   *See* n-octyl n-decyl adipate.

nonacosane $C_{29}H_{60}$ Colorless hydrocarbon, melting at 63°C; found in beeswax and the fat of cabbage leaves.

nonadecane $CH_3(CH_2)_{17}CH_3$ Flammable crystals, soluble in ether and alcohol, insoluble in water; melts at 32°C; used as a chemical intermediate.

nonanal $C_8H_{17}CHO$ A colorless liquid with an orange rose odor; used in perfumes and for flavoring. Also known as aldehyde C–9; n-nonyl aldehyde; pelargonic aldehyde.

nonane $CH_3(CH_2)_7CH_3$ Flammable, colorless liquid, boiling at 151°C; soluble in alcohol, insoluble in water; used as a chemical intermediate. Also known as nonyl hydride.

nonanedioic acid   *See* azelaic acid.

n-nonanoic acid   *See* pelargonic acid.

nonanol   *See* n-nonyl alcohol.

nonaqueous   Pertaining to a liquid or solution containing no water.

nonbenzenoid aromatic compound   A compound exhibiting aromatic character but not containing a benzene nucleus, or having one or more rings in a fused ring system that are not benzene rings.

noncompetitive inhibition   Enzyme inhibition in which the inhibitor can combine with either the free enzyme or the enzyme-substrate complex so that the inhibitor does not compete with the substrate for the enzyme.

noncrossing rule   The rule that when the potential energies of two electronic states of a diatomic molecule are plotted as a function of distance between the nuclei, the resulting curves do not cross, unless the states have different symmetry.

nonene   *See* 1-nonylene.

nonessential amino acid   An amino acid which can be synthesized by an organism and thus need not be supplied in the diet.

nonfaradaic path One of the two available paths for transfer of energy across an electrolyte-metal interface, in which energy is carried by capacitive transfer, that is, by charging and discharging the double-layer capacitance.

nonhypergolic Not capable of igniting spontaneously upon contact; used especially with reference to rocket fuels.

***n*-nonic acid** *See* pelargonic acid.

nonideal solution A solution whose behavior does not conform to that of an ideal solution; that is, the behavior is not predictable over a wide range of concentrations and temperatures by the use of Raoult's law.

nonine *See* nonyne.

nonlinear molecule A branched-chain molecule, that is, one whose atoms do not all lie along a straight line. Also known as isomolecule.

nonlinear spectroscopy The study of energy levels not normally accessible with optical spectroscopy, through the use of nonlinear effects such as multiphoton absorption and ionization.

nonoic acid $C_8H_{17}COOH$ Any of a family of acids which are mixed isomers produced in the Fischer-Tropsch process; pelargonic acid is the straight-chain member; used as a chemical intermediate.

***n*-nonoic acid** *See* pelargonic acid.

nonpolar Pertaining to an element or compound which has no permanent electric dipole moment.

nonprotein nitrogen The nitrogen fraction in the body tissues, excretions, and secretions, not precipitated by protein precipitants.

nonprotic solvent A solvent that does not contain a hydrogen ion source.

nonspherical nucleus A nucleus which appears to have a permanent ellipsoidal shape in its ground state, as suggested by a large electric quadrupole moment.

nonyl acetate $C_9H_{19}OOCCH_3$ Any of a family of isomers, such as *n*-nonyl acetate and diisobutyl carbinyl acetate, which are products of Fischer-Tropsch and oxo syntheses.

***n*-nonyl acetate** $CH_3COO(CH_2)_8CH_3$ Alcohol-soluble, colorless liquid with pungent odor; boiling point 208–212°C; used in perfumery.

***n*-nonyl alcohol** $CH_3(CH_2)_7CH_2OH$ One of a family of $C_9H_{19}OH$ isomers; a colorless liquid with rose aroma; boils at 215°C; insoluble in water, soluble in alcohol; used in perfumery and flavorings. Also known as alcohol C-9; octyl carbinol; nonanol; pelargonic alcohol.

***n*-nonyl aldehyde** *See* nonanal.

nonyl benzene $C_9H_{19}C_6H_5$ Liquid boiling at 245–252°C; straw-colored with aromatic aroma; used to make surface-active agents.

nonyl carbinol *See* decyl alcohol.

1-nonylene C_9H_{18} Colorless liquid boiling at 150°C; soluble in alcohol, insoluble in water; used as a chemical intermediate. Also known as nonene.

nonyl hydride *See* nonane.

***n*-nonylic acid**   *See* pelargonic acid.

nonyl phenol $C_9H_{19}C_6H_4OH$ Pale-yellow liquid boiling at 283–302°C; soluble in organic solvents, insoluble in water; a mixture of monoalkyl phenol isomers, mostly para-substituted; used to make surface-active agents, resins, and plasticizers.

nonylphenoxyacetic acid $C_9H_{19}C_6H_4OCH_2COOH$ A viscous, amber-colored liquid, soluble in alkali; used in turbine oils, lubricants, greases, and other materials as a corrosion inhibitor.

nonyne $CH_3(CH_2)_6{\equiv}CCH$ Water-insoluble, colorless liquid boiling at 160°C. Also known as *n*-heptylacetylene; nonine.

nopinene   *See* pinene.

nor-   Chemical formula prefix for normal; indicates a parent for another compound to be formed by removal of one or more carbons and associated hydrogens.

noradrenaline   *See* norepinephrine.

nordihydroguaiaretic acid $C_{18}H_{22}O_4$ A crystalline compound with a melting point of 184–185°C; soluble in alcohols, ether, acetone, glycerol, and propyleneglycol; used as an antioxidant in fats and oils. Abbreviated NDGA.

norepinephrine $C_8H_{11}O_3N$ A hormone produced by chromaffin cells of the adrenal medulla; acts as a vasoconstrictor and mediates transmission of sympathetic nerve impulses. Also known as noradrenaline.

norflurazon $C_{12}H_9ClF_3N_3O$ An off-white powder with a melting point of 175–178°C; slightly soluble in water; used as a preemergence herbicide for cotton, cranberries, and fruit trees. Also known as 4-chloro-5-(methylamino)-2(α,α,α-trifluoro-*meta*-tolyl)-3-(2*H*)-pyridazine.

normal bonded-phase chromatography   A technique of bonded-phase chromatography in which the stationary phase is polar and the mobile phase is nonpolar.

normality   Measure of the number of gram-equivalent weights of a compound per liter of solution. Abbreviated N.

normal mass shift   The portion of the mass shift that corresponds to the variation of reduced mass, and is thus easily calculated for all transitions.

normal potassium pyrophosphate   *See* potassium pyrophosphate.

normal salt   A salt in which all of the acid hydrogen atoms have been replaced by a metal, or the hydroxide radicals of a base are replaced by an acid radical; for example, Na_2CO_3.

normal silver sulfate   *See* silver sulfate.

normal solution   An aqueous solution containing one equivalent of the active reagent in grams in 1 liter of the solution.

normal state   A term sometimes used for ground state.

normal thorium sulfate   *See* thorium sulfate.

normal twin   A twin crystal whose twin axis is perpendicular to the composition surface.

normal uranium   *See* native uranium.

norphytane *See* pristane.

Np *See* neptunium.

NPA *See* naptalam.

NRS *See* nuclear reaction spectrometry.

N shell The fourth layer of electrons about the nucleus of an atom, having electrons characterized by the principal quantum number 4.

NTA *See* nitrilotriacetic acid.

nuclear 1. Pertaining to a group of atoms joined directly to the central group of atoms or central ring of a molecule. 2. Pertaining to the atomic nucleus.

nuclear absorption Absorption of energy by the nucleus of an atom.

nuclear angular momentum *See* nuclear spin.

nuclear atom An atomic structure consisting of dense, positively charged nucleus (neutrons and protons) surrounded by a corresponding set of negatively charged electrons.

nuclear binding energy The energy required to separate an atom into its constituent protons, neutrons, and electrons.

nuclear capture Any process in which a particle, such as a neutron, proton, electron, muon, or alpha particle, combines with a nucleus.

nuclear chemistry Study of the atomic nucleus, including fission and fusion reactions and their products.

nuclear collision A collision between an atomic nucleus and another nucleus or particle.

nuclear cross section A measure of the probability for a reaction to occur between a nucleus and a particle; it is an area such that the number of reactions which occur in a sample exposed to a beam of particles equals the product of the number of nuclei in the sample and the number of incident particles which would pass through this area if their velocities were perpendicular to it.

nuclear decay mode One of the ways in which a nucleus can undergo radioactive decay, distinguished from other decay modes by the resulting isotope and the particles emitted.

nuclear density The mass per unit volume of a nucleus as a function of distance from the center of the nucleus, as determined by a number of different types of experiments which are in reasonably good agreement.

nuclear fission *See* fission.

nuclear force That part of the force between nucleons which is not electromagnetic; it is much stronger than electromagnetic forces, but drops off very rapidly at distances greater than about 10^{-13} centimeter; it is responsible for holding the nucleus together.

nuclear fusion *See* fusion.

nuclear ground state The stationary state of lowest energy of an isotope.

nuclear magnetic moment The magnetic dipole moment of an atomic nucleus; a vector whose scalar product with the magnetic flux density gives the negative of the energy of interaction of a nucleus with a magnetic field.

nuclear magnetic resonance spectrometer A spectrometer in which nuclear magnetic resonance is used for the analysis of protons and nuclei and for the study of changes in chemical and physical quantities over wide frequency ranges.

nuclear magneton A unit of magnetic dipole moment used to express magnetic moments of nuclei and baryons; equal to the electron charge times Planck's constant divided by the product of 4π, the proton mass, and the speed of light.

nuclear mass The mass of an atomic nucleus, which is usually measured in atomic mass units; it is less than the sum of the masses of its constituent protons and neutrons by the binding energy of the nucleus divided by the square of the speed of light.

nuclear moment One of the various static electric or magnetic multipole moments of a nucleus.

nuclear number *See* mass number.

nuclear polarization For a nucleus in a mixed state, with spin I and probability $p(I_z)$ that the I_z substate is populated, the polarization is the sum over allowed values of I_z of $I_z p(I_z)/I$.

nuclear potential The potential energy of a nuclear particle as a function of its position in the field of a nucleus or of another nuclear particle.

nuclear potential energy The average total potential energy of all the protons and neutrons in a nucleus due to the nuclear forces between them, excluding the electrostatic potential energy.

nuclear potential scattering That part of elastic scattering of particles by a nucleus which may be treated by studying the scattering of a wave which obeys the Schrödinger equation with a potential determined by the properties of the nucleus.

nuclear quadrupole moment The electric quadrupole moment of an atomic nucleus.

nuclear radiation A term used to denote alpha particles, neutrons, electrons, photons, and other particles which emanate from the atomic nucleus as a result of radioactive decay and nuclear reactions.

nuclear radiation spectroscopy Study of the distribution of energies or momenta of particles emitted by nuclei.

nuclear radius The radius of a sphere within which the nuclear density is large, and at the surface of which it falls off sharply.

nuclear reaction A reaction involving a change in an atomic nucleus, such as fission, fusion, neutron capture, or radioactive decay, as distinct from a chemical reaction, which is limited to changes in the electron structure surrounding the nucleus. Also known as reaction.

nuclear reaction spectrometry A method of determining the concentration of a given element as a function of depth beneath the surface of a sample, by measuring the yield of characteristic gamma rays from a resonance reaction occurring when the surface is bombarded by a beam of ions. Abbreviated NRS.

nuclear resonance 1. An unstable excited state formed in the collision of a nucleus and a bombarding particle, and associated with a peak in a plot of cross section versus energy. **2.** The absorption of energy by nuclei from radio-frequency fields at certain frequencies when these nuclei are also subjected to certain types of static fields, as in magnetic resonance and nuclear quadrupole resonance.

nuclear scattering The change in directions of particles as a result of collisions with nuclei.

nuclear spallation *See* spallation.

nuclear species *See* nuclide.

nuclear spectrum **1.** The relative number of particles emitted by atomic nuclei as a function of energy or momenta of these particles. **2.** The graphical display of data from devices used to measure these quantities.

nuclear spin The total angular momentum of an atomic nucleus, resulting from the coupled spin and orbital angular momenta of its constituent nuclei. Also known as nuclear angular momentum. Symbolized I.

nuclear spontaneous reaction *See* radioactive decay.

nuclear stability The ability of an isotope to resist decay or fission.

nuclear transformation *See* transmutation.

nuclear Zeeman effect A splitting of atomic spectral lines resulting from the interaction of the magnetic moment of the nucleus with an applied magnetic field.

nuclease An enzyme that catalyzes the splitting of nucleic acids to nucleotides, nucleosides, or the components of the latter.

nucleate boiling Boiling in which bubble formation is at the liquid-solid interface rather than from external or mechanical devices; occurs in kettle-type and natural-circulation heaters or reboilers.

nucleation In crystallization processes, the formation of new crystal nuclei in supersaturated solutions.

nucleic acid A large, acidic, chainlike molecule containing phosphoric acid, sugar, and purine and pyrimidine bases; two types are ribonucleic acid and deoxyribonucleic acid.

nuclein Any of a poorly defined group of nucleic acid protein complexes occurring in cell nuclei.

nucleonium A bound state of a nucleus and an antinucleus.

nucleon number *See* mass number.

nucleoprotein Any member of a class of conjugated proteins in which molecules of nucleic acid are closely associated with molecules of protein.

nucleosidase An enzyme that catalyzes the hydrolysis of a nucleoside to its component pentose and its purine or pyrimidine base.

nucleoside The glycoside resulting from removal of the phosphate group from a nucleotide; consists of a pentose sugar linked to a purine or pyrimidine base.

nucleotidase Any of a group of enzymes which split phosphoric acid from nucleotides, leaving nucleosides.

nucleotide An ester of a nucleoside and phosphoric acid; the structural unit of a nucleic acid.

nucleus The central, positively charged, dense portion of an atom. Also known as atomic nucleus.

nuclide A species of atom characterized by the number of protons, number of neutrons, and energy content in the nucleus, or alternatively by the atomic number, mass number, and atomic mass; to be regarded as a distinct nuclide, the atom must be capable of existing for a measurable lifetime, generally greater than 10^{-10} second. Also known as nuclear species; species.

Nylander reagent A solution of Rochelle salt (potassium sodium tartrate), potassium or sodium hydroxide, and bismuth subnitrate in water; used to test for sugar in urine.

O

O *See* oxygen.

Obermayer's reagent A 0.4% solution of ferric chloride in concentrated hydrochloric acid; used to test for indican in urine, with a pale-blue or deep-violet color indicating positive.

obtuse bisectrix The bisectrix of the obtuse angle between the axes of a biaxial crystal.

ochratoxin A $C_{20}H_{18}ClNO_6$ A toxic metabolite from *Aspergillus ochraceus*; a crystalline compound which exhibits green fluorescence; melting point is 169°C; inhibits phosphorylase and mitochondrial respiration in rat liver.

9,12-octadecadienoic acid *See* linoleic acid.

octadecanamide *See* stearamide.

n-octadecane $C_{18}H_{38}$ Colorless liquid boiling at 318°C; soluble in alcohol, acetone, ether, and petroleum, insoluble in water; used as a solvent and chemical intermediate.

n-octadecanoic acid *See* stearic acid.

1-octadecanol *See* stearyl alcohol.

octadeca-9,11,13-trienoic acid *See* eleostearic acid.

9,12,15-octadecatrienoic acid *See* linolenic acid.

octadecatrienol *See* linolenyl alcohol.

9-octadecen-1,12-diol *See* ricinoleyl alcohol.

1-octadecene $C_{18}H_{36}$ Colorless liquid boiling at 180°C; soluble in alcohol, acetone, ether, and petroleum, insoluble in water; used as a chemical intermediate.

octadecenol *See* oleyl alcohol.

octadecenyl aldehyde $C_{17}H_{35}CHO$ A flammable liquid with a boiling point of 167°C; used in the manufacture of vulcanization accelerators, rubber antioxidants, and pesticides. Also known as oleyl aldehyde.

octadecyl alcohol *See* stearyl alcohol.

octafluorocyclobutane C_4F_8 A colorless gas or liquid with a boiling point of −4°C and a freezing point of −41.4°C; soluble in ether; used as a dielectric, refrigerant, and aerosol propellant. Also known as perfluorocyclobutane.

octafluoropropane C_3F_8 A colorless gas with a boiling point of −36.7°C and a freezing point of approximately −160°C; used as a refrigerant and gaseous insulator. Also known as perfluoropropane.

octahedral cleavage   Crystal cleavage in the four planes parallel to the face of the octahedron.

octahedral plane   The plane in a cubic lattice having three numerically equal Miller indices.

octanal   *See* octyl aldehyde.

n-octane C_8H_{18} Colorless liquid boiling at 126°C; soluble in alcohol, acetone, and ether, insoluble in water; used as a solvent and chemical intermediate.

octanedioic acid   *See* suberic acid.

n-octanoic acid   *See* caprylic acid.

octanone-3   *See* ethyl amyl ketone.

2-octanone   *See* methyl hexyl ketone.

1-octene $CH_3(CH_2)_5CHCH_2$ A colorless, flammable liquid; used as a plasticizer and in synthesis of organic compounds. Also known as 1-caprylene; 1-octylene.

2-octene $CH_3(CH_2)_4CHCHCH_3$ A colorless, flammable liquid, with trans and cis forms; used to manufacture lubricants and to synthesize organic materials.

octet   A collection of eight valence electrons in an atom or ion, which form the most stable configuration of the outermost, or valence, electron shell.

octine   *See* octyne.

octomethylene   *See* cyclooctane.

octyl-   Prefix indicating the eight-carbon hydrocarbon radical (C_8H_{17}—).

n-octyl acetate $CH_3COO(CH_2)_7CH_3$ A colorless liquid with a fruity odor and a boiling point of 199°C; soluble in alcohol and other organic liquids; used for perfumes and flavoring. Also known as acetate C-8; caprylyl acetate.

octylacetic acid   *See* capric acid.

octyl alcohol   *See* 2-ethylhexyl alcohol.

octyl aldehyde $C_8H_{16}O$ A liquid aldehyde boiling at 172°C; found in essential oils of many plants; used in perfume compositions. Also known as caprylaldehyde; octanal.

n-octyl bromide   *See* 1-bromooctane.

octyl carbinol   *See* n-nonyl alcohol.

n-octyl n-decyl adipate   A liquid with a boiling range of 250–254°C; used as a low-temperature plasticizer. Abbreviated NODA.

n-octyl n-decyl phthalate   A clear liquid with a boiling range of 232–267°C; used as a plasticizer for vinyl resins.

1-octylene   *See* 1-octene.

octylene glycol   *See* ethohexadiol.

octyl formate $C_8H_{17}OOCH$ A colorless liquid with a fruity odor; soluble in mineral oil; used for flavoring.

octylic acid   *See* caprylic acid.

_n_-octyl mercaptan $C_8H_{17}SH$ Clear, colorless liquid boiling at 199°C; used as a chemical intermediate and polymerization conditioner.

octyl phenol $C_8H_{17}C_6H_4OH$ White flakes, congealing at 73°C; soluble in organic solvents, insoluble in water; used to make surfactants, plasticizers, and antioxidants. Also known as diisobutyl phenol.

octyne $CHC(CH_2)_5CH_3$ Colorless hydrocarbon liquid, boiling at 125°C. Also known as caprilydene; hexylacetylene; octine.

odd-even nucleus   A nucleus which has an odd number of protons and an even number of neutrons.

odd-odd nucleus   A nucleus that has an odd number of protons and an odd number of neutrons.

odd term   A term of an atom or molecule for which the sum of the angular-momentum quantum numbers of all the electrons is odd, so that the states have odd parity; designated by a superscript o or u.

odorize   To add an unpleasant odor as a safety measure to an odorless material such as fuel gas.

O electron   An electron in the fifth (O) shell of electrons surrounding the atomic nucleus, having the principal quantum number 5.

off-site facility   In a chemical process plant, any supporting facility that is not a direct part of the reaction train, such as utilities, steam, and waste-treatment facilities.

ogdosymmetric class   A merohedral crystal class whose general form has one-eighth the number of equivalent faces of the corresponding holohedral form.

-oic   A suffix indicating the presence of a —COOH group, as in ethyloic (—CH_2—COOH).

oil blue   Violet-blue copper sulfide pigment used in varnishes.

oil garlic   _See_ allyl sulfide.

oil-gas process   Process to manufacture high-caloric-value fuel gas by the destructive distillation of high-boiling petroleum oils.

oil of mirbane   _See_ nitrobenzene.

oil of vitriol   _See_ sulfuric acid.

-ol   Chemical suffix for an —OH group in organic compounds, such as phenol (C_6H_5OH).

oleate   Salt made up of a metal or alkaloid with oleic acid; used for external medicines and in soaps and paints.

olefiant gas   _See_ ethylene.

olefin C_nH_{2n} A family of unsaturated, chemically active hydrocarbons with one carbon-carbon double bond; includes ethylene and propylene.

olefin copolymer   Polymer made by the interreaction of two or more kinds of olefin monomers, such as butylene and propylene.

olefin resin   Long-chain polymeric material produced by the chain reaction of olefinic monomers, such as polyethylene from ethylene, or polypropylene from propylene.

oleic acid $C_{17}H_{33}COOH$ Yellowish, unsaturated fatty acid with lardlike aroma; soluble in organic solvents, slightly soluble in water; boils at 286°C (100 mm Hg); the main component of olive and cooking oils; used in soaps, ointments, cosmetics, and ore beneficiation. Also known as red oil.

olein $(C_{17}H_{33}COO)_3C_3H_5$ Oleic acid triglyceride; yellow liquid melting at -5°C; slightly soluble in alcohol, soluble in chloroform, ether, and carbon tetrachloride; found in most fats and oils; used in textile lubrication.

oleum 1. Latin name for oil. 2. *See* fuming sulfuric acid.

oleyl alcohol $C_{18}H_{35}OH$ Clear liquid, boiling at 282–349°C; fatty alcohol derived from oleic acid; commercial grade 80–90% pure; used to make resins and surface-active agents, and as a chemical intermediate. Also known as octadecenol.

oleyl aldehyde *See* octadecenyl aldehyde.

oligomer A polymer made up of two, three, or four monomer units.

oligomeric protein A protein composed of two or more polypeptide chains.

oligonucleotide A polynucleotide of low molecular weight, consisting of less than 20 nucleotide polymers.

oligopeptide A peptide composed of no more than 10 amino acids.

oligosaccharide A sugar composed of two to eight monosaccharide units joined by glycosidic bonds. Also known as compound sugar.

omission solid solution A crystal with certain atomic sites incompletely filled.

ONB *See* ortho-nitrobiphenyl.

-one Chemical suffix indicating a ketone, a substance related to starches and sugars, or an alkone.

one-dimensional lattice A simplified model of a crystal lattice consisting of particles lying along a straight line at either equal or periodically repeating distances.

one-face-centered lattice A crystal lattice in which there are lattice points at the centers of one pair of faces in each unit cell as well as at the corners.

-onium Chemical suffix indicating a complex cation, as for oxonium, $(H_3O)^+$.

Onsager equation An equation which relates the measured equivalent conductance of a solution at a certain concentration to that of the pure solvent.

on stream Of plant or process-operations, unit, being in operation.

on-stream time In plant or process operations, the actual time that a unit is operating and producing product.

open-circle deoxyribonucleic acid *See* relaxed circular deoxyribonucleic acid.

open-circuit potential The steady-state or equilibrium potential of an electrode in absence of external current flow to or from the electrode.

open form A crystal form in which the crystal faces do not entirely enclose a space.

open-packed structure A crystal structure corresponding to the stacking of spheres in an orthogonal arrangement so that each sphere is in contact with six others.

operating line In the graphical solution of equilibrium processes (such as distillation absorption extraction), the actual liquid-vapor relationship of a key component, in contrast to a true equilibrium relationship.

opine A type of amino acid usually not found in nature, such as that secreted by a crown gall.

Oppenauer oxidation The oxidation of a primary or secondary hydroxyl compound to form the corresponding carbonyl compound; aluminum alkoxide and an excess amount of a carbonyl hydrogen acceptor, such as benzophenone or acetone, are required.

Oppenheimer-Phillips reaction A type of stripping reaction which can occur when a deuteron passes near a nucleus, in which the proton in the deuteron experiences Coulomb repulsion from the nucleus while the neutron is attracted to the nucleus by nuclear forces, with the result that the neutron-proton bond in the deuteron is broken, the neutron is absorbed into the nucleus, and the proton is repelled.

optical anomaly The phenomenon in which an organic compound has a molar refraction which does not agree with the value calculated from the equivalents of atoms and other structural units composing it.

optical antipode *See* enantiomorph.

optical crystal Any natural or synthetic crystal, such as sodium chloride, calcium fluoride, silver chloride, potassium iodide, or stilbene, that is used in infrared and ultraviolet optics and for its piezoelectric effects.

optical cytopherometry *See* microelectrophoresis.

optical exaltation Optical anomaly in which the observed molar refraction exceeds the calculated one; most cases of optical anomaly are in this category.

optical isomer *See* enantiomorph.

optical isomerism Existence of two forms of a molecule such that one is a mirror image of the other; the two molecules differ in that rotation of light is equal but in opposite directions.

optical model *See* cloudy crystal-ball model.

optical monochromator A monochromator used to observe the intensity of radiation at wavelengths in the visible, infrared, or ultraviolet regions.

optical null method In infrared spectrometry, the adjustment of a reference beam's energy transmission to match that of a beam that has been passed through a sample being analyzed.

optical spectra Electromagnetic spectra for wavelengths in the ultraviolet, visible and infrared regions, ranging from about 10 nanometers to 1 millimeter, associated with excitations of valence electrons of atoms and molecules, and vibrations and rotations of molecules.

optical spectrograph An optical spectroscope provided with a photographic camera or other device for recording the spectrum made by the spectroscope.

optical spectrometer An optical spectroscope that is provided with a calibrated scale either for measurement of wavelength or for measurement of refractive indices of transparent prism materials.

optical spectroscope An optical instrument, consisting of a slit, collimator lens, prism or grating, and a telescope or objective lens, which produces an optical spectrum arising from emission or absorption of radiant energy by a substance, for visual observation.

optical spectroscopy The production, measurement, and interpretation of optical spectra arising from either emission or absorption of radiant energy by various substances.

optical twinning Growing together of two crystals which are the same except that the structure of one is the mirror image of the structure of the other. Also known as chiral twinning.

optic angle *See* axial angle.

optic-axial angle *See* axial angle.

optimum cure The degree of vulcanization at which maximum desired property is reached.

optoacoustic detection method A method of detecting trace impurities in a gas, in which the absorption of a sample of the gas at various light frequencies is measured by directing a periodically interrupted laser beam through the sample in a spectrophone and measuring the sound generated by the optoacoustic effect at the frequency of interruption of the beam.

orange cadmium *See* cadmium sulfide.

orange oxide *See* uranium trioxide.

orange spectrometer A type of beta-ray spectrometer that consists of a number of modified double-focusing spectrometers employing a common source and a common detector, and has exceptionally high transmission.

orange toner A diazo dyestuff coupled to diacetoacetic acid anhydride; contains no sulfonic or carboxylic groups; used for printing inks.

orbital The space-dependent part of the Schrödinger wave function of an electron in an atom or molecule in an approximation such that each electron has a definite wave function, independent of the other electrons.

orbital decay A change of an atom from one energy state to another of lower energy in which the orbital of one of the electrons changes.

orbital electron An electron which has a high probability of being in the vicinity (at distances on the order of 10^{-10} meter or less) of a particular nucleus, but has only a very small probability of being within the nucleus itself. Also known as planetary electron.

orbital symmetry The property of certain molecular orbitals of being carried into themselves or into the negative of themselves by certain geometrical operations, such as a rotation of 180° about an axis in the plane of the molecule, or reflection through this plane.

orcin $CH_3C_6H_3(OH)_2 \cdot H_2O$ White crystals with strong, sweet, unpleasant taste; soluble in water, alcohol, and ether; extracted from lichens; used in medicine and as an analytical reagent. Also known as 3,5-dihydroxytoluene; orcinol.

order A classification of chemical reactions, in which the order is described as first, second, third, or higher, according to the number of molecules (one, two, three, or more) which appear to enter into the reaction; decomposition of H_2O_2 to form water and oxygen is a first-order reaction.

organic Of chemical compounds, based on carbon chains or rings and also containing hydrogen with or without oxygen, nitrogen, or other elements.

organic acid   A chemical compound with one or more carboxyl radicals (COOH) in its structure; examples are butyric acid, $CH_3(CH_2)_2COOH$, maleic acid, HOOCCHCHCOOH, and benzoic acid, C_6H_5COOH.

organic chemistry   The study of the composition, reactions, and properties of carbon-chain or carbon-ring compounds or mixtures thereof.

organic geochemistry   A branch of geochemistry which deals with naturally occurring carbonaceous and biologically derived substances which are of geological interest.

organic pigment   Any of the materials with organic-chemical bases used to add color to dyes, plastics, linoleum, tones, and lakes.

organic quantitative analysis   Quantitative determination of elements, functional groups, or molecules in organic materials.

organic reaction mechanism   A pathway of chemical states traversed by an organic chemical system in its passage from reactants to products.

organic salt   The reaction product of an organic acid and an inorganic base, for example, sodium acetate (CH_3COONa) from the reaction of acetic acid (CH_3COOH) and sodium hydroxide (NaOH).

organic solvent   Liquid organic compound with the power to dissolve solids, gases, or liquids (miscibility); examples are methanol (methyl alcohol), CH_3OH, and benzene, C_6H_6.

organized ferment   *See* intracellular enzyme.

organometallic compound   Molecules containing carbon-metal linkage; a compound containing an alkyl or aryl radical bonded to a metal, such as tetraethyllead, $Pb(C_2H_5)_4$.

organophosphate   A soluble fertilizer material made up of organic phosphate esters such as glucose, glycol, or sorbitol; useful for providing phosphorus to deep-root systems.

organosulfur compound   One of a group of substances which contain both carbon and sulfur.

orientation   **1.** The directions of the axes of a crystal lattice relative to the surfaces of the crystal, to applied fields, or to some other planes or directions of interest. **2.** The arrangement of radicals in an organic compound in relation to each other and to the parent compound.

orientation effect   A method of determining attractive forces among molecules, or components of these forces, from the interaction energy associated with the relative orientation of molecular dipoles.

orientation force   A type of van der Waals force, resulting from interaction of the dipole moments of two polar molecules. Also known as dipole-dipole force; Keesom force.

ornithine $C_5H_{12}O_2N_2$ An amino acid occurring in the urine of some birds, but not found in native proteins.

orotic acid $C_4H_4O_4N_2$ A crystalline acid which is a growth factor for certain bacteria and is also a pyrimidine precursor.

Orsat analyzer   Gas analysis apparatus in which various gases are absorbed selectively (volumetric basis) by passing them through a series of preselected solvents.

ortho acid   1. Aromatic acid with a carboxyl group in the ortho position (1,2 position). **2.** Organic acid with one added molecule of water in chemical combination; for example, $HC(OH)_3$, orthoformic acid, in contrast to $HCOOH$, formic acid; $H_3PO_4(P_2O_5 \cdot 3H_2O)$, orthophosphoric acid, in contrast to the less hydrated form, metaphosphoric acid, $HPO_3(P_2O_5 \cdot H_2O)$.

orthoarsenic acid   *See* arsenic acid.

orthoaxis   The diagonal or lateral axis perpendicular to the vertical axis in the monoclinic system.

orthoboric acid   *See* boric acid.

orthochem   A precipitate formed within a depositional basin or within the sediment itself by direct chemical action.

orthogonal crystal   A crystal whose axes are mutually perpendicular.

orthographic projection   A projection for displaying the poles of a crystal in which the poles are projected from a reference sphere onto an equatorial plane by dropping perpendiculars from the poles to the plane.

orthohelium   Those states of helium atoms in which the spins of the two electrons are parallel.

orthohexagonal axes   A set of crystallographic axes, two of which have a fixed ratio, as in hexagonal or trigonal crystals.

orthohydrogen   Those states of hydrogen molecules in which the spins of the two nuclei are parallel.

orthophosphate   One of the possible salts of orthophosphoric acid; the general formula is M_3PO_4, where M may be potassium as in potassium orthophosphate, K_3PO_4.

orthophosphoric acid   *See* phosphoric acid.

orthopinacoid   *See* front pinacoid.

orthorhombic lattice   A crystal lattice in which the three axes of a unit cell are mutually perpendicular, and no two have the same length. Also known as rhombic lattice.

orthorhombic system   A crystal system characterized by three axes of symmetry that are mutually perpendicular and of unequal length. Also known as rhombic system.

orthosymmetric crystal   A crystal that has orthorhombic symmetry.

orthotungstic acid   *See* tungstic acid.

Os   *See* osmium.

osazone   Any of the compounds that contain two phenylhydrazine residues and are produced by a reaction between a reducing sugar and phenylhydrazine.

oscillator strength   A quantum-mechanical analog of the number of dispersion electrons having a given natural frequency in an atom, used in an equation for the absorption coefficient of a spectral line; it need not be a whole number. Also known as f value; Ladenburg f value.

oscillatory reaction   A chemical reaction in which a variable of a chemical system exhibits regular periodic changes in time or in space.

oscillatory twinning Repeated, parallel twinning.

oscillographic polarography A type of voltammetry using a dropping mercury electrode with oscillographic scanning of the applied potential; used to measure the concentration of electroactive species in solutions.

oscillometric titration Radio-frequency technique used for conductometric and dielectrometric titrations; the changes in conductance or dielectric properties changes the solution capacity and thus the frequency of the connected oscillator circuit.

oscillometry Electrode measurement of oscillation-frequency changes to detect the progress of a titration of electrolytic solutions.

oscine *See* scopoline.

O shell The fifth layer of electrons about the nucleus of an atom, having electrons characterized by the principal quantum number 5.

osmate A salt or ester of osmic acid, containing the osmate radical, $OsO_4{}^{2-}$; for example, potassium osmate (K_2OsO_4).

osmic acid anhydride OsO_4 Poisonous yellow crystals with disagreeable odor; melts at 40°C; soluble in water, alcohol, and ether; used in medicine, photography, and catalysis. Also known as osmium oxide; osmium tetroxide.

osmium A chemical element, symbol Os, atomic number 76, atomic weight 190.2.

osmium oxide *See* osmic acid anhydride.

osmium tetroxide *See* osmic acid anhydride.

osmolality The molality of an ideal solution of a nondissociating substance that exerts the same osmotic pressure as the solution being considered.

osmolarity The molarity of an ideal solution of a nondissociating substance that exerts the same osmotic pressure as the solution being considered.

osmole 1. The unit of osmolarity equal to the osmolarity of a solution that exerts an osmotic pressure equal to that of an ideal solution of a nondissociating substance that has a concentration of 1 mole of solute per liter of solution. 2. The unit of osmolality equal to the osmolality of a solution that exerts an osmotic pressure equal to that of an ideal solution of a nondissociating substance that has a concentration of 1 mole of solute per kilogram of solvent.

osmometer A device for measuring molecular weights by measuring the osmotic pressure exerted by solvent molecules diffusing through a semipermeable membrane.

osmosis The transport of a solvent through a semipermeable membrane separating two solutions of different solute concentration, from the solution that is dilute in solute to the solution that is concentrated.

osmotic gradient *See* osmotic pressure.

osmotic pressure 1. The applied pressure required to prevent the flow of a solvent across a membrane which offers no obstruction to passage of the solvent, but does not allow passage of the solute, and which separates a solution from the pure solvent. 2. The applied pressure required to prevent passage of a solvent across a membrane which separates solutions of different concentration, and which allows passage of the solute, but may also allow limited passage of the solvent. Also known as osmotic gradient.

Ostwald dilution law   The law that for a sufficiently dilute solution of univalent electrolyte, the dissociation constant approximates $a^2c/(1-a)$, where c is the concentration of electrolyte and a is the degree of dissociation.

Ostwald process   An industrial preparation of nitric acid by the oxidation of ammonia; the oxidation takes place in successive stages to nitric oxide, nitrogen dioxide, and nitric acid; a catalyst of platinum gauze is used and high temperatures are needed.

Ostwald ripening   Solution-crystallizer phenomenon in which small crystals, more soluble than large ones, dissolve and reprecipitate onto larger particles.

Oswald diagram   Diagram used in fuel Orsat analyses by plotting percent by volume CO_2 maximum in the fuel (ordinate) versus percent by volume O_2 in air (abscissa); O_2 and CO_2 Orsat readings should fall on a line connecting these maximum values if the analysis is proceeding properly.

Othmer process   Process to produce acetic acid from pyroligneous acid by azeotropic distillation.

ouabain $C_{29}H_{44}O_{12} \cdot 8H_2O$ White crystals that melt with decomposition at 190°C, soluble in water and ethanol; used in medicine.

Oudeman law   The law that the molecular rotations of the various salts of an acid or base tend toward an identical limiting value as the concentration of the solution is reduced to zero.

outer orbital complex   A metal coordination compound in which the d orbital used in forming the coordinate bond is at the same energy level as the s and p orbitals.

outflow   Flow of fluid product out of a process facility.

ovalbumin   The major, conjugated protein of eggwhite.

overall plate efficiency   For a specified liquid-mixture separation in a fractionation (or distillation) tower, the ratio of actual to theoretical plates (or trays) required.

overall stability constant   Reaction equilibrium constant for the reaction forming soluble complexes during compleximetric titration.

overcritical binding   A binding energy for electrons in atoms which is so large that a vacancy in the bound state results in the spontaneous formation of an electron-positron pair; predicted to occur when the atomic number exceeds 173.

overcritical electric field   An electric field so strong that an electron-positron pair is created spontaneously; quantum electrodynamics predicts that this will happen near a nucleus having more than approximately 173 protons.

overcuring   A condition resulting from vulcanizing longer than necessary to achieve full development of physical strength; causes softness or brittleness and impaired age-resisting quality of the material.

overgrowth   A crystal growth in optical and crystallographic continuity around another crystal of different composition.

Overhauser effect   The effect whereby, if a radio frequency field is applied to a substance in an external magnetic field, whose nuclei have spin ½ and which has unpaired electrons, at the electron spin resonance frequency, the resulting polarization of the nuclei is as great as if the nuclei had the much larger electron magnetic moment.

overhead Pertaining to fluid (gas or liquid) effluent from the top of a process vessel, such as a distillation column.

overlapping orbitals Two orbitals (usually of electrons associated with different atoms in a molecule) for which there is a region of space where both are of appreciable magnitude.

overpoint The initial boiling point in a distillation process; specifically, the temperature at which the first drop falls from the tip of the condenser into the condensate flask.

overpotential *See* overvoltage.

overprint A complete or partial disturbance of an isolated radioactive system by thermal, igneous, or tectonic activities which results in loss or gain of radioactive or radiogenic isotopes and, hence, a change in the radiometric age that will be given the disturbed system.

overtone band The spectral band associated with transitions of a molecule in which the vibrational quantum number changes by 2 or more.

overvoltage The difference between electrode potential under electrolysis conditions and the thermodynamic value of the electrode potential in the absence of electrolysis for the same experimental conditions. Also known as overpotential.

ovex $ClC_6H_4OSO_2C_6H_4Cl$ A white, crystalline solid with a melting point of 86.5°C; soluble in acetone and aromatic solvents; used as an insecticide and acaricide.

oxadiazon $C_{13}H_{18}Cl_2N_2O_3$ A white solid with a melting point of 88–90°C; slight solubility in water; used as a pre- and postemergence herbicide to control weeds in rice, turf, soybeans, peanuts, and orchards. Also known as 2-*tert*-butyl-4-(2,4-dichloro-5-isopropoxyphenyl)-δ-2-1-1,3,4 oxadiazalin-5-one.

oxalate Salt of oxalic acid; contains the $(COO)_2$ radical; examples are sodium oxalate, $Na_2C_2O_4$, ammonium oxalate, $(NH_4)_2C_2O_4 \cdot H_2O$, and ethyl oxalate, $C_2H_5(C_2O_4)C_2H_5$.

oxaldehydic acid *See* glyoxalic acid.

oxalic acid $HOOCCOOH \cdot 2H_2O$ Poisonous, transparent, colorless crystals melting at 187°C; soluble in water, alcohol, and ether; used as a chemical intermediate and a bleach, and in polishes and rust removers.

oxalyl chloride $(COCl)_2$ Toxic, colorless liquid boiling at 64°C; soluble in ether, benzene, and chloroform; used as a chlorinating agent and for military poison gas. Also known as ethanedioyl chloride.

oxalylurea *See* parabanic acid.

oxamic acid hydrazine *See* semioxamazide.

oxamide $NH_2COCONH_2$ Water-insoluble white powder, melting at 419°C; used as a stabilizer for nitrocellulose products.

oxammonium *See* ortho-hydroxylaniline.

oxammonium hydrochloride *See* hydroxylamine hydrochloride.

oxamyl $C_7H_{13}N_3O_3S$ A white, crystalline compound with a melting point of 100–102°C; used to control pests of tobacco, ornamentals, fruits, and crops. Also known as methyl N',N'-dimethyl N[(methylcarbamoyl)oxy]-1-thiooxamimidate.

oxazole C_3H_3ON A structure that consists of a five-membered ring containing oxygen and nitrogen in the 1 and 3 position; a colorless liquid (boiling point 69–70°C) that is miscible with organic solvents and water; used to prepare other organic compounds.

oxidase An enzyme that catalyzes oxidation reactions by the utilization of molecular oxygen as an electron acceptor.

oxidation 1. A chemical reaction that increases the oxygen content of a compound. 2. A chemical reaction in which a compound or radical loses electrons, that is in which the positive valence is increased.

oxidation number 1. Numerical charge on the ions of an element. 2. *See* oxidation state.

oxidation potential The difference in potential between an atom or ion and the state in which an electron has been removed to an infinite distance from this atom or ion.

oxidation-reduction indicator A compound whose color in the oxidized state differs from that in the reduced state.

oxidation-reduction reaction An oxidizing chemical change, where an element's positive valence is increased (electron loss), accompanied by a simultaneous reduction of an associated element (electron gain).

oxidation state The number of electrons to be added (or subtracted) from an atom in a combined state to convert it to elemental form. Also known as oxidation number.

oxidative phosphorylation Conversion of inorganic phosphate to the energy-rich phosphate of adenosinetriphosphatase by reactions associated with the electron transfer system.

oxide Binary chemical compound in which oxygen is combined with a metal (such as Na_2O; basic) or nonmetal (such as NO_2; acidic).

oxidizing agent Compound that gives up oxygen easily, removes hydrogen from another compound, or attracts negative electrons.

oxidizing atmosphere Gaseous atmosphere in which an oxidation reaction occurs; usually refers to the oxidation of solids.

oxidizing flame A flame, or the portion of it, that contains an excess of oxygen.

oxidoreductase An enzyme catalyzing a reaction in which two molecules of a compound interact so that one molecule is oxidized and the other reduced, with a molecule of water entering the reaction.

oxime Compound containing the CH(:NOH) radical; condensation product of hydroxylamine with aldehydes or ketones.

oximido *See* nitroso.

oxine C_9H_6NOH White powder that darkens when exposed to light; slightly soluble in water, dissolves in ethanol, acetone, and benzene; used to prepare fungicides and to separate metals by precipitation. Also known as 8-hydroxyquinoline; oxyquinoline; 8-quinolinol.

oxine sulfate   *See* 8-hydroxyquinoline sulfate.

oxo-   Chemical prefix designating the keto group, C:O.

oxoethanoic acid   *See* glyoxalic acid.

para-oxon $(C_2H_5O)_2P(O)C_6H_4NO_2$ A reddish-yellow oil with a boiling point of 148–151°C; soluble in most organic solvents; used as an insecticide. Also known as diethyl *para*-nitrophenyl phosphate.

oxonium ion   *See* hydronium ion.

oxo process   Catalytic process for production of alcohols, aldehydes, and other oxygenated organic compounds by reaction of olefin vapors with carbon monoxide and hydrogen.

oxosilane   *See* siloxane.

oxoxanthone   *See* genicide.

oxy-   1. Prefix indicating the oxygen radical (—O—) in a chemical compound. 2. Prefix incorrectly used as a substitute for hydroxy-.

oxyacanthine $C_{37}H_{40}N_2O_6$ An alkaloid obtained from the root of *Berberis vulgaris*; a white, crystalline powder with a melting point of 202–214°C; soluble in water, chloroform, benzene, alcohol, and ether; used in medicine. Also known as vinetine.

oxyamination   *See* ammoxidation.

oxybenzone $C_{14}H_{12}O_3$ A crystalline substance with a melting point of 66°C; used as a sunscreen agent. Also known as 4-methoxy-2-hydroxybenzophenone.

oxycarboxin $C_{12}H_{13}NO_4S$ An off-white, crystalline compound with a melting point of 127.5–130°C; used to control rust disease in greenhouse carnations. Also known as 5,6-dihydro-2-methyl-1,4-oxathiin-3-carboxanilide-4,4-dioxide.

oxydemeton methyl   *See* demeton-S-methyl sulfoxide.

oxyethylsulfonic acid   *See* isethionic acid.

oxygen   A gaseous chemical element, symbol O, atomic number 8, and atomic weight 15.9994; an essential element in cellular respiration and in combustion processes; the most abundant element in the earth's crust, and about 20% of the air by volume.

oxygen-18   Oxygen isotope with atomic weight 18; found 8 parts to 10,000 of oxygen-16 in water, air, and rocks; used in tracer experiments. Also known as heavy oxygen.

oxygen absorbent   Any material that will absorb (dissolve) oxygen into its body without reacting with it.

oxygenase   An oxidoreductase that catalyzes the direct incorporation of oxygen into its substrate.

oxygenate   To treat, infuse, or combine with oxygen.

oxygen burning   The synthesis of nuclei in stars through reactions involving the fusion of two oxygen-16 nuclei at temperatures of about 10^9 K.

oxygen cell   *See* aeration cell.

oxygen deficit   The difference between the actual amount of dissolved oxygen in lake or sea water and the saturation concentration at the temperature of the water mass sampled.

oxygen-flask method   Technique to determine the presence of combustible elements; the sample is burned with oxygen in a closed flask, and combustion products are absorbed in water of dilute alkali with subsequent analysis of the solution.

oxygen isotope fractionation   The use of temperature-dependent variations of the oxygen-18/oxygen-16 ratio in the carbonate shells of marine organisms, to measure water temperature at the time of deposition.

oxygen ratio   *See* acidity coefficient.

oxyhemoglobin   The red crystalline pigment formed in blood by the combination of oxygen and hemoglobin, without the oxidation of iron.

oxyhydrogen flame   A flame obtained from the combustion of a mixture of oxygen and hydrogen.

oxyl process   Modified Fischer-Tropsch process used to make alcohols, other oxygenated compounds, paraffins, and olefin hydrocarbons from carbon monoxide and hydrogen.

β-oxynaphthoic acid   *See* β-hydroxynaphthoic acid.

oxyneurine   *See* betaine.

4-oxypentanoic acid   *See* levulinic acid.

oxyphile   *See* lithophile.

oxyquinoline   *See* oxine.

oxyquinoline sulfate   *See* 8-hydroxyquinoline sulfate.

oxythioquinox $C_{10}H_6N_2OS_2$ A yellow, crystalline powder with a melting point of 172°C; insoluble in water; used as an insecticide, fungicide, and miticide for fruit and nut trees, for ornamentals, and in greenhouses. Also known as 6-methyl-2,3-quinoxalinedithiol cyclic-S,S-dithiocarbonate.

oxytocin $C_{43}H_{66}O_{12}N_{12}S_2$ A polypeptide hormone secreted by the neurohypophysis that stimulates contraction of the uterine muscles.

ozone O_3 Unstable blue gas with pungent odor; an allotropic form of oxygen; a powerful oxidant boiling at -112°C; used as an oxidant, bleach, and water purifier, and to treat industrial wastes.

ozonide   Any of the oily, thick, unstable compounds formed by reaction of ozone with unsaturated compounds; an example is oleic ozonide from the reaction of oleic acid and ozone.

ozonization   The process of treating, impregnating, or combining with ozone.

ozonolysis   **1.** Oxidation of an organic substance by means of ozone. **2.** The use of ozone to locate double bonds.

P

p- *See* para-; peta-.

P *See* phosphorus.

Pa *See* protactinium.

Paal-Knorr synthesis A method of converting a 1,4-dicarbonyl compound by cyclization with ammonia or a primary amine to a pyrrole.

Paar turbidimeter A visual-extinction device for measurement of solution turbidity; the length of the column of liquid suspension is adjusted until the light filament can no longer be seen.

PABA *See para*-aminobenzoic acid.

PABA sodium *See* sodium *para*-aminobenzoate.

Pachuca tank Air-agitated, solid-liquid mixing vessel in which the air is injected into the bottom of a center draft tube; air and solids rise through the tube, with solids exiting the top of the tube and falling through the bulk of the liquid.

package power reactor A small nuclear power plant designed to be crated in packages small enough for transportation to remote locations.

packed bed A fixed layer of small particles or objects arranged in a vessel to promote intimate contact between gases, vapors, liquids, solids, or various combinations thereof; used in catalysis, ion exchange, sand filtration, distillation, absorption, and mixing.

packed tower A fractionating or absorber tower filled with small objects (packing) to bring about intimate contact between rising fluid (vapor or liquid) and falling liquid.

packed tube A pipe or tube filled with high-heat-capacity granular material; used to heat gases when tubes are externally heated.

packing Arrangement of atoms or ions in a crystal lattice.

packing fraction The quantity $(M-A)/A$, where M is the mass of an atom in atomic mass units and A is its atomic number.

packing index The volume of ion divided by the volume of the unit cell in a crystal.

packing radius One-half the smallest approach distance of atoms or ions.

pairing isomer An excited nuclear state which has an unusually long lifetime because the microscopic motions of its constituent nucleons differ sharply from those of states of lower energy into which it is permitted to decay.

palladium   A chemical element, symbol Pd, atomic number 46, atomic weight 106.4.

palladium bichloride   *See* palladium chloride.

palladium chloride $PdCl_2$ or $PdCl_2 \cdot 2H_2O$ Dark-brown, deliquescent powder that decomposes at 501°C; soluble in water, alcohol, acetone, and hydrochloric acid; used in medicine, analytical chemistry, photographic chemicals, and indelible inks. Also known as palladium bichloride; palladous chloride.

palladium iodide PdI_2 Black powder that decomposes above 100°C; soluble in potassium iodide solution, insoluble in water and alcohol. Also known as palladous iodide.

palladium monoxide   *See* palladium oxide.

palladium nitrate $Pd(NO_3)_2$ Brown, water-soluble, deliquescent salt; used as an analytical reagent. Also known as palladous nitrate.

palladium oxide PdO Amber or black-green powder that decomposes at 750°C; soluble in dilute acids; used in chemical synthesis as a reduction catalyst. Also known as palladium monoxide.

palladous chloride   *See* palladium chloride.

palladous iodide   *See* palladium iodide.

palladous nitrate   *See* palladium nitrate.

pall ring   A specially shaped steel ring used as packing for distillation columns.

palmitate   A derivative ester or salt of palmitic acid.

palmitic acid $C_{15}H_{31}COOH$ A fatty acid; white crystals, soluble in alcohol and ether, insoluble in water; melts at 63.4°C, boils at 271.5°C (100 mm Hg); derived from spermaceti; used to make metallic palmitates and in soaps, waterproofing, and lube oils. Also known as cetylic acid; palmitinic acid.

palmitin   *See* tripalmitin.

palmitinic acid   *See* palmitic acid.

palmitoleic acid $C_{16}H_{30}O_2$ An unsaturated fatty acid, found in marine animal oils; it is a clear liquid used as a standard in chromatography. Also known as *cis*-9-hexadecenoic acid.

palytoxin   A water-soluble toxin produced by several species of *Palythoa;* considered to be one of the most poisonous substances known.

pancreatic lipase   *See* steapsin.

pancreatin   A cream-colored, amorphous powder obtained from the fresh pancreas of a hog; contains amylopsin, trypsin, steapsin, and other enzymes.

pancreozymin   A crude extract of the intestinal mucosa that stimulates secretion of pancreatic juice.

Paneth's adsorption rule   The rule that an element is strongly absorbed on a precipitate which has a surface charge opposite in sign to that carried by the element, provided that the resulting adsorbed compound is very sparingly soluble in the solvent.

pan tank   *See* rundown tank.

pantothenate   A salt or ester of pantothenic acid.

pantothenic acid $C_9H_{17}O_5N$ A member of the vitamin B complex that is essential for nutrition of some animal species. Also known as vitamin B_3.

papain An enzyme preparation obtained from the juice of the fruit and leaves of the papaya (*Carica papaya*); contains proteolytic enzymes.

Papanicolaou's stains A group of stains used on exfoliated cells, particularly those from the vagina, for examination and diagnosis.

papaverine $C_{20}H_{21}O_4N$ A white, crystalline alkaloid, melting at 147°C; soluble in acetone and chloroform, insoluble in water; used as a smooth muscle relaxant and weak analgesic, usually as the water-soluble hydrochloride salt. Also known as 6,7-dimethoxy-1-veratrylisoquinoline.

paper chromatography Procedure for analysis of complex chemical mixtures by the progressive absorption of the components of the unknown sample (in a solvent) on a special grade of paper.

paper electrochromatography Variation of paper electrophoresis in which the electrolyte-impregnated absorbent paper is suspended vertically and the electrodes are connected to the sides of the paper, producing a current at right angles to the downward movement of the unknown sample.

paper electrophoresis A variation of paper chromatography in which an electric current is applied to the ends of the electrolyte-impregnated absorbent paper, thus moving chargeable molecules of the unknown sample toward the appropriate electrode.

paper-tape chemical analyzer Chemically treated paper tape that is continuously unreeled, exposed to the sample, and viewed by a phototube to measure the color change that is empirically related to changes in the sample's chemical composition.

para- Chemical prefix designating the positions of substituting radicals on the opposite ends of a benzene nucleus, for example, paraxylene, $CH_3C_6H_4CH_3$. Abbreviated *p-*.

parabanic acid $C_3H_2O_3N_2$ A water-soluble cyclic compound that decomposes when heated to about 227°C; used in organic synthesis. Also known as imidazoletrione; oxalylurea.

paracetaldehyde *See* paraldehyde.

parachlorometaxylenol *See* 4-chloro-3,5-xylenol.

paracyanic acid *See* fulminic acid.

paracyanogen $(CN)_x$ A white solid produced by polymerization of cyanogen gas when heated to 400°C.

paraffin Any of the saturated aliphatic hydrocarbons of the methane series C_nH_{2n+2}.

paraffinicity The paraffinic nature or composition of crude petroleum or its products.

paraform *See* paraformaldehyde.

paraformaldehyde $(HCHO)_n$ Polymer of formaldehyde where n is greater than 6; white, alkali-soluble solid, insoluble in alcohol, ether, and water; used as a disinfectant, fumigant, and fungicide, and to make resins. Also known as paraform.

parahelium Those states of helium in which the spins of the two electrons are antiparallel, in contrast to orthohelium. Also spelled parhelium.

parahydrogen Those states of hydrogen molecules in which the spins of the two nuclei are antiparallel; known as spin isomers.

paraldehyde $C_6H_{12}O_3$ Acetaldehyde polymer; colorless, flammable, toxic liquid, miscible with most organic solvents, soluble in water; melts at 12.6°C, boils at 124.5°C; used as a chemical intermediate, in medicine, and as a solvent. Also known as *para*-acetaldehyde; paracetaldehyde.

paraldol $(CH_3CHOHCH_2CHO)_2$ Water-soluble, white crystals, boiling at 90-100°C; used as a chemical intermediate, to make resins, and in cadmium plating baths.

parallel growth *See* parallel intergrowth.

parallel intergrowth Intergrowth of two or more crystals in such a way that one or more axes in each crystal are approximately parallel. Also known as parallel growth.

parallel twin A twinned crystal whose twin axis is parallel to the composition surface.

paramagnetic analytical methods Analysis of fluid mixtures by measurement of the paramagnetic (versus diamagnetic) susceptibilities of materials when exposed to a magnetic field.

paramagnetic spectra Spectra associated with the coupling of the electronic magnetic moments of atoms or ions in paramagnetic substances, or in paramagnetic centers of diamagnetic substances, to the surrounding liquid or crystal environment, generally at microwave frequencies.

parameter Any of the axial lengths or interaxial angles that define a unit cell.

paramylum A reserve, starchlike carbohydrate of various protozoans and algae.

paramyosin A type of fibrous protein found in the adductor muscles of bivalves and thought to form the core of a filament with myosin molecules at the surface.

paranitraniline red *See* para red.

paraoxon $C_{10}H_{14}NO_6P$ A reddish-yellow oil, used by veterinarians as an insecticide. Also known as O,O-diethyl-*ortho-para*-nitrophenyl phosphate.

paraoxypropiophenone *See* para-hydroxypropiophenone.

paraquat $[CH_3(C_5H_4N)_2CH_3]\cdot 2CH_3SO_4$ A yellow, water-soluble solid, used as an herbicide.

para red $C_{10}H_6(OH)NNC_6H_4NO_2$ Red pigment derived from the coupling of β-naphthol with diazotized paranitroaniline. Also known as paranitraniline red.

pararosaniline $HOC(C_6H_4NH_2)_3$ Red to colorless crystals, melting at 205°C; soluble in ethanol, in the hydrochloride salt; used as a dye.

pararosolic acid *See* aurin.

parastate A state of a diatomic molecule in which the spins of the nuclei are antiparallel.

parathion $(C_2H_5O)_2PSOC_6H_4NO_2$ Poisonous, deep-brown to yellow liquid boiling at 157–162°C; soluble in most organic solvents; used as an insecticide and acaricide. Also known as O,O-diethyl-*ortho-para*-nitrophenyl phosphorothioate.

parathormone A polypeptide hormone that functions in regulating calcium and phosphate metabolism. Also known as parathyroid hormone.

parent A radionuclide that upon disintegration yields a specified nuclide, the daughter, either directly, or indirectly as a later member of a radioactive series.

parent compound A chemical compound that is the basis for one or more derivatives; for example, ethane is the parent compound for ethyl alcohol and ethyl acetate.

parent name That part of a chemical compound's name from which the name of a derivative comes; for example, ethane is the parent name for ethanol.

parhelium *See* parahelium.

parinol $C_{18}H_{12}Cl_2NO$ A white to pale yellow solid with a melting point of 169–170°C; used as a fungicide to control powdery mildew on flowers, non-bearing apple trees, and grape vines. Also known as α,α-bis(*para*-chlorophenyl)-3-pyridine methanol.

paroxypropione *See para*-hydroxypropiophenone.

partial condensation The cooling (or pressurization) of a saturated vapor until a part of it is condensed out as liquid.

partial dislocation The line at the edge of an extended dislocation where a slip through a fraction of a lattice constant has occurred.

partial molal quantity The molal concentration of one component of a mixture of components as related to total molal concentration for all components in the mixture.

partial molar volume That portion of the volume of a solution or mixture related to the molar content of one of the components within the solution or mixture.

particle counting Microscopic or photomicrographic technique for the visual counting of the numbers of particles in a known quantity of a solid-liquid suspension.

particle electrophoresis Electrophoresis in which the particles undergoing analysis are of sufficient size to be viewed either with the naked eye or with the assistance of an optical microscope.

particle emission The ejection of a particle other than a photon from a nucleus, in contrast to gamma emission.

particle-induced x-ray emission A method of trace analysis in which a beam of ions is directed at a thin foil on which the sample to be analyzed has been deposited, and the energy spectrum of the resulting x-rays is measured.

particle-scattering factor Factor in light-scattering equations used to compensate for the loss in scattered light intensity caused by destructive interference during the analysis of macromolecular compounds.

particle-thickness technique Microscopic technique for visual measurement of the thickness of a fine particle (in the 3–100 micrometer range).

partition chromatography Chromatographic procedure in which the stationary phase is a high-boiling liquid spread as a thin film on an inert support, and the mobile phase is a vaporous mixture of the components to be separated in an inert carrier gas.

partition coefficient In the equilibrium distribution of a solute between two liquid phases, the constant ratio of the solute's concentration in the upper phase to its concentration in the lower phase. Symbolized K.

parylene Polyparaxylylene, used in ultrathin plastic films for capacitor dielectrics, and as a pore-free coating.

PAS *See* photoacoustic spectroscopy.

Pascal rules Rules which give the diamagnetic susceptiblity of a complex molecule in terms of the sum of the susceptibilities of its constituent atoms, and a correction factor which depends on the type of bonds linking the atoms.

Paschen-Back effect An effect on spectral lines obtained when the light source is placed in a very strong magnetic field; the anomalous Zeeman effect obtained with weaker fields changes over to what is, in a first approximation, the normal Zeeman effect.

Paschen-Runge mounting A diffraction grating mounting in which the slit and grating are fixed, and photographic plates are clamped to a fixed track running along the corresponding Rowland circle.

Paschen series A series of lines in the infrared spectrum of atomic hydrogen whose wave numbers are given by $R_H [(\frac{1}{9}) - (1/n^2)]$, where R_H is the Rydberg constant for hydrogen, and n is any integer greater than 3.

passivation potential The potential corresponding to the critical anodic current density of an electrode which behaves in an active-passive manner.

passivity A state of chemical inactivity, especially of a metal that is relatively resistant to corrosion due to loss of chemical activity.

Pasteur's salt solution Laboratory reagent consisting of potassium phosphate and calcium phosphate, magnesium sulfate, and ammonium tartrate in distilled water.

Pauli g-permanence rule For given L, S, and M_J in LS coupling, the sum, over J, of the weak-field g-factors is equal to the sum of the strong-field factors.

Pauli g-sum rule For all the states arising from a given electron configuration, the sum of the g-factors for levels with the same J value is a constant, independent of the coupling scheme.

Pavy's solution Laboratory reagent used to determine the concentration of sugars in solution by color titration; contains copper sulfate, sodium potassium tartrate, sodium hydroxide, and ammonia in water solution.

Pb *See* lead.

PBI *See* protein-bound iodine.

PBI test *See* protein-bound iodine test.

PC *See* phosphocreatine.

PCNB *See* pentachloronitrobenzene.

Pd *See* palladium.

PDMS   *See* plasma desorption mass spectrometry.

peacock blue $HSO_3C_6H_4COH$ $[C_6H_4N(C_2H_5)CH_2C_6H_4SO_3Na]_2$ Blue pigment used in inks for multicolor printing.

peak enthalpimetry   A thermochemical analytical procedure applicable to biochemical and chemical analyses; the salient feature is rapid mixing of a reagent stream with an isothermal solvent stream into which discrete samples are intermittently injected; peak enthalpograms result which exhibit the response characteristics of genuine differential detectors.

peak width   In a gas chromatogram (plot of eluent rise and fall versus time), the width of the base (time duration) of a symmetrical peak (rise and fall) of eluent.

pearl hardening   Commercial name for a crystallized grade of calcium sulfate; used as a paper filler.

peat formation   Decomposition of vegetation in stagnant water with small amounts of oxygen, under conditions intermediate between those of putrefaction and those of moldering.

pebble heater   Gas-heating device (for air, hydrogen, methane, and steam) in which heat is transferred to the gas via a countercurrent movement of preheated pebbles.

pebulate $C_{10}IH_{21}NOS$ An amber to yellow liquid, used to control weeds in preemergence sugarbeets, tobacco, and tomatoes. Also known as *S*-propyl butylethylthiocarbamate.

Peclet number   Dimensionless group used to determine the chemical reaction similitude for the scale-up from pilot-plant data to commercial-sized units; incorporates heat capacity, density, fluid velocity, and other pertinent physical parameters.

pectic acid   A complex acid, partially demethylated, obtained from the pectin of fruits.

pectin   A purified carbohydrate obtained from the inner portion of the rind of citrus fruits, or from apple pomace; consists chiefly of partially methoxylated polygalacturonic acids.

pectinase   An enzyme that catalyzes the transformation of pectin into sugars and galacturonic acid.

pectinesterase   An enzyme that catalyzes the hydrolytic breakdown of pectins to pectic acids.

pedial class   That class in the triclinic system which has no symmetry.

pedion   A crystal form with only one face; member of the asymmetric class of the triclinic system.

pedogeochemical survey   A geochemical prospecting survey in which the materials sampled are soil and till.

pelargonic acid $CH_3(CH_2)_7CO_2H$ A colorless or yellowish oil, boiling at 254°C; soluble in ether and alcohol, insoluble in water; used as a chemical intermediate and flotation agent, in lacquers, pharmaceuticals, synthetic flavors and aromas, and plastics. Also known as *n*-nonanoic acid; *n*-nonoic acid; *n*-nonylic acid.

pelargonic alcohol   *See* n-nonyl alcohol.

pelargonic aldehyde   *See* nonanal.

pelargonidin   An anthocyanidin pigment obtained by hydrolysis of pelargonin in the form of its red-brown crystalline chloride, $C_{15}H_{11}ClO_5$.

pelargonin   An anthocyanin obtained from the dried petals of red pelargoniums or blue cornflowers in the form of its red crystalline chloride, $C_{27}H_{31}ClO_{15}$.

p electron   In the approximation that each electron has a definite central-field wave function, an atomic electron that has an orbital angular momentum quantum number of unity.

pellet cooler   Gas-cooled, gravity-bed device for the cooling and drying of extruded pellets and briquets.

pellet technique   *See* potassium bromide–disk technique.

pellicular resins   Glass spheres coated with a thin layer of ion-exchange resin, used in liquid chromatography.

pellotine $C_{13}H_{19}O_3N$ A colorless, crystalline alkaloid, derived from the dried cactus pellote, *Lophophora williamsi* (Mexico), slightly soluble in water; used as a hypnotic.

penetrating shower   A cosmic-ray shower, consisting mainly of muons, that can penetrate 15 to 20 centimeters of lead.

penetration twin   *See* interpenetration twin.

Penex process   A continuous, nonregenerative petroleum-refinery process for isomerization of C_5 or C_6 fractions in the presence of hydrogen and a platinum catalyst.

penicillinase   A bacterial enzyme that hydrolyzes and inactivates penicillin.

Penning ionization   The ionization of gas atoms or molecules in collisions with metastable atoms.

Pensky-Martens closed tester   Device to determine the ASTM flash point of fuel oils and cutback asphalts and other viscous materials and suspensions of solids.

pentabasic   A description of a molecule that has five hydrogen atoms that may be replaced by metals or bases.

pentaborane B_5H_9 Flammable liquid boiling at 48°C; ignites spontaneously in air; proposed as high-energy fuel for aircraft and missiles.

pentachloride   A molecule containing five atoms of chlorine in its structure.

pentachloroethane $CHCl_2CCl_3$ Colorless, water-insoluble liquid, boiling at 159°C; used as a solvent to degrease metals. Also known as pentalin.

pentachloronitrobenzene $C_6Cl_5NO_2$ Cream-colored crystals with a melting point of 142–145°C; slightly soluble in alcohols; used as a fungicide and herbicide. Abbreviated PCNB. Also known as quintozene; terrachlor.

pentachlorophenol C_6Cl_5OH A toxic white powder, decomposing at 310°C, melting at 190°C; soluble in alcohol, acetone, ether, and benzene; used as a fungicide, bactericide, algicide, herbicide, and chemical intermediate.

pentacosane $C_{25}H_{52}$ A water-insoluble hydrocarbon derived from beeswax.

pentadecalactone *See* pentadecanolide.

pentadecane $C_{15}H_{32}$ A colorless, water-insoluble liquid, boiling at 270.5°C; soluble in alcohol; used as a chemical intermediate.

pentadecanolide $C_{15}H_{18}O_2$ A colorless liquid with a musky odor extracted from angelica oil; soluble in 90% ethyl alcohol in equal volume; used in perfumes. Also known as 15-hydroxypentadecanoic acid lactone; pentadecalactone.

pentadentate ligand A chelating agent having five groups capable of attachment to a metal ion. Also known as quinquidentate ligand.

pentadiene C_5H_8 Any of several straight-chain liquid diolefins: 1,2-CH_2=C=CHCH$_3$, a colorless liquid boiling at 45°C, also known as ethylallene; 1,3-CH_3=CHCH=CHCH$_3$, a colorless liquid boiling at 43°C, also known as α-methyl bivinyl, piperylene; 1,4-$(CH_2$=CH)$_2$CH, a colorless liquid boiling at 26°C.

pentaerythritol $(CH_2OH)_4C$ A white crystalline solid, melting at 261-262°C; moderately soluble in cold water, freely soluble in hot water; used to make the explosive pentaerythritol tetranitrate (PETN) and in the manufacture of alkyol resins and other coating compounds. Also known as tetramethylolmethane.

pentaerythritol tetranitrate $C(CH_2ONO_2)_4$ A white crystalline compound, melting at 139°C; explodes at 205–215°C; soluble in acetone, insoluble in water; used in medicines and explosives. Also known as penthrite; PETN.

pentaerythritol tetrastearate $C(CH_2OOCC_{17}H_{35})_4$ A hard, ivory-colored wax with a softening point of 67°C; used in polishes and textile finishes.

pentafluoride A chemical compound onto which five fluoride atoms are bonded.

pentahydroxycyclohexane *See* quercitol.

pentahydroxyhexoic acid *See* galactonic acid.

pentalin *See* pentachloroethane.

pentamethine *See* dicarbocyanine.

pentamethylene glycol *See* 1,5-pentanediol.

pentanamide *See* valeramide.

n-pentane $CH_3(CH_2)_3CH_3$ A colorless, flammable, water-insoluble hydrocarbon liquid, freezing at −130°C, boiling at 36°C; soluble in hydrocarbons and ethers; used as a chemical intermediate, solvent, and anesthetic.

1,5-pentanediol $HOCH_2(CH_2)_3CH_2OH$ Colorless, water-miscible liquid boiling at 242.5°C; used as a hydraulic fluid, lube-oil additive, and antifreeze, and in manufacture of polyester and polyurethane resins. Also known as pentamethylene glycol.

2,3-pentanedione *See* acetyl propionyl.

2,4-pentanedione *See* acetylacetone.

pentane insolubles Insoluble matter that can be separated from used lubricating oil in solution in *n*-pentane; may include resinous bitumens produced from the oxidation of oil and fuel; used in an ASTM test.

pentanethiol *See* amyl mercaptan.

n-pentanoic acid   *See* valeric acid.

pentanol $C_5H_{11}OH$ A toxic organic alcohol; 1-pentanol is *n*-amyl alcohol, primary; 2-pentanol is methylpropylcarbinol; 3-pentanol is diethylcarbinol; *tert*-pentanol is *tert*-amyl alcohol; pentanols are used in pharmaceuticals, as chemical intermediates, and as solvents.

2-pentanol   *See* methyl propyl carbinol.

pentanone   Either of two isomeric ketones derived from pentane: $CH_3COC_3H_7$ is a flammable, colorless, clear liquid, a mixture of methyl propyl and diethyl ketones; insoluble in water, soluble in ether and alcohol; used as a solvent; also known as ethyl acetone, methyl propyl ketone (MPK), 2-pentanone. $C_2H_5COC_2H_5$ is a colorless, flammable liquid with acetone aroma, boiling at 101°C; soluble in alcohol and ether; used in medicine and organic synthesis; also known as diethyl ketone, 3-pentanone, propione.

2-pentanone   *See* pentanone.

3-pentanone   *See* pentanone.

pentasodium triphosphate   *See* sodium tripolyphosphate.

pentavalent   An atom or radical that exhibits a valency of 5.

pentene C_5H_{12} Colorless, flammable liquids derived from natural gasoline; isomeric forms are α-*n*-amylene and β-*n*-amylene.

penthrite   *See* pentaerythritol tetranitrate.

pentine   *See* pentyne.

2-pentine   *See* pentyne.

pentoglycerine   *See* trimethylolethane.

pentosan   A hemicellulose present in cereal, straws, brans, and other woody plant tissues; yields five-carbon-atom sugars.

pentose   Any one of a class of carbohydrates containing five atoms of carbon.

pentose phosphate pathway   A pathway by which glucose is metabolized or transformed in plants and microorganisms; glucose-6-phosphate is oxidized to 6-phosphogluconic acid, which then undergoes oxidative decarboxylation to form ribulose-5-phosphate, which is ultimately transformed to fructose-6-phosphate.

pentoxide   A compound that is binary and has five atoms of oxygen; for example, phosphorus pentoxide, P_2O_5.

para-pentoxyphenol   *See para*-pentyloxyphenol.

pentyl   *See* amyl.

pentylamine   *See n*-amylamine.

pentylformic acid   *See* caproic acid.

para-pentyloxyphenol $C_{11}H_{16}O_2$ Compound melting at 49–50°C; used as a bactericide. Also known as *para*-amoxyphenol; hydroquinone monopentyl ether; *para*-pentoxyphenol.

pentyne C_5H_8 Either of two normal isometric acetylene hydrocarbons: $HC\equiv C(CH_2)_2CH_3$, colorless liquid boiling at 40°C, also known as pentine, propylacetylene; $CH_3C\equiv CC_2H_5$, liquid boiling at 56°C, also known as ethylmethylacetylene, 2-pentine.

peppermint camphor *See* menthol.

PeP reaction *See* proton-electron-proton reaction.

pepsin A proteolytic enzyme found in the gastric juice of mammals, birds, reptiles, and fishes.

pepsinogen The precursor of pepsin, found in the stomach mucosa.

peptidase An enzyme that catalyzes the hydrolysis of peptides to amino acids.

peptide A compound of two or more amino acids joined by peptide bonds.

peptide bond A bond in which the carboxyl group of one amino acid is condensed with the amino group of another to form a $—CO \cdot NH—$ linkage. Also known as peptide linkage.

peptide synthetase A ribosomal synthetase that catalyzes the formation of peptide bonds during protein synthesis.

peptization 1. Aggregation in which a hydrophobic colloidal sol is stabilized by the addition of electrolytes (peptizing agents) which are adsorbed on the particle surfaces. 2. Liquefaction of a substance by trace amounts of another substance.

peptone A water-soluble mixture of proteoses and amino acids derived from albumin, meat, or milk; used as a nutrient and to prepare nutrient media for bacteriology.

per- Chemical prefix meaning: 1. Complete, as in hydrogen peroxide. 2. Extreme, or the presence of the peroxy ($—O—O—$) group. 3. Exhaustive (complete) substitution, as in perchloroethylene.

peracetic acid CH_3COOOH A toxic, colorless liquid with strong aroma; boils at 105°C; explodes at 110°C; miscible with water, alcohol, glycerin, and ether; used as an oxidizer, bleach, catalyst, bactericide, fungicide, epoxy-resin precursor, and chemical intermediate. Also known as peroxyacetic acid.

peracid Acid containing the peroxy ($—O—O—$) group, such as peracetic acid or perchloric acid.

peralcohol Chemical compound containing the peroxy group ($—O—O—$), such as peracetic acid and perchromic acid.

perbenzoic acid $C_6H_5CO_2OH$ A crystalline compound forming leaflets from benzene solution, melting at 41–45°C, freely soluble in organic solvents; used in analysis of unsaturated compounds and to change ethylinic compounds into oxides. Also known as benzylhydroperoxide; peroxybenzoic acid.

perchlorate A salt of perchloric acid containing the ClO_4^- radical; for example, potassium perchlorate, $KClO_4$.

perchloric acid $HClO_4$ Strongly oxidizing, corrosive, colorless, hygroscopic liquid, boiling at 16°C (8 mmHg, or 1067 newtons per square meter); soluble in water; unstable in pure form, but stable when diluted in water; used in medicine, electrolytic baths, electropolishing, explosives, and analytical chemistry, and as a chemical intermediate. Also known as Fraude's reagent.

perchlorobenzene *See* hexachlorobenzene.

perchloroethane *See* hexachloroethane.

perchloroethylene CCl_2CCl_2 Stable, colorless liquid, boiling at 121°C; nonflammable and nonexplosive, with low toxicity; used as a dry-cleaning and industrial solvent, in pharmaceuticals and medicine, and for metal cleaning. Also known as tetrachloroethylene.

perchloromethyl mercaptan $ClSCCl_3$ Poisonous, yellow oil with disagreeable aroma; decomposes at 148°C; used as a chemical intermediary, granary fumigant, and military poison gas.

perchloryl fluoride $ClFO_3$ A colorless gas with a sweet odor; boiling point is $-46.8°C$ and melting point is $-146°C$; used as an oxidant in rocket fuels.

percolation filtration A continuous petroleum-refinery process in which lubricating oils and waxes are percolated through a clay bed to improve color, odor, and stability.

percussion figure Radiating lines on a crystal section produced by a sharp blow.

perfect crystal A crystal without lattice defects; it is an unattained ideal or standard.

perfect fractionation path On a phase diagram, a line or a path representing a crystallization sequence in which any crystal that has been formed remains inert, that is, its composition is not altered.

perfectly mobile component A component whose quantity in a system is determined by its externally imposed chemical potential rather than by its initial quantity in the system. Also known as boundary value component.

perfect solution A solution that is ideal throughout its entire compositional range.

perfect vacuum *See* absolute vacuum.

perfluorochemical A hydrocarbon in which all the hydrogen atoms have been replaced by fluorine.

perfluorocyclobutane *See* octafluorocyclobutane.

perfluoroethylene *See* tetrafluoroethylene.

perfluoropropane *See* octafluoropropane.

perforated-pipe distributor Liquid distribution device consisting of a length of piping or tubing with holes at spaced intervals along the length; used in spray columns, liquid-vapor contactors, and spray driers. Also known as sparger.

perforated plate Flat plate with series of holes used to control fluid distribution, as in a perforated-plate (distillation) column.

perforated-plate column Distillation column in which vapor-liquid contact is provided by perforated plates instead of bubble-cap trays.

perforated-plate distributor 1. A perforated plate or screen used to even out liquid-flow fluctuations through flow channels. 2. A perforated plate as used in a distillation column or liquid-liquid extraction column.

perforated-plate extractor A liquid-liquid extraction vessel in which perforated plates are used to bring about contact between the two or more liquid phases.

perhydro- Prefix designating a completely saturated aromatic compound, as for decalin ($C_{10}H_{18}$), also known as perhydronaphthalene.

pericline twin law A parallel twin law in triclinic feldspars, in which the b-axis is the twinning axis and the composition surface is a rhombic section.

pericondensed polycyclic Referring to an aromatic compound in which three or more rings share common carbon atoms.

period A family of elements with consecutive atomic numbers in the periodic table and with closely related properties; for example, chromium through copper.

periodate A salt of periodic acid, HIO_4, for example, potassium periodate, KIO_4.

periodic acid $HIO_4 \cdot 2H_2O$ Water- and alcohol-soluble white crystals; loses water at 100°C; used as an oxidant.

periodic lattice *See* lattice.

periodic law The law that the properties of the chemical elements and their compounds are a periodic function of their atomic weights.

periodic table A table of the elements, written in sequence in the order of atomic number or atomic weight and arranged in horizontal rows (periods) and vertical columns (groups) to illustrate the occurrence of similarities in the properties of the elements as a periodic function of the sequence.

peritectic An isothermal reversible reaction in which a liquid phase reacts with a solid phase during cooling to produce a second solid phase.

peritectic point In a binary two-phase heteroazeotropic system at constant pressure, that point up to which the boiling point has remained constant until one of the phases has boiled away.

peritectoid An isothermal reversible reaction in which a solid phase on cooling reacts with another solid phase to form a third solid phase.

Perkins' reaction The formation of unsaturated cinnamic-type acids by the condensation of aromatic aldehydes (Ar) with fatty acids in the presence of acetic anhydride; for example, $ArCHO + CH_3COONa \rightarrow ArCH{=}CHCOONa + H_2O$.

permanent hardness The hardness of water persisting after boiling.

permanent-press resin A thermosetting resin, based on chemicals such as formaldehyde and maleic anhydride, which is used to impart crease resistance to textiles and fibers. Also known as durable-press resin.

permanganate A purple salt of permanganic acid containing the $MnO_4{}^-$ radical; used as an oxidizing agent and a disinfectant.

permanganic acid $HMnO_4$ An unstable acid that exists only in dilute solutions; decomposes to manganese dioxide and oxygen.

permeable membrane A thin sheet or membrane of material through which selected liquid or gas molecules or ions will pass, either through capillary pores in the membrane or by ion exchange; used in dialysis, electrodialysis, and reverse osmosis.

permeametry Determination of the average size of fine particles in a fluid (gas or liquid) by passing the mixture through a powder bed of known dimensions and recording the pressure drop and flow rate through the bed.

permease Any of a group of enzymes which mediate the phenomenon of active transport.

permeation The movement of atoms, molecules, or ions into or through a porous or permeable substance (such as zeolite or a membrane).

permeator A membrane assembly that performs an ion-exchange function, for example, desalting in a membrane water-desalting process.

permselective membrane An ion-exchange material that allows ions of one electrical sign to enter and pass through.

peroxidase An enzyme that catalyzes reactions in which hydrogen peroxide is an electron acceptor.

peroxide 1. A compound containing the peroxy (—O—O—) group, as in hydrogen peroxide. 2. *See* hydrogen peroxide.

peroxide number Measure of millimoles of peroxide (or milliequivalents of oxygen) taken up by 1000 grams of fat or oil; used to measure rancidity. Also known as peroxide value.

peroxide value *See* peroxide number.

peroxyacetic acid *See* peracetic acid.

peroxybenzoic acid *See* perbenzoic acid.

peroxydol *See* sodium perborate.

peroxymonosulfuric acid *See* Caro's acid.

Persian red Red pigment made from basic lead chromate or ferric oxide.

persulfate Salt derived from persulfuric acid and containing the radical $S_2O_8^{2-}$; made by electrolysis of sulfate solutions.

persulfuric acid $H_2S_2O_8$ Acid formed in lead-cell batteries by electrolyzing sulfuric acid; strong oxidizing agent.

PETN *See* pentaerythritol tetranitrate.

petrochemicals Chemicals made from feedstocks derived from petroleum or natural gas; examples are ethylene, butadiene, most large-scale plastics and resins, and petrochemical sulfur. Also known as petroleum chemicals.

petrochemistry 1. An aspect of geochemistry that deals with the study of the chemical composition of rocks. 2. The chemistry and reactions of materials derived from petroleum, natural gas, or asphalt deposits.

petroleum chemicals *See* petrochemicals.

petroleum isomerization process A fixed-bed, vapor-phase petroleum-refinery process using a precious-metal catalyst and external hydrogen; feedstocks include natural gas, pentane, and hexane cuts; the product is high-octane blending stock.

petroleum processing The recovery and processing of various usable fractions from the complex crude oils; usable fractions include gasoline, kerosine, diesel oil, fuel oil, and asphalt. Also known as petroleum refining.

petroleum refining *See* petroleum processing.

Pfund series A series of lines in the infrared spectrum of atomic hydrogen whose wave numbers are given by $R_H[(1/25) - (1/n^2)]$, where R_H is the Rydberg constant for hydrogen, and n is any integer greater than 5.

PGA *See* folic acid.

pH A term used to describe the hydrogen-ion activity of a system; it is equal to $-\log a_{H}{}^{+}$; here $a_{H}{}^{+}$ is the activity of the hydrogen ion; in dilute solution, activity is essentially equal to concentration and pH is defined as $-\log_{10}[H^{+}]$, where $[H^{+}]$ is hydrogen-ion concentration in moles per liter; a solution of pH 0 to 7 is acid, pH of 7 is neutral, pH over 7 to 14 is alkaline.

phagocytin A type of bactericidal agent present within phagocytic cells.

phallotoxin One of a group of toxic peptides produced by the mushroom *Amanita phalloides*.

phantom crystal A crystal containing an earlier stage of crystallization outlined by dust, minute inclusions, or bubbles. Also known as ghost crystal.

pharmaceutical chemistry The chemistry of drugs and of medicinal and pharmaceutical products.

pharmacology The science dealing with the nature and properties of drugs, particularly their actions.

phase Portion of a physical system (liquid, gas, solid) that is homogeneous throughout, has definable boundaries, and can be separated physically from other phases.

phase diagram A graphical representation of the equilibrium relationships between phases (such as vapor-liquid, liquid-solid) of a chemical compound, mixture of compounds, or solution.

phase equilibria The equilibrium relationships between phases (such as vapor, liquid, solid) of a chemical compound or mixture under various conditions of temperature, pressure, and composition.

phase solubility The different solubilities of a sample's solid constituents (phases) in a selected solvent.

phase-solubility analysis Solvent technique used to determine the amount and number of components in a solid substance; the weight of sample added to the solvent is plotted against the weight of sample dissolved, with breakpoints in the curve occurring with each progressive saturation of the solvent with respect to each of the components; can be combined with extraction and recrystallization procedures.

phase titration Analysis of a binary mixture of miscible liquids by titrating with a third liquid that is miscible with only one of the components, using the ternary phase diagram to determine the end point.

pH electrode Membrane-type glass electrode used as the hydrogen-ion sensor of most pH meters; the pH-response electrode surface is a thin membrane made of a special glass.

α-phellandrene $C_{10}H_{16}$ A colorless oil soluble in ether; boiling point of d-optical isomer is 66–68°C, of l-optical isomer is 58–59°C; used in flavoring and perfumes.

phenacetin *See* acetophenetidin.

phenacyl chloride *See* chloroacetophenone.

phenanthrene $C_{14}H_{10}$ A colorless, crystalline hydrocarbon; melts at about 100°C; the nucleus is produced by the degradation of certain alkaloids; used in the synthesis of dyes and drugs.

phenanthroline $C_{12}H_8N_2$ Any of three nitrogen bases related to phenanthrene; the ortho form is an oxidation-reduction indicator, turning faint blue when oxidized.

phenanthroline indicator A sensitive, red-colored specific reagent for iron. Also known as bathophenanthroline; 4, 7-diphenyl-1,10-phenanthroline.

phenarsazine chloride $C_{12}H_9AsClN$ A yellow, crystalline compound obtained as a precipitate from carbon tetrachloride solutions; it sublimes readily, and is slightly soluble in xylene, benzene, and carbon tetrachloride; used as a war gas. Also known as adamsite; 10-chloro-5,10-dihydrophenarsazine.

phenazine $C_6H_4N_2C_6H_4$ Yellow crystals, melting at 170°C; slightly soluble in water, soluble in alcohol and ether; used as chemical intermediate and to make dyes. Also known as azophenylene.

phencapton $C_{11}H_{15}Cl_2PS_3$ An amber oil that decomposes on boiling; insoluble in water; used as a miticide. Also known as O,O-diethyl S-(2,5-dichloro-phenylthiomethyl)phosphorodithioate.

phenethyl acetate $C_6H_5CH_2CH_2OOCCH_3$ A colorless liquid with a peachlike odor and a boiling point of 226°C; soluble in alcohol, ether, and fixed oils; used in perfumes. Also known as phenylethyl acetate.

phenethyl alcohol $C_8H_{10}O$ A liquid with a floral odor found in many natural essential oils; soluble in 50% alcohol; used in perfumes and flavors, and in medicine as an antibacterial agent in diseases of the eye. Also known as benzyl carbinol; 2-phenylethanol; β-phenylethyl alcohol.

phenethyl isobutyrate $(CH_3)2CHCOOC_2H_4C_6H_5$ A colorless liquid, soluble in alcohol and ether; used in perfumes and flavoring. Also known as phenyl-ethyl isobutyrate.

phenetidine $NH_2C_6H_4OC_2H_5$ Either of two toxic, oily liquids that darken when exposed to light and air; soluble in alcohol, insoluble in water; the ortho form boils at 228–230°C, is used to make dyes, and is also known at 2-amino-phenetole; the para form boils at 253–255°C, is used to make dyes and in pharmaceuticals, and is also known as 4-aminophenetole.

phenmedipham $C_{16}H_{16}O_4N_2$ A colorless solid that melts at 143–144°C; used in an emulsifiable concentrate form as a preemergent herbicide for weeds in beet crops. Also known as *meta*-methylcarbanilate; methyl-*meta*-hydroxy-carbanilate.

phenol 1. C_6H_5OH White, poisonous, corrosive crystals with sharp, burning taste; melts at 43°C, boils at 182°C; soluble in alcohol, water, ether, carbon disulfide, and other solvents; used to make resins and weed killers, and as a solvent and chemical intermediate. Also known as carbolic acid; phenylic acid. 2. A chemical compound based on the substitution product of phenol, for example, ethylphenol ($C_2H_4C_4H_5OH$), the ethyl substitute of phenol.

phenolate process A process which employs sodium phenolate to remove hydrogen sulfide from gas.

phenol coefficient Number scale for comparison of antiseptics, using the efficacy of phenol as unity.

phenol-coefficient method A method for evaluating water-miscible disinfectants in which a test organism is added to a series of dilutions of the disinfectant; the phenol coefficient is the number obtained by dividing the greatest dilution of the disinfectant killing the test organism by the greatest dilution of phenol showing the same result.

β-phenoldiazine *See* phthalazine.

phenol extraction Petroleum-refinery solvent-extraction process using phenol as the solvent to remove aromatic, unsaturated and naphthenic constituents from lubricating-oil stocks.

phenol-formaldehyde resin Thermosetting resin made by the reaction of phenol and formaldehyde; has good strength and chemical resistance and low cost; used as a molding material for mechanical and electrical parts. Originally known as Bakelite.

phenol-furfural resin A phenolic resin characterized by the ability to be fabricated by injection molding since it hardens after curing conditions are reached.

phenolphthalein $(C_6H_4OH)_2COC_6H_4CO$ Pale-yellow crystals; soluble in alcohol, ether, and alkalies, insoluble in water; used as an acid-base indicator (carmine-colored to alkalies, colorless to acids) for titrations, as a laxative and dye, and in medicine.

phenol process A single-solvent petroleum-refining process in which phenol is the selective solvent.

phenol red *See* phenolsulfonephthalein.

phenolsulfonephthalein $C_{19}H_{14}O_5S$ A bright-red, crystalline compound, soluble in water, alcohol, and acetone; used as a pH indicator, and to test for kidney function in dogs. Also known as phenol red.

phenolsulfonic acid $C_6H_5SO_3H$ Water- and alcohol-soluble mixture of *ortho-* and *para-*phenolsulfonic acids; yellowish liquid that turns brown when exposed to air; used as a chemical intermediate and in water analysis. Also known as sulfocarbolic acid.

phenothiazine $C_{12}H_9N$ A yellow, crystalline compound, forming rhomboid leaflets or diamond-shaped plates, obtained from toluene or butanol solution; soluble in hot acetic acid, benzene, and ether; used as an insecticide and in pharmaceutical manufacture. Also known as dibenzothiazine; thiophenylamine.

phenotole $C_6H_5OC_2H_5$ Combustible, colorless liquid, boiling at 172°C; soluble in alcohol and ether, insoluble in water.

phenoxyacetic acid $C_6H_5OCH_2COOH$ A light tan powder with a melting point of 98°C; soluble in ether, water, carbon disulfide, methanol, and glacial acetic acid; used in the manufacture of pharmaceuticals, pesticides, fungicides, and dyes.

phenoxybenzamine hydrochloride $C_{18}H_{22}ONCl\cdot HCl$ White crystals, slightly soluble in water, melting at 139°C; used in medicine.

2-phenoxyethanol $C_6H_5OCH_2CH_2OH$ An oily liquid with a faint aromatic odor; melting point is 14°C; soluble in water; used in perfumes as a fixative, in organic synthesis, as an insect repellent, and as a topical anesthetic. Also known as ethylene glycol monophenyl ether; β-hydroxyethylphenyl ether; 1-hydroxy-2-phenoxyethane.

phenoxypropanediol $C_9H_{12}O_3$ A white, crystalline solid with a melting point of 53°C; soluble in water, alcohol, glycerin, and carbon tetrachloride; used in medicine and as a plasticizer.

phenoxy resin A high-molecular-weight thermoplastic polyether resin based on bisphenol-A and epichlorohydrin with bisphenol-A terminal groups; used for injection molding, extrusion, coatings, and adhesives.

phenthoate $C_{12}H_{17}O_4PS_2$ A reddish yellow liquid; melting point is 17.5°C; used to control pests on rice, cotton, vegetables, and fruit, and to control mosquitoes. Also known as O,O-dimethyl-S-(α-ethoxycarbonylbenzyl)phosphorodithioate.

phentolamine hydrochloride $C_{17}H_{19}ON_2 \cdot HCl$ White, water-soluble crystals, melting at 240°C; a sympatholytic; used in medicine.

phentriazophos $C_{12}H_{15}N_3O_3PS$ A pale yellow liquid that decomposes before boiling; used as an insecticide, miticide, and nematicide for vegetable and food crops. Also known as 1-phenyl-3-(O,O-diethylthionophosphoryl)-1,2,4-triazole; triazophos.

phenyl C_6H_5— A functional group consisting of a benzene ring from which a hydrogen has been removed.

phenylacetaldehyde C_8H_8O A colorless liquid with a boiling point of 193–194°C; soluble in ether and fixed oils; used in perfumes and flavoring. Also known as α-toluic aldehyde.

N-**phenylacetamide** *See* acetanilide.

phenylacetic acid $C_8H_8O_2$ White crystals with a boiling point of 262°C; soluble in alcohol and ether; used in perfumes, medicine, and flavoring and in the manufacture of penicillin. Also known as α-toluic acid.

phenylacetonitrile *See* benzyl cyanide.

phenylacetylurea *See* phenacemide.

phenylacrolein *See* cinnamaldehyde.

β-phenylacrylic acid *See* cinnamic acid.

phenylacrylyl chloride *See* cinnamoyl chloride.

β-phenylalanine *See* phenylalanine.

phenylalanine $C_9H_{11}O_2N$ An essential amino acid, obtained in the levo form by hydrolysis of proteins (as lactalbumin); converted to tyrosine in the normal body. Also known as α-aminohydrocinnamic acid; α-amino-β-phenylpropionic acid; β-phenylalanine.

phenylallylic alcohol *See* cinnamic alcohol.

phenylaniline *See* diphenylamine.

N-**phenylanthranilic acid** $(C_6H_5NH)C_6H_4COOH$ A crystalline compound, soluble in hot alcohol; decomposes at 183–184°C; used to detect vanadium in steel. Also known as 2-anilinobenzoic acid; diphenylamine-2-carboxylic acid.

phenylbenzene *See* biphenyl.

2-phenyl-3,1-benzoxazinone-(4) $C_{14}H_9NO_2$ A white solid with a melting point of 123–124°C; solubility in water is 5–6 parts per million at 20°C; used as a postemergence herbicide. Also known as bentranil.

phenyl benzoyl carbinol *See* benzoin.

phenylbutazone $C_{19}H_{20}O_2N_2$ White or light-yellow powder with aromatic aroma and bitter taste; melts at 107°C; slightly soluble in water, soluble in acetone;

used in medicine as an analgesic and antipyretic. Also known as butazolidine; 4-butyl-1,2-diphenyl-3,5-pyrazolindinedione.

phenylchloroform   *See* benzotrichloride.

phenyl cyanide   *See* benzonitrile.

phenylcyclohexane $C_{12}H_{16}$ A colorless, oily liquid with a boiling point of 237.5°C; soluble in alcohol, benzene, castor oil, carbon tetrachloride, xylene, and hexane; used as a high-boiling solvent and a penetrating agent. Also known as cyclohexylbenzene.

phenyldichloroarsine $C_6H_5AsCl_2$ A liquid which becomes a microcrystalline mass at −20°C (melting point) and decomposes in water; soluble in alcohol, ether, and benzene; used as a poison gas.

phenyl 5,6-dichloro-2-trifluoromethyl-1-benzimidazole carboxylate   *See* fenazaflor.

phenyl diglycol carbonate $C_{18}H_{18}O_7$ A colorless solid with a melting point of 40°C; soluble in organic solvents; used as a plasticizer.

3-phenyl-1,1-dimethylurea   *See* fenuron.

3-phenyl-1,1-dimethylurea trichloroacetate   *See* fenuron-TCA.

phenylene blue   *See* indamine.

phenylenediamine $C_6H_4(NH_2)_2$ Also known as diaminobenzene. Any of three toxic isomeric crystalline compounds that are diamino derivatives of benzene; the ortho form, toxic colorless crystals melting at 102–104°C and soluble in alcohol, ether, water, and chloroform, is used to manufacture dyes, in photographic developers, and as a chemical intermediate, and is also known as *ortho*-diaminobenzene, orthamine; the meta form, colorless crystals unstable in air, melting at 63°C, and soluble in alcohol, ether, and water, is used to manufacture dyes, in textile dyeing, and as a nitrous acid detector; the para form, white to purple crystals melting at 147°C, soluble in alcohol and ether, and irritating to the skin, is used to manufacture dyes, in chemical analysis, and in photographic developers, and is also known as *para*-diaminobenzene.

2-phenylethanol   *See* phenethyl alcohol.

phenyl ether   *See* diphenyl oxide.

phenylethyl acetate   *See* phenethyl acetate.

β-phenylethyl alcohol   *See* phenethyl alcohol.

phenylethyl carbinol   *See* phenylpropyl alcohol.

phenylethylene   *See* styrene.

phenylethyl isobutyrate   *See* phenethyl isobutyrate.

phenylethylmalonylurea   *See* phenobarbital.

phenyl fluoride   *See* fluorobenzene.

phenylglycolic acid   *See* mandelic acid.

***N*-phenylglycine** $C_6H_5NHCH_2COOH$ A crystalline compound, moderately soluble in water, melting at 127–128°C; used in dye manufacture (indigo). Also known as anilinoacetic acid.

phenylglyoxylonitriloxime *O,O*-diethyl phosphorothioate $(H_5C_2O)_2PSONCCNC_6H_5$ A yellow liquid with a boiling point of 102°C at 0.01 mmHg (1.333 newtons per square meter); solubility in water is 7 parts per million at 20°C; used as an insecticide for stored products. Also known as phoxim.

phenylhydrazine $C_6H_5NHNH_2$ Poisonous, oily liquid, boiling at 244°C; soluble in alcohol, ether, chloroform, and benzene, slightly soluble in water; used in analytical chemistry to detect sugars and aldehydes, and as a chemical intermediate. Also known as hydrazinobenzene.

α-phenylhydroxyacetic acid   *See* mandelic acid.

2-phenyl-1,3-indanedione   *See* phenindione.

phenyl isothiocyanate   *See* phenyl mustard oil.

phenyl ketone   *See* benzophenone.

phenyl mercaptan   *See* thiophenol.

phenylmercuric acetate $C_8H_8O_2Hg$ White to cream-colored prisms with a melting point of 148–150°C; soluble in alcohol, benzene, and glacial acetic acid; used as an antiseptic, fungicide, herbicide, and mildewcide.

phenylmercuric chloride C_6H_5HgCl White crystals with a melting point of 251°C; soluble in benzene and ether; used as an antiseptic and fungicide.

phenylmercuric hydroxide C_6H_5HgOH White to cream-colored crystals with a melting point of 197–205°C; soluble in acetic acid and alcohol; used as a fungicide, germicide, and alcohol denaturant.

phenylmercuric oleate $C_{41}H_{21}O_2Hg$ A white, crystalline powder with a melting point of 45°C; soluble in organic solvents; used in paints as a mildewproofing agent, and as a fungicide.

phenylmercuric propionate $C_9H_{10}O_2Hg$ A white, waxlike powder with a melting point of 65–70°C; used in paints as a fungicide and bactericide.

phenylmercuriethanolammonium acetate $C_{10}H_{15}O_3NHg$ A white, water-soluble, crystalline solid; used as an insecticide and fungicide.

phenylmethane   *See* toluene.

phenylmethanol   *See* benzyl alcohol.

phenylmethyl acetate   *See* benzyl acetate.

phenylmethylcarbinol   *See* α-methylbenzyl alcohol.

phenylmethylcarbinyl acetate   *See* α-methylbenzyl acetate.

phenyl methyl ketone   *See* acetophenone.

***N*-phenylmorpholine** $C_{10}H_{13}NO$ A white, water-soluble solid with a melting point of 57°C; used in the manufacture of dyestuffs, corrosion inhibitors, and photographic developers, and as an insecticide.

phenyl mustard oil C_6H_5NCS A pale yellow or colorless liquid with a boiling point of 221°C; soluble in alcohol and ether; used in medicine. Also known as phenyl isothiocyanate; phenylthiocarbonimide; thiocarbanil.

α-phenyl phenacyl   *See* desyl.

phenylphenol $C_6H_5C_6H_4OH$ Almost white crystals, soluble in alcohol, insoluble in water; the ortho form, melting at 56–58°C, is used to manufacture dyes, as germicide and fungicide, and in the rubber industry, and is also known as 2-hydroxybiphenyl, *ortho*-xenol; the para form, melting at 164–165°C, is used to manufacture dyes, resins, and rubber chemicals, and as a fungicide, and is also known as *para*-hydroxydiphenyl, *para*-xenol.

phenylphosphoric acid *See* benzenephosphoric acid.

N-phenylpiperazine $C_{10}H_{14}N_2$ A pale yellow oil with a boiling point of 286.5°C; soluble in alcohol and ether; used for pharmaceuticals and in the manufacture of synthetic fibers.

phenylpropane *See* propyl benzene.

3-phenyl-1-propanol *See* phenylpropyl alcohol.

1-phenyl-1-propanol $C_6H_5CH(OH)CH_2CH_3$ An oily liquid that has a weak esterlike odor; miscible with methanol, ethanol, ether, benzene, and toluene; used in industry as a heat transfer medium, in the manufacture of perfumes, and as a choleretic in medicine. Also known as α-ethylbenzyl alcohol; ethyl phenyl carbinol; α-hydroxypropylbenzene; 1-phenylpropyl alcohol.

3-phenylpropenal *See* cinnamic aldehyde.

3-phenyl-2-propen-1-ol *See* cinnamic alcohol.

3-phenylpropionaldehyde *See* phenylpropyl aldehyde.

3-phenylpropionic acid *See* hydrocinnamic acid.

phenylpropyl alcohol $C_9H_{12}O$ A colorless liquid with a floral odor and a boiling point of 219°C; soluble in 70% alcohol; used in perfumes and flavoring. Also known as hydrocinnamic alcohol; phenylethyl carbinol; 3-phenyl-1-propanol.

1-phenylpropyl alcohol *See* 1-phenyl-1-propanol.

phenylpropyl aldehyde $C_9H_{10}O$ A colorless liquid with a floral odor; soluble in 50% alcohol; used in perfumes and flavoring. Also known as hydrocinnamic aldehyde; 3-phenylpropionaldehyde.

1-phenyl-3-pyrazolidinone $C_9H_{10}N_2O$ A crystalline compound soluble in dilute aqueous solutions of acids and alkalies; melting point is 121°C; used as a high-contrast photographic developer. Also known as 1-phenyl-3-pyrazolidone.

1-phenyl-3-pyrazolidone *See* 1-phenyl-3-pyrazolidinone.

phenylpyruvic acid $C_6H_5CH_2·CO·COOH$ A keto acid, occurring as a metabolic product of phenylalanine.

phenyl salicylate *See* salol.

N-phenylsalicylamide *See* salicylanilide.

phenylthiocarbamide *See* phenylthiourea.

phenylthiocarbonimide *See* phenyl mustard oil.

phenylthiourea $C_6H_5NHCSNH_2$ A crystalline compound that has either a bitter taste or is tasteless, depending on the heredity of the taster; used in human genetics studies. Also known as phenylthiocarbamide.

phenyl *ortho*-tolyl ketone *See ortho*-phenyl tolyl ketone.

***ortho*-phenyl tolyl ketone** $CH_3C_6H_4COC_6H_5$ An oily liquid with a boiling point of 309–311°C; soluble in alcohol, oils, and organic solvents; used as a fixative in perfumery. Also known as 2-methyl-benzophenone; phenyl *ortho*-tolyl ketone.

1-phenyl-3-(*O,O*-diethylthionophosphoryl)-1,2,4-triazole   *See* phentriazophos.

philosopher's wool   *See* zinc oxide.

phloridzin $C_{21}H_{24}O_{19} \cdot 2H_2O$ A glycoside extracted from the root bark of apple, plum, and pear trees; white needles with a melting point of 109°C; soluble in alcohol and hot water; used in medicine. Also known as phlorizin.

phlorizin   *See* phloridzin.

phloroglucine   *See* phloroglucinol.

phloroglucinol $C_6H_3(OH)_3 \cdot 2H_2O$ White to yellow crystals with a melting point of 212–217°C when heated rapidly and 200–209°C when heated slowly; soluble in alcohol and ether; used as a bone decalcifying agent, as a floral preservative, and in the manufacture of pharmaceuticals. Also known as phloroglucine; 1,3,5-trihydroxy benzene.

pH measurement   Determination of the hydrogen-ion concentration in an ionized solution by means of an indicator solution (such as phenolphthalein) or a pH meter.

phorate $C_7H_{17}O_2PS_2$ A clear liquid with slight solubility in water; used as an insecticide for a wide range of insects on a wide range of crops. Also known as *O,O*-diethyl *S*-(ethylthio)-methyl phosphorodithioate.

phosalone $C_{12}H_{15}ClNO_4PS_2$ A white solid, insoluble in water, with a melting point of 43–45°C; used as an insecticide and miticide on fruits. Also known as *O,O*-diethyl *S*[(6-chloro-2-oxobenzoxazolin-3-yl)methyl]phosphorodithioate.

phosazetim $C_{14}H_{13}N_2PO_2Cl_2$ A white, crystalline compound with a melting point of 104–106°C; used as a rodenticide. Also known as *O,O*-bis(*para*-chlorophenyl)acetimidoylphosphoramidothioate.

phosgene $COCl_2$ A highly toxic, colorless gas that condenses at 0°C to a fuming liquid; used as a war gas and in manufacture of organic compounds. Also known as carbonyl chloride; chloroformyl chloride.

phosmet $C_{11}H_{11}NO_4PS_2$ A white solid with a melting point of 71.9°C; used as an insecticide for fruit tree pests and alfalfa weevil. Also known as *N*-(mercaptomethyl)phthalimide *S*-(*O,O*-dimethylphosphorodithioate).

phosphamidon $C_{10}H_{19}NO_5P$ A colorless liquid, miscible with water; used as a systemic and contact insecticide to control mites, aphids, beetles, and plant bugs. Also known as 2-chloro-2-diethylcarbamoyl-1-methylvinyl dimethylphosphate.

phosphatase   An enzyme that catalyzes the hydrolysis and synthesis of phosphoric acid esters and the transfer of phosphate groups from phosphoric acid to other compounds.

phosphate   1. Generic term for any compound containing a phosphate group (PO_4^{3-}), such as potassium phosphate, K_3PO_4. 2. Generic term for a phosphate-containing fertilizer material.

phosphate anion PO_4^{3-} The negative ion of phosphoric acid.

phosphate buffer   Laboratory pH reference solution made of KH_2PO_4 and Na_2HPO_4; when 0.025 molal (equimolal of the K and Na salts), the pH is 6.865 at 25°C.

phosphate desulfurization A continuous, regenerative petroleum-refinery process using a tripotassium phosphate solution to remove hydrogen sulfide from natural gas, refinery gas, or liquid hydrocarbons.

phosphatide *See* phospholipid.

phosphatization Conversion to a phosphate or phosphates; for example, the diagenetic replacement of limestone, mudstone, or shale by phosphate-bearing solutions, producing phosphates of calcium, aluminum, or iron.

phosphatizing *See* phosphating.

phosphide Binary compound of trivalent phosphorus, as in Na_3P.

phosphine PH_3 Poisonous, colorless, spontaneously flammable gas with garlic aroma; soluble in alcohol, slightly soluble in cold water; boils at $-85°C$; used in organic reactions. Also known as hydrogen phosphide; phosphoretted hydrogen.

phosphinic acid Organic derivative of hypophosphorous acid; contains the radical $-H_2PO_2$ or $=HPO_2$; examples are methylphosphinic acid, CH_3HPOOH, and dimethyl phosphinic acid, $(CH_3)_2POOH$.

phosphite Salt of phosphorous acid; contains the radical $PO_3{}^{3-}$; an example is normal sodium phosphite, Na_3PO_3

phosphocreatine $C_4H_{10}N_3O_5P$ Creatine phosphate, a phosphoric acid derivative of creatine which contains an energy-rich phosphate bond; it is present in muscle and other tissues, and during the anaerobic phase of muscular contraction it hydrolyzes to creatine and phosphate and makes energy available. Abbreviated PC.

phosphoenolpyruvic acid $CH_2=O(OPO_3H_2)COOH$ A high-energy phosphate formed by dehydration of 2-phosphoglyceric acid; it reacts with adenosinediphosphate to form adenosinetriphosphate and enolpyruvic acid.

phosphoglucoisomerase An enzyme that catalyzes the conversion of galactose-1-phosphate to glucose-1-phosphate.

phosphoglucomutase An enzyme that catalyzes the conversion of glucose-1-phosphate to glucose-6-phosphate.

phospholan $C_6H_{14}O_3PNS_2$ A colorless to yellow solid with a melting point of $37–45°C$; used as an insecticide and miticide for cotton. Also known as 2-(diethyoxyphosphenylimino)-1,3-dithiolane.

phospholipase An enzyme that catalyzes a hydrolysis of a phospholipid, especially a lecithinase that acts in this manner on a lecithin.

phospholipid Any of a class of esters of phosphoric acid containing one or two molecules of fatty acid, an alcohol, and a nitrogenous base. Also known as phosphatide.

phosphomolybdic acid $H_3PO_4 \cdot 12MoO_3 \cdot xH_2O$ Yellow crystals; soluble in alcohol, ether, and water; used as an alkaloid reagent and a pigment. Abbreviated PMA.

phosphomonoesterase An enzyme catalyzing hydrolysis of phosphoric acid esters containing one ester linkage.

phosphonic acid $ROP(OH)_2$, where R is an organic radical such as $C_6H_5{}^-$, as in phenylphosphonic acid.

N-(phosphonomethyl)glycine *See* glyphosate.

N,N-bis(phosphonomethyl)glycine *See* glyphosine.

phosphorescence 1. Luminescence that persists after removal of the exciting source. Also known as afterglow. 2. Luminescence whose decay, upon removal of the exciting source, is temperature-dependent.

phosphoretted hydrogen *See* phosphine.

phosphoric acid H_3PO_4 Water-soluble, transparent crystals, melting at 42°C; used as a fertilizer, in soft drinks and flavor syrups, pharmaceuticals, water treatment, and animal feeds and to pickle and rust-proof metals. Also known as orthophosphoric acid.

phosphoric acid polymerization A petroleum-refinery process using phosphoric acid catalyst to convert propylene, butylene, or both, into high-octane gasoline or petrochemical polymers.

phosphoric anhydride P_2O_5 A flammable, dangerous, soft-white deliquescent powder; used as a dehydrating agent, in medicine and sugar refining, and as a chemical intermediate and analytical reagent. Also known as anhydrous phosphoric acid; phosphoric oxide; phosphorus pentoxide.

phosphoric bromide *See* phosphorus pentabromide.

phosphoric chloride *See* phosphorus pentachloride.

phosphoric oxide *See* phosphoric anhydride.

phosphoric perbromide *See* phosphorus pentabromide.

phosphoric perchloride *See* phosphorus pentachloride.

phosphoric sulfide *See* phosphorus pentasulfide.

phosphorimetry Low-temperature, analytical procedure related to fluorometry; based on the nature and intensity of the phosphorescent light emitted by an appropriately excited molecule.

phosphorization Impregnation or combination with phosphorus or a compound of phosphorus; for example, the diagenetic process of phosphatization.

phosphorolysis A reaction by which elements of phosphoric acid are incorporated into the molecule of a compound.

phosphorous acid H_3PO_3 Alcohol- and water-soluble deliquescent white or yellowish crystals; decomposes at 200°C; used as an analytical reagent and reducing agent.

phosphorus A nonmetallic element, symbol P, atomic number 15, atomic weight 30.98; used to manufacture phosphoric acid, in phosphor bronzes, incendiaries, pyrotechnics, matches, and rat poisons; the white (or yellow) allotrope is a soft waxy solid melting at 44.5°C, is soluble in carbon disulfide, insoluble in water and alcohol, and is poisonous and self-igniting in air; the red allotrope is an amorphous powder subliming at 416°C, igniting at 260°C, is insoluble in all solvents, and is nonpoisonous; the black allotrope comprises lustrous crystals similar to graphite, and is insoluble in most solvents.

phosphorus bromide *See* phosphorus pentabromide.

phosphorus nitride P_3N_5 Amorphous white solid that decomposes in hot water; insoluble in cold water, soluble in organic solvents; used to dope semiconductors.

phosphorus oxide An oxygen compound of phosphorus; examples are phosphorus monoxide (P_2O), phosphorus trioxide (P_2O_3), phosphorus suboxide (P_4O).

phosphorus oxychloride $POCl_3$ Toxic, colorless, fuming liquid with pungent aroma; boils at 107°C; decomposes in water or alcohol; causes skin burns; used as a catalyst, chlorinating agent, and in manufacture of various anhydrides. Also known as phosphoryl chloride.

phosphorus pentabromide PBr_5 Yellow crystals, decomposing at 106°C and in water; used in organic synthesis. Also known as phosphoric bromide; phosphoric perbromide; phosphorus bromide.

phosphorus pentachloride PCl_5 Toxic, yellowish crystals with irritating aroma; an eye irritant; sublimes on heating, but will melt at 148°C under pressure; soluble in carbon disulfide; decomposes in water; used as a catalyst and chlorinating agent. Also known as phosphoric chloride; phosphoric perchloride.

phosphorus pentasulfide P_2S_5 Flammable, hygroscopic, yellow crystals, melting at 281°C; decomposes in moist air; soluble in alkali hydroxides; used to make lube-oil additives, rubber additives, and flotation agents. Also known as phosphoric sulfide; phosphorus persulfide; thiophosphoric anhydride.

phosphorus pentoxide *See* phosphoric anhydride.

phosphorus persulfide *See* phosphorus pentasulfide.

phosphorus sesquisulfide P_4S_3 Flammable, yellow crystals, melting at 172°C; decomposed by hot water, insoluble in water, soluble in carbon disulfide; used as chemical intermediate and to make matches. Also known as tetraphosphorus trisulfide.

phosphorus sulfide *See* phosphorus trisulfide.

phosphorus thiochloride $PSCl_3$ Yellow liquid, boiling at 125°C; used to make insecticides and oil additives.

phosphorus tribromide PBr_3 A corrosive, fuming, colorless liquid with penetrating aroma; soluble in acetone, alcohol, carbon disulfide, and hydrogen sulfide; decomposes in water; used as an analytical reagent to test for sugar and oxygen.

phosphorus trichloride PCl_3 A colorless, fuming liquid that decomposes rapidly in moist air and water; soluble in ether, benzene, carbon disulfide, and carbon tetrachloride; boils at 76°C; used as a chlorinating agent, phosphorus solvent, and in saccharin manufacture.

phosphorus triiodide PI_3 Hygroscopic, red crystals, melting at 61°C; soluble in alcohol and carbon disulfide; decomposes in water; used in organic syntheses.

phosphorus trisulfide P_2S_3 or P_4S_6 Grayish-yellow, tasteless, odorless solid that burns in air; soluble in alcohol, carbon disulfide, and ether; melts at 290°C; used as an analytical reagent. Also known as phosphorus sulfide.

phosphorylase An enzyme that catalyzes the formation of glucose-1-phosphate (Cori ester) from glycogen and inorganic phosphate; it is widely distributed in animals, plants, and microorganisms.

phosphorylation The esterification of compounds with phosphoric acid.

phosphoryl chloride *See* phosphorus oxychloride.

phosphotransacetylase   An enzyme that catalyzes the reversible transfer of an acetyl group from acetyl coenzyme A to a phosphate, with formation of acetyl phosphate.

phosphotungstic acid $H_3PO_4 \cdot 12WO_3 \cdot xH_2O$ Heavy-greenish, water- and alcohol-soluble crystals; used as an analytical reagent and in the manufacture of organic pigments. Also known as heavy acid; phosphowolframic acid; PTA.

phosphotungstic pigment   A green or blue pigment prepared by precipitating solutions of phosphotungstic or phosphomolybdic acid with malachite green, Victoria blue, and other basic dyestuffs; used in printing inks, paints, and enamels. Also known as tungsten lake.

phosphowolframic acid   *See* phosphotungstic acid.

photoacoustic spectroscopy   A spectroscopic technique for investigating solid and semisolid materials, in which the sample is placed in a closed chamber filled with a gas such as air and illuminated with monochromatic radiation of any desired wavelength, with intensity modulated at some suitable acoustic frequency; absorption of radiation results in a periodic heat flow from the sample, which generates sound that is detected by a sensitive microphone attached to the chamber. Abbreviated PAS.

photoaddition   A bimolecular photochemical process in which a single product is formed by electronically excited unsaturated molecules.

photochemical oxidant   Any of the chemicals which enter into oxidation reactions in the presence of light or other radiant energy.

photochemical reaction   A chemical reaction influenced or initiated by light, particularly ultraviolet light, as in the chlorination of benzene to produce benzene hexachloride.

photochemistry   The study of the effects of light on chemical reactions.

photochromic compound   A chemical compound that changes in color when exposed to visible or near-visible radiant energy; the effect is reversible; used to produce very-high-density microimages.

photochromic reaction   A chemical reaction that produces a color change.

photochromism   The ability of a chemically treated plastic or other transparent material to darken reversibly in strong light.

photocurrent   An electric current induced at an electrode by radiant energy.

photodegradation   Decomposition of a compound by radiant energy.

photodetachment   The removal of an electron from a negative ion by absorption of a photon, resulting in a neutral atom or molecule.

photodimerization   A bimolecular photochemical process involving an electronically excited unsaturated molecule that undergoes addition with an unexcited molecule of the same species.

photodisintegration   The breakup of an atomic nucleus into two or more fragments as a result of bombardment by gamma radiation. Also known as Chadwick-Goldhaber effect.

photodissociation   The removal of one or more atoms from a molecule by the absorption of a quantum of electromagnetic energy.

photoelectric absorption analysis Type of activation analysis in which the γ-photon gives all of its energy to an electron in the crystal under analysis, generating a maximum-sized pulse for that particular γ-energy.

photoelectric colorimetry Measurement of the colorant concentration in a solution by means of the tristimulus values of three primary light filter-photocell combinations.

photoelectrolysis The process of using optical energy to assist or effect electrolytic processes that ordinarily require the use of electrical energy.

photoelectron spectroscopy The branch of electron spectroscopy concerned with the energy analysis of photoelectrons ejected from a substance as the direct result of bombardment by ultraviolet radiation or x-radiation.

photofission Fission of an atomic nucleus that results from absorption by the nucleus of a high-energy photon.

photoglycine *See* glycin.

photographic photometry The use of a comparator-densitometer to analyze a photographed spectrograph spectrum by emulsion density measurements.

photohomolysis A homolysis reaction in which bond breaking is caused by radiant energy.

photoionization The removal of one or more electrons from an atom or molecule by absorption of a photon of visible or ultraviolet light. Also known as atomic photoelectric effect.

photoisomer An isomer produced by photolysis.

photoluminescence Luminescence stimulated by visible, infrared, or ultraviolet radiation.

photolysis The use of radiant energy to produce chemical changes.

photomechanochemistry A branch of polymer sciences that deals with photochemical conversion of chemical energy into mechanical energy.

photometric titration A titration in which the titrant and solution cause the formation of a metal complex accompanied by an observable change in light absorbance by the titrated solution.

photoneutron A neutron released from a nucleus in a photonuclear reaction.

photonuclear reaction A nuclear reaction resulting from the collision of a photon with a nucleus.

photophosphorylase An enzyme that is associated with the surface of a thylakoid membrane and is involved in the final stages of adenosinetriphosphate production by photosynthetic phosphorylation.

photophosphorylation Phosphorylation induced by light energy in photosynthesis.

photopigment A pigment that is unstable in the presence of light of appropriate wavelengths, such as the chromophore pigment which combines with opsins to form rhodopsin in the rods and cones of the vertebrate eye.

photopolymer Any of a class of light-sensitive polymers which undergo a spontaneous and permanent change in physical properties on exposure to light.

photoproton A proton released from a nucleus in a photonuclear reaction.

photoreactive chlorophyll Chlorophyll molecules which receive light quanta from antenna chlorophyll and constitute a photoreaction center where light energy conversion occurs.

photorespiration Respiratory activity taking place in plants during the light period; CO_2 is released and O_2 is taken up, but no useful form of energy, such as adenosinetriphosphate, is derived.

photosynthesis Synthesis of chemical compounds in light, especially the manufacture of organic compounds (primarily carbohydrates) from carbon dioxide and a hydrogen source (such as water), with simultaneous liberation of oxygen, by chlorophyll-containing plant cells.

photosystem I One of two reaction sequences of the light phase of photosynthesis in green plants that involves a pigment system which is excited by wavelengths shorter than 700 nanometers and which transfers this energy to energy carriers such as NADPH that are subsequently utilized in carbon dioxide fixation.

photosystem II One of two reaction sequences of the light phase of photosynthesis in green plants which involves a pigment system excited by wavelengths shorter than 685 nanometers and which is directly involved in the splitting or photolysis of water.

phoxim *See* phenylglyoxylonitriloxime *O,O*-diethyl phosphorothioate.

pH(S) *See* pH standard.

pH standard Five standard laboratory solutions available from the U. S. National Bureau of Standards, each solution having a known pH value; the standards cover pH ranges from 3.557 to 8.833. Abbreviated pH(S).

phthalate A salt of phthalic acid; contains the radical $C_6H_4(COO)_2^{2-}$; an example is dibutylphthalate, $C_{16}H_{22}O_4$; used as a plasticizer in plastics, and as a buffer in standard laboratory solutions.

phthalate buffer Laboratory pH reference solution made of potassium hydrogen phthalate, $KHC_8H_4O_4$; at 0.05 molal, the pH is 4.008 at 25°C.

phthalate ester Any of a group of plastics plasticizers made by the direct action of alcohol on phthalic anhydride; generally characterized by moderate cost, good stability, and good general properties.

phthalazine $C_6H_4CHN_2CH$ Colorless crystals, melting at 91°C; soluble in alcohol. Also known as 2, 3-benzodiazine; β-benzo-*o*-diazine; β-phenoldiazine.

***para*-phthalic acid** *See* terephthalic acid.

phthalic acid $C_6H_4(CO_2H)_2$ Any of three isomeric benzene dicarboxylic acids; the ortho form is usually called phthalic acid, comprises alcohol-soluble, colorless crystals decomposing at 191°C, slightly soluble in water and ether, is used to make dyes, medicine, and synthetic perfumes, and as a chemical intermediate, and is also known as benzene orthodicarboxylic acid; the para form, known as terephthalic acid, is used to make polyester resins (Dacron) and as poultry feed additives; the meta form is isophthalic acid.

***meta*-phthalic acid** *See* isophthalic acid.

***ortho*-phthalic acid** *See* phthalic acid.

phthalic anhydride $C_6H_4(CO)_2O$ White crystals, melting at 131°C; sublimes when heated; slightly soluble in ether and hot water, soluble in alcohol; used to make dyes, resins, plasticizers, and insect repellents. Also known as acid phthalic anhydride.

phthalocyanine pigments A group of light-fast organic pigments with four isoindole groups, $(C_6H_4)C_2N$, linked by four nitrogen atoms to form a conjugated chain; included are phthalocyanine (blue-green), copper phthalocyanine (blue), chlorinated copper phthalocyanine (green), and sulfonated copper phthalocyanine (green); used in enamels, plastics, linoleum, inks, wallpaper, and rubber goods.

phthalonitrile $C_6H_4(CN)_2$ Buff-colored crystals with a melting point of 138°C; soluble in acetone and benzene; used in organic synthesis and as an insecticide. Also known as O-dicyanobenzene.

phycite *See* erythritol.

phycobilin Any of various protein-bound pigments which are open-chain tetrapyrroles and occur in some groups of algae.

phycocyanin A blue phycobilin.

phycocyanobilin $C_{31}H_{38}O_2N_4$ Phycobilin with an ethylidene side chain ($=CH— CH_3$) and only one asymmetric carbon atom (C_1).

phycoerythrin A red phycobilin.

phycoerythrobilin $C_{31}H_{38}O_2N_4$ Phycobilin with seven conjugated double bonds, an ethylidine side chain ($=CH—CH_3$), and two asymmetric carbon atoms (C_1 and C_7).

physical adsorption Reversible adsorption in which the adsorbate is held by weak physical forces.

physical chemistry The branch of chemistry that deals with the interpretation of chemical phenomena and properties in terms of the underlying physical processes, and with the development of techniques for their investigation.

physical property Property of a compound that can change without involving a change in chemical composition; examples are the melting point and boiling point.

physisorption A physical adsorption process in which there are van der Waals forces of interaction between gas or liquid molecules and a solid surface.

physostigmine $C_{15}H_{21}O_2N_3$ An alkaloid; poisonous, colorless-to-pinkish crystals; soluble in alcohol and dilute acids; melts at 86°C; used as a source of salicylate and sulfate forms. Also known as calabarine; eserine.

physostigmine salicylate $C_{15}H_{21}O_2N_3 \cdot C_7H_6O_3$ Poisonous, colorless-to-yellow crystals; soluble in water, alcohol, and chloroform; melts at 182°C; used for medicines.

physostigmine sulfate $(C_{15}H_{21}O_2N_3)_2 \cdot H_2SO_4$ Poisonous, white crystals; soluble in water, alcohol, and chloroform; melts at 150°C; used for medicines.

phytase An enzyme occurring in plants, especially cereals, which catalyzes hydrolysis of phytic acid to inositol and phosphoric acid.

phytic acid $C_6H_6[(OPO(OH)_2]_6$ An acid found in seeds of plants as the insoluble calcium magnesium salt (phytin); derived from corn steep liquor; inhibits calcium absorption in intestine; used to treat hard water, to remove iron and copper from wines, and to inactivate trace-metal contaminants in animal and vegetable oils.

phytochrome A protein plant pigment which serves to direct the course of plant growth and development in response variously to the presence or absence of light, to photoperiod, and to light quality.

phytohemagglutinin *See* phytolectin.

phytohormone *See* plant hormone.

phytol $C_{20}H_{40}O$ A liquid with a boiling point of 202–204°C; soluble in organic solvents; used in the synthesis of vitamins E and K.

phytolectin A lectin found in plants. Also known as phytohemagglutinin.

phytonadione $C_{31}H_{46}O_2$ A yellow, viscous liquid soluble in benzene, chloroform, and vegetable oils; used in medicine and as a food supplement. Also known as 2-methyl-3-phytyl-1,4-napthoquinone; vitamin K_1.

phytosterol Any of various sterols obtained from plants, including ergosterol and stigmasterol.

phytotoxin 1. A substance toxic to plants. 2. A toxin produced by plants.

pi bonding Covalent bonding in which the greatest overlap between atomic orbitals is along a plane perpendicular to the line joining the nuclei of the two atoms.

pickling A method of preparing hides for tanning by immersion in a salt solution with a pH of 2.5 or less.

pickling acid Any of the acids used in pickling solutions, such as hydrochloric, sulfuric, nitric, phosphoric, or hydrofluoric acid.

pickup A type of nuclear reaction in which the incident particle takes a nucleon from the target nucleus and proceeds with this nucleon bound to itself.

picloram $C_6H_3Cl_3N_2$ A colorless powder that decomposes at 215°C; used as an herbicide to control brush for rights-of-way, pasture, and rangeland. Also known as 4-amino-3,5,6-trichloropicolinic acid.

picoline $C_5H_4N(CH_3)$ Family of colorless liquid isomers, soluble in water and alcohol; the alpha form, boiling at 129°C, is used as a solvent and chemical intermediate, and is also known as 2-methyl pyridine; the beta form, boiling at 143.5°C, is used as a solvent for chemical synthesis reactions, to make nicotinic acid, and in fabric waterproofing, and is also known as 3-methyl pyridine; the gamma form, boiling at 143.1°C, is used as a solvent for chemical synthesis reactions and in fabric waterproofing.

picolinic acid $C_{10}H_8N_4O_5$ An alcohol-soluble crystalline compound, forming yellow leaflets that melt at 116–117°C; used as a reagent in phenylalanine, tryptophan, and alkaloids production, and for the quantitative detection of calcium. Also known as 3-methyl-4-nitro-1-(*para*-nitrophenyl)-2-pyrazoline-5-one.

picramic acid $C_6H_5N_3O_5$ A crystalline acid, forming dark red needles from alcohol solutions, melting at 169–170°C; used in dye manufacture and as a reagent in tests for albumin. Also known as 2-amino-4,6-dinitrophenol; 4,6-dinitro-2-aminophenol.

picric acid $C_6H_2(NO_2)_3OH$ Poisonous, explosive, highly oxidative yellow crystals with bitter taste; soluble in water, alcohol, chloroform, benzene, and ether; melts at 122°C; used in explosives, in external medicines; to make dyes, matches, and batteries, and to etch copper. Also known as carbazotic acid; nitroxanthic acid; picronitric acid; trinitrophenol.

picronitric acid *See* picric acid.

picrotoxin $C_{30}H_{34}O_{13}$ A poisonous, crystalline plant alkaloid found primarily in *Cocculus indicus*; used as a stimulant and convulsant drug. Also known as cocculin.

pi electron An electron which participates in pi bonding.

piezochemistry The field of chemical reactions under high pressures.

pigment Any coloring matter in plant or animal cells.

pilocarpine $C_{11}H_{16}N_2O_2$ An alkaloid, in either oil or crystal form, melting at 34°C; soluble in chloroform, water, and alcohol; used in medicine.

pilot-scale chemical reaction Small-scale chemical reaction used to test operating conditions and product yields; used as a pilot for design of large-scale reaction systems.

pimaricin $C_{33}H_{47}NO_{13}$ A compound crystallizing from a methanol-water solution, decomposing at about 200°C; soluble in water and organic solvents; used in medicine as an antifungal agent for *Candida albicans* vaginitis. Also known as tennecetin.

pimelic acid $HOOC(CH_2)_5COOH$ Crystals melting at 105°C; slightly soluble in water, soluble in alcohol and ether; used in biochemical research. Also known as heptanedioic acid.

pimelic ketone *See* cyclohexanone.

pinacoid An open crystal form that comprises two parallel faces.

pinacoidal class That crystal class in the triclinic system having only a center of symmetry.

pinacoidal cleavage A type of crystal cleavage that is parallel to one of the crystal's pinacoidal surfaces.

pindone $C_{14}H_{14}O_3$ A yellow, crystalline compound with a melting point of 108.5–110.5°C; slight solubility in water; used as a rodenticide. Also known as 2-pivalyl-1,3-indandione.

pinene $C_{10}H_{16}$ Either of two colorless isomeric unsaturated bicyclic terpene hydrocarbon liquids derived from sulfate wood turpentine; 95% of the alpha form boils in the range 156–160°C, and of the beta form boils in the range 164–169°C; used as solvents for coatings and wax formulations, as chemical intermediates for resins, and as lube-oil additives. Also known as nopinene.

pinene hydrochloride *See* terpene hydrochloride.

pinic acid $C_9H_{14}O_4$ A crystalline dicarboxylic acid derived from α-pinene; used to make diesters for plasticizers and lubricants.

pion condensate A state of nuclear matter compressed to abnormally high densities, in which great numbers of pairs of particles, each consisting of a positive pion and a negative pion, are generated, and interact strongly with the nucleons, causing them to form a coherent spin-isospin structure.

piotine *See* saponite.

piperalin $C_{16}H_{21}O_2NCl_2$ An amber liquid with a boiling point of 156–157°C; used as a fungicide. Also known as 3-(2-methylpiperidino)propyl 3,4-dichlorobenzoate.

piperazidine *See* piperazine.

piperazine $C_4H_{10}N_2$ A cyclic compound; colorless, deliquescent crystals, melting at 104–107°C; soluble in water, alcohol, glycerol, and glycols; absorbs carbon dioxide from air; used in medicine. Also known as diethylenediamine; ethyleneamine; piperazidine.

piperazine dihydrochloride $C_4H_{10}N_2\cdot 2HCl$ White, water-soluble needles; used for insecticides and pharmaceuticals.

2,5-piperazine-dione *See* diketopiperazine.

N,N′-[1,4-piperazinediyl-bis(2,2,2-trichloroethylidine)]-bis[formamide] *See* triforine.

piperazine hexahydrate $C_4H_{10}N_2\cdot 6H_2O$ White crystals with a melting point of 44°C; soluble in alcohol and water; used for pharmaceuticals and insecticides.

piperidic acid *See* γ-aminobutyric acid.

piperidine $C_5H_{11}N$ A cyclic compound, and strong base; colorless liquid with pepper aroma; boils at 106°C; soluble in water, alcohol, and ether; used as a chemical intermediate and rubber accelerator, and in medicine.

piperine $C_{17}H_{19}NO_3$ A crystalline compound that is found in black pepper; melting point is 130°C; soluble in benzene and acetic acid; used to give a pungent taste to brandy and as an insecticide.

piperocaine hydrochloride $C_{16}H_{23}NO_2\cdot HCl$ A white, crystalline powder with a bitter taste and a melting point of 172–175°C; soluble in water, chloroform, and alcohol; used in medicine. Also known as 3-(2-methyl-1-piperidyl)propyl benzoate hydrochloride.

piperonal $C_8H_6O_3$ White crystals with a floral odor and a melting point of 35.5–37°C; soluble in alcohol and ether; used in medicine, perfumes, suntan preparations, and mosquito repellents. Also known as heliotropin; 3,4-methylenedioxybenzaldehyde; piperonyl aldehyde.

piperonal bis[2-(2-butoxyethoxy)ethyl]acetal *See* piprotal.

piperonyl aldehyde *See* piperonal.

piperylene *See* pentadiene.

pipe still A petroleum-refinery still in which heat is applied to the oil while it is being pumped through a coil or pipe arranged in a firebox, the oil then running to a fractionator with continuous removal of overhead vapor and liquid bottoms.

pipet Graduated or calibrated tube which may have a center reservoir (bulb); used to transfer known volumes of liquids from one vessel to another; types are volumetric or transfer, graduated, and micro.

piprotal $C_{16}H_{31}O_8$ An amber liquid, insoluble in water; used as a synergist with pyrethrum and carbamate insecticides for control of wood borers. Also known as piperonal bis[2-(2-butoxyethoxy)ethyl]acetal.

pirimiphosethyl $C_{13}H_{24}N_3O_3PS$ A straw-colored liquid which decomposes at 130°C; used as an insecticide for the control of soil insects in vegetables and other crops. Also known as 2-diethylamino-6-methylpyrimidin-4-yl diethyl phosphorothioate.

pirimiphosmethyl *See* 2-diethylamino-6-methylpyrimidin-4-yl dimethyl phosphorothionate.

Pitzer equation Equation for the approximation of data for heats of vaporization for organic and simple inorganic compounds; derived from temperature and reduced temperature relationships.

2-pivalyl-1,3-indandione *See* pindone.

PIXE *See* proton-induced x-ray emission.

pK The logarithm (to base 10) of the reciprocal of the equilibrium constant for a specified reaction under specified conditions.

plagiohedral Pertaining to obliquely arranged spiral faces; in particular, to a member of a group in the isometric system with 13 axes but no center or planes.

plait point Composition conditions in which the three coexisting phases of partially soluble components of a three-phase liquid system approach each other in composition.

plane defect A type of crystal defect that occurs along the boundary plane of two regions of a crystal, or between two grains.

plane-dendritic crystal An ice crystal exhibiting an elaborately branched (dendritic) structure of hexagonal symmetry, with its much larger dimension lying perpendicular to the principal (c-axis) of the crystal. Also known as plane dendrite; stellar crystal.

plane of mirror symmetry In certain crystals, a symmetry element whereby reflection of the crystal through a certain plane leaves the crystal unchanged. Also known as mirror plane of symmetry; plane of reflection; plane of symmetry; reflection plane; symmetry plane.

plane of reflection *See* plane of mirror symmetry.

plane of symmetry *See* plane of mirror symmetry.

planetary electron *See* orbital electron.

planocaine base *See* procaine base.

plant fermentation A form of plant metabolism in which carbohydrates are partially degraded without the consumption of molecular oxygen.

plant hormone An organic compound that is synthesized in minute quantities by one part of a plant and translocated to another part, where it influences physiological processes. Also known as phytohormone.

plasma desorption mass spectrometry A technique for analysis of nonvolatile molecules, particularly heavy molecules with atomic weight over 2000, in which heavy ions with energies on the order of 100 MeV penetrate and deposit energy in thin films, giving rise to chemical reactions that result in the formation of molecular ions and shock waves that result in the ejection of these ions from the surface; the ions are then analyzed in a mass spectrometer. Abbreviated PDMS.

plasma-jet excitation The use of a high-temperature plasma jet to excite an element to provide measurable spectra with many ion lines similar to those from spark-excited spectra.

plasma thromboplastin antecedent *See* factor XI.

plasma thromboplastin component *See* Christmas factor.

plasmin A proteolytic enzyme in plasma which can digest many proteins through the process of hydrolysis. Also known as fibrinolysin.

plasminogen The inert precursor, or zymogen, of plasmin. Also known as profibrinolysin.

plaster of paris White powder consisting essentially of the hemihydrate of calcium sulfate ($CaSO_4 \cdot \frac{1}{2}H_2O$ or $2CaSO_4 \cdot H_2O$), produced by calcining gypsum until it is partially dehydrated; forms with water a paste that quickly sets; used for casts and molds, building materials, and surgical bandages. Also known as calcined gypsum.

plate efficiency The equilibrium produced by an actual plate of a distillation column or countercurrent tower extractor compared with that of a perfect plate, expressed as a ratio.

plate theory In gas chromatography, the theory that the column operates similarly to a distillation column; for example, chromatographic columns are considered as consisting of a number of theoretical plates, each performing a partial separation of components.

plate tower A distillation tower along the internal height of which is a series of transverse plates (bubble-cap or sieve) to force intimate contact between downward flowing liquid and upward flowing vapor.

platinic chloride *See* chloroplatinic acid.

platinic sodium chloride *See* sodium chloroplatinate.

platinic sulfate *See* platinum sulfate.

platinite *See* platynite.

platinochloride *See* chloroplatinate.

platinocyanide A double salt of platinous cyanide and another cyanide, such as $K_2Pt(CN)_4$; used in photography and fluorescent x-ray screens. Also known as cyanoplatinate.

platinous chloride *See* platinum dichloride.

platinous iodide *See* platinum iodide.

platinum A chemical element, symbol Pt, atomic number 78, atomic weight 195.09.

platinum bichloride *See* platinum dichloride.

platinum chloride $PtCl_4$ or $PtCl_4 \cdot 5H_2O$ A brown solid or red crystals; soluble in alcohol and water; decomposes when heated (loses $4H_2O$ at 100°C); used as an analytical reagent.

platinum dichloride $PtCl_2$ Water-insoluble, green-gray powder; decomposes to platinum at red heat; used to make platinum salts. Also known as platinous chloride; platinum bichloride.

platinum diiodide *See* platinum iodide.

platinum electrode A solid platinum wire electrode used during voltammetric analyses of electrolytes.

platinum iodide PtI_2 Water- and alkali-insoluble black powder; slightly soluble in hydrochloric acid; decomposes at 300–350°C. Also known as platinous iodide; platinum diiodide.

platinum oxide An oxide of platinum; examples are platinum monoxide (or platinous oxide), PtO, and platinum dioxide (or platinic oxide), PtO_2.

platinum potassium chloride *See* potassium chloroplatinate.

platinum sodium chloride *See* sodium chloroplatinate.

platinum sulfate $Pt(SO_4)_2$ A hygroscopic, dark mass; soluble in alcohol, ether, water, and dilute acids; used in microanalysis for halogens. Also known as platinic sulfate.

pleomorphism *See* polymorphism.

Plessy's green $CrPO_4 \cdot xH_2O$ Deep-green pigment made of chromium phosphate mixed with chromium oxide and calcium phosphate.

plumbous oxide *See* lead monoxide.

plumbous sulfide *See* lead sulfide.

plumbum Latin name for lead; source of the element symbol Pb.

plutonium A reactive metallic element, symbol Pu, atomic number 94, in the transuranium series of elements; the first isotope to be identified was plutonium-239; used as a nuclear fuel, to produce radioactive isotopes for research, and as the fissile agent in nuclear weapons.

plutonium-238 The first synthetic isomer made of plutonium; similar chemically to uranium and neptunium; atomic number 94; formed by bombardment of uranium with deuterons.

plutonium-239 A synthetic element chemically similar to uranium and neptunium; atomic number 94; made by bombardment of uranium-238 with slow electrons in a nuclear reactor; used as nuclear reactor fuel and an ingredient for nuclear weapons.

plutonium oxide PuO_2 A radioactively poisonous pyrophoric oxide of plutonium; particles may be easily airborne.

Pm *See* promethium.

PMA *See* phosphomolybdic acid; pyromellitic acid.

PMDA *See* pyromellitic dianhydride.

pNa Logarithm of the sodium-ion concentration in a solution; that is, $pNa = -\log a_{Na+}$, where a_{Na+} is the sodium-ion concentration.

pnicogen Any member of the nitrogen family of elements, group V in the periodic table.

pnictide A simple compound of a pnicogen and an electropositive element.

POD analysis A precision laboratory distillation procedure used to separate low-boiling hydrocarbon fractions quantitatively for analytical purposes. Also known as Podbielniak analysis.

Podbielniak analysis *See* POD analysis.

Podbielniak extractor A solvent-extraction device in which centrifugal action enhances liquid-liquid contact and increases resultant separation efficiency.

point defect   A departure from crystal symmetry which affects only one, or, in some cases, two lattice sites.

point group   A group consisting of the symmetry elements of an object having a single fixed point; 32 such groups are possible.

poison   1. A substance which reduces the phosphorescence of a luminescent material.   2. A substance that exerts inhibitive effects on catalysts, even when present only in small amounts; for example, traces of sulfur or lead will poison platinum-based catalysts.

polar axis   An axis of crystal symmetry which does not have a plane of symmetry perpendicular to it.

polar compound   Molecules which contain polar covalent bonds; they can ionize when dissolved or fused; polar compounds include inorganic acids, bases, and salts.

polar covalent bond   A bond in which a pair of electrons is shared in common between two atoms, but the pair is held more closely by one of the atoms.

polarimetric analysis   A method of chemical analysis based on the optical activity of the substance being determined; the measurement of the extent of the optical rotation of the substance is used to identify the substance or determine its quantity.

polarization potential   The reverse potential of an electrolytic cell which opposes the direct electrolytic potential of the cell.

polarization spectroscopy   A type of saturation spectroscopy in which a circularly polarized saturating laser beam depletes molecules with a certain orientation preferentially, leaving the remaining ones polarized; the latter are detected through their induction of elliptical polarization in a probe beam, allowing the beam to pass through crossed linear polarizers.

polarized scattering   In a quasielastic light scattering experiment performed with polarizers, the type of scattering produced when the polarizers select both the incident and final polarizations perpendicular to the scattering plane.

polar molecule   A molecule having a permanent electric dipole moment.

polarogram   Plotted output (current versus electrode voltage) for polarographic analysis of an electrolyte.

polarographic analysis   An electroanalytical technique in which the current through an electrolysis cell is measured as a function of the applied potential; the apparatus consists of a potentiometer for adjusting the potential, a galvanometer for measuring current, and a cell which contains two electrodes, a reference electrode whose potential is constant and an indicator electrode which is commonly the dropping mercury electrode. Also known as polarography.

polarographic cell   Device for polarographic (voltammetric) analysis of an electrolyte solution; a known voltage is applied to the solution, and the ensuing current that passes through the cell (to an electrode) is measured.

polarographic maximum   A deceptively high voltage buildup on an electrode during polarographic analysis of an electrolyte; caused by a reduction or oxidation process at the electrode.

polarography   *See* polarographic analysis.

polar symmetry A type of crystal symmetry in which the two ends of the central crystallographic axis are not symmetrical.

pole A direction perpendicular to one of the faces of a crystal.

polishing In petroleum refining, removal of final traces of impurities, as for a lubricant, by clay adsorption or mild hydrogen treating.

polonium A chemical element, symbol Po, atomic number 84; all polonium isotopes are radioactive; polonium-210 is the naturally occurring isotope found in pitchblende.

polonium-210 Radioactive isotope of polonium; mass 210, half-life 140 days, α-radiation; used to calibrate radiation counters, and in oil well logging and atomic batteries. Also known as radium F.

poly- A chemical prefix meaning many; for example, a polymer is made of a number of single molecules known as monomers, as polyethylene is made from ethylene.

polyacetals *See* acetal resins.

polyacrylamide $(CH_2CHCONH_2)_x$ A white, water-soluble high polymer based on acrylamide; used as a thickening or suspending agent in water-base formulations.

polyacrylate A polymer of an ester or salt of acrylic acid.

polyacrylic acid $(CH_2CHCOOH)_x$ An acrylic or acrylate resin formed by the polymerization of acrylic acid; water-soluble; used as a suspending and textile-sizing agent, and in adhesives, paints, and hydraulic fluids.

polyacrylic fiber Continuous-strand fiber extruded from an acrylate resin.

polyacrylonitrile Polymer of acrylonitrile; semiconductive; used like an inorganic oxide catalyst to dehydrogenate *tert*-butyl alcohol to produce isobutylene and water.

polyalcohol *See* polyhydric alcohol.

polyallomer A copolymer of propylene with other olefins.

polyamide *See* polyamide resin.

polyamide resin Product of polymerization of amino acid or the condensation of a polyamine with a polycarboxylic acid; an example is the nylons. Also known as polyamide.

polyatomic molecule A chemical molecule with three or more atoms.

polybasic A chemical compound in solution that yields two or more H^+ ions per molecule, such as sulfuric acid, H_2SO_4.

polybutadiene Oil-extendable synthetic elastomer polymer made from butadiene; resilience is similar to natural rubber; it is blended with natural rubber for use in tire and other rubber products. Also known as butadiene rubber.

polybutene A polymer of isobutene, $(CH_3)_2CCH_2$; made in varying chain lengths to give a wide range of properties from oily to solid; used as a lube-oil additive, in adhesives, and in rubber products.

polybutylene A polymer of one or more butylenes whose consistency ranges from a viscous liquid to a rubbery solid.

polycarbonate $[OC_6H_4C(CH_3)_2C_6H_4OCO]_x$ A linear polymer of carbonic acid which is a thermoplastic synthetic resin made from bisphenol and phosgene; used in emulsion coatings with glass fiber reinforcement.

polycarboxylic Prefix for a compound containing two or more carboxyl (—COOH) groups.

polychlorinated biphenyl A colorless liquid, used as an insulating fluid in electrical equipment.

polycistronic messenger In ribonucleic acid viruses, messenger ribonucleic acid that contains the amino acid sequence for several proteins.

polyclonal mixed cryoglobulin A cryoglobulin made of heterogeneous immunoglobin molecules belonging to two or more different classes, and sometimes additional serum proteins.

polycondensation A chemical condensation leading to the formation of a polymer by the linking together of molecules of a monomer and the releasing of water or a similar simple substance.

polycyclic A molecule that contains two or more closed atomic rings; can be aromatic (such as DDT), aliphatic (bianthryl), or mixed (dicarbazyl).

polycyclic hydrocarbon *See* polynuclear hydrocarbon.

polydispersity Molecular-weight nonhomogeneity in a polymer system; that is, there is some molecular-weight distribution throughout the body of the polymer.

polyelectrolyte A natural or synthetic electrolyte with high molecular weight, such as proteins, polysaccharides, and alkyl addition products of polyvinyl pyridine; can be a weak or strong electrolyte; when dissociated in solution, it does not give uniform distribution of positive and negative ions (the ions of one sign are bound to the polymer chain while the ions of the other sign diffuse through the solution).

polyene Compound containing many double bonds, such as the carotenoids.

polyester resin A thermosetting or thermoplastic synthetic resin made by esterification of polybasic organic acids with polyhydric acids; examples are Dacron and Mylar; the resin has high strength and excellent resistance to moisture and chemicals when cured.

polyester rubber *See* polyurethane rubber.

polyether resin A polymer that contains —$(CH_2CHRO—)_x$ in the main-chain or side-chain linkage.

polyethylene *See* ethylene resin.

polyethylene glycol Any of a family of colorless, water-soluble liquids with molecular weights from 200 to 6000; soluble also in aromatic hydrocarbons (not aliphatics) and many organic solvents; used to make emulsifying agents and detergents, and as plasticizers, humectants, and water-soluble textile lubricants.

polyethylene glycol distearate *See* polyglycol distearate.

polyethylene resin *See* ethylene resin.

polyethylene terephthalate A thermoplastic polyester resin made from ethylene glycol and terephthalic acid; melts at 265°C; used to make films or fibers.

polyforming A noncatalytic, petroleum-refinery process charging C_3 and C_4 gases with naphtha or gas oil at high temperature to produce high-quality gasoline and fuel oil; mostly replaced by catalytic reforming; the product is known as polyformdistillate.

polygalacturonase An enzyme that catalyzes the hydrolysis of glycosidic linkage of polymerized galacturonic acids.

polygen *See* polyvalent.

polyglycol A dihydroxy ether derived from the dehydration (removal of a water molecule) of two or more glycol molecules; an example is diethylene glycol, $CH_2OHCH_2OCH_2CH_2OH$.

polyglycol distearate $(C_{17}H_{35})_2CO_2CO(CH_2CH_2O)_x$ An off-white, soft solid with a melting point of $43°C$; soluble in chlorinated solvents, acetone, and light esters; used as a resin plasticizer. Also known as polyethylene glycol distearate.

polyhydric alcohol An alcohol with many hydroxyl (—OH) radicals, such as glycerol, $C_3H_5(OH_3)$. Also known as polyalcohol; polyol.

polyhydric phenol A phenolic compound containing two or more hydroxyl groups, such as diphenol, $C_6H_4(OH)_2$.

polyimide resin An aromatic polyimide made by reacting pyromellitic dianhydride with an aromatic diamine; has high resistance to thermal stresses; used to make components of internal combustion engines.

polyisoprene $(C_5H_8)_x$ The basis of natural rubber, balata, gutta-percha, and other rubberlike materials; can also be made synthetically; the stereospecific forms are *cis*-1,4- and *trans*-1,4-polyisoprene; the polymer is thermoplastic.

polylactic resin A soft, elastic resin made by the heat reaction of lactic acid with castor oil or other fatty oils; used to produce tough, water-resistant coatings.

polymer Substance made of giant molecules formed by the union of simple molecules (monomers); for example polymerization of ethylene forms a polyethylene chain, or condensation of phenol and formaldehyde (with production of water) forms phenol-formaldehyde resins.

polymerase An enzyme that links nucleotides together to form polynucleotide chains.

polymerization 1. The bonding of two or more monomers to produce a polymer. 2. Any chemical reaction that produces such a bonding.

polymethyl methacrylate A thermoplastic polymer derived from methyl methacrylate, $CH_2=C(CH_3)COOCH_3$; transparent solid with excellent optical qualities and water resistance; used for aircraft domes, lighting fixtures, optical instruments, and surgical appliances.

polymorph One of the crystal forms of a substance displaying polymorphism. Also known as polymorphic modification.

polymorphic modification *See* polymorph.

polymorphism The property of a chemical substance crystallizing into two or more forms having different structures, such as diamond and graphite. Also known as pleomorphism.

polynuclear hydrocarbon Hydrocarbon molecule with two or more closed rings; examples are naphthalene, $C_{10}H_8$, with two benzene rings side by side, or diphenyl, $(C_6H_5)_2$, with two bond-connected benzene rings. Also known as polycyclic hydrocarbon.

polynucleotide A linear sequence of nucleotides.

polyol *See* polyhydric alcohol.

polyolefin A resinous material made by the polymerization of olefins, such as polyethylene from ethylene, polypropylene from propylene, or polybutene from butylene.

polyorganosiloxane *See* polysiloxane.

polyoxyalkylene resin Condensation polymer produced from an oxyalkene, such as polyethylene glycol from oxyethylene or ethylene glycol.

polyoxyethylene (8) stearate *See* polyoxyl (8) stearate.

polyoxyl (8) stearate A cream-colored, soft, waxy solid at 25°C; soluble in toluene, acetone, ether, and ethanol; used in bakery products as an emulsifier. Also known as polyoxyethylene (8) stearate.

polypeptide A chain of amino acids linked together by peptide bonds but with a lower molecular weight than a protein; obtained by synthesis, or by partial hydrolysis of protein.

polyphenol oxidase A copper-containing enzyme that catalyzes the oxidation of phenol derivatives to quinones.

polyphenyl Any of a group of direct colors used to dye cotton and wool.

polyphenylene oxide A polyether resin of 2, 6-dimethylphenol, $(CH_3)_2C_6H_3OH$; useful temperature range is -275 to $375°F$ (-168 to $191°C$), with intermittent use possible up to $400°F$ ($204°C$).

polyphosphoric acid $H_6P_4O_{13}$ Viscous, water-soluble, hygroscopic, water-white liquid; used wherever concentrated phosphoric acid is needed.

polypropylene (C_3H_{6x}) A crystalline, thermoplastic resin made by the polymerization of propylene, C_3H_6; the product is hard and tough, resists moisture, oils, and solvents, and withstands temperatures up to 170°C; used to make molded articles, fibers, film, rope, printing plates, and toys.

polypropylene glycol $CH_3CHOH(CH_2OCHCH_3)_xCH_2OH$ Polymeric material similar to polyethylene glycol, but with greater oil solubility and less water solubility; used as a solvent for vegetable oils, waxes, and resins, in hydraulic fluids and as a chemical intermediate.

polysaccharide A carbohydrate composed of many monosaccharides.

polysiloxane $(R_2SiO)_n$ A polymer in which the chain contains alternate silicon and oxygen atoms; in the formula, R can be H or an alkyl or aryl group; commercially, the R is usually CH_3 (the methylsiloxanes); properties vary with molecular weight, from oils to greases to rubbers to plastics. Also known as polyorganosiloxane.

polystyrene $(C_6H_5CHCH_2)_x$ A water-white, tough synthetic resin made by polymerization of styrene; soluble in aromatic and chlorinated hydrocarbon solvents; used for injection molding, extrusion or casting for electrical insulation, fabric lamination, and molding of plastic objects.

polysulfide rubber A synthetic polymer made by the reaction of sodium polysulfide with an organic dichloride; resistant to light, oxygen, oils, and solvents; impermeable to gases; poor tensile strength and abrasion resistance.

polysulfide treating   A petroleum-refinery process used to remove elemental sulfur from refinery liquids by contacting them with a nonregenerable solution of sodium polysulfide.

polysynthetic twinning   Repeated twinning that involves three or more individual crystals according to the same twin law and on parallel twin planes.

polyterpene resin   A thermoplastic resin or viscous liquid from polymerization of turpentine; used in paints, polishes, and rubber plasticizers, and to cure concrete and impregnate paper.

polythene   Common name for polyethlylene in Great Britain.

polytrifluorochloroethylene resin   *See* chlorotrifluoroethylene polymer.

polytype   A type of polymorph whose different forms are due to more than one possible mode of atomic packing.

polytypism   Referring to the ability of a mineral to crystallize into more than one form, because of more than one possible mode of atomic packing.

polyunsaturated acid   A fatty acid with two or more double bonds per molecule, such as linoleic or linolenic acid.

polyurethane resin   Any resins resulting from the reaction of diisocyanates (such as toluene diisocyanate) with a phenol, amine, or hydroxylic or carboxylic compound to produce a polymer with free isocyanate groups; used as protective coatings, potting or casting resins, adhesives, rubbers, and foams, and in paints, varnishes, and adhesives.

polyurethane rubber   A synthetic polyurethane-resin elastomer made by the reaction of a diisocyanate to a polyester (such as the glycol–adipic acid ester); has high resistance to abrasion, oil, ozone, and high temperatures. Also known as polyester rubber.

polyvalent   An ion or radical with more than one valency, such as the sulfate ion, SO_4^{2-}. Also known as multivalent; polygen.

polyvinyl acetal resin   *See* vinyl acetal resin.

polyvinyl acetate $(H_2CCHOOCCH_3)_x$ A thermoplastic polymer; insoluble in water, gasoline, oils, and fats, soluble in ketones, alcohols, benzene, esters, and chlorinated hydrocarbons; used in adhesives, films, lacquers, inks, latex paints, and paper sizes. Abbreviated PVA; pVAc.

polyvinyl alcohol   Water-soluble polymer made by hydrolysis of a polyvinyl ester (such as polyvinyl acetate); used in adhesives, as textile and paper sizes, and for emulsifying, suspending, and thickening of solutions. Abbreviated PVA.

polyvinyl carbazole   Thermoplastic resin made by reaction of acetylene with carbazole; softens at 150°C; has good electrical properties and heat and chemical stabilities; used as a paper-capacitor impregnant and as a substitute for electrical mica.

polyvinyl chloride $(H_2CCHCl)_x$ Polymer of vinyl chloride; tasteless, odorless; insoluble in most organic solvents; a member of the family of vinyl resins; used in soft flexible films for food packaging and in molded rigid products such as pipes, fibers, upholstery, and bristles. Abbreviated PVC.

polyvinyl chloride acetate   Thermoplastic copolymer of vinyl chloride, CH_2CHCl, and vinyl acetate, $CH_3COOCH=CH_2$; colorless solid with good resistance to water, concentrated acids, and alkalies; compounded with plasticizers, it yields a flexible material superior to rubber in aging properties; used for cable and wire coverings and protective garments.

polyvinyl dichloride   A high-strength polymer of chlorinated polyvinyl chloride; it is self-extinguishing and has superior chemical resistance; used for pipes carrying hot, corrosive materials. Abbreviated PVDC.

polyvinyl ether   *See* polyvinyl ethyl ether.

polyvinyl ethyl ether   [—CH(OC$_2$H$_5$)CH$_2$—]$_x$ A viscous gum to rubbery solid, soluble in organic solvents; used for pressure-sensitive tape. Also known as polyvinyl ether.

polyvinyl fluoride   (—H$_2$CCHF—)$_x$ Vinyl fluoride polymer; has superior resistance to weather, chemicals, oils, and stains, and has high strength; used for packaging (but not of food) and electrical equipment.

polyvinyl formate resin   (CH$_2$=CHOOCH)$_x$ Clear-colored resin that is hard and solvent-resistant; used to make clear, hard plastics.

polyvinylidene chloride   Thermoplastic polymer of vinylidene chloride, H$_2$C=CCl$_2$; white powder softening at 185–200°C; used to make soft-flexible to rigid products.

polyvinylidene fluoride   H$_2$C=CF$_2$ Fluorocarbon polymer made from vinylidene fluoride; has good tensile and compressive strength and high impact strength; used in chemical equipment for gaskets, impellers, and other pump parts, and for drum linings and protective coatings.

polyvinylidene resin   *See* vinylidene resin.

polyvinyl isobutyl ether   [—CH$_2$CHOCH$_2$CH(CH$_3$)$_2$—]$_x$ An odorless synthetic resin; elastomer to viscous liquid depending on molecular weight; soluble in hydrocarbons, esters, ethers, and ketones, insoluble in water; used in adhesives, waxes, plasticizers, lubricating oils, and surface coatings. Abbreviated PVI.

polyvinyl methyl ether   (—CH$_2$CHOCH$_3$—)$_x$ A colorless, tacky liquid, soluble in organic solvents, except aliphatic hydrocarbons, and in water below 32°C; used for pressure-sensitive adhesives, as a heat sensitizer for rubber latex, and as a pigment binder in inks and textile finishing. Abbreviated PVM.

polyvinyl pyrrolidone   (C$_6$H$_9$NO)$_x$ A water-soluble, white, resinous solid; used in pharmaceuticals, cosmetics, detergents, and foods, and as a synthetic blood plasma. Abbreviated PVP.

polyvinyl resin   Any resin or polymer derived from vinyl monomers. Also known as vinyl plastic.

Ponchon-Savarit method   Graphical solution on an enthalpy-concentration diagram of liquid-vapor equilibrium values between trays of a distillation column.

pontocaine   *See* tetracaine hydrochloride.

population inversion   The condition in which a higher energy state in an atomic system is more heavily populated with electrons than a lower energy state of the same system.

p orbital   The orbital of an atomic electron with an orbital angular momentum quantum number of unity.

porous alum   *See* aluminum sodium sulfate.

porphin   A heterocyclic ring consisting of four pyrrole rings linked by methine (—CH=) bridges; the basic structure of chlorophyll, hemoglobin, the cytochromes, and certain other related substances.

porphobilinogen $C_{10}H_{14}O_4N_2$ Dicarboxylic acid derived from pyrrole; a product of hemoglobin breakdown that gives the urine a Burgundy-red color.

porphyrin   A class of red-pigmented compounds with a cyclic tetrapyrrolic structure in which the four pyrrole rings are joined through their α-carbon atoms by four methene bridges (=C—); the porphyrins form the active nucleus of chlorophylls and hemoglobin.

positional isomer   **1.** One of a set of structural isomers which differ only in the point at which a side-chain group is attached. **2.** Constitutional isomer having the same functional group located in different positions along a chain or in a ring.

positive ion   An atom or group of atoms which by loss of one or more electrons has acquired a positive electric charge; occurs on ionization of chemical compounds as H^+ from ionization of hydrochloric acid, HCl.

positron emission   A β-decay process in which a nucleus ejects a positron and a neutrino.

postignition   Surface ignition after the passage of the normal spark.

postprecipitation   Precipitation of an impurity from a supersaturated solution onto the surface of an already present precipitate; used for analytical laboratory separations.

potash   *See* potassium carbonate.

potash blue   A pigment made by oxidizing ferrous ferrocyanide; used in making carbon paper.

potassium   A chemical element, symbol K, atomic number 19, atomic weight 39.102; an alkali metal. Also known as kalium.

potassium-40   A radioactive isotope of potassium having a mass number of 40, a half-life of approximately 1.31×10^9 years, and an atomic abundance of 0.000122 gram per gram of potassium.

potassium-42   Radioactive isotope with mass number of 42; half-life is 12.4 hours, with β- and γ-radiation; radiotoxic; used as radiotracer in medicine.

potassium acetate $KC_2H_3O_2$ White, deliquescent solid; soluble in water and alcohol, insoluble in ether; melts at 292°C; used as analytical reagent, dehydrating agent, in medicine, and in crystal glass manufacture.

potassium acid carbonate   *See* potassium bicarbonate.

potassium acid fluoride   *See* potassium bifluoride.

potassium acid oxalate   *See* potassium binoxalate.

potassium acid phosphate   *See* potassium phosphate.

potassium acid phthalate   *See* potassium biphthalate.

potassium acid saccharate $HOOC(CHOH)_4COOK$ An off-white powder, soluble in hot water, acid, or alkaline solutions; used in rubber formulations, soaps, and detergents, and for metal plating.

488   potassium acid sulfate

potassium acid sulfate *See* potassium bisulfate.

potassium acid sulfite *See* potassium bisulfite.

potassium acid tartrate *See* potassium bitartrate.

potassium alginate $(C_6H_7O_6K)_n$ A hydrophilic colloid occurring as filaments, grains, granules, and powder; used in food processing as a thickener and stabilizer. Also known as potassium polymannuronate.

potassium alum *See* potassium aluminum sulfate.

potassium aluminate $K_2Al_2O_4 \cdot 3H_2O$ Water-soluble, alcohol-insoluble, lustrous crystals; used as a dyeing and printing mordant, and as a paper sizing.

potassium aluminum fluoride K_3AlF_6 A toxic, white powder used as an insecticide.

potassium aluminum sulfate $KAl(SO_4)_2 \cdot 12H_2O$ White, odorless crystals that are soluble in water; used in medicines and baking powder, in dyeing, papermaking, and tanning. Also known as alum; aluminum potassium sulfate; potassium alum.

potassium antimonate $KSbO_3$ White, water-soluble crystals. Also known as potassium stibnate.

potassium antimonyl tartrate *See* tartar emetic.

potassium argentocyanide *See* silver potassium cyanide.

potassium arsenate K_3AsO_4 Poisonous, colorless crystals; soluble in water, insoluble in alcohol; used as an insecticide, analytical reagent, and in hide preservation and textile printing. Also known as Macquer's salt.

potassium arsenite $KH(AsO_2)_2$ Poisonous, hygroscopic, white powder; soluble in alcohol; decomposes slowly in air; used in medicine, on mirrors, and as an analytical reagent. Also known as potassium metarsenite.

potassium aurichloride *See* potassium gold chloride.

potassium bicarbonate $KHCO_3$ A white powder or granules, or transparent colorless crystals; used in baking powder and in medicine as an antacid. Also known as potassium acid carbonate.

potassium bichromate *See* potassium dichromate.

potassium bifluoride KHF_2 Colorless, corrosive, poisonous crystals; soluble in water and dilute alcohol; used to etch glass and as a metallurgy flux. Also known as Fremy's salt; potassium acid fluoride.

potassium binoxalate $KHC_2O_4 \cdot H_2O$ A poisonous, white, odorless, crystalline compound; used to clean wood and remove ink stains, as a mordant in dyeing, and in photography. Also known as potassium acid oxalate; sal acetosella; salt of sorrel.

potassium biphthalate $HOOCC_6H_4COOK$ A crystalline compound, soluble in 12 parts of water; used as a buffer in pH determinations and as a primary standard for preparation of volumetric alkali solutions. Also known as acid potassium phthalate; potassium acid phthalate; potassium hydrogen phthalate.

potassium bismuth tartrate A white, granular powder with a sweet taste; soluble in water; used in medicine. Also known as bismuth potassium tartrate.

potassium bisulfate $KHSO_4$ Water-soluble, colorless crystals, melting at 214°C; used in winemaking, fertilizer manufacture, and as a flux and food preservative. Also known as acid potassium sulfate; potassium acid sulfate.

potassium bisulfite $KHSO_3$ White, water-soluble powder with sulfur dioxide aroma; insoluble in alcohol; decomposes when heated; used as an antiseptic and reducing chemical, and in analytical chemistry, tanning, and bleaching. Also known as potassium acid sulfite.

potassium bitartrate $KHC_4H_4O_6$ White, water-soluble crystals or powder; used in baking powder, for medicine, and as an acid and buffer in foods. Also known as cream of tartar; potassium acid tartrate.

potassium borohydride KBH_4 A white, crystalline powder, soluble in water, alcohol, and ammonia; used as a hydrogen source and a reducing agent for aldehydes and ketones.

potassium bromate $KBrO_3$ Water-soluble, white crystals, melting at 434°C; insoluble in alcohol; strong oxidizer and a fire hazard; used in analytical chemistry and as an additive for permanent-wave compounds.

potassium bromide KBr White, hygroscopic crystals with bitter taste; soluble in water and glycerin, slightly soluble in alcohol and ether; melts at 730°C; used in medicine, soaps, photography, and lithography.

potassium bromide–disk technique Method of preparing an infrared spectrometry sample by grinding it and mixing it with a dry powdered alkali halide (such as KBr), then compressing the mixture into a tablet or pellet. Also known as pellet technique; pressed-disk technique.

potassium cadmium iodide *See* potassium tetraiodocadmate.

potassium carbonate K_2CO_3 White, water-soluble, deliquescent powder, melting at 891°C; insoluble in alcohol; used in brewing, ceramics, explosives, fertilizers, and as a chemical intermediate. Also known as potash; salt of tartar.

potassium chlorate $KClO_3$ Transparent, colorless crystals or a white powder with a melting point of 356°C; soluble in water, alcohol, and alkalies; used as an oxidizing agent, for explosives and matches, and in textile printing and paper manufacture.

potassium chloride KCl Colorless crystals with saline taste; soluble in water, insoluble in alcohol; melts at 776°C; used as a fertilizer and in photography and pharmaceutical preparations. Also known as potassium muriate.

potassium chloroaurate *See* potassium gold chloride.

potassium chloroplatinate K_2PtCl_6 Orange-yellow crystals or powder which decomposes when heated (250°C); used in photography. Also known as platinum potassium chloride; potassium platinichloride.

potassium chromate K_2CrO_4 Yellow crystals, melting at 971°C; soluble in water, insoluble in alcohol; used as an analytical reagent and textile mordant, in enamels, inks, and medicines, and as a chemical intermediate.

potassium chromium sulfate *See* chrome alum.

potassium citrate $K_3C_6H_5O_7 \cdot H_2O$ Odorless crystals with saline taste; soluble in water and glycerol, deliquescent and insoluble in alcohol; decomposes about 230°C; used in medicine.

potassium cobaltinitrite *See* cobalt potassium nitrite.

potassium cyanate KOCN Colorless, water-soluble crystals; used as an herbicide and for the manufacture of drugs and organic chemicals.

potassium cyanide KCN Poisonous, white, deliquescent crystals with bitter almond taste; soluble in water, alcohol, and glycerol; used for metal extraction, for electroplating, for heat-treating steel, and as an analytical reagent and insecticide.

potassium cyanoargentate *See* silver potassium cyanide.

potassium cyanoaurite *See* potassium gold cyanide.

potassium dichloroisocyanurate White, crystalline powder or granules; strong oxidant used in dry household bleaches, detergents, and scouring powders.

potassium dichromate $K_2Cr_2O_7$ Poisonous, yellowish-red crystals with metallic taste; soluble in water, insoluble in alcohol; melts at 396°C, decomposes at 500°C; used as an oxidizing agent and analytical reagent, and in explosives, matches, and electroplating. Also known as potassium bichromate; red potassium chromate.

potassium dihydrogen phosphate *See* potassium phosphate.

potassium diphosphate *See* potassium phosphate.

potassium ethyldithiocarbonate *See* potassium xanthate.

potassium ethylxanthogenate *See* potassium xanthate.

potassium ferric oxalate $K_3Fe(C_2O_4)_3 \cdot 3H_2O$ Green crystals decomposing at 230°C, soluble in water and acetic acid; used in photography and blueprinting.

potassium ferricyanide $K_3Fe(CN)_6$ Poisonous, water-soluble, bright-red crystals; decomposes when heated; used in calico printing and wool dyeing. Also known as red potassium prussiate; red prussiate of potash.

potassium ferrocyanide $K_4Fe(CN)_6 \cdot 3H_2O$ Yellow crystals with saline taste; soluble in water, insoluble in alcohol; loses water at 60°C; used in medicine, dry colors, explosives, and as an analytical reagent. Also known as yellow prussiate of potash.

potassium fluoborate KBF_4 White powder or gelatinous crystals that decompose at high temperatures; slightly soluble in water and hot alcohol; used as a sand agent to cast magnesium and aluminum, and in electrochemical processes.

potassium fluoride KF or $KF \cdot 2H_2O$ Poisonous, white, deliquescent crystals with saline taste; soluble in water and hydrofluoric acid, insoluble in alcohol; melts at 846°C; used to etch glass and as a preservative and insecticide.

potassium fluosilicate K_2SiF_6 An odorless, white crystalline compound; slightly soluble in water; used in vitreous frits, synthetic mica, metallurgy, and ceramics. Also known as potassium silicofluoride.

potassium gluconate $KC_6H_{11}O_7$ An odorless, white crystalline compound with salty taste; soluble in water, insoluble in alcohol and benzene; used in medicine.

potassium glutamate $KOOC(CH_2)_2CH(NH_2)COOH \cdot H_2O$ White, hygroscopic, water-soluble powder; used as a flavor enhancer and salt substitute. Also known as monopotassium L-glutamate.

potassium glycerinophosphate *See* potassium glycerophosphate.

potassium glycerophosphate $K_2C_3H_5O_2 \cdot H_2PO_4 \cdot 3H_2O$ Pale yellow, syrupy liquid, soluble in alcohol; used in medicine and as a dietary supplement. Also known as potassium glycerinophosphate.

potassium gold chloride $KAuCl_4 \cdot 2H_2O$ Yellow crystals, soluble in water, ether, and alcohol; used in photography and medicine. Also known as gold potassium chloride; potassium aurichloride; potassium chloroaurate.

potassium gold cyanide $KAu(CN)_2$ A white, water-soluble, crystalline powder; used in medicine and for gold plating. Also known as gold potassium cyanide; potassium cyanoaurite.

potassium hexafluoroarsenate *See* hexaflurate.

potassium hydrate *See* potassium hydroxide.

potassium hydrogen phosphate *See* potassium phosphate.

potassium hydrogen phthalate *See* potassium biphthalate.

potassium hydroxide KOH Toxic, corrosive, water-soluble, white solid, melting at 360°C; used to make soap and matches, and as an analytical reagent and chemical intermediate. Also known as caustic potash; potassium hydrate.

potassium N-hydroxy-methyl-N-methyldithiocarbamate $C_3H_6OS_2K$ A fungicide, nematicide, and bactericide used on foliage and soil.

potassium hyperchlorate *See* potassium perchlorate.

potassium hypophosphite KH_2PO_2 White, opaque crystals or powder, soluble in water and alcohol; used in medicine.

potassium iodate KIO_3 Odorless, white crystals; soluble in water, insoluble in alcohol; melts at 560°C; used as an analytical reagent and in medicine.

potassium iodide KI Water- and alcohol-soluble, white crystals with saline taste; melts at 686°C; used in medicine and photography, and as an analytical reagent.

potassium linoleate $C_{17}H_{31}COOK$ Light-tan, water-soluble paste; used as an emulsifying agent.

potassium manganate K_2MnO_4 Water-soluble dark-green crystals, decomposing at 190°C; used as an analytical reagent, bleach, oxidizing agent, disinfectant, mordant for dyeing wool and in photography, printing, and water purification.

potassium metabisulfite $K_2S_2O_5$ White granules or powder, decomposing at 150–190°C; used as an antiseptic, for winemaking, food preservation, and process engraving, and as a source for sulfurous acid. Also known as potassium pyrosulfite.

potassium metarsenite *See* potassium arsenite.

potassium monophosphate *See* potassium phosphate.

potassium muriate *See* potassium chloride.

potassium nitrate KNO_3 Flammable, water-soluble, white crystals with saline taste; melts at 337°C; used in pyrotechnics, explosives, and matches, as a fertilizer, and as an analytical reagent. Also known as niter.

potassium nitrite KNO_2 White, deliquescent prisms, melting at 297–450°C; soluble in water, insoluble in alcohol; strong oxidizer, exploding at over 550°C; used as an analytical reagent, in medicine, organic synthesis, pyrotechnics, and explosives.

potassium oxalate $K_2C_2O_4 \cdot H_2O$ Odorless, efflorescent, water-soluble, colorless crystals; decomposes when heated; used in analytical chemistry and photography and as a bleach and oxalic acid source.

potassium oxide K_2O Gray, water-soluble crystals; melts at red heat; forms potassium hydroxide in water.

potassium penicillin G₁ See benzyl penicillin potassium.

potassium percarbonate $K_2C_2O_6 \cdot H_2O$ White, granular, water-soluble mass with a melting point of 200–300°C; used in microscopy, photography, and textile printing.

potassium perchlorate $KClO_4$ Explosive, oxidative, colorless crystals; soluble in water, insoluble in alcohol; decomposes at 400°C; used in explosives, medicine, pyrotechnics, analysis, and as a reagent and oxidizing agent. Also known as potassium hyperchlorate.

potassium permanganate $KMnO_4$ Highly oxidative, water-soluble, purple crystals with sweet taste; decomposes at 240°C; and explodes in contact with oxidizable materials; used as a disinfectant and analytical reagent, in dyes, bleaches, and medicines, and as a chemical intermediate. Also known as purple salt.

potassium peroxide K_2O_2 Yellow mass with a melting point of 490°C; decomposes with oxygen evolution in water; used as an oxidizing and bleaching agent.

potassium peroxydisulfate See potassium persulfate.

potassium persulfate $K_2S_2O_8$ White, water-soluble crystals, decomposing below 100°C; used for bleaching and textile desizing, as an oxidizing agent and antiseptic, and in the manufacture of soap and pharmaceuticals. Also known as potassium peroxydisulfate.

potassium phosphate Any one of three orthophosphates of potassium. The monobasic form, KH_2PO_4, consists of colorless, water-soluble crystals melting at 253°C; used in sonar transducers, optical modulation, medicine, baking powders, and nutrient solutions; also known as potassium acid phosphate, potassium dihydrogen phosphate (KDP), potassium diphosphate, potassium orthophosphate. The dibasic form, K_2HOP_4, consists of white, water-soluble crystals; used in medicine, fermentation, and nutrient solutions; also known as potassium hydrogen phosphate, potassium monophosphate. The tribasic form, K_3PO_4, is a water-soluble, hygroscopic white powder, melting at 1340°C; used to purify gasoline, to soften water, and to make liquid soaps and fertilizers; also known as neutral potassium phosphate, tripotassium orthophosphate.

potassium platinichloride See potassium chloroplatinate.

potassium polymannuronate See potassium alginate.

potassium polymetaphosphate $(KPO_3)_n$ White powder with a molecular weight up to 500,000; used in foods as a fat emulsifier and moisture-retaining agent.

potassium pyrophosphate $K_4P_2O_7 \cdot 3H_2O$ Water-soluble, colorless crystals; dehydrates below 300°C, melts at 1090°C; used in tin plating, china-clay purification, dyeing, oil-drilling muds, and synthetic rubber production. Also known as normal potassium pyrophosphate; tetrapotassium pyrophosphate.

potassium pyrosulfite *See* potassium metabisulfite.

potassium rhodanide *See* potassium thiocyanate.

potassium silicate $SiO_2=K_2O$ A compound existing in two forms, solution and solid (glass); as a solution, it is colorless to turgid in water, and is used in paints and coatings, as an arc-electrode binder and catalyst and in detergents; as a solid, it is colorless and water-soluble solid, and is used in glass manufacture and for dyeing and bleaching.

potassium silicofluoride *See* potassium fluosilicate.

potassium sodium ferricyanide $K_2NaFe(CN)_6$ Red, water-soluble crystals; used for blueprint paper and in photography.

potassium sodium tartrate $KNaC_4H_4O_6 \cdot 4H_2O$ Colorless, water-soluble, efflorescent crystals or white powder with a melting point of 70–80°C; used in medicine and as a buffer and sequestrant in foods. Also known as Rochelle salt; sodium potassium tartrate.

potassium sorbate $C_6H_7KO_2$ A crystalline compound, more soluble in water than in alcohol; decomposes above 270°C; used to inhibit mold and yeast growth in food. Also known as sorbic acid potassium salt.

potassium stannate $K_2SnO_3 \cdot 3H_2O$ White crystals; soluble in water, insoluble in alcohol; used in textile printing and dyeing, and in tin-plating baths.

potassium stibnate *See* potassium antimonate.

potassium sulfate K_2SO_4 Colorless crystals with bitter taste; soluble in water, insoluble in alcohol; melts at 1072°C; used as an analytical reagent, medicine, and fertilizer, and in aluminum and glass manufacture. Also known as salt of Lemery.

potassium sulfide K_2S Moderately flammable, water-soluble, deliquescent red crystals; melts at 840°C; used in analytical chemistry, medicine, and depilatories. Also known as fused potassium sulfide; hepar sulfuris; potassium sulfuret.

potassium sulfite $K_2SO_3 \cdot 2H_2O$ Water-soluble, white crystals; used in medicine and photography.

potassium sulfocyanate *See* potassium thiocyanate.

potassium sulfocyanide *See* potassium thiocyanate.

potassium sulfuret *See* potassium sulfide.

potassium tetraiodocadmate $K_2(CdI_4) \cdot 2H_2O$ A crystalline compound; used in analytical chemistry for alkaloids, amines, and other compounds. Also known as cadmium potassium iodide; potassium cadmium iodide.

potassium thiocyanate KCNS Water- and alcohol-soluble, colorless, odorless hygroscopic crystals with saline taste; decomposes at 500°C; used as an analytical reagent and in freezing mixtures, chemicals manufacture, textile printing and dyeing, and photographic chemicals. Also known as potassium rhodanide; potassium sulfocyanate; potassium sulfocyanide.

potassium undecylenate $CH_2:CH(CH_2)_8COOK$ A white, water-soluble powder, decomposing at about 250°C; used in pharmaceuticals and cosmetics as a fungistat and bacteriostat.

potassium xanthate　KC_2H_5OCSS　Water- and alcohol-soluble, yellow crystals; used as an analytical reagent and soil-treatment fungicide. Also known as potassium ethyldithiocarbonate; potassium ethylxanthogenate; potassium xanthogenate.

potassium xanthogenate　*See* potassium xanthate.

potentiometric cell　Container for the two electrodes and the electrolytic solution being titrated potentiometrically.

potentiometric titration　Solution titration in which the end point is read from the electrode-potential variations with the concentrations of potential-determining ions, following the Nernst concept. Also known as constant-current titration.

powder diffraction camera　A metal cylinder having a window through which an x-ray beam of known wavelength is sent by an x-ray tube to strike a finely ground powder sample mounted in the center of the cylinder; crystal planes in this powder sample diffract the x-ray beam at different angles to expose a photographic film that lines the inside of the cylinder; used to study crystal structure. Also known as x-ray powder diffractometer.

powder pattern　In the powder method of x-ray diffraction analysis, the display of lines made on film by the Debye-Scherrer method or on paper by a recording diffractometer.

Powerforming　A fixed-bed petroleum-refinery process to catalytically reform petroleum naphtha to produce high-octane motor gasoline.

p-process　The synthesis of certain nuclides in stars through capture of protons or ejection of neutrons by gamma rays.

Pr　*See* praseodymium.

praseodymium　A chemical element, symbol Pr, atomic number 59, atomic weight 140.91; a metallic element of the rare-earth group.

precession camera　An x-ray diffraction camera used in the Buerger precession method for recording the diffractions of an individual crystal.

precipitant　A chemical or chemicals that cause a precipitate to form when added to a solution.

precipitate　1. A substance separating, in solid particles, from a liquid as the result of a chemical or physical change. 2. To form a precipitate.

precipitation　The process of producing a separable solid phase within a liquid medium; represents the formation of a new condensed phase, such as a vapor or gas condensing to liquid droplets; a new solid phase gradually precipitates within a solid alloy as a result of slow, inner chemical reaction; in analytical chemistry, precipitation is used to separate a solid phase in an aqueous solution.

precipitation number　The number of milliliters of asphaltic precipitate formed when 10 milliliters of petroleum-lubricating oil is mixed with 90 milliliters of a special-quality petroleum naphtha, then centrifuged according to ASTM test conditions; used to determine the quantity of asphalt in petroleum-lubricating oil.

precipitation titration　Amperometric titration in which the potential of a suitable indicator electrode is measured during the titration.

predissociation The dissociation of a molecule that has absorbed energy before it can lose energy by radiation.

prednisolone $C_{21}H_{28}O_5$ A glucocorticoid that is a dehydrogenated analog of hydrocortisone.

Pregl procedure Microanalysis technique in which the sample is decomposed thermally, with subsequent oxidation of decomposition products.

pregnanediol $C_{21}H_{36}O_2$ A metabolite of progesterone, present in urine during the progestational phase of the menstrual cycle and also during pregnancy.

pregnenolone $C_{21}H_{32}O_2$ A steroid ketone that is formed as an oxidation product of cholesterol, stigmasterol, and certain other steroids.

preparing salt *See* sodium stannate.

prepolymer A plastic or resin intermediate whose molecular weight is between that of the original monomer or monomers and that of the final, cured polymer or resin.

pressed-disk technique *See* potassium bromide–disk technique.

pressure-base factor Factor used in orifice pressure-drop calculations to allow for conditions where the pressure base used for calculating the orifice factor is not 14.73 psia (101.56 megapascals); calculated as $F_{pb} = 14.73$/pressure base (absolute).

pressure broadening A spreading of spectral lines when pressure is increased, due to an increase in collision broadening.

pressure effect The effect of changes in pressure on spectral lines in the radiation emitted or absorbed by a substance; namely, pressure broadening and pressure shift.

pressure process Treatment of timber to prevent decay by forcing a preservative such as creosote and zinc chloride into the cells of the wood.

pressure shift An increase in the wavelength at which a spectral line has maximum intensity, which takes place when pressure is increased.

pressure still A continuous-flow, petroleum-refinery still in which heated oil (liquid and vapor) is kept under pressure so that it will crack (decompose into smaller molecules) to produce lower-boiling products (pressure distillate or pressure naphtha).

pressure tank A pressurized tank into which timber is inserted for impregnation with preservative.

pressure treater Any chemical treating device operated at higher-than-atmospheric pressure, as in the chemical and petroleum industries.

prill To form pellet-sized crystals or agglomerates of material by the action of upward-blowing air on falling hot solution; used in the manufacture of ammonium nitrate and urea fertilizers.

primary A term used to distinguish basic compounds from similar or isomeric forms; in organic compounds, for example, RCH_2OH is a primary alcohol, R_1R_2CHOH is a secondary alcohol, and $R_1R_2R_3COH$ is a tertiary alcohol; in inorganic compounds, for example, NaH_2PO_4 is primary sodium phosphate, Na_2HPO_4 is the secondary form, and Na_3PO_4 is the tertiary form.

primary alcohol An alcohol whose molecular structure may be written as RCH_2OH, rather than as R_1R_2CHOH (secondary) or $R_1R_2R_3COH$ (tertiary).

primary amine   An amine whose molecular structure may be written as RNH_2, instead of R_1R_2NH (secondary) or $R_1R_2R_3N$ (tertiary).

primary carbon atom   A carbon atom in a molecule that is singly bonded to only one other carbon atom.

primary cosmic rays   *See* cosmic rays.

primary hydrogen atom   A hydrogen atom that is bonded to a primary carbon atom.

primary isoamyl alcohol   *See* isobutyl carbinol.

primitive cell   A parallelepiped whose edges are defined by the primitive translations of a crystal lattice; it is a unit cell of minimum volume.

primitive lattice   A crystal lattice in which there are lattice points only at its corners. Also known as simple lattice.

primitive translation   For a space lattice, one of three translations which can be repeatedly applied to generate any translation which leaves the lattice unchanged.

principal axis   The longest axis in a crystal.

principal line   That spectral line which is most easily excited or observed.

principal moments   The three moments of inertia of a rigid molecule calculated with respect to the principal axes.

principal quantum number   A quantum number for orbital electrons, which, together with the orbital angular momentum and spin quantum numbers, labels the electron wave function; the energy level and the average distance of an electron from the nucleus depend mainly upon this quantum number.

principal series   A series occurring in the line spectra of many atoms and ions with one, two, or three electrons in the outer shell, in which the total orbital angular momentum quantum number changes from 1 to 0.

priscol   *See* tolazoline hydrochloride.

prism   A crystal which has three, four, six, eight, or twelve faces, with the face intersection edges parallel, and which is open only at the two ends of the axis parallel to the intersection edges.

prismatic cleavage   A type of crystal cleavage that occurs parallel to the faces of a prism.

prism spectrograph   Analysis device in which a prism is used to give two different but simultaneous light wavelengths derived from a common light source; used for the analysis of materials by flame photometry.

pristane $C_{19}H_{40}$ A liquid soluble in such organic solvents as ether, petroleum ether, benzene, chloroform, and carbon tetrachloride; used as a lubricant, as an oil in transformers, and as an anticorrosion agent. Also known as norphytane.

proaccelerin   A labile procoagulant in normal plasma but deficient in the blood of patients with parahemophilia; essential for rapid conversion of prothrombin to thrombin. Also known as factor V; labile factor.

procaine   *See* procaine base.

procaine base $C_6H_4NH_2COOCH_2CH_2N(C_2H_5)_2$ Water-insoluble, light-sensitive, odorless, white powder, melting at 60°C; soluble in alcohol, ether, chloroform, and benzene; used in medicine as a local anesthetic. Also known as *para*-aminobenzoyldiethylaminoethanol base; planocaine base; procaine.

procaine penicillin G $C_{29}H_{38}N_4O_6S{\cdot}H_2O$ White crystals or powder, fairly soluble in chloroform; used as an antibiotic in animal feed.

process analyzer   An instrument for determining the chemical composition of the substances involved in a chemical process directly, or for measuring the physical parameters indicative of composition.

process furnace   Furnace used to heat process-stream materials (liquids, gases, or solids) in a chemical-plant operation; types are direct-fired, indirect-fired, and pebble heaters.

process heater   Equipment for the heating of chemical process streams (gases, liquids, or solids); usually refers to furnaces, in contrast to heat exchangers.

process monitoring   The observation of chemical process variables by means of pressure, temperature, flow, and other types of indicators; usually occurs in a central control room.

process variable   Any of those varying operational and physical conditions associated with a chemical processing operation, such as temperature, pressure, flowrate, density, pH, viscosity, or chemical composition.

prochirality   The property displayed by a molecule or atom which contains (or is bonded to) two constitutionally identical ligands. Also known as prostereoisomerism.

procoagulant   Any of blood clotting factors V to VIII; accelerates the conversion of prothrombin to thrombin in the presence of thromboplastin and calcium.

procollagen   A high-molecular-weight form of collagen that is found in intracellular spaces and is believed to be the precursor of collagen.

product   *See* discharge liquor.

product water   Fresh water that is produced by a desalination process. Also known as converted water.

proenzyme   *See* zymogen.

profibrinolysin   *See* plasminogen.

proflavine   *See* proflavine sulfate.

proflavine sulfate $C_{13}H_{11}N_3{\cdot}H_2SO_4$ A reddish-brown, crystalline powder, soluble in alcohol and water; used in medicine. Also known as 6-diaminoacridinium hydrogen sulfate; proflavine.

profluralin $C_{14}H_{16}F_3N_3O_4$ An orange to yellow, crystalline compound; melting point is 33–36°C; used to control grasses and weeds in cotton, soybeans, turf, and ornamentals. Also known as N-(cyclopropylmethy)-α,α,α-trifluoro-2,6-dinitro-N-propyl-*para*-toluidine.

progestational hormone   **1.** The natural hormone progesterone, which induces progestational changes of the uterine mucosa. **2.** Any derivative or modification or progesterone having similar actions.

progesterone $C_{21}H_{30}O_2$ A steroid hormone produced in the corpus luteum, placenta, testes, and adrenals; plays an important physiological role in the luteal phase of the menstrual cycle and in the maintenance of pregnancy; it is an intermediate in the biosynthesis of androgens, estrogens, and the corticoids.

prolactin A protein hormone produced by the adenohypophysis; stimulates lactation and promotes functional activity of the corpus luteum. Also known as lactogenic hormone; luteotropic hormone; mammary-stimulating hormone; mammogen; mammogenic hormone; mammotropin.

prolamin Any of the simple proteins, such as zein, found in plants; soluble in strong alcohol, insoluble in absolute alcohol and water.

proline $C_5H_9O_2$ A heterocyclic amino acid occurring in essentially all proteins, and as a major constituent in collagen proteins.

promazine hydrochloride $C_{17}H_{20}N_2S \cdot HCl$ A white to slightly yellow, crystalline powder, melting at 172–182°C; used in medicine and as a food additive.

promecarb $C_{12}H_{17}C_2N$ A colorless, crystalline compound with a melting point of 87–88°C; slight solubility in water; used as an insecticide for potatoes and fruits. Also known as *meta*-cym-*s*-yl-methylcarbamate.

promethium A chemical element, symbol Pm, atomic number 61; atomic weight of the most abundant isotope is 147; a member of the rare-earth group of metals.

promethium-147 Artificially produced rare-earth element with atomic number 61 and mass 147; produced during fission of ^{235}U. Also known as florentium; illinium.

prometon $C_{10}H_{19}N_5O$ A white, crystalline solid with a melting point of 91–92°C; used as a postemergence herbicide on noncrop land. Also known as 2,4-bis(isopropylamino)-6-methoxy-*S*-triazine.

prometryn $C_{10}H_{19}N_5S$ A white, crystalline solid with a melting point of 118–120°C; used as an herbicide to control weeds in cotton and celery. Also known as 2,4-bis(isopropylamino)-6-(methylthio)-*S*-triazine.

promoter A chemical which itself is a feeble catalyst, but greatly increases the activity of a given catalyst.

prompt neutron A neutron released coincident with the fission process, as opposed to neutrons subsequently released.

prompt radiation Radiation emitted within a time too short for measurement, including γ-rays, characteristic x-rays, conversion and Auger electrons, prompt neutrons, and annihilation radiation.

pronamide $C_{12}H_{11}Cl_2NO$ An off-white solid with a melting point of 154–156°C; slightly soluble in water; used as a pre- or postemergence herbicide on vegetable crops. Also known as *N*-(1,1-dimethyl-2-propynyl)-3,5-dichlorobenzamide.

propachlor $C_{11}H_{14}ClNO$ A tan solid with a melting point of 67–76°C; used as a preemergence herbicide for treatment of sweet corn, field corn, sorghum, and seed soybeans. Also known as 2-chloro-*N*-isopropylacetanilide.

propadiene *See* allene.

propagation rate The speed at which a flame front progresses through the body of a flammable fuel-oxidizer mixture, such as gas and air.

propane $CH_3CH_2CH_3$ A heavy, colorless, gaseous petroleum hydrocarbon gas of the paraffin series; boils at $-44.5°C$; used as a solvent, refrigerant, and chemical intermediate. Also known as dimethylmethane.

propane deasphalting Petroleum-refinery solvent process using propane to remove and precipitate asphalt from petroleum stocks, such as for lubricating oils.

propane decarbonizing Petroleum-refinery solvent process using propane to recover catalytic-cracking feedstock from heavy-fuel residues; when butane or butane-propane solvent is used, the process is called solvent decarbonizing.

propane dewaxing Petroleum-refinery solvent process using propane to remove waxes from lubricating oils to lower the lube-oil pour point.

1,2-propanediol See propylene glycol.

propane fractionation Continuous, petroleum-refinery solvent process using liquid propane to segregate long-vacuum residue into two or more grades of lube-oil stock (such as heavy neutral stock or bright stock) and asphalt.

1-propanethiol See n-propyl mercaptan.

propanil $C_9H_9Cl_2NO$ A light-colored solid with a melting point of 90.6–91.6°C; slightly soluble in water; used as an herbicide for control of weeds in rice. Also known as 3,4-dichloropropionanilide.

propanoic acid See propionic acid.

propanol See propyl alcohol.

2-propanol See isopropyl alcohol.

2-propanone See acetone.

2-propanone oxime See acetoxime.

propargite $C_{19}H_{26}O_4S$ A light to dark brown, viscous liquid, insoluble in water; used to control many types of mites on almonds, field corn, hops, mint, potatoes, strawberries, and walnuts. Also known as 2-(para-tert-butyl phenoxy)cyclohexyl 2-propynyl sulfite.

propargyl alcohol $HCCCH_2OH$ Colorless, water- and alcohol-soluble liquid, boiling at 114°C; used as a chemical intermediate, stabilizer, and corrosion inhibitor. Also known as 2-propyn-1-ol.

propargyl bromide C_3H_3Cl A flammable liquid with a boiling point range of 56.0–57.1°C; used as a soil fumigant. Also known as 3-chloro-1-propyne.

propargyl chloride C_3H_3Cl A liquid miscible with benzene, carbon tetrachloride, ethanol, and ethylene glycol; used as an intermediate in organic synthesis.

propazine $C_9H_{16}ClN_5$ A colorless solid with a melting point of 212–214°C; used as a preemergence herbicide for control of weeds in milo and sweet sorghum. Also known as 2-chloro-4,6-bis(isopropylamino)-S-triazine.

propellant 23 See fluoroform.

2-propenal See acrolein.

2-propen-1-amine See allylamine.

propene See propylene.

2-propene-1-thiol *See* allyl mercaptan.

2-propen-1-ol *See* allyl alcohol.

1-propen-2-yl-acetate *See* isopropenyl acetate.

para-**propenylanisole** *See* anethole.

α-**propenyldichlorohydrin** *See* α-dichlorohydrin.

propenyl guaethol $C_{11}H_{14}O_2$ A white powder with a vanilla flavor and a melting point of 85–86°C; soluble in fats, essential oils, and edible solvents; used for artificial vanilla flavoring. Also known as 1-ethoxy-2-hydroxy-4-propenyl-benzene.

propham $C_{10}H_{13}NO_2$ A light brown solid with a melting point of 87–88°C; slightly soluble in water; used as a pre- and postemergence herbicide for vegetable crops. Also known as isopropyl-N-phenylcarbamate (IPC).

prophenpyridamine maleate *See* pheniramine maleate.

propine *See* allylene.

propineb *See* zinc-[N,N'-propylene-1,2-bis(dithiocarbamate)].

β-**propiolactone** $C_3H_4O_2$ Water-soluble liquid that decomposes rapidly at boiling point (155°C); miscible with ethanol, acetone, chloroform, and ether; reacts with alcohol; used as a chemical intermediate.

propionaldehyde C_2H_5CHO Flammable, water-soluble, water-white liquid, with suffocating aroma; boils at 48.8°C; used to manufacture acetals, plastics, and rubber chemicals, and as a disinfectant and preservative.

propionate A salt of propionic acid, CH_3CH_2COOH; an example is sodium propionate, CH_3CH_2COONa.

propione *See* pentanone.

propionic acid CH_3CH_2COOH Water- and alcohol-soluble, clear, colorless liquid with pungent aroma; boils at 140.7°C; used to manufacture various propionates, in nickel-electroplating solutions, for perfume esters and artificial flavors, for pharmaceuticals, and as a cellulosics solvent. Also known as methylacetic acid; propanoic acid.

propionic anhydride $(CH_3CH_2CO)_2O$ A colorless liquid with a boiling point of 167–169°C; soluble in ether, alcohol, and chloroform; used as an esterifying agent and for dyestuffs and pharmaceuticals.

propionic ether *See* ethyl propionate.

propionitrile *See* ethyl cyanide.

para-**propionylphenol** *See* para-hydroxypropiophenone.

propyl- The $CH_3CH_2CH_2$— radical, derived from propane; found, for example, in 1-propanol.

n-**propyl acetate** $C_3H_7OOCCH_3$ Colorless liquid with pleasant aroma; miscible with alcohols, ketones, esters, and hydrocarbons; boils at 96–102°C; used for flavors and perfumes, in organic synthesis, and as a solvent.

propylacetone *See* methyl butyl ketone.

propylacetylene *See* pentyne.

propyl alcohol $CH_3CH_2CH_2OH$ A colorless liquid made by oxidation of aliphatic hydrocarbons; boils at 97°C; used as a solvent and chemical intermediate. Also known as ethyl carbinol; propanol.

sec-propyl alcohol *See* isopropyl alcohol.

n-propylamine $C_3H_7NH_2$ Colorless, flammable liquid, boiling at 46–51°C; used as a sedative.

propyl benzene $C_6H_5C_3H_7$ Water-insoluble, colorless liquid, boiling at 158°C. Also known as phenylpropane.

S-propyl butylethylthiocarbamate *See* pebulate.

propyl cyanide *See* butyronitrile.

propylene $CH_3CH{=}CH_2$ Colorless unsaturated hydrocarbon gas, with boiling point of -47°C; used to manufacture plastics and as a chemical intermediate. Also known as methyl ethylene; propene.

S-propyldipropylthiocarbamate *See* vernolate.

propylene aldehyde *See* crotonaldehyde.

propylene carbonate $C_3H_6CO_3$ Odorless, colorless liquid, boiling at 242°C; miscible with acetone, benzene, and ether; used as a solvent, extractant, plasticizer, and chemical intermediate.

propylene chloride *See* propylene dichloride.

propylene dichloride $CH_3CHClCH_2Cl$ Water-insoluble, colorless, moderately flammable liquid, with chloroform aroma; boils at 96.3°C; miscible with most common solvents; used as a solvent, dry-cleaning fluid, metal degreaser, and fumigant. Also known as 1,2-dichloropropane; propylene chloride.

propylene glycol $CH_3CHOHCH_2OH$ A viscous, colorless liquid, miscible with water, alcohol, and many solvents; boils at 188°C; used as a chemical intermediate, antifreeze, solvent, lubricant, plasticizer, and bactericide. Also known as 1,2-dihydroxypropane; methyl ethylene glycol; methyl glycol; 1,2-propanediol.

propylene glycol alginate $C_9H_{14}O_7$ A white, water-soluble powder; used as a stabilizer, thickener, and emulsifier. Also known as hydroxypropyl alginate.

propylene glycol methyl ether *See* propylene glycol monoethyl ether.

propylene glycol monomethyl ether $C_4H_{10}O_2$ A colorless liquid with a boiling point of 120.1°C; soluble in water, methanol, and ether; used as a solvent for cellulose, dyes, and inks. Also known as propylene glycol methyl ether.

propylene glycol monoricinoleate $C_{21}H_{30}O_4$ A pale yellow, moderately viscous oily liquid, soluble in organic solvents; used as a plasticizer and lubricant and in dye solvents and cosmetics.

propyleneimine C_3H_7N A clear, colorless liquid with a boiling point of 66–67°C; soluble in water and organic solvents; used as an intermediate in organic synthesis. Also known as 2-methylaziridine.

propylene oxide C_3H_6O Colorless, flammable liquid, with etherlike aroma; soluble in water, alcohol, and ether; boils at 33.9°C; used as a solvent and fumigant, in lacquers, coatings, and plastics, and as a petrochemical intermediate.

propylene tetramer   *See* dodecane.

propyl formate $C_4H_8O_2$ A flammable liquid with a boiling point of 81.3°C; used for flavoring.

propylformic acid   *See* butyric acid.

***n*-propyl furoate** $C_8H_{10}O_3$ A colorless, fragrant liquid with a boiling point of 210.9°C; soluble in alcohol and ether; used for flavoring.

propyl gallate $C_3H_7OOCC_6H_2(OH)_3$ Colorless crystals with a melting point of 150°C; used to prevent or retard rancidity in edible fats and oils.

propyl *para*-hydroxybenzoate   *See* propylparaben.

propyliodone $C_{10}H_{11}O_3NI_2$ A white, crystalline powder with a melting point of 187–190°C; soluble in alcohol, acetone, and ether; used in medicine as a radiopaque medium.

***n*-propyl mercaptan** C_3H_7SH A liquid with an offensive odor and a boiling range of 67–73°C; used as an herbicide. Also known as 1-propanethiol.

***N*-propyl nitrate** $C_3H_7NO_3$ A white to straw-colored liquid with a boiling range of 104–127°C; used as a monopropellant rocket fuel.

propylparaben $C_{10}H_{12}O_3$ Colorless crystals or white powder with a melting point of 95–98°C; soluble in acetone, ether, and alcohol; used in medicine and as a food preservative and fungicide. Also known as propyl *para*-hydroxybenzoate.

1-propylphosphonic acid $C_3H_9O_3P$ A white solid with a melting point of 68–69°C; soluble in water; used as a growth regulator for herbaceous and woody species.

propylpiperidine   *See* coniine.

propylthiopyrophosphate $C_{12}H_{28}P_2S_2O$ A straw-colored to dark amber liquid with a boiling point of 148°C; used as an insecticide for chinch bugs in lawns and turf.

propyne   *See* allylene.

2-propyn-1-ol   *See* propargyl alcohol.

prorennin   *See* renninogen.

prostaglandin   Any of various physiologically active compounds containing 20 carbon atoms and formed from essential fatty acids; found in highest concentrations in normal human semen; activities affect the nervous system, circulation, female reproductive organs, and metabolism.

prostereoisomerism   *See* prochirality.

prosthetic group   A characteristic nonamino acid substance that is strongly bound to a protein and necessary for the protein portion of an enzyme to function; often used to describe the function, as in hemeprotein for hemoglobin.

protactinium   A chemical element, symbol Pa, atomic number 91; the third member of the actinide group of elements; all the isotopes are radioactive; the longest-lived isotope is protactinium-231.

protamine   Any of the simple proteins that are combined with nucleic acid in the sperm of certain fish, and that upon hydrolysis yield basic amino acids;

used in medicine to control hemorrhage, and in the preparation of an insulin form to control diabetes.

protease An enzyme that digests proteins.

protein Any of a class of high-molecular-weight polymer compounds composed of a variety of α-amino acids joined by peptide linkages.

proteinase A type of protease which acts directly on native proteins in the first step of their conversion to simpler substances.

protein-bound iodine Iodine bound to blood protein. Abbreviated PBI.

proteinometer *See* hand sugar refractometer.

proteoglycan A high-molecular-weight polyanionic substance covalently linked by numerous heteropolysaccharide side chains to a polypeptide chain backbone.

proteolysin A lysin that produces proteolysis.

proteolysis Fragmentation of a protein molecule by addition of water to the peptide bonds.

proteolytic enzyme Any enzyme that catalyzes the breakdown of protein.

proteose One of a group of derived proteins intermediate between native proteins and peptones; soluble in water, not coagulable by heat, but precipitated by saturation with ammonium or zinc sulfate.

prothiocarb $C_8H_{18}N_2O \cdot HCl$ A colorless, crystalline compound with a melting point of 120–121°C; soluble in water; used as a fungicide for seed dressing. Also known as ethyl-N-(3-dimethylamino-propyl)-thiolcarbamate hydrochloride.

prothoate $C_9H_{20}NO_3PS_2$ A white, crystalline compound with a melting point of 23–24°C; used as an insecticide and miticide on fruit trees. Also known as O,O-diethyl-S-(N-isopropylcarbamoylmethyl)phosphorodithioate.

prothrombin An inactive plasma protein precursor of thrombin. Also known as factor II; thrombinogen.

prothrombin factor *See* vitamin K.

proticity In oxidative phosphorylation, the flowing of protons in the proton circuit from high to low protic potential.

protium The lightest hydrogen isotope, having a mass number of 1 and consisting of a single proton and electron. Also known as light hydrogen.

protogenic Strongly acidic.

protomer One of the polypeptide chains composing an oligomeric protein. Also known as subunit.

proton acid *See* Brönsted acid.

protonate To add protons to a base by a proton source.

proton capture A nuclear reaction in which a proton combines with a nucleus.

proton-electron-proton reaction A nuclear reaction in which two protons and an electron react to form a deuteron and a neutrino; it is an important source of detectable neutrinos from the sun. Abbreviated PeP reaction.

protonic acid *See* Brönsted acid.

proton-induced x-ray emission A method of elemental analysis in which the energy of the characteristic x-rays emitted when a sample is bombarded with a beam of energetic protons is used to identify the elements present in the sample. Abbreviated PIXE.

protonium A bound state of a proton and an antiproton.

proton moment The magnetic dipole moment of the proton, a physical constant equal to $(1.41062 \pm 0.00001) \times 10^{-123}$ erg per gauss.

proton-proton chain An energy-releasing nuclear reaction chain which is believed to be of major importance in energy production in hydrogen-rich stars. Also known as deuterium cycle.

proton resonance A phenomenon in which protons absorb energy from an alternating magnetic field at certain characteristic frequencies when they are also subjected to a static magnetic field; this phenomenon is used in nuclear magnetic resonance quantitative analysis technique.

proton-stability constant The reciprocal of the dissociation constant of a weak base in solution.

protophilic Strongly basic.

prototropy A reversible interconversion of structural isomers that involves the transfer of a proton.

protropic Pertaining to chemical reactions that are influenced by protons.

protrypsin *See* trypsinogen.

Prout's hypothesis The hypothesis that all atoms are built up from hydrogen atoms.

provitamin A vitamin precursor; assumes vitamin activity upon activation or chemical change.

proximate analysis A technique that separates and identifies categories of compounds in a mixture; reported are moisture and ash content, the extracts of the mixture made with alcohol, petroleum ether, water, hydrochloric acid and resins, starches, reducing sugars, proteins, fats, esters, free acids, and so on; this type of analysis of solid fuels allows a prediction to be made as to how the fuel will behave in a furnace.

Prussian blue $Fe_4[Fe(CN)_6]_3$ Ferric ferrocyanide, used as a blue pigment and in the removal of hydrogen sulfide from gases.

prussic acid *See* hydrocyanic acid.

prynachlor $C_{12}H_{12}ClN$ A white, crystalline compound with a melting point of 40–47°C; used as a preemergence herbicide on sorghum, soybeans, corn, and onions. Also known as 2-chloro-*N*-(1-methyl-2-propynyl)acetanilide.

pryrrolidine C_4H_9N A colorless to pale yellow liquid with a boiling point of 87°C; soluble in water and alcohol; used in the manufacture of pharmaceuticals, insecticides, and fungicides.

pseudocritical properties Effective (empirical) values for the critical properties (such as temperature, pressure, and volume) of a multicomponent chemical system.

pseudocrystal A substance that appears to be crystalline but does not have a true crystalline diffraction pattern.

pseudocumene C_9H_{12} Water-insoluble, hydrocarbon liquid, boiling at 168°C; soluble in alcohol, benzene, and ether; used to manufacture perfumes and dyes, and as a catgut sterilant. Also known as pseudocumol; uns-trimethyl-benzene.

pseudocumol *See* pseudocumene.

pseudoionone $C_{13}H_{20}O$ A pale yellow liquid with a boiling point of 143–145°C; soluble in alcohol and ether; used for perfumes and cosmetics.

pseudoreduced compressibility The compressibility factor for a multicomponent gaseous system, calculated at reduced conditions using the pseudoreduced properties of the mixture.

pseudoreduced properties Reduced-state relationships (such as reduced pressure, reduced temperature, and reduced volume) calculated for multicomponent chemical systems by using pseudocritical properties.

pseudosymmetry Apparent symmetry of a crystal, resembling that of another system; generally due to twinning.

P shell The sixth layer of electrons about the nucleus of an atom, having electrons whose principal quantum number is 6.

Pt *See* platinum.

PTA *See* factor XI; phosphotungstic acid.

pteroic acid $C_{14}H_{12}N_6O_3$ A crystalline amino acid formed by hydrolysis of folic acid or other pteroylglutamic acids.

pteroylglutamic acid *See* folic acid.

ptyalase *See* ptyalin.

ptyalin A diastatic enzyme found in saliva which catalyzes the hydrolysis of starch to dextrin, maltose, and glucose, and the hydrolysis of sucrose to glucose and fructose. Also known as ptyalase; salivary amylase; salivary diastase.

Pu *See* plutonium.

puberulic acid $(HO)_3(C_7H_2O)COOH$ A keto acid formed as a metabolic product of certain species of *Penicillium;* has some germicidal activity against gram-positive bacteria.

puking In a distillation column, the foaming and rising of liquid so that part of it is driven out of the vessel through the vapor line.

pulse column Continuous-phase process column (such as liquid only or gas only) in which the flow-through is pulsating; used to increase mass-transfer rates, as in a liquid-liquid extraction operation.

pulsed-bed sorption Solid-liquid countercurrent adsorption process (such as an ion-exchange process) in which the granulated solids bed and the solution flow alternately, in opposite directions.

pulser Device used to create a pulsating fluid flow through a process vessel, such as a liquid-liquid or vapor-liquid extraction tower; used to increase contact and mass transfer rates.

pulse radiolysis   A method of studying fast chemical reactions in which a sample is subjected to a pulse of ionizing radiation, and the products formed by the resulting reactions are studied spectroscopically.

pumparound   A system or process vessel that moves liquid out of and back into the vessel at a new location; for example, in a bubble tower, the withdrawing of liquid from a plate or tray, followed by cooling, and returning to another plate to induce condensation of vapors.

purine   A heterocyclic compound containing fused pyrimidine and imidazole rings; adenine and guanine are the purine components of nucleic acids and coenzymes.

purity   The state of a chemical compound when no impurity can be detected by any experimental method; absolute purity is never reached in practice.

purple of Cassius   *See* gold tin purple.

purple salt   *See* potassium permanganate.

purpurin $C_{14}H_8O_5$ A compound crystallizing as long orange needles from dilute alcohol solutions; used in the manufacture of dyes, and as a reagent for the detection of boron. Also known as natural red; 1,2,4-trihydroxyanthraquinone.

purpurin red   *See* anthrapurpurin.

purpurogallin $C_{11}H_8O_5$ A red, crystalline compound, the aglycone of several glycosides from nutgalls; decomposes at 274–275°C; soluble in boiling alcohol, methanol, and acetone; used as an antioxidant or to retard metal contamination in hydrocarbon fuels or lubricants.

putrefaction   Decomposition of organic matter, particularly the anaerobic breakdown of proteins by bacteria, with the production of foul-smelling compounds.

PVA   *See* polyvinyl acetate; polyvinyl alcohol.

PVAc   *See* polyvinyl acetate.

PVC   *See* polyvinyl chloride.

PVDC   *See* polyvinyl dichloride.

PVI   *See* polyvinyl isobutyl ether.

PVM   *See* polyvinyl methyl ether.

PVP   *See* polyvinyl pyrrolidone.

pyracarbolid $C_{13}H_{15}NO_2$ A gray solid with a melting point of 106–107°C; used as a fungicide to treat seed and foliage to control diseases of cereals, ornamentals, coffee, tea, vegetables, and cotton. Also known as 2-methyl-5,6-dihydro-4*H*-pyran-3-carboxylic acid anilide.

pyracetic acid   *See* pyroligneous acid.

pyramid   An open crystal having three, four, six, eight, or twelve nonparallel faces that meet at a point.

pyramidal cleavage   A type of crystal cleavage that occurs parallel to the faces of a pyramid.

pyrazine carboxylamide   *See* pyrazinamide.

pyrazinoic acid amide   *See* pyrazinamide.

pyrazolone dye An acid dye containing both —N=N— and =C=C= chromophore groups, such as tartrazine; used for silk and wool.

pyrazon $C_{10}H_8ClN_3O$ A tan solid with a melting point of 185–195°C; used as pre- and postemergence herbicide for weed control in sugarbeets, red beets, and fodder beets. Also known as 5-amino-4-chloro-2-phenyl-3(2H)-pyridazinone.

pyrazophos *See* 2-(O,O-diethyl thionophosphoryl)-5-methyl-6-carbethoxy-pyrazolo-(1,5a)pyrimidine.

pyridine C_5H_5N Organic base; flammable, toxic yellowish liquid, with penetrating aroma and burning taste; soluble in water, alcohol, ether, benzene, and fatty oils; boils at 116°C; used as an alcohol denaturant, solvent, in paints, medicine, and textile dyeing.

pyridine-4-carboxylic acid *See* isonicotinic acid.

2,3-pyridinedicarboxylic acid $C_7H_5NO_4$ An odorless, crystalline compound with a melting point of 190°C; soluble in water; inhibits glucose synthesis. Also known as quinolinic acid.

pyridoxal hydrochloride *See* pyridoxine hydrochloride.

pyridoxal phosphate *See* codecarboxylase.

pyridoxine hydrochloride $C_8H_{11}NO_3 \cdot HCl$ A crystalline compound, decomposing at about 208°C; used in medicine in vitamin therapy. Also known as pyridoxal hydrochloride; vitamin B_6 hydrochloride.

pyrimidine $C_4H_4N_2$ A heterocyclic organic compound containing nitrogen atoms at positions 1 and 3; naturally occurring derivatives are components of nucleic acids and coenzymes.

pyritohedron A dodecahedral crystal with 12 irregular pentagonal faces; it is characteristic of pyrite. Also known as pentagonal dodecahedron; pyritoid; regular dodecahedron.

pyro- A chemical prefix for compounds formed by heat, such as pyrophosphoric acid, an inorganic acid formed by the loss of one water molecule from two molecules of an ortho acid.

pyrocarbonic acid diethyl ester *See* diethyl pyrocarbonate.

pyrocatechol $C_6H_4(OH)_2$ Colorless crystals with a melting point of 104°C; used as a fungicide for potatoes in storage. Also known as catechol; O-dihydroxybenzene.

pyrocatechuic acid *See* catechol.

pyrocellulose Highly nitrated cellulose; used to make explosives; originally called guncotton in the United States, cordite in England.

pyrogallic acid $C_6H_3(OH)_3$ Lustrous, light-sensitive white crystals, melting at 133°C; soluble in alcohol, ether, and water; used for photography, dyes, drugs, medicines, and process engravings, and as an analytical reagent and protective colloid. Also known as pyrogallol; 1,2,3,-trihydroxybenzene.

pyrogallol *See* pyrogallic acid.

pyrogallolphthalein *See* gallein.

pyrogen A group of substances thought to be polysaccharides of microbial origin that produce an increase in body temperature when injected into humans and some animals.

pyrogenic distillation A cracking process that runs at high temperatures, high pressures, or both, resulting in greater yields of the light hydrocarbon components of gasoline.

pyroligneous Referring to a substance obtained by the destructive distillation of wood.

pyroligneous acid An impure acetic acid derived from destructive distillation of wood or pine tar. Also known as pyracetic acid; wood vinegar.

pyrolithic acid $HOC(NCOH)_2N \cdot 2H_2O$ Colorless monoclinic crystals, slightly soluble in water. Also known as cyanuric acid.

pyrolysis The breaking apart of complex molecules into simpler units by the use of heat, as in the pyrolysis of heavy oil to make gasoline.

pyromellitic acid $C_6H_2(COOH)_4$ A white powder with a melting point of 257–265°C; used as an intermediate for polyesters and polyamides. Abbreviated PMA. Also known as 1,2,3,4,5-benzenetetracarboxylic acid.

pyromellitic dianhydride $C_6H_2(C_2O_3)_2$ A white powder with a melting point of 286°C; soluble in some organic solvents; used for curing epoxy resins. Abbreviated PMDA.

pyrophosphatase An enzyme catalyzing hydrolysis of esters containing two or more molecules of phosphoric acid to form a simpler phosphate ester.

pyrophosphoric acid $H_4P_2O_7$ Water-soluble, syrupy liquid melting at 61°C; used as a catalyst and to make organic phosphate esters.

pyroracemic acid *See* pyruvic acid.

pyrosin *See* tetraiodofluorescein.

pyrouric acid *See* cyanuric acid.

pyroxylin $[C_{12}H_{16}O_6(NO_3)_4]_x$ Any member of the group of commercially available nitrocelluloses that are used for properties other than their combustibility; the term is commonly used to identify products that are principally made from nitrocellulose, such as pyroxylin plastic or pyroxylin lacquer. Also known as collodion cotton; soluble guncotton; soluble nitrocellulose.

pyrrole C_4H_5N Water-insoluble, yellowish oil, with pungent taste; soluble in alcohol, ether, and dilute acids; boils at 130°C; polymerizes in light; used to make drugs.

2-pyrrolidone C_4H_7ON Combustible, light-yellow liquid, boiling at 245°C; soluble in ethyl alcohol, water, chloroform, and carbon disulfide; used as a plasticizer and polymer solvent, in insecticides and specialty inks, and as a nylon-4 precursor.

pyrrone A polyimidazopyrrolone synthesized from dianhydrides and tetramines; soluble in sulfuric acid; resists temperatures to 600°C.

pyruric acid *See* cyanuric acid.

pyruvaldehyde bis(aminohydrazone) *See* methyl glyoxal bis(guanylhydrazone).

pyruvate Salt of pyruvic acid, such as sodium pyruvate, $NaOOCCOCH_3$.

pyruvic acid Important intermediate in protein and carbohydrate metabolism; liquid with acetic-acid aroma; melts at 11.8°C; miscible with alcohol, ether, and water; used in biochemical research. Also known as acetyl formic acid; α-ketopropionic acid; pyroracemic acid.

Q

Q *See* disintegration energy.

quadratic Stark effect A splitting of spectral lines of atoms in an electric field in which the energy levels shift by an amount proportional to the square of the electric field, and all levels shift to lower energies; observed in lines resulting from the lower energy states of many-electron atoms.

quadratic Zeeman effect A splitting of spectral lines of atoms in a magnetic field in which the energy levels shift by an amount proportional to the square of the magnetic field.

quadridentate ligand A group which forms a chelate and has four points of attachment.

quadruple point Temperature at which four phases are in equilibrium, such as a saturated solution containing an excess of solute.

quadrupole spectrometer A type of mass spectroscope in which ions pass along a line of symmetry between four parallel cylindrical rods; an alternating potential superimposed on a steady potential between pairs of rods filters out all ions except those of a predetermined mass. Also known as Massenfilter.

qualitative analysis The analysis of a gas, liquid, or solid sample or mixture to identify the elements, radicals, or compounds composing the sample.

quantitative analysis The analysis of a gas, liquid, or solid sample or mixture to determine the precise percentage composition of the sample in terms of elements, radicals, or compounds.

quantum chemistry A branch of physical chemistry concerned with the explanation of chemical phenomena by means of the laws of quantum mechanics.

quantum theory of valence The theory of valence based on quantum mechanics; it accounts for many experimental facts, explains the stability of a chemical bond, and allows the correlation and prediction of many different properties of molecules not possible in earlier theories.

quantum yield For a photochemical reaction, the number of moles of a stated reactant disappearing, or the number of moles of a stated product produced, per einstein of light of the stated wavelength absorbed.

quasi-atom A system formed by two colliding atoms whose nuclei approach each other so closely that, for a very short time, the atomic electrons arrange themselves as if they belonged to a single atom whose atomic number equals the sum of the atomic numbers of the colliding atoms.

quasi-fission A nuclear reaction induced by heavy ions in which the two product nuclei have kinetic energies typical of fission products, but have masses close to those of the target and projectile, individually. Also known as deep inelastic transfer; incomplete fusion; relaxed peak process; strongly damped collision.

quasi-molecule The structure formed by two colliding atoms when their nuclei are close enough for the atoms to interact, but not so close as to form a quasiatom.

quaternary ammonium base Ammonium hydroxide (NH_4OH) with the ammonium hydrogens replaced by organic radicals, such as $(CH_3)_4NOH$.

quaternary ammonium salt A nitrogen compound in which a central nitrogen atom is joined to four organic radicals and one acid radical, for example, hexamethonium chloride; used as an emulsifying agent, corrosion inhibitor, and antiseptic.

quaternary carbon atom A carbon atom bonded to four other carbon atoms with single bonds.

quaternary phase equilibria The solubility relationships in any liquid system with four nonreactive components with varying degrees of mutual solubility.

quaternary system An equilibrium relationship between a mixture of four (four phases, four components, and so on).

quenching Phenomenon in which a very strong electric field, such as a crystal field, causes the orbit of an electron in an atom to precess rapidly so that the average magnetic moment associated with its orbital angular momentum is reduced to zero.

quercetin $C_{15}H_5O_2(OH)_5$ A yellow, crystalline flavonol obtained from oak bark and Douglas fir bark; used as an antioxidant and absorber of ultraviolet rays, and in rubber, plastics, and vegetable oils.

quercimelin *See* quercitrin.

quercitrin $C_{21}H_{20}O_{11}$ The 3-rhamnoside of quercitin, forming yellow crystals from dilute ethanol or methanol solution, melting at 176–179°C, soluble in alcohol; used as a textile dye. Also known as quercimelin; quercitroside.

quercitroside *See* quercitrin.

Quevenne scale Arbitrary scale used with hydrometers or lactometers in the determination of the specific gravity of milk; degrees Quevenne = 1000 (specific gravity − 1).

quicklime *See* calcium oxide.

quicksilver *See* mercury.

quicksilver vermilion *See* mercuric sulfide.

quinalbarbitone *See* secobarbital.

quinaldine $C_9H_6NCH_3$ A colorless, oily liquid with a boiling point of 246–247°C; soluble in alcohol, chloroform, and ether; used in medicine as an antimalarial. Also known as chinaldine; α-methylquinoline.

quinalizarin $C_{14}H_8O_6$ A red, crystalline compound, soluble in water solutions of alkalies, and in acetic and sulfuric acid; used to dye cottons.

quinhydrone $C_6H_4O_2 \cdot C_6H_4(OH)_2$ Green, water-soluble powder, subliming at 171°C; a compound of quinone and hydroquinone dissociating in solution.

quinhydrone electrode   A platinum wire in a saturated solution of quinhydrone; used as a reversible electrode standard in pH determinations.

quinic acid $C_6H_7(OH)_4COOH\cdot H_2O$ Ether-insoluble, white crystals with acid taste; melts at 162°C; soluble in alcohol, water, and glacial acetic acid; used in medicine. Also known as chinic acid; kinic acid.

quinidine $C_{20}H_{24}N_2O_2$ A crystalline alkaloid that melts at 171.5°C and that may be derived from the bark of cinchona; used as the salt in medicine. Also known as chinidine; β-quinine.

quinine $C_{20}H_{24}N_2O_2\cdot 3H_2O$ White powder or crystals, soluble in alcohol, ether, carbon disulfide, chloroform, and glycerol; an alkaloid derived from cinchona bark; used as an antimalarial drug and in beverages.

β-quinine   *See* quinidine.

quinine-β-naphthol α-sulfonate   *See* quinaphthol.

quinoidine   A brownish-black mass consisting of a mixture of alkaloids which remain in solution after extracting crystallized alkaloids from cinchona bark; soluble in dilute acids, alcohol, and chloroform; used in medicine. Also known as chinoidine.

quinol   *See* hydroquinone.

quinoline C_9H_7N Water-soluble, aromatic nitrogen compound; colorless, hygroscopic liquid; also soluble in alcohol, ether, and carbon disulfide; boils at 238°C; used in medicine and as a chemical intermediate. Also known as chinoline; leucoline; leukol.

quinoline blue   *See* cyanine dye.

quinolinic acid   *See* 2,3-pyradine dicarboxylic acid.

8-quinolinol   *See* 8-hydroxyquinoline; oxine.

quinone $CO(CHCH)_2CO$ Yellow crystalline compound with irritating aroma; melts at 116°C; soluble in alcohol, alkalies, and ether; used to make dyes and hydroquinone. Also known as benzoquinone; chinone.

quinoxaline $C_8H_6N_2$ Bicyclic organic base; colorless powder, soluble in water and organic solvents; melts at 30°C; used in organic synthesis. Also known as 1,4-benzodiazine; benzo-*para*-diazine.

N′-2-quinoxalysulfanilimide $C_{14}H_{12}N_4SO_2$ Crystals with a melting point of 247°C; almost insoluble in water; used as a rodenticide. Also known as sulfaquinoxaline.

quinquidentate ligand   *See* pentadentate ligand.

quintozene   *See* pentachloronitrabenzene.

Q value   *See* disintegration energy.

Ra *See* radium.

rabicide $C_8H_2Cl_4O_2$ A colorless, crystalline compound with a melting point of 209–210°C; slightly soluble in water; used as a fungicide in rice crops. Also known as 4,5,6,7–tetrachlorophthalide.

racemase Any of a group of enzymes that catalyze racemization reactions.

racemic acid $C_2H_4O_2(COOH)_2 \cdot H_2O$ Colorless crystals, melting at 205°C; soluble in water, slightly soluble in alcohol; used as a chemical intermediate. Also known as inactive tartaric acid.

racemic mixture A compound which is a mixture of equal quantities of dextrorotatory and levorotatory isomers of the same compound, and therefore is optically inactive.

radial chromatography A circular disk of absorbent paper which has a strip (wick) cut from edge to center to dip into a solvent; the solvent climbs the wick, touches the sample, and resolves it into concentric rings (the chromatogram). Also known as circular chromatography; radial-paper chromatography.

radial distribution function A function $\rho(r)$ equal to the average over all directions of the number density of molecules at distance r from a given molecule in a liquid.

radial-paper chromatography *See* radial chromatography.

radiation biochemistry The study of the response of the constituents of living matter to radiation.

radiation catalysis The use of radiation (such as gamma, neutron, proton, electron, or x-ray) to activate or speed up a chemical or physical change; for example, radiation alone can initiate polymerization without heat, pressure, or chemical catalysts.

radiative capture A nuclear capture process whose prompt result is the emission of electromagnetic radiation only.

radical *See* free radical.

radicofunctional name A name for an organic compound that uses two key words; the first word corresponds to the group or groups involved and the second word indicates the functional group—for example, alkyl halide.

radioactinium Conventional name for the isotope of thorium which has mass number 227 and is in the actinium series. Symbolized RdAc.

radioactive Exhibiting radioactivity or pertaining to radioactivity.

radioactive chain *See* radioactive series.

radioactive clock A radioactive isotope such as potassium-40 which spontaneously decays to a stable end product at a constant rate, allowing absolute geologic age to be determined.

radioactive cobalt Radioactive form of cobalt, such as cobalt-60 with a half-life of 5.3 years.

radioactive decay The spontaneous transformation of a nuclide into one or more different nuclides, accompanied by either the emission of particles from the nucleus, nuclear capture or ejection of orbital electrons, or fission. Also known as decay; nuclear spontaneous reaction; radioactive disintegration; radioactive transformation; radioactivity.

radioactive decay series *See* radioactive series.

radioactive disintegration *See* radioactive decay.

radioactive displacement law The statement of the changes in mass number A and atomic number Z that take place during various nuclear transformations. Also known as displacement law.

radioactive element An element all of whose isotopes spontaneously transform into one or more different nuclides, giving off various types of radiation; examples include promethium, radium, thorium, and uranium.

radioactive emanation A radioactive gas given off by certain radioactive elements; all of these gases are isotopes of the element radon. Also known as emanation.

radioactive equilibrium In radioactivity, the condition of equilibrium in which the rate of decay of the parent isotope is exactly matched by the rate of decay of every intermediate daughter isotope.

radioactive metal A luminous metallic element, such as actinium, radium, or uranium, that spontaneously and continuously emits radiation capable in some degree of penetrating matter impervious to ordinary light.

radioactive series A succession of nuclides, each of which transforms by radioactive disintegration into the next until a stable nuclide results. Also known as decay chain; decay family; decay series; disintegration chain; disintegration family; disintegration series; radioactive chain; radioactive decay series; series decay; transformation series.

radioactive transformation *See* radioactive decay.

radioactivity 1. A particular type of radiation emitted by a radioactive substance, such as α-radioactivity. 2. *See* radioactive decay. 3. *See* activity.

radioactivity equilibrium A condition which may arise in the decay of a radioactive parent with short-lived descendants, in which the ratio of the activity of a parent to that of a descendant remains constant.

radioassay An assay procedure involving the measurement of the radiation intensity of a radioactive sample.

radiocarbon *See* carbon-14.

radiocesium *See* cesium-137.

radiochemical laboratory A specially equipped and shielded chemical laboratory designed for conducting radiochemical studies without danger to the laboratory personnel.

radiochemistry That area of chemistry concerned with the study of radioactive substances.

radio element A radioactive isotope of an element, or a sample consisting of one or more radioactive isotopes of an element.

radio-frequency spectrometer An instrument which measures the intensity of radiation emitted or absorbed by atoms or molecules as a function of frequency at frequencies from 10^5 to 10^9 hertz; examples include the atomic-beam apparatus, and instruments for detecting magnetic resonance.

radio-frequency spectroscopy The branch of spectroscopy concerned with the measurement of the intervals between atomic or molecular energy levels that are separated by frequencies from about 10^5 to 10^9 hertz, as compared to the frequencies that separate optical energy levels of about 6×10^{14} hertz.

radiogenic Pertaining to a material produced by radioactive decay, as the production of lead from uranium decay.

radiogenic argon Argon occurring in rocks and minerals that is the result of in-place decay of potassium-40 since the formation of the earth.

radiogenic isotope An isotope which was produced by the decay of a radioisotope, but which itself may or may not be radioactive.

radiogenic lead Stable, end-product lead (Pb-206, Pb-207, and Pb-208) occurring in rocks and minerals that is the result of in-place decay of uranium and thorium since the formation of the earth.

radiogenic strontium Strontium-87 occurring in rocks and minerals that is the direct result of in-place decay of rubidium-87 since the formation of the earth.

radiogeology The study of the distribution patterns of radioactive elements in the earth's crust and the role of radioactive processes in geologic phenomena.

radioiodine Any radioactive isotope of iodine, especially iodine-131; used as a tracer to determine the activity and size of the thyroid gland, and experimentally, to destroy the thyroid glands of animals.

radioisotope An isotope which exhibits radioactivity. Also known as radioactive isotope; unstable isotope.

radioisotope assay An analytical technique including procedures for separating and reproducibly measuring a radioactive tracer.

radiolysis The dissociation of molecules by radiation; for example, a small amount of water in a reactor core dissociates into hydrogen and oxygen during operation.

radiometric analysis Quantitative chemical analysis that is based on measurement of the absolute disintegration rate of a radioactive component having a known specific activity.

radiometric titration Use of radioactive indicator to track the transfer of material between two liquid phases in equilibrium, such as titration of $^{110}AgNO_3$ (silver nitrate, with the silver atom having mass number 110) against potassium chloride.

radiomimetic substances Chemical substances which cause biological effects similar to those caused by ionizing radiation.

radionuclide A nuclide that exhibits radioactivity.

radiothorium   Conventional name of the isotope of thorium which has mass number 228. Symbolized RdTh.

radium   A radioactive member of group IIA, symbol Ra, atomic number 88; the most abundant naturally occurring isotope has mass number 226 and a half-life of 1620 years. A highly toxic solid that forms water-soluble compounds; decays by emission of α, β, and γ-radiation; melts at 700°C, boils at 1140°C; turns black in air; used in medicine, in industrial radiography, and as a source of neutrons and radon.

radium bromide $RaBr_2$ Water-soluble, poisonous, radioactive white powder, corrosive to skin or flesh; melts at 728°C; used in medicine, physical research, and luminous paint.

radium carbonate $RaCO_3$ Water-insoluble, poisonous, radioactive, white powder; used in medicine.

radium chloride $RaCl_2$ Water- and alcohol-soluble, poisonous, radioactive, yellow-white crystals; corrosive effect on skin and flesh; melts at 1000°C; used in medicine, physical research, and luminous paint.

radium F   *See* polonium-210.

radium sulfate $RaSO_4$ Water-insoluble, radioactive, poisonous, white crystals; used in medicine.

radius ratio   The ratio of the radius of a cation to the radius of an ion; relative ionic radii are pertinent to crystal lattice structure, particularly the determination of coordination number.

radon   A chemical element, symbol Rn, atomic number 86; all isotopes are radioactive, the longest half-life being 3.82 days for mass number 222; it is the heaviest element of the noble-gas group, produced as a gaseous emanation from the radioactive decay of radium.

radon-220   The isotope of radon having mass number 220, symbol ^{220}Ra, which is a radioactive member of the thorium series with a half-life of 56 seconds.

radon-222   The isotope of radon having mass number 222, symbol ^{222}Ra, which is a radioactive member of the uranium series with a half-life of 3.82 days.

raffinase   An enzyme that hydrolyzes raffinose, yielding fructose in the reaction.

raffinate   In solvent refining, that portion of the treated liquid mixture that remains undissolved and is not removed by the selective solvent. Also known as good oil to petroleum-refinery operators.

raffinose $C_{18}H_{32}O_{16} \cdot 5H_2O$ A white, crystalline trisaccharide found in sugar-beets, cottonseed meal, and molasses; yields glucose, fructose, and galactose on complete hydrolysis. Also known as gossypose; melitose; melitriose.

Raman spectrophotometry   The study of spectral-line patterns on a photograph taken at right angles through a substance illuminated with a quartz mercury lamp.

Raman spectroscopy   Analysis of the intensity of Raman scattering of monochromatic light as a function of frequency of the scattered light.

Raman spectrum   A display, record, or graph of the intensity of Raman scattering of monochromatic light as a function of frequency of the scattered light.

Ramsauer effect The vanishing of the scattering cross section of electrons from atoms of a noble gas at some value of the electron energy, always below 25 electronvolts.

Ramsbottom coke test A laboratory test for carbon residue in petroleum products.

random coil Any of various irregularly coiled polymers that can occur in solution. Also known as cyclic coil.

random copolymer Resin copolymer in which the molecules of each monomer are randomly arranged in the polymer backbone.

Raoult's law The law that the vapor pressure of a solution equals the product of the vapor pressure of the pure solvent and the mole fraction of solvent.

rare-earth element The name given to any of the group of chemical elements with atomic numbers 58 to 71; the name is a misnomer since they are neither rare nor earths; examples are cerium, erbium, and gadolinium.

rare-earth salts Salts derived from monazite, and with rare earths in similar proportions as in monazite; contains La, Ce, Pr, Nd, Sm, Gd, and Y as acetates, carbonates, chlorides, fluorides, nitrates, sulfates, and so on.

rare gas *See* noble gas.

Raschig process A method for production of phenol that begins with a first-stage chlorination of benzene, using an air–hydrochloric acid mixture.

Raschig ring A type of packing in the shape of a short pipe; used in columns for absorption operations, and to a limited extent for distillation operations.

Rast method The melting-point depression method often used for the determination of the molecular weight of organic compounds.

rat distillate A refinery designation for gasoline and other fuels as they come from the condenser, before undesirable substances are removed by further processing.

rate constant Numerical constant in a rate-of-reaction equation; for example, $r_A = kC_A{}^aC_B{}^bC_C{}^c$, where C_A, C_B, and C_C are reactant concentrations, k is the rate constant (specific reaction rate constant), and a, b, and c are empirical constants.

ratio of specific heats The ratio of specific heat at constant pressure to specific heat at constant volume, $\gamma = C_p/C_v$.

Rayleigh line Spectrum line in scattered radiation which has the same frequency as the corresponding incident radiation.

Rb *See* rubidium.

RdAc *See* radioactinium.

RDC extractor *See* rotary-disk contactor.

RDGE *See* resorcinol diglycidyl ether.

RdTh *See* radiothorium.

Re *See* rhenium.

reactants The molecules that act upon one another to produce a new set of molecules (products); for example, in the reaction $HCl + NaOH \rightarrow NaCl + H_2O$, the HCl and NaOH are the reactants.

reaction *See* nuclear reaction.

reaction boundary *See* reaction line.

reaction curve *See* reaction line.

reaction energy *See* disintegration energy.

reaction enthalpy number A dimensionless number used in the study of interphase transfer in chemical reactions, equal to the enthalpy of reaction per unit mass of a specified compound produced in a reaction, times the mass fraction of that compound, divided by the product of the specific heat at constant pressure and the temperature change during the reaction.

reaction kinetics *See* chemical kinetics.

reaction line In a ternary system, a special case of the boundary line along which one of the two crystalline phases present reacts with the liquid, as the temperature is decreased, to form the other crystalline phase. Also known as reaction boundary; reaction curve.

reaction zone In a catalytic reactor vessel, the location or zone within the vessel where the bulk of the chemical reaction takes place.

reactive bond A bond between atoms that is easily invaded (reacted to) by another atom or radical; for example, the double bond in $CH_2\!\!=\!\!CH_2$ (ethylene) is highly reactive to other ethylene molecules in the reaction known as polymerization to form polyethylene.

reactivity The relative capacity of an atom, molecule, or radical to combine chemically with another atom, molecule, or radical.

reactor Device or process vessel in which chemical reactions (catalyzed or noncatalyzed) take place during a chemical conversion type of process.

reagent 1. The compound that supplies the molecule, ion, or free radical which is arbitrarily considered as the attacking species in a chemical reaction. 2. A substance, chemical, or solution used in the laboratory to detect, measure, or otherwise examine other substances, chemicals, or solutions; grades include ACS (American Chemical Society standards), reagent (for analytical reagents), CP (chemically pure), USP (U.S. Pharmacopeia standards), NF (National Formulary standards), and purified, technical (for industrial use).

reagent chemicals High-purity chemicals used for analytical reactions, for testing of new reactions where the effect of impurities are unknown, and, in general, for chemical work where impurities must either be absent or at a known concentration.

rearrangement reaction A nuclear reaction in which nucleons are exchanged between nuclei.

reboiler An auxiliary heating unit for a fractionating tower designed to supply additional heat to the lower portion of the tower; liquid withdrawn from the side or bottom of the tower is reheated by heat exchange, then reintroduced into the tower.

receiver Vessel, container, or tank used to receive and collect liquid material from a process unit, such as the distillate receiver from the overhead condenser of a distillation column.

receiving house A building where liquid streams from petroleum-refining-process condensers are observed through a look box, and samples are taken for testing, and also where products are diverted to storage tanks or to other processing units.

receiving tank   *See* rundown tank.

receptor   A site or structure in a cell which combines with a drug or other biological to produce a specific alteration of cell function.

reciprocal lattice   A lattice array of points formed by drawing perpendiculars to each plane *(hkl)* in a crystal lattice through a common point as origin; the distance from each point to the origin is inversely proportional to spacing of the specific lattice planes; the axes of the reciprocal lattice are perpendicular to those of the crystal lattice.

reciprocal vectors   For a set of three vectors forming the primitive translations of a lattice, the vectors that form the primitive translations of the reciprocal lattice.

reciprocal velocity region   The energy region in which the capture cross section for neutrons by a given element is inversely proportional to neutron velocity.

reciprocating-plate extractor   A liquid-liquid contactor in which equally spaced perforated plates (as in a distillation column) move up and down rapidly over a short distance to cause liquid agitation and mixing. Also known as reciprocating-plate column.

recoil ion spectroscopy   A method of studying highly ionized and highly excited atomic states, in which relatively light atoms in a gaseous target are bombarded by highly ionized, fast, heavy projectiles, resulting in single collisions in which the target atoms are raised to very high states of ionization and excitation while incurring relatively small recoil velocities.

reconstructive processing   The spinning of an inorganic compound of an organic support or binder subsequently removed by oxidation or volatilization to form an inorganic polymer.

reconstructive transformation   A type of crystal transformation that involves the breaking of either first- or second-order coordination bonds.

recording balance   An analytical balance equipped to record weight results by electromagnetic or servomotor-driven accessories.

recrystallization   **1.** Repeated crystallization of a material from fresh solvent to obtain an increasingly pure product. **2.** A change in the structure of a crystal without a chemical alteration.

rectification distillation   A distillation technique in which a rectifying column is used.

rectifying column   Portion of a distillation column above the feed tray in which rising vapor is enriched by interaction with a countercurrent falling stream of condensed vapor; contrasted to the stripping column section below the column feed tray.

recycle mixing   The mixing of a portion of a product stream (fluid or solid) from a processing unit with incoming raw feed.

recycle ratio   In a continuous chemical process, the ratio of recycle stock to fresh feed.

recycle stock   That portion of a feedstock that has passed through a processing unit and is recirculated (recycled) back through the process.

red lead   *See* lead tetroxide.

red mercury sulfide   *See* mercuric sulfide.

red ocher   *See* ferric oxide.

redoxomorphic stage   The earliest geochemical stage of diagenesis characterized by mineral changes primarily due to oxidation and reduction reactions.

redox potential   Voltage difference at an inert electrode immersed in a reversible oxidation-reduction system; measurement of the state of oxidation of the system. Also known as oxidation-reduction potential.

redox potentiometry   Use of neutral electrode probes to measure the solution potential developed as the result of an oxidation or reduction reaction.

redox system   A chemical system in which reduction and oxidation (redox) reactions occur.

redox titration   A titration characterized by the transfer of electrons from one substance to another (from the reductant to the oxidant) with the end point determined colorimetrically or potentiometrically.

red phosphorus   An allotropic form of the element phosphorus; violet-red, amorphous powder subliming at 416°C, igniting at 260°; insoluble in all solvents; nonpoisonous.

red potassium chromate   *See* potassium dichromate.

red potassium prussiate   *See* potassium ferricyanide.

red precipitate   *See* mercuric oxide.

red prussiate of potash   *See* potassium ferricyanide.

red prussiate of soda   *See* sodium ferricyanide.

red tetrazolium   *See* triphenyltetrazolium chloride.

reducer   *See* reducing agent.

reducing agent   Also known as reducer. **1.** A material that adds hydrogen to an element or compound. **2.** A material that adds an electron to an element or compound, that is, decreases the positiveness of its valence.

reducing atmosphere   An atmosphere of hydrogen (or other substance that readily provides electrons) surrounding a chemical reaction or physical device; the effect is the opposite to that of an oxidizing atmosphere.

reducing flame   A flame having excess fuel and being capable of chemical reduction, such as extracting oxygen from a metallic oxide.

reducing sugar   Any of the sugars that because of their free or potentially free aldehyde or ketone groups, possess the property of readily reducing alkaline solutions of many metallic salts such as copper, silver, or bismuth; examples are the monosaccharides and most of the disaccharides, including maltose and lactose.

reduction   **1.** Reaction of hydrogen with another substance. **2.** Chemical reaction in which an element gains an electron (has a decrease in positive valence).

reduction cell   A vessel in which aqueous solutions of salts or fused salts are reduced electrolytically.

reduction potential   The potential drop involved in the reduction of a positively charged ion to a neutral form or to a less highly charged ion, or of a neutral atom to a negatively charged ion.

reentrant angle   The angle between two plane surfaces on a crystalline solid, in which the external angle is less than 180°.

reference electrode   A nonpolarizable electrode that generates highly reproducible potentials; used for pH measurements and polarographic analyses; examples are the calomel electrode, silver-silver chloride electrode, and mercury pool.

refinery   System of process units used to convert crude petroleum into fuels, lubricants, and other petroleum-derived products.

reflectance spectrophotometry   Measurement of the ratio of spectral radiant flux reflected from a light-diffusing specimen to that reflected from a light-diffusing standard substituted for the specimen.

reflection plane   *See* plane of mirror symmetry; plane of reflection.

reflection twin   A crystal twin whose symmetry is formed by an apparent mirror image across a plane.

reflux   In a chemical process, that part of the product stream that may be returned to the process to assist in giving increased conversion or recovery, as in distillation or liquid-liquid extraction.

reflux condenser   An auxiliary vessel for a distillation column that constantly condenses vapors and returns liquid to the column.

reflux ratio   The quantity of liquid reflux per unit quantity of product removed from the process unit, such as a distillation tower or extraction column.

Reformatsky reaction   A condensation-type reaction between ketones and α-bromoaliphatic acids in the presence of zinc or magnesium, such as R_2CO + $BrCH_2 \cdot COOR$ + $Zn \rightarrow (ZnO \cdot HBr)$ + $R_2C(OH)CH_2COOR$.

reforming   The thermal or catalytic conversion of petroleum naphtha into more volatile products of higher octane number; represents the total effect of numerous simultaneous reactions, such as cracking, polymerization, dehydrogenation and isomerization.

refractory hard metals   True chemical compounds composed of two or more metals in the crystalline form, and having a very high melting point and high hardness.

refrigerant 23   *See* fluoroform.

regenerant   A solution whose purpose is to restore the activity of an ion-exchange bed.

regenerate   To clean of impurities and make reusable as in regeneration of a catalytic cracking catalyst by burning off carbon residue, regeneration of clay adsorbent by washing free of adherents, or regeneration of a filtration system by cleaning off the filter media.

regenerator   Device or system used to return a system or a component of it to full strength in a chemical process; examples are a furnace to burn carbon from a catalyst, a tower to wash impurities from clay, and a flush system to clean off the surface of filter media.

regioselective Pertaining to a chemical reaction which favors a single positional or structural isomer, leading to its yield being greater than that of the other products in the reaction. Sometimes known as regiospecific.

regiospecific 1. Referring to a chemical reaction which has the potential of yielding two or more structural isomers, but actually produces only one. 2. *See* regioselective.

regular dodecahedron *See* pyritohedron.

Reichert-Meissl number An indicator of the measure of volatile soluble fatty acids.

Reich process Process to purify carbon dioxide produced during fermentation; organic impurities in the gas are oxidized and absorbed, then the gas is dehydrated.

Reimer-Tiemann reaction Formation of phenolic aldehydes by reaction of phenol with chloroform in the presence of an alkali.

Reinecke's salt $[(NH_3)_2Cr(SCN)_4]NH_4 \cdot H_2O$ A reagent to detect mercury (gives a red color or a precipitate), and to isolate organic bases (such as proline or histidine).

Reinsch test A test for detecting small amounts of arsenic, silver, bismuth, and mercury.

relative fugacity The ratio of the fugacity in a given state to the fugacity in a defined standard state.

relative stability test A color test using methylene blue that indicates when the oxygen present in a sewage plant's effluent or polluted water is exhausted.

relative volatility The volatility of a standard material whose relative volatility is by definition equal to unity.

relaxed circular deoxyribonucleic acid A form of circular deoxyribonucleic acid in which the circle of one strand is broken. Also known as open-circle deoxyribonucleic acid.

relaxed peak process *See* quasi-fission.

relaxin A hormone found in the serum of humans and certain other animals during pregnancy; probably acting with progesterone and estrogen, it causes relaxation of pelvic ligaments in the guinea pig.

relief The apparent topography exhibited by minerals in thin section as a consequence of refractive index.

remedial operation In a chemical process operation, the revision of operating conditions so as to correct the overall operation and bring the product into desired rote or specification limits. Also known as corrective operation.

renaturation The process of restoring denatured proteins to their original condition.

renin A proteolytic enzyme produced in the afferent glomerular arteriole which reacts with the plasma component hypertensinogen to produce angiotensin II.

rennin An enzyme found in the gastric juice of the fourth stomach of calves; used for coagulating milk casein in cheesemaking. Also known as chymosin.

renninogen   The zymogen of rennin. Also known as prorennin.

repeated twinning   Crystal twinning that involves more than two simple crystals.

repellency   Ability to repel water, or being hydrophobic; opposite to water wettability.

replication   The formation of a faithful mold or replica of a solid that is thin enough for penetration by an electron microscope beam; can use plastic (such as collodion) or vacuum deposition (such as of carbon or metals) to make the mold.

Reppe process   A family of high-pressure, catalytic acetylene-reaction processes yielding (depending upon what the acetylene reacts with) butadiene, allyl alcohol, acrylonitrile, vinyl ethers and derivatives, acrylic acid esters, cyclooctatraene, and resins.

repressor   An end product of metabolism which represses the synthesis of enzymes in the metabolic pathway.

rerunning   The distillation of a liquid material that has already been distilled; usually implies taking a large proportion of the charge stock overhead.

resacetophenone   *See* 2,4'-dihydroxyacetophenone.

resbenzophenone   *See* benzoresorcinol.

residence time   The average length of time a particle of reactant spends within a process vessel or in contact with a catalyst.

residue   1. The substance left after distilling off all but the heaviest components from crude oil in petroleum refinery operations. Also known as bottoms; residuum. 2. Solids deposited onto the filter medium during filtration. Also known as cake; discharged solids.

residuum   *See* residual oil; residue.

resin   Any of a class of solid or semisolid organic products of natural or synthetic origin with no definite melting point, generally of high molecular weight; most resins are polymers.

resin-in-pulp ion exchange   Combination of coarse anion-exchange resin with a slurry of finely ground uranium ore in an acid-leach liquor.

resinography   Science of resins, polymers, plastics, and their products; includes study of morphology, structure, and other characteristics relatable to composition or treatment.

resinoid   A thermosetting synthetic resin either in its initial (temporarily fusible) or in its final (infusible) state.

resistivity factor   *See* formation factor.

resite   *See* C-stage resin.

resolution   *See* resolving power.

resolving power   A measure of the ability of a spectroscope or interferometer to separate spectral lines of nearly equal wavelength, equal to the average wavelength of two equally strong spectral lines whose images can barely be separated, divided by the difference in wavelengths; for spectroscopes, the lines must be resolved according to the Rayleigh criterion; for interferometers, the wavelengths at which the lines have half of maximum intensity must be equal. Also known as resolution.

resonance capture The combination of an incident particle and a nucleus in a resonance level of the resulting compound nucleus, characterized by having a large cross section at and very near the corresponding resonance energy.

resonance fluorescence 1. Resonant scattering from an atomic nucleus. 2. *See* resonance radiation.

resonance hybrid A molecule that may be considered an intermediate between two or more valence bond structures.

resonance ionization spectroscopy A technique capable of detecting single atoms or molecules of a given element or compound in a gas, in which an atom or molecule in its ground state is excited to a bound state when a photon is absorbed from a laser beam at a very well-controlled wavelength that is resonant with the excitation energy; a second photon removes the excited electron from the atom or molecule, and this electron is then accelerated by an electric field and collides with the gas molecules, creating additional ionization which is detected by a proportional counter. Abbreviated RIS.

resonance lamp An evacuated quartz bulb containing mercury, which acts as a source of radiation at the wavelength of the pure resonance line of mercury when irradiated by a mercury-arc lamp.

resonance luminescence *See* resonance radiation.

resonance radiation The emission of radiation by a gas or vapor as a result of excitation of atoms to higher energy levels by incident photons at the resonance frequency of the gas or vapor; the radiation is characteristic of the particular gas or vapor atom but is not necessarily the same frequency as the absorbed radiation. Also known as resonance fluorescence; resonance luminescence.

resonance reaction A nuclear reaction that takes place only when the energy of the incident particles is at or very close to a characteristic value.

resonance scattering A peak in the cross section of a nucleus for elastic scattering of neutrons at energies near a resonance level, accompanied by an anomalous phase shift in the scattered neutrons.

resonance spectrum An emission spectrum resulting from illumination of a substance (usually a molecular gas) by radiation of a definite frequency or definite frequencies.

resonance structure Any of two or more possible structures of the same compound that have identical geometry but different arrangements of their paired electrons; none of the structures has physical reality or adequately accounts for the properties of the compound, which exists as an intermediate form.

resorcin *See* resorcinol.

resorcinol $C_6H_4(OH)_2$ Sweet-tasting, white, toxic crystals; soluble in water, alcohol, ether, benzene, and glycerol; melts at 111°C; used for resins, dyes, pharmaceuticals, and adhesives, and as a chemical intermediate. Also known as *meta*-dihydroxybenzene; resorcin.

resorcinol acetate $HOC_6H_4OCOCH_3$ A viscous, combustible, yellow to amber liquid with burning taste; soluble in alcohol and solvent; boils at 283°C; used in cosmetics and medicine. Also known as resorcinol monoacetate.

resorcinol diglycidyl ether $C_{12}H_{14}O_2$ A straw yellow liquid with a boiling point of 172°C (at 0.8 mmHg, or 106.66 newtons per square meter); used for epoxy resins. Abbreviated RDGE.

resorcinol-formaldehyde resin A phenol-formaldehyde resin, soluble in water, ketones, and alchol; used to make fast-curing adhesives for wood gluing.

resorcinol monoacetate *See* resorcinol acetate.

β-resorcylic acid $(OH)_2C_6H_3COOH$ Combustible, white needles; soluble in alcohol and ether, very slightly soluble in water; decomposes at 220°C; used as a dyestuff and a pharmaceutical intermediate, and in the manufacture of fine chemicals. Also known as BRA; 4-carboxy-resorcinol; 2,4-dihydroxybenzene carboxylic acid; 2,4-dihydroxybenzoic acid; 4-hydroxysalicylic acid.

respiratory pigment Any of various conjugated proteins that function in living organisms to transfer oxygen in cellular respiration.

restricted internal rotation Restrictions on the rotational motion of molecules or parts of molecules in some substances, such as solid methane, at certain temperatures.

ret The reduction or digestion of fibers (usually linen) by enzymes.

retene $C_{18}H_{18}$ A cyclic hydrocarbon, melting at 100.5–101°C, soluble in benzene and hot ethanol; used in organic syntheses. Also known as 7-isopropyl-1-methyl phenanthrene; methylisopropylphenanthrene.

retention index In gas chromatography, the relationship of retention volume with arbitrarily assigned numbers to the compound being analyzed; used to indicate the volume retention behavior during analysis.

retention time In gas chromatography, the time at which the center, or maximum, of a symmetrical peak occurs on a gas chromatogram.

retention volume In gas chromatography, the product of retention time and flow rate.

reticulin A protein isolated from reticular fibers.

retinal A carotenoid, produced as an intermediate in the bleaching of rhodopsin and decomposition to vitamin A. Also known as vitamin A aldehyde.

retinal pigment *See* rhodopsin.

retinene A pigment extracted from the retina, which turns yellow by the action of light; the chief carotenoid of the retina.

retinoid The set of molecules composing vitamin A and its synthetic analogs, such as retinal or retinyl acetate.

retinol *See* vitamin A.

retort 1. A closed refractory chamber in which coal is carbonized for manufacture of coal gas. 2. A vessel for the distillation or decomposition of a substance.

retrogradation 1. Generally, a process of deterioration; a reversal or retrogression to a simpler physical form. 2. A chemical reaction involving vegetable adhesives, which revert to a simpler molecular structure.

retrograde condensation Phenomenon associated with the behavior of a hydrocarbon mixture in the critical as diaphthoresis; retrogressive metamorphism.

retting Soaking vegetable stalks to decompose the gummy material and release the fibers.

reversal spectrum A spectrum which may be observed in intense white light which has traversed luminous gas, in which there are dark lines where there were bright lines in the emission spectrum of the gas.

reversal temperature The temperature of a blackbody source such that, when light from this source is passed through a luminous gas and analyzed in a spectroscope, a given spectral line of the gas disappears, whereas it appears as a bright line at lower blackbody temperatures, and a dark line at higher temperatures.

reverse bonded-phase chromatography A technique of bonded-phase chromatography in which the stationary phase is nonpolar and the mobile phase is polar.

reversed air-blast process A gasmaking process in which, after a short period of the ordinary blow, the air blast is reversed so as to enter the top of the superheater, and passes back to the top of the generator and down.

reverse deionization A process in which an ion-exchange unit and a cation-exchange unit are used in sequence to remove all ions from a solution.

reversed-phase partition chromatography Paper chromatography in which the low-polarity phase (such as paraffin, paraffin jelly, or grease) is put onto the support (paper) and the high-polarity phase (such as water, acids, or organic solvents) is allowed to flow over it.

reverse osmosis A technique used in desalination and waste-water treatment; pressure is applied to the surface of a saline (or waste) solution, forcing pure water to pass from the solution through a membrane (hollow fibers of cellulose acetate or nylon) that will not pass sodium or chloride ions.

reversible chemical reaction A chemical reaction that can be made to proceed in either direction by suitable variations in the temperature, volume, pressure, or quantities of reactants or products.

reversible electrode An electrode that owes its potential to unit charges of a reversible nature, in contrast to electrodes used in electroplating and destroyed during their use.

Rexforming A proprietary petroleum-refinery process combining Platforming with aromatics extraction, wherein low-octane raffinate is recycled to the Platformer.

Reychler's acid *See* d-camphorsulfonic acid.

Reynolds analogy Relationship showing the similarity between the transfer of mass, heat, and momentum.

Reynolds number A dimensionless number which is significant in the design of a model of any system in which the effect of viscosity is important in controlling the velocities or the flow pattern of a fluid; equal to the density of a fluid, times its velocity, times a characteristic length, divided by the fluid viscosity. Symbolized N_{Re}. Also known as Damköhler number V (DaV).

Rh *See* rhodium.

rhamnose $C_6H_{12}O_5$ A deoxysugar occurring free in poison sumac, and in glycoside combination in many plants. Also known as isodulcitol.

rhenium   A metallic element, symbol Re, atomic number 75, atomic weight 186.2; a transition element.

rhenium halide   Halogen compound of rhenium; examples are $ReCl_3$, $ReCl_4$, ReF_4, and ReF_6.

rheopexy   A property of certain sols, having particles shaped like rods or plates, which set to gel form more quickly when mechanical means are used to hasten the orientation of the particles.

rheum emodin   *See* emodin.

rhizoctol   *See* methylarsinic sulfide.

rhodamine B $C_{28}H_{31}ClN_2O_3$ Red, green, or reddish-violet powder, soluble in alcohol and water; forms bluish red, fluorescent solution in water; used as red dye for paper, wool, and silk, and as an analytical reagent and biological stain.

rhodanate   *See* thiocyanate.

rhodanic acid   *See* thiocyanic acid.

rhodanide   *See* thiocyanate.

rhodanine $C_3H_3NOS_2$ A pale-yellow crystalline compound that may decompose violently when heated, giving off toxic by-products; used in organic synthesis. Also known as 2-thioxo-4-thiazolidinone.

L-rhodeose   *See* L-fucose.

rhodium   A chemical element, symbol Rh, atomic number 45, atomic weight 102.905.

rhodium chloride $RhCl_3$ Water-insoluble, brown-red powder, soluble in cyanides and alkalies; decomposes at 450–500°C. Also known as rhodium trichloride.

rhodium trichloride   *See* rhodium chloride.

rhodopsin   A deep-red photosensitive pigment contained in the rods of the retina of marine fishes and most higher vertebrates. Also known as retinal pigment; visual purple.

rhodotorulic acid   A diketopiperazine composed of two molecules of N^δ-acetyl-N^δ-hydroxy-ornithine; a potent growth factor for microorganisms that require hemin or iron transport compounds, for example, *Arthrobacter* species.

rhodoxanthin $C_{40}H_{50}O_2$ A xanthophyll carotenoid pigment.

rhomb   *See* rhombohedron.

rhombic dodecahedron   A crystal form in the cubic system that is a dodecahedron whose faces are equal rhombuses.

rhombic lattice   *See* orthorhombic lattice.

rhombic sulfur   Crystalline sulfur with three unequal axes, all at right angles.

rhombic system   *See* orthorhombic system.

rhombohedral   1. Of or pertaining to the rhombohedral system. 2. Of or pertaining to crystal cleavage in or a centered lattice of the hexagonal system.

rhombohedral lattice   A crystal lattice in which the three axes of a unit cell are of equal length, and the three angles between axes are the same, and are not right angles. Also known as trigonal lattice.

rhombohedral packing   The tightest manner of systematic arrangement of uniform solid spheres in a clastic sediment or crystal lattice, characterized by a unit cell of six planes passed through eight sphere centers situated at the corners of a regular rhombohedron.

rhombohedral system   A division of the trigonal crystal system in which the rhombohedron is the basic unit cell.

rhombohedron   A trigonal crystal form that is a parallelepiped, the six identical faces being rhombs. Also known as rhomb.

ribitol   *See* adonitol.

riboflavin $C_{17}H_{20}N_4O_6$ A water-soluble, yellow-orange fluorescent pigment that is essential to human nutrition as a component of the coenzymes flavin mononucleotide and flavin adenine dinucleotide. Also known as lactoflavin; vitamin B_2; vitamin G.

riboflavin 5′-phosphate $C_{17}H_{21}N_4O_9P$ The phosphoric acid ester of riboflavin. Also known as flavin phosphate; flavin mononucleotide; FMN; isoalloxazine mononucleotide; vitamin B_2 phosphate.

D-ribo-2-ketohexose   *See* allulose.

D-riboketose   *See* ribulose.

ribonuclease $C_{587}H_{909}N_{171}O_{197}S_{12}$ An enzyme that catalyzes the depolymerization of ribonucleic acid.

ribonucleic acid   A long-chain, usually single-stranded nucleic acid consisting of repeating nucleotide units containing four kinds of heterocyclic, organic bases: adenine, cytosine, quanine, and uracil; they are conjugated to the pentose sugar ribose and held in sequence by phosphodiester bonds; involved intracellularly in protein synthesis. Abbreviated RNA.

ribonucleoprotein   Any of a large group of conjugated proteins in which molecules of ribonucleic acid are closely associated with molecules of protein.

ribose $C_5H_{10}O_5$ A pentose sugar occurring as a component of various nucleotides, including ribonucleic acid.

riboside   Any glycoside containing ribose as the sugar component.

ribosomal ribonucleic acid   Any of three large types of ribonucleic acid found in ribosomes: 5S RNA, with molecular weight 40,000; 14–16S RNA, with molecular weight 600,000; and 18–22S RNA with molecular weight 1,200,000. Abbreviated rRNA.

ribulose $C_5H_{10}O_5$ A pentose sugar that exists only as a syrup; synthesized from arabinose by isomerization with pyridine; important in carbohydrate metabolism. Also known as D-erythropentose; D-riboketose.

ribulose diphosphate $C_5H_{12}O_{11}P_2$ The phosphate ester of ribulose.

Rice's bromine solution   Analytical reagent for the quantitative analysis of urea; has 12.5% bromine and sodium bromide in aqueous solution.

rich mixture   An air-fuel mixture that is high in its concentration of combustible component.

ricinoleic acid $C_{18}H_{34}O_3$ Unsaturated fatty acid; a combustible, water-insoluble, viscous liquid; soluble in most organic solvents; boils at 226°C (10 mm Hg); used as a chemical intermediate, in soaps and Turkey red oils, and for textile finishing. Also known as castor oil acid; *cis*-12-hydroxyocta-dec-9-enoic acid; 12-hydroxyoleic acid.

ricinoleyl alcohol $C_{18}H_{36}O_2$ Fatty alcohol of ricinoleic acid; a combustible, colorless, nondrying liquid, boiling at 170–328°C; used as a chemical intermediate, in protective coatings, surface-active agents, pharmaceuticals, and plasticizers. Also known as 9-octadecen-1,12-diol.

Riegler's test Analytical technique for nitrous acid; uses sodium naphthionate and β-naphthol.

right-handed Having a crystal structure with a mirror-image relationship to a left-handed structure.

rim gypsum Gypsum in thin films between anhydrite crystals, believed to have been introduced in solution rather than produced by replacement.

ring and ball test A test for determining the melting point of asphalt, waxes, and paraffins in which a small ring is fitted with a test sample upon which a small ball is then placed; the melting point is that temperature at which the sample softens sufficiently to allow the ball to fall through the ring. Also known as ball and ring method.

ring deoxyribonucleic acid *See* circular deoxyribonucleic acid.

Ringer's solution A solution of 0.86 gram sodium chloride, 0.03 gram potassium chloride, and 0.033 gram calcium chloride in boiled, purified water, used topically as a physiological salt solution.

ring isomerism A type of geometrical isomerism in which bond lengths and bond angles prevent the existence of the trans structure if substituents are attached to alkenic carbons which are part of a cyclic system, the ring of which contains fewer than eight members; for example, 1,2-dichlorocyclohexane.

ring system Arbitrary designation of certain compounds as closed, circular structures, as in the six-carbon benzene ring; common rings have four, five, and six members, either carbon or some combination of carbon, nitrogen, oxygen, sulfur, or other elements.

ring whizzer A fluxional molecule frequently encountered in organometallic chemistry in which rapid rearrangements occur by migrations about unsaturated organic rings.

RIS *See* resonance ionization spectroscopy.

risalin $C_{12}H_{18}N_4O_6S$ An orange, crystalline compound with a melting point of 137–138°C; used as a preemergence herbicide for grasses. Also known as 3,5-dinitro-*N,N*-di(*n*-propyl)sulfanilamide.

riser That portion of a bubble-cap assembly in a distillation tower that channels the rising vapor and causes it to flow downward to pass through the liquid held on the bubble plate.

Ritter reaction A procedure for the preparation of amides by reacting alkenes or tertiary alcohols with nitriles in an acidic medium.

Ritz formula A particular expansion of an equation used in studying the spectra of atoms.

Ritz's combination principle The empirical rule that sums and differences of the frequencies of spectral lines often equal other observed frequencies. Also known as combination principle.

Rn *See* radon.

RNA *See* ribonucleic acid.

roasting regeneration Regeneration of a processing (treating) clay by heating or burning it in contact with air to remove combustible impurities adsorbed onto the surface.

Roberts evaporator *See* short-tube vertical evaporator.

Robisonester *See* glucose-6-phosphate.

roentgen spectrometry *See* x-ray spectrometry.

Roese-Gottlieb method A solvent extraction method used to obtain an accurate determination of the fat content of milk.

roller drying A method used to dry milk for purposes other than human consumption; concentrated milk is fed between two heated and narrowly spaced stainless steel rollers, the adhering thin film of milk dries as the rollers turn and is scraped off the roller by a doctor blade.

ronnel $C_8H_8Cl_3O_3PS$ A tan solid with a melting point of 37–38°C; slightly soluble in water; used as an insecticide and miticide for flies and cockroaches. Also known as O,O-dimethyl-O-(2,4,5-trichlorophenyl)phosphorothioate.

Rosenmund reaction Catalytic hydrogenation of an acid chloride to form an aldehyde; the reaction is in the presence of sulfur to prevent the subsequent hydrogenation of the aldehyde.

rosolic acid *See* aurin.

rotary-disk contactor Liquid-liquid contactor, having a vertical cylindrical shell with vertical rotating shaft upon which are mounted a spaced series of flat disks; spinning of the disks forces liquid into shell-mounted baffles, causing mixing; used for liquid-liquid extraction processes. Also known as RDC extractor.

rotating-drum heat transfer Procedure for solidifying layers of solids onto the outside surface of an inside-cooled drum that is partly immersed in a melt of the solids material.

rotating platinum electrode Platinum wire sealed in a soft-glass tubing and rotated by a constant-speed motor; used as the electrode in amperometric titrations. Abbreviated RPE.

rotational constant That constant inversely proportioned to the moment of inertia of a linear molecule; used in calculations of microwave spectroscopy quantums.

rotational energy For a diatomic molecule, the difference between the energy of the actual molecule and that of an idealized molecule which is obtained by the hypothetical process of gradually stopping the relative rotation of the nuclei without placing any new constraint on their vibration, or on motions of electrons.

rotational level An energy level of a diatomic or polyatomic molecule characterized by a particular value of the rotational energy and of the angular momentum associated with the motion of the nuclei.

rotational quantum number A quantum number J characterizing the angular momentum associated with the motion of the nuclei of a molecule; the angular momentum is $(h/2\pi)\sqrt{J(J+1)}$ and the largest component is $(h/2\pi)J$, where h is Planck's constant.

rotational spectrum The molecular spectrum resulting from transitions between rotational levels of a molecule which behaves as the quantum-mechanical analog of a rotating rigid body.

rotational sum rule The rule that, for a molecule which behaves as a symmetric top, the sum of the line strengths corresponding to transitions to or from a given rotational level is proportional to the statistical weight of that level, that is, to $2J + 1$, where J is the total angular momentum quantum number of the level.

rotational transformation A type of crystal transformation that is a change from an ordered phase to a partially disordered phase by rotation of groups of atoms.

rotation axis A symmetry element of certain crystals in which the crystal can be brought into a position physically indistinguishable from its original position by a rotation through an angle of $360°/n$ about the axis, where n is the multiplicity of the axis, equal to 2, 3, 4, or 6. Also known as symmetry axis.

rotation-inversion axis A symmetry element of certain crystals in which a crystal can be brought into a position physically indistinguishable from its original position by a rotation through an angle of $360°/n$ about the axis followed by an inversion, where n is the multiplicity of the axis, equal to 1, 2, 3, 4, or 6. Also known as inversion axis.

rotation-reflection axis A symmetry element of certain crystals in which a crystal can be brought into a position physically indistinguishable from its original position by a rotation through an angle of $360°/n$ about the axis followed by a reflection in the plane perpendicular to the axis, where n is the multiplicity of the axis, equal to 1, 2, 3, 4, or 6.

rotation spectrum Absorption-spectrum (absorbed electromagnetic energy) wavelengths produced if only the rotational energy of a molecule is affected during excitation.

rotation twin A twin crystal in which the parts will coincide if one part is rotated 180° (sometimes 30, 60, or 120°).

rotation-vibration spectrum Absorption-spectrum (absorbed electromagnetic energy) wavelengths produced when both the energy of vibration and energy of rotation of a molecule are affected by excitation.

rotatory power The product of the specific rotation of an element or compound and its atomic or molecular weight.

rotatory reflection axis *See* rotoreflection axis.

rotaxane A compound with two or more independent portions not bonded to each other but linked by a linear portion threaded through a ring and maintained in this position by bulky end groups.

rotenone $C_{23}H_{22}O_6$ White crystals with a melting point of 163°C; soluble in ether and acetone; used as an insecticide and in flea powders and fly sprays. Also known as tubatoxin.

rotoinversion axis A type of crystal symmetry element that combines a rotation of 60, 90, 120, or 180° with inversion across the center. Also known as symmetry axis of rotary inversion; symmetry axis of rotoinversion.

rotoreflection axis A type of symmetry element that combines a rotation of 60, 90, 120, or 180° with reflection across the plane perpendicular to the axis. Also known as rotatory reflection axis.

rowland A unit of length, formerly used in spectroscopy, equal to 999.81/999.94 angstrom, or approximately 0.99987 × 10^{-10} meter.

Rowland circle A circle drawn tangent to the face of a concave diffraction grating at its midpoint, having a diameter equal to the radius of curvature of a grating surface; the slit and camera for the grating should lie on this circle.

Rowland ghost A false spectral line produced by a diffraction grating, arising from periodic errors in groove position.

Rowland grating *See* concave grating.

Rowland mounting A mounting for a concave grating spectrograph in which camera and grating are connected by a bar forming a diameter of the Rowland circle, and the two run on perpendicular tracks with the slit placed at their junction.

roxarsone *See* 4-hydroxy-3-nitrobenzenearsonic acid.

RPE *See* rotating platinum electrode.

r-process The synthesis of elements and nuclides in supernovas through rapid captures of neutrons in a matter of seconds, followed by beta decay.

rRNA *See* ribosomal ribonucleic acid.

Ru *See* ruthenium.

rubber A natural, synthetic, or modified high polymer with elastic properties and, after vulcanization, elastic recovery; the generic term is elastomer.

rubber accelerator A substance that increases the speed of curing of rubber, such as thiocarbanilide.

rubber hydrochloride White, thermoplastic hydrochloric acid derivative of rubber; water-insoluble powder or clear film, soluble in aromatic hydrocarbons; softens at 110–120°C; used for protective coverings, food packaging, shower curtains, and rainwear.

rubeanic acid *See* dithiooxamide.

rubidium A chemical element, symbol Rb, atomic number 37, atomic weight 85.47; a reactive alkali metal; salts of the metal may be used in glass and ceramic manufacture.

rubidium bromide RbBr Colorless, regular crystals, melting at 683°C; soluble in water; used as a nerve sedative.

rubidium chloride RbCl A water-soluble, white, lustrous powder melting at 715°C; used as a source for rubidium metal, and as a laboratory reagent.

rubidium halide Any of the halogen compounds of rubidium; examples are RbBr, RbCl, RbF, RbIBrCl, $RbBr_2Cl$, and $RbIBr_2$.

rubidium halometallate Halogen-metal-containing compounds of rubidium; examples are Rb_2GeF_6 (rubidium hexafluorogermanate), Rb_2PtCl_6 (rubidium chloroplatinate), and Rb_2PdCl_5 (rubidium palladium chloride).

rubidium sulfate Rb_2SO_4 Colorless, water-soluble rhomboid crystals, melting at 1060°C; used as a cathartic.

ruling engine A machine operated by a long micrometer screw which rules equally spaced lines on an optical diffraction grating.

run 1. The amount of feedstock processed by a petroleum refinery unit during a given time; often used colloquially in relation to the type of stock being processed, as in crude run or naphtha run. 2. A processing-cycle or batch-treatment operation.

runback A pipe through which all or part of a distillation column's overhead condensate can be run back into the column, instead of being drawn off as product.

rundown line A line from a process unit that connects the look box in the receiving house with the tank in which the product is temporarily stored.

rundown tank A tank in which the product from a still, agitator, or other processing equipment is received, and from which the product is pumped to larger storage tanks. Also known as pan tank; receiving tank.

Rushton-Oldshue column A mixing unit used for continuous pipeline blending in which two-phase contacting is desired; it is a column containing separation plates, baffles, and mixing impellers.

Russell movable-wall oven An oven for coal carbonization which cokes a 400-pound (180-kilogram) charge in a horizontal, 12-inch-wide (30-centimeter-wide) chamber, heated from both sides, but with one side floating and balanced against scales.

ruthenic chloride *See* ruthenium chloride.

ruthenium A chemical element, symbol Ru, atomic number 44, atomic weight 101.07.

ruthenium chloride $RuCl_3$ Black, deliquescent, water-insoluble solid that decomposes in hot water and above 500°C; used as a laboratory reagent. Also known as ruthenic chloride; ruthenium sesquichloride.

ruthenium halide Halogen compound of ruthenium; examples are $RuCl_2$, $RuCl_3$, $RuCl_4$, $RuBr_3$, and RuF_5.

ruthenium red $Ru_2(OH)_2Cl_4 \cdot 7NH_3 \cdot 3H_2O$ A water-soluble, brownish-red powder; used as an analytical reagent and stain. Also known as ammoniated ruthenium oxychloride.

ruthenium sesquichloride *See* ruthenium chloride.

ruthenium tetroxide RuO_4 A yellow, toxic solid, melting at 25°C; used as an oxidizing agent.

Rutherford backscattering spectrometry A method of determining the concentrations of various elements as a function of depth beneath the surface of a sample, by measuring the energy spectrum of ions which are backscattered out of a beam directed at the surface.

Rutherford nuclear atom A theory of atomic structure in which nearly all the mass is concentrated in a small nucleus, electrons surrounding the nucleus fill nearly all the atom's volume, the number of these electrons equals the atomic number, and the positive charge on the nucleus is equal in magnitude to the negative charge of the electrons.

Rutherford scattering Scattering of heavy charged particles by the Coulomb field of an atomic nucleus.

rutin $C_{27}H_{32}O_{16}$ A hydroxyflavone glucorhamnoside derived from cowslip and other plants; yellow needles melting at 190°C; used to treat capillary disorders.

ry *See* rydberg.

rydberg **1.** A unit of energy used in atomic physics, equal to the square of the charge of the electron divided by twice the Bohr radius; equal to 13.60583 ± 0.00004 electronvolts. Symbolized ry. **2.** *See* kayser.

Rydberg atom An atom whose outer electron has been excited to very high energy states, far from the nucleus.

Rydberg constant **1.** An atomic constant which enters into the formulas for wave numbers of atomic spectra, equal to $2\pi^2 me^4/ch^3$, where m and e are the rest mass and charge of the electron, c is the speed of light, and h is Planck's constant; equal to $109,737.31 \pm 0.01$ inverse centimeters. Symbolized R_∞. **2.** For any atom, the Rydberg constant (first definition) divided by $1 + m/M$, where m and M are the masses of an electron and of the nucleus.

Rydberg correction A term inserted into a formula for the energy of a single electron in the outermost shell of an atom to take into account the failure of the inner electron shells to screen the nuclear charge completely.

Rydberg series formula An empirical formula for the wave numbers of various lines of certain spectral series such as neutral hydrogen and alkali metals; it states that the wave number of the nth member of the series is $\lambda_\infty - R/(n+a)^2$, where λ_∞ is the series limit, R is the Rydberg constant of the atom, and a is an empirical constant.

S

S *See* sulfur.

S$_m$ *See* Semenov number.

saccharase An enzyme that catalyzes the hydrolysis of disaccharide to monosaccharides, specifically of sucrose to dextrose and levulose. Also known as invertase; invertin; sucrase.

saccharin $C_6H_4COSO_2NH$ A sweet-tasting, white powder, soluble in acetates, benzene, and alcohol; slightly soluble in water and ether; melts at 228°C; used as a sugar substitute for syrups, and in medicines, foods, and beverages. Also known as benzosulfimide; gluside.

saccharolactic acid *See* mucic acid.

saccharose *See* sucrose.

sacrificial metal A metal that can be used for a sacrificial anode.

saddle-point azeotrope A rarely occurring azeotrope which is formed in ternary systems and for which the boiling point is intermediate between the highest and lowest boiling mixture in the system.

safranine Any of a group of phenazine-based dyes; some are used as biological stains.

safrole $C_3H_5C_6H_3O_2CH_2$ A toxic, water-insoluble, colorless oil that boils at 233°C; found in sassafras and camphorwood oils; used in medicine, perfumes, insecticides, and soaps, and as a chemical intermediate. Also known as 3,4-methylenedioxy-1-allyl benzene.

sal acetosella *See* potassium binoxalate.

salazosulfadimidine $C_{19}H_{17}N_5O_5S$ A brown crystalline compound that melts at 207°C; used in medicine in cases of ulcerative colitis. Also known as salicylazosulfadimidine; salicylazosulfamethazine.

sal ethyl *See* ethyl salicylate.

salicin $C_{13}H_{18}O_7$ A glucoside; colorless crystals, soluble in water, alcohol, alkalies, and glacial acetic acid; melts at 199°C; used in medicine and as an analytical reagent.

salicylal *See* salicylaldehyde.

salicyl alcohol $C_7H_8O_2$ A crystalline alcohol that forms plates or powder, melting at 86–87°C; used in medicine as a local anesthetic. Also known as *ortho*-hydroxybenzyl alcohol; saligenol.

salicylaldehyde C_6H_4OHCHO Clear to dark-red oily liquid, with burning taste and almond aroma; soluble in alcohol, benzene, and ether, very slightly soluble in water; boils at 196°C; used in analytical chemistry, in perfumery, and for synthesis of chemicals. Also known as helicin; *ortho*-hydroxybenzaldehyde; salicylal; salicylic aldehyde.

salicylamide $C_6H_4(OH)CONH_2$ Pinkish or white crystals; soluble in alcohol, ether, chloroform, and hot water; melts at 193°C; used in medicine as an analgesic, antipyretic, and antirheumatic drug. Also known as ethbenzamide; 2-ethoxybenzene carbonamide; *ortho*-ethyl ether; *ortho*-hydroxybenzamide.

salicylate A salt of salicylic acid with the formula $C_6H_4(OH)COOM$, where M is a monovalent metal; for example, $NaC_7H_5O_3$, sodium salicylate.

salicylated mercury *See* mercuric salicylate.

salicylazosulfadimidine *See* salazosulfadimidine.

salicylazosulfamethazine *See* salazosulfadimidine.

salicylic acid $C_6H_4(OH)(COOH)$ White crystals with sweetish taste; soluble in alcohol, acetone, ether, benzene, and turpentine, slightly soluble in water; discolored by light; melts at 158°C; used as a chemical intermediate and in medicine, dyes, perfumes, and preservatives. Also known as *ortho*-hydroxybenzoic acid.

salicylic acid ethyl ether *See* ethyl salicylate.

salicylic acid phenyl ester *See* salol.

salicylic aldehyde *See* salicylaldehyde.

salicylic ether *See* ethyl salicylate.

salicylsulfonic acid *See* sulfosalicylic acid.

saligenol *See* salicyl alcohol.

saline-water reclamation Purification and removal of salts from brine or brackish water by ion exchange, crystallization, distillation, evaporation, and reverse osmosis.

salivary amylase *See* ptyalin.

salivary diastase *See* ptyalin.

salol $C_6H_4OHCOOC_6H_5$ White powder with aromatic taste and aroma; soluble in alcohol, ether, chloroform, and benzene; slightly soluble in water; melts at 42°C; used in medicinals and as a preservative. Also known as phenyl salicylate; salicylic acid phenyl ester.

sal soda $Na_2CO_3 \cdot 10H_2O$ White, water-soluble crystals, insoluble in alcohol; melts and loses water at about 33°C; mild irritant to mucous membrane; used in cleansers and for washing textiles and bleaching linen and cotton. Also known as sodium carbonate decahydrate; washing soda.

salt The reaction product when a metal displaces the hydrogen of an acid; for example, $H_2SO_4 + 2NaOH \rightarrow Na_2SO_4$ (a salt) + $2H_2O$.

sal tartar *See* sodium tartrate.

salt bridge A bridge of a salt solution, usually potassium chloride, placed between the two half-cells of a galvanic cell, either to reduce to a minimum the potential of the liquid junction between the solutions of the two half-cells or to isolate a solution under study from a reference half-cell and prevent chemical precipitations.

salt cake Impure sodium sulfate; used in soaps, paper pulping, detergents, glass, ceramic glaze, and dyes.

salt error An error introduced in an analytical determination of a saline liquid such as sea water; caused by the effect of the neutral ions in the solution on the color of the pH indicator, and hence upon the apparent pH.

salt grainer Type of evaporative crystallizer in which the solution is kept hot, and supersaturation is developed by evaporation rather than by cooling.

salting-out effect The growth of crystals of a substance on heated, liquid-holding surfaces of a crystallizing evaporator as a result of the decrease in solubility of the substance with increase in temperature.

salt of Lemery *See* potassium sulfate.

salt of sorrel *See* potassium binoxalate.

salt of tartar *See* potassium carbonate.

salt pan A pool used for obtaining salt by the natural evaporation of sea water.

saltpeter *See* potassium nitrate.

samarium Group III rare-earth metal, atomic number 62, symbol Sm; melts at 1350°C, tarnishes in air, ignites at 200–400°C.

samarium oxide Sm_2O_3 A cream-colored powder with a melting point of 2300°C; soluble in acids; used for infrared-absorbing glass and as a neutron absorber.

SAN *See* styrene-acrylonitrile resin.

Sandmeyer's reaction Conversion of diazo compounds (in the presence of cuprous halogen salts) into halogen compounds.

Sanger's reagent *See* 1-fluoro-2,4-dinitrobenzene.

santalol $C_{15}H_{24}O$ A colorless liquid with a boiling point of 300°C; derived from sandalwood oil and used for perfumes.

santonin $C_{15}H_{18}O_3$ A white powder with a melting point of 170–173°C; soluble in chloroform and alcohol; used in medicine.

saponification The process of converting chemicals into soap; involves the alkaline hydrolysis of a fat or oil, or the neutralization of a fatty acid.

saponification equivalent The quantity of fat in grams that can be saponified by 1 liter of normal alkalies.

saponification number Milligrams of potassium hydroxide required to saponify the fat, oil, or wax in a 1-gram sample of a given material, using a specific ASTM test method.

saponin Any of numerous plant glycosides characterized by foaming in water and by producing hemolysis when water solutions are injected into the bloodstream; used as beverage foam producer, textile detergent and sizing, soap substitute, and emulsifier.

sarcosine CH_3NHCH_2COOH Sweet-tasting, deliquescent crystals; soluble in water, slightly soluble in alcohol; decomposes at 210–215°C; used in toothpaste manufacture. Also known as methyl aminoacetic acid; methyl glycocol.

Sargent curve A graph of logarithms of decay constants of radioisotopes subject to beta-decay against logarithms of the corresponding maximum beta-particle energies; most of the points fall on two straight lines.

sarmentocymarin A cardioactive, steroid glycoside from the seeds of *Strophanthus sarmentosus;* on hydrolysis it yields sarmentogenin and sarmentose.

sarmentogenin $C_{23}H_{34}O_5$ The steroid aglycon of sarmentocymarin; isometric with digitoxigenin, and characterized by a hydroxyl group at carbon number 11.

SAS *See* aluminum sodium sulfate

satellite infrared spectrometer A spectrometer carried aboard satellites in the Nimbus series which measures the radiation from carbon dioxide in the atmosphere at several different wavelengths in the infrared region, giving the vertical temperature structure of the atmosphere over a large part of the earth. Abbreviated SIRS.

saturated ammonia **1.** Liquid ammonia in a state in which adding heat at constant pressure causes the liquid to vaporize at constant temperature, and in which removing heat at constant pressure causes the temperature of the liquid to drop immediately. **2.** Ammonia vapor in a state in which adding heat at constant pressure causes an immediate temperature rise (superheating) and in which removing heat at constant pressure starts immediate condensation at constant temperature.

saturated calomel electrode A reference electrode of mercury topped by a layer of mercury (I) chloride paste with potassium chloride solution placed above; easier to assemble than the normal and the one-tenth normal (referring to the concentration of KCl) calomel electrodes.

saturated compound An organic compound with all carbon bonds satisfied; it does not contain double or triple bonds and thus cannot add elements or compounds.

saturated hydrocarbon A saturated carbon-hydrogen compound with all carbon bonds filled; that is, there are no double or triple bonds as in olefins and acetylenics.

saturated interference spectroscopy A version of saturation spectroscopy in which the gas sample is placed inside an interferometer that splits a probe laser beam into parallel components in such a way that they cancel on recombination; intensity changes in the recombined probe beam resulting from changes in absorption or refractive index induced by a laser saturating beam are then measured.

saturated liquid A solution that contains enough of a dissolved solid, liquid, or gas so that no more will dissolve into the solution at a given temperature and pressure.

saturation The condition in which the partial pressure of any fluid constituent is equal to its maximum possible partial pressure under the existing environmental conditions, such that any increase in the amount of that constituent will initiate within it a change to a more condensed state.

saturation spectroscopy A branch of spectroscopy in which the intense, monochromatic beam produced by a laser is used to alter the energy-level populations of a resonant medium over a narrow range of particle velocities, giving rise to extremely narrow spectral lines that are free from Doppler broadening; used to study atomic, molecular, and nuclear structure, and to establish accurate values for fundamental physical constants.

saxitoxin A nonprotein toxin produced by the dinoflagellate *Gonyaulax catenella*.

Sb *See* antimony.

SBK catalytic reforming A proprietary petroleum refinery process using SBK (Sinclair-Baker-Kellogg) regenerable platinum catalyst; used for catalytic reforming of gasoline-range hydrocarbons to produce high octanes.

Sc *See* scandium.

Sc₃ *See* Schmidt number 3.

scalenohedron A closed crystal form whose faces are scalene triangles.

scandia *See* scandium oxide.

scandium A metallic group III element, symbol Sc, atomic number 21; melts at 1200°C; found associated with rare-earth elements.

scandium halide A compound of scandium and a halogen; for example, scandium chloride, $ScCl_3$.

scandium oxide Sc_2O_3 White powder, soluble in hot acids; used to prepare scandium. Also known as scandia.

scandium sulfate $Sc_2(SO_4)_3$ Water-soluble, colorless crystals.

scandium sulfide Sc_3S_3 Yellowish powder; decomposes in dilute acids and boiling water to give off hydrogen sulfide.

scarlet red $CH_3C_6H_4H{:}NC_6H_3CH_3N{:}NC_{10}H_{15}OH$ A brown, water-insoluble powder, used as a dye in ointments. Also known as Biebrich red; scarlet.

scattering length A parameter used in analyzing nuclear scattering at low energies; as the energy of the bombarding particle becomes very small, the scattering cross section approaches that of an impenetrable sphere whose radius equals this length. Also known as scattering power.

scattering plane In a quasielastic light-scattering experiment performed with the use of polarizers, the plane containing the incident and scattered beams.

scattering power *See* scattering length.

scavenger A substance added to a mixture or other system to remove or inactivate impurities.

Schaeffer's salt $HOC_{10}SO_3Na$ A light-yellow to pink, water-soluble powder; the sodium salt formed from 2-naphthol-6-sulfonic acid; used as an intermediate in synthesis of organic compounds.

Schardinger dextrin Any of several water-soluble, nonreducing, dextrorotatory polysaccharides that have a low molecular weight and that are obtained by the action of *Bacillus macerans* on a starch solution.

Scheele's green *See* copper arsenite.

Scheibel column *See* Scheibel extractor.

Scheibel extractor Liquid-liquid contact vessel used in liquid-liquid extraction processes: a vertical cylinder with interspersed open spaces and wire-mesh packing along its height, with liquid agitators in the open spaces, or a vertical cylinder fully filled with wire-mesh packing. Also known as Scheibel column; Scheibel-York extractor; York-Scheibel column.

Scheibel-York extractor   *See* Scheibel extractor.

Schiff base $RR'C{=}NR''$ Any of a class of derivatives of the condensation of aldehydes or ketones with primary amines; colorless crystals, weakly basic; hydrolyzed by water and strong acids to form carbonyl compounds and amines; used as chemical intermediates and perfume bases, in dyes and rubber accelerators, and in liquid crystals for electronics.

Schiff's reagent   An aqueous solution of rosaniline and sulfurous acid; used in the Schiff test.

Schiff test   A test for aldehydes by using an aqueous solution of rosaniline and sulfurous acid.

Schmidt lines   Two lines, on a graph of nuclear magnetic moment versus nuclear spin, on which points describing all nuclides should lie, according to the independent particle model; experimentally, however, points describing nuclides are scattered between the lines.

Schmidt number 2   *See* Semenov number 1.

Schmidt number 3   A dimensionless number used in electrochemistry, equal to the product of the dielectric susceptibility and the dynamic viscosity of a fluid divided by the product of the fluid density, electrical conductivity, and the square of a characteristic length. Symbolized Sc_3.

Schoelkopf's acid   A dye of the following types: 1-naphthol-4,8-disulfonic acid, 1-naphthylamine-4,8-disulfonic acid, and 1-naphthylamine-8-sulfonic acid; may be toxic.

Schoenherr-Hessberger process   A nitrogen-fixation process used in Norway; employs a very long (7 meters) alternating-current arc around which air moves in a helical path in a 746-kilowatt furnace.

Schönflies crystal symbols   Symbols denoting the 32 crystal point groups or symmetry classes; capital letters indicate the general type of class, and subscripts the multiplicity of rotation axes and the existence of additional symmetries.

Schotten-Baumann reaction   An acylation reaction that uses an acid chloride in the presence of dilute alkali to acylate the hydroxyl and amino group of organic compounds.

Schuermann series   A list of metals so arranged that the sulfide of one is precipitated at the expense of the sulfide of any lower metal in the series.

Schultz-Hardy rule   The sensitivity of lyophobic colloids to coagulating electrolytes is governed by the charge of the ion opposite that of the colloid, and the sensitivity increases more rapidly than the charge of the ion.

Schulze's reagent   An oxidizing mixture consisting of a saturated aqueous solution of $KClO_3$ and varying amounts of concentrated HNO_3; commonly used in palynologic macerations.

Schuster method   A method for focusing a prism spectroscope without using a distant object or a Gauss eyepiece.

Schweitzer's reagent   An ammoniacal solution of cupric hydroxide; used to dissolve cellulose, silk, and linen, and to test for wool.

scleroprotein Any one of a class of proteins, such as keratin, fibroin, and the collagens, which occur in hard parts of the animal body and serve to support or protect. Also known as albuminoid.

scopoline $C_8H_{13}O_2H$ A white crystalline alkaloid that melts at 108–109°C, soluble in water and ethanol; used in medicine. Also known as oscine.

scorching Premature vulcanization caused by heat during the processing of rubber.

scraped-surface exchanger A liquid-liquid heat-exchange device that has a rotating element with spring-loaded scraper blades to wipe the process-fluid exchange surfaces clean of crystals or other foulants; used in paraffin-wax processing.

screen dryer *See* traveling-screen dryer.

screening The reduction of the electric field about a nucleus by the space charge of the surrounding electrons.

screening agent A nonchelating dye used to improve the colorimetric end point of a complexometric titration; a dye addition forms a complementary pair of colors with the metalized and unmetalized forms of the end-point indicator.

screening constant The difference between the atomic number of an element and the apparent atomic number for a given process; this difference results from screening.

screw axis A symmetry element of some crystal lattices, in which the lattice is unaltered by a rotation about the axis combined with a translation parallel to the axis and equal to a fraction of the unit lattice distance in this direction.

screw dislocation A dislocation in which atomic planes form a spiral ramp winding around the line of the dislocation.

Se *See* selenium.

seasonal balancing A seasonal adjustment of the front-end boiling range (volatility) of a motor gasoline to control engine starting characteristics by compensating for seasonal temperature changes.

sebacic acid $COOH(CH_2)_8COOH$ Combustible, white crystals; slightly soluble in water, soluble in alcohol and ether; melts at 133°C; used in perfumes, paints, and hydraulic fluids and to stabilize synthetic resins. Also known as decanedioic acid; sebacylic acid.

sebacylic acid *See* sebacic acid.

secalose A polysaccharide consisting of fructose units; occurs in green rye and oats, and in rye flour.

secbumeton $C_{10}H_{19}N_5O$ A white solid with a melting point of 86–88°C; used as an herbicide for alfalfa. Also known as 2-*sec*-butylamino-4-ethylamino-6-methoxy-S-triazine.

secondary alcohol An organic alcohol with molecular structure R_1R_2CHOH, where R_1 and R_2 designate either identical or different groups.

secondary amine An organic compound that may be written R_1R_2NH, where R_1 and R_2 designate either identical or different groups.

secondary carbon atom A carbon atom that is singly bonded to two other carbon atoms.

secondary hydrogen atom   A hydrogen atom that is bonded to a secondary carbon atom.

secondary manganous phosphate   *See* acid manganous phosphate.

secondary twinning   Twinning of a crystal caused by an external influence, such as pressure in rock.

second boiling point   In certain mixtures, the temperature at which a gas phase develops from a liquid phase upon cooling.

second-order reaction   A reaction whose rate of reaction is determined by the concentration of two chemical species.

secretin   A basic polypeptide hormone produced by the duodenum in response to the presence of acid; acts to excite the pancreas to activity.

sedimentation balance   A device to measure and record the weight of sediment (solid particles settled out of a liquid) versus time; used to determine particle sizes of fine solids.

sedimentation coefficient   In the sedimentation of molecules in an accelerating field, such as that of a centrifuge, the velocity of the boundary between the solution containing the molecules and the solvent divided by the accelerating field. (In the case of a centrifuge, the accelerating field equals the distance of the boundary from the axis of rotation multiplied by the square of the angular velocity in radians per second.)

sedimentation constant   A quantity used in studying the behavior of colloidal particles subject to forces, especially centrifugal forces; it is equal to $2r^2(\rho - \rho')/9\eta$, where r is the particle's radius, ρ and ρ' are reciprocals of partial specific volumes of particle and medium respectively, and η is the medium's viscosity.

sedimentation equilibrium   The equilibrium between the forward movement of a sample's liquid-sediment boundary and reverse diffusion during centrifugation; used in molecular-weight determinations.

sedimentation velocity   The rate of movement of the liquid-sediment boundary in the sample holder during centrifugation; used in molecular-weight determinations.

sedoheptulose   A seven-carbon ketose sugar widely distributed in plants of the Crassulaceae family; a significant intermediary compound in the cyclic regeneration of D-ribulose.

seed   A small, single crystal of a desired substance added to a solution to induce crystallization.

seed charge   A small amount of material added to a supersaturated solution to initiate precipitation.

seeding   The adding of a seed charge to a supersaturated solution, or a single crystal of a desired substance to a solution of the substance to induce crystallization.

Segas process   A process for the production of low-Btu gas by the catalytic method using a fixed bed catalyst, lime-bauxite mixture bonded with bentonite.

Segrè chart   A chart of the nuclides which is laid off in squares, each square displaying data about a nuclide; each column contains nuclides with a given neutron number and each row contains nuclides with a given atomic number;

successive columns and rows represent successively higher numbers of neutrons and protons.

selective adsorbent Material that will selectively adsorb (or reject) one or more specific components from a multicomponent mixture of gases or liquids; common adsorbents are silica gel, carbon and activated carbon, activated alumina, and synthetic or natural zeolites (molecular sieves).

selective cracking A refinery process in which recycled stock is distilled in equipment kept separate from that used for distillation of original stock.

selective polymerization The polymerization of a single type of molecule in a mixture of monomers; for example, the production of diisobutylene from a mixture of butylenes.

selective solubility diffusion The transmission of fluids through a nonporous, polymeric barrier (membrane) by an adsorption-solution-diffusion-desorption sequence.

selective solvent A solvent that, at certain temperatures and ratios with other materials, preferentially dissolves more of one component of a liquid or solids mixture than of another, thereby permitting partial separation.

selectivity coefficient Ion equilibria relationship formula for ion-exchange-resin systems.

selectivity diagram A triangular plot of solubilities in a ternary liquid system; used to calculate the ability of a solvent to extract a component from a mixture (its selectivity) at various concentration combinations.

s-electron An atomic electron that is described by a wave function with orbital angular momentum quantum number of zero in the independent particle approximation.

selenic acid H_2SeO_4 A highly toxic, water-soluble, white solid, melting point 58°C, decomposing at 260°C.

selenide 1. M_2Se A binary compound of divalent selenium, such as Ag_2Se, silver selenide. 2. An organic compound containing divalent selenium, such as $(C_2H_5)_2Se$, ethyl selenide.

selenious acid *See* selenous acid.

selenium A highly toxic, nonmetallic element in group VI, symbol Se, atomic number 34; steel-gray color; soluble in carbon disulfide, insoluble in water and alcohol; melts at 217°C; and boils at 690°C; used in analytical chemistry, metallurgy, and photoelectric cells, and as a lube-oil stabilizer and chemicals intermediate.

selenium bromide Any of three compounds of selenium and bromine: Se_2Br_2, a red liquid that melts at −46°C, also known as selenium monobromide; $SeBr_2$, a brown liquid, also known as selenium dibromide; and $SeBr_4$, orange, carbon-disulfide-soluble crystals, also known as selenium tetrabromide.

selenium dibromide *See* selenium bromide.

selenium dioxide SeO_2 Water- and alcohol-soluble, white to reddish, lustrous crystals; melts at 340°C; used in medicine, and as an oxidizing agent and catalyst. Also known as selenous acid anhydride; selenous anhydride; selenium oxide.

selenium disulfide *See* selenium sulfide.

selenium halide A compound of selenium and a halogen, for example, Se_2Br_2, $SeBr_2$, $SeBr_4$; Se_2Cl_2, $SeCl_2$, $SeCl_4$; Se_2I_2, SeI_4.

selenium monobromide *See* selenium bromide.

selenium nitride Se_2N_2 A water-insoluble, yellow solid that explodes at 200°C.

selenium oxide *See* selenium dioxide.

selenium tetrabromide *See* selenium bromide.

selenone A group of organic selenium compounds with the general formula R_2SeO_2.

selenonic acid Any organic acid containing the radical —SeO_3H; analogous to a sulfonic acid.

selenous acid H_2SeO_3 Colorless, transparent crystals; soluble in water and alcohol, insoluble in ammonia; decomposes when heated; used as an analytical reagent. Also spelled selenious acid.

selenous acid anhydride *See* selenium dioxide.

selenous anhydride *See* selenium dioxide.

selenoxide A group of organic selenium compounds with the general formula R_2SeO.

self-absorption Reduction of the intensity of the center of an emission line caused by selective absorption by the cooler portions of the source of radiation. Also known as self-reduction; self-reversal.

self-reduction *See* self-absorption.

self-reversal *See* self-absorption.

Seliwanoff's test A color test helpful in the identification of ketoses, which develop a red color with resorcinol in hydrochloric acid.

sellite A solution of sodium sulfate that is used during the manufacture of trinitrotoluene.

Semenov number 1 A dimensionless number used in reaction kinetics, equal to a mass transfer constant divided by a reaction rate constant. Symbolized S_m. Formerly known as Schmidt number 2.

semibatch chemical reactor A reactor in which a constant liquid volume is maintained without any overflow, and with the continuous addition of one reactant, usually a gas.

semicarbazide hydrochloride $CH_5ON_3 \cdot HCl$ Colorless prisms, soluble in water, decomposing at 175°C; used as an analytical reagent for aldehydes and ketones, and to recover constituents of essential oils. Also known as aminourea hydrochloride; carbamylhydrazine hydrochloride.

semicarbazone $R_2C:N_2HCONH_2$ A condensation product of an aldehyde or ketone with semicarbazide.

semichemical pulping A method of producing wood-fiber products in which the wood chips are merely softened by chemical treatment (neutral sodium sulfite solution), while the remainder of the pulping action is supplied by a disk attrition mill or by some similar mechanical device for separating the fibers.

semimicroanalysis A chemical analysis procedure in which the weight of the sample is between 10 and 100 milligrams.

semioxamazide $H_2NCOCONHNH_2$ A crystalline compound that decomposes at 220°C; soluble in hot water, acids, and alkalies; used as a reagent for aldehydes and ketones. Also known as aminooxamide; oxamic acid hydrazide.

sensing zone technique Particle-size measurement in a dilute solution, with fine particles passed through a small zone (opening) so that individual particles may be observed and measured by electrolytic, photic, or sonic methods.

separating power The measure of the ability of a system (such as a rectifying system) to separate the components of a mixture, when the components have increasingly close boiling points.

separation The separation of liquids or gases in a mixture, as by distillation or extraction.

separation energy The energy needed to remove a proton, neutron, or alpha particle from a nucleus.

separative work unit A fundamental measure of work required to separate a quantity of isotopic mixture into two component parts, one having a higher percentage of concentration of the desired isotope and one having a lower percentage.

Separator-Nobel dewaxing *See* S-N dewaxing.

separatory funnel A funnel-shaped device used for the careful and accurate separation of two immiscible liquids; a stopcock on the funnel stem controls the rate and amount of outflow of the lower liquid.

sequestering agent A substance that removes a metal ion from a solution system by forming a complex ion that does not have the chemical reactions of the ion that is removed; can be a chelating or a complexing agent.

Serber potential A potential between nucleons, equal to $\frac{1}{2}(1 + M)V(r)$, where $V(r)$ is a function of the distance between the nucleons, and M is an operator which exchanges the spatial coordinates of the particles but not their spins (corresponding to the Majorana force).

series decay *See* radioactive series.

series disintegration The successive radioactive transformations in a radioactive series. Also known as chain decay; chain disintegration.

series of lines A collection of spectral lines of an atom or ion for a set of transitions, with the same selection rules, to a single final state; often the frequencies have a general formula of the form $[R/(a + c_1)^2] - [R/(n + c_2)^2]$, where R is the Rydberg constant for the atom, a and c_1 and c_2 are constants, and n takes on the values of the integers greater than a for the various lines in the series.

serine $C_3H_7O_3N$ An amino acid obtained by hydrolysis of many proteins; a biosynthetic precursor of several metabolites, including cysteine, glycine, and choline.

serotonin $C_{10}H_{12}ON_2$ A compound derived from tryptophan which functions as a local vasoconstrictor, plays a role in neurotransmission, and has pharmacologic properties. Also known as 5-hydroxytryptamine.

serum albumin The principal protein fraction of blood serum and serous fluids.

serum globulin The globulin fraction of blood serum.

sesquioxide A compound composed of a metal and oxygen in the ratio 2:3; for example, Al_2O_3.

sesquiterpene Any terpene with the formula $C_{15}H_{24}$; that is, 1½ times the terpene formula.

set The hardening or solidifying of a plastic or liquid substance.

severity factor A measure of the severeness or intensity of overall reaction conditions in a chemical reaction; for example, the temperature, pressure, or conversion in a catalytic cracker or reformer.

sexadentate ligand *See* hexadentate ligand.

sex hormone Any hormone secreted by a gonad, but also found in other tissues.

SFS *See* sodium formaldehyde sulfoxylate.

shaking-out A procedure in which a sample of crude oil is centrifuged at high speed to separate its components; used to determine sediment and water content.

shape isomer An excited nuclear state which has an unusually long lifetime because of its deformed shape, which differs drastically from that of the lower energy states into which it is permitted to decay.

sharp series A series occurring in the line spectra of many atoms and ions with one, two, or three electrons in the outer shell, in which the total orbital angular momentum quantum number changes from 0 to 1.

shed A unit of cross section, used in studying collisions of nuclei and particles, equal to 10^{-24} barn, or 10^{-48} square centimeter.

Shell fluid catalytic cracking A two-stage, proprietary petroleum-refining process for catalytic cracking of gas oils; the catalyst is regenerated in one of the stages.

shell model A model of the nucleus in which the shell structure is either postulated or is a consequence of other postulates; especially the model in which the nucleons act as independent particles filling a preassigned set of energy levels as permitted by the quantum numbers and Pauli principle.

shell still A distillation device formerly used in petroleum refineries; oil was charged into a closed, cylindrical shell and heat was applied to the outside of the bottom by a firebox.

shell structure Structure of the nucleus in which nucleons of each kind occupy quantum states which are in groups of approximately the same energy, called shells, the number of nucleons in each shell being limited by the Pauli exclusion principle.

shift of spectral line A small change in the position of a spectral line that is due to a corresponding change in frequency which, in turn, results from one or more of several causes, such as the Doppler effect.

shikimic acid $C_7H_{10}O_5$ A crystalline acid that is a plant constituent, and an intermediate in the biochemical pathway from phosphoenolpyruvic acid to tyrosine.

short residuum A petroleum refinery term for residual oil from crude-oil distillation operations in which neutral oils are taken overhead with the distillate.

short stop A substance added during a polymerization process to terminate the reaction.

short-tube vertical evaporator   A liquid evaporation process unit with a vertical bundle of tubes 2–3 inches (5–8 centimeters) in diameter and 4–6 feet (1.2–1.8 meters) long; the heating fluid is inside the tubes, and the liquid to be evaporated is in the shell area outside the tubes; used mainly to evaporate cane-sugar juice. Also known as calandria evaporator; Roberts evaporator; standard evaporator.

shower   *See* cosmic-ray shower.

Si   *See* silicon.

sialic acid   Any of a family of amino sugars, containing nine or more carbon atoms, that are nitrogen- and oxygen-substituted acyl derivatives of neuraminic acid; as components of lipids, polysaccharides, and mucoproteins, they are widely distributed in bacteria and in animal tissues.

sialomucin   An acid mucopolysaccharide containing sialic acid as the acid component.

side-centered lattice   A type of centered lattice that is centered on the side faces only.

side chain   A grouping of similar atoms (two or more, generally carbons, as in the ethyl radical, C_2H_5—) that branches off from a straight-chain or cyclic (for example, benzene) molecule.

side pinacoid   A pinacoid with Miller indices (010) in an orthorhombic, monoclinic, or triclinic crystal.

side reaction   A secondary or subsidiary reaction that takes place simultaneously with the reaction of primary interest.

siderophile element   An element with a weak affinity for oxygen and sulfur and that is readily soluble in molten iron; includes iron, nickel, cobalt, platinum, gold, tin, and tantalum.

sidestream   A liquid stream taken from an intermediate point of a liquids-processing unit, for example, a distillation or extraction tower.

sidestream stripper   A device used to perform further distillation on a liquid stream (sidestream) from any one of the plates of a bubble tower, usually with the use of steam.

siduron $C_{14}H_{20}N_2O$ A colorless, crystalline solid with a melting point of 133–138°C; used as a preemergence control of undesirable grasses in turf. Also known as 1-(2-methylcyclohexyl)-3-phenylurea.

siegbahn   A unit of length, formerly used to express wavelengths of x-rays, equal to exactly 1/3029.45 of the spacing of the (200) planes of calcite at 18°C, or to $(1.00202 \pm 0.00003) \times 10^{-13}$ meter. Also known as x-ray unit; X-unit. Symbolized X; XU.

sieve plate   A distillation-tower tray that is perforated so that the vapor emerges vertically through the tray, passing through the liquid holdup on top of the tray; used as a replacement for bubble-cap trays in distillation. Also known as sieve tray.

sieve shaker   A device used to shake a stacked column of standard sieve-test trays to cause solids to sift progressively from the top (large openings) to the bottom (small openings and a final pan), according to particle size.

sieve tray   *See* sieve plate.

sigma bond   The chemical bond resulting from the formation of a molecular orbital by the end-on overlap of atomic orbitals.

sigma-minus hyperonic atom   An atom consisting of a negatively charged sigma hyperon orbiting around an ordinary nucleus. Designated Σ^- hyperonic atom.

silane Si_nH_{2n+2} A class of silicon-based compounds analogous to alkanes, that is, straight-chain, saturated paraffin hydrocarbons; they can be gaseous or liquid. Also known as silicon hydride.

silica gel   A colloidal, highly absorbent silica used as a dehumidifying and dehydrating agent, as a catalyst carrier, and sometimes as a catalyst.

silicate   The generic term for a compound that contains silicon, oxygen, and one or more metals, and may contain hydrogen.

silicate of soda   *See* sodium silicate.

silicic acid $SiO_2 \cdot nH_2O$ A white, amorphous precipitate; used to bleach fats, waxes, and oils. Also known as hydrated silica.

silicon   A group IV nonmetallic element, symbol Si, with atomic number 14, atomic weight 28.086; dark-brown crystals that burn in air when ignited; soluble in hydrofluoric acid and alkalies; melts at 1410°C; used to make silicon-containing alloys, as an intermediate for silicon-containing compounds, and in rectifiers and transistors.

silicon bromide   *See* silicon tetrabromide.

silicon burning   The synthesis, in stars, of elements, chiefly in the iron group, resulting from the photodisintegration of silicon-28 and other intermediate-mass nuclei; copious supplies of protons, alpha particles, and neutrons are produced, followed by the capture of these particles by other intermediate-mass nuclei.

silicon carbide   SiC   Water-insoluble, bluish-black crystals, very hard and iridescent; soluble in fused alkalies; sublimes at 2210°C; used as an abrasive and a heat refractory, and in light-emitting diodes to produce green or yellow light.

silicon chloride   *See* silicon tetrachloride.

silicon dioxide SiO_2 Colorless, transparent crystals, soluble in molten alkalies and hydrofluoric acid; melts at 1710°C; used to make glass, ceramic products, abrasives, foundry molds, and concrete.

silicon fluoride   *See* silicon tetrafluoride.

silicon halide   A compound of silicon and a halogen; for example, $SiBr_4$, Si_2Br_6, $SiCl_4$, Si_2Cl_6, Si_3Cl_8, SiF_4, Si_2F_6, SiI_4, and Si_2F_6.

silicon hydride   *See* silane.

silicon monoxide   SiO   A hard, abrasive, amorphous solid used as thin surface films to protect optical parts, mirrors, and aluminum coatings.

silicon nitride Si_3N_4 A white, water-insoluble powder, resistant to thermal shock and to chemical reagents; used as a catalyst support and for stator blades of high-temperature gas turbines.

silicon tetrabromide $SiBr_4$ A fuming, colorless liquid that yellows in air; disagreeable aroma; boils at 153°C. Also known as silicon bromide.

silicon tetrachloride $SiCl_4$ A clear, corrosive, fuming liquid with suffocating aroma; decomposes in water and alcohol; boils at 57.6°C; used in warfare smoke screens, to make ethyl silicate and silicones, and as a source of pure silicon and silica. Also known as silicon chloride.

silicon tetrafluoride SiF_4 A colorless, suffocating gas absorbed readily by water, in which it decomposes; boiling point, −86°C; used in chemical analysis and to make fluosilicic acid. Also known as silicon fluoride.

siloxane R_2SiO Any of a family of silica-based polymers in which R is an alkyl group, usually methyl; these polymers exist as oily liquids, greases, rubbers, resins, or plastics. Also known as oxosilane.

silver A white metallic element in group I, symbol Ag, with atomic number 47; soluble in acids and alkalies, insoluble in water; melts at 961°C, boils at 2212°C; used in photographic chemicals, alloys, conductors, and plating.

silver acetate CH_3COOAg A white powder, moderately soluble in water and nitric acid; used in medicine.

silver acetylide Ag_2C_2 A white explosive powder used in detonators.

silver arsenite Ag_3AsO_3 A poisonous, light-sensitive, yellow powder; soluble in acids and alkalies, insoluble in water and alcohol; decomposes at 150°C; used in medicine.

silver bromate $AgBrO_3$ A poisonous, light- and heat-sensitive, white powder; soluble in ammonium hydroxide, slightly soluble in hot water; decomposed by heat.

silver bromide $AgBr$ Yellowish, light-sensitive crystals; soluble in potassium bromide and potassium cyanide, slightly soluble in water; melts at 432°C; used in photographic films and plates.

silver carbonate Ag_2CO_3 Yellowish, light-sensitive crystals; insoluble in water and alcohol, soluble in alkalies and acids; decomposes at 220°C; used as a reagent.

silver chloride $AgCl$ A white, poisonous, light-sensitive powder; slightly soluble in water, soluble in alkalies and acids; melts at 445°C; used in photography, photometry, silver plating, and medicine.

silver chromate Ag_2CrO_4 Dark-colored crystals insoluble in water, soluble in acids and in solutions of alkali chromates; used as an analytical reagent.

silver cyanide $AgCN$ A poisonous, white, light-sensitive powder; insoluble in water, soluble in alkalies and acids; decomposes at 320°C; used in medicine and in silver plating.

silver fluoride $AgF \cdot H_2O$ A light-sensitive, yellow or brownish solid, soluble in water; dehydrated form melts at 435°C; used in medicine. Also known as tachiol.

silver halide A compound of silver and a halogen; for example, silver bromide (AgBr), silver chloride (AgCl), silver fluoride (AgF), and silver iodide (AgI).

silver iodate $AgIO_3$ A white powder, soluble in ammonium hydroxide and nitric acid, slightly soluble in water; melts above 200°C; used in medicine.

silver iodide AgI A pale-yellow powder, insoluble in water, soluble in potassium iodide–sodium chloride solutions and ammonium hydroxide; melts at 556°C; used in medicine, photography, and artificial rainmaking.

silver lactate $CH_3CHOHCOOAg \cdot H_2O$ Gray-to-white, light-sensitive crystals; slightly soluble in water and in alcohol; used in medicine. Also known as actol.

silver methylarsonate $CH_3AsO_3Ag_2$ A toxic compound used in algicides. Also known as disilver salt; methanearsonic acid.

silver nitrate $AgNO_3$ Poisonous, corrosive, colorless crystals; soluble in glycerol, water, and hot alcohol; melts at 212°C; used in external medicine, photography, hair dyeing, silver plating, ink manufacture, and mirror silvering, and as a chemical reagent.

silver nitrite $AgNO_2$ Yellow or grayish-yellow needles which decompose at 140°C; soluble in hot water; used in organic synthesis and in testing for alcohols.

silver orthophosphate *See* silver phosphate.

silver oxide Ag_2O An odorless, dark-brown powder with a metallic taste; soluble in nitric acid and ammonium hydroxide, insoluble in alcohol; decomposes above 300°C; used in medicine and in glass polishing and coloring, as a catalyst, and to purify drinking water.

silver permanganate $AgMnO_4$ Water-soluble, violet crystals that decompose in alcohol; used in medicine and in gas masks.

silver phosphate Ag_3PO_4 A poisonous, yellow powder; darkens when heated or exposed to light; soluble in acids and in ammonium carbonate, very slightly soluble in water; melts at 849°C; used in photographic emulsions and in pharmaceuticals, and as a catalyst. Also known as silver orthophosphate.

silver picrate $C_6H_2O(NO_2)_3Ag \cdot H_2O$ Yellow crystals, soluble in water, insoluble in ether and chloroform; used in medicine.

silver potassium cyanide $KAg(CN)_2$ Toxic, white crystals soluble in water and alcohol; used in silver plating and as a bactericide and antiseptic. Also known as potassium argentocyanide; potassium cyanoargentate.

silver protein A brown, hygroscopic powder containing 7.5–8.5% silver; made by reaction of a silver compound with gelatin in the presence of an alkali; used as an antibacterial.

silver selenide Ag_2Se A gray powder, insoluble in water, soluble in ammonium hydroxide; melts at 880°C.

silver suboxide AgO A charcoal-gray powder that crystallizes in the cubic or orthorhombic system, and has diamagnetic properties; used in making silver oxide-zinc alkali batteries. Also known as argentic oxide.

silver sulfate Ag_2SO_4 Light-sensitive, colorless, lustrous crystals; soluble in alkalies and acids, insoluble in alcohol; melts at 652°C; used as an analytical reagent. Also known as normal silver sulfate.

silver sulfide Ag_2S A dark, heavy powder, insoluble in water, soluble in concentrated sulfuric and nitric acids; melts at 825°C; used in ceramics and in inlay metalwork.

silvex $C_9H_7Cl_3O_3$ A colorless powder with a melting point of 178.2°C; used to promote ripening in apples and to control turf weeds, aquatic plant pests, and weeds in food crops. Also known as 2-(2,4,5-trichlorophenoxy)propionic acid.

simazine $C_7H_{12}ClN_5$ A colorless, crystalline compound with a melting point of 225–227°C; slight solubility in water; used as an herbicide to control weeds in corn, alfalfa, and fruit trees, and in industrial areas. Also known as 2-chloro-4,6-bis(ethylamino)-S-triazine.

simple continuous distillation *See* equilibrium flash vaporization.

simple cubic lattice A crystal lattice whose unit cell is a cube, and whose lattice points are located at the vertices of the cube.

simple lattice *See* primitive lattice.

simple protein One of a group of proteins which, upon hydrolysis, yield exclusively amino acids; included are globulins, glutelins, histones, prolamines, and protamines.

simple salt One of four classes of salts in a classification system that depends on the character of completeness of the ionization; examples are NaCl, $NaHCO_3$, and Pb(OH)Cl.

simple twin A twinned crystal composed of only two individuals in twin relation.

Sinclair-Baker reforming A proprietary petroleum-refinery catalytic-reforming process, using platinum-on-alumina as the catalyst.

Sinclair hydrogen treating A proprietary petroleum-refinery hydrodesulfurization process, using cobalt-molybdenum-on-alumina as the catalyst.

single crystal A crystal, usually grown artificially, in which all parts have the same crystallographic orientation.

single-effect evaporation An evaporation process completed entirely in one vessel or by means of a single heating unit.

single-phase flow The flow of a material, as a gas, single-phase liquid, or a solid, but not in any combination of the three.

SIRS *See* satellite infrared spectrometer.

size enlargement Making large particles out of small ones by crystallization, particle cementation, tableting, briquetting, agglomeration, flocculation, melting, casting, compaction and extrusion, and sintering or nodulizing.

skatole C_9H_9N A white, crystalline compound that melts at 93–95°C, dissolves in hot water, and has an unpleasant feceslike odor. Also known as 3-methylindole.

skeleton crystal A crystal formed in microscopic outline with incomplete filling in of the faces.

skimming plant A petroleum refinery designed to remove and finish only the lighter constituents of crude oil, such as gasoline and kerosine; the heavy ends are sold as fuel oil or for further processing elsewhere.

Skraup synthesis A method for the preparation of commercial synthetic quinoline by heating aniline and glycerol in the presence of sulfuric acid and an oxidizing agent to form pyridine unsubstituted quinolines.

slaked lime *See* calcium hydroxide.

sliding-vane compressor A rotary-element gas compressor in which spring-loaded sliding vanes (evenly spaced around a cylinder off-center in a surrounding chamber) pick up, compress, and discharge gas as the cylinder revolves.

slip The movement of one atomic plane over another in a crystal; it is one of the ways that plastic deformation occurs in a solid. Also known as glide.

slipband One of the microscopic parallel lines (Lüders' lines) on the surface of a crystalline material stretched beyond its elastic limit, located at the intersection of the surface with intracrystalline slip planes in the grains of the material. Also known as slip line.

slip direction The crystallographic direction in which the translation of slip occurs.

slip line *See* slipband.

slip plane *See* glide plane.

slop A petroleum-refinery term for odds and ends of oil produced in the refinery; the slop must be rerun or further processed to make it suitable for use. Also known as slop oil.

slop oil *See* slop.

slow neutron 1. A neutron having low-kinetic energy, up to about 100 electronvolts. 2. *See* thermal neutron.

sludge 1. Residue left after acid treatment of petroleum oils. 2. Any semisolid waste from a chemical process.

sludge coking The recovery of sulfuric acid from dry acid sludge.

slurry bed reactor *See* ebullating-bed reactor.

Sm *See* samarium.

smectic-A A subclass of smectic liquid crystals in which molecules are free to move within layers and are oriented perpendicular to the layers.

smectic-B A subclass of smectic liquid crystals in which molecules in each layer are arranged in a close-packed lattice and are oriented perpendicular to the layers.

smectic-C A subclass of smectic liquid crystals in which molecules are free to move within layers and are oriented with their axes tilted with respect to the normal to the layers.

smectic phase A form of the liquid crystal (mesomorphic) state in which molecules are arranged in layers that are free to glide over each other with relatively small viscosity.

smectogenic solid A solid which will form a smectic liquid crystal when heated.

Sn *See* tin.

S-N dewaxing A proprietary, petroleum-refinery solvent process for removal of wax from petroleum oils. Also known as Separator-Nobel dewaxing.

snow point Referring to a gas mixture, the temperature at which the vapor pressure of the sublimable component is equal to the actual partial pressure of that component in the gas mixture; analogous to dew point.

soaking drum A heated petroleum-refinery process vessel used in connection with petroleum thermal-cracking coils to furnish the residence time needed to complete the cracking reaction.

soda *See* sodium carbonate.

soda alum *See* aluminum sodium sulfate.

soda ash Na_2CO_3 The commercial grade of sodium carbonate; a powder soluble in water, insoluble in alcohol; used in glass manufacture and petroleum refining, and for soaps and detergents. Also known as anhydrous sodium carbonate; calcined soda.

soda crystals *See* metahydrate sodium carbonate.

sodamide *See* sodium amide.

soda pulping process The digestion of wood chips by caustic soda; used to manufacture pulp for paper products.

Soddy's displacement law The atomic number of a nuclide decreases by 2 in alpha decay, increases by 1 in beta negatron decay, and decreases by 1 in beta positron decay and electron capture.

sodide A member of a class of alkalides in which the metal anion is sodium (Na^-).

sodium A metallic element of group I, symbol Na, with atomic number 11, atomic weight 22.9898; silver-white, soft, and malleable; oxidizes in air; melts at 97.6°C; used as a chemical intermediate and in pharmaceuticals, petroleum refining, and metallurgy; the source of the symbol Na is natrium.

sodium-24 A radioactive isotope of sodium, mass 24, half-life 15.5 hours; formed by deuteron bombardment of sodium; decomposes to magnesium with emission of beta rays.

sodium acetate $NaC_2H_3O_2$ Colorless, efflorescent crystals, soluble in water and ether; melts at 324°C; used as a chemical intermediate and for pharmaceuticals, dyes, and dry colors.

sodium-acetone bisulfite *See* acetone-sodium bisulfite.

sodium acid carbonate *See* sodium bicarbonate.

sodium acid chromate *See* sodium dichromate.

sodium acid fluoride *See* sodium bifluoride.

sodium acid methanearsonate *See* sodium methanearsonate.

sodium acid sulfate *See* sodium bisulfate.

sodium acid sulfite *See* sodium bisulfite.

sodium alginate $C_6H_7O_6Na$ Colorless or light yellow filaments, granules, or powder which forms a viscous colloid in water; used in food thickeners and stabilizers, in medicine and textile printing, and for paper coating and water-base paint. Also known as alginic acid sodium salt; sodium polymannuronate.

sodium alphanaphthylamine sulfonate *See* sodium naphthionate.

sodium aluminate $Na_2Al_2O_4$ A white powder soluble in water, insoluble in alcohol; melts at 1800°C; used as a zeolite-type of material and a mordant, and in water purification, milkglass manufacture, and cleaning compounds.

sodium aluminosilicate White, amorphous powder or beads, partially soluble in strong acids and alkali hydroxide solutions between 80 and 100°C; used in food as an anticaking agent. Also known as sodium silicoaluminate.

sodium aluminum phosphate $NaAl_3H_{14}(PO_4)_8 \cdot 4H_2O$ or $Na_3Al_2H_{15}(PO_4)_8$ White powder, soluble in hydrochloric acid; used as a food additive for baked products. Also known as acidic sodium aluminum phosphate.

sodium aluminum silicofluoride $Na_5Al(SiF_6)_4$ A toxic, white powder, used for mothproofing and in insecticides.

sodium aluminum sulfate   *See* aluminum sodium sulfate.

sodium amalgam Na_xHg_x A fire-hazardous, silver-white crystal mass that decomposes in water; used to make hydrogen and as an analytical reagent.

sodium amide $NaNH_2$ White crystals that decompose in water; melts at 210°C; a fire hazard; used to make sodium cyanide. Also known as sodamide.

sodium aminophenylarsonate   *See* sodium arsanilate.

sodium anilinearsonate   *See* sodium arsanilate.

sodium antimonate $NaSbO_3$ A white, granular powder, used as an enamel opacifier and high-temperature oxidizing agent. Also known as antimony sodiate.

sodium antimony bis(pyrocatechol-2,4-disulfonate)   *See* stibophen.

sodium arsanilate $C_6H_4NH_2(AsO \cdot OH \cdot ONa)$ A white, water-soluble, poisonous powder with a faint saline taste, used in medicine and as a chemical intermediate. Also known as sodium aminophenylarsonate; sodium anilinearsonate.

sodium arsenate $Na_3AsO_4 \cdot 12H_2O$ Water-soluble, poisonous, clear, colorless crystals with a mild alkaline taste; melts at 86°C; used in medicine, insecticides, dry colors, and textiles, and as a germicide and a chemical intermediate.

sodium arsenite $NaAsO_2$ A poisonous, water-soluble, grayish powder; used in antiseptics, dyeing, insecticides, and soaps for taxidermy. Also known as sodium metaarsenite.

sodium ascorbate $CH_2OH(CHOH)_2COHCOHCOONa$ White, odorless crystals; soluble in water, insoluble in alcohol; decomposes at 218°C; used in therapy for vitamin C deficiency.

sodium aurichloride   *See* sodium gold chloride.

sodium aurocyanide   *See* sodium gold cyanide.

sodium aurothiomalate   *See* gold sodium thiomalate.

sodium azide NaN_3 Poisonous, colorless crystals; soluble in water and liquid ammonia; decomposes at 300°C; used in medicine and to make lead azide explosives.

sodium barbiturate $C_4H_3N_2O_3Na$ White to slightly yellow powder, soluble in water and dilute mineral acid; used in wood-impregnating solutions.

sodium benzoate $NaC_7H_5O_2$ Water- and alcohol-soluble, white, amorphous crystals with a sweetish taste; used as a food preservative and an antiseptic and in tobacco, pharmaceuticals, and medicine.

sodium benzosulfimide   *See* sodium saccharine.

sodium benzoylacetone dihydrate   A metal chelate with low melting point (115°C) and slight solubility in acetone.

sodium benzyl penicillinate *See* benzyl penicillin sodium.

sodium bicarbonate $NaHCO_3$ White, water-soluble crystals with an alkaline taste; loses carbon dioxide at 270°C; used as a medicine and a butter preservative, in food preparation, in effervescent salts and beverages, in ceramics, and to prevent timber mold. Also known as baking soda; sodium acid carbonate.

sodium bichromate *See* sodium dichromate.

sodium bifluoride $NaHF_2$ Poisonous, water-soluble, white crystals; decomposes when heated; used as a laundry-rinse neutralizer, preservative, and antiseptic, and in glass etching and tinplating. Also known as sodium acid fluoride.

sodium bismuthate $NaBiO_3$ A yellow to brown amorphous powder; used as an analytical reagent and in pharmaceuticals.

sodium bisulfate $NaHSO_4$ Colorless crystals, soluble in water; the aqueous solution is strongly acidic; decomposes at 315°C; used for flux to decompose minerals, as a disinfectant, and in dyeing and manufacture of magnesia, cements, perfumes, brick, and glue. Also known as niter cake; sodium acid sulfate.

sodium bisulfide *See* sodium hydrosulfide.

sodium bisulfite $NaHSO_3$ A colorless, water-soluble solid; decomposes when heated. Also known as sodium acid sulfite.

sodium bisulfite test A test for aldehydes in which aldehydes form a crystalline salt upon addition of a 40% aqueous solution of sodium bisulfite.

sodium bitartrate $NaHC_4H_5O_6 \cdot H_2O$ A white, combustible, water-soluble powder that loses water at 100°C, decomposes at 219°C; used in effervescing mixtures and as an analytical reagent. Also known as acid sodium tartrate.

sodium borate $Na_2B_4O_7 \cdot 10H_2O$ A water-soluble, odorless, white powder; melts between 75 and 200°C; used in glass, ceramics, starch and adhesives, detergents, agricultural chemicals, pharmaceuticals, and photography; the impure form is known as borax. Also known as sodium pyroborate; sodium tetraborate.

sodium boroformate $NaH_2BO_3 \cdot 2HCOOH \cdot 2H_2O$ Water-soluble, white crystals, melting at 110°C; used in textile treating and in tanning, and as a buffering agent.

sodium borohydride $NaBH_4$ A flammable, hygroscopic, white to gray powder; soluble in water, insoluble in ether and hydrocarbons; decomposes in damp air; used as a hydrogen source, a chemical reagent, and a rubber foaming agent.

sodium bromate $NaBrO_3$ Odorless, white crystals; soluble in water, insoluble in alcohol; decomposes at 381°C; a fire hazard, used as an analytical reagent.

sodium bromide $NaBr$ White, water-soluble, crystals with a bitter, saline taste; absorbs moisture from air; melts at 758°C; used in photography and medicine, as a chemical intermediate, and to make bromides.

sodium cacodylate $(CH_3)_2AsOONa \cdot 3H_2O$ A white crystalline compound that melts at approximately 60°C and dissolves in ethanol and water; used as a herbicide. Also known as sodium dimethylarsenate.

sodium carbolate *See* sodium phenate.

sodium carbonate Na_2CO_3 A white, water-soluble powder that decomposes when heated to about 852°C; used as a reagent; forms a monohydrate compound, $Na_2CO_3 \cdot H_2O$, and a decahydrate compound, $Na_2CO_3 \cdot 10H_2O$. Also known as soda.

sodium carbonate decahydrate *See* sal soda.

sodium carbonate peroxide $2Na_2CO_3 \cdot 3H_2O$ A white, crystalline powder; used in household detergents, in dental cleansers, and for bleaching and dyeing.

sodium carboxymethylcellulose *See* carboxymethyl cellulose.

sodium caseinate A tasteless, odorless, water-soluble, white powder; used in medicine, foods, emulsification, and stabilization; formed by dissolving casein in sodium hydroxide and then evaporating. Also known as casein sodium; nutrose.

sodium cellulose glycolate *See* carboxymethyl cellulose.

sodium chlorate $NaClO_3$ Water- and alcohol-soluble, colorless crystals with a saline taste; melts at 255°C; used as a medicine, weed killer, defoliant, and oxidizing agent, and in matches, explosives, and bleaching.

sodium chloride $NaCl$ Colorless or white crystals; soluble in water and glycerol, slightly soluble in alcohol; melts at 804°C; used in foods and as a chemical intermediate and an analytical reagent. Also known as common salt; table salt.

sodium chlorite $NaClO_2$ An explosive, white, mildly hygroscopic, water-soluble powder; decomposes at 175°C; used as an analytical reagent and oxidizing agent.

sodium chloroacetate $ClCH_2COONa$ A white, water-soluble powder; used as a defoliant and in the manufacture of weed killers, dyes, and pharmaceuticals.

sodium chloroaurate *See* sodium gold chloride.

sodium chloroplatinate $Na_2PtCl_6 \cdot 4H_2O$ A yellow powder, soluble in alcohol and water; used for zinc etching, indelible ink, plating, and mirrors, and in photography and medicine. Also known as platinic sodium chloride; platinum sodium chloride; sodium platinichloride.

sodium chromate $Na_2CrO_4 \cdot 10H_2O$ Water-soluble, translucent, yellow, efflorescent crystals that melt at 20°C; used as a rust preventive and in inks, dyeing, and leather tanning.

sodium citrate $C_6H_5Na_3O_7 \cdot 2H_2O$ A white powder with the taste of salt; soluble in water, slightly soluble in alcohol; has an acid taste; loses water at 150°C; decomposes at red heat; used in medicine as an anticoagulant, in soft drinks, cheesemaking, and electroplating. Also known as trisodium citrate.

sodium cobaltinitrite $Na_3Co(NO_2)_6 \cdot \frac{1}{2}H_2O$ Purple, water-soluble, hygroscopic crystals; used as a reagent for analysis of potassium.

sodium cyanate $NaOCN$ A poisonous, white powder; soluble in water, insoluble in alcohol and ether; used as a chemical intermediate and for the manufacture of medicine and the heat-treating of steels.

sodium cyanide $NaCN$ A poisonous, water-soluble, white powder melting at 563°C; decomposes rapidly when standing; used to manufacture pigments, in heat treatment of metals, and as a silver- and gold-ore extractant.

sodium cyanoaurite *See* sodium gold cyanide.

sodium cyclamate $C_6H_{11}NHSO_3Na$ White, water-soluble crystals; sweetness 30 times that of sucrose; formerly used as an artificial sweetener for foods, but now prohibited. Also known as sodium cyclohexylsulfamate.

sodium cyclohexylsulfomate *See* sodium cyclamate.

sodium dehydroacetate $C_8H_7NaO_4\cdot H_2O$ A tasteless, white powder, soluble in water and propylene glycol; used as a fungicide and plasticizer, in toothpaste, and for pharmaceuticals.

sodium 3,5-diacetamido-2,4,6-triiodobenzoate *See* sodium diatrizoate.

sodium diacetate $CH_3COONa\cdot x(CH_3COOH)$ Combustible, white, water-soluble crystals with an acetic acid aroma; decomposes above 150°C; used to inhibit mold, and as a buffer, varnish hardener, sequestrant, and food preservative, and in mordants.

sodium diatrizoate $C_{11}H_8NO_4I_3Na$ White, water-soluble crystals which give a radiopaque solution; used in medicine as a radiopaque medium. Also known as sodium 3,5-diacetamido-2,4,6-triiodobenzoate.

sodium dichloroisocyanate $HC_3N_3O_3NaCl$ A white, crystalline compound, soluble in water; used as a bactericide and algicide in swimming pools.

sodium dichloroisocyanurate $C_3N_3O_3Cl_2Na$ A white, crystalline powder; used in dry bleaches, detergents, and cleaning compounds, and for water and sewage treatment.

sodium dichromate $Na_2Cr_2O_7\cdot 2H_2O$ Poisonous, red to orange deliquescent crystals; soluble in water, insoluble in alcohol; melts at 320°C; loses water of hydration upon prolonged heating at 105°C; used as a chemical intermediate and corrosion inhibitor and in the manufacture of pigments, leather tanning, and electroplating. Also known as bichromate of soda; sodium acid chromate; sodium bichromate.

sodium diethyldithiocarbamate $(C_2H_5)_2NCS_2Na$ A solid that is soluble in water and in alcohol; the trihydrate is used to determine small amounts of copper and to separate copper from other metals.

sodium dimethylarsenate *See* sodium cacodylate.

sodium dimethyldithiocarbamate $(CH_3)_2NCS_2Na$ Amber to light green liquid; used as a fungicide, corrosion inhibitor, and rubber accelerator. Abbreviated SDDC.

sodium dinitro-*ortho*-cresylate $CH_3C_6H_2(NO_2)_2ONa$ A toxic, orange-yellow dye, used as an herbicide and fungicide.

sodium *para*-diphenylaminoazobenzenesulfonate *See* tropeoline 00.

sodium dithionite *See* sodium hydrosulfite.

sodium diuranate $Na_2U_2O_7\cdot 6H_2O$ A yellow-orange solid, soluble in dilute acids; used for colored glazes on ceramics and in the manufacture of fluorescent uranium glass. Also known as uranium yellow.

sodium dodecylbenzenesulfanate $C_{18}H_{29}SO_3Na$ Biodegradable, white to yellow flakes, granules, or powder, used as a synthetic detergent.

sodium ethoxide *See* sodium ethylate.

sodium ethylate C_2H_5ONa A white powder formed from ethanol by replacement of the hydroxyl groups' hydrogen by monovalent sodium; used in organic synthesis. Also known as caustic alcohol; sodium ethoxide.

sodium 2-ethylhexyl sulfoacetate $C_{10}H_{19}O_2SO_3Na$ Cream-colored, water-soluble flakes, used as a stabilizing agent in soapless shampoos.

sodium ethylmercurithiosalicylate   *See* thimerosal.

sodium ethylxanthate $C_2H_5OC(S)SNa$ A yellowish powder, soluble in water and alcohol; used as an ore flotation agent. Also known as sodium xanthate; sodium xanthogenate.

sodium ferricyanide $Na_3Fe(CN)_6 \cdot H_2O$ A poisonous, deliquescent, red powder; soluble in water, insoluble in alcohol; used in printing and for the manufacture of pigments. Also known as red prussiate of soda.

sodium ferrocyanide $Na_4Fe(CN)_6 \cdot 10H_2O$ Semitransparent crystals, soluble in water; insoluble in alcohol; used in photography, dyes, tanning, and blueprint paper. Also known as yellow prussiate of soda.

sodium fluoborate $NaBF_4$ A white powder with a bitter taste; soluble in water, slightly soluble in alcohol; decomposes when heated, fuses below 500°C; used in electrochemical processes, as flux for nonferrous metals refining, and as an oxidation inhibitor.

sodium fluorescein   *See* uranine.

sodium fluoride NaF A poisonous, water-soluble, white powder, melting at 988°C; used as an insecticide and a wood and adhesive preservative, and in fungicides, vitreous enamels, and dentistry.

sodium fluoroacetate $C_2H_2FO_2Na$ A white powder, hygroscopic and nonvolatile; decomposes at 200°C; very soluble in water; used as a repellent for rodents and predatory animals.

sodium fluosilicate Na_2SiF_6 A poisonous, white, amorphous powder; slightly soluble in water; decomposes at red heat; used to fluoridate drinking water and to kill rodents and insects. Also known as sodium silicofluoride.

sodium folate $C_{19}H_{18}N_7NaO_6$ A yellow to yellow-orange liquid; used in medicine for folic acid deficiency. Also known as folic acid sodium salt; sodium pteroylglutamate.

sodium formaldehyde bisulfite   *See* formaldehyde sodium bisulfite.

sodium formaldehyde sulfoxylate $HCHO \cdot HSO_2Na \cdot 2H_2O$ A white solid with a melting point of 64°C; soluble in water and alcohol; used as a textile stripping agent and a bleaching agent for soap and molasses. Abbreviated SFS. Also known as sodium sulfoxylate; sodium sulfoxylate formaldehyde.

sodium formate $HCOONa$ A mildly hygroscopic, white powder, soluble in water; has a formic acid aroma; melts at 245°C; used in medicine and as a chemical intermediate and reducing agent.

sodium glucoheptonate $C_7H_{13}O_8Na$ A light tan, crystalline powder; used for cleaning metal, mercerizing, paint stripping, and aluminum etching.

sodium gluconate $C_6H_{11}NaO_7$ A water-soluble, yellow to white, crystalline powder, produced by fermentation; used in food and pharmaceutical industries, and as a metal cleaner. Also known as gluconic acid sodium salt.

sodium glutamate $COOH(CH_2)_2CH(NH_2)COONa$ A salt of an amino acid; a white powder, soluble in water and alcohol; used as a taste enhancer. Also known as monosodium glutamate (MSG).

sodium gold chloride $NaAuCl_4 \cdot 2H_2O$ Yellow crystals, soluble in water and alcohol; used in photography, fine glass staining, porcelain decorating, and medicine. Also known as gold salt; gold sodium chloride; sodium aurichloride; sodium chloroaurate.

sodium gold cyanide $NaAu(CN)_2$ A yellow, water-soluble powder; used for gold plating radar and electric parts, jewelry, and tableware. Also known as gold sodium cyanide; sodium aurocyanide; sodium cyanoaurite.

sodium halide   A compound of sodium with a halogen; for example, sodium bromide ($NaBr$), sodium chloride ($NaCl$), sodium iodide (NaI), and sodium fluoride (NaF).

sodium halometallate   A compound of sodium with halogen and a metal; for example, sodium platinichloride, $Na_2PtCl_6 \cdot 6H_2O$.

sodium hexylene glycol monoborate $C_6H_{12}O_3BNa$ An amorphous, white solid with a melting point of 426°C; used as a corrosion inhibitor, flame retardant, and lubricating-oil additive.

sodium hydrate   *See* sodium hydroxide.

sodium hydride NaH A white powder, decomposed by water, and igniting in moist air; used to make sodium borohydride and as a drying agent and a reagent.

sodium hydrogen phosphate $NaH_2PO_4 \cdot H_2O$ Hygroscopic, transparent, water-soluble crystals; used as a purgative, reagent, and buffer.

sodium hydrogen sulfide   *See* sodium hydrosulfide.

sodium hydrosulfide $NaSH \cdot 2H_2O$ Toxic, colorless, water-soluble needles, melting at 55°C; used in pulping of paper, processing dyestuffs, hide dehairing, and bleaching. Also known as sodium bisulfide; sodium hydrogen sulfide; sodium sulfhydrate.

sodium hydrosulfite $Na_2S_2O_4$ A fire-hazardous, lemon to whitish-gray powder; soluble in water, insoluble in alcohol; melts at 55°C; used as a chemical intermediate and catalyst and in ore flotation. Also known as sodium dithionite.

sodium hydroxide $NaOH$ White, deliquescent crystals; absorbs carbon dioxide and water from air; soluble in water, alcohol, and glycerol; melts at 318°C; used as an analytical reagent and chemical intermediate, in rubber reclaiming and petroleum refining, and in detergents. Also known as caustic soda; sodium hydrate.

sodium hypochlorite $NaOCl$ Air-unstable, pale-green crystals with sweet aroma; soluble in cold water, decomposes in hot water; used as a bleaching agent for paper pulp and textiles, as a chemical intermediate, and in medicine.

sodium hypophosphite $NaH_2PO_2 \cdot H_2O$ Colorless, pearly, water-soluble crystalline plates or a white, granular powder; used in medicine and electroless nickel plating of plastic and metal.

sodium hyposulfite   *See* sodium thiosulfate.

sodium indigotin disulfonate   *See* indigo carmine.

sodium iodate $NaIO_3$ A white, water- and acetone-soluble powder; used as a disinfectant and in medicine.

sodium iodide   NaI   A white, air-sensitive powder, deliquescent, with bitter taste; soluble in water, alcohol, and glycerin; melts at 653°C; used in photography and in medicine and as an analytical reagent.

sodium iodomethanesulfonate   *See* sodium methiodal.

sodium isopropylxanthate $C_5H_7ONaS_2$ Light yellow, crystalline compound that decomposes at 150°C; soluble in water; used as a postemergence herbicide and as a flotation agent for ores.

sodium lactate $CH_3CHOHCOONa$ A water-soluble, hygroscopic, yellow to colorless, syrupy liquid; solidifies at 17°C; used in medicine, as a corrosion inhibitor in antifreeze, and a hygroscopic agent.

sodium lauryl sulfate $CH_3(CH_2)_{10}CH_2OSO_3Na$ A water-soluble salt, produced as a white or cream powder, crystals, or flakes; used in the textile industry as a wetting agent and detergent. Also known as dodecyl sodium sulfate.

sodium lead hyposulfate   *See* lead sodium thiosulfate.

sodium lead thiosulfate   *See* lead sodium thiosulfate.

sodium mercaptoacetate   *See* sodium thioglycolate.

sodium metaarsenite   *See* sodium arsenite.

sodium metaborate $NaBO_2$ Water-soluble, white crystals, melting at 966°C; the aqueous solution is alkaline; made by fusing sodium carbonate with borax; used as an herbicide.

sodium metaphosphate $(NaPO_3)_x$ Sodium phosphate groupings; cyclic forms range from $x = 3$ for the trimetaphosphate, to $x = 10$ for the decametaphosphate; sodium hexametaphosphate with $x = 10$ to 20 is probably a polymer; used for dental polishing, building detergents, and water softening, and as a sequestrant, emulsifier, and food additive.

sodium metasilicate   *See* sodium silicate.

sodium metavanadate $NaVO_3$ Colorless crystals or a pale green, crystalline powder with a melting point of 630°C; soluble in water; used in inks, fur dyeing, and photography, and as a corrosion inhibitor in gas scrubbers.

sodium methanearsonate $CH_3AsO(OH)(ONa)$ White, water-soluble solid with a melting point of 130–140°C; used as an herbicide. Also known as monosodium methanearsonate; sodium acid methanearsonate; sodium methylarsonate.

sodium methiodal ICH_2SO_3Na A white, crystalline powder, soluble in water and methanol; used in medicine as a radiopaque medium. Also known as sodium iodomethanesulfonate.

sodium methoxide CH_3ONa A salt produced as a free-flowing powder, soluble in methanol and ethanol; used as an intermediate in organic synthesis. Also known as sodium methylate.

sodium methylarsonate   *See* sodium methanearsonate.

sodium methylate   *See* sodium methoxide.

sodium N-methyldithiocarbamate dihydrate $CH_3NHC(S)SNa \cdot 2H_2O$ A white, water-soluble, crystalline solid; used as a fungicide, insecticide, nematicide, and weed killer.

sodium molybdate Na_2MoO_4 Water-soluble crystals, melting at 687°C; used as an analytical reagent, corrosion inhibitor, catalyst, and zinc-plating brightening agent, and in medicine.

sodium 12-molybdophosphate $Na_3PMo_{12}O_{40}$ Yellow, water-soluble crystals; used in neuromicroscopy and photography, and as a water-resisting agent in plastic adhesives and cements.

sodium monoxide Na_2O A strong basic white powder soluble in molten caustic soda; forms sodium hydroxide in water; used as a dehydrating and polymerization agent. Also known as sodium oxide.

sodium naphthalenesulfonate $C_{10}H_7SO_3Na$ Yellow, water-soluble crystalline plates or white scales; used as a liquefying agent in animal glue.

sodium naphthionate $NaC_{10}H_6(NH_2)SO_3 \cdot 4H_2O$ White, light-sensitive crystals, soluble in water and insoluble in ether; used in analysis (Riegler's reagent) for nitrous acid. Also known as sodium alphanaphthylamine sulfonate.

sodium 1,2-naphthoquinone-4-sulfonate *See* sodium-β-naphthoquinone-4-sulfonate.

sodium β-naphthoquinone-4-sulfonate $C_{10}H_5NaO_5S$ A yellow, crystalline compound, soluble in water; used for the determination of amines and amino acids. Also known as 1,2-naphthoquinone-4-sulfonic acid sodium salt; β-naphthoquinone-4-sulfonic acid sodium salt; sodium 1,2-naphthoquinone-4-sulfonate.

sodium nitrate $NaNO_3$ Fire-hazardous, transparent, colorless crystals with bitter taste; soluble in glycerol and water; melts at 308°C; decomposes when heated; used in manufacture of glass and pottery enamel and as a fertilizer and food preservative.

sodium nitrite $NaNO_2$ A fire-hazardous, air-sensitive, yellowish powder, soluble in water; decomposes above 320°C; used as an intermediate for dyestuffs and for pickling meat, textiles dyeing, and rust-proofing, and in medicine.

sodium nitroferricyanide $Na_2Fe(CN)_5NO \cdot 2H_2O$ Water-soluble, transparent, reddish crystals; slowly decomposes in water; used as an analytical reagent. Also known as sodium nitroprussiate; sodium nitroprusside.

sodium nitroprussiate *See* sodium nitroferricyanide.

sodium nitroprusside *See* sodium nitroferricyanide.

sodium oleate $C_{17}H_{33}COONa$ A white powder with a tallow aroma; soluble in alcohol and water, with partial decomposition; used in medicine and textile waterproofing.

sodium oxalate $Na_2C_2O_4$ A poisonous, white powder; soluble in water, insoluble in alcohol; used for leather tanning and as an analytical reagent.

sodium oxide *See* sodium monoxide.

sodium paraperiodate $Na_3H_2IO_6$ White, crystalline solid, soluble in concentrated sodium hydroxide solutions; used to wet-strengthen paper and to aid in tobacco combustion. Also known as sodium triparaperiodate.

sodium pentaborate $Na_2B_{10}O_{16} \cdot 10H_2O$ A white, water-soluble powder; used in glassmaking, weed killers, and fireproofing compositions.

sodium pentachlorophenate C_6Cl_5ONa A white or tan powder, soluble in water, ethanol, and acetone; used as a fungicide and herbicide.

sodium perborate $NaBO_2 \cdot H_2O_2 \cdot 3H_2O$ A white powder with a saline taste; slightly soluble in water, decomposes in moist air; used in deodorants, in dental compositions and as a germicide. Also known as peroxydol.

sodium perchlorate $NaClO_4$ Fire-hazardous, white, deliquescent crystals; soluble in water and alcohol; melts at 482°C; explosive when in contact with concentrated sulfuric acid; used in jet fuel, as an analytical reagent, and for explosives.

sodium permanganate $NaMnO_4 \cdot 3H_2O$ A fire-hazardous, water-soluble, purple powder; decomposes when heated; used to make saccharin, as a disinfectant, and as an oxidizing agent.

sodium peroxide Na_2O_2 A fire-hazardous, white powder that yellows with heating; decomposes when heated; causes ignition when in contact with water; used as an oxidizing agent and a bleach, and in medicinal soap.

sodium peroxydisulfate *See* sodium persulfate.

sodium persulfate $Na_2S_2O_8$ A white, water-soluble, crystalline powder; used as a bleaching agent and in medicine. Also known as sodium peroxydisulfate.

sodium phenate C_6H_5ONa White, deliquescent crystals, soluble in water and alcohol; decomposed by carbon dioxide in air; used as a chemical intermediate, antiseptic, and military gas absorbent. Also known as sodium carbolate; sodium phenolate.

sodium phenolate *See* sodium phenate.

sodium phenylacetate $C_6H_5CH_2 \cdot COONa$ Pale yellow, 50% aqueous solution which crystallizes at 15°C; used in the manufacture of penicillin G. Also known as sodium α-toluene.

sodium *ortho*-phenylphenate $C_6H_4(C_6H_5)ONa \cdot 4H_2O$ White flakes, soluble in water, methanol, and acetone; used as a bactericide and fungicide. Also known as sodium *ortho*-phenylphenolate.

sodium *ortho*-phenylphenolate *See* sodium *ortho*-phenylphenate.

sodium phenylphosphinate $C_6H_5PH(O)(ONa)$ Crystals with a melting point of 355°C; used as an antioxidant and as a heat and light stabilizer.

sodium phosphate A general term encompassing the following compounds: sodium hexametaphosphate, sodium metaphosphate, dibasic sodium phosphate, hemibasic sodium phosphate, monobasic sodium phosphate, tribasic sodium phosphate, sodium pyrophosphate, and acid sodium pyrophosphate.

sodium phosphite $Na_2HPO_3 \cdot 5H_2O$ White, hygroscopic crystals, melting at 53°C; soluble in water, insoluble in alcohol; used in medicine.

sodium phosphotungstate *See* sodium 12-tungstophosphate.

sodium phytate $C_6H_9O_{24}P_6Na_9$ A hygroscopic, water-soluble powder; used as a chelating agent for trace metals and in medicine.

sodium picramate $NaOC_6H_2(NO_2)_2NH_2$ A yellow salt, soluble in water; used in the manufacture of dye intermediates.

sodium platinichloride *See* sodium chloroplatinate.

sodium plumbite $Na_2PbO_2 \cdot 3H_2O$ A toxic, corrosive solution of lead oxide (litharge) in sodium hydroxide; used (as doctor solution) to sweeten gasoline.

sodium polymannuronate *See* sodium alginate.

sodium polysulfide Na_2S_x Yellow-brown granules, used to make dyes and colors, and insecticides, as a petroleum additive, and in electroplating.

sodium-potassium tartrate *See* potassium-sodium tartrate.

sodium propionate CH_3CH_2COONa Deliquescent, transparent crystals; soluble in water, slightly soluble in alcohol; used as a fungicide, and mold preventive.

sodium pteroylglutamate *See* sodium folate.

sodium pyroborate *See* sodium borate.

sodium pyrophosphate $Na_4P_2O_7$ A white powder; soluble in water, insoluble in alcohol and ammonia; melts at 880°C; used as a water softener and newsprint deinker, and to control drilling-mud viscosity. Also known as normal sodium pyrophosphate; tetrasodium pyrophosphate (TSPP).

sodium saccharin $C_7H_4NNaO_3S \cdot 2H_2O$ White crystals or a crystalline powder, soluble in water and slightly soluble in alcohol; used in medicine and as a nonnutritive food sweetener. Also known as sodium benzosulfimide; soluble gluside; soluble saccharin.

sodium salicylate HOC_6H_4COONa A shiny, white powder with sweetish taste and mild aromatic aroma; soluble in water, glycerol, and alcohol; used in medicine and as a preservative.

sodium selenate $Na_2SeO_4 \cdot 10H_2O$ White, poisonous, water-soluble crystals; used as an insecticide.

sodium selenite $Na_2SeO_3 \cdot 5H_2O$ White, water-soluble crystals; used in glass manufacture, as a bacteriological reagent, and for decorating porcelain.

sodium sesquicarbonate $Na_2CO_3 \cdot NaHCO_3 \cdot 2H_2O$ White, water-soluble, needle-shaped crystals; used as a detergent, an alkaline agent for water softening and leather tanning, and a food additive.

sodium sesquisilicate $Na_6Si_2O_7$ A white, water-soluble powder; used for metals cleaning and textile processing.

sodium silicate Na_2SiO_3 A gray-white powder; soluble in alkalies and water, insoluble in alcohol and acids; used to fireproof textiles, in petroleum refining and corrugated paperboard manufacture, and as an egg preservative. Also known as liquid glass; silicate of soda; sodium metasilicate; soluble glass; water glass.

sodium silicoaluminate *See* sodium aluminosilicate.

sodium silicofluoride *See* sodium fluosilicate.

sodium stannate $Na_2SnO_3 \cdot 3H_2O$ Water- and alcohol-insoluble, whitish crystals; used in ceramics, dyeing, and textile fireproofing, and as a mordant. Also known as preparing salt.

sodium stearate $NaC_{18}H_{35}O_2$ A white powder with a fatty aroma; soluble in hot water and alcohol; used in medicine and toothpaste and as a waterproofing agent.

sodium subsulfite *See* sodium thiosulfate.

sodium succinate $Na_2C_4H_4O_4 \cdot 6H_2O$ Water-soluble, white crystals; loses water at 120°C; used in medicine.

sodium sulfate Na_2SO_4 Crystalline compound, melts at 888°C, soluble in water; used to make paperboard, kraft paper, glass, and freezing mixtures.

sodium sulfhydrate   *See* sodium hydrosulfide.

sodium sulfide Na_2S An irritating, water-soluble, yellow to red, deliquescent powder; melts at 1180°C; used as a chemical intermediate, solvent, photographic reagent, and analytical reagent. Also known as sodium sulfuret.

sodium sulfite Na_2SO_3 White, water-soluble, crystals with a sulfurous, salty taste; decomposes when heated; used as a chemical intermediate and food preservative, in medicine and paper manufacturing, and for dyes and photographic developing.

sodium sulfite process   A process for the digestion of wood chips in a solution of magnesium, ammonium, or calcium disulfite containing free sulfur dioxide; used in papermaking.

sodium sulfocyanate   *See* sodium thiocyanate.

sodium sulfoxylate   *See* sodium formaldehyde sulfoxylate.

sodium sulfoxylate formaldehyde   *See* sodium formaldehyde sulfoxylate.

sodium sulfuret   *See* sodium sulfide.

sodium tartrate $Na_2C_4H_4O_6\cdot2H_2O$ White, water-soluble crystals or granules; loses water at 150°C; used in medicine and as a food stabilizer and sequestrant. Also known as disodium tartrate; sal tartar.

sodium TCA   *See* sodium trichloroacetate.

sodium tetraborate   *See* sodium borate.

sodium tetrafluorescein   *See* easin.

sodium tetraphenylborate $[(C_6H_5)_4B]Na$ A snow-white, crystalline compound, soluble in water and acetone; used as a reagent in the determination of the following ions: potassium, ammonium, rubidium, and cesium. Also known as tetraphenylboron sodium.

sodium tetrasulfide Na_2S_4 Hygroscopic, yellow or dark-red crystals, melting at 275°C; used for insecticides and fungicides, ore flotation, and dye manufacture, and as a reducing agent.

sodium thiocyanate   NaSCN   A poisonous, water- and alcohol-soluble, deliquescent, white powder; melts at 287°C; used as an analytical reagent, solvent, and chemical intermediate, and for rubber treatment and textile dyeing and printing. Also known as sodium sulfocyanate.

sodium thioglycolate $C_2H_3NaO_3S$ A water-soluble compound produced as hygroscopic crystals; used as an ingredient in bacteriology mediums, and in hair-waving solutions. Also known as sodium mercaptoacetate.

sodium thiosulfate $Na_2S_2O_3\cdot5H_2O$ White, translucent crystals or powder with a melting point of 48°C; soluble in water and oil of turpentine; used as a fixing agent in photography, for extracting silver from ore, in medicine, and as a sequestrant in food. Also known as hypo; sodium hyposulfite; sodium subsulfite.

sodium α-toluene   *See* sodium phenylacetate.

sodium trichloroacetate CCl_3COONa A toxic material, used in herbicides and pesticides. Abbreviated sodium TCA.

sodium 2,4,5-trichlorophenate $C_6H_2Cl_3ONa\cdot1\frac{1}{2}H_2O$ Buff to light brown flakes, soluble in water, methanol, and acetone; used as a bactericide and fungicide.

sodium triparaperiodate *See* sodium paraperiodate.

sodium triphosphate *See* sodium tripolyphosphate.

sodium tripolyphosphate $Na_5P_3O_{10}$ A white powder with a melting point of 622°C; used for water softening and as a food additive and texturizer. Abbreviated STPP. Also known as pentasodium triphosphate; sodium triphosphate.

sodium tungstate $Na_2WO_4\cdot2H_2O$ Water-soluble, colorless crystals; lose water at 100°C, melts at 692°C; used as a chemical intermediate analytical reagent, and for fireproofing. Also known as sodium wolframate.

sodium 12-tungstophosphate $Na_3[P(W_3O_{10})_4]\cdot xH_2O$ A yellowish-white powder, soluble in water and alcohols; used to manufacture organic pigments, as an antistatic agent for textiles, in leather tanning, and as a water-resistant agent in plastic films, adhesives, and cements. Also known as sodium phosphotungstate.

sodium undecylenate $C_{11}H_{19}O_2Na$ A white, water-soluble powder that decomposes above 200°C; used in cosmetics and pharmaceuticals as a bacteriostat and fungistat.

sodium wolframate *See* sodium tungstate.

sodium xanthate *See* sodium ethylxanthate.

sodium xanthogenate *See* sodium ethylxanthate.

soft shower A cosmic-ray shower that cannot penetrate 15 to 20 centimeters of lead; consists mainly of electrons and positrons.

soft water Water that is free of magnesium or calcium salts.

soft x-ray absorption spectroscopy A spectroscopic technique which is used to get information about unoccupied states above the Fermi level in a metal or about empty conduction bands in an inoculator.

soft x-ray appearance potential spectroscopy A branch of electron spectroscopy in which a solid surface is bombarded with monochromatic electrons, and small but abrupt changes in the resulting total x-ray emission intensity are detected as the energy of the electrons is varied. Abbreviated SXAPS.

soil chemistry The study and analysis of the inorganic and organic components and the life cycles within soils.

sol A colloidal solution consisting of a suitable dispersion medium, which may be gas, liquid, or solid, and the colloidal substance, the disperse phase, which is distributed throughout the dispersion medium.

solar distillation A procedure in which the sun's heat is used to evaporate seawater in order to produce sodium chloride and other salts or potable water.

solar still A device for evaporating seawater, in which water is confined in one or more shallow pools, over which is placed a roof-shaped transparent cover made of glass or plastic film; the sun's heat evaporates the water, leaving behind a residue of salt; the vapor from the evaporated water condenses on the surface of the cover and trickles down into gutters, which thus collect fresh water.

solation The change of a substance from a gel to a sol.

solid-liquid equilibrium   1. The interrelation of a solid material and its melt at constant vapor pressure. 2. The concentration relationship of a solid with a solvent liquid other than its melt. Also known as liquid-solid equilibrium.

solidus   In a constitution or equilibrium diagram, the locus of points representing the temperature below which the various compositions finish freezing on cooling, or begin to melt on heating.

solidus curve   A curve on the phase diagram of a system with two components which represents the equilibrium between the liquid phase and the solid phase.

soliquid   A system in which solid particles are dispersed in a liquid.

solubility   The ability of a substance to form a solution with another substance.

solubility coefficient   The volume of a gas that can be dissolved by a unit volume of solvent at a specified pressure and temperature.

solubility curve   A graph showing the concentration of a substance in its saturated solution in a solvent as a function of temperature.

solubility product constant   A type of simplified equilibrium constant, K_{sp}, defined for and useful for equilibria between solids and their respective ions in solution; for example, the equilibrium

$$AgCl(s) \rightleftarrows Ag^+ + Cl^-, [Ag^+][Cl^-] \cong K_{sp}$$

where $[Ag^+]$ and $[Cl^-]$ are molar concentrations of silver ions and chloride ions.

solubility test   1. A test for the degree of solubility of asphalts and other bituminous materials in solvents, such as carbon tetrachloride, carbon disulfide, or petroleum ether. 2. Any test made to show the solubility of one material in another (such as liquid-liquid, solid-liquid, gas-liquid, or solid-solid).

soluble   Capable of being dissolved.

soluble barbital   *See* sodium barbital.

soluble glass   *See* sodium silicate.

soluble gluside   *See* sodium saccharine.

soluble guncotton   *See* pyroxylin.

soluble indigo blue   *See* indigo carmine.

soluble nitrocellulose   *See* pyroxylin.

soluble saccharin   *See* sodium saccharin.

solute   The substance dissolved in a solvent.

solution   A single, homogeneous liquid, solid, or gas phase that is a mixture in which the components (liquid, gas, solid, or combinations thereof) are uniformly distributed throughout the mixture.

solution pressure   1. A measure of the tendency of molecules or atoms to cross a bounding surface between phases and to enter into a solution. 2. A measure of the tendency of hydrogen, metals, and certain nonmetals to pass into solution as ions.

solution process An oil-refining process for separating mercaptans from gasoline by washing with a caustic solution containing organic compounds in which the mercaptans are soluble.

solutizer-air regenerative process A petroleum refinery process that is identical to the solutizer-steam regeneration process, except for the regeneration step; the newer units use uncatalyzed air regeneration.

solutizer-steam regenerative process A petroleum refinery process used to extract mercaptans from gasoline or naphtha; uses solutizers (potassium isobutyrate or potassium alkyl phenolate) in strong potassium hydroxide solution as the selective solvent.

solutizer-tannin process A petroleum refinery process that is an early variation of the solutizer-air regenerative process for extraction of mercaptans from gasoline; uses tannin-catalyzed oxidation for the regeneration step.

solutrope A ternary mixture with two liquid phases and a third component distributed between the phases, or selectively dissolved in one or the other of the phases; analogous to an azeotrope.

solvation The process of swelling, gelling, or dissolving of a material by a solvent; for resins, the solvent can be a plasticizer.

Solvay process The process to make sodium carbonate and calcium chloride by treating sodium chloride with ammonia and carbon dioxide.

solvent That part of a solution that is present in the largest amount, or the compound that is normally liquid in the pure state (as for solutions of solids or gases in liquids).

solvent deasphalting A petroleum refinery process used to remove asphaltic and resinous materials from reduced crude oils, lubricating oil stocks, gas oils, or middle distillates through the extractive or precipitant action of solvents. Also known as solvent deresining.

solvent deresining *See* solvent deasphalting.

solvent dewaxing A petroleum refinery process for solvent removal of wax from oils; the mixture of waxy oil and solvent is chilled, then filtered or centrifuged to remove the precipitated oil; the solvent is recovered for reuse.

solvent extraction The separation of materials of different chemical types and solubilities by selective solvent action; that is, some materials are more soluble in one solvent than in another, hence there is a preferential extractive action; used to refine petroleum products, chemicals, vegetable oils, and vitamins.

solvent recovery For reuse purposes, the catching and recovery of solvent vapors from vent lines, process vessels, or other sources of evaporative loss, usually with a solid adsorbent material.

solvent-refined Pertaining to any product material whose final quality and condition is in part the result of a solvent treatment during processing of the feedstock material.

solvent refining The process of treating a mixed material with a solvent that preferentially dissolves and removes certain minor constituents (usually the undesired ones); common in the petroleum refining industry.

solvolysis A reaction in which a solvent reacts with the solute to form a new substance.

solvus   In a phase or equilibrium diagram, the locus of points representing the solid-solubility temperatures of various compositions of the solid phase.

somatotropin   The growth hormone of the pituitary gland.

Sommelet process   The preparation of thiophene aldehydes by treatment of thiophene with hexamethylenetetramine.

Sommerfeld law for doublets   According to the Bohr-Sommerfeld theory, the splitting in frequency of regular or relativistic doublets is $\alpha^2 R(Z - \sigma)^4/n^3(l + 1)$, where α is the fine structure constant, R is the Rydberg constant of the atom, Z is the atomic number, σ is a screening constant, n is the principal quantum number, and l is the orbital angular momentum quantum number.

sonic nucleation   In supersaturated solutions, the use of sonic or ultrasonic radiation to help bring about nucleation and corresponding crystallization of substances otherwise difficult to crystallize.

Sonnenschein's reagent   A solution of phosphomolybdic acid that forms a yellow precipitate with alkaloid sulfates.

sorbic acid $CH_3CH{=}CHCH{=}CHCOOH$ A white, crystalline compound; soluble in most organic solvents, slightly soluble in water; melts at 135°C; used as a fungicide and food preservative, and in the manufacture of plasticizers and lubricants. Also known as 2,4-hexadienoic acid.

sorbic acid potassium salt   *See* potassium sorbate.

sorbide   The generic term for anhydrides derived from sorbitol.

sorbin   *See* sorbose.

D-sorbite   *See* sorbitol.

sorbitol $C_6H_8(OH)_6$ Combustible, white, water-soluble, hygroscopic crystals with a sweet taste; melt at 93 to 97.5°C (depending on the form); used in cosmetic creams and lotions, toothpaste, and resins; as a food additive; and for ascorbic acid fermentation. Also known as D-sorbite; D-sorbitol.

D-sorbitol   *See* sorbitol.

sorbose $C_6H_{12}O_6$ A carbohydrate prepared by fermentation; produced as water-soluble crystals that melt at 165°C; used in the production of vitamin C. Also known as sorbin; L-sorbose.

L-sorbose   *See* sorbose.

Sörensen titration   Titration with one of the Sörensen hydrogen-ion-concentration indicators.

sorption   A general term used to encompass the processes of adsorption, absorption, desorption, ion exchange, ion exclusion, ion retardation, chemisorption, and dialysis.

sosoloid   A system consisting of particles of a solid dispersed in another solid.

sour   The condition of containing large amounts of sulfur or sulfur compounds (such as mercaptans or hydrogen sulfide), as in crude oils, naphthas, or gasoline.

source   The arc or spark that supplies light for a spectroscope.

Sovafining   A proprietary petroleum refinery hydrogenation process for pretreatment of straight-run and thermal naphtha for feed to catalytic reforming units, and for desulfurization of fuel oils.

Sovaforming A proprietary petroleum refinery process using a platinum catalyst in a multiple-reactor, fixed-bed catalytic system; used to catalytically reform gasoline-range hydrocarbons to produce high octanes.

Soxhlet extractor A flask and condenser device for the continuous extraction of alcohol- or ether-soluble materials.

space cloth Woven cloth or wire used for solids screening, and for which the openings between the fibers or strands are designated in terms of space or clear opening.

space group A group of operations which leave the infinitely extended, regularly repeating pattern of a crystal unchanged; there are 230 such groups.

space group extinction The absence of certain classes of reflections in the x-ray diffraction pattern of a crystal due to the existence of symmetry elements in the space group of the crystal which are not present in its point group.

space lattice *See* crystal lattice; lattice.

space velocity The relationship between feed rate and reactor volume in a flow process; defined as the volume or weight of feed (measured at standard conditions) per unit time per unit volume of reactor (or per unit weight of catalyst).

spallation A nuclear reaction in which the energy of each incident particle is so high that more than two or three particles are ejected from the target nucleus and both its mass number and atomic number are changed. Also known as nuclear spallation.

spallation reaction A high-energy nuclear reaction which results in the release of large numbers of nucleons as reaction products.

Spanish white *See* bismuth subnitrate.

sparger *See* perforated-pipe distributor.

spark excitation The use of an electric spark (10,000 to 30,000 volts) to excite spectral line emissions from otherwise hard-to-excite samples; used in emission spectroscopy.

spark explosion method A technique for the analysis of hydrogen; the sample is mixed with an oxidant and exploded by a spark or hot wire, and the combustion products are then analyzed.

spark spectrum The spectrum produced by a spark discharging through a gas or vapor; with metal electrodes, a spectrum of the metallic vapor is obtained.

sparteine $C_{15}H_{26}N_2$ A poisonous, colorless, oily alkaloid; soluble in alcohol and ether, slightly soluble in water; boils at 173°C; used in medicine. Also known as lupinidine.

species 1. A chemical entity or molecular particle, such as a radical, ion, molecule, or atom. 2. *See* nuclide.

specific mass shift The portion of the mass shift that is produced by the correlated motion of different pairs of atomic electrons and is therefore absent in one-electron systems.

specific retention volume The relationship among retention volume, void volume, and adsorbent weight, used to standardize gas chromatography adsorbents by the elution of a standard solute by a standard eluent from the adsorbent under test.

specific surface The surface area per unit weight or volume of a particulate solid; used in size-reduction (crushing and grinding) calculations.

specific susceptibility *See* mass susceptibility.

spectral bandwidth The minimum radiant-energy bandwidth to which a spectrophotometer is accurate; that is, 1–5 nanometers for better models.

spectral directional reflectance factor In spectrophotometric colorimetry, the ratio of the energy diffused in any desired direction by the object under analysis to that energy diffused in the same direction by an ideal perfect (energy) diffuser.

spectral line A discrete value of a quantity, such as frequency, wavelength, energy, or mass, whose spectrum is being investigated; one may observe a finite spread of values resulting from such factors as level width, Doppler broadening, and instrument imperfections. Also known as spectrum line.

spectral radiance factor A situation when the desired directions for analysis of energy diffused from (reflected from) an object under spectrophotometric colorimetric analysis are all substantially the same (a solid angle of nearly zero steradians).

spectral reflectance Situation when the desired directions for analysis of energy from (reflected from) an object under spectrophotometric colorimetric analysis is diffused in all directions (not directed as a single beam).

spectral regions Arbitrary ranges of wavelength, some of them overlapping, into which the electromagnetic spectrum is divided, according to the types of sources that are required to produce and detect the various wavelengths, such as x-ray, ultraviolet, visible, infrared, or radio-frequency.

spectral series Spectral lines or groups of lines that occur in sequence.

spectrobolometer An instrument that measures radiation from stars; measurement can be made in a narrow band of wavelengths in the electromagnetic spectrum; the instrument itself is a combination spectrometer and bolometer.

spectrofluorometer A device used in fluorescence spectroscopy to increase the selectivity of fluorometry by passing emitted fluorescent light through a monochromator to record the fluorescence emission spectrum.

spectrogram The record of a spectrum produced by a spectrograph.

spectrograph A spectroscope provided with a photographic camera or other device for recording the spectrum.

spectrography The use of photography to record the electromagnetic spectrum displayed in a spectroscope.

spectrometer **1.** A spectroscope that is provided with a calibrated scale either for measurement of wavelength or for measurement of refractive indices of transparent prism materials. **2.** A spectroscope equipped with a photoelectric photometer to measure radiant intensities at various wavelengths.

spectrometry The use of spectrographic techniques for deriving the physical constants of materials.

spectrophone A cell containing the sample in the optoacoustic detection method; equipped with windows through which the laser beam enters the cell and a microphone for detecting sound.

spectrophotometer An instrument that measures transmission or apparent reflectance of visible light as a function of wavelength, permitting accurate analysis of color or accurate comparison of luminous intensities of two sources or specific wavelengths.

spectrophotometric titration An analytical method in which the radiant-energy absorption of a solution is measured spectrophotometrically after each increment of titrant is added.

spectrophotometry A procedure to measure photometrically the wavelength range of radiant energy absorbed by a sample under analysis; can be by visible light, ultraviolet light, or x-rays.

spectropyrheliometer An astronomical instrument used to measure distribution of radiant energy from the sun in the ultraviolet and visible wavelengths.

spectroscope An optical instrument consisting of a slit, collimator lens, prism or grating, and a telescope or objective lens which produces a spectrum for visual observation.

spectroscopic displacement law The spectrum of an un-ionized atom resembles that of a singly ionized atom of the element one place higher in the periodic table, and that of a doubly ionized atom two places higher in the table, and so forth.

spectroscopic splitting factor *See* Landé g factor.

spectrum line *See* spectral line.

spermatin An albuminoid material occurring in semen.

spermidine $H_2N(CH_2)_3NH(CH_2)_4NH_2$ The triamine found in semen and other animal tissues.

spermine $C_{10}H_{26}N_4$ A tetramine found in semen, blood serum, and other body tissues.

Sperry process The electrolytic manufacture of basic lead carbonate (white lead) from desilverized lead that contains some bismuth; impure lead collects at the anode, and carbon dioxide is passed into the solution to convert the lead to carbonate.

sphenoid An open crystal, occurring in monoclinic crystals of the sphenoidal class, and characterized by two nonparallel faces symmetrical with an axis of twofold symmetry.

sphere of attraction The distance within which the potential energy arising from mutual attraction of two molecules is not negligible with respect to the molecules' average thermal energy at room temperature.

spheroidal group A group in the tetragonal symmetry system; the sphenoid is the typical form.

sphingolipid Any lipid, such as a sphingomyelin, that yields sphingosine or one of its derivatives as a product of hydrolysis.

sphingomyelin A phospholipid consisting of choline, sphingosine, phosphoric acid, and a fatty acid.

sphingosine $C_{18}H_{37}O_2N$ A moiety of sphingomyelin, cerebrosides, and certain other phosphatides.

spiculin An organic material making up a portion of a spicule.

spinacene *See* squalene.

spin echo technique A variation of the nuclear magnetic resonance technique in which the radio frequency field is applied in two pulses, separated by a time interval t, and a strong nuclear induction signal is observed at a time t after the second pulse.

spin isomer 1. An excited nuclear state which has an unusually long lifetime because of the large difference between the spin of the state and the spins of the states of lower energy into which it is permitted to decay. 2. *See* parahydrogen.

spin label A molecule which contains an atom or group of atoms exhibiting an unpaired electron spin that can be detected by electron spin resonance (ESR) spectroscopy and can be bonded to another molecule.

spinning band column An analytical distillation column inside of which is a series of driven, spinning bands; centrifugal action of the bands throws a layer of liquid onto the inner surface of the column; used as an aid in liquid-vapor contact.

spinochrome A type of echinochrome; an organic pigment that is known only from sea urchins and certain homopteran insects.

spiral plate exchanger A heat-transfer device made from a pair of plates rolled in a spiral to provide two relatively long, rectangular passages for heat-transfer between fluids in countercurrent flow.

spiral wire column An analytical rectification (distillation) column with a wire spiral the length of the inside of the column to serve as a liquid-vapor contact surface.

spiran A polycyclic compound containing a carbon atom which is a member of two rings.

spiro ring system A molecular structure with two ring structures having one atom in common; for example, spiropentane.

split interstitial A crystal defect in which a displaced atom forms a bond with a normal atom in such a way that neither atom is on the normal site but the two are symmetrically displaced from it.

splitter A petroleum-refinery term for a fractionating tower that produces only an overhead and bottom stream.

sponge Wood shavings coated with iron oxide and used as a catalyst in processes for removing hydrogen sulfide from industrial gases.

spongin A scleroprotein, occurring as the principal component of skeletal fibers in many sponges.

spontaneous fission Nuclear fission in which no particles or photons enter the nucleus from the outside.

spontaneous heating The slow reaction of material with atmospheric oxygen at ambient temperatures; liberated heat, if undissipated, accumulates so that in the presence of combustible substances a fire will result.

spontaneous ignition Ignition which can occur when certain materials such as tung oil are stored in bulk, resulting from the generation of heat, which cannot be readily dissipated; often heat is generated by microbial action.

sporopollenin   A substance related to suberin and cutin but more resistant to decay that is found in the exine of pollen grains.

spot test   The addition of a drop of reagent to a drop or two of sample solution to obtain distinctive colors or precipitates; used in qualitative analysis.

spray tower   A vertical column, at the top of which is a liquid spray device; used to contact liquids with gas streams for absorption, humidification, or drying.

spreading factor   *See* hyaluronidase.

s-process   The synthesis of elements, predominantly in the iron group, over long periods of time through the capture of slow neutrons which are produced mainly by the reactions of α-particles with carbon-13 and neon-21.

squalene $C_{30}H_{50}$ A liquid triterpene which is found in large quantities in shark liver oil, and which appears to play a role in the biosynthesis of sterols and polycyclic terpenes; used as a bactericide and as an intermediate in the synthesis of pharmaceuticals. Also known as spinacene.

Sr   *See* strontium.

SSD   *See* steady-state distribution.

stability   The property of a chemical compound which is not readily decomposed and does not react with other compounds.

stability constant   Refers to the equilibrium reaction of a metal cation and a ligand to form a chelating mononuclear complex; the absolute-stability constant is expressed by the product of the concentration of products divided by the product of the concentrations of the reactants; the apparent-stability constant (also known as the conditional- or effective-stability constant) allows for the nonideality of the system because of the combination of the ligand with other complexing agents present in the solution.

stabilization   A petroleum-refinery process for separating light gases from petroleum or gasoline, thus leaving a stable (less volatile) liquid so that it can be handled or stored with less change in composition.

stabilizer   The fractionation column in a petroleum refinery used to stabilize (remove fractions from) hydrocarbon mixtures.

stable factor   *See* factor VII.

stable isobar   One of two or more stable nuclides which have the same mass number but differ in atomic number.

stable isotope   An isotope which does not spontaneously undergo radioactive decay.

stable nucleus   A nucleus which does not spontaneously undergo radioactive decay.

stack   In gas works, a row of benches containing retorts.

stacking fault   A defect in a face-centered cubic or hexagonal close-packed crystal in which there is a change from the regular sequence of positions of atomic planes.

standard calomel electrode   A mercury-mercurous chloride electrode used as a reference (standard) measurement in polarographic determinations.

standard electrode potential   The reversible or equilibrium potential of an electrode in an environment where reactants and products are at unit activity.

standard evaporator *See* short-tube vertical evaporator.

standard plane The crystal plane whose Miller indices are (111), that is, whose intercepts on the crystal axes are proportional to the corresponding sides of a unit cell.

standard potential The potential of an electrode composed of a substance in its standard state, in equilibrium with ions in their standard states compared to a hydrogen electrode.

stannane *See* tin hydride.

stannic acid *See* stannic oxide.

stannic anhydride *See* stannic oxide.

stannic bromide $SnBr_4$ Water- and alcohol-soluble, white crystals that fume when exposed to air, and melt at 31°C; used in mineral separations. Also known as tin bromide; tin tetrabromide.

stannic chloride $SnCl_4$ A colorless, fuming liquid; soluble in cold water, alcohol, carbon disulfide, and oil of turpentine; decomposed by hot water; boils at 114°C; used as a conductive coating and a sugar bleach, and in drugs, ceramics, soaps, and blueprinting. Also known as tin chloride; tin tetrachloride.

stannic chromate $Sn(CrO_4)_2$ Toxic, brownish-yellow crystals, slightly soluble in water; used to decorate porcelain and china. Also known as tin chromate.

stannic iodide SnI_4 Yellow-reddish crystals; insoluble in water, soluble in alcohol, ether, chloroform, carbon disulfide, and benzene; decomposed by water, melt at 144°C, sublime at 180°C. Also known as tin iodide; tin tetraiodide.

stannic oxide SnO_2 A white powder; insoluble in water, soluble in concentrated sulfuric acid; melts at 1127°C; used in ceramic glazes and colors, special glasses, putty, and cosmetics, and as a catalyst. Also known as flowers of tin; stannic acid; stannic anhydride; tin dioxide; tin oxide; tin peroxide.

stannic sulfide SnS_2 A yellow-brown powder; insoluble in water, soluble in alkaline sulfides; decomposes at red heat; used as a pigment and for imitation gilding. Also known as artificial gold; mosaic gold; tin bisulfide.

stannous bromide $SnBr_2$ A yellow powder; soluble in water, alcohol, acetone, ether, and dilute hydrochloric acid; browns in air; melts at 215°C. Also known as tin bromide.

stannous chloride $SnCl_2$ White crystals; soluble in water, alcohol, and alkalies; oxidized in air to the oxychloride; melt at 247°C; used as a chemical intermediate, reducing agent, and ink-stain remover, and for silvering mirrors. Also known as tin chloride; tin crystals; tin dichloride; tin salts.

stannous chromate $SnCrO_4$ A brown powder; very slightly soluble in water; used to decorate porcelain. Also known as tin chromate.

stannous 2-ethylhexoate $Sn(C_8H_{15}O_2)_2$ A light yellow liquid, soluble in benzene, toluene, and petroleum ether; used as a lubricant, a vulcanizing agent, and a stabilizer for transformer oil. Also known as stannous octoate; tin octoate.

stannous fluoride SnF_2 A white, lustrous powder; slightly soluble in water; used to fluoridate toothpaste and as a medicine. Also known as tin difluoride; tin fluoride.

stannous octoate *See* stannous 2-ethylhexoate.

stannous oxalate SnC_2O_4 A white, crystalline powder that decomposes at about 280°C; soluble in acids; used in textile dyeing and printing. Also known as tin oxalate.

stannous oxide SnO An air-unstable, brown to black powder; insoluble in water, soluble in acids and strong bases; decomposes when heated; used as a reducing agent and chemical intermediate, and for glass plating. Also known as tin oxide; tin protoxide.

stannous sulfate $SnSO_4$ Heavy light-colored crystals; decomposes rapidly in water, loses SO_2 at 360°C; used for dyeing and tin plating. Also known as tin sulfate.

stannous sulfide SnS Dark crystals; insoluble in water, soluble (with decomposition) in concentrated hydrochloric acid; melts at 880°C; used as an analytical reagent and catalyst, and in bearing material. Also known as tin monosulfide; tin protosulfide; tin sulfide.

stannum The Latin name for tin, thus the symbol Sn for the element.

staphylotoxin Any of the various toxins elaborated by strains of *Staphylococcus aureus,* including hemolysins, enterotoxins, and leukocidin.

starch Any one of a group of carbohydrates or polysaccharides, of the general composition $(C_6H_{10}O_5)_n$, occurring as organized or structural granules of varying size and markings in many plant cells; it hydrolyzes to several forms of dextrin and glucose; its chemical structure is not completely known, but the granules consist of concentric shells containing at least two fractions: an inner portion called amylose, and an outer portion called amylopectin.

Stark effect The effect on spectrum lines of an electric field which is either externally applied or is an internal field caused by the presence of neighboring ions or atoms in a gas, liquid, or solid. Also known as electric field effect.

Stark-Einstein law *See* Einstein photochemical equivalence law.

static bed Refers to a layer of solids in a process vessel (absorber, catalytic reactor, packed distillation column, or granular filter bed) in which the particles rest upon one another at essentially the settled bulk density of the solids phase; contrasted to moving-solids or fluidized-solids beds.

steady-state distribution The equilibrium condition between phases in each step of a multistage, countercurrent liquid-liquid extraction. Abbreviated SSD.

steady-state flow Fluid flow without any change in composition or phase equilibria relationships.

steam cracking High-temperature cracking of petroleum hydrocarbons in the presence of steam.

steam distillation A distillation in which vaporization of the volatile constituents of a liquid mixture takes place at a lower temperature by the introduction of steam directly into the charge; steam used in this manner is known as open steam. Also known as steam stripping.

steam refining A petroleum refinery distillation process, in which the only heat used comes from steam in open and closed coils near the bottom of the still; used to produce gasoline and naphthas where odor and color are of prime importance; where open steam is used, it is known as steam distillation.

steam still　A still in which steam provides most of the heat; distillation requires a lower temperature than in standard equipment (except for a vacuum distillation unit).

steam stripping　*See* steam distillation.

steapsin　An enzyme in pancreatic juice that catalyzes the hydrolysis of fats. Also known as pancreatic lipase.

stearamide　$CH_3(CH_2)_{16}CONH_2$　Colorless leaflets with a melting point of 109°C; used as a corrosion inhibitor in oil wells. Also known as octadecanamide.

stearate　$C_{17}H_{35}COOM$　A salt or ester of stearic acid where M is a monovalent radical, for example, sodium stearate, $C_{17}H_{35}COONa$.

stearic acid　$CH_3(CH_2)_{16}COOH$　Nature's most common fatty acid, derived from natural animal and vegetable fats; colorless, waxlike solid, insoluble in water, soluble in alcohol, ether, and chloroform; melts at 70°C; used as a lubricant and in pharmaceuticals, cosmetics, and food packaging. Also known as *n*-octadecanoic acid.

stearin　$C_3H_5(C_{18}H_{35}O_2)_3$　A colorless combustible powder; insoluble in water, soluble in alcohol, chloroform, and carbon disulfide; melts at 72°C; used in metal polishes, pastes, candies, candles, and soap, and to waterproof paper. Also known as glyceryl tristearate; tristearin.

stearyl alcohol　$CH_3(CH_2)_{16}CH_2OH$　Oily white, combustible flakes; insoluble in water, soluble in alcohol, acetone, and ether; melt at 59°C; used in lubricants, resins, perfumes, and cosmetics, and as a surface-active agent. Also known as 1-octadecanol; octadecyl alcohol.

Stengel process　Manufacture of ammonium nitrate fertilizer from nitric acid and anhydrous ammonia.

step　*See* elementary reaction.

sterane　A cycloalkane derived from a sterol.

stercobilin　Urobilin as a component of the brown fecal pigment.

stercobilinogen　A colorless reduction product of stercobilin found in feces.

stereochemistry　The study of the spatial arrangement of atoms in molecules and the chemical and physical consequences of such arrangement.

stereognomogram　The projection resulting from the superposition of the projection planes of a stereogram and a gnomogram.

stereoisomers　Compounds whose molecules have the same number and kind of atoms and the same atomic arrangement, but differ in their spatial relationship.

stereorubber　Synthetic rubber, *cis*-polyisoprene, a polymer with stereospecificity.

stereoselective reaction　A chemical reaction in which one stereoisomer is produced or decomposed more rapidly than another.

stereospecificity　The condition of a polymer whose molecular structure has a fixed spatial (geometric) arrangement of its constituent atoms, thus having crystalline properties; for example, synthetic natural rubber, *cis*-polyisoprene.

stereospecific polymer A polymer with specific or definite order of arrangement of molecules in space, as in isotactic polypropylene; permits close packing of molecules and leads to a high degree of polymer crystallinity. Also known as stereoregular polymer.

stereospecific synthesis Catalytic polymerization of monomer molecules to produce stereospecific polymers, as with Ziegler or Natta catalysts (derived from a transition metal halide and a metal alkyl).

steric effect The influence of the spatial configuration of reacting substances upon the rate, nature, and extent of reaction.

steric hindrance The prevention or retardation of chemical reaction because of neighboring groups on the same molecule; for example, ortho-substituted aromatic acids are more difficult to esterify than are the meta and para substitutions.

Stern-Gerlach effect The splitting of a beam of atoms passing through a strong, inhomogeneous magnetic field into several beams.

stern layer One of two electrically charged layers of electrolyte ions, the layer of ions immediately adjacent to the surface, in the neighborhood of a negatively charged surface.

steroid A member of a group of compounds, occurring in plants and animals, that are considered to be derivatives of a fused, reduced ring system, cyclopenta[α]-phenanthrene, which consists of three fused cyclohexane rings in a nonlinear or phenanthrene arrangement.

sterol Any of the natural products derived from the steroid nucleus; all are waxy, colorless solids soluble in most organic solvents but not in water, and contain one alcohol functional group.

sterone A ketone derived from a steroid.

stibide *See* antimonide.

stibium The Latin name for antimony, thus the symbol Sb for the element.

stibnate *See* potassium antimonate.

Stickland reaction An amino acid fermentation involving the coupled decomposition of two or more substrates.

stilbene $C_6H_5CH:CHC_6H_5$ Colorless crystals soluble in ether and benzene, insoluble in water; melts at 124°C; used to make dyes and bleaches and as phosphors. Also known as diphenylethylene; toluylene.

stilbesterol *See* diethylstilbesterol.

still A device used to evaporate liquids; heat is applied to the liquid, and the resulting vapor is condensed to a liquid state.

Stobbe reaction A type of aldol condensation reaction represented by the reaction of benzophenone with dimethyl succinate and sodium methoxide to form monoesters of an α-alkylidene (or arylidene) succinic acid.

stoichiometry The numerical relationship of elements and compounds as reactants and products in chemical reactions.

Stokes' law The wavelength of luminescence excited by radiation is always greater than that of the exciting radiation.

Stokes line A spectrum line in luminescent radiation whose wavelength is greater than that of the radiation which excited the luminescence, and thus obeys Stokes' law.

Stokes shift The displacement of spectral lines or bands of luminescent radiation toward longer wavelengths than those of the absorption lines or bands.

STPP *See* sodium tripolyphosphate.

straight-run Petroleum fractions derived from the straight distillation of crude oil without chemical reaction or molecular modification. Also known as virgin.

straight-run distillation Continuous nonreactive distillation of petroleum oil to separate it into products in the order of their boiling points.

Stratco alkylation A proprietary petroleum refinery process for the manufacture of alkylate gasoline; uses sulfuric acid as catalyst and refrigeration of reactor effluent to control isobutane-olefin ratio in the reactor feedstream. Also known as effluent refrigeration alkylation.

stream day Denoting a 24-hour actual operation of a processing unit, in contrast to the hours actually operated during a calendar (24-hour) day.

strepogenin A factor, possibly a peptide derivative of glutamic acid, reported to exist in certain proteins, acting as a growth stimulant in bacteria and mice in the presence of completely hydrolyzed protein. Also known as streptogenin.

streptobiosamine $C_{13}H_{23}NO_9$ A nitrogen-containing disaccharide, obtained when streptomycin undergoes acid hydrolysis; in the streptomycin molecule it is glycosidally linked to streptidine.

streptogenin *See* strepogenin.

streptokinase An enzyme occurring as a component of fibrinolysin in cultures of certain hemolytic streptococci.

streptolysin Any of a group of hemolysins elaborated by *Streptococcus pyogenes*.

stripped atom An ionized atom which has appreciably fewer electrons than it has protons in the nucleus.

stripper An evaporative device for the removal of vapors from liquids; can be in a bubble-tray distillation tower, a vacuum vessel, or an evaporator; if it is a part of a distillation column below the feed tray, it is called the stripping section.

stripping In petroleum refining, the removal (by flash evaporation or steam-induced vaporation) of the more volatile components from a cut or fraction; used to raise the flash point of kerosine, gas oil, or lubricating oil.

stripping analysis An analytic process of solutions or concentrations containing ions, in which the ions are electrodeposited onto an electrode, stripped (dissolved) from the material from the electrode, and weighed.

stripping reaction A nuclear reaction in which part of the incident nucleus combines with the target nucleus, and the other part proceeds with most of its original momentum in practically its original direction; especially the reaction in which the incident nucleus is a deuteron and only a proton emerges from the target.

strong acid   An acid with a high degree of dissociation in solution, for example, mineral acids, such as hydrochloric acid, HCl, sulfuric acid, H_2SO_4, or nitric acid, HNO_3.

strong base   A base with a high degree of dissociation in solution, for example, sodium hydroxide, NaOH, potassium hydroxide, KOH.

strongly damped collision   *See* quasi-fission.

strontia   *See* strontium oxide.

strontium   A metallic element in group IIA, symbol Sr, with atomic number 38, atomic weight 87.62; flammable, soft, pale-yellow solid; soluble in alcohol and acids, decomposes in water; melts at 770°C, boils at 1380°C; chemistry is similar to that of calcium; used as electron-tube getter.

strontium-90   A poisonous, radioactive isotope of strontium; 28-year half life with β radiation; derived from reactor-fuel fission products; used in thickness gages, medical treatment, phosphor activation, and atomic batteries.

strontium acetate $Sr(C_2H_3O_2)_2 \cdot \frac{1}{2}H_2O$ White, water-soluble crystals, loses water at 150°C; used for catalysts, as a chemical intermediate, and in medicine.

strontium bromide $SrBr_2 \cdot 6H_2O$ A white, hygroscopic powder soluble in water and alcohol; loses water at 180°C, melts at 643°C; used in medicine and as an analytical reagent.

strontium carbonate $SrCO_3$ A white powder slightly soluble in water, decomposes at 1340°C; used to make TV-tube glass, strontium salts, and ceramic ferrites, and in pyrotechnics.

strontium chlorate $Sr(ClO_3)_2$ Shock-sensitive, highly combustible, white, water-soluble crystals that decompose at 120°C; used in pyrotechnics and tracer bullets.

strontium chloride $SrCl_2$ Water- and alcohol-soluble white crystals, melts at 872°C; used in medicine and pyrotechnics and to make strontium salts.

strontium chromate $SrCrO_4$ A light yellow, rust and corrosion-resistant pigment used in metal coatings and for pyrotechnics.

strontium dioxide   *See* strontium peroxide.

strontium fluoride SrF_2 A white powder, soluble in hydrochloric acid and hydrofluoric acid; used in medicine and for single crystals for lasers.

strontium hydrate   *See* strontium hydroxide.

strontium hydroxide $Sr(OH)_2$ Colorless deliquescent crystals that absorb carbon dioxide from air, soluble in hot water and acids, melts at 375°C; used by the sugar industry, in lubricants and soaps, and as a plastic stabilizer. Also known as strontium hydrate.

strontium iodide SrI_2 Air-yellowing, white crystals that decompose in moist air, melts at 515°C; used in medicine and as a chemicals intermediate.

strontium monosulfide   *See* strontium sulfide.

strontium nitrate $Sr(NO_3)_2$ A white, water-soluble powder melting at 570°C; used in pyrotechnics, signals and flares, medicine, and matches, and as a chemicals intermediate.

strontium oxalate $SrC_2O_4 \cdot H_2O$ A white powder that loses water at 150°C; used in pyrotechnics and tanning.

strontium oxide SrO A grayish powder, melts at 2430°C, becomes the hydroxide in water; used in medicine, pyrotechnics, pigments, greases, soaps, and as a chemicals intermediate. Also known as strontia.

strontium peroxide SrO_2 A strongly oxidizing, fire-hazardous, white, alcohol-soluble powder that decomposes in hot water; used in medicine, bleaching, and fireworks. Also known as strontium dioxide.

strontium salicylate $Sr(C_7H_5O_3)_2 \cdot 2H_2O$ White crystals or powder with a sweet saline taste; soluble in water and alcohol; used in medicine and manufacture of pharmaceuticals.

strontium sulfate $SrSO_4$ White crystals insoluble in alcohol, slightly soluble in water and concentrated acids, melts at 1605°C; used in paper manufacture, pyrotechnics, ceramics, and glass.

strontium sulfide SrS A gray powder with a hydrogen sulfide aroma in moist air, slightly soluble in water, soluble (with decomposition) in acids, melts above 2000°C; used in depilatories and luminous paints and as a chemicals intermediate. Also known as strontium monosulfide.

strontium titanate $SrTiO_3$ A solid material, insoluble in water and melting at 2060°C; used in electronics and electrical insulation.

structural formula A system of notation used for organic compounds in which the exact structure, if it is known, is given in schematic representation.

structure cell See unit cell.

structure type The structural arrangement of a crystal, regardless of the atomic elements present; it corresponds to the crystal's space group.

strychnine $C_{21}H_{22}O_2N_2$ An alkaloid obtained primarily from the plant nux vomica, formerly used for therapeutic stimulation of the central nervous system.

Stuart factor A procoagulant in normal plasma but deficient in the blood of patients with a hereditary bleeding disorder; may be closely related to pro-thrombin since both are formed in the liver by action of vitamin K. Also known as factor X; Stuart-Power factor.

Stuart-Power factor See Stuart factor.

styphnic acid $C_6H(OH)_2(HO_2)_3$ An explosive, yellow, crystalline compound, melting at 179–180°C, slightly soluble in water; used in explosives as a priming agent. Also known as 2,4,6-trinitroresorcinol.

styralyl acetate See α-methylbenzyl acetate.

styralyl alcohol See α-methylbenzyl alcohol.

styrene $C_6H_5CH:CH_2$ A colorless, toxic liquid with a strong aromatic aroma; insoluble in water, soluble in alcohol and ether; polymerizes rapidly, can become explosive; boils at 145°C; used to make polymers and copolymers, polystyrene plastics, and rubbers. Also known as phenylethylene; styrene monomer; vinylbenzene.

styrene-acrylonitrile resin A thermoplastic copolymer of styrene and acrylonitrile with good stiffness and resistance to scratching, chemicals, and stress. Also known as SAN.

styrene monomer See styrene.

styrene oxide C_8H_8O A moderately toxic, combustible, colorless or straw-colored liquid miscible in acetone, ether, and benzene, and melts at 195°C; used as a chemical intermediate.

styrene plastic A plastic made by the polymerization of styrene or the copolymerization of styrene with other unsaturated compounds.

styryl carbinol *See* cinnamic alcohol.

styrylformic acid *See* cinnamic acid.

subalkaline Pertaining to a soil in which the pH is 8.0 to 8.5, usually in a limestone or salt-marsh region.

subcompound A compound, generally in the vapor phase, in which an element exhibits a valency lower than that exhibited in its ordinary compounds.

suberic acid $HOOC(CH_2)_6COOH$ A colorless, crystalline compound that melts at 143°C, and dissolves slightly in cold water; used in organic synthesis. Also known as octanedioic acid.

suberin A fatty substance found in many plant cell walls, especially cork.

sublevel *See* subshell.

sublimatography A procedure of fractional sublimation in which a solid mixture is separated into bands along a condensing tube with a temperature gradient.

sublimator Device used for the heating of solids (usually under vacuum) to the temperature at which the solid sublimes.

sublimed sulfur *See* flowers of sulfur.

subshell Electrons of an atom within the same shell (energy level) and having the same azimuthal quantum numbers. Also known as sublevel.

subsolvus A range of conditions in which two or more solid phases can form by exsolution from an original homogeneous phase.

substance P An undecapeptide widely distributed in the central nervous system and found in highest concentrations in superficial layers of the dorsal horn of the spinal cord, in the trigeminal nerve nucleus, and in the substantia nigra; acts as a neurotransmitter.

substitution reaction Replacement of an atom or radical by another one in a chemical compound.

substitutive nomenclature A system in which the name of a compound is derived by using the functional group (the substituent) as a prefix or suffix to the name of the parent compound to which it is attached; for example, in 2-chloropropane a chlorine atom has replaced a hydrogen atom on the central carbon of the propane chain.

substrate 1. The substance with which an enzyme reacts. 2. A compound with which a reagent reacts.

subunit *See* protomer.

succinamide $H_2NCOCH_2CONH_2$ The amide of succinic acid.

succinate A salt or ester of succinic acid; for example, sodium succinate, $Na_2C_4H_4O_4 \cdot 6H_2O$, the reaction product of succinic acid and sodium hydroxide.

succinbromimide   *See* N-bromosuccinamide.

succinic acid $CO_2H(CH_2)_2CO_2H$ Water-soluble, colorless crystals with an acid taste; melts at 185°C; used as a chemical intermediate, in medicine, and to make perfume esters. Also known as butanedioic acid.

succinic acid dehydrogenase   An enzyme that catalyzes the dehydrogenation of succinic acid to fumaric acid in the presence of a hydrogen acceptor. Also known as succinic dehydrogenase.

succinic acid 2,2-dimethylhydrazide $C_6H_{12}O_3N_2$ White crystals with a melting point of 154–156°C; soluble in water; used as a growth regulator for many crops and ornamentals. Also known as aminocide.

succinic anhydride $C_4H_4O_3$ Colorless or pale needles soluble in alcohol and chloroform; converts to succinic acid in water; melts at 120°C; used as a chemical and pharmaceutical intermediate and a resin hardener. Also known as butanedioic anhydride; 2,5-diketotetrahydrofurane; succinyl oxide.

succinic dehydrogenase   *See* succinic acid dehydrogenase.

succinimide $C_4H_5O_2N \cdot H_2O$ Colorless or tannish water-soluble crystals with a sweet taste; melts at 126°C; used to make plant growth stimulants and as a chemical intermediate. Also known as 2,5-diketopyrrolidine.

succinonitrite   *See* ethylene cyanide.

succinoxidase   A complex enzyme system containing succinic dehydrogenase and cytochromes that catalyzes the conversion of succinate ion and molecular oxygen to fumarate ion.

succinylcholine chloride $[Cl(CH_3)_3N(CH_2)_2OOCH_2]_2 \cdot 2H_2O$ Water-soluble white crystals with a bitter taste, melts at 162°C; used in medicine. Also known as choline succinate dichloride dihydrate.

succinyl oxide   *See* succinic anhydride.

sucrase   *See* saccharase.

sucrose $C_{12}H_{22}O_{11}$ Combustible, white crystals soluble in water, decomposes at 160 to 186°C; derived from sugarcane or sugarbeet; used as a sweetener in drinks and foods and to make syrups, preserves, and jams. Also known as saccharose; table sugar.

sucrose octaacetate $C_{28}H_{38}O_{19}$ A bitter crystalline compound that forms needles from alcohol solution, melts at 89°C, and breaks down at 286°C or above; used as an adhesive, to impregnate and insulate paper, and in lacquers and plastics.

sugar   A generic term for a class of carbohydrates usually crystalline, sweet, and water soluble; examples are glucose and fructose.

sugar alcohol   Any of the acyclic linear polyhydric alcohols; may be considered sugars in which the aldehydic group of the first carbon atom is reduced to a primary alcohol; classified according to the number of hydroxyl groups in the molecule; sorbitol (D-glucitol, sorbite) is one of the most widespread of all the naturally occurring sugar alcohols.

sulfallate $C_8H_{14}NS_2Cl$ An oily liquid, used as a preemergence herbicide for vegetable crops and ornamentals. Also known as 2-chloroallyl diethyldithiocarbamate.

sulfamate A salt of sulfamic acid; for example, calcium sulfamate, $Ca(SO_3NH_2)_2 \cdot 4H_2O$.

sulfamic acid HSO_3NH_2 White, nonvolatile crystals slightly soluble in water and organic solvents, decomposes at 205°C; used to clean metals and ceramics, and as a plasticizer, fire retardant, chemical intermediate, and textile and paper bleach.

sulfamidyl *See* sulfanilamide.

sulfanilic acid $C_6H_4NH_2 \cdot SO_3H \cdot H_2O$ Combustible, grayish-white crystals slightly soluble in water, alcohol, and ether, soluble in fuming hydrochloric acid; chars at 280–300°C; used in medicine and dyestuffs and as a chemical intermediate. Also known as *para*-aminobenzenesulfonic acid; *para*-anilinesulfonic acid.

***meta*-sulfanilic acid** *See* metanillic acid.

sulfaquinoxaline *See* N'-2-quinoxalysulfanilimide.

sulfatase Any of a group of esterases that catalyze the hydrolysis of sulfuric esters.

sulfate 1. A compound containing the $-SO_4$ group, as in sodium sulfate Na_2SO_4. 2. A salt of sulfuric acid.

sulfate pulping A wood-pulping process in which sodium sulfate is used in the caustic soda pulp-digestion liquor. Also known as kraft process; kraft pulping.

sulfation The conversion of a compound into a sulfate by the oxidation of sulfur, as in sodium sulfide, Na_2S, oxidized to sodium sulfate, Na_2SO_4; or the addition of a sulfate group, as in the reaction of sodium and sulfuric acid to form Na_2SO_4.

sulfenic acid An oxy acid of sulfur with the general formula RSOH, where R is an alkyl or aryl group such as CH_3; known as the esters and halides.

sulfenyl chloride Any of a group of well-known organosulfur compounds with the general formula RSCl; although highly reactive compounds, they can generally be synthesized and isolated; examples are trichloromethanesulfenyl chloride and 2,4-dinitrobenzenesulfenyl chloride.

sulfhemoglobin A greenish substance derived from hemoglobin by the action of hydrogen sulfide; it may appear in the blood following the ingestion of sulfanilamide and other substances.

sulfhydryl compound A compound with a $-SH$ group. Also known as a mercapto compound.

sulfidation The chemical insertion of a sulfur atom into a compound.

sulfide Any compound with one or more sulfur atoms in which the sulfur is connected directly to a carbon, metal, or other nonoxygen atom; for example, sodium sulfide, Na_2S.

sulfide dye A dye containing sulfur and soluble in a 0.25–0.50% sodium sulfide solution, and used to dye cotton; the dyes are manufactured from aromatic polyamines or hydroxy amines; the amine group is primary, secondary, or tertiary, or may be an equivalent nitro, nitroso, or imino group; an example is the dye sulfur blue. Also known as sulfur dye.

sulfinate 1. A compound containing the R_2SX_2 grouping, where X is a halide. **2.** A salt of sulfinic acid having the general formula R·OH·S:O.

sulfinic acid Any of the monobasic organic acids of sulfur with the general formula RS:O(OH); for example, ethanesulfinic acid, $C_2H_5SO_2H$.

sulfinyl bromide *See* thionyl bromide.

sulfite M_2SO_3 A salt of sulfurous acid, for example, sodium sulfite, Na_2SO_3.

sulfo- Prefix for a compound with either a divalent sulfur atom, or the presence of —SO_3H, the sulfo group in a compound. Also spelled sulpho-.

sulfocarbanilide *See* thiocarbanilide.

sulfocarbimide *See* isothiocyanate.

sulfocyanate *See* thiocyanate.

sulfocyanic acid *See* thiocyanic acid.

sulfocyanide *See* thiocyanate.

sulfofication Oxidation of sulfur and sulfur compounds into sulfates, occurring in soils by the agency of bacteria.

sulfolane $C_4H_8SO_2$ A liquid with a boiling point of 285°C and outstanding solvent properties; used for extraction of aromatic hydrocarbons, fractionation of fatty acids, and textile finishing, and as a solvent and plasticizer. Also known as tetrahydrothiophene-1,1-dioxide; tetramethylene sulfone.

sulfonamide One of a group of organosulfur compounds, RSO_2NH_2, prepared by the reaction of sulfonyl chloride and ammonia; used for sulfa drugs.

sulfonamide P *See* sulfanilamide.

sulfonate 1. A sulfuric acid derivative or a sulfonic acid ester containing a SO_4^{2-} group. **2.** Any of a group of petroleum hydrocarbons derived from sulfuric-acid treatment of oils, used as synthetic detergents, emulsifying and wetting agents, and chemical intermediates.

sulfonation Substitution of —SO_3H groups (from sulfuric acid) for hydrogen atoms, for example, conversion of benzene, C_6H_6, into benzenesulfonic acid, $C_6H_5SO_3H$.

sulfone R_2SO_2 (or RSOOR) A compound formed by the oxidation of sulfides, for example, ethyl sulfone, $C_4H_{10}SO_2$, from ethyl sulfide, $C_4H_{10}S$; the use of sulfones, particularly 4,4'-sulfonyldianiline (dapsone) in the treatment of leprosy leads to apparent improvement; relapses associated with sulfone-resistant strains have been encountered.

sulfonic acid A compound with the radical —SO_2OH, derived by the sulfuric acid replacement of a hydrogen atom; for example, conversion of benzene, C_6H_6, to the water-soluble benzenesulfonic acid, $C_6H_5SO_3H$, by treatment with sulfuric acid; used to make dyes and drugs.

3-sulfonic acid *See* gamma acid.

6-sulfonic acid *See* gamma acid.

sulfonyl Also known as sulfuryl. **1.** A compound containing the radical —SO_2—. **2.** A prefix denoting the presence of a sulfone group.

sulfonyl chloride *See* sulfuryl chloride.

sulfoparaldehyde *See* trithioacetaldehyde.

sulfophile element An element occurring preferentially in an oxygen-free mineral. Also known as thiophile element.

sulfosalicylic acid $C_7H_6O_6S$ A trifunctional aromatic compound whose dihydrate is in the form of white crystals or crystalline powder; soluble in water and alcohol; melting point is 120°C; used as an indicator for albumin in urine and as a reagent for the determination of ferric ion; it also has industrial uses. Also known as 2-hydroxybenzoic-5-sulfonic acid; salicylsulfonic acid.

sulfotepp $C_8H_{20}O_5P_2S_2$ A yellow liquid, slightly soluble in water; used in a smoke generator as an insecticide, and as a miticide in a greenhouse. Also known as *O,O,O,O*-tetraethyldithiopyrophosphate.

sulfoxide R_2SO A compound with the radical $=SO$; derived from oxidation of sulfides, the proportion of oxidant, such as hydrogen peroxide, and temperature being set to avoid excessive oxidation; an example is dimethyl sulfoxide, $(CH_3)_2SO$.

sulfur A nonmetallic element in group VIa, symbol S, atomic number 16, atomic weight 32.064, existing in a crystalline or amorphous form and in four stable isotopes; used as a chemical intermediate and fungicide, and in rubber vulcanization.

sulfur-35 Radioactive sulfur with mass number 35; radiotoxic, with 87.1-day half-life, β radiation; derived from pile irradiation; used as a tracer to study chemical reactions, engine wear, and protein metabolism.

sulfurated lime *See* calcium sulfide.

sulfuration The chemical act of combining an element or compound with sulfur.

sulfur bichloride *See* sulfur dichloride.

sulfur bromide S_2Br_2 A toxic, irritating, yellow liquid that reddens in air, soluble in carbon disulfide, decomposes in water, boils at 54°C. Also known as sulfur monobromide.

sulfur chloride S_2Cl_2 A combustible, water-soluble, oily, fuming, amber to yellow-red liquid with an irritating effect on the eyes and lungs, boils at 138°C; used to make military gas and insecticides, in rubber substitutes and cements, to purify sugar juices, and as a chemical intermediate. Also known as sulfur monochloride; sulfur subchloride.

sulfur dichloride SCl_2 A red-brown liquid boiling (when heated rapidly) at 60°C, decomposes in water; used to make insecticides, for rubber vulcanization, and as a chemical intermediate and a solvent. Also known as sulfur bichloride.

sulfur dioxide SO_2 A toxic, irritating, colorless gas soluble in water, alcohol, and ether, boils at -10°C; used as a chemical intermediate, in artificial ice, paper pulping, and ore refining, and as a solvent. Also known as sulfurous acid anhydride.

sulfur dye *See* sulfide dye.

sulfur hexafluoride SF_6 A colorless gas soluble in alcohol and ether, slightly soluble in water, sublimes at -64°C; used as a dielectric in electronics.

sulfuric acid H_2SO_4 A toxic, corrosive, strongly acid, colorless liquid that is miscible with water and dissolves most metals, and melts at 10°C; used in industry in the manufacture of chemicals, fertilizers, and explosives, and in petroleum refining. Also known as dipping acid.

sulfuric acid alkylation   A petroleum refinery alkylation process in which three-carbon, four-carbon, and five-carbon olefins combine with isobutane in the presence of a sulfuric acid catalyst to form high-octane, branched-chain hydrocarbons; used in motor gasoline.

sulfuric chloride   *See* sulfuryl chloride.

sulfuric chlorohydrin   *See* chlorosulfonic acid.

sulfur iodide   *See* sulfur iodine.

sulfur iodine I_2S_2 A gray-black brittle mass with an iodine aroma and a metallic luster, insoluble in water, soluble in carbon disulfide; used in medicine. Also known as iodine bisulfide; iodine disulfide; sulfur iodide.

sulfur monobromide   *See* sulfur bromide.

sulfur monochloride   *See* sulfur chloride.

sulfur monoxide   SO   A gas at ordinary temperatures; produces an orange-red deposit when cooled to temperatures of liquid air; prepared by passing an electric discharge through a mixture of sulfur vapor and sulfur dioxide at low temperature.

sulfur number   The number of milligrams of sulfur per 100 milliliters of sample, determined by electrometric titration; used in the petroleum industry for oils.

sulfurous acid H_2SO_3 An unstable, water-soluble, colorless liquid with a strong sulfur aroma; derived from absorption of sulfur dioxide in water; used in the synthesis of medicine and chemicals, manufacture of paper and wine, brewing, metallurgy, and ore flotation, as a bleach and analytic reagent, and to refine petroleum products.

sulfurous acid anhydride   *See* sulfur dioxide.

sulfurous oxychloride   *See* thionyl chloride.

sulfur oxide   An oxide of sulfur, such as sulfur dioxide, SO_2, and sulfur trioxide, SO_3.

sulfur oxychloride   *See* thionyl chloride.

sulfur subchloride   *See* sulfur chloride.

sulfur test   1. Method to determine the sulfur content of a petroleum material by combustion in a bomb. 2. Analysis of sulfur in petroleum products by lamp combustion in which combustion of the sample is controlled by varying the flow of carbon dioxide and oxygen to the burner.

sulfur trioxide SO_3 A toxic, irritating liquid in three forms, α, β, γ, with respective melting points of 62°C, 33°C, and 17°C; a strong oxidizing agent and fire hazard; used for sulfonation of organic chemicals.

sulfuryl   *See* sulfonyl.

sulfuryl chloride SO_2Cl_2 A colorless liquid with a pungent aroma, boils at 69°C, decomposed by hot water and alkalies; used as a chlorinating agent and solvent and for pharmaceuticals, dyestuffs, rayon, and poison gas. Also known as chlorosulfuric acid; sulfonyl chloride; sulfuric chloride.

sulfuryl fluoride SO_2F_2 A colorless gas with a melting point of $-136.7°C$ and a boiling point of 55.4°C; used as an insecticide and fumigant.

Sullivan reaction The formation of a red-brown color when cysteine is reacted with 1,2-naphthoquinone-4-sodium sulfate in a highly alkaline reducing medium.

sulpho- *See* sulfo-.

superacid **1.** An acidic medium that has a proton-donating ability equal to or greater than 100% sulfuric acid. **2.** A solution of acetic or phosphoric acid.

superfluorescence The process of spontaneous emission of electromagnetic radiation from a collection of excited atoms.

superheavy element A chemical element with an atomic number of 110 or greater.

superhelix A macromolecular structure consisting of a number of alpha-helical polypeptide strands which are twisted together.

superinsulation A multilayer insulation for cryogenic systems, composed of many floating radiation shields in an evacuated double-wall annulus, closely spaced but thermally separated by a poor-conducting fiber.

supermolecule A single quantum-mechanical entity presumably formed by two reacting molecules and in existence only during the collision process; a concept in the hard-sphere collision theory of chemical kinetics.

supersaturation The condition existing in a solution when it contains more solute than is needed to cause saturation. Also known as supersolubility.

supersolubility *See* supersaturation.

surface chemistry The study and measurement of the forces and processes that act on the surfaces of fluids (gases and liquids) and solids, or at an interface separating two phases; for example, surface tension.

surface orientation Arrangement of molecules on the surface of a liquid with one part of the molecule turned toward the liquid.

surface phase A thin rock layer differing in geochemical properties from those of the volume phases on either side. Also known as volume phase.

surface reaction A chemical reaction carried out on a surface as on an adsorbent or solid catalyst.

suspension A mixture of fine, nonsettling particles of any solid within a liquid or gas, the particles being the dispersed phase, while the suspending medium is the continuous phase. Also known as suspended solids.

svedberg A unit of sedimentation coefficient, equal to 10^{-13} second.

Swarts reaction The reaction of chlorinated hydrocarbons with metallic fluorides to form chlorofluorohydrocarbons, such as CCl_2F_2, which is quite inert and nontoxic.

sweat Exudation of nitroglycerin from dynamite due to separation of nitroglycerin from its adsorbent.

sweating Separation of paraffin oil from low-melting petroleum wax obtained from paraffin wax in a chamber (sweater) by first cooling the mixture until it is a solid cake, then warming gradually to cause partial fusion of the mixture to allow drainage of liquid from the cake. Also known as exudation.

sweetening Improvement of a petroleum-product color and odor by converting sulfur compounds into disulfides with sodium plumbite (doctor treating), or by removing them by contacting the petroleum stream with alkalies or other sweetening agents.

sweet spirits of niter *See* ethyl nitrite.

swep $C_8H_7Cl_2NO_2$ A white, crystalline compound with a melting point of 112–114°C; insoluble in water; used as a pre- and postemergence herbicide for rice, carrots, potatoes, and cotton. Also known as methyl-N-(3,4-dichlorophenyl)carbamate.

SXAPS *See* soft x-ray appearance potential spectroscopy.

sym- A chemical prefix; denotes structure of a compound in which substituents are symmetrical with respect to a functional group or to the carbon skeleton.

symbol Letter or combination of letters and numbers that represent various conditions or properties of an element, for example, a normal atom, O (oxygen); with its atomic weight, ^{16}O; its atomic number, $_8$O; as a molecule, O_2; as an ion, O^{2+}; in excited state, O*; or as an isotope, ^{18}O.

symmetric top molecule A nonlinear molecule which has one and only one axis of threefold or higher symmetry.

symmetry axis *See* rotation axis.

symmetry axis of rotary inversion *See* rotoinversion axis.

symmetry axis of rotoinversion *See* rotoinversion axis.

symmetry class *See* crystal class.

symmetry element 1. Some combination of rotations and reflections and translations which brings a crystal into a position that cannot be distinguished from its original position. Also known as symmetry operation; symmetry transformation. 2. The rotational axes, mirror planes, and center of symmetry characteristic of a given crystal.

symmetry number The number of indistinguishable orientations that a molecule can exhibit by being rotated around symmetry axes.

symmetry plane *See* plane of mirror symmetry.

symmetry transformation *See* symmetry element.

syndiotactic polymer A vinyl polymer in which the side chains alternate regularly above and below the plane of the backbone.

syneresis Spontaneous separation of a liquid from a gel or colloidal suspension due to contraction of the gel.

synthesis Any process or reaction for building up a complex compound by the union of simpler compounds or elements.

synthesis gas A mixture of gases prepared as feedstock for a chemical reaction, for example, carbon monoxide and hydrogen to make hydrocarbons or organic chemicals, or hydrogen and nitrogen to make ammonia.

synthetase *See* ligase.

synthetic resin Amorphous, organic, semisolid, or solid material derived from the polymerization of unsaturated monomers such as ethylene, butylene, propylene, and styrene.

synthol process A reaction of carbon monoxide and hydrogen with an iron and sodium carbonate catalyst; produces a mixture of higher alcohols, aldehydes, ketones, higher fatty acids, and aliphatic hydrocarbons, usable as a synthetic gasoline.

Szechtman cell An electrolytic process for manufacture of chlorine that is a variation of both the mercury cell and molten salt cell.

T

T *See* tritium.

2,4,5-T *See* 2,4,5-trichlorophenoxyacetic acid.

2,4,6-T *See* 2,4,6-trichlorophenol.

Ta *See* tantalum.

table salt *See* sodium chloride.

table sugar *See* sucrose.

tabular crystal A crystal that appears broad and flat due to two prominent parallel faces.

tabun $(CH_3)_2NP(O)(C_2H_5O)(CN)$ A toxic liquid with a boiling point of 240°C; soluble in organic solvents; used as a nerve gas.

tachiol *See* silver fluoride.

tachysterol The precursor of calciferol in the irradiation of ergosterol; an isomer of ergosterol.

tactic polymer A polymer with regularity or symmetry in the structural arrangement of its molecules, as in a stereospecific polymer such as some types of polypropylene.

tactoid A particle that appears as a spindle-shaped body under the polarizing microscope and occurs in mosaic virus, fibrin, and myosin.

tag *See* isotopic tracer.

Tag closed-cup tester A laboratory device used to determine the flash point of mobile petroleum liquids flashing below 175°F (79.4°C). Also known as Tagliabue closed tester.

tagged molecule A molecule having one or more atoms which are either radioactive or have a mass which differs from that of the atoms which normally make up the molecule.

tail house An installation in a refinery containing a look box, facilities for sampling, and controls for diverting the products to storage tanks or to other locations in the refinery for further processing.

tank bottom The liquid material in a tank below the level of the outlet pipe; often a mixture of the stored liquid with rust and other sediment.

tannase An enzyme that catalyzes the hydrolysis of tannic acid to gallic acid; found in cultures of *Aspergillus* and *Penicillium*.

tannic acid   1. $C_{14}H_{10}O_9$ A yellowish powder with an astringent taste; soluble in water and alcohol, insoluble in acetone and ether; derived from nutgalls; decomposes at 210°C; used as an alcohol denaturant and a chemical intermediate, and in tanning and textiles. Also known as digallic acid; gallotannic acid; gallotannin; tannin.   2. $C_{76}H_{52}O_{46}$ Yellowish-white to light-brown amorphous powder or flakes; decomposes at 210–215°C; very soluble in alcohol and acetone; used as a mordant in dyeing, in photography, as a reagent, and in clarifying wine or beer. Also known as pentadigalloylglucose.

tannin   See tannic acid.

tantalic acid anhydride   See tantalum oxide.

tantalic chloride   See tantalum chloride.

tantalum   Metallic element in group V, symbol Ta, atomic number 73, atomic weight 180.948; black powder or steel-blue solid soluble in fused alkalies, insoluble in acids (except hydrofluoric and fuming sulfuric); melts about 3000°C.

tantalum carbide   TaC   Hard, chemical-resistant crystals melting at 3875°C; used in cutting tools and dies.

tantalum chloride   TaCl$_5$ A highly reactive, pale-yellow powder decomposing in moist air; soluble in alcohol and potassium hydroxide; melts at 221°C; used to produce tantalum and as a chemical intermediate. Also known as tantalic chloride; tantalum pentachloride.

tantalum nitride   TaN   A very hard, black, water-insoluble solid, melting at 3360°C.

tantalum oxide   Ta$_2$O$_5$ Prisms insoluble in water and acids (except for hydrofluoric); melts at 1800°C; used to make tantalum, in optical glass and electronic equipment, and as a chemical intermediate. Also known as tantalic acid anhydride; tantalum pentoxide.

tantalum pentachloride   See tantalum chloride.

tantalum pentoxide   See tantalum oxide.

tar base   A basic nitrogen compound found in coal tar, for example, pyridine and quinoline.

target   The atom or nucleus in an atomic or nuclear reaction which is initially stationary.

targusic acid   See lapachol.

tartar emetic   K(SbO)C$_4$H$_4$O$_6$·½H$_2$O   A transparent crystalline compound, soluble in water; used to attract and kill moths, wasps, and yellow jackets. Also known as antimony potassium tartrate; potassium antimonyl tartrate.

tartaric acid   HOOC(CHOH)$_2$COOH   Water- and alcohol-soluble colorless crystals with an acid taste, melts at 170°C; used as a chemical intermediate and a sequestrant and in tanning, effervescent beverages, baking power, ceramics, photography, textile processing, mirror silvering, and metal coloring. Also known as dihydroxysuccinic acid.

tartrate   M$_2$C$_4$H$_4$O$_6$ A salt or ester of tartaric acid, for example, sodium tartrate, Na$_2$C$_4$H$_4$O$_6$.

tartrazine   C$_{16}$H$_9$N$_4$O$_9$S$_2$ A bright orange-yellow, water-soluble powder, used as a food, drug, and cosmetic dye.

Tauber test   A color test for identification of pentose sugars; the sugars produce a cherry-red color when heated with a solution of benzidine in glacial acetic acid.

taurine $NH_2CH_2CH_2SO_3H$ A crystalline compound that decomposes at about 300°C; present in bile combined with cholic acid. Also known as 2-aminoethanesulfonic acid.

taurocholic acid $C_{26}H_{45}NO_7S$ A common bile acid with a five-carbon chain; it is the product of the conjugation of taurine with cholic acid; crystallizes from an alcohol ether solution, and decomposes at about 125°C. Also known as cholaic acid; cholytaurine.

tautomerism   The reversible interconversion of structural isomers of organic chemical compounds; such interconversions usually involve transfer of a proton.

Taylor-Orowan dislocation   *See* edge dislocation.

Tb   *See* terbium.

TBH   *See* 1,2,3,4,5,6-hexachlorocyclohexane.

TBP   *See* tributyl phosphate.

TBT   *See* tetrabutyl titanate.

Tc   *See* technetium.

TCA   *See* trichloroacetic acid.

TCP   *See* tricresyl phosphate.

TCTP   *See* tetrachlorothiophene.

TDE   *See* 2,2-bis(*para*-chlorophenyl)-1,1-dichloroethane.

Te   *See* tellurium.

TEA chloride   *See* tetraethylammonium chloride.

tebuthiuron $C_9H_{16}N_4OS$ A colorless solid with a melting point of 161.5–164°C; used as an herbicide in noncropland areas. Also known as N-[5-(1,1-dimethyl)1,3,4–thiadiazol-2-yl]-N,N'-dimethylurea.

technetium   A member of group VII, symbol Tc, atomic number 43; derived from uranium and plutonium fission products; chemically similar to rhenium and manganese; isotope ^{99}Tc has a half-life of 2×10^5 years; used to absorb slow neutrons in reactor technology.

technical chlorinated camphene   *See* toxaphene.

tectoquinone $C_{15}H_{10}O_2$ A white compound with needlelike crystals; sublimes at 177°C; insoluble in water; used as an insecticide to treat wood. Also known as 2-methyl anthraquinone.

TEDP   *See* tetraethyl dithionopyrophosphate.

TEG   *See* tetraethylene glycol; triethylene glycol.

teichoic acid   A polymer of ribitol or glycerol phosphate with additional compounds such as glucose linked to the backbone of the polymer; found in the cell walls of some bacteria.

TEL   *See* tetraethyllead.

Tellerette A type of inert packing with the appearance of a circular-wound spiral, used to create a large surface area to increase contact between falling liquid and rising vapor; used in gas-absorption operations.

Teller-Redlich rule For two isotopic molecules, the product of the frequency ratio values of all vibrations of a given symmetry type depends only on the geometrical structure of the molecule and the masses of the atoms, and not on the potential constants.

telluric acid H_6TeO_6 Toxic white crystals, slightly soluble in cold water, soluble in hot water and alkalies; melts at 136°C; used as an analytical reagent. Also known as hydrogen tellurate.

telluric line Any of the spectral bands and lines in the spectrum of the sun and stars produced by the absorption of their light in the atmosphere of the earth.

tellurinic acid A compound of tellurium with the general formula R_2TeOOH; an example is methanetellurinic acid, C_6H_5TeOOH.

tellurium A member of group VI, symbol Te, atomic number 52, atomic weight 127.60; dark-gray crystals, insoluble in water, soluble in nitric and sulfuric acids and potassium hydroxide; melts at 452°C, boils at 1390°C; used in alloys (with lead or steel), glass, and ceramics.

tellurium bromide *See* tellurium dibromide.

tellurium chloride *See* tellurium dichloride.

tellurium dibromide $TeBr_2$ Toxic, hygroscopic, green- or gray-black crystals with violet vapor, soluble in ether, decomposes in water, and melts at 210°C. Also known as tellurium bromide; tellurous bromide.

tellurium dichloride $TeCl_2$ A toxic, amorphous, black or green-yellow powder decomposing in water, melting at 209°C. Also known as tellurium chloride; tellurous chloride.

tellurium dioxide TeO_2 The most stable oxide of tellurium, formed when tellurium is burned in oxygen or air or by oxidation of tellurium with cold nitric acid; crystallizes as colorless, tetragonal, hexagonlike crystals that melt at 452°C.

tellurium disulfide TeS_2 A toxic, red powder, insoluble in water and acids. Also known as tellurium sulfide.

tellurium hexafluoride TeF_6 A colorless gas which is formed from the elements tellurium and fluorine; it is slowly hydrolyzed by water.

tellurium monoxide TeO A black, amorphous powder, stable in cold dry air; formed by heating the mixed oxide $TeSO_3$.

tellurium sulfide *See* tellurium disulfide.

telluroketone One of a group of compounds with the general formula R_2CTe.

telluromercaptan One of a group of compounds with the general formula $RTeH$.

tellurous acid H_2TeO_3 Toxic, white crystals, soluble in alkalies and acids, slightly soluble in water and alcohol; decomposes at 40°C.

tellurous bromide *See* tellurium dibromide.

tellurous chloride *See* tellurium dichloride.

telvar   The common name for the herbicide 3-(*para*-chlorophenyl)-1,1-di-methylurea; used as a soil sterilant.

TEM   *See* triethylenemelamine.

TEMA standard   Shell-and-tube heat-exchange standard designed to supplement the ASME code for unfired pressure vessels.

temporary hardness   The portion of the total hardness of water that can be removed by boiling whereby the soluble calcium and magnesium bicarbonate are precipitated as insoluble carbonates.

tennecetin   *See* pimaricin.

tensor force   A spin-dependent force between nucleons, having the same form as the interaction between magnetic dipoles; it is introduced to account for the observed values of the magnetic dipole moment and electric quadrupole moment of the deuteron.

TEP   *See* triethyl phosphate.

terbacil $C_9H_{13}ClN_2O_2$ A colorless, crystalline compound with a melting point of 175–177°C; used as an herbicide to control weeds in sugarcane, apples, peaches, citrus, and mints. Also known as 3-*tert*-butyl-5-chloro-6-methyl-uracil.

terbia   *See* terbium oxide.

terbium   A rare-earth element, symbol Tb, in the yttrium subgroup of group III, atomic number 65, atomic weight 158.924.

terbium chloride $TbCl_3 \cdot 6H_2O$ Water- and alcohol-soluble, hygroscopic, colorless, transparent prisms; anhydrous form melts at 588°C.

terbium nitrate $Tb(NO_3)_3 \cdot 6H_2O$ A colorless, fire-hazardous (strong oxidant) powder, soluble in water; melts at 89°C.

terbium oxide Tb_2O_3 A slightly hygroscopic, dark-brown powder soluble in dilute acids, absorbs carbon dioxide from air. Also known as terbia.

terbumeton $C_{10}H_{19}N_5$ A colorless, crystalline compound with a melting point of 123–124°C; slightly soluble in water; used as a postemergence herbicide for citrus and apple orchards, vineyards, and forests. Also known as 2-*tert*-butylamino-4-ethylamino-6-methoxy-*S*-triazine.

terbutol   The common name for the herbicide 2,6-di-*tert*-butyl-*p*-tolylmethyl-carbamate; used as a selective preemergence crabgrass herbicide for turf.

terbutryn $C_{13}H_{19}N_5S$ A colorless powder with a melting point of 104–105°C; used for weed control for wheat, barley, and grain sorghum. Also known as 2-*tert*-butylamino-4-ethylamino-6-methylthio-*S*-triazine.

terbutylhylazine $C_9H_{16}N_5Cl$ A white solid with a melting point of 177–179°C; used as a preemergence herbicide. Also known as 2-*tert*-butylamino-4-chloro-6-ethylamino-*S*-triazine.

terdentate ligand   *See* tridentate ligand.

terephthalic acid $C_6H_4(COOH)_2$ A combustible white powder, insoluble in water, soluble in alkalies, sublimes above 300°C; used to make polyester resins for fibers and films and as an analytical reagent and poultry-feed additive. Also known as benzene-*para*-dicarboxylic acid; *para*-phthalic acid; TPA.

terephthaloyl chloride $C_6H_4(COCl)_2$ Colorless needles with a melting point of 82–84°C; soluble in ether; used in the manufacture of dyes, synthetic fibers, resins, and pharmaceuticals. Also known as 1,4-benzenedicarbonyl chloride.

term   A set of $(2S + 1)(2L + 1)$ atomic states belonging to a definite configuration and to definite spin and orbital angular momentum quantum numbers S and L.

ternary compound   A molecule consisting of three different types of atoms; for example, sulfuric acid, H_2SO_4.

ternary system   Any system with three nonreactive components; in liquid systems, the components may or may not be partially soluble.

terpane   *See* menthane.

terpene $C_{10}H_{16}$ A moderately toxic, flammable, unsaturated hydrocarbon liquid found in essential oils and plant oleoresins; used as an intermediate for camphor, menthol, and terpineol.

terpene alcohol   A generic name for an alcohol related to or derived from a terpene hydrocarbon, such as terpineol or borneol.

terpene hydrochloride $C_{10}H_{16} \cdot HCl$ A solid, water-insoluble material melting at 125°C; used as an antiseptic. Also known as artificial camphor; dipentene hydrochloride; pinene hydrochloride; turpentine camphor.

terpenoid   Any compound with an isoprenoid structure similar to that of the terpene hydrocarbons.

***para*-terphenyl** $(C_6H_5)_2C_6H_4$ A combustible, toxic liquid boiling at 405°C; crystals are used for scintillation counters; polymerized with styrene to make plastic phosphor. Also known as 1,4-diphenylbenzene.

terpineol $C_{10}H_{17}OH$ A combustible, colorless liquid with a lilac scent, derived from pine oil, soluble in alcohol, slightly soluble in water, boils at 214–224°C; used in medicine, perfumes, soaps, and disinfectants, and as an antioxidant, a flavoring agent, and a solvent; isomeric forms are alpha-, beta- and gamma-.

terpin hydrate $CH_3(OH)C_6H_9C(CH_3)_2OH \cdot H_2O$ Combustible, efflorescent, lustrous white prisms soluble in alcohol and ether, slightly soluble in water; melts at 116°C; used for pharmaceuticals and to make terpineol. Also known as dipentene glycol.

terpinolene $C_{10}H_{16}$ A flammable, water-white liquid insoluble in water, soluble in alcohol, ether, and glycols, boils at 184°C; used as a solvent and as a chemical intermediate for resins and essential oils.

terpinyl acetate $C_{10}H_{17}OOCCH_3$ A combustible, colorless, liquid slightly soluble in water and glycerol, soluble in water, boils at 220°C; used in perfumes and flavors.

terpolymer   A polymer that contains three distinct monomers; for example, acrylonitrile-butadiene-styrene terpolymer, ABS.

terrachlor   *See* pentachloronitrabenzene.

tert- $(R_1R_2R_3C—)$ Abbreviation for tertiary; trisubstituted methyl radical with the central carbon attached to three other carbons.

tertiary alcohol   A trisubstituted alcohol in which the hydroxyl group is attached to a carbon that is joined to three carbons; for example, *tert*-butyl alcohol.

tertiary amine R_3N A trisubstituted amine in which the hydroxyl group is attached to a carbon that is joined to three carbons; for example, trimethylamine, $(CH_3)_3N$.

tertiary carbon atom A carbon atom bonded to three other carbon atoms with single bonds.

tertiary hydrogen atom A hydrogen atom that is bonded to a tertiary carbon atom.

tertiary mercuric phosphate *See* mercuric phosphate.

tertiary mercurous phosphate *See* mercurous phosphate.

tertiary sodium phosphate *See* trisodium phosphate.

tertiary structure The characteristic three-dimensional folding of the polypeptide chains in a protein molecule.

testicular hormone Any of various hormones secreted by the testes.

testosterone $C_{19}H_{28}O_2$ The principal androgenic hormone released by the human testis; may be synthesized from cholesterol and certain other sterols.

tetanolysin A hemolysin produced by *Clostridium tetani*.

tetanospasmin A neurotoxin elaborated by the bacterium *Clostridium tetani* and which is responsible for the manifestations of tetanus.

TETD *See* tetraethylthiuram disulfide.

tetraamylbenzene $(C_5H_{11})_4C_6H_2$ A colorless liquid with a boiling range of 320–350°C; used as a solvent.

tetrabromobisphenol A $(C_6H_2Br_2OH)_2C(CH_3)_2$ An off-white powder with a melting point of 180–184°C; soluble in methanol and ether; used as a flame retardant for plastics, paper, and textiles.

***sym*-tetrabromoethane** *See* acetylene tetrabromide.

tetrabromophthalic anhydride $C_6Br_4C_2O_3$ A pale yellow, crystalline solid with a melting point of 280°C; used as a flame retardant for paper, plastics, and textiles.

tetrabutylthiuram monosulfide $[(C_4H_9)_2NCS]_2S$ A brown liquid, soluble in acetone, benzene, gasoline, and ethylene dichloride; used as a rubber accelerator.

tetrabutyltin $(C_4H_9)_4Sn$ A colorless or slightly yellow, oily liquid with a boiling point of 145°C; soluble in most organic solvents; used as a stabilizing agent and rust inhibitor for silicones, and as a lubricant and fuel additive.

tetrabutyl titanate $Ti(OC_4H_9)_4$ A combustible, colorless to yellowish liquid soluble in many solvents, boils at 312°C, decomposes in water; used in paints, surface coatings, and heat-resistant paints. Abbreviated TBT. Also known as butyl titanate; titanium butylate.

tetrabutyl urea $(C_4H_9)_4N_2CO$ A liquid with a boiling point of 305°C; used as a plasticizer.

tetracaine hydrochloride $C_{15}H_{24}O_2N_2$·HCl Bitter-tasting, water-soluble crystals melting at 148°C; used as a local anesthetic. Also known as amethocaine hydrochloride; decicaine; pontocaine.

tetracene   *See* naphthacene.

tetrachlorobenzene $C_6H_2Cl_4$ Water-insoluble, combustible white crystals that appear in two forms: 1,2,3,4-tetrachlorobenzene which melts at 47°C and is used in chemical synthesis and in dielectric fluids; and 1,2,4,5-tetrachlorobenzene which melts at 138°C and is used to make herbicides, defoliants, and electrical insulation.

***sym*-tetrachlorodifluoroethane** CCl_2FCCl_2F A white, toxic liquid with a camphor aroma, soluble in alcohol, insoluble in water, boils at 93°C; used for metal degreasing.

2,4,4′,5-tetrachlorodiphenyl sulfide $C_{12}H_6Cl_4S$ White crystals with a melting point of 87°C; slightly soluble in water; used as a miticide for mite eggs.

***sym*-tetrachloroethane** $CHCl_2CHCl_2$ A colorless, corrosive, toxic liquid with a chloroform scent, soluble in alcohol and ether, slightly soluble in water, boils at 147°C; used as a solvent, metal cleaner, paint remover, and weed killer. Also known as acetylene tetrachloride.

tetrachloroethylene   *See* perchloroethylene.

2,4,5,6-tetrachloroisophthalonitrile   *See* chlorothalonil.

tetrachloromethane   *See* carbon tetrachloride.

tetrachlorophenol C_6HCl_4OH Either of two toxic compounds: 2,3,4,6-tetrachlorophenol comprises brown flakes, soluble in common solvents, melting at 70°C, and is used as a fungicide; 2,4,5,6-tetrachlorophenol is a brown solid, insoluble in water, soluble in sodium hydroxide, has a phenol scent, melts at about 50°C, and is used as a fungicide and for wood preservatives.

tetrachlorophthalic acid $C_6Cl_4(CO_2H)_2$ Colorless plates, soluble in hot water; used in making dyes.

tetrachlorophthalic anhydride $C_6Cl_4(CO)_2O$ A white powder with a melting point of 254–255°C; slightly soluble in water; used in the manufacture of dyes and pharmaceuticals and as a flame retardant for epoxy resins.

4,5,6,7-tetrachlorophthalide   *See* rabicide.

tetrachloroquinone   *See* chloranil.

tetrachlorothiophene C_4Cl_4S A yellow-white solid with a melting point of 29°C; insoluble in water; used as a nematicide and fumigant for tobacco. Abbreviated TCTP.

tetrachlorvinphos $C_{10}H_9Cl_4O_4P$ An off-white, crystalline compound with a melting point of 94–97°C; used as an insecticide to control houseflies, chicken mites, lice, screwworms, and hornflies. Also known as 2-chloro-1-(2,4,5-trichlorophenyl)vinyl dimethylphosphate.

tetracosane $C_{24}H_{50}$ Combustible crystals insoluble in water, soluble in alcohol, melts at 52°C; used as a chemical intermediate.

tetracyanoethylene $(CN)_2C:C(CN)_2$ A member of the cyanocarbon compounds; colorless crystals with a melting point of 198–200°C; used in dye manufacture.

***n*-tetradecane** $C_{14}H_{30}$ A combustible, colorless, water-insoluble liquid boiling at 254°C; used as a solvent and distillation chaser and in organic synthesis.

1-tetradecanol   *See* myristyl alcohol.

1-tetradecene CH_2:$CH(CH_2)_{11}CH_3$ A combustible, colorless, water-insoluble liquid boiling at 256°C; used as a solvent for perfumes and flavors and in medicine. Also known as α-tetradecylene.

tetradecylamine $C_{14}H_{29}NH_2$ A white solid with a melting point of 37°C; soluble in alcohol and ether; used in making germicides.

α-tetradecylene *See* 1-tetradecene.

tetradecyl mercaptan $CH_3(CH_2)_{13}SH$ A combustible liquid with a boiling point of 176–180°C; used for processing synthetic rubber. Also known as myristyl mercaptan.

tetradentate ligand A chelating agent which has four groups capable of attachment to a metal ion. Also known as quadridentate ligand.

tetradifon $C_{12}H_6Cl_4O_2S$ A colorless, crystalline compound with a melting point of 145–146°C; used to control mites on fruits, vegetables, cotton, and ornamentals. Also known as S-*para*-chlorophenyl-2,4,5-trichlorophenyl sulfone.

tetraethanolammonium hydroxide $(HOCH_2CH_2)_4NOH$ A white, water-soluble, crystalline solid with a melting point of 123°C; used as a dye solvent and in metal-plating solutions.

tetraethylammonium chloride $(C_2H_5)_4NCl$ Colorless, hygroscopic crystals with a melting point of 37.5°C; soluble in water, alcohol, acetone, and chloroform; used in medicine. Abbreviated TEAC. Also known as TEA chloride.

tetraethylammonium hexafluorophosphate $(C_2H_5)_4NPF_6$ A water-soluble solid with a melting point of 255°C; used for bactericides and fungicides.

tetra-(2-ethylbutyl)silicate $[(C_2H_5)C_4H_8O]_4Si$ A colorless liquid with a boiling point of 238°C at 50 mmHg (6660 newtons per square meter); used as a lubricant and hydraulic fluid.

tetraethyl dithionopyrophosphate $(C_2H_5O)_4P_2OS_2$ A toxic, combustible liquid with a boiling point of 138°C, soluble in alcohol; used as an insecticide. Abbreviated TEDP.

***O,O,O,O*-tetraethyldithiopyrophosphate** *See* sulfotepp.

tetraethylene glycol $HO(C_2H_4O)_3C_2H_4OH$ A combustible, hygroscopic, colorless, water-soluble liquid, boils at 327°C; used as a nitrocellulose solvent and plasticizer and in lacquers and coatings. Abbreviated TEG.

tetraethylene glycol dimethacrylate A colorless to pale straw-colored liquid with a boiling point of 200°C at 1 mmHg (133.32 newtons per square meter); soluble in styrene and some esters and aromatics; used as a plasticizer.

tetraethylenepentamine $C_8H_{23}N_2$ A toxic, viscous liquid with a boiling point of 333°C and a freezing point of -30°C; soluble in water and organic solvents; used as a motor oil additive, in the manufacture of synthetic rubber, and as a solvent for dyes, acid gases, and sulfur.

tetraethyllead $Pb(C_2H_5)_4$ A highly toxic lead compound that, when added in small proportions to gasoline, increases the fuel's antiknock quality. Abbreviated TEL.

***O,O,O',O'*-tetraethyl *S,S'*-methylene bisphosphorodithioate** *See* ethion.

tetraethylpyrophosphate $C_8H_{20}O_7P_2$ A hygroscopic corrosive liquid miscible with although decomposed by water, and miscible with many organic solvents;

inhibits the enzyme acetylcholinesterase; used as an insecticide in place of nicotine sulfate.

tetraethylthiuram disulfide $[(C_2H_5)_2NCS]_2S_2$ A light gray solid with a melting range of 65–70°C; soluble in carbon disulfide, benzene, and chloroform; used as a fungicide and insecticide. Abbreviated TETD; TTD. Also known as bis(diethylthiocarbamyl)disulfide; disulfram.

tetraethylthiuram sulfide $[(C_2H_5)_2NCS]_2S$ A dark brown solid with a boiling range of 225–240°C; used in pharmaceutical ointments and as a fungicide and insecticide. Also known as bis(diethylthiocarbamyl)sulfide.

tetrafluoroethylene $F_2C{:}CF_2$ A flammable, colorless, heavy gas, insoluble in water, boils at 78°C; used as a monomer to make polytetrafluoroethylene polymers, for example, Teflon. Also known as perfluoroethylene; TFE.

tetrafluorohydrazine F_2NNF_2 A colorless liquid or gas with a calculated boiling point of −73°C; used as an oxidizer in rocket fuels.

tetrafluoromethane *See* carbon tetrafluoride.

tetragonal lattice A crystal lattice in which the axes of a unit cell are perpendicular, and two of them are equal in length to each other, but not to the third axis.

tetragonal trisoctahedron *See* trapezohedron.

tetragonal tristetrahedron *See* deltohedron.

tetrahedron An isometric crystal form in cubic crystals, in the shape of a four-faced polyhedron, each face of which is a triangle.

tetrahexahedron A form of regular crystal system with four triangular isosceles faces on each side of a cube; there are altogether 24 congruent faces.

1,2,3,4-tetrahydrobenzene *See* cyclohexene.

tetrahydro-3,5-dimethyl-2H-1,3,5-thiadiazine-6-thione *See* dazomet.

tetrahydrofuran C_4H_8O A clear, colorless liquid with a boiling point of 66°C; soluble in water and organic solvents; used as a solvent for resins and in adhesives, printing inks, and polymerizations. Abbreviated THF.

tetrahydrofurfuryl acetate $C_7H_{12}O_3$ A colorless liquid with a boiling point of 194–195°C; soluble in water, alcohol, ether, and chloroform; used for flavoring.

tetrahydrofurfuryl alcohol $C_4H_7OCH_2OH$ A hygroscopic, colorless liquid, miscible with water, boils at 178°C; used as a solvent for resins, in leather dyes, and in nylon. Also known as tetrahydrofurfuryl carbinol.

tetrahydrofurfurylamine $C_4H_7OCH_2NH_2$ A colorless to light yellow liquid with a distillation range of −150 to 156°C; used for fine-grain photographic development and to accelerate vulcanization.

tetrahydrofurfuryl carbinol *See* tetrahydrofurfuryl alcohol.

tetrahydrofurfuryl oleate $C_{23}H_{42}O_3$ A colorless liquid with a boiling point of 240°C at 2 mmHg (266.64 newtons per square meter); used as a plasticizer.

tetrahydrofurfuryl phthalate $C_6H_4(COOCH_2C_4H_7O)_2$ A colorless liquid with a melting point below 15°C; used as a plasticizer.

tetrahydrolinalool $C_{10}H_{21}OH$ A colorless liquid with a floral odor, used in perfumery and flavoring. Also known as 3,7-dimethyl-3-octanol.

tetrahydro-*para*-methoxyquinoline *See* thalline.

tetrahydronaphthalene $C_{10}H_{12}$ A colorless, oily liquid that boils at 206°C, and is miscible with organic solvents; used as an intermediate in chemical synthesis and as a solvent.

tetrahydro-1,4-oxazine *See* morpholine.

N-(tetrahydro-2-oxo-3-thienyl)acetamide *See* N-acetylhomocysteinethiolactone.

tetrahydro-*para*-quinanisol *See* thalline.

tetrahydrothiophene-1,1-dioxide *See* sulfolane.

tetrahydroxyadipic acid *See* mucic acid.

4,6,3′,4′-tetrahydroxyaurone *See* aureusidin.

tetrahydroxy butane *See* erythritol.

3′,4′,5,7-tetrahydroxyflavanone *See* eriodictyol.

tetraiodoethylene $I_2C:CI_2$ Light yellow crystals with a melting point of 187°C; soluble in organic solvents; used in surgical dusting powder and antiseptic ointments, and as a fungicide. Also known as iodoethylene.

tetraiodofluorescein $C_{20}H_8O_5I_4$ A yellow, water-insoluble, crystalline compound; used as a dye. Also known as pyrosin.

tetraisopropylthiuram disulfide $[(CH_3CH_3CH)_2NCS]_2S_2$ A tan powder with a melting range of 95–99°C; soluble in benzene, chloroform, and gasoline; used as a rubber accelerator.

tetrakis(hydroxymethyl)phosphonium chloride $(HOCH_2)_4PCl$ A crystalline compound made from phosphine, formaldehyde, and hydrochloric acid; used as a flame retardant for cotton fabrics. Abbreviated THPC.

tetralite *See* tetryl.

tetramer A polymer that results from the union of four identical monomers; for example, the tetramer C_8H_8 forms from union of four molecules of C_2H_2.

tetramethrin $C_{20}H_{25}NO_4$ A light yellow powder with a melting point between 65 and 80°C; used to control houseflies, cattle insects, pests in stored products, and garden pests. Also known as 1-cyclohexene-1,2-dicarboximidomethyl-2,2-dimethyl-3-(2-methylpropenyl) cyclopropanecarboxylate.

tetramethyldiaminobenzophenone $[(CH_3)_2NC_6H_4]_2CO$ White to greenish, crystalline leaflets with a melting point of 172°C; soluble in alcohol, ether, water, and warm benzene; used in the manufacture of dyes. Also known as Michler's ketone.

tetramethyldiene *See* cyclobutadiene.

tetramethylene *See* cyclobutane.

Δ¹,³-tetramethylene *See* cyclobutadiene.

tetramethylene bis(methanesulfonate) *See* busulfan.

tetramethylethylenediamine $(CH_3)_4N_2(CH_2)_2$ A colorless liquid with a boiling point of 121–122°C; soluble in organic solvents and water; used in the formation of polyurethane, as a corrosion inhibitor, and for textile finishing agents.

tetramethylene dichloride   *See* 1,4-dichlorobutane.

tetramethylene glycol   *See* 1,4-butylene glycol.

tetramethylene sulfone   ¯*See* sulfolane.

tetramethyllead $Pb(CH_3)_4$ An organic compound of lead that, when added in small amounts to motor gasoline, increases the antiknock quality of the fuel; not widely used.

tetramethylolmethane   *See* pentaerythritol.

tetramethylsilane $(CH_3)_4Si$ A colorless, volatile, toxic liquid with a boiling point of 26.5°C; soluble in organic solvents; used as an aviation fuel.

***O,O,O′,O′*-tetramethyl *O,O′*-thiodi-*para*-phenylene phosphorothioate** $C_{16}H_{20}O_6S_3P_2$ A brown, viscous liquid with a melting point of 30°C; insoluble in water; used as a larvicide for mosquitos, black flies, and midges.

tetramethylthiuram monosulfide $[(CH_3)_2NCS]_2S$ A yellow powder with a melting point of 104–107°C; soluble in acetone, benzene, and ethylene dichloride; used as a rubber accelerator, fungicide, and insecticide. Also known as bis(dimethylthiocarbamyl)sulfide.

tetramethylurea $C_5H_{12}N_2O$ A liquid that boils at 176.5°C, and is miscible in water and organic solvents; used as a reagent and solvent.

tetranitromethane $C(NO_2)_4$ A powerful oxidant; toxic, colorless liquid with a pungent aroma, insoluble in water, soluble in alcohol and ether, boils at 126°C; used in rocket fuels and as an analytical reagent.

tetraphenylboron sodium   *See* sodium tetraphenylborate.

tetraphenyltin $(C_6H_5)_4Sn$ A white powder with a melting point of 225–228°C; soluble in hot benzene, toluene, and xylene; used for mothproofing.

tetraphosphorus trisulfide   *See* phosphorus sesquisulfide.

tetrapotassium pyrophosphate   *See* potassium pyrophosphate.

tetrapropylene   *See* dodecane.

tetrasodium pyrophosphate   *See* sodium pyrophosphate.

tetraterpene   A class of terpene compounds that contain isoprene units; best known are the carotenoid pigments from plants and animals, such as lycopene, the red coloring matter in tomatoes.

tetratohedral crystal   A crystal which has one quarter of the maximum number of faces allowed by the crystal system to which the crystal belongs.

tetrazene $H_2NC(NH)_3N_2C(NH)_3NO$ An explosive, colorless to yellowish solid practically insoluble in water and alcohol; used as an explosive initiator and in detonators.

tetrazolium blue   *See* blue tetrazolium.

tetrodotoxin $C_{11}H_{17}N_3O_8$ A toxin that blocks the action potential in the nerve impulse.

tetrol   *See* furan.

tetrose Any of a group of monosaccharides that have a four-carbon chain; an example is erythrose, $CH_2OH \cdot (CHOH)_2 \cdot CHO$.

tetryl $(NO_2)_3C_6H_2N(NO_2)CH_3$ A yellow, water-insoluble, crystalline explosive material melting at 130°C; used in explosives and ammunition. Also known as N-methyl-N,2,4,6-tetranitroaniline; methylpicrylnitramine; tetralite.

TFE *See* tetrafluoroethylene.

th *See* thermie.

Th *See* thorium.

thalassophile element An element that is relatively more abundant in sea water than in normal continental waters, such as sodium and chlorine.

thalline $C_9H_6N(OCH_3)H_4$ Colorless rhomboids soluble in water and melting at 40°C. Also known as tetrahydro-*para*-methoxyquinoline; tetrahydro-*para*-quinanisol.

thallium A metallic element in group III, symbol Tl, atomic number 81, atomic weight 204.37; insoluble in water, soluble in nitric and sulfuric acids, melts at 302°C, boils at 1457°C.

thallium acetate $TlOCOCH_3$ Toxic, white, deliquescent crystals, soluble in water and alcohol, melts at 131°C; used as an ore-flotation solvent and in medicine.

thallium bromide $TlBr$ A toxic, yellowish powder soluble in alcohol, slightly soluble in water, melts at 460°C; used in infrared radiation transmitters and detectors. Also known as thallous bromide.

thallium carbonate Tl_2CO_3 Toxic, shiny, colorless needles soluble in water, insoluble in alcohol, melts at 272°C; used as an analytical reagent and in artificial gems. Also known as thallous carbonate.

thallium chloride $TlCl$ A white, toxic, light-sensitive powder, slightly soluble in water, insoluble in alcohol, melts at 430°C; used as a chlorination catalyst and in medicine and suntan lamps. Also known as thallous chloride.

thallium hydroxide $TlOH \cdot H_2O$ Toxic yellow, water- and alcohol-soluble needles, decomposes at 139°C; used as an analytical reagent. Also known as thallous hydroxide.

thallium iodide TlI A toxic, yellow powder, insoluble in alcohol, slightly soluble in water, melts at 440°C; used in infrared radiation transmitters and in medicine. Also known as thallous iodide.

thallium monoxide Tl_2O A black, toxic, water- and alcohol-soluble powder, melts at 300°C; used as an analytical reagent and in artificial gems and optical glass. Also known as thallium oxide; thallous oxide.

thallium nitrate $TlNO_3$ Colorless, toxic, fire-hazardous crystals soluble in hot water, insoluble in alcohol, melts at 206°C, decomposes at 450°C; used as an analytical reagent and in pyrotechnics. Also known as thallous nitrate.

thallium oxide *See* thallium monoxide.

thallium sulfate Tl_2SO_4 Toxic, water-soluble, colorless crystals melting at 632°C; used as an analytical reagent and in medicine, rodenticides, and pesticides. Also known as thallous sulfate.

thallium sulfide Tl_2S Lustrous, toxic, blue-black crystals insoluble in water, alcohol, and ether, soluble in mineral acids, melts at 448°C; used in infrared-sensitive devices. Also known as thallous sulfide.

thallous bromide   *See* thallium bromide.

thallous carbonate   *See* thallium carbonate.

thallous chloride   *See* thallium chloride.

thallous hydroxide   *See* thallium hydroxide.

thallous iodide   *See* thallium iodide.

thallous nitrate   *See* thallium nitrate.

thallous oxide   *See* thallium monoxide.

thallous sulfate   *See* thallium sulfate.

thallous sulfide   *See* thallium sulfide.

THAM   *See* tromethamine.

thenyl $C_4H_3SCH_2$— An organic radical based on methylthiophene; thus thenyl alcohol is also known as thiophenemethanol.

theobromine $C_7H_8N_4O_2$ A toxic alkaloid found in cocoa, chocolate products, tea, and cola nuts; closely related to caffeine. Also known as 3,7-dimethylxanthine.

theophylline $C_7H_8N_4O_2 \cdot H_2O$ Alkaloid from tea leaves; bitter-tasting white crystals slightly soluble in water and alcohol, melts at 272°C; used in medicine.

theoretical plate   A distillation column plate or tray that produces perfect distillation (that is, produces the same difference in composition as that existing between a liquid mixture and the vapor in equilibrium with it); the packed-column equivalent of a theoretical plate is the HETP, or height (of packing) equivalent to a theoretical plate.

thermal analysis   Any analysis of physical or thermodynamic properties of materials in which heat (or its removal) is directly involved; for example, boiling, freezing, solidification-point determinations, heat of fusion and heat of vaporization measurements, distillation, calorimetry, and differential thermal, thermogravimetric, thermometric, and thermometric titration analyses. Also known as thermoanalysis.

thermal black   A type of carbon black made by a thermal process using natural gas; used in the rubber industry.

thermal cracking   A petroleum refining process that decomposes, rearranges, or combines hydrocarbon molecules by the application of heat, without the aid of catalysts.

thermal degradation   Molecular deterioration of materials (usually organics) because of overheat; can be avoided by low-temperature or vacuum processing, as for foods and pharmaceuticals.

thermal diffusion   A phenomenon in which a temperature gradient in a mixture of fluids gives rise to a flow of one constituent relative to the mixture as a whole. Also known as thermodiffusion.

thermal efficiency   In a tube-and-shell heat-exchange system, the ratio of the actual temperature range of the tube-side fluid (inlet versus outlet temperature) to the maximum possible temperature range.

thermal excitation The process in which atoms or molecules acquire internal energy in collisions with other particles.

thermal-liquid system A system with a special liquid that acts as a heat sink or heat source (for example, steam, hot water, mercury, Dowtherm, molten salts, or mineral oils); used for process heating and cooling.

thermal polymerization A thermal, petroleum refining process used to convert light hydrocarbon gases into liquid fuels; paraffinic hydrocarbons are cracked to produce olefinic material which is concurrently polymerized by heat and pressure to form liquids, the product being known as polymer gasoline.

thermal process Any process that utilizes heat, without the aid of a catalyst, to accomplish chemical change; for example, thermal cracking, thermal reforming, or thermal polymerization.

thermal reactor A device, system, or vessel in which chemical reactions take place because of heat (no catalysis); for example, thermal cracking, thermal reforming, or thermal polymerization.

thermal reforming A petroleum refining process using heat (but no catalyst) to effect molecular rearrangement of a low-octane naphtha to form high-octane motor gasoline.

thermal titration *See* thermometric titration.

thermoanalysis *See* thermal analysis.

thermobalance An analytical balance modified for thermogravimetric analysis, involving the measurement of weight changes associated with the transformations of matter when heated.

thermochemistry The measurement, interpretation, and analysis of heat changes accompanying chemical reactions and changes in state.

thermodiffusion *See* thermal diffusion.

Thermofor catalytic cracking A proprietary, petroleum refining process with a continuously moving bed of catalyst; used to crack heavy petroleum fractions into lighter, gasoline-range liquids, and the product is known as crackate or cat crackate.

Thermofor catalytic reforming A proprietary, petroleum refining process with a continuously moving bed of catalyst; used to rearrange low-octane naphtha molecules to produce high-octane motor gasoline, and the product is known as reformate.

Thermofor continuous percolation A proprietary petroleum refinery clay-treating adsorbent process, used to stabilize and decolorize lubricants and waxes.

thermogravimetric analysis Chemical analysis by the measurement of weight changes of a system or compound as a function of increasing temperature.

thermogravitational column A device in which thermal diffusion results from the countercurrent flow of hot and cold material, thus increasing the separation of materials in a solution by the formation of a concentration gradient (difference). Also known as Clausius-Dickel column.

thermokinetic analysis A type of enthalpimetric analysis which uses kinetic titrimetry; involves rapid and continuous automatic delivery of a suitable titrant, under judiciously controlled experimental conditions with temperature measurement; the end points obtained are converted by mathematical procedures into valid stoichiometric equivalence points and used for determining reaction rate constants.

thermoluminescence 1. Broadly, any luminescence appearing in a material due to application of heat. 2. Specifically, the luminescence appearing as the temperature of a material is steadily increased; it is usually caused by a process in which electrons receiving increasing amounts of thermal energy escape from a center in a solid where they have been trapped and go over to a luminescent center, giving it energy and causing it to luminesce.

thermometric analysis A method for determination of the transformations a substance undergoes while being heated or cooled at an essentially constant rate, for example, freezing-point determinations.

thermometric titration A titration in an adiabatic system, yielding a plot of temperature versus volume of titrant; used for neutralization, precipitation, redox, organic condensation, and complex-formation reactions. Also known as calorimetric titration; enthalpy titration; thermal titration.

thermonuclear reaction A nuclear fusion reaction which occurs between various nuclei of light elements when they are constituents of a gas at very high temperature.

thermosiphon reboiler A liquid reheater (as for distillation-column bottoms) in which natural circulation of the boiling liquid is obtained by maintaining a sufficient liquid head.

thermotropic liquid crystal A liquid crystal prepared by heating the substance.

THF *See* tetrahydrofuran.

thiabendazole $C_{10}H_7N_3S$ A white powder with a melting point of 304–305°C; controls fungi on citrus fruits, sugarbeets, turf, and ornamentals, and roundworms of cattle and other animals. Also known as 2-(4-thiazolyl)benzimidazole.

thiacetic acid *See* thioacetic acid.

thiadiazin *See* milneb.

thiamine $C_{12}H_{17}ClN_4OS$ A member of the vitamin B complex that occurs in many natural sources, frequently in the form of cocarboxylase. Also known as aneurine; vitamin B_1.

thiamine hydrochloride $C_{12}H_{17} \cdot ON_4S \cdot HCl$ White, hygroscopic crystals soluble in water, insoluble in ether, with a yeasty aroma and a salty, nutlike taste, decomposes at 247°C; the form in which thiamine is generally employed.

thiamine pyrophosphate The coenzyme or prosthetic component of carboxylase; catalyzes decarboxylation of various α-keto acids. Also known as cocarboxylase.

thianaphthene C_8H_6S A crystalline compound with a melting point of 32°C; soluble in organic solvents; used in the production of pharmaceuticals. Also known as benzothiofuran.

thiazole C_3H_3NS A colorless to yellowish liquid with a pyridinelike aroma, slightly soluble in water, soluble in alcohol and ether; used as an intermediate for fungicides, dyes, and rubber accelerators.

thiazole dye One of a family of dyes in which the chromophore groups are =C=N—, —S—C=, and used mainly for cotton; an example is primuline.

2-(4-thiazolyl)benzimidazole *See* thiabendazole.

thickening The concentration of the solids in a suspension in order to recover a fraction with a higher concentration of solids than in the original suspension.

Thiele coordinates A graphical method for calculating the solvent-free composition of two components being separated by solvent extraction.

Thiele-Geddes method A method for the prediction of the product distribution from a multicomponent distillation system.

Thiele melting point apparatus A stirred, specially shaped test-tube device used for the determination of the melting point of a crystalline chemical.

thimerosal $C_9H_9HgNaO_2S$ A light cream-colored, crystalline powder, soluble in water and alcohol; used as a topical antiseptic. Also known as sodium ethylmercurithiosalicylate.

thin-layer chromatography Chromatographing on thin layers of adsorbents rather than in columns; adsorbent can be alumina, silica gel, silicates, charcoals, or cellulose.

thio- A chemical prefix derived from the Greek *theion,* meaning sulfur; indicates the replacement of an oxygen in an acid radical by sulfur with a negative valence of 2.

thioacetamide C_2H_5NS A crystalline compound with a melting point of 113–114°C; soluble in water and ethanol; used in laboratories in place of hydrogen sulfide.

thioacetic acid CH_3COSH A toxic, clear-yellow liquid with an unpleasant aroma, soluble in water, alcohol, and ether, boils at 82°C; used as an analytical reagent and a lacrimator. Also known as ethanethiolic acid; thiacetic acid.

thioaldehyde An organic compound that contains the —CHS radical and has the suffix -thial; for example, ethanethial, CH_3CHS.

thioallyl ether *See* allyl sulfide.

thiobarbituric acid $C_6H_4N_2O_2S$ Malonyl thiourea, the parent compound of the thiobarbiturates; represents barbituric acid in which the oxygen atom of the urea component has been replaced by sulfur.

thiocarbamide *See* thiourea.

thiocarbamisin *See* thiocarbamazine.

thiocarbanil *See* phenyl mustard oil.

thiocarbanilide $CS(NHC_6H_5)_2$ A gray powder with a melting point of 148°C; soluble in alcohol and ether; used for making dyes, and as a vulcanization accelerator and ore flotation agent. Also known as N,N'-diphenylthiourea; sulfocarbanilide.

thiocyanate A salt of thiocyanic acid that contains the —SCN radical; for example, sodium thiocyanate, NaSCN. Also known as rhodanate; rhodanide; sulfocyanate; sulfocyanide; thiocyanide.

thiocyanic acid HSC:N A colorless, water-soluble liquid decomposing at 200°C; used to inhibit paper deterioration due to the action of light, and (in the form of organic esters) as an insecticide. Also known as rhodanic acid; sulfocyanic acid.

thiocyanide *See* thiocyanate.

thiocyanogen NCSSCN White, light-unstable rhombic crystals melting at $-2°C$.

2-(thiocyanomethylthio)benzothiazole $C_9H_7N_2S_2$ A fungicide used on barley, corn, oats, rice, wheat, and sorghum.

thiodiethylene glycol *See* thiodiglycol.

thiodiglycol $(CH_2CH_2OH)_2S$ A combustible, colorless, syrupy liquid soluble in water, alcohol, acetone, and chloroform, boils at 283°C; used as a chemical intermediate, textile-dyeing solvent, and antioxidant. Also known as dihydroxyethylsulfide; thiodiethylene glycol.

thiodiglycolic acid $HOOCCH_2SCH_2COOH$ Combustible, colorless, water- and alcohol-soluble crystals melting at 128°C; used as an analytical reagent.

thiodihydracrylic acid *See* 3,3′-thiodiproprionic acid.

thiodipropionic acid *See* dilauryl thiodipropionate.

3,3′-thiodiproprionic acid $(CH_2CH_2COOH)_2S$ A crystalline compound with a melting point of 134°C; soluble in hot water, acetone, and alcohol; used as an antioxidant for soap products and polymers of ethylene. Also known as diethyl sulfide 2,2′-dicarboxylic acid; thiodihydracrylic acid.

thioethanolamine *See* 2-aminoethanethiol.

thioether RSR A general formula for colorless, volatile organic compounds obtained from alkyl halides and alkali sulfides; the R groups can be the same, or different as in methylthioethane $(CH_3SC_2H_5)$.

thioethyl alcohol *See* ethyl mercaptan.

thioflavine T $C_{16}H_{17}N_2Cl$ A yellow basic dye, used for textile dyeing and fluorescent sign paints.

thiofuran *See* thiophene.

thioglycolic acid $HSCH_2COOH$ A liquid with a strong unpleasant odor; used as a reagent for metals such as iron, molybdenum, silver, and tin, and in bacteriology. Also known as mercaptoacetic acid.

2-thiohydantoin $NHC(S)NHC(O)CH_2$ Crystals or a tan powder with a melting point of 230°C; used in the manufacture of pharmaceuticals, rubber accelerators, and copper-plating brighteners. Also known as glycolythiourea.

thiol *See* mercaptan.

thiolactic acid $CH_3CH(SH)COOH$ An oil with a disagreeable odor; used in toiletry preparation. Also known as 2-mercaptopropionic acid; 2-thiolpropionic acid.

2-thiolpropionic acid *See* thiolactic acid.

thiomalic acid $C_4H_6O_4S$ White crystals or powder with a melting point of 149–150°C; soluble in water, alcohol, and acetone; used as a sealer for fuel cells and machine and electrical parts, for caulking compounds, and as a propellant binder. Also known as mercaptosuccinic acid.

thionazin $C_8H_{13}N_2O_3PS$ A liquid used as an insecticide and miticide for soil. Also known as O,O-diethyl-O-(2-pyrazinyl)phosphorothioate.

thionic acid **1.** $H_2S_xO_6$, where x varies from 2 to 6. **2.** An organic acid with the radical —CSOH.

thionyl bromide $SOBr_2$ A red liquid boiling at 68°C (40 mmHg). Also known as sulfinyl bromide.

thionyl chloride $SOCl_2$ A toxic, yellowish to red liquid with a pungent aroma, soluble in benzene, decomposes in water and at 140°C; boils at 79°C; used as a chemical intermediate and catalyst. Also known as sulfur oxychloride; sulfurous oxychloride.

thiopental sodium $C_{11}H_{17}O_2N_2NaS$ Yellow, water-soluble crystals with a characteristic aroma; used in medicine as a short-acting anesthetic. Also known as thiopentone sodium.

thiopentone sodium *See* thiopental sodium.

thiophanate $C_{14}H_{18}N_4O_4S_2$ A tan to colorless solid that decomposes at 195°C; slightly soluble in water; used to control fungus diseases of turf. Also known as diethyl 4,4'-O-phenylene-bis(3-thioallophanate).

thiophene C_4H_4S A toxic, flammable, highly reactive, colorless liquid, insoluble in water, soluble in alcohol and ether, boils at 84°C; used as a chemical intermediate and to make condensation copolymers. Also known as thiofuran.

thiophenol C_6H_5SH A toxic, fire-hazardous, water-white liquid with a disagreeable aroma, insoluble in water, soluble in alcohol and ether, boils at 168°C; used to make pharmaceuticals. Also known as phenyl mercaptan.

thiophenylamine *See* phenothiazine.

thiophile element *See* sulfophile element.

thiophosphoric anhydride *See* phosphorus pentasulfide.

thiosalicylic acid $HOOCC_6H_4SH$ A yellow solid with a melting point of 164–165°C; soluble in alcohol, ether, and acetic acid; used for making dyes. Also known as 2-mercaptobenzoic acid.

thiosemicarbazide $NH_2CSNHNH_2$ A white, water- and alcohol-soluble powder melting at 182°C; used as an analytical reagent and in photography and rodenticides. Also known as aminothiourea.

thiosinamine *See* allylthiourea.

thiosulfate $M_2S_2O_3$ A salt of thiosulfuric acid and a base; for example, reaction of sodium hydroxide and thiosulfuric acid to produce sodium thiosulfate.

thiosulfonic acid Name for a group of oxy acids of sulfur, with the general formula RS_2O_2H; they are known as esters and salts.

thiosulfuric acid $H_2S_2O_3$ An unstable acid that decomposes readily to form sulfur and sulfurous acid.

thiourea $(NH_2)_2CS$ Bitter-tasting, white crystals with a melting point of 180–182°C; soluble in cold water and alcohol; used in photography and photocopying, as a rubber accelerator, and as an antithyroid drug in treating hyperthyroidism. Also known as thiocarbamide.

β-thiovaline *See* penicillamine.

2-thioxo-4-thiazolidinone *See* rhodanine.

thiram $C_6H_{12}N_2S_4$ A colorless, crystalline solid with a melting point of 155–156°C; used to control fungus diseases of fruits, turf, and vegetables, and to protect fruit and other trees from grazers. Also known as bis(dimethylthiocarbamoyl)disulfide.

third-order reaction A chemical reaction in which the rate of reaction is determined by the concentration of three reactants.

thiuram A chemical compound containing a R_2NCS radical; occurs mainly in disulfide compounds; the most common monosulfide compound is tetramethylthiuram monosulfide.

thixotropy Property of certain gels which liquefy when subjected to vibratory forces, such as ultrasonic waves or even simple shaking, and then solidify again when left standing.

Thomas-Fermi atom model A method of approximating the electrostatic potential and the electron density in an atom in its ground state, in which these two quantities are related by the Poisson equation on the one hand, and on the other hand by a semiclassical formula for the density of quantum states in phase space.

Thomas-Fermi equation The differential equation $x^{1/2}(d^2y/dx^2) = y^{3/2}$ that arises in calculating the potential in the Thomas-Fermi atom model; the physically meaningful solution satisfies the boundary conditions $y(0) = 1$ and $y(\infty) = 0$.

Thomas-Reiche-Kuhn sum rule *See* f-sum rule.

Thomson-Berthelot principle The assumption that the heat released in a chemical reaction is directly related to the chemical affinity, and that, in the absence of the application of external energy, that chemical reaction which releases the greatest heat is favored over others; the principle is in general incorrect, but applies in certain special cases.

thoria *See* thorium dioxide.

thorium An element of the actinium series, symbol Th, atomic number 90, atomic weight 232; soft, radioactive, insoluble in water and alkalies, soluble in acids, melts at 1750°C, boils at 4500°C.

thorium anhydride *See* thorium dioxide.

thorium carbide ThC_2 A yellow solid melting at above 2630°C, decomposes in water; used in nuclear fuel.

thorium chloride $ThCl_4$ Hygroscopic, toxic colorless crystal needles soluble in alcohol, melts at 820°C, decomposes at 928°C; used in incandescent lighting. Also known as thorium tetrachloride.

thorium dioxide ThO_2 A heavy, white powder soluble in sulfuric acid, insoluble in water, melts at 3300°C; used in medicine, ceramics, flame spraying, and electrodes. Also known as thoria; thorium anhydride; thorium oxide.

thorium fluoride ThF_4 A white, toxic powder, melts at 1111°C; used to make thorium metal and magnesium-thorium alloys and in high-temperature ceramics.

thorium nitrate $Th(NO_3)_4 \cdot 4H_2O$ Explosive white crystals soluble in water and alcohol, strong oxidizer; the anhydrous form decomposes at 500°C; used in medicine and as an analytical reagent.

thorium oxalate $Th(C_2O_4)_2 \cdot 2H_2O$ A white, toxic powder soluble in alkalies and ammonium oxalate, insoluble in water and most acids, decomposes to thorium dioxide, ThO_2, above 300–400°C; used in ceramics.

thorium oxide   *See* thorium dioxide.

thorium sulfate $Th(SO_4)_2 \cdot 8H_2O$ A white powder soluble in ice water, loses water at 42° and 400°C. Also known as normal thorium sulfate.

thorium tetrachloride   *See* thorium chloride.

thoron   The conventional name for radon-220. Symbolized Tn.

Thorpe reaction   The reaction by which, in presence of lithium amides, α,ω-dinitriles undergo base-catalyzed condensation to cyclic iminonitriles, which can be hydrolyzed and decarboxylated to cyclic ketones.

THPC   *See* tetrakis(hydroxymethyl)phosphonium chloride.

threeling   *See* trilling.

threonine $CH_3CHOHCH(NH_2)COOH$ A crystalline α-amino acid considered essential for normal growth of animals; it is biosynthesized from aspartic acid and is a precursor of isoleucine in microorganisms.

threshold detector   An element or isotope in which radioactivity is induced only by the capture of neutrons having energies in excess of a certain characteristic threshold value; used to determine the neutron spectrum from a nuclear explosion.

threshold treatment   The process of stopping a precipitation-type reaction at the threshold of precipitate formation; used in water-treatment reactions.

thrombin   An enzyme elaborated from prothrombin in shed blood which induces clotting by converting fibrinogen to fibrin.

thrombokinase   A proteolytic enzyme in blood plasma that, together with thromboplastin, calcium, and factor V, converts prothrombin to thrombin.

thromboplastin   Any of a group of lipid and protein complexes in blood that accelerate the conversion of prothrombin to thrombin. Also known as factor III; plasma thromboplastin component (PTC).

thromboplastinogen   *See* antihemophilic factor.

throughput   The volume of feedstock charged to a process equipment unit during a specified time.

thulia   *See* thulium oxide.

thulium   A rare-earth element, symbol Tm, group IIIB, of the lanthanide group, atomic number 69, atomic weight 168.934; reacts slowly with water, soluble in dilute acids, melts at 1550°C, boils at 1727°C; the dust is a fire hazard; used as x-ray source and to make ferrites.

thulium-170   The radioactive isotope of thulium, with mass number 170; used as a portable x-ray source.

thulium chloride $TmCl_3 \cdot 7H_2O$ Green, deliquescent crystals soluble in water and alcohol, melts at 824°C.

thulium oxalate $Tm_2(C_2O_4)_3 \cdot 6H_2O$ A toxic, greenish-white solid, soluble in aqueous alkali oxalates, loses one water at 50°C; used for analytical separation of thulium from common metals.

thulium oxide Tm_2O_3 A white, slightly hygroscopic powder that absorbs water and carbon dioxide from the air, and is slowly soluble in strong acids; used to make thulium metal. Also known as thulia.

Thunburg technique A technique used to study oxidation of a substrate occurring by dehydrogenation reactions; methylene blue, a reversibly oxidizable indicator, substitutes for molecular oxygen as the ultimate hydrogen acceptor (oxidant), becoming reduced to the colorless leuco form.

Thylox process The conversion of coke-oven-gas, hydrogen sulfide, into elemental sulfur by absorption in a solution of arsenious oxide and soda ash in water, then air-blowing to liberate the sulfur which can be skimmed off the solution.

thyme camphor *See* thymol.

thymidine $C_{10}H_{14}N_2O_5$ A nucleoside derived from deoxyribonucleic acid; essential growth factor for certain microorganisms in mediums lacking vitamin B_{12} and folic acid.

thymidylic acid $C_{10}H_{15}N_2O_8P$ A mononucleotide component of deoxyribonucleic acid which yields thymine, D-ribose, and phosphoric acid on complete hydrolysis.

thymine $C_5H_6N_2O_2$ A pyrimidine component of nucleic acid, first isolated from the thymus.

thymol $C_{10}H_{14}O$ A naturally occurring crystalline phenol obtained from thyme or thyme oil, melting at 515°C; used to kill parasites in herbaria, to preserve anatomical specimens, and in medicine as a topical antifungal agent. Also known as isopropyl-*meta*-cresol; thyme camphor.

thymol blue $C_6H_4SO_2OC[C_6H_2(CH_3)(OH)CH(CH_3)_2]_2$ Brown-green crystals soluble in alcohol and dilute alkalies, insoluble in water, decomposes at 223°C; used as acid-base pH indicator. Also known as thymolsulfonphthalein.

thymol iodide $[C_6H_2(CH_3)(OI)(C_3H_7)]_2$ A red-brown, light-sensitive powder with an aromatic aroma, soluble in ether and chloroform, insoluble in water; used in medicine and as a feed additive.

thymolphthalein $C_6H_4COOC[C_6H_2(CH_3)(OH)CH(CH_3)_2]_2$ A white powder insoluble in water, soluble in alcohol and acetone, melts at 245°C; used in medicine and as an acid-base titration indicator.

thymolsulfonphthalein *See* thymol blue.

para-thymoquinone $C_{10}H_{12}O_2$ Bright yellow crystals with a melting point of 45.5°C; soluble in alcohol and ether; used as a fungicide. Also known as 2-isopropyl-5-methylbenzoquinone.

thyrocalcitonin *See* calcitonin.

thyroglobulin An iodinated protein found as the storage form of the iodinated hormones in the thyroid follicular lumen and epithelial cells.

thyroid hormone Commonly, thyroxine or triiodothyronine, or both; a metabolically active compound formed and stored in the thyroid gland which functions to regulate the rate of metabolism.

thyroprotein A protein secreted in the thyroid gland, such as thyroxine.

thyrotropic hormone A hormone produced by the adenohypophysis which regulates thyroid gland function. Also known as thyroid-stimulating hormone (TSH).

thyrotropin A thyroid-stimulating hormone produced by the adenohypophysis.

thyroxine $C_{15}H_{11}I_4NO_4$ The active physiologic principle of the thyroid gland; used in the form of the sodium salt for replacement therapy in states of hypothyroidism or absent thyroid function.

Ti *See* titanium.

tiba $C_7H_3I_3O_2$ A colorless solid with a melting point of 226–228°C; insoluble in water; used as a growth regulator for fruit. Also known as 2,3,5-triiodobenzoic acid.

tie line A line on a phase diagram joining the two points which represent the composition of systems in equilibrium. Also known as conode.

tight ion pair An ion pair composed of individual ions which keep their stereochemical configuration; no solvent molecules separate the cation and anion. Also known as contact ion pair; intimate ion pair.

time-of-flight mass spectrometer A mass spectrometer in which all the positive ions of the material being analyzed are ejected into the drift region of the spectrometer tube with essentially the same energies, and spread out in accordance with their masses as they reach the cathode of a magnetic electron multiplier at the other end of the tube.

time-resolved laser spectroscopy A method of studying transient phenomena in the interaction of light with matter through the exposure of samples to extremely short and intense pulses of laser light, down to subnanosecond or subpicosecond duration.

tin Metallic element in group IV, symbol Sn, atomic number 50, atomic weight 118.69; insoluble in water, soluble in acids and hot potassium hydroxide solution; melts at 232°C, boils at 2260°C.

tin bisulfide *See* stannic sulfide.

tin bromide *See* stannic bromide; stannous bromide.

tin chloride *See* stannic chloride; stannous chloride.

tin chromate *See* stannic chromate; stannous chromate.

tin crystals *See* stannous chloride.

tin dichloride *See* stannous chloride.

tin difluoride *See* stannous fluoride.

tin dioxide *See* stannic oxide.

tin fluoride *See* stannous fluoride.

tin hydride SnH_4 A gas boiling at -52°C. Also known as stannane.

tin iodide *See* stannic iodide.

tin monosulfide *See* stannous sulfide.

tin octoate *See* stannous 2-ethylhexoate.

tin oxalate *See* stannous oxalate.

tin oxide *See* stannic oxide; stannous oxide.

tin peroxide *See* stannic oxide.

tin protosulfide *See* stannous sulfide.

tin protoxide   *See* stannous oxide.

tin salts   *See* stannous chloride.

tin sulfate   *See* stannous sulfate.

tin sulfide   *See* stannous sulfide.

tin tetrabromide   *See* stannic bromide.

tin tetrachloride   *See* stannic chloride.

tin tetraiodide   *See* stannic iodide.

Tischenko reaction   The formation of an ester by the condensation of two molecules of aldehyde utilizing a catalyst of aluminum alkoxides in the presence of a halide.

titanate   A salt of titanic acid; titanates of the M_2TiO_3 type are called metatitanates, those of the M_4TiO_4 type are called orthotitanates; an example is sodium titanate, $(Na_2O)_2Ti_2O_5$.

titanellow   *See* titanium trioxide.

titania   *See* titanium dioxide.

titanic acid H_2TiO_3 A white, water-insoluble powder; used as a dyeing mordant. Also known as metatitanic acid; titanic hydroxide.

titanic anhydride   *See* titanium dioxide.

titanic chloride   *See* titanium tetrachloride.

titanic hydroxide   *See* titanic acid.

titanic sulfate   *See* titanium sulfate.

titanium   A metallic element in group IV, symbol Ti, atomic number 22, atomic weight 47.90; ninth most abundant element in the earth's crust; insoluble in water, melts at 1660°C, boils above 3000°C.

titanium boride TiB_2 A hard solid that resists oxidation at elevated temperatures and melts at 2980°C; used as a refractory and in alloys, high-temperature electrical conductors, and cermets.

titanium butylate   *See* tetrabutyl titanate.

titanium carbide   TiC   Very hard gray crystals insoluble in water, soluble in nitric acid and aqua regia, melts at about 3140°C; used in cermets, arc-melting electrodes, and tungsten-carbide tools.

titanium chloride   *See* titanium dichloride.

titanium dichloride $TiCl_2$ A flammable, alcohol-soluble, black powder that decomposes in water, and in vacuum at 475°C, and burns in air. Also known as titanium chloride.

titanium dioxide TiO_2 A white, water-insoluble powder that melts at 1560°C, and which is produced commercially from the titanium dioxide minerals ilmenite and rutile; used in paints and cosmetics. Also known as titania; titanic anhydride; titanium oxide; titanium white.

titanium hydride TiH_2 A black metallic powder whose dust is an explosion hazard and which dissociates above 288°C; used in powder metallurgy, hydrogen production, foamed metals, glass solder, and refractories, and as an electronic gas getter.

titanium nitride TiN Golden-brown brittle crystals melting at 2927°C; used in refractories, alloys, cermets, and semiconductors.

titanium oxalate $Ti_2(C_2O_4)_3 \cdot 10H_2O$ Toxic, yellow prisms soluble in water, insoluble in alcohol; used to make titanic acid and titanium metal. Also known as titanous oxalate.

titanium oxide *See* titanium dioxide; titanium trioxide.

titanium peroxide *See* titanium trioxide.

titanium sesquisulfate *See* titanous sulfate.

titanium sulfate $Ti(SO_4)_2 \cdot 9H_2O$ Caked solid, soluble in water, toxic, highly acidic; used as a dye stripper, reducing agent, laundry chemical, and in treatment of chrome yellow colors. Also known as basic titanium sulfate; titanic sulfate; titanyl sulfate.

titanium tetrachloride $TiCl_4$ A colorless, toxic liquid soluble in water, fumes when exposed to moist air, boils at 136°C; used to make titanium and titanium salts, as a dye mordant and polymerization catalyst, and in smoke screens and pigments. Also known as titanic chloride.

titanium trichloride $TiCl_3$ Toxic, dark-violet, deliquescent crystals soluble in alcohol and some amines, decomposes in water with heat evolution, decomposes above 440°C; used as a reducing agent, chemical intermediate, polymerization catalyst, and laundry stripping agent. Also known as titanous chloride.

titanium trioxide TiO_3 Yellow titanium oxide used to make ivory shades in ceramics. Also known as titanellow; titanium oxide; titanium peroxide.

titanium white *See* titanium dioxide.

titanous chloride *See* titanium trichloride.

titanous oxalate *See* titanium oxalate.

titanous sulfate $Ti_2(SO_4)_3$ Green crystals soluble in dilute hydrochloric and sulfuric acids, insoluble in water and alcohol; used as a textile reducing agent. Also known as titanium sesquisulfate.

titanyl sulfate *See* titanium sulfate.

titer 1. The concentration in a solution of a dissolved substance as shown by titration. 2. The least amount or volume needed to give a desired result in titration. 3. The solidification point of hydrolyzed fatty acids.

titrand The substance that is analyzed in a titration procedure.

titrant A standard solution of known concentration and composition used for analytical titrations.

titration A method of analyzing the composition of a solution by adding known amounts of a standardized solution until a given reaction (color change, precipitation, or conductivity change) is produced.

Tl *See* thallium.

Tm *See* thulium.

TMA   *See* trimethylamine.

Tn   *See* thoron.

TNF   *See* 2,4,7-trinitrofluorene.

TNT   *See* 2,4,6-trinitrotoluene.

tocopherol $C_{29}H_{50}O_2$ Any of several substances having vitamin E activity that occur naturally in certain oils; alpha tocopherol is the most potent.

tolazoline hydrochloride $C_{10}H_{12}N_2 \cdot HCl$ Water-soluble white crystals, and melting at 173°C; used as a sympatholytic and vasodilator. Also known as priscol.

***ortho*-tolidine** $[C_6H_3(CH_3)NH_2]_2$ Light-sensitive, combustible white to reddish crystals soluble in alcohol and ether, slightly soluble in water, melts at 130°C; used as an anlytical reagent and a curing agent for urethane resins. Also known as diaminoditolyl.

Tollen's aldehyde test   A test that uses an ammoniacal solution of silver oxides to test for aldehydes and ketones.

α-toluamide   *See* α-phenylacetamide.

toluene $C_6H_5CH_3$ A colorless, aromatic liquid derived from coal tar or from the catalytic reforming of petroleum naphthas; insoluble in water, soluble in alcohol and ether, boils at 111°C; used as a chemical intermediate, for explosives, and in high-octane gasolines. Also known as methylbenzene; phenylmethane; toluol.

toluene 2,4-diisocyanate $CH_3C_6H_3(NCO)_2$ A liquid (at room temperature) with a sharp, pungent odor; miscible with ether, acetone, and benzene; used to make polyurethane foams and other elastomers, and also as a protein cross-linking agent. Also known as 2,4-diisocyanotototoluene; 2,4-tolylene diisocyanate.

***para*-toluenesulfonate**   *See para*-toluenesulfonic acid.

***para*-toluenesulfonic acid** $C_6H_4(SO_3H)(CH_3)$ Toxic, colorless, combustible crystals soluble in water, alcohol, and ether; melts at 107°C; used in dyes and as a chemical intermediate and organic catalyst. Also known as *para*-toluenesulfonate.

toluenethiol   *See* thiocresol.

toluene trichloride   *See* benzotrichloride.

toluene trifluoride   *See* benzotrifluoride.

α-toluic acid   *See* phenylacetic acid.

***meta*-toluic acid** $C_6H_4CH_3COOH$ White to yellow, combustible crystals soluble in alcohol and ether, slightly soluble in water, melts at 109°C; used as a chemical intermediate and base for insect repellants. Also known as 3-methylbenzoic acid; *meta*-toluylic acid.

***ortho*-toluic acid** $C_6H_4CH_3COOH$ White, combustible crystals soluble in alcohol and chloroform, slightly soluble in water, melts at 104°C; used as a bacteriostat. Also known as 2-methylbenzoic acid; *ortho*-toluylic acid.

α-toluic aldehyde   *See* phenylacetaldehyde.

***meta*-toluidine** $CH_3C_6H_4NH_2$ A combustible, colorless, toxic liquid soluble in alcohol and ether, slightly soluble in water, boils at 203°C; used for dyes and as a chemical intermediate. Also known as *meta*-aminotoluene.

ortho-toluidine $CH_3C_6H_4NH_2$ A light-green, light-sensitive, combustible, toxic liquid soluble in alcohol and ether, very slightly soluble in water, boils at 200°C; used for dyes and textile printing and as a chemical intermediate. Also known as *ortho*-aminotoluene.

para-toluidine $CH_3C_6H_4NH_2$ Toxic, combustible, white leaflets soluble in alcohol and ether, very slightly soluble in water, boils at 200°C; used as an analytical reagent and in dyes. Also known as *para*-aminotoluene.

α-tolunitrile *See* benzyl cyanide.

toluol *See* toluene.

toluylene *See* stilbene.

toluylene red *See* neutral red.

ortho-toluylic acid *See ortho*-toluic acid.

meta-toluylic acid *See meta*-toluic acid.

para-toluylic acid *See para*-toluic acid.

2,4-tolylene diisocyanate *See* toluene 2,4-diisocyanate.

meta-tolyl-N-methylcarbamate $C_9H_{11}NO_2$ A white, crystalline compound, insoluble in water; used to control pests on rice, citrus, fruits, and cotton. Abbreviated MTMC.

para-tolylsulfonylmethylnitrosamide $C_8H_{10}N_2O_3S$ Yellow crystals with a melting point of 62°C; soluble in ether, petroleum ether, benzene, carbon tetrachloride, and chloroform; a precursor to diazomethane; a useful reagent for the preparation of a wide range of biologically active compounds for gas chromatography analysis. Also known as N-methyl-N-nitroso-*para*-toluenesulfonamide.

tomatine $C_{50}H_{83}NO_{21}$ A glycosidal alkaloid obtained from the leaves and stems from the tomato plant; the crude extract is known as tomatin: white, toxic crystals; used as a plant fungicide and as a precipitating agent for cholesterol.

topochemical control In a chemical reaction, product formation that is determined by the orientation of molecules in the crystal.

topping The distillation of crude petroleum to remove the light fractions only; the unrefined distillate is called tops.

top steam Steam admitted near the top of a shell still to purge the still, and to prevent a vacuum from forming when pumping out the liquid contents.

total heat of dilution *See* heat of dilution.

total heat of solution *See* heat of solution.

total solids The total content of suspended and dissolved solids in water.

tower A vertical, cylindrical vessel used in chemical and petroleum processing to increase the degree of separation of liquid mixtures by distillation or extraction. Also known as column.

toxaphene $C_{10}H_{10}Cl_8$ A toxic, waxy, amber solid with a mild chlorine-camphor aroma, soluble in organic solvents, melts at 65–90°C; used as an insecticide. Also known as technical chlorinated camphene.

toxin Any of various poisonous substances produced by certain plant and animal cells, including bacterial toxins, phytotoxins, and zootoxins.

toxoplasmin The *Toxoplasma* antigen; used in a skin test to demonstrate delayed hypersensitivity to toxoplasmosis.

TPA *See* terephthalic acid.

TPN *See* triphosphopyridine nucleotide.

trace analysis Analysis of a very small quantity of material of a sample by such techniques as polarography or spectroscopy.

trace element A nonessential element found in small quantities (usually less than 1.0%) in a mineral. Also known as accessory element; guest element.

tracer A foreign substance, usually radioactive, that is mixed with or attached to a given substance so the distribution or location of the latter can later be determined; used to trace chemical behavior of a natural element in an organism. Also known as tracer element.

tracer element *See* tracer.

transaminase One of a group of enzymes that catalyze the transfer of the amino group of an amino acid to a keto acid to form another amino acid. Also known as aminotransferase.

transamination 1. The transfer of one or more amino groups from one compound to another. 2. The transposition of an amino group within a single compound.

transesterification Conversion of an organic acid ester into another ester of that same acid.

transferase Any of various enzymes that catalyze the transfer of a chemical group from one molecule to another.

transference number The portion of the total electrical current carried by any ion species in a fluid-state electrolyte.

transfer reaction A nuclear reaction in which one or more nucleons are exchanged between the target nucleus and an incident projectile.

transferrin Any of various beta globulins in blood serum which bind and transport iron to the bone marrow and storage areas.

transfer unit The relationship between the overall rate coefficient (for whatever transfer operation is being calculated), column volume, and fluid volumetric flow rate in fixed-bed sorption operations.

transformation *See* inversion.

transformation series *See* radioactive series.

transformation twin A crystal twin developed by a growth transformation from a higher to a lower symmetry.

transition element One of a group of metallic elements in which the members have the filling of the outermost shell to 8 electrons interrupted to bring the penultimate shell from 8 to 18 or 32 electrons; includes elements 21 through 29 (scandium through copper), 39 through 47 (yttrium through silver), 57 through 79 (lanthanum through gold), and all known elements from 89 (actinium) on.

transition time The time interval needed for a working (nonreference) electrode to become polarized during chronopotentiometry (time-measurement electrolysis of a sample).

transketolase An enzyme that cleaves a substrate at the position of the carbonyl carbon and transports a two-carbon fragment to an acceptor compound to form a new compound.

translation gliding *See* crystal gliding.

translation group The collection of all translation operations which carry a crystal lattice into itself.

transmethylase A transferase enzyme involved in catalyzing chemical reactions in which methyl groups are transferred from a substrate to a new compound.

transmethylation A metabolic reaction in which a methyl group is transferred from one compound to another; methionine and choline are important donors of methyl groups.

transmission diffraction A type of electron diffraction analysis in which the electron beam is transmitted through a thin film or powder whose smallest dimension is no greater than a few tenths of a micrometer.

transmittance During absorption spectroscopy, the amount of radiant energy transmitted by the solution under analysis.

transmutation A nuclear process in which one nuclide is transformed into the nuclide of a different element. Also known as nuclear transformation.

transpassive region That portion of an anodic polarization curve in which metal dissolution increases as the potential becomes noble.

transplutonium element An element having an atomic number greater than that of plutonium (94).

transport number The fraction of the total current carried by a given ion in an electrolyte.

transuranic elements Elements that have atomic numbers greater than 92; all are radioactive, are products of artificial nuclear changes, and are members of the actinide group. Also known as transuranium elements.

transuranium elements *See* transuranic elements.

trapezohedron An isometric crystal form of 24 faces, each face of which is an irregular four-sided figure. Also known as icositetrahedron; leucitohedron; tetragonal trisoctahedron.

Traube's rule In dilute solutions, the concentration of a member of a homologous series at which a given lowering of surface tension is observed decreases threefold for each additional methylene group in a given series.

traveling-screen dryer A moving screen belt on which damp material is conveyed through a heated drying zone. Also known as screen dryer.

tray tower A vertical process tower for liquid-vapor contacting (as in distillation, absorption, stripping, evaporation, spray drying, dehumidification, humidification, flashing, rectification, dephlegmation), along the height of which is a series of trays designed to cause intimate contact between the falling liquid and the rising vapor.

treater A vessel or system for the contacting of a process stream with reagent (treating) chemicals; for example, acid treating or caustic treating.

treating Usually, the contacting of a fluid stream (for example, water, sewage, petroleum products, or mixed gases) with chemicals to improve the fluid properties by removing, sequestering, or converting undesirable impurities.

tretamine *See* triethylenemelamine.

triacetin $C_3H_5(CO_2CH_3)_3$ A colorless, combustible oil with a bitter taste and a fatty aroma; found in cod liver and butter; soluble in alcohol and ether, slightly soluble in water; boils at 259°C; used in plasticizers, perfumery, cosmetics, and external medicine and as a solvent and food additive. Also known as glyceryl triacetate.

triallate *See* S(2,3,3-trichlorallyl)diisopropylthiocarbamate.

triamcinolone $C_{21}H_{27}FO_6$ White, toxic crystals; insoluble in water, soluble in dimethylformamide; melts at 266°C; used as an intermediate for ion-exchange resin, wetting and frothing agent, and photographic developer.

2,4,6-triamino-*s*-triazine *See* melamine.

triamylamine $(C_5H_{11})_3N$ A combustible, colorless, toxic liquid; soluble in gasoline, insoluble in water; used to inhibit corrosion and in insecticides.

triamyl borate $(C_5H_{11})_3BO_3$ A combustible, colorless liquid with an alcoholic aroma; soluble in alcohol and ether; boils at 220–280°C; used in varnishes.

***s*-triazinetriol** *See* cyanuric acid.

triazole A five-membered chemical ring compound with three nitrogens in the ring; for example, $C_2H_3N_3$; proposed for use as a photoconductor and for copying systems.

triazophos *See* phentriazophos.

tribasic calcium phosphate *See* calcium phosphate.

tribasic zinc phosphate *See* zinc phosphate.

triboluminescence Luminescence produced by friction between two materials.

tribonate $C_{17}H_{14}N_4O_{10}$ A postcontact herbicide. Also known as 2,4-dinitrophenyl-2-(*sec*-butyl)-4,6-dinitrophenyl carbonate.

tribromoethyl alcohol *See* tribromoethanol.

tribromomethane *See* bromoform.

tribromonitromethane *See* bromopicrin.

tributoxyethyl phosphate $_3[(CH_2)_3O(CH_2)_2O]PO$ A light yellow, oily liquid with a boiling range of 215–228°C; soluble in organic solvents; used as a plasticizer and flame retardant, and in floor waxes.

tributyl borate $(C_4H_9)_3BO_3$ A combustible, water-white liquid miscible with common organic liquids; boils at 232°C; used in welding fluxes and as a chemical intermediate and textile flame-retardant. Also known as butyl borate.

tri-*n*-butylchlorostannane *See* tributyltin chloride.

tributyl[(5-chloro-2-thienyl)methyl]phosphonium chloride $C_{14}H_{25}PSCl_2$ An amorphous, tan solid with a melting point of 157–159°C; soluble in water; used as a growth regulator for soybeans, peas, snap beans, and elm trees.

tributyl citrate *See* butyl citrate.

tributyl phosphate $(C_4H_9)_3PO_4$ A combustible, toxic, stable liquid; soluble in most solvents, and very slightly soluble in water; boils at 292°C; used as a heat-exchange medium, pigment-grinding assistant, antifoam agent, and solvent. Abbreviated TBP.

S,S,S,-tributylphosphorotrithioate $(CH_3)_3(CH_2)_9S_3P$ A pale yellow liquid, used as a defoliant for cotton.

tributyltin acetate $(C_4H_9)_3Sn-OOCCH_3$ An organic compound of tin, used as an antimicrobial agent in the paper, wood, plastics, leather, and textile industries.

tributyltin chloride $(C_4H_9)_3SnCl$ A colorless liquid with a boiling point of 145–147°C; soluble in alcohol, benzene, and other organic solvents; used as a rodenticide. Also known as tri-*n*-butylchlorostannane.

tricaine $C_{10}H_{15}NO_5S$ Fine, needlelike crystals, soluble in water; used as an anesthetic for fish. Also known as *meta*-aminobenzoate methanesulfonate.

tricalcium phosphate *See* calcium phosphate.

tricarboxylic acid cycle *See* Krebs cycle.

S(2,3,3-trichlorallyl)diisopropylthiocarbamate $C_{10}H_{16}NSCl_3$ An oily liquid with a boiling point of 117°C at 0.6 mmHg (80 newtons per square meter); decomposes at 200°C; solubility in water is 4 parts per million at 25°C; used as a preemergence herbicide for wild oats, peas, barley, and fruit crops. Also known as triallate.

trichlorfon $C_4H_8Cl_3O_4P$ A colorless, crystalline compound with a melting point of 83–84°C; used as an insecticide in agriculture. Also known as O,O-dimethyl(2,2,2-trichloro-1-hydroxyethyl)phosphonate.

trichloroacetic acid CCl_3COOH Toxic, deliquescent, colorless crystals with a pungent aroma; soluble in water, alcohol, and ether; boils at 198°C; used as a chemical intermediate and laboratory reagent, and in medicine, pharmacy, and herbicides. Abbreviated TCA.

trichloroacetic aldehyde *See* chloral.

trichlorobenzene $C_6H_3Cl_3$ Either of two toxic compounds: 1,2,3-trichlorobenzene forms white crystals, soluble in ether, insoluble in water, boiling at 221°C, and is used as a chemical intermediate; 1,2,4-trichlorobenzene is a combustible, colorless liquid, soluble in most organic solvents and oils, insoluble in water, boiling at 213°C, and is used as a solvent and in dielectric fluids, synthetic transformer oils, lubricants, and insecticides.

trichloro-*tert*-butyl alcohol *See* chlorobutanol.

trichloroethanal *See* chloral.

trichloroethane $C_2H_3Cl_3$ Either of two nonflammable, irritating liquid isomeric compounds: 1,1,1-trichloroethane (CH_3CCl_3) is toxic, soluble in alcohol and ether, insoluble in water, and boils at 75°C; it is used as a solvent, aerosol propellant, and pesticide and for metal degreasing, and is also known as methyl chloroform; 1,1,2-trichloroethane ($CHCl_2CH_2Cl$) is clear and colorless, is soluble in alcohols, ethers, esters, and ketones, insoluble in water, has a sweet aroma, and boils at 114°C; it is used as a chemical intermediate and solvent, and is also known as vinyl trichloride.

trichloroethyl alcohol *See* 2,2,2-trichloroethanol.

trichloroethylene $CHCl:CCl_2$ A heavy, stable, toxic liquid with a chloroform aroma; slightly soluble in water, soluble with greases and common organic solvents; boils at 87°C; used for metal degreasing, solvent extraction, and dry cleaning and as a fumigant and chemical intermediate.

trichlorofluoromethane CCl_3F A toxic, noncombustible, colorless liquid boiling at 24°C; used as a chemical intermediate, solvent, refrigerant, aerosol propellant, and blowing agent (plastic foams) and in fire extinguishers. Also known as fluorocarbon-11; fluorotrichloromethane.

trichloroiminocyanuric acid *See* trichloroisocyanuric acid.

trichloroisocyanuric acid $C_3Cl_3N_3O_3$ A crystalline substance that releases hypochlorous acid on contact with water; melting point is 246–247°C; soluble in chlorinated and highly polar solvents; used as a chlorinating agent, disinfectant, and industrial deodorant. Also known as symclosene; trichloroiminocyanuric acid.

trichloromethane *See* chloroform.

trichloromethyl chloroformate $ClCOOCCl_3$ A toxic, colorless liquid with a boiling point of 127–128°C; soluble in alcohol, ether, and benzene; used in organic synthesis, and as a military poison gas during World War I. Also known as diphosgene.

***cis*-*N*-(trichloromethylthio)-4-cyclohexene-1,2-dicarboximide** *See* captan.

***N*-(trichloromethylthio)phthalimide** *See* folpet.

trichloronate *See* ortho-ethyl[O-2,4,5-trichlorophenyl]ethylphosphonothioate.

trichloronitromethane *See* chloropicrin.

trichlorophenol $C_6H_2Cl_3OH$ Either of two toxic nonflammable compounds with a phenol aroma: 2,4,5-trichlorophenol is a gray solid, is soluble in alcohol, acetone, and ether, melts at 69°C, and is used as a fungicide and bactericide; 2,4,6-trichlorophenol forms yellow flakes, is soluble in alcohol, acetone, and ether, boils at 248°C, and is used as a fungicide, defoliant, and herbicide; it is also known as 2,4,6-T.

2,4,5-trichlorophenoxyacetic acid $C_6H_2Cl_3OCH_2CO_2H$ A toxic, light-tan solid; soluble in alcohol, insoluble in water; melts at 152°C; used as a defoliant, plant hormone, and herbicide. Also known as 2,4,5-T.

2-(2,4,5-trichlorophenoxy)ethyl 2,2-dichloropropionate *See* erbon.

2-(2,4,5-trichlorophenoxy)propionic acid *See* silvex.

2,3,6-trichlorophenylacetic acid *See* chlorofenac.

1,2,3-trichloropropane $CH_2ClCHClCH_2Cl$ A toxic, colorless liquid with a boiling point of 156.17°C; used as a paint and varnish remover and degreasing agent.

1,1,2-trichloro-1,2,2-trifluoroethane CCl_2FCClF_2 A colorless, volatile liquid with a boiling point of 47.6°C; used as a solvent for dry cleaning, as a refrigerant, and in fire extinguishers. Also known as trifluorotrichloroethane.

trickle hydrodesulfurization A fixed-bed, petroleum refining process for desulfurization of middle distillates and gas oils; catalyst is cobalt molybdenum on alumina.

triclinic crystal A crystal whose unit cell has axes which are not at right angles, and are unequal. Also known as anorthic crystal.

triclinic system The most general and least symmetric crystal system, referred to by three axes of different length which are not at right angles to one another.

tricosane $CH_3(CH_2)_{21}CH_3$ Combustible, glittering crystals; soluble in alcohol, insoluble in water; melts at 48°C; used as a chemical intermediate.

tricresyl phosphate $(CH_3C_6H_4O)_3PO$ A combustible, colorless liquid; insoluble in water, soluble in common solvents and vegetable oils; boils at 420°C; used as a plasticizer, plastics fire retardant, air-filter medium, and gasoline and lubricant additive. Abbreviated TCP.

tricyclic dibenzopyran See xanthene.

tricyclohexyltin hydroxide See cyhexatin.

n-tridecane $CH_3(CH_2)_{11}CH_3$ A combustible liquid; soluble in alcohol, insoluble in water; boils at 226°C; used as a distillation chaser and chemical intermediate.

tridecanol See tridecyl alcohol.

tridecyl alcohol $C_{12}H_{25}CH_2OH$ An isomer mixture; a white, combustible solid with a pleasant aroma; melts at 31°C; used in detergents and perfumery and to make synthetic lubricants. Also known as tridecanol.

N-tridecyl-2,6-dimethyl morpholine See tridemorph.

tridemorph $C_{19}H_{39}NO$ A colorless liquid with a boiling point of 134°C; used as a fungicide for powdery mildew in cereals. Also known as N-tridecyl-2,6-dimethyl morpholine.

tridentate ligand A chelating agent having three groups capable of attachment to a metal ion. Also known as terdentate ligand.

triethanolamine $(HOCH_2CH_2)_3N$ A viscous, hygroscopic liquid with an ammonia aroma, soluble in chloroform, water, and alcohol, and boiling at 335°C; used in dry-cleaning soaps, cosmetics, household detergents, and textile processing, for wool scouring, and as a corrosion inhibitor.

triethanolamine stearate See trihydroxyethylamine stearate.

triethylamine $(C_2H_5)_3N$ A colorless, toxic, flammable liquid with an ammonia aroma; soluble in water and alcohol; boils at 90°C; used as a solvent, rubber-accelerator activator, corrosion inhibitor, and propellant, and in penetrating and waterproofing agents.

triethylborane $(C_2H_5)_3B$ A colorless liquid with a boiling point of 95°C; used as a jet fuel or igniter for jet engines and as a fuel additive. Also known as boron triethyl; triethylborine.

triethylborine See triethylborane.

triethylene glycol $HO(C_2H_4O)_3H$ A colorless, combustible, hygroscopic, water-soluble liquid; boils at 287°C; used as a chemical intermediate, solvent, bactericide, humectant, and fungicide. Abbreviated TEG.

triethylenemelamine $NC[N(CH_2)_2]NC[N(CH_2)_2]NC[N(CH_2)_2]$ White crystals, soluble in water, alcohol, acetone, chloroform, and methanol; polymerizes at 160°C; used in medicine and insecticides and as a chemosterilant. Abbreviated TEM. Also known as tretamine.

triethylenetetramine $NH_2(C_2H_4NH)_2C_2H_4NH_2$ A yellow, water-soluble liquid with a boiling point of 277.5°C; used in detergents and in the manufacture of dyes and pharmaceuticals.

triethylic borate *See* ethyl borate.

triethyl phosphate $(C_2H_5)_3PO_4$ A toxic, colorless liquid that acts as a cholinesterase inhibitor; boiling point is 216°C; soluble in organic solvents; used as a solvent and plasticizer and for pesticides manufacture. Abbreviated TEP.

trifluorochlorethylene resin A fluorocarbon used as a base for polychlorotrifluoroethylene resin, marketed as Kel-F.

trifluorochloromethane *See* chlorotrifluoromethane.

α,α,α-trifluoro-2,6-dinitro-*N,N*-dipropyl-*para*-toluidine *See* trifluralin.

trifluoromethane *See* fluoroform.

trifluoromethyl benzene *See* benzotrifluoride.

3′-trifluoromethyldiphenylamine-2-carboxylic acid *See* flufenamic acid.

trifluorotrichloroethane *See* 1,1,2-trichloro-1,2,2-trifluoroethane.

trifluralin $C_{13}H_{16}F_3N_3O_4$ A yellow-orange, crystalline compound with a melting point of 48.5–49°C; used as a preemergence herbicide for use on cotton, beans, and vegetables. Also known as α,α,α-trifluoro-2,6-dinitro-*N,N*-dipropyl-*para*-toluidine.

triforine $C_{10}H_{10}Cl_6N_4O_2$ A colorless, crystalline compound with a melting point of 155°C; used to control fungus diseases of ornamentals, cereals, fruits, and vegetables. Also known as *N,N*′-[1,4-piperazinediyl-bis(2,2,2-trichloroethylidine)]bis[formamide].

triformal *See* *sym*-trioxane.

triglyceride $CH_2(OOCR_1)CH(OOCR_2)CH_2(OOCR_3)$ A naturally occurring ester of normal, fatty acids and glycerol; used in the manufacture of edible oils, fats, and monoglycerides.

trigonal lattice *See* rhombohedral lattice.

trigonal system A crystal system which is characterized by threefold symmetry, and which is usually considered as part of the hexagonal system since the lattice may be either hexagonal or rhombohedral.

1,2,4-trihydroxyanthraquinone *See* purpurin.

1,2,3-trihydroxybenzene *See* pyrogallic acid.

1,3,5-trihydroxybenzene *See* phloroglucinol.

trihydroxycyanidine *See* cyanuric acid.

trihydroxyethylamine stearate $C_{24}H_{49}NO_5$ A cream-colored, waxy solid with a melting point of 42–44°C; soluble in methanol, ethanol, mineral oil, and vegetable oil; used as an emulsifier in cosmetics and pharmaceuticals. Also known as triethanolamine stearate.

2,3,5-triiodobenzoic acid *See* tiba.

triiodomethane *See* iodoform.

tri-iron dodecacarbonyl *See* iron tetracarbonyl.

triisobutylene $(C_4H_8)_3$ A mixture of isomers; combustible liquid boiling at 348–354°C; used as a chemical and resin intermediate, lubricating oil additive, and motor-fuel alkylation feedstock.

triketohydrindene hydrate *See* ninhydrin.

trill *See* trilling.

trilling A cyclic crystal twin consisting of three individual crystals. Also known as threeling; trill.

trimellitic acid *See* 1,2,4-benzenetricarboxylic acid.

trimer A condensation product of three monomer molecules; C_6H_6 is a trimer of C_2H_2.

trimercuric orthophosphate *See* mercuric phosphate.

trimercurous orthophosphate *See* mercurous phosphate.

trimethylamine $(CH_3)_3N$ A colorless, liquefied gas with a fishy odor and a boiling point of $-4°C$; soluble in water, ether, and alcohol; used as a warning agent for natural gas, a flotation agent, and insect attractant. Abbreviated TMA.

***uns*-trimethylbenzene** *See* pseudocumene.

3,7,7-trimethylbicyclo-[4.1.0]-hept-3-ene *See* δ-3-carene.

trimethyl borate $B(OCH_3)_3$ A water-white liquid, boiling at 67–68°C; used as a solvent for resins, waxes, and oils, and as a catalyst and a reagent in analysis of paint and varnish. Also known as methyl borate.

2,2,3-trimethylbutane *See* triptane.

trimethylchlorosilane $(CH_3)_3SiCl$ A colorless liquid with a boiling point of 57°C; soluble in ether and benzene; used as a water-repelling agent.

trimethylethylene *See* methyl butene.

trimethylol aminomethane *See* tromethamine.

trimethylolethane $CH_3C(CH_2OH)_3$ Colorless crystals, soluble in alcohol and water; used in the manufacture of varnishes and drying oils. Also known as methyltrimethylolmethane; pentoglycerine.

2,2,4-trimethylpentane *See* isooctane.

2,4,6-trimethylpyridine *See* 2,4,6-collidine.

trimethylvinylammonium hydroxide *See* neurine.

trinickelous orthophosphate *See* nickel phosphate.

trinitrobenzene $C_6H_3(NO_2)_3$ A yellow crystalline compound, soluble in alcohol and ether; used as an explosive.

2,4,6-trinitro-1,3-dimethyl-5-tert-butylbenzene *See* musk xylol.

2,4,7-trinitrofluorenone $C_{13}H_5N_3O_7$ A yellow, crystalline compound with a melting point of 175.2–176°C; forms crystalline complexes with indoles for identification by mass spectroscopy. Abbreviated TNF.

trinitromethane $CH(NO_2)_3$ A crystalline compound, melting at 150°C, decomposing above 25°C; used to make explosives. Also known as nitroform.

trinitrophenol   *See* picric acid.

2,4,6-trinitroresorcinol   *See* styphnic acid.

2,4,6-trinitrotoluene $CH_3C_6H_2(NO_2)_3$ Toxic, flammable, explosive, yellow crystals; soluble in alcohol and ether, insoluble in water; melts at 81°C; used as an explosive and chemical intermediate and in photographic chemicals. Also known as methyltrinitrobenzene; TNT.

triose   A group of monsaccharide compounds that have a three-carbon chain length.

***sym*-trioxane** $(CH_2O)_3$ White, flammable, explosive crystals; soluble in water, alcohol, and ether; melts at 62°C; used as a chemical intermediate, disinfectant, and fuel. Also known as metaformaldehyde; triformol; trioxin.

trioxin   *See sym*-trioxane.

trioxycyanidine   *See* cyanuric acid.

tripalmitin $C_3H_5(OOCC_{15}H_{31})_3$ A white, water-insoluble powder that melts at 65.5°C; used in the preparation of leather dressings and soaps. Also known as glyceryl tripalmitate; palmitin.

triphenylmethane dye   A family of dyes with a molecular structure derived from $(C_6H_5)_3CH$, usually by NH_2, OH, or HSO_3 substitution for one of the C_6H_5 hydrogens; includes many coal tar dyes, for example, rosaniline and fuchsin.

triphenylmethyl radical   A free radical in which three phenyl rings are bonded to a single carbon. Also known as trityl radical.

triphenyl phosphate $(C_6H_5O)_3PO$ A crystalline compound with a melting point of 49–50°C; soluble in benzene, chloroform, ether, and acetone; used as a substitute for camphor in celluloid, as a plasticizer in lacquers and varnishes, and to impregnate roofing paper.

triphenylphosphine $(C_6H_5)_3P$ A crystalline compound with a melting point of 80.5°C; soluble in ether, benzene, chloroform, and glacial acetic acid; used as an initiator of polymerization and in organic synthesis.

triphenyltetrazolium chloride $C_{19}H_{15}ClN_4$ A crystalline compound, soluble in water, alcohol, and acetone; used as a sensitive reagent for reducing sugars. Also known as red tetrazolium.

triphenyltinacetate   *See* fentinacetate.

triphosphopyridine dinucleotide   *See* triphosphopyridine nucleotide.

triphosphopyridine nucleotide $C_{12}H_{28}N_7O_{17}P_3$ A grayish-white powder, soluble in methanol and in water; a coenzyme and an important component of enzymatic systems concerned with biological oxidation-reduction systems. Abbreviated TPN. Also known as codehydrogenase II; coenzyme II; triphosphopyridine dinucleotide.

triple point   A particular temperature and pressure at which three different phases of one substance can coexist in equilibrium.

triplet state   Electronic state of an atom or molecule whose total spin angular momentum quantum number is equal to 1.

tripotassium orthophosphate   *See* potassium phosphate.

triprene $C_{18}H_{32}O_2S$ An amber liquid used as a growth regulator for crops. Also known as ethyl (2*E*,4*E*)-11-methoxy-3,7,11-trimethyl-2,4-dodecadienethiolate.

triptane C_7H_{16} A hydrocarbon compound made commercially in small quantities, but having one of the highest antiknock ratings known. Also known as 2,2,3-trimethylbutane.

TRIS *See* tromethamine.

trisaccharide A carbohydrate which, on hydrolysis, yields three molecules of monosaccharides.

trisamine *See* tromethamine.

tris buffer *See* tromethamine.

tris(hydroxymethyl)aminomethane *See* tromethamine.

trisodium citrate *See* sodium citrate.

trisodium orthophosphate *See* trisodium phosphate.

trisodium phosphate Na_3PO_4 A water-soluble crystalline compound; used as a cleaning compound and as a water softener. Abbreviated TSP. Also known as tertiary sodium phosphate; trisodium orthophosphate.

tristazine $C_9H_{16}ClN_5$ A solid compound, insoluble in water; used to control weeds in potatoes and peas. Also known as 2-chloro-4-(methylamino)-6-(ethylamino)-S-triazine.

tristearin *See* stearin.

trisulfide A binary chemical compound that contains three sulfur atoms in its molecule, for example, iron trisulfide, Fe_2S_3.

tris[2-(2,4-dichlorophenoxy)ethyl]phosphite $C_{24}H_{21}Cl_6O_6P$ A dark liquid that boils above 200°C; used as a preemergence herbicide for corn, peanuts, and strawberries. Abbreviated 2,4-DEP.

triterpene One of a class of compounds having molecular skeletons containing 30 carbon atoms, and theoretically composed of six isoprene units; numerous and widely distributed in nature, occurring principally in plant resins and sap; an example is ambrein.

trithioacetaldehyde $(C_4H_4S_2)_3$ A colorless, water-insoluble, crystalline compound; used as a hypnotic. Also known as sulfoparaldehyde.

tritium The hydrogen isotope having mass number 3; it is one form of heavy hydrogen, the other being deuterium. Symbolized H^3; T.

triton The nucleus of tritium.

tritopine *See* laudanidine.

trityl radical *See* triphenylmethyl radical.

triuranium octoxide U_3O_8 Olive green to black crystals or granules, soluble in nitric acid and sulfuric acid; decomposes at 1300°C; used in nuclear technology and in the preparation of other uranium compounds. Also known as uranous-uranic oxide; uranyl uranate.

trivial name Unsystematic nomenclature, being the name of a chemical compound derived from the names of the natural source of the compound at the time of its isolation and before anything is known about its molecular structure.

tromethamine $C_4H_{11}NO_3$ A crystalline compound with a melting point of 171–172°C; soluble in water, ethylene glycol, methanol, and ethanol; used to make surface-active agents, vulcanization accelerators, and pharmaceuticals, and as a titrimetric standard. Also known as THAM; trimethylol aminomethane; TRIS; trisamine; tris buffers; tris(hydroxymethyl)aminomethane.

Trona process A process for separation and purification of a mixture of soda ash, anhydrous sodium sulfate, bromine, and potassium chloride from brine (Searles Lake, California).

tropeoline 00 $NaSO_3C_6H_4NNC_6H_4NHC_6H_5$ An acid-base indicator with a pH range of 1.4–3.0, color change (from acid to base) red to yellow; used as a biological stain. Also known as sodium *para*-diphenylaminoazobenzene sulfonate.

tropocollagen The fundamental units of collagen fibrils.

tropomyosin A muscle protein similar to myosin and implicated as being part of the structure of the Z bands separating sarcomeres from each other.

troponin A protein species located at specific stations every 36.5 nanometers on the actin helix in muscle sarcomere.

Trouton's rule An approximation rule for the derivation of molar heats of vaporization of normal liquids at their boiling points.

true-boiling-point analysis A standard laboratory technique used to predict the refining qualities of crude petroleum; gives distillation cuts for gasoline, kerosine, distillate (diesel) fuel, cracking, and lube distillate stocks. Also known as true-boiling-point distillation.

true-boiling-point distillation *See* true-boiling-point analysis.

true condensing point *See* critical condensation temperature.

true freezing point The temperature at which the liquid and solid forms of a substance exist in equilibrium at a given pressure (usually 1 standard atmosphere, or 101,325 newtons per square meter).

trypsin A proteolytic enzyme which catalyzes the hydrolysis of peptide linkages in proteins and partially hydrolyzed proteins; derived from trypsinogen by the action of enterokinase in intestinal juice.

trypsinogen The zymogen of trypsin, secreted in the pancreatic juice. Also known as protrypsin.

tryptophan $C_{11}H_{12}O_2N_2$ An amino acid obtained from casein, fibrin, and certain other proteins; it is a precursor of indoleacetic acid, serotonin, and nicotinic acid.

TSP *See* trisodium phosphate.

TSPP *See* sodium pyrophosphate.

TTD *See* tetraethylthiuram disulfide.

tubatoxin *See* rotenone.

tube-still heater A firebox containing a pipe coil through which oil for a tube still (pipe still) is pumped.

tubulin A globular protein containing two subunits; 10–14 molecules are arranged to form a microtubule.

tungstate M_2WO_4 A salt of tungstic acid; for example, sodium tungstate, Na_2WO_4.

tungstate white *See* barium tungstate.

tungsten Also known as wolfram. A metallic element in group VI, symbol W, atomic number 74, atomic weight 183.85; soluble in mixed nitric and hydrofluoric acids; melts at 3400°C.

tungsten boride WB_2 A silvery solid; insoluble in water, soluble in aqua regia and concentrated acids; melts at 2900°C; used as a refractory for furnaces and chemical process equipment.

tungsten carbide WC A hard, gray powder; insoluble in water; readily attacked by nitric-hydrofluoric acid mixture; melts at 2780°C; used in tools, dies, ceramics, cermets, and wear-resistant mechanical parts, and as an abrasive.

tungsten carbonyl *See* tungsten hexacarbonyl.

tungsten disulfide WS_2 A grayish-black solid with a melting point above 1480°C; used as a lubricant and aerosol.

tungsten hexacarbonyl $W(CO)_6$ A white, refractive, crystalline solid which decomposes at 150°C; used for tungsten coatings on base metals. Also known as tungsten carbonyl.

tungsten hexachloride WCl_6 Dark blue or violet crystals with a melting point of 275°C; soluble in organic solvents; used for tungsten coatings on base metals and as a catalyst for olefin polymers.

tungsten lake *See* phosphotungstic pigment.

tungsten oxychloride $WOCl_4$ Dark red crystals with a melting point of approximately 211°C; soluble in carbon disulfide; used for incandescent lamps.

tungstic acid H_2WO_4 A yellow powder; insoluble in water, soluble in alkalies; used as a color-resist mordant for textiles, as an ingredient in plastics, and for the manufacture of tungsten metal products. Also known as orthotungstic acid; wolframic acid.

tungstic acid anhydride *See* tungstic oxide.

tungstic anhydride *See* tungstic oxide.

tungstic oxide WO_3 A heavy, canary-yellow powder; soluble in caustic, insoluble in water; melts at 1473°C; used in alloys, in fabric fireproofing, for ceramic pigments, and for the manufacture of tungsten metal. Also known as anhydrous wolframic acid; tungstic acid anhydride; tungstic anhydride; tungstic trioxide.

tungstic trioxide *See* tungstic oxide.

turbidimetric analysis A scattered-light procedure for the determination of the weight concentration of particles in cloudy, dull, or muddy solutions; uses a device that measures the loss in intensity of a light beam as it passes through the solution. Also known as turbidimetry.

turbidimetric titration Titration in which the end point is indicated by the developing turbidity of the titrated solution.

turbidimetry *See* tubidimetric analysis.

turbidity 1. Measure of the clarity (using APHA or colorimetric scales) of an otherwise clear liquid. 2. Cloudy or hazy appearance in a naturally clear liquid caused by a suspension of colloidal liquid droplets or fine solids.

turbogrid plate A tray for distillation columns that consists of a flat grid of parallel slots extending over the entire cross-sectional area of the column; the liquid level on each tray is maintained by a dynamic balance between down-flowing liquid and up-flowing vapor.

turnaround In petroleum refining, the shutdown of a unit after a normal run for maintenance and repair work, then putting the unit back into operation.

Turnbull's blue A blue pigment that precipitates from the reaction of potassium ferricyanide with a ferrous salt.

turnover number 1. The number of molecules of a substrate acted upon in a period of 1 minute by a single enzyme molecule, with the enzyme working at a maximum rate. 2. In an industrial catalytic process, a value that indicates the amount of feed or substrate converted per a measured amount of catalyst.

turnover rate In an industrial catalytic process, a value corresponding to the turnover number per specified unit of time.

turpentine camphor See terpene hydrochloride.

twin See twin crystal.

twin axis The crystal axis about which one individual of a twin crystal may be rotated (usually 180°) to bring it into coincidence with the other individual.

twin boundary A grain boundary whose lattice structures are mirror images of each other in the plane of the boundary.

twin crystal A compound crystal which has one or more parts whose lattice structure is the mirror image of that in the other parts of the crystal. Also known as twin.

twin law A statement relating two or more individuals of a twin to one another in terms of their crystallography (twin plane, twin axis, and so on).

twinning The development of a twin crystal by growth, translation, or gliding.

twinning plane See twin plane.

twin plane The plane common to and across which the individual crystals or components of a crystal twin are symmetrically arranged or reflected. Also known as twinning plane.

Twitchell reagent A catalyst for the acid hydrolysis of fats; a sulfonated addition product of naphthalene and oleic acid, that is, a naphthalenestearosulfonic acid.

two-dimensional chromatography A paper chromatography technique in which the sample is resolved by standard procedures (ascending, descending, or horizontal solvent movement) and then turned at right angles in a second solvent and re-resolved.

two-fluid cell Cell having different electrolytes at the positive and negative electrodes.

Tyler screen A screen standard for the openings in screen-type mediums based on meshes per linear inch; convertible to the U.S. Sieve Series.

tyrosinase An enzyme found in plants, molds, crustaceans, mollusks, and some bacteria which, in the presence of oxygen, catalyzes the oxidation of mono-phenols and polyphenols with the introduction of —OH groups and the formation of quinones.

tyrosine $C_9H_{11}NO_3$ A phenolic alpha amino acid found in many proteins; a precursor of the hormones epinephrine, norepinephrine, thyroxine, and triiodothyronine, and of the black pigment melanin.

tyvelose A dideoxy sugar found in bacterial lipopolysaccharides.

U

U center The color-center type of point lattice defect in ionic crystals created by the incorporation of an impurity such as hydrogen into alkali halides.

Udex process A proprietary petroleum refining process used to extract aromatic hydrocarbons from a complex hydrocarbon mixture; uses a glycol-water solvent.

UDMH *See uns*-dimethylhydrazine.

UDP *See* uridine diphosphate.

UDPG *See* uridine diphosphoglucose.

Ullmann reaction A variation of the Fittig synthesis, using copper powder instead of sodium.

ultimate analysis The determination of the percentage of elements contained in a chemical substance.

ultrafiltration Separation of colloidal or very fine solid materials by filtration through microporous or semipermeable mediums.

Ultrafining A proprietary fixed-bed catalytic hydrogenation process used in petroleum refineries to desulfurize and upgrade naphtha distillates.

Ultraforming A proprietary catalytic reforming process used to increase the octane ratings of petroleum naphthas.

ultramarine blue A blue pigment; a powder with heat resistance, used for enamels on toys and machinery, white baking enamels, printing inks, and cosmetics, and in textile printing.

ultraviolet absorption spectrophotometry The study of the spectra produced by the absorption of ultraviolet radiant energy during the transformation of an electron from the ground state to an excited state as a function of the wavelength causing the transformation.

ultraviolet densitometry An ultraviolet-spectrophotometry technique for measurement of the colors on thin-layer chromatography absorbents following elution.

ultraviolet photoemission spectroscopy A spectroscopic technique in which photons in the energy range 10–200 electronvolts bombard a surface and the energy spectrum of the emitted electrons gives information about the states of electrons in atoms and chemical bonding. Abbreviated UPS.

ultraviolet spectrometer A device which produces a spectrum of ultraviolet light and is provided with a calibrated scale for measurement of wavelength.

ultraviolet spectrophotometry Determination of the spectra of ultraviolet absorption by specific molecules in gases or liquids (for example, Cl_2, SO_2, NO_2, CS_2, ozone, mercury vapor, and various unsaturated compounds).

ultraviolet spectroscopy Absorption spectroscopy involving electromagnetic wavelengths in the range 4–400 nanometers.

ultraviolet stabilizer *See* UV stabilizer.

UMP *See* uridylic acid.

uncharged species A chemical entity with no net electric charge. Also known as neutral species.

uncompetitive enzyme inhibition The prevention of an enzymic process as a result of the interaction of an inhibitor with the enzyme-substrate complex or a subsequent intermediate form of the enzyme, but not with the free enzyme.

uncoupling phenomena Deviations of observed spectra from those predicted in a diatomic molecule as the magnitude of the angular momentum increases, caused by interactions which could be neglected at low angular momenta.

undecanal $CH_3(CH_2)_9CHO$ A sweet-smelling, colorless liquid, soluble in oils and alcohol; used in perfumes and flavoring. Also known as hendecanal; *n*-undecyclic aldehyde.

n-undecane $CH_3(CH_2)_9CH_3$ A colorless, combustible liquid, boiling at 367°F (196°C), flash point at 149°F (65°C); used as a chemical intermediate and in petroleum research.

undecanoic acid $CH_3(CH_2)_9COOH$ Colorless crystals, soluble in alcohol and ether, insoluble in water; melts at 29°C; used as a chemical intermediate. Also known as hendecanoic acid; *n*-undecylic acid.

2-undecanone *See* methyl nonyl ketone.

10-undecen-1-ol *See* undecylenic alcohol.

undecyl $C_{11}H_{23}$ The radical of undecane. Also known as hendecyl.

undecylenic acid $CH_2 \cdot CH(CH_2)_8COOH$ A light-colored, combustible liquid with a fruity aroma; soluble in alcohol, ether, chloroform, and benzene, almost insoluble in water; used in medicine, perfumes, flavors, and plastics.

n-undecylenic acid *See* undecylenic alcohol.

undecylenic alcohol $C_{11}H_{22}O$ A colorless liquid with a citrus odor, soluble in 70% alcohol; used in perfumes. Also known as alcohol C-11; 10-undecen-1-ol; *n*-undecylenic acid.

undecylenyl acetate $C_{13}H_{24}O_2$ A colorless liquid with a floral-fruity odor, soluble in 80% alcohol; used in perfumes and for flavoring. Also known as 10-hendecenyl acetate.

n-undecylic acid *See* undecanoic acid.

n-undecylic aldehyde *See* undecanal.

undercutting In distillation, the technique of taking the products coming off the distillation tower at a temperature below the desired ultimate boiling point range to prevent contaminating the products with the compound that would distill just beyond the ultimate boiling point range.

undersaturated fluid   Any fluid (liquid or gas) capable of holding additional vapor or liquid components in solution at specified conditions of pressure and temperature.

Underwood chart   A graphical solution of mass balances for a single equilibrium stage in the calculation of a solvent-extraction operation.

Underwood distillation method   A method for calculation of liquid separations from binary distillation systems operated at partial reflux.

unfired pressure vessel   A pressure vessel that is not in direct contact with a heating flame.

unimolecular reaction   A chemical reaction involving only one molecular species as a reactant; for example, $2H_2O \rightarrow 2H_2 + O_2$, as in the electrolytic dissociation of water.

Unisol process   A proprietary solvent-extraction process used in petroleum refineries to extract mercaptan sulfur and certain nitrogen compounds from sour gasolines or distillates.

unit cell   A parallelepiped which will fill all space under the action of translations which leave the crystal lattice unchanged. Also known as structure cell.

unit operations   The basic physical operations of chemical engineering in a chemical process plant, that is, distillation, fluid transport, heat and mass transfer, evaporation, extraction, drying, crystallization, filtration, mixing, size separation, crushing and grinding, and conveying.

unloading   **1.** The release downstream of a trapped contaminant. **2.** A filter medium failure and release of system pressure. **3.** The depressuring or emptying of a process unit.

unsaturated compound   Any chemical compound with more than one bond between adjacent atoms, usually carbon, and thus reactive toward the addition of other atoms at that point; for example, olefins, diolefins, and unsaturated fatty acids.

unsaturated hydrocarbon   One of a class of hydrocarbons that have at least one double or triple carbon-to-carbon bond that is not in an aromatic ring; examples are ethylene, propadiene, and acetylene.

unsaturation   A state in which the atomic bonds of an organic compound's chain or ring are not completely satisfied (that is, not saturated); usually applies to carbon, but can include other ring or chain atoms; unsaturation usually results in a double bond (as for olefins) or a triple bond (as for the acetylenes).

uns-, unsym-   A chemical prefix denoting that the substituents of an organic compound are structurally unsymmetrical with respect to the carbon skeleton, or with respect to a function group (for example, double or triple bond).

upflow   In an ion-exchange unit, an operation in which solutions enter at the bottom of the unit and leave at the top.

UPS   *See* ultraviolet photoemission spectroscopy.

upstream   That portion of a process stream that has not yet entered the system or unit under consideration; for example, upstream to a refinery or to a distillation column.

uracil $C_4H_4N_2O_2$ A pyrimidine base important as a component of ribonucleic acid.

uranate   A salt of uranic acid; for example, sodium uranate, Na_2UO_4.

urania   *See* uranium dioxide.

uranic chloride   *See* uranium tetrachloride.

uranic oxide   *See* uranium dioxide.

uranin   *See* uranine.

uranine $Na_2C_{20}H_{10}O_5$ A brown or orange-red hygroscopic powder soluble in water; used as a yellow dye for silk and wool, a marker in the ocean to facilitate air and sea rescues, and as an analytical reagent. Also known as sodium fluorescein; uranin; uranine yellow.

uranine yellow   *See* uranine.

uranium   A metallic element in the actinide series, symbol U, atomic number 92, atomic weight 238.03; highly toxic and radioactive; ignites spontaneously in air and reacts with nearly all nonmetals; melts at 1132°C, boils at 3818°C; used in nuclear fuel and as the source of uranium-235 and plutonium.

uranium acetate   *See* uranyl acetate.

uranium carbide   One of the carbides of uranium, such as uranium monocarbide; used chiefly as a nuclear fuel.

uranium decay series   *See* uranium series.

uranium dioxide UO_2 Black, highly toxic, spontaneously flammable, radioactive crystals; insoluble in water, soluble in nitric and sulfuric acids; melts at approximately 3000°C; used to pack nuclear fuel rods and in ceramics, pigments, and photographic chemicals. Also known as urania; uranic oxide; uranium oxide.

uranium hexafluoride UF_6 Highly toxic, radioactive, corrosive, colorless crystals; soluble in carbon tetrachloride, fluorocarbons, and liquid halogens; it reacts vigorously with alcohol, water, ether, and most metals, and it sublimes; used to separate uranium isotopes in the gaseous-diffusion process.

uranium hydride UH_3 A highly toxic, gray to black powder that ignites spontaneously in air, and that conducts electricity; used for making powdered uranium metal, for hydrogen-isotope separation, and as a reducing agent.

uranium nitrate   *See* uranyl nitrate.

uranium oxide   *See* uranium dioxide; uranium trioxide.

uranium-radium series   *See* uranium series.

uranium series   The series of nuclides resulting from the decay of uranium-238, including uranium I, II, X_1, X_2, Y, and Z, and radium A, B, C, C′, C″, D, E, E″, F, and G. Also known as uranium decay series; uranium-radium series.

uranium sulfate   *See* uranyl sulfate.

uranium tetrachloride UCl_4 Poisonous, radioactive, hygroscopic, dark-green crystals; soluble in alcohol and water; melts at 590°C, boils at 792°C. Also known as uranic chloride.

uranium tetrafluoride UF_4 Toxic, radioactive, corrosive green crystals; insoluble in water; melts at 1036°C; used in the manufacture of uranium metal. Also known as green salt.

uranium trioxide UO_3 A poisonous, radioactive, red to yellow powder; soluble in nitric acid, insoluble in water; decomposes when heated; used in ceramics and pigments and for uranium refining. Also known as orange oxide; uranium oxide.

uranous-uranic oxide *See* triuranium octoxide.

uranyl acetate $UO_2(C_2H_3O_2)_2 \cdot 2H_2O$ Poisonous, radioactive yellow crystals, decomposed by light; soluble in cold water, decomposes in hot water; loses water of crystallization at 110°C, decomposes at 275°C; used in medicine and as an analytical reagent and bacterial oxidant. Also known as uranium acetate.

uranyl nitrate $UO_2(NO_3)_2 \cdot 6H_2O$ Toxic, explosive, unstable yellow crystals; soluble in water, alcohol, and ether; melts at 60°C and boils at 118°C; used in photography, in medicine, and for uranium extraction and uranium glaze. Also known as uranium nitrate; yellow salt.

uranyl salts Salts of UO_3 that ionize to form UO_2^{2+} and that are yellow in solution; for example, uranyl chloride, UO_2Cl_2.

uranyl sulfate $UO_2SO_4 \cdot 3\frac{1}{2}H_2O$ and $UO_2SO_4 \cdot 3H_2O$ Poisonous, radioactive yellow crystals; soluble in water and concentrated hydrochloric acid; used as an analytical reagent. Also known as uranium sulfate.

uranyl uranate *See* triuranium octoxide.

urbacid $C_7H_{15}AsN_2S_3$ A colorless, crystalline compound with a melting point of 144°C; insoluble in water; used to control apple scale and diseases of coffee trees. Also known as bis(dimethylthiocarbamoylthio)methylarsine.

urea $CO(HN_2)_2$ A natural product of protein metabolism found in urine; synthesized as white crystals or powder with a melting point of 132.7°C; soluble in water, alcohol, and benzene; used as a fertilizer, in plastics, adhesives, and flameproofing agents, and in medicine. Also known as carbamide.

urea anhydride *See* cyanamide.

urea dewaxing A continuous, petroleum refinery process used to produce low-pour-point oils; urea forms a filterable solid complex (adduct) with the straight-chain wax paraffins in the stock.

urea-formaldehyde resin A synthetic thermoset resin derived by the reaction of urea (carbamide) with formaldehyde or its polymers. Also known as urea resin.

urea nitrate $CO(NH_2)_2 \cdot HNO_3$ Colorless, explosive, fire-hazardous crystals; soluble in alcohol, slightly soluble in water; decomposes at 152°C; used in explosives and to make urethane.

urea peroxide $CO(NH_2)_2 \cdot H_2O_2$ An unstable, fire-hazardous white powder; soluble in water, alcohol, and ethylene glycol; decomposes at 75–85°C or by moisture; used as a source of water-free hydrogen peroxide, as a disinfectant, in cosmetics and pharmaceuticals, and for bleaching. Also known as carbamide peroxide.

urease An enzyme that catalyzes the degradation of urea to ammonia and carbon dioxide; obtained from the seed of jack bean.

(*para*-ureidophenylarsylenedithio)diacetic acid *See* thiocarbarsone.

(*para*-ureidophenylarsylenedithio)di-*ortho*-benzoic acid *See* thiocarbamazine.

urethane $CO(NH_2)OC_2H_5$ A combustible, toxic, colorless powder; soluble in water and alcohol; melts at 49°C; used as a solvent and chemical intermediate and in biochemical research and veterinary medicine. Also known as ethyl carbamate; ethyl urethane.

uric acid $C_5H_4N_4O_3$ A white, crystalline compound, the excretory end product in amino acid metabolism by uricotelic species. Also known as 8-hydroxyxanthine.

uricase An enzyme present in the liver, spleen, and kidney of most mammals except man; converts uric acid to allantoin in the presence of gaseous oxygen.

uridine $C_9H_{12}N_2O_6$ A crystalline nucleoside composed of one molecule of uracil and one molecule of D-ribose; a component of ribonucleic acid.

uridine diphosphate The chief transferring coenzyme in carbohydrate metabolism. Abbreviated UDP.

uridine diphosphoglucose A compound in which α-glucopyranose is esterified, at carbon atom 1, with the terminal phosphate group of uridine-5'-pyrophosphate (that is, uridine diphosphate); occurs in animal, plant, and microbial cells; functions as a key in the transformation of glucose to other sugars. Abbreviated UDPG.

uridine monophosphate *See* uridylic acid.

uridine phosphoric acid *See* uridylic acid.

uridylic acid $C_9H_{13}N_2O_9P$ Water- and alcohol-soluble crystals, melting at 202°C; used in biochemical research. Also known as uridine monophosphate (UMP); uridine phosphoric acid.

urobilin A bile pigment produced by reduction of bilirubin by intestinal bacteria and excreted by the kidneys or removed by the liver.

urobilinogen A chromogen, formed in feces and present in urine, from which urobilin is formed by oxidation.

urocanic acid $C_6H_6N_2O_2$ A crystalline compound formed as an intermediate in the degradative pathway of histidine.

urochrome $C_{43}H_{51}O_{26}N$ Yellow pigment found in normal urine.

urokinase An enzyme, present in human urine, that catalyzes the conversion of plasminogen to plasmin.

uronic acid One of the compounds that are similar to sugars, except that the terminal carbon has been oxidized from the alcohol to a carboxyl group; for example, galacturonic acid and glucuronic acid.

uropepsin The end product of the secretion of pepsinogen into the blood by gastric cells; occurs in urine.

uroporphyrin Any of several isomeric, metal-free porphyrins, occurring in small quantities in normal urine and feces; molecule has four acetic acid ($—CH_2COOH$) and four propionic acid ($—CH_2CH_2COOH$) groups.

urotropin *See* cystamine.

ursolic acid $C_{30}H_{48}O_3$ A pentacyclic terpene that crystallizes from absolute alcohol solution, found in leaves and berries of plants; used in pharmaceutical and food industries as an emulsifying agent.

usnic acid $C_{18}H_{16}O_7$ Yellow crystals, insoluble in water, slightly soluble in alcohol and ether, melts about 198°C; found in lichens; used as an antibiotic. Also known as usninic acid.

usninic acid *See* usnic acid.

USP acid test A United States Pharmacopoeia test to determine the carbonizable substances present in petroleum white oils.

U-tube heat exchanger A heat-exchanger system consisting of a bundle of U tubes (hairpin tubes) surrounded by a shell (outer vessel); one fluid flows through the tubes, and the other fluid flows through the shell, around the tubes.

UV stabilizer Any chemical compound that, admixed with a thermoplastic resin, selectively absorbs ultraviolet rays; used to prevent ultraviolet degradation of polymers. Also known as ultraviolet stabilizer.

V

vacuum condensing point Temperature at which the sublimate (vaporized solid) condenses in a vacuum. Abbreviated vcp.

vacuum crystallizer Crystallizer in which a warm saturated solution is fed to a lagged, closed vessel maintained under vacuum; the solution evaporates and cools adiabatically, resulting in crystallization.

vacuum distillation Liquid distillation under reduced (less than atmospheric) pressure; used to lower boiling temperatures and lessen the risk of thermal degradation during distillation. Also known as reduced-pressure distillation.

vacuum flashing The heating of a liquid that, upon release to a lower pressure (vacuum), undergoes considerable vaporization (flashing). Also known as flash vaporization.

vacuum pan salt A salt made from salt brine boiled at reduced pressure in a triple-effect evaporator.

vacuum thermobalance An instrument used in thermogravimetry consisting of a precision balance and furnace that have been adapted for continuously measuring or recording changes in weight of a substance as a function of temperature; used in many types of physicochemical reactions where rates of reaction and energies of activation for vaporization, sublimation, and chemical reaction can be obtained.

vacuum-type insulation Highly reflective double-wall structure with high vacuum between the walls; used as insulation for cryogenic systems; Dewar flasks have vacuum-type insulation.

vacuum ultraviolet spectroscopy Absorption spectroscopy involving electromagnetic wavelengths shorter than 200 nanometers; so called because the interference of the high absorption of most gases necessitates work with evacuated equipment.

vagotonine An endocrine substance which is thought to be elaborated by cells of the pancreas and which regulates autonomic tonus.

valence 1. The relative ability of a biological substance to react or combine. 2. A positive number that characterizes the combining power of an element for other elements, as measured by the number of bonds to other atoms which one atom of the given element forms upon chemical combination; hydrogen is assigned valence 1, and the valence is the number of hydrogen atoms, or their equivalent, with which an atom of the given element combines.

valence angle *See* bond angle.

valence bond The bond formed between the electrons of two or more atoms.

valence-bond method   A method of calculating binding energies and other parameters of molecules by taking linear combinations of electronic wave functions, some of which represent covalent structures, others ionic structures; the coefficients in the linear combination are calculated by the variational method. Also known as valence-bond resonance method.

valence-bond resonance method   *See* valence-bond method.

valence-bond theory   A theory of the structure of chemical compounds according to which the principal requirements for the formation of a covalent bond are a pair of electrons and suitably oriented electron orbitals on each of the atoms being bonded; the geometry of the atoms in the resulting coordination polyhedron is coordinated with the orientation of the orbitals on the central atom.

valence electron   An electron that belongs to the outermost shell of an atom.

valence number   A number that is equal to the valence of an atom or ion multiplied by $+1$ or -1, depending on whether the ion is positive or negative, or equivalently on whether the atom in the molecule under consideration has lost or gained electrons from its free state.

valence shell   The electrons that form the outermost shell of an atom.

valence transition   A change in the electronic occupation of the $4f$ or $5f$ orbitals of the rare-earth or actinide atoms in certain substances at a certain temperature, pressure, or composition.

valeral   *See* n-valeraldehyde.

n-valeraldehyde $CH_3(CH_2)_3CHO$ A flammable liquid, soluble in ether and alcohol, slightly soluble in water; boils at 102°C; used in flavors and as a rubber accelerator. Also known as amyl aldehyde; valeral; valeric aldehyde.

valeramide $CH_3(CH_2)_3CONH_2$ Water-soluble, colorless crystals, melting at 127°C. Also known as pentanamide; valeric amide.

valerianic acid   *See* valeric acid.

valeric acid $CH_3(CH_2)_3COOH$ A combustible, toxic, colorless liquid with a penetrating aroma; soluble in water, alcohol, and ether; boils at 185°C; used to make flavors, perfumes, lubricants, plasticizers, and pharmaceuticals. Also known as n-pentanoic acid; valerianic acid.

valeric aldehyde   *See* n-valeraldehyde.

valeric amide   *See* valeramide.

γ-valerolactone $C_5H_8O_2$ A combustible, mostly immiscible, colorless liquid, boiling at 205°C; used as a dye-bath coupling agent, in brake fluids and cutting oils, and as a solvent for adhesives, lacquers, and insecticides.

valine $C_5H_{11}NO_2$ An amino acid considered essential for normal growth of animals, and biosynthesized from pyruvic acid. Also known as 2-aminoisovaleric acid; α-aminoisovaleric acid; 2-amino-3-methylbutyric acid.

value of isotope mixture   A measure of the effort required to prepare a quantity of an isotope mixture; it is proportional to the amount of the mixture, and also depends on the composition of the mixture to be prepared and the composition of the original mixture.

vamidothion $C_7H_{16}NO_4PS_2$ A white wax with a melting point of 40°C; very soluble in water; used to control pests in orchards, vineyards, rice, cotton, and ornamentals. Also known as O,O-dimethyl-S-{2-[(1-methylcarbamoyl-ethyl)thio]ethyl}phosphorodithioate.

vanadic acid   Any of various acids that do not exist in a pure state and are found in various alkali and other metal vanadates; forms are meta- (HVO_3), ortho- (H_3VO_4), and pyro- ($H_4V_2O_7$).

vanadic acid anhydride   *See* vanadium pentoxide.

vanadic sulfate   *See* vanadyl sulfate.

vanadic sulfide   *See* vanadium sulfide.

vanadium   A metal in group Vb, symbol V, atomic number 23; soluble in strong acids and alkalies; melts at 1900°C, boils about 3000°C; used as a catalyst.

vanadium carbide   VC   Hard, black crystals, melting at 2800°C, boiling at 3900°C; insoluble in acids, except nitric acid; used in cutting-tool alloys and as a steel additive.

vanadium dichloride VCl_2 Toxic, green crystals, soluble in alcohol and ether; decomposes in hot water; used as a reducing agent. Also known as vanadous chloride.

vanadium oxide   A compound of vanadium with oxygen, for example, vanadium tetroxide (V_2O_4), vanadium trioxide or sesquioxide (V_2O_3), vanadium oxide (VO), and vanadium pentoxide (V_2O_5).

vanadium oxydichloride   *See* vanadyl chloride.

vanadium oxytrichloride $VOCl_3$ A toxic, yellow liquid that dissolves or reacts with many organic substances; hydrolyzes in moisture; boils at 126°C; used as an olefin-polymerization catalyst and in organovanadium synthesis.

vanadium pentasulfide   *See* vanadium sulfide.

vanadium pentoxide V_2O_5 A toxic, yellow to red powder, soluble in alkalies and acids, slightly soluble in water; melts at 690°C; used in medicine, as a catalyst, as a ceramics coloring, for ultraviolet-resistant glass, photographic developers, textiles dyeing, and nuclear reactors. Also known as vanadic acid anhydride.

vanadium sesquioxide   *See* vanadium trioxide.

vanadium sulfate   *See* vanadyl sulfate.

vanadium sulfide V_2S_5 A toxic, black-green powder; insoluble in water, soluble in alkalies and acids; decomposes when heated; used to make vanadium compounds. Also known as vanadic sulfide; vanadium pentasulfide.

vanadium tetrachloride VCl_4 A toxic, red liquid; soluble in ether and absolute alcohol; boils at 154°C; used in medicine and to manufacture vanadium and organovanadium compounds.

vanadium tetraoxide V_2O_4 A toxic blue-black powder; insoluble in water, soluble in alkalies and acids; melts at 1967°C; used as a catalyst.

vanadium trichloride VCl_3 Toxic, deliquescent, pink crystals; soluble in ether and absolute alcohol; decomposes in water and when heated; used to prepare vanadium and organovanadium compounds.

vanadium trioxide V_2O_3 Toxic, black crystals; soluble in alkalies and hydrofluoric acid, slightly soluble in water; melts at 1970°C; used as a catalyst. Also known as vanadium sesquioxide.

vanadous chloride *See* vanadium dichloride.

vanadyl chloride $V_2O_2Cl_4 \cdot 5H_2O$ Toxic, deliquescent, water- and alcohol-soluble green crystals; used to mordant textiles. Also known as divanadyl tetrachloride; vanadium oxydichloride; vanadyl dichloride.

vanadyl dichloride *See* vanadyl chloride.

vanadyl sulfate $VOSO_4 \cdot 2H_2O$ Blue, toxic, water-soluble crystals; used as a reducing agent, catalyst, glass and ceramics colorant, and mordant. Also known as vanadic sulfate; vanadium sulfate.

Van Deemter rate theory A theory that the sample phase in gas chromatography flows continuously, not stepwise.

van der Waals adsorption Adsorption in which the cohesion between gas and solid arises from van der Waals forces.

van der Waals attraction *See* van der Waals force.

van der Waals covolume The constant b in the Van der Waals equation, which is approximately four times the volume of an atom of the gas in question multiplied by Avogadro's number.

van der Waals equation An empirical equation of state which takes into account the finite size of the molecules and the attractive forces between them: $p = [RT/ (v - b)] - (a/v^2)$, where p is the pressure, v is the volume per mole, T is the absolute temperature, R is the gas constant, and a and b are constants.

van der Waals force An attractive force between two atoms or nonpolar molecules, which arises because a fluctuating dipole moment in one molecule induces a dipole moment in the other, and the two dipole moments then interact. Also known as dispersion force; London dispersion force; van der Waals attraction.

van der Waals–London interactions The interaction associated with the van der Waals force.

van der Waals structure The structure of a molecular crystal.

vanillin $C_8H_8O_3$ A combustible solid, soluble in water, alcohol, ether, and chloroform; melts at 82°C; used in pharmaceuticals, perfumes, and flavors, and as an analytical reagent. Also known as vanillic aldehyde.

van't Hoff equation An equation for the variation with temperature T of the equilibrium constant K of a gaseous reaction in terms of the heat of reaction at constant pressure, ΔH: $d(\ln K)/dT = \Delta H/RT^2$, where R is the gas constant. Also known as van't Hoff isochore.

van't Hoff formula The expression that the number of stereoisomers of a sugar molecule is equal to 2^n, where n is the number of asymmetric carbon atoms.

van't Hoff isochore *See* van't Hoff equation.

van't Hoff isotherm An equation for the change in free energy during a chemical reaction in terms of the reaction, the temperature, and the concentration and number of molecules of the reactants.

vaporizer A process vessel in which a liquid is heated until it vaporizes; heat can be indirect (steam or heat-transfer fluid) or direct (hot gases or submerged combustion).

vapor-liquid equilibrium *See* liquid-vapor equilibrium.

vapor-liquid separation   The removal of liquid droplets from a flowing stream of gas or vapor; accomplished by impingement, cyclonic action, and absorption or adsorption operations.

vapor-phase reactor   A heavy steel vessel for carrying out chemical reactions on an industrial scale where efficient control over a vapor phase is needed, for example, in an oxidation process.

vapor-pressure osmometer   A device for the determination of molecular weights by the decrease of vapor pressure of a solvent upon addition of a soluble sample.

vapor rate   In distillation, the upward flow rate of vapor through a distillation column.

vasopressin   A peptide hormone which is elaborated by the posterior pituitary and which has a pressor effect; used medicinally as an antidiuretic. Also known as antidiuretic hormone (ADH).

vasotocin   A hormone from the neurosecretory cells of the posterior pituitary of lower vertebrates; increases permeability to water in amphibian skin and in bladder.

V band   Absorption bands that appear in the ultraviolet part of the spectrum due to color centers produced in potassium bromide by exposure of the crystal at temperature of liquid nitrogen (81 K) to intense penetrating x-rays.

vcp   *See* vacuum condensing point.

vector model of atomic structure   A model of atomic structure in which spin and orbital angular momenta of the electrons are represented by vectors, with special rules for their addition imposed by underlying quantum-mechanical considerations.

vegetable diastase   *See* diastase.

Venetian red   A pigment with a true red hue; contains 15–40% ferric oxide and 60–80% calcium sulfate.

venturi scrubber   A gas-cleaning device in which liquid injected at the throat of a venturi is used to scrub dust and mist from the gas flowing through the venturi.

verdigris   *See* cupric acetate.

vermilion   *See* mercuric sulfide.

vernine   *See* guanosine.

vernolate $C_{10}H_{21}NOS$ An amber liquid, used to control weeds in sweet potatoes, peanuts, soybeans, and tobacco. Also known as S-propyldipropylthiocarbamate.

VI   *See* viscosity index.

vibrational energy   For a diatomic molecule, the difference between the energy of the molecule idealized by setting the rotational energy equal to zero, and that of a further idealized molecule which is obtained by gradually stopping the vibration of the nuclei without placing any new constraint on the motions of electrons.

vibrational level   An energy level of a diatomic or polyatomic molecule characterized by a particular value of the vibrational energy.

vibrational quantum number   A quantum number v characterizing the vibrational motion of nuclei in a molecule; in the approximation that the molecule behaves as a quantum-mechanical harmonic oscillator, the vibrational energy is $h(v + \frac{1}{2})f$, where h is Planck's constant and f is the vibration frequency.

vibrational spectrum   The molecular spectrum resulting from transitions between vibrational levels of a molecule which behaves like the quantum-mechanical harmonic oscillator.

vibrational sum rule   **1.** The rule that the sums of the band strengths of all emission bands with the same upper state is proportional to the number of molecules in the upper state, where the band strength is the emission intensity divided by the fourth power of the frequency. **2.** The sums of the band strengths of all absorption bands with the same lower state is proportional to the number of molecules in the lower state, where the band strength is the absorption intensity divided by the frequency.

vic-   A chemical prefix indicating vicinal (neighboring or adjoining) positions on a carbon structure (ring or chain); used to identify the location of substituting groups when naming derivatives.

vicinal   Referring to neighboring or adjoining positions on a carbon structure (ring or chain).

vicinal faces   Macroscopic crystal faces which are inclined only a few minutes of arc to crystal faces with low Miller indices, and which therefore must have high Miller indices themselves.

Victoria blue $C_{33}H_{31}N_3 \cdot HCl$ Bronze crystals, soluble in hot water, alcohol, and ether; used as a dye for silk, wool, and cotton, as a biological stain, and to make pigment toners.

Vigreaux column   An obsolete apparatus used in laboratory fractional distillation; it is a long glass tube with indentation in its walls; a thermometer is placed at the top of the tube and a side arm is attached to a condenser.

vinetine   *See* oxyacanthine.

vinyl acetal resin $[CH_2CH(OC_2H_5)]_x$ A colorless, odorless, light-stable thermoplastic that is unaffected by water, gasoline, or oils; soluble in lower alcohols, benzene, and chlorinated hydrocarbons; used in lacquers, coatings, and molded objects. Also known as polyvinyl acetal resin.

vinyl acetate $CH_3COOCH:CH_2$ A colorless, water-insoluble, flammable liquid that boils at 73°C; used as a chemical intermediate and in the production of polymers and copolymers (for example, the polyvinyl resins).

vinyl acetate resin $(CH_2:CHOOCCH_3)_x$ An odorless thermoplastic formed by the polymerization of vinyl acetate; resists attack by water, gasoline, and oils; soluble in lower alcohols, benzene, and chlorinated hydrocarbons; used in lacquers, coatings, and molded products.

vinylacetonitrile   *See* allyl cyanide.

vinylacetylene $H_2CCHCCH$ A combustible dimer of acetylene, boiling at 5°C; used for the manufacture of neoprene rubber and as a chemical intermediate.

vinyl alcohol $CH_2:CHOH$ A flammable, unstable liquid found only in ester or polymer form. Also known as ethenol.

vinylation   Formation of a vinyl-derived product by reaction with acetylene; for example, vinylation of alcohols gives vinyl ethers, such as vinyl ethyl ether.

vinylbenzene *See* styrene.

vinyl chloride CH_2:CHCl A flammable, explosive gas with an ethereal aroma; soluble in alcohol and ether, slightly soluble in water; boils at $-14°C$; an important monomer for polyvinyl chloride and its copolymers; used in organic synthesis and in adhesives. Also known as chloroethene; chloroethylene.

vinyl chloride resin $(CH_2CHCl)_x$ A white-power polymer made by the polymerization of vinyl chloride; used to make chemical-resistant pipe (when unplasticized) or bottles and parts (when plasticized).

vinylcyanide *See* acrylonitrile.

vinyl ether CH_2:CHOCH:CH_2 A colorless, light-sensitive, flammable, explosive liquid; soluble in alcohol, acetone, ether, and chloroform, slightly soluble in water; boils at 39°C; used as an anesthetic and a comonomer in polyvinyl chloride polymers. Also known as divinyl ether; divinyl oxide.

vinyl ether resin Any of a group of vinyl ether polymers; for example, polyvinyl methyl ether, polyvinyl ethyl ether, and polyvinyl butyl ether.

vinyl group CH_2=CH— A group of atoms derived when one hydrogen atom is removed from ethylene.

vinylidene chloride CH_2:CCl_2 A colorless, flammable, explosive liquid, insoluble in water; boils at 37°C; used to make polymers copolymerized with vinyl chloride or acrylonitrile (Saran).

vinylidene resin A polymer made up of the (—H_2CCX_2—) unit, with X usually a chloride, fluoride, or cyanide radical. Also known as polyvinylidene resin.

vinylog Any of the organic compounds that differ from each other by a vinylene linkage (—CH=CH—); for example, ethyl crotonate is a vinylog of ethyl acetate and of the next higher vinylog, ethyl sorbate.

vinyl plastic *See* polyvinyl resin.

vinyl polymerization Addition polymerization where the unsaturated monomer contains a CH_2=C— group.

vinylpyridine C_5H_4NCH:CH_2 A toxic, combustible liquid; soluble in water, alcohol, hydrocarbons, esters, ketones, and dilute acids; used to manufacture elastomers and pharmaceuticals.

***N*-vinyl-2-pyrrolidone** C_6H_9ON A colorless, toxic, combustible liquid, boiling at 148°C (100 mm Hg); used as a chemical intermediate and to make polyvinyl pyrrolidone.

vinylstyrene *See* divinylbenzene.

vinyltoluene CH_2:$CHC_6H_4CH_3$ A colorless, flammable, moderately toxic liquid; soluble in ether and methanol, slightly soluble in water; boils at 170°C; used as a chemical intermediate and solvent. Also known as methyl styrene.

vinyl trichloride *See* trichloroethane.

vinyl trichlorosilone $CH_2CH_3SiCl_3$ A liquid that boils at 90.6°C and is soluble in organic solvents; used in silicones and adhesives.

virgin *See* straight-run.

virotoxin One of a group of toxins present in the mushroom *Amanita virosa*.

virtual level The energy of a virtual state.

virtual orbital An orbital that is either empty or unoccupied while in the ground state.

virtual state An unstable state of a compound nucleus which has a lifetime many times longer than the time it takes a nucleon, with the same energy as it has in the virtual state, to cross the nucleus.

visbreaking *See* viscosity breaking.

viscose process A process for the manufacture of rayon by treating cellulose with caustic soda, and with carbon disulfide to form cellulose xanthate, which is then dissolved in a weak caustic solution to form the viscose; fibers are used as silk substitutes.

viscosity blending chart A graphical means for estimating the viscosity at a given temperature of a blend of petroleum products.

viscosity breaking A petroleum refinery process used to lower or break the viscosity of high-viscosity residuum by thermal cracking of molecules at relatively low temperatures. Also known as visbreaking.

viscosity conversion table A table or chart with which kinematic viscosity, in centistokes, can be converted to Saybolt viscosity, in seconds, at the same temperature.

viscosity-gravity constant An index of the chemical composition of crude oil; defined as the general relation between specific gravity and Saybolt Universal viscosity; the constant is low for paraffinic crude oils, high for naphthenic crude oils. Abbreviated VGC.

viscosity index An arbitrary scale used to show the magnitude of viscosity changes in lubricating oils with changes in temperature. Abbreviated VI.

viscosity-temperature chart A chart with which the kinematic or Saybolt viscosity of a petroleum oil at any temperature within a limited range may be ascertained, provided viscosities at two temperatures are known.

visible absorption spectrophotometry Study of the spectra produced by the absorption of visible-light energy during the transformation of an electron from the ground state to an excited state as a function of the wavelength causing the transformation.

visible spectrophotometry In spectrophotometric analysis, the use of a spectrophotometer with a tungsten lamp that has an electromagnetic spectrum of 380–780 nanometers as a light source, glass or quartz prisms or gratings in the monochromator, and a photomultiplier cell as a detector.

visible spectrum 1. The range of wavelengths of visible radiation. 2. A display or graph of the intensity of visible radiation emitted or absorbed by a material as a function of wavelength or some related parameter.

visual colorimetry A procedure for the determination of the color of an unknown solution by visual comparison to color standards (solutions or color-tinted disks).

visual pigment Any of various photosensitive pigments of vertebrate and invertebrate photoreceptors.

visual purple *See* rhodopsin.

visual yellow   An intermediary substance formed from rhodopsin in the retina after exposure to light; it is ultimately broken down to retinene and vitamin A.

vitamin   An organic compound present in variable, minute quantities in natural foodstuffs and essential for the normal processes of growth and maintenance of the body; vitamins do not furnish energy, but are essential for energy transformation and regulation of metabolism.

vitamin A $C_{20}H_{29}OH$ A pale-yellow alcohol that is soluble in fat and insoluble in water; found in liver oils and carotenoids, and produced synthetically; it is a component of visual pigments and is essential for normal growth and maintenance of epithelial tissue. Also known as antiinfective vitamin; antixerophthalmic vitamin; retinol.

vitamin A aldehyde   *See* retinal.

vitamin B₁   *See* thiamine.

vitamin B₂   *See* riboflavin.

vitamin B₃   *See* pantothenic acid.

vitamin B₆   A vitamin which exists as three chemically related and water-soluble forms found in food: pyridoxine, pyridoxal, and pyridoxamine; dietary requirements and physiological activities are uncertain.

vitamin B₁₂   A group of closely related polypyrrole compounds containing trivalent cobalt; the antipernicious anemia factor, essential for normal hemopoiesis. Also known as cobalamin; cyanocobalamin; extrinsic factor.

vitamin B complex   A group of water-soluble vitamins that include thiamine, riboflavin, nicotinic acid, pyridoxine, pantothenic acid, inositol, *p*-aminobenzoic acid, biotin, folic acid, and vitamin B_{12}.

vitamin B₆ hydrochloride   *See* pyridoxine hydrochloride.

vitamin B₂ phosphate   *See* riboflavin 5′-phosphate.

vitamin C   *See* ascorbic acid.

vitamin D   Either of two fat-soluble, sterol-like compounds, calciferol or ergocalciferol (vitamin D_2) and cholecalciferol (vitamin D_3); occurs in fish liver oils and is essential for normal calcium and phosphorus deposition in bones and teeth. Also known as antirachitic vitamin.

vitamin E $C_{29}H_{50}O_2$ Any of a series of eight related compounds called tocopherols, α-tocopherol having the highest biological activity; occurs in wheat germ and other oils and is believed to be needed in certain human physiological processes.

vitamin G   *See* riboflavin.

vitamin K   Any of three yellowish oils which are fat-soluble, nonsteroid, and nonsaponifiable; it is essential for formation of prothrombin. Also known as antihemorrhagic vitamin; prothrombin factor.

vitamin K₁   *See* phytonadione.

vitamin P   A substance, such as citrin or one or more of its components, believed to be concerned with maintenance of the normal state of the walls of small blood vessels.

vitamin P complex *See* bioflavanoid.

volatile Readily passing off by evaporation.

volatile fluid A liquid with the tendency to become vapor at specified conditions of temperature and pressure.

volatility product The product of the concentrations of two or more molecules or ions that react to form a volatile substance.

Volhard titration Determination of the halogen content of a solution by titration with a standard thiocyanate solution.

voltametry Any electrochemical technique in which a faradaic current passing through the electrolysis solution is measured while an appropriate potential is applied to the polarizable or indicator electrode; for example, polarography.

Volta series *See* displacement series.

volume phase *See* surface phase.

volume shift *See* field shift.

volume susceptibility The magnetic susceptibility of a specified volume (for example, 1 cubic centimeter) of a magnetically susceptible material.

volumetric analysis Quantitative analysis of solutions of known volume but unknown strength by adding reagents of known concentration until a reaction end point (color change or precipitation) is reached; the most common technique is by titration.

volumetric flask A laboratory flask primarily intended for the preparation of definite, fixed volumes of solutions, and therefore calibrated for a single volume only.

volumetric pipet A graduated glass tubing used to measure quantities of a solution; the tube is open at the top and bottom, and a slight vacuum (suction) at the top pulls liquid into the calibrated section; breaking the vacuum allows liquid to leave the tube.

volutin A basophilic substance, thought to be a nucleic acid, occurring as granules in the cytoplasm and vacuoles of algae and other microorganisms.

Vorce diaphragm cell A cylindrical cell with graphite anodes and asbestos-covered cathode, used in the electrolytic process for the manufacture of chlorine.

vulcanization A chemical reaction of sulfur (or other vulcanizing agent) with rubber or plastic to cause cross-linking of the polymer chains; it increases strength and resiliency of the polymer.

Wacker process A process for the oxidation of ethylene to acetaldehyde by oxygen in the presence of palladium chloride and cupric chloride.

Wagner's reagent An aqueous solution of iodine and potassium iodide; used for microchemical analysis of alkaloids. Also known as Wagner's solution.

Wagner's solution *See* Wagner's reagent.

Walden's rule A rule which states that the product of the viscosity and the equivalent ionic conductance at infinite dilution in electrolytic solutions is a constant, independent of the solvent; it is only approximately correct.

Wallach transformation By the use of concentrated sulfuric acid, an azoxybenzene is converted into a *para*-hydroxyazobenzene.

warfarin $C_{19}H_{16}O_4$ A white, crystalline compound with a melting point of 161°C; insoluble in water; used as a rodenticide. Also known as 3-(α-acetonylbenzyl)-4-hydroxy-coumarin; coumafene.

washing **1.** In the purification of a laboratory sample, the cleaning of residual liquid impurities from precipitates by adding washing solution to the precipitates, mixing, then decanting, and repeating the operation as often as needed. **2.** The removal of soluble components from a mixture of solids by using the effect of differential solubility. **3.** In a process operation, cleaning of a solids bed (settler) or cake (filter) with a liquid in which the solid is not soluble.

washing soda *See* sal soda.

wash water Water contacted with process streams (liquid or gas), packed beds, or filter cakes to flush or dissolve out impurities.

waste-heat boiler A heat-retrieval unit using hot by-product gas or oil from chemical processes; used to produce steam in a boiler-type system. Also known as gas-tube boiler.

water H_2O Clear, odorless, tasteless liquid that is essential for most animal and plant life and is an excellent solvent for many substances; melting point 0°C (32°F), boiling point 100°C (212°F); the chemical compound may be termed hydrogen oxide.

water absorption tube A glass tube filled with a solid absorbent (calcium chloride or silica gel) to remove water from gaseous streams during or after chemical analyses.

watercolor pigment A permanent pigment used in watercolor painting, for example, titanium oxide (white).

water demineralizing The removal of minerals (for example, compounds of Ca, Mg, and Na) from water by chemical, ion-exchange, or distillation procedures.

water gas reaction A method used to prepare carbon monoxide by passing steam over hot coke or coal at 600–1000°C.

water glass *See* sodium silicate.

water of hydration Water present in a definite amount and attached to a compound to form a hydrate; can be removed, as by heating, without altering the composition of the compound.

water saturation 1. A solid adsorbent that holds the maximum possible amount of water under specified conditions. 2. A liquid solution in which additional water will cause the appearance of a second liquid phase. 3. A gas that is at or just under its dew point because of its water content.

water scrubber A device or system in which gases are contacted with water (either by spray or bubbling through) to wash out traces of water-soluble components of the gas stream.

water softening Removal of scale-forming calcium and magnesium ions from hard water, or replacing them by the more soluble sodium ions; can be done by chemicals or ion exchange.

water-wettable Denoting the capability of a material to accept water, or of being hydrophilic or hydrophoric.

water white A grade of color for liquids that has the appearance of clear water; for petroleum products, a plus 21 in the scale of the Saybolt chromometer.

Watson equation Calculation method to extend heat of vaporization data for organic compounds to within 10 or 15°C of the critical temperature; uses known latent heats of vaporization and reduced temperature data.

wavelength standards Accurately measured lengths of waves emitted by specified light sources for the purpose of obtaining the wavelengths in other spectra by interpolating between the standards.

wax fractionation A continuous solvent-recovery/crystallization petroleum-refinery process for the production of waxes with low oil content from wax concentrates; for example, MEK (methyl ethyl ketone) deoiling.

wax manufacturing A petroleum refinery process similar to wax fractionation for the manufacture of oil-free waxes by chilling and crystallization from a solvent.

weak acid An acid that does not ionize greatly; for example, acetic acid or carbonic acid.

Weber number 3 A dimensionless number used in interfacial area determination in distillation equipment, equal to the surface tension divided by the product of the liquid density, the acceleration of gravity, and the depth of liquid on the tray under consideration. Symbolized N_{We3}.

wedge spectrograph A spectrograph in which the intensity of the radiation passing through the entrance slit is varied by moving an optical wedge.

Weiss magneton A unit of magnetic moment, equal to 1.853×10^{-21} erg/oersted, about one-fifth of the Bohr magneton; it is experimentally derived, the magnetic moments of certain molecules being close to integral multiples of this quantity.

Weisz ring oven   A device for vaporization of solvent from filter paper, leaving the solute in a ring (circular) shape; used for qualitative analysis of very small samples.

Werner band   A band in the ultraviolet spectrum of molecular hydrogen extending from 116 to 125 nanometers.

Werner complex   *See* coordination compound.

wet ashing   The conversion of an organic compound into ash (decomposition) by treating the compound with nitric or sulfuric acid.

wettability   The ability of any solid surface to be wetted when in contact with a liquid; that is, the surface tension of the liquid is reduced so that the liquid spreads over the surface.

wetted   Pertaining to material that has accepted water or other liquid, either on its surface or within its pore structure.

wetted-wall column   A vertical column that operates with the inner walls wetted by the liquid being processed; used in theoretical studies of mass transfer rates and in analytical distillations; an example is a spinning-band column.

white copperas   *See* zinc sulfate.

white lead   Basic lead carbonate of variable composition, the oldest and most important lead paint pigment; also used in putty and ceramics.

white phosphorus   The element phosphorus in its allotropic form, a soft, waxy, poisonous solid melting at 44.5°C; soluble in carbon disulfide, insoluble in water and alcohol; self-igniting in air. Also known as yellow phosphorus.

white vitriol   *See* zinc sulfate.

Wiedemann's additivity law   The law that the mass (or specific) magnetic susceptibility of a mixture or solution of components is the sum of the proportionate (by weight fraction) susceptibilities of each component in the mixture.

Wien effect   An increase in the conductance of an electrolyte at very high potential gradients.

Wigner nuclides   The most important class of mirror nuclides, comprising pairs of odd-mass-number isobars for which the atomic number and the neutron number differ by 1.

Wigner-Seitz cell   A polyhedron about an atom in a face-centered cubic structure, made by drawing planes which perpendicularly bisect the lines to the nearest neighbors; in a body-centered cubic structure, bisecting planes of lines to nearest neighbors and next-nearest neighbors are used; such polyhedra fill space.

Wigner supermultiplet   A set of quantum-mechanical states of a collection of nucleons which form the basis of a representation of SU(4), especially appropriate when spin and isospin dependence of the nuclear interaction may be disregarded; several combinations of spin and isospin multiplets may occur in a supermultiplet.

Wijs' iodine monochloride solution   A solution in glacial acetic acid of iodine monochloride; used to determine iodine numbers. Also known as Wijs' special solution.

Wijs' special solution *See* Wijs' iodine monochloride solution.

Williamson synthesis The synthesis of ethers utilizing an alkyl iodide and sodium alcoholate.

Winkler titration A chemical method for estimating the dissolved oxygen in seawater: manganous hydroxide is added to the sample and reacts with oxygen to produce a manganese compound which in the presence of acid potassium iodide liberates an equivalent quantity of iodine that can be titrated with standard sodium thiosulfate.

wintergreen oil *See* methyl salicylate.

Wittig ether rearrangement The rearrangement of benzyl and alkyl ethers when reacted with a methylating agent, producing secondary and tertiary alcohols.

Witt theory A theory of the mechanism of dyeing stating that all colored organic compounds (called chromogens) contain certain unsaturated chromophoric groups which are responsible for the color, and if these compounds also contain certain auxochromic groups, they possess dyeing properties.

Wolf-Kishner reduction Conversion of aldehydes and ketones to corresponding hydrocarbons by heating their semicarbazones, phenylhydrazones, and hydrazones with sodium ethoxide or by heating the carbonyl compound with excess sodium ethoxide and hydrazine sulfate.

wolfram *See* tungsten.

wolframic acid *See* tungstic acid.

wolfram white *See* barium tungstate.

wood alcohol *See* methyl alcohol.

wood ether *See* dimethyl ether.

wood sugar *See* xylose.

wood vinegar *See* pyroligneous acid.

Woodward-Hoffmann rule A concept which can predict or explain the stereochemistry of certain types of reactions in organic chemistry; it is also described as the conservation of orbital symmetry.

Woodward's Reagent K *See* N-ethyl-5-phenylisoxazolium-3'-sulfonate.

working electrode The electrode used in corrosion testing by an electrochemical cell.

work life The period of time a resin or an adhesive will remain usable after it is mixed with a catalyst and other ingredients.

Wulff process A chemical process to make acetylene and ethylene by cracking a hydrocarbon gas (for example, butane) with high-temperature steam in a regenerative furnace.

Wurtz-Fittig reaction A modified Wurtz reaction in which an aromatic halide reacts with an aklyl halide in the presence of sodium and an anhydrous solvent to form alkylated aromatic hydrocarbons.

Wurtz reaction Synthesis of hydrocarbons by treating alkyl iodides in ethereal solution with sodium according to the reaction $2CH_3I + 2Na \rightarrow CH_3CH_3 + 2NaI$.

X *See* siegbahn.

xanthan gum A high-molecular-weight (5–10 million) water-soluble natural gum; a heteropolysaccharide made up of building blocks of D-glucose, D-mannose, and D-glucuronic acid residues; produced by pure culture fermentation of glucose with *Xanthomonas campestris*.

xanthate A water-soluble salt of xanthic acid, usually potassium or sodium; used as an ore-flotation collector.

xanthene $CH_2(C_6H_4)_2O$ Yellowish crystals that are soluble in ether, slightly soluble in water and alcohol; melts at 100°C; used as a fungicide and chemical intermediate. Also known as tricyclic dibenzopyran.

xanthene dye Any of a family of dyes related to the xanthenes; the chromophore groups are (C_6H_4).

xanthene ketone *See* xanthone.

xanthine $C_5H_4N_4O_2$ A toxic yellow-white purine base that is found in blood and urine, and occasionally in plants; it is a powder, insoluble in water and acids, soluble in caustic soda; sublimes when heated; used in medicine and as a chemical intermediate. Also known as dioxopurine.

xanthine oxidase A flavoprotein enzyme catalyzing the oxidation of certain purines.

xanthone $CO(C_6H_4)_2O$ White needle crystals that are found in some plant pigments; insoluble in water, soluble in alcohol, chloroform, and benzene; melts at 173°C, sublimes at 350°C; used as a larvicide, as a dye intermediate, and in perfumes and pharmaceuticals. Also known as benzophenone oxide; dibenzopyrone; genicide; xanthene ketone.

xanthophyll $C_{40}H_{56}O_2$ Any of a group of yellow, alcohol-soluble carotenoid pigments that are oxygen derivatives of the carotenes, and are found in certain flowers, fruits, and leaves. Also known as lutein.

xanthurenic acid $C_{10}H_7NO_4$ Sulfur yellow crystals with a melting point of 286°C; soluble in aqueous alkali hydroxides and carbonates; excreted by pyridoxine-deficient animals after ingestion of tryptophan.

x axis A reference axis within a quartz crystal.

x-cut A quartz-crystal cut made in such a manner that the x axis is perpendicular to the faces of the resulting slab.

Xe *See* xenon.

xenon   An element, symbol Xe, member of the noble gas family, group O, atomic number 54, atomic weight 131.30; colorless, boiling point −108°C (1 atmosphere, or 101,325 newtons per square meter), noncombustible, nontoxic, and nonreactive; used in photographic flash lamps, luminescent tubes, and lasers, and as an anesthetic.

xenon-135   A radioactive isotope of xenon produced in nuclear reactors; readily absorbs neutrons; half-life is 9.2 hours.

xenyl   The chemical radical $C_6H_5C_6H_4$—.

xenylamine   *See para*-biphenylamine.

XPS   *See* x-ray photoelectron spectroscopy.

x-ray crystallography   The study of crystal structure by x-ray diffraction techniques. Also known as roentgen diffractometry.

x-ray crystal spectrometer   An instrument designed to produce an x-ray spectrum and measure the wavelengths of its components, by diffracting x-rays from a crystal with known lattice spacing.

x-ray diffraction analysis   Analysis of the crystal structure of materials by passing x-rays through them and registering the diffraction (scattering) image of the rays.

x-ray emission   *See* x-ray fluorescence.

x-ray fluorescence   Emission by a substance of its characteristic x-ray line spectrum upon exposure to x-rays. Also known as x-ray emission.

x-ray fluorescence analysis   A nondestructive physical method used for chemical analyses of solids and liquids; the specimen is irradiated by an intense x-ray beam and the lines in the spectrum of the resulting x-ray fluorescence are diffracted at various angles by a crystal with known lattice spacing; the elements in the specimen are identified by the wavelengths of their spectral lines, and their concentrations are determined by the intensities of these lines. Also known as x-ray fluorimetry.

x-ray fluorescent emission spectrometer   An x-ray crystal spectrometer used to measure wavelengths of x-ray fluorescence; in order to concentrate beams of low intensity, it has bent reflecting or transmitting crystals arranged so that the theoretical curvature required can be varied with the diffraction angle of a spectrum line.

x-ray fluorimetry   *See* x-ray fluorescence analysis.

x-ray image spectrography   A modification of x-ray fluorescence analysis in which x-rays irradiate a cylindrically bent crystal, and Bragg diffraction of the resulting emissions produces a slightly enlarged image with a resolution of about 50 micrometers.

x-ray photoelectron spectroscopy   A form of electron spectroscopy in which a sample is irradiated with a beam of monochromatic x-rays and the energies of the resulting photoelectrons are measured. Abbreviated XPS. Also known as electron spectroscopy for chemical analysis (ESCA).

x-ray powder diffractometer   *See* powder diffraction camera.

x-ray spectrograph   An x-ray spectrometer equipped with photographic or other recording apparatus; one application is fluorescence analysis.

x-ray spectrometer An instrument for producing the x-ray spectrum of a material and measuring the wavelengths of the various components.

x-ray spectrometry The measure of wavelengths of x-rays by observing their diffraction by crystals of known lattice spacing. Also known as roentgen spectrometry; x-ray spectroscopy.

x-ray spectroscopy *See* x-ray spectrometry.

x-ray spectrum A display or graph of the intensity of x-rays, produced when electrons strike a solid object, as a function of wavelengths or some related parameter; it consists of a continuous bremsstrahlung spectrum on which are superimposed groups of sharp lines characteristic of the elements in the target.

x-ray unit *See* siegbahn.

XU *See* siegbahn.

X unit *See* siegbahn.

xylene $C_6H_4(CH_3)_2$ Any one of the family of isomeric, colorless aromatic hydrocarbon liquids, produced by the destructive distillation of coal or by the catalytic reforming of petroleum naphthenic fractions; used for high-octane and aviation gasolines, solvents, chemical intermediates, and the manufacture of polyester resins. Also known as dimethylbenzene; xylol.

***meta*-xylene** $1,3\text{-}C_6H_4(CH_3)_2$ A flammable, toxic liquid; insoluble in water, soluble in alcohol and ether; boils at 139°C; used as an intermediate for dyes, a chemical intermediate, and a solvent, and in insecticides and aviation fuel. Also known as 1,3-dimethylbenzene.

***ortho*-xylene** $1,2\text{-}C_6H_4(CH_3)_2$ A flammable, moderately toxic liquid; insoluble in water, soluble in alcohol and ether; boils at 144°C; used to make phthalic anhydride, vitamins, pharmaceuticals, and dyes, and in insecticides and motor fuels. Also known as 1,2-dimethylbenzene.

***para*-xylene** $1,4\text{-}C_6H_4(CH_3)_2$ A toxic, combustible liquid; insoluble in water, soluble in alcohol and ether; boils at 139°C; used as a chemical intermediate, and to synthesize terephthalic acid, vitamins, and pharmaceuticals, and in insecticides. Also known as 1,4-dimethylbenzene.

xylenol $(CH_3)_2C_6H_3OH$ Highly toxic, combustible crystals; slightly soluble in water, soluble in most organic solvents; melts at 20–76°C; used as a chemical intermediate, disinfectant, solvent, and fungicide, and for pharmaceuticals and dyestuffs. Also known as dimethylhydroxybenzene; dimethylphenol; hydroxydimethylbenzene.

xylidine $(CH_3)_2C_6H_3NH_2$ A toxic, combustible liquid; soluble in alcohol and ether, slightly soluble in water; boils about 220°C; used as a chemical intermediate and to make dyes and pharmaceuticals. Also known as aminodimethylbenzene; aminoxylene.

xylitol $CH_2OH(CHOH)_3CH_2OH$ Pentahydric alcohols derived from xylose. Also known as xylite.

xylol *See* xylene.

xylose $C_5H_{10}O_5$ A pentose sugar found in many woody materials; combustible, white crystals with a sweet taste; soluble in water and alcohol; melts about 148°C; used as a nonnutritive sweetener and in dyeing and tanning. Also known as wood sugar.

Y

Y *See* yttrium.

yacca gum *See* acaroid resin.

y axis A line perpendicular to two opposite parallel faces of a quartz crystal.

Yb *See* ytterbium.

Y cut A quartz-crystal cut such that the y axis is perpendicular to the faces of the resulting slab.

yellow lead oxide *See* lead monoxide.

yellow phosphorus *See* white phosphorus.

yellow precipitate *See* mercuric oxide.

yellow prussiate of potash *See* potassium ferrocyanide.

yellow prussiate of soda *See* sodium ferrocyanide.

yellow pyoktanin *See* auramine hydrochloride.

yellow salt *See* uranyl nitrate.

ylium ion *See* enium ion.

York-Scheibel column *See* Scheibel extractor.

yrast state An energy state of a nucleus whose energy is less than that of any other state with the same spin.

ytterbia *See* ytterbium oxide.

ytterbium A rare-earth metal of the yttrium subgroup, symbol Yb, atomic number 70, atomic weight 173.04; lustrous, malleable, soluble in dilute acids and liquid ammonia, reacts slowly with water; melts at 824°C, boils at 1427°C; used in chemical research, lasers, garnet doping, and x-ray tubes.

ytterbium oxide Yb_2O_3 A colorless compound, melts at 2346°C, dissolves in hot dilute acids; used to prepare alloys, ceramics, and special glasses. Also known as ytterbia.

yttria *See* yttrium oxide.

yttrium A rare-earth metal, symbol Y, atomic number 39, atomic weight 88.905; dark-gray, flammable (as powder), soluble in dilute acids and potassium hydroxide solution, and decomposes in water; melts at 1500°C, boils at 2927°C; used in alloys and nuclear technology and as a metal deoxidizer.

yttrium acetate $Y(C_2H_3O_2)_3 \cdot 8H_2O$ Colorless, water-soluble crystals used as an analytical reagent.

yttrium chloride $YCl_3 \cdot 6H_2O$ Reddish, transparent, water- and alcohol-soluble prisms; decomposes at 100°C; used as an analytical reagent.

yttrium oxide Y_2O_3 A yellowish powder, insoluble in water, soluble in dilute acids; used as television tube phosphor and microwave filters. Also known as yttria.

yttrium sulfate $Y_2(SO_4)_3 \cdot 8H_2O$ Reddish crystals that are soluble in concentrated sulfuric acid, slightly soluble in water; decomposes at 700°C; used as an analytical reagent.

Yukawa force The strong, short-range force between nucleons, as calculated on the assumption that this force is due to the exchange of a particle of finite mass (Yukawa meson), just as electrostatic forces are interpreted in quantum electrodynamics as being due to the exchange of photons.

Yukawa potential The potential function that is associated with the Yukawa force, with the form $V(r) = -V_0(b/r) \exp(-r/b)$, where r is the distance between the nucleons and V_0 and b are constants, giving measures of the strength and range of the force respectively.

Z

z axis The optical axis of a quartz crystal, perpendicular to both the x and y axes.

Zeeman displacement The separation, in wave numbers, of adjacent spectral lines in the normal Zeeman effect in a unit magnetic field, equal (in centimeter-gram-second Gaussian units) to $e/4\pi mc^2$, where e and m are the charge and mass of the electron, or to approximately 4.67×10^{-15} (centimeter)$^{-1}$(gauss)$^{-1}$.

Zeeman effect A splitting of spectral lines in the radiation emitted by atoms or molecules in a static magnetic field.

Zeeman energy The energy of interaction between an atomic or molecular magnetic moment and an applied magnetic field.

zeolite catalyst Hydrated aluminum and calcium (or sodium) silicates (for example, $CaO\cdot2Al_2O_3\cdot5SiO_2$ or $Na_2O\cdot2Al_2O_3\cdot5SiO_2$) made with controlled porosity; used as a catalytic cracking catalyst in petroleum refineries, or loaded with catalyst for other chemical reactions.

zeotrope A nonazeotropic liquid mixture which may be separated by distillation, and in which the components are miscible in all proportions (homogeneous zeotrope or homozeotrope) or not miscible in all proportions (heterogeneous zeotrope or heterozeotrope).

Zerewitinoff reagent A light-colored methylmagnesium iodide–n-butyl ether solution that reacts rapidly with moisture and oxygen; used to determine water, alcohols, and amines in inert solvents.

zero branch A spectral band whose Fortrat parabola lies between two other Fortrat parabolas, with its vertex almost on the wave number axis.

zerogel A gel which has dried until apparently solid; sometimes it will swell or redisperse to form a sol when treated with a suitable solvent.

zero-order reaction A reaction for which reaction rate is independent of the concentrations of the reactants; for example, a photochemical reaction in which the rate is determined by the intensity of light.

Ziegler process A process for the low-pressure linear polymerization of ethylene and stereospecific polymerization of propylene; the product is a high-density polymer or elastomer.

Zimm plot A graphical determination of the root-square-mean end-to-end distances of coillike polymer molecules during scattered-light photometric analyses.

zinc A metal of group IIb, symbol Zn, atomic number 30, atomic weight 65.37; explosive as powder; soluble in acids and alkalies, insoluble in water; strongly electropositive; melts at 419°C, boils at 907°C.

zinc-65 A radioactive isotope of zinc, which has a 250-day half-life with beta and gamma radiation; used in alloy-wear tracer studies and body metabolism studies.

zinc acetate $Zn(C_2H_3O_2)_2 \cdot 2H_2O$ Pearly-white crystals with an astringent taste; soluble in water and alcohol; decomposes at 200°C; used to preserve wood in textile dyeing, and as an analytical reagent, a feed additive, and a polymer cross-linking agent.

zinc arsenate $ZnHAsO_4$ A toxic white powder that is insoluble in water, soluble in alkalies; used as an insecticide. Also known as zinc orthoarsenate.

zinc arsenite $Zn(AsO_2)_2$ A toxic white powder that is insoluble in water, soluble in alkalies; used as an insecticide and timber preservative. Also known as zinc metaarsenite.

zincate A reaction product of zinc with an alkali metal or with ammonia; for example, sodium zincate, Na_2ZnO_2.

zinc borate $3ZnO \cdot 2B_2O_3$ A white, amorphous powder that is soluble in dilute acids, slightly soluble in water; melts at 980°C; used in medicine, as a ceramics flux, as an inhibitor for mildew, and to fireproof textiles.

zinc bromide $ZnBr_2$ Water- and alcohol-soluble, white crystals that melt at 294°C; used in medicine, manufacture of rayon, and photography, and as a radiation viewing screen.

zinc carbonate $ZnCO_3$ White crystals that are insoluble in water, soluble in alkalies and acids; used in ceramics and ointments, and as a fireproofing agent and feed additive.

zinc chloride $ZnCl_2$ Water- and alcohol-soluble, white, fire-hazardous crystals that melt at 290°C, and are irritating to the skin; used as a catalyst and in electroplating, wood preservation, textile processing, petroleum refining, medicine, and feed additives.

zinc chromate $ZnCrO_4$ A toxic, yellow powder that is insoluble in water, soluble in acids; used as a pigment in paints (artists', automotive, primer), varnishes, linoleum, and epoxy laminates.

zinc cyanide $Zn(CN)_2$ A toxic, white powder that is insoluble in water and alcohol, soluble in alkalies and dilute acids; melts at 800°C; used as an analytical reagent and insecticide, and in medicine and metal plating.

zinc dimethyldithiocarbamate *See* ziram.

zinc ethylenebisdithiocarbamate *See* zineb.

zinc fluoride ZnF_2 A toxic white powder that is slightly soluble in water and melts at 872°C; used in enamels, ceramic glazes, and galvanizing.

zinc formate $Zn(CHO_2)_2 \cdot 2H_2O$ Toxic, white crystals that are soluble in water, insoluble in alcohol; used as a catalyst, weatherproofing agent, and wood preservative.

zinc halide A binary compound of zinc and a halogen; for example, $ZnBr_2$, $ZnCl_2$, ZnF_2, and ZnI_2.

zinc hydroxide $Zn(OH)_2$ Colorless, water-soluble crystals that decompose at 125°C; used as a chemical intermediate and in rubber compounding and surgical dressings.

zinc metaarsenite *See* zinc arsenite.

zinc naphthenate $Zn(C_6H_5COO)_2$ A combustible, viscous, acetone-soluble solid; used in paints, varnishes, and resins, and as a drier and wetting agent, insecticide, fungicide, and mildewstat.

zinc orthoarsenate *See* zinc arsenate.

zinc orthophosphate *See* zinc phosphate.

zinc oxide ZnO A bitter-tasting, white to gray powder that is insoluble in water, soluble in alkalies and acids; melts at 1978°C; used as a pigment, mold-growth inhibitor, and dietary supplement, and in cosmetics, electronics, and color photography.

zinc phosphate $Zn_3(PO_4)_2$ A white powder that is insoluble in water, soluble in acids and ammonium hydroxide; melts at 900°C; used in coatings for steel, aluminum, and other metals, and in dental cements and phosphors. Also known as tribasic zinc phosphate; zinc orthophosphate.

zinc phosphide Zn_3P_2 A toxic, alcohol-insoluble, gray gritty powder that reacts violently with oxidizing agents; melts at over 420°C, decomposes in water; used as a rat poison and in medicine.

zinc selenide $ZnSe$ A water-insoluble, moderately toxic, yellow to reddish solid that is a fire hazard when in contact with water and acids; melts above 1100°C; used as infrared optical windows.

zinc sulfate $ZnSO_4 \cdot 7H_2O$ Efflorescent, water-soluble, colorless crystals with an astringent taste; used to preserve skins and wood and as a paper bleach, analytical reagent, feed additive, and fungicide. Also known as white copperas; white vitriol; zinc vitriol.

zinc sulfide ZnS A yellowish powder that is insoluble in water, soluble in acids; exists in two crystalline forms (alpha, or wartzite, and beta, or sphalerite); beta becomes alpha at 1020°C, and sublimes at 1180°C; used as a pigment for paints and linoleum, in opaque glass, rubber, and plastics, for hydrosulfite dyeing process, as x-ray and television screen phosphor, and as a fungicide.

zinc telluride $ZnTe$ Moderately toxic, reddish crystals that melt at 1238°C and decompose in water.

zinc vitriol *See* zinc sulfate.

zinc white *See* Chinese white.

zinc-[N,N'-propylene-1,2-bis(dithiocarbamate)]ₓ A yellow powder which decomposes at 160°C; insoluble in water; used as a fungicide for potatoes, tobacco, grapevines, and bananas. Also known as propineb.

zineb $C_4H_6N_2S_4Zn$ An off-white solid compound, used as a fungicide for fruits and vegetables. Also known as zinc ethylenebisdithiocarbamate.

ziram $C_6H_{12}N_2S_4Zn$ A colorless solid; melting point is 246°C; used as a fungicide for vegetables and some fruit crops. Also known as zinc dimethyldithiocarbamate.

zirconia *See* zirconium oxide.

zirconic anhydride *See* zirconium oxide.

zirconium A metallic element of group IVb, symbol Zr, atomic number 40, atomic weight 91.22; occurs as crystals, flammable as powder; insoluble in water, soluble in hot, concentrated acids; melts at 1850°C, boils at 4377°C.

zirconium-95 A radioactive isotope of zirconium; half-life of 63 days with beta and gamma radiation; used to trace petroleum-pipeline flows and in the circulation of a catalyst in a cracking plant.

zirconium boride ZrB_2 A hard, toxic, gray powder that melts at 3000°C; used as an aerospace refractory, in cutting tools, and to protect thermocouple tubes. Also known as zirconium diboride.

zirconium carbide ZrC Hard, gray crystals that are soluble in water, soluble in acids; as powder, it ignites spontaneously in air; melts at 3400°C, boils at 5100°C; used as an abrasive, refractory, and metal cladding, and in cermets, incandescent filaments, and cutting tools.

zirconium chloride *See* zirconium tetrachloride.

zirconium diboride *See* zirconium boride.

zirconium dioxide *See* zirconium oxide.

zirconium halide A compound of zirconium with a halogen; for example, $ZrBr_2$, $ZrCl_2$, $ZrCl_3$, $ZrCl_4$, $ZrBr_2$, $ZrBr_3$, ZrF_4, and ZrI_4.

zirconium hydride ZrH_2 A flammable, gray-black powder; used in powder metallurgy and nuclear moderators, and as a reducing agent, vacuum-tube getter, and metal-foaming agent.

zirconium hydroxide $Zr(OH)_4$ A toxic, amorphous white powder; insoluble in water, soluble in dilute mineral acids; decomposes at 550°C; used in pigments, glass, and dyes, and to make zirconium compounds.

zirconium nitride ZrN A hard, brassy powder that is soluble in concentrated acids; melts at 2930°C; used in refractories, cermets, and laboratory crucibles.

zirconium orthophosphate *See* zirconium phosphate.

zirconium oxide ZrO_2 A toxic, heavy white powder that is insoluble in water, soluble in mineral acids; melts at 2700°C; used in ceramic glazes, special glasses, and medicine, and to make piezoelectric crystals. Also known as zirconia; zirconic anhydride; zirconium dioxide.

zirconium oxychloride $ZrOCl_2 \cdot 8H_2O$ White crystals that are soluble in water, insoluble in organic solvents, and acidic in aqueous solution; used for textile dyeing and oil-field acidizing, in cosmetics and greases, and for antiperspirants and water repellents. Also known as basic zirconium chloride; zirconyl chloride.

zirconium phosphate $ZrO(H_2PO_4)_2 \cdot 3H_2O$ A toxic, dense white powder that is insoluble in water, soluble in acids and organic solvents; decomposes on heating; used as an analytical reagent, coagulant, and radioactive-phosphor carrier. Also known as basic zirconium phosphate; zirconium orthophosphate.

zirconium tetrachloride $ZrCl_4$ Toxic, alcohol-soluble, white lustrous crystals; sublimes above 300°C and decomposes in water; used to make pure zirconium and for water-repellent textiles and as an analytical reagent. Also known as zirconium chloride.

zirconyl chloride *See* zirconium oxychloride.

Zn *See* zinc.

zone 1. A set of crystal faces which intersect (or would intersect, if extended) along edges which are all parallel. 2. *See* band.

zone axis A line through the center of a crystal which is parallel to all the faces of a zone.

zone indices Three integers identifying a zone of a crystal; they are the crystallographic coordinates of a point joined to the origin by a line parallel to the zone axis.

zone law A law which states that the Miller indices (h, k, l) of any crystal plane lying in a zone with zone indices (u, v, w) satisfy the equation $hu + lv + kw = 0$.

zone melting crystallization A method for purification of crystalline solids; the sample, packed in a narrow column, is heated so that a molten zone passes down through the sample, carrying impurities with it.

zoning A variation in the composition of a crystal from core to margin due to a separation of the crystal phases during its growth by loss of equilibrium in a continuous reaction series.

Zr *See* zirconium.

Zsigmondy gold number The number of milligrams of protective colloid necessary to prevent 10 milliliters of gold sol from coagulating when 0.5 milliliter of 10% sodium chloride solution is added.

zymase A complex of enzymes that catalyze glycosis.

zymogen The inactive precursor of an enzyme; liberates an active enzyme on reaction with an appropriate kinose. Also known as proenzyme.

zymogen granules Granules of zymogen in gland cells, particularly those of the pancreatic acini and of the gastric chief cells.

zymophore The active portion of an enzyme.

zymosterol $C_{27}H_{43}OH$ An unsaturated sterol obtained from yeast fat; yields cholesterol on hydrogenation.